Plant Biosystematics

Academic Press Rapid Manuscript Reproduction

Based on the Symposium of the
International Organization of Plant Biosystematists,
McGill University, Montreal, Canada,
July 17 - 21, 1983.

Plant Biosystematics

Edited by

William F. Grant

Department of Plant Science
Macdonald College of McGill University
Ste. Anne de Bellevue, Quebec, Canada

1984

ACADEMIC PRESS
(Harcourt Brace Jovanovich, Publishers)
Toronto   Orlando   San Diego   New York   London
Montreal   San Francisco   Sydney   Tokyo   São Paulo

ACADEMIC PRESS CANADA
55 Barber Greene Road, Don Mills, Ontario   M3C 2A1

United States Edition published by
ACADEMIC PRESS, INC.
Orlando, Florida 32887

United Kingdom Edition published by
ACADEMIC PRESS, INC. (LONDON) LTD.
24/28 Oval Road, London NW1 7DX

Library of Congress Cataloging in Publication Data

Main entry under title:
Plant biosystematics.

 Proceedings of a symposium sponsored by the Inter-
national Organization of Plant Biosystematists and held
at McGill University in July 1983.
   Includes index.
   1. Botany—Classification—Congresses. I. Grant,
William F.   II. International Organization of Plant
Biosystematists.
QK95.P545 1984   580'.12   84-6490
ISBN 0-12-295680-X (alk. paper)

Canadian Cataloguing in Publication Data

Main entry under title:
Plant biosystematics

Based on a symposium sponsored by the International
Organization of Plant Biosystematists, held at McGill
University, Montreal, Quebec, July 17-21, 1983.
Includes index.
ISBN 0-12-295680-X

I. Botany—Classification—Congresses. I. Grant,
W.F., 1924-   II. International Organization of
Plant Biosystematists.

QK95.P42 1984   580   C84-098685-8

Contents

Contributors   ix

Preface   xiii

INTRODUCTION

Biosystematics 1983   1
  Robert K. Vickery, Jr.

CYTOLOGY AND CYTOGENETICS

Cytology and Biosystematics: 1983   25
  Keith Jones

The Genome, the Natural Karyotype, and Biosystematics   41
  Michael D. Bennett

Chromosome Pairing in Species and Hybrids   67
  R. C. Jackson

Nuclear DNA Variation and the Homology of Chromosomes   87
  H. Rees

Chromosome Banding and Biosystematics   97
  Ichiro Fukuda

Chromosome Evolution and Adaptation in Mistletoes   117
  B. A. Barlow and Nicole J. Martin

Differentiation and Evolution of the Genus *Campanula* in the
Mediterranean Region   141
  J. Contandriopoulos

BREEDING SYSTEMS AND HYBRIDIZATION

The Biological Species Concept Reexamined   159
  Bengt Jonsell

The Pursuit of Hybridity and Population Divergence in *Isotoma petraea*   169
  S. H. James

The Role of Hybridization in the Evolution of *Bidens* on the Hawaiian Islands   179
  Fred R. Ganders and *Kenneth M. Nagata*

Hybridization in the Domesticated-Weed-Wild Complex   195
  Ernest Small

PLANT REPRODUCTION AND REPRODUCTION ISOLATION

Plant Reproductive Strategies   211
  Krystyna M. Urbanska

The Relationships between Self-incompatibility, Pseudo-compatibility, and Self-compatibility   229
  David L. Mulcahy

Apomixis and Biosystematics   237
  Sven Asker

Constraints on the Evolution of Plant Breeding Systems and Their Relevance to Systematics   249
  C. J. Webb

Pollination by Animals and Angiosperm Biosystematics   271
  Peter G. Kevan

METHODOLOGIES IN BIOSYSTEMATICS

The Biosystematic Importance of Phenotypic Plasticity   293
  Pierre Morisset and Céline Boutin

A Biosystematic and Phylogenetic Study of the Dipsacaceae   307
  R. Verlaque

Evolution of rDNA in *Claytonia* Polyploid Complexes   321
  J. J. Doyle, R. N. Beachy and W. H. Lewis

Isozyme Evidence and Problem Solving in Plant Systematics   343
  L. D. Gottlieb

Phytochemical Approaches to Biosystematics   359
  K. E. Denford

Pollen Morphology and Biosystematics of the Subfamily Papilionoideae
(Leguminosae) 377
   I. K. Ferguson

Numerical Taxonomy and Biosystematics 395
   J. McNeill

Problems of Hybridity in the Cladistics of *Crataegus* (Rosaceae) 417
   J. B. Phipps

POPULATION BIOLOGY AND BIOGEOGRAPHY

Population Biology and Biosystematics: Current Experimental Approaches 439
   Barbara A. Schaal

Cytogeography and Biosystematics 453
   C. Favarger

Biosystematics of Macaronesian Flowering Plants 477
   Liv Borgen

Biosystematics of Tropical Forest Plants: A Problem of Rare Species 497
   P. S. Ashton

BIOSYSTEMATICS OF MOSSES AND FERNS

Biosystematics of Bryophytes: An Overview 519
   Robert Wyatt and Ann Stoneburner

Biosystematic Studies on Pteridophytes in Canada: Progress and Problems 543
   D. M. Britton

BIOSYSTEMATICS AND ITS PRACTICAL APPLICATIONS

Biosystematics and Medicine 561
   Walter H. Lewis

Modes of Evolution in Plants Under Domestication 579
   Daniel Zohary

Zea—A Biosystematical Odyssey 587
   Hugh H. Iltis and John F. Doebley

Biosystematics and Hybridization in Horticultural Plants 617
   Willem A. Brandenburg

Biosystematics and Conservation 633
   David Bramwell

OVERVIEWS

A Comparison of Taxonomic Methods in Biosystematics 643
  Warren H. Wagner, Jr.

Observations on IOPB 1983 and Notes on the Discussions among Participants 655
  John C. Semple

Index 659

Contributors

Numbers in parentheses indicate the pages on which the authors' contributions begin.

P. S. Ashton (497), *The Arnold Arboretum, Harvard University, Cambridge, Massachusetts 02138, U.S.A.*

Sven Asker (237), *Institute of Genetics, University of Lund, S-223 62 Lund, Sweden*

B. A. Barlow (117), *Division of Plant Industry, C.S.I.R.O., G.P.O. Box 1600, Canberra City, ACT 2601, Australia*

R. N. Beachy (321), *Department of Biology, Washington University, St. Louis, Missouri 63130, U.S.A.*

Michael D. Bennett (41), *Plant Breeding Institute, Maris Lane, Trumpington, Cambridge CB2 2LQ, England*

Liv Borgen (477), *Botanical Garden and Museum, University of Oslo, Trondheimsveien 23B, Oslo 5, Norway*

Céline Boutin[1] (293), *Département de biologie, Faculté des sciences, Université Laval, Québec G1K 7P4, Canada*

David Bramwell (633), *Jardin Botanico "Viera y Clavijo," Aparto de Correos 14, Tafira Alta, Las Palmas de Gran Canaria, Canary Islands, Spain*

Willem A. Brandenburg (617), *Department of Taxonomy of Cultivated Plants and Weeds, The Agricultural University, Haagsteg 3, Wageningen NL-6708 PM, The Netherlands*

D. M. Britton (543), *Department of Botany, University of Guelph, Guelph, Ontario N1G 2W1, Canada*

J. Contandriopoulos (141), *Laboratoire de Cytotaxinomie végétale, Université de Provence, Centre de Saint-Charles, 3 Place Victor-Hugo, 13331 Marseille, Cedex 3, France*

K. E. Denford (359), *Department of Botany, University of Alberta, Edmonton, Alberta T6G 2E9, Canada*

---

[1]Present address: School of Plant Biology, University College of North Wales, Bangor, Gwynned LL 57 2UW, U.K.

John F. Doebley (587), *Department of Genetics, North Carolina State University, Raleigh, North Carolina 27650, U.S.A.*

J. J. Doyle (321), *Department of Biology, Washington University, St. Louis, Missouri 63130, U.S.A.*

C. Favarger (453), *Institut de Botanique, Université de Neuchâtel, 2000 Neuchâtel, Switzerland*

I. K. Ferguson (377), *The Royal Botanic Gardens, Kew, Richmond, Surrey TW9 3DS, England*

Ichiro Fukuda (97), *Division of Biology, Tokyo Woman's Christian University, Zempukuji, Suginami, Tokyo 167, Japan*

Fred R. Ganders (179), *Department of Botany, University of British Columbia, 2075 Wesbrook Mall, Vancouver, British Columbia V6T 1W5, Canada*

L. D. Gottlieb (343), *Department of Genetics, University of California, Davis, California 95616, U.S.A.*

Hugh H. Iltis (587), *Herbarium and Department of Botany, University of Wisconsin, Madison, Wisconsin 53706, U.S.A.*

R. C. Jackson (67), *Department of Biology, Texas Technical University, Lubbock, Texas 79409, U.S.A.*

S. H. James (169), *Department of Botany, University of Western Australia, Nedlands, WA 6009, Australia*

Keith Jones (25), *Jodrell Laboratory, The Royal Botanic Gardens, Kew, Richmond, Surrey TW9 3DS, England*

Bengt Jonsell (159), *Bergius Botanic Garden, Stockholm University, Box 50017, S-104 05 Stockholm, Sweden*

Peter G. Kevan (271), *Department of Environmental Biology, University of Guelph, Guelph, Ontario N1G 2W1, Canada*

Walter H. Lewis (321, 561), *Department of Biology, Washington University, St. Louis, Missouri 63130, U.S.A.*

Nicole J. Martin (117), *Cytogenetics Unit, Royal Brisbane Hospital, Brisbane, Queensland 4029, Australia*

J. McNeill (395), *Department of Biology, University of Ottawa, Ottawa, Ontario K1N 6N5, Canada*

Pierre Morisset (293), *Département de biologie, Faculté des sciences, Université Laval, Québec, Québec G1K 7P4, Canada*

David L. Mulcahy (229), *Department of Botany, University of Massachusetts, Amherst, Massachusetts 01002, U.S.A.*

Kenneth M. Nagata (179), *Harold L. Lyon Arboretum, Honolulu, Hawaii, U.S.A.*

J. B. Phipps (417), *Department of Plant Science, University of Western Ontario, London, Ontario N6A 3K7, Canada*

H. Rees (87), *Department of Agricultural Botany, University College of Wales, Penglais, Aberystwyth, Cardiganshire, Wales SY23 3DD, U.K.*

Barbara A. Schaal (439), *Department of Biology, Washington University, St. Louis, Missouri 63130, U.S.A.*

John C. Semple (655), *Department of Biology, University of Waterloo, Waterloo, Ontario N2L 3G1, Canada*

Ernest Small (195), *Biosystematics Research Institute, Saunders Building, Central Experimental Farm, Agriculture Canada, Ottawa, Ontario K1A 0C6, Canada*

Ann Stoneburner (519), *Department of Botany, University of Georgia, Athens, Georgia 30602, U.S.A.*

Krystyna M. Urbanska (211), *Geobotanisches Institut ETH, Stiftung Rübel, 38 Zürichbergstrasse, CH-8044 Zürich, Switzerland*

R. Verlaque (307), *Laboratoire de Cytotaxinomie végétale, Centre de Saint-Charles, 3 Place Victor-Hugo, 13331 Marseille, Cedex 3, France*

Robert K. Vickery, Jr. (1), *Department of Biological Sciences, University of Utah, Salt Lake City, Utah 84112, U.S.A.*

Warren H. Wagner, Jr. (643), *Department of Botany, University of Michigan, Ann Arbor, Michigan 48109, U.S.A.*

C. J. Webb (249), *Botany Division, D.S.I.R., Private Bag, Christchurch, New Zealand*

Robert Wyatt (519), *Department of Botany, University of Georgia, Athens, Georgia 30602, U.S.A.*

Daniel Zohary (579), *Department of Botany, The Hebrew University, Jerusalem, Israel*

Preface

Plant Biosystematics is the proceedings of a four-day symposium held at McGill University, Montreal, in July, 1983. The symposium was sponsored by the International Organization of Plant Biosystematists in their continuing role to foster international cooperation in plant biosystematics. Delegates from 16 countries participated. The theme of the symposium was "Plant Biosystematics: Forty Years Later," in recognition of the introduction of the term "biosystematy" by Camp and Gilly in 1943. Camp and Gilly considered "biosystematy" as a descriptive term to cover newer approaches to "(1) delimit the natural biotic units and (2) apply to these units a system of nomenclature adequate to the task of conveying precise information regarding their defined limits, relationships, variability, and dynamic structure." The types of studies they advocated were to provide a system of classification which was to form a basis for the organization of information of all types concerning living organisms, and ultimately to provide solutions to questions concerning biological diversity within the framework of evolution.

Until the 1960s biosystematics was viewed as the "alpha and omega" of biology and as an "unending synthesis." It was considered that the "processes of microevolution and the origin of higher plant diversity" would be readily resolved and that these would be able to be expressed "satisfactorily in our systematic arrangements."

In the 1960s chemistry and statistics became increasingly important and often very powerful analytical tools. But by 1970 there was a growing awareness among biosystematists that they were not in a position to offer unequivocal solutions to the quandaries faced by classical taxonomy, and that it was not possible to set up a system of classification which reflected all types of data with equal fidelity.

Plant systematics in the 1970s matured to the point where it is generally understood that taxonomy provides a general-purpose classification, incorporating and summing up characters of many kinds, which can serve as a framework for many types of investigation, including the biology of plant populations. Biosystematics has placed more emphasis on an understanding of processes at the level of the population, and of evolutionary forces, especially those of microevolution, and on all aspects of the biology of extant populations and the individual organisms of which they consist.

In the brief history of this discipline, biosystematists have used a wide variety of techniques, in addition to those traditional to classical taxonomy, in an attempt to resolve these issues. Two of these, cytology and hybridization, have been significant contributors to biosystematic studies. Karyological studies have from the beginning had a central role in biosystematics. The importance of chromosomes has been twofold: they have been used as a diagnostic character in classification and to provide insights into genetic phenomena and evolutionary processes leading to speciation. Three characteristics of chromosomes have been found to have especially high information content, namely, chromosome number, chromosome morphology, and chromosome behavior at meiosis. Likewise, hybridization, observed either in wild populations, or carried out artificially, has played a central role in biosystematic studies. It has been used both to test for reproductive barriers and to explain the transfer of genetic information from one taxon to another. Since species differ from each other genetically, the consequences of hybridization have also been much linked to the biological species concept.

The program was planned to consider biosystematics in its broadest sense in which not only traditional themes of speciation, karyology, polyploidy, hybridization, and specialized systems (apomixis, incompatibility, pollination) would be considered, but also some of the current methodologies and future trends, as well as some of the practical aspects (agriculture, conservation, horticulture, and medicine).

The volume has provided a review of the current field of biosystematics from which botanists, geneticists, agriculturalists, evolutionists, and biologists in general, who are interested in the evolution of natural biota, may obtain some perception of the present status of plant biosystematics and how this discipline contributes to human well-being. Of the 38 papers presented at the meeting, 37 are reproduced herein. Those by J. Contandriopoulos, C. Favarger, P. Morisset, and R. Verlaque were presented in French but were translated into English for publication with the authors' consent. Although discussions were not recorded, J. Semple has summarized in a final chapter some of the flavor of the meeting.

Acknowledgments

Although the symposium was organized by the Executive and Council of the International Organization of Plant Biosystematists, many other individuals were responsible for suggestions and help with arrangements. The active and enthusiastic participation during the initial organizational phase of two of my former students, C. W. Crompton and A. E. Stahevitch of the Biosystematics Research Institute, Ottawa, was most fruitful. Walter Lewis not only contributed to the program but delivered the post-banquet talk on ''Recent Contributions of Plants to Medicine and Dentistry.'' Hugh Iltis supplied the Camp and Gilly cartoon which decorated the cover of the program. Luc Brouillet made all the arrangements for the tour of the Montreal Botanical Gardens, and the reception by the City of Montreal. Pierre Morisset translated the abstracts for the program into French. Paulette Lachance served most efficiently as Executive Secretary and headed the registration desk. Others who helped during the symposium included Lawrence Goldstein, Pierre St. Marseille, John Raelson, Kate Merlin and Nancy MacLean. Finally, financial contributions, which made the meeting possible, are gratefully acknowledged from McGill University, the Genetics Society of Canada, l'Association Canadienne-Française pour l'Avancement des Sciences, and the Natural Sciences and Engineering Research Council of Canada.

William F. Grant

Biosystematics 1983

Robert K. Vickery, Jr.
Department of Biological Sciences
University of Utah
Salt Lake City, Utah, U.S.A.

INTRODUCTION

1983 finds biosystematics alive, well and thriving! However, as it has developed and grown over the last several decades, it has faced several challenges and new opportunities. But first, let us see what the roots of biosystematics are and how it has developed before we assess its status, challenges and opportunities today.

Biosystematics or biosystematy as it was originally defined by Camp and Gilly in 1943 "seeks (1) to delimit the natural biotic units and (2) to apply to these units a system of nomenclature adequate to the task of conveying precise information regarding their defined limits, relationships, variability and dynamic structure." Purists may argue that Camp and Gilly should have used the term *biotaxonomy* inasmuch as their stated emphasis falls heavily on the philosophic consideration of "what is a species?" or, more accurately, "what are species?". In practice, their approach consisted of using the methods of cytology and genetics to experimentally ascertain species relationships, that is, it was, in fact, a systematic approach, but to a taxonomic problem. What they were doing was attempting to define and formalize the growing emphasis on the use of experimental studies as the foundation for a better, more biological classification.

HISTORICAL BACKGROUND

Historically, classification had been highly artificial from the days of the "Father of Botany," Theophrastus (3rd and 4th Century B.C.), and the celebrated physician, Dioscorides (1st Century A.D.) down to the 16th century. Theophrastus grouped plants by form: herbs, undershrubs, shrubs and trees (Lawrence 1951). Trees were plants with a woody main stem, e.g., the olive, fig and vine. Dioscorides grouped plants by use: pot herbs, sharp herbs, etc. (Gunther 1959). Artificial classification culminated in the well-known, comprehensive, sexual system of Linneaus (Linné 1737). His system provided a highly expedient way of coping with the flood of new plants being brought to Europe from all over the planet by the voyages of discovery (Eisley 1961).

However, well before Linneaus' time, classification was starting to become more natural in the great Herbals, e.g., that of Gerard (Gerard 1597) in which natural groups such as *Narcisus, Geranium,* etc. were recognized. Actually, Linneaus, too, had begun working out natural groups, e.g. Palmae, Gramina, Orchideae, etc. (Linné 1778). Linneaus' successors, such as the De Candolles (A. De Candolle 1824-1873) and Bentham and Hooker (1862-1883), brought the natural system of plant classification to a high level.

Philosophically, the natural systems of classification were based on the static, Great Scale of Nature (Eisley 1961). The Darwinian revolution (Darwin 1859) necessitated remarkably few changes in the systems (Bessey 1915; Engler and Prantl 1897-1915), a testimony to how carefully worked out the old systems were. However, the new philosophical base, *evolutionary change,* raised intriguing and complex questions of phylogenies. Were groups related by common descent? closely? distantly? What about convergent evolution? etc. Initially, these questions were probed using data from comparative morphology, the growing knowledge of biogeography and, when possible, the fossil record.

RISE OF EXPERIMENTALLY BASED TAXONOMY

The need for better, more objective methods of ascertaining relationships was being increasingly felt by the early years of this century. For example, Mez (1926) applied the techniques of immunology to classification. He assessed the similarity of taxa by observing the comparative antibody responses they illicited. In general, the results were in line with those of the classical, natural systems. The method is effective, but cumbersome and only occasionally used, for example, Petersen and Fairbrothers (1983) recent use of it to resolve the placement of *Amphipterygium* and *Leitneria* in the *Rutiflorae*.

However, at the time of Mez — the early 20's — the time was ripe for developing experimental approaches to classification. Soon, a surge of studies started that used the observational, investigative techniques of cytology, others that employed the experimental hybridizations of genetics and still others that were based on the methods of ecology, transplant experiments, etc.

These cytogenetic-ecologic experimental studies began with the very active Scandinavian school. This diverse, but enthusiastic group — they would go back and forth for evening seminars in Lund or Copenhagen — included such figures as Turreson (1922) developing his ecologic-genetic concepts of the ecotype, ecospecies, etc., Nilson-Ehle (Briggs and Knowles 1967) working out the genetics of grain color in polyploid series in wheat,

Clausen (1926) elucidating the intricate genetics of *Viola* flower colors, Muntzing (1932) elegantly demonstrating the allotetraploid origin of *Galeopsis tetrahit*, Winge, DuRietz, Danser and others.

Another group of taxonomic experimentalists was developing at the same time in the San Francisco Bay area. And, they, too, would go back and forth, that is, from Stanford to Berkeley for evening Biosystematic seminars. This group included Babcock (1920, 1931) with his cytogenetic-evolutionary-taxonomic studies of *Crepis*, Stebbins (Stebbins and Babcock 1939; Stebbins 1950) with his wide-ranging cytogenetic-evolutionary interests in the Compositae, Goodspeed (1923, 1938, 1954) with his extensive field studies coupled with laboratory and experimental garden cytogenetic investigations of *Nicotiana*, Hall (Hall and Clements 1923; Hall 1926) with his transplant studies and Clausen, Keck and Hiesey (1934, 1939, 1940) who continued Hall's transplant studies and initiated their own, well-known cytogenetic-taxonomic studies. There were other, scattered workers such as Anderson (1949) with his studies of introgressive hybridization, Gregor (1939) with his work on clines, Chetverikov (1927) with his interest in the genetics of wild populations and Cleland (1954) with his long-continued studies of *Oenothera* and their unusual rings of chromosomes. This tide of experimental studies continued to grow and swell during the 20's and 30's. It added whole new dimensions to our understanding of the biology of species!

ESTABLISHMENT OF BIOSYSTEMATICS AS A SCIENCE

By the early 1940's the time was ripe for the next step – to synthesize and emphasize the new, experimental approaches to systematics and taxonomy. This, Clausen, Keck and Hiesey (1940) did with their studies and the term "Experimental Taxonomy" plus the use of Turesson's (1922) earlier terms of *biotype, ecotype, ecospecies, coenospecies* and *comparium* to describe experimentally delimited entities. This, Camp and Gilly (1943) did also with their studies and their term *Biosystematy* plus terms to characterize various specific products of evolution: *homogeneon, phenon, parageneon, dysploidion, euploidion, alloploidion, micton, rheogameon, cleistogameon, heterogameon, apogameon* and *agameon*.

Both groups, as practicing taxonomists – Camp (1942) on *Vaccinium*, Gilly (Camp and Gilly 1943) on *Carex* and Keck (1936) on *Penstemon* – tended to combine the new experimental approaches and terms with those of classical taxonomy. Indeed, this has been the usual practice ever since.

In time, the "natural selection" of usage has established the term *Biosystematics* rather than Biosystematy or Experimental Taxonomy as the name for our rapidly developing new scientific field of experimental, biologically based systematics.

REVIEW OF FOUR DECADES OF BIOSYSTEMATICS

How has Biosystematics developed and matured from its coming of
age and naming in the early 1940's to adulthood today in 1983?
Clearly it has flourished, expanded and spread worldwide from its
Scandinavian and Californian main "centers of origin," but, has it
changed directions? methods? emphasis? or, the questions being
asked?

   Let's recall thoughtfully, if very briefly, a few
representative investigations of recent decades in order to give
us a feel for these questions and a good assessment of where
Biosystematics actually is in 1983.

   The Scandinavian "center of origin" has continued to flourish
with, for example, Nordenskiold's (1951) investigation of
Luzula species with their unusual diffuse centromeres and
consequent unusual mixoploid series, Ellerström's (Ellerström and
Sjödin 1966) study of aneuploids in Red Clover, Westergaard's
(1946) study of sex determination and polyploidy in *Melandrium*,
Nygren's (1962) work on hybridization in *Calamagrostis* and
Böcher's (1949, 1966, 1972) extensive cytological and experimental
studies on varietal and racial variations in *Prunella*, *Campanula*
and *Clinopodium*.

   The Scandinavian center has had its "adaptive radiations." For
example, Linhart (1973) with his pollen dispersal studies of
Heliconia and the incredibly extensive cytotaxonomic studies
of the Löves of so many groups ranging from the whole Icelandic
flora (A. and D. Löve 1956) to arctic polyploidy (A. and D. Löve
1957) to the Iberian Cistaceae (A. Löve and Kjellquist 1964) to
genera such as *Phyllitis* (A. and D. Löve 1973) and
Acanthoxanthium (D. Löve 1975) and the very useful and
extensive IOPB Documented Chromosome Number Reports starting with
number 1 in 1964 (A. Löve and Solbrig 1964) and continuing to the
present – number 79 appeared this April (A. Löve 1983).

   Biosystematics, has continued to thrive in the San Francisco
Bay Area "center of origin", also. For example, there are the
continuing transplant, cytologic, genetic and sophisticated ecophysiological studies of the the Carnegie (Stanford) group on the
Madiinae, *Achillea*, *Potentilla*, *Mimulus* and *Poa* (Clausen
et al. 1945, 1948; Clausen and Hiesey 1958; Hiesey *et al*.
1971; Hiesey and Nobs 1982). At Berkeley there are the broad
guage biosystematic studies of Constance (1964), of Baker (earlier
at Leeds) on *Armeria* (Baker 1954), on colonizing species
(Baker and Stebbins 1965) and on nectar composition in taxonomic
and phylogenetic contexts (H. and I. Baker 1977), of Ornduff on a
wide variety of plants such as *Hypericum* (Ornduff 1975),
Jepsonia (Ornduff and Weller 1975), *Lasthenia* (Ornduff 1976)
and *Amsinckia* (Weller and Ornduff 1977) often from the point
of view of their reproductive biology (Ornduff 1969), of Stebbins
(later at Davis) on *Bromus*, *Cryptantha*, and many other groups
(oral commun.) and of Gottlieb (1982) on allozymes particularly

but also on such studies as the origin of a new species of
Stephanomeria (Gottlieb and Bennett 1983), and more specialized
studies, e.g. that of Millar (1983) on clines in *Pinus muricata*.
"Adaptive radiations" from the Bay Area center include the
elegant studies on speciation in *Clarkia* by H. Lewis and his
students (Lewis and Roberts 1956; Lewis and Raven 1958; Lewis
1962, 1966), the extensive cytological studies of Bell (Bell and
Constance 1957) the wide-ranging - United States to South America
to South Africa to New Zealand to China - studies of Raven and
coworkers (Ehrlich and Raven 1965; Moore and Raven 1970; Raven
et al. 1971; Bartholomew *et al.* 1973; Raven 1975, 1977; Seavey
and Raven 1977; P. and T. Raven 1976) as well as V. Grant's
detailed analyses of the semi-species, species, super species (or
syngamia) of *Gilia* (Grant 1950, 1957; V. and A. Grant 1960)
and more recently of the clonal microspecies, semi-species, etc.
of *Opuntia* (V. and K. Grant 1980, 1982).

Other examples of diverse biosystematic investigations from the
United States are the studies of Ownbey (1950) on natural
hybridization in *Tragopogon,* of Rollins (1957, 1963) on
Lesquerella and *Leavenworthia,* of W. Lewis (Lewis, 1959, 1980,
oral commun.) on polyploidy in *Rosa, Claytonia,* etc., of
Solbrig on *Viola* using a many-faceted population biology
approach (Cook and Solbrig oral commun.) and of biosystematics in
general (Solbrig 1971), of Smith (1983) on *Coreopsis,* of
Wagner (Wagner *et al.* 1983) on long-term cytotaxonomic and
evolutionary studies of ferns, of Uhl (1982) on ploidy in
Echeveria and of Carr and Carr (1983) on chromosome races in
Calycadenia.

In Canada there is the thorough work of W. Grant on *Lotus*
and *Betula* in which he and his coworkers have stressed
cytological studies (Somaroo and Grant 1972) including the amounts
of DNA present (Taper and Grant 1973; Cheng and Grant 1973) as
well as crossing relationships (Alam and Grant 1972; Somaroo and
Grant 1972).

From Australia comes the cyto-biogeographic studies of the
mistletoes by Barlow and Wiens (1971).

In England there are diverse examples of biosystematic
investigations. Consider Manton's (1950) meticulous analysis of
the ploidy relationships and evolution of the British ferns which,
early on, set a very high standard and model for biosystematic
studies. Another example is Valentine's (1950) thoughtful study
of the differing role of interspecific compatibility in
differentiating species in different groups such as *Primula*
and *Viola* which raised troublesome questions of objectivity
vs. subjectivity. Recall also, how Bradshaw's (Alston and
Bradshaw 1966) studies on rapid evolution challenged conventional
ideas on the speed of speciation. And, Harbred's (Harbred and
McArthur 1980) studies of the Brassicaceae which show the
continuing effective use of cytological observations and genetic
experiments on biosystematic studies.

In Europe, there are the studies of Vida, for example, his

resynthesis of fern allopolyploid species (Lovis and Vida 1969),
the detailed cytotaxonomic studies of Gadella, e.g., on *Campanula*
and *Symphytum* (Gadella 1964; Gadella and Kliphuis 1969) of
Negrul (1968) on the evolution of *Vitis* and the remarkably
extensive investigations of Ehrendorfer on cytology, hybridization
and evolution in groups such as *Galium, Achillea* and *Anacyclus*
(Ehrendorfer, 1953 1961; Schweizer and Ehrendorfer 1976).

From Argentina comes the earlier work of Solbrig (1964) on
highly polymorphic *Gutierrezia sarothrae* which he found to
exhibit neither clines nor ecotypes but to consist of diploid and
tetraploid sibling species. Also from Argentina comes the work of
Hunziker (Hunziker *et al.* 1973; Wells and Hunziker 1976) on
the evolutionary and intercontinental relationships in
Hordeum and *Larrea*.

From Japan come the careful genetic studies and extensive field
work in eastern Asia, Australia and Africa of Morishima (Morishima
and Oka 1970), Oka (Oka *et al.* 1977; Oka 1978) and their
coworkers on the evolution of rice, particularly of the cultivated
varieties. Also from Japan come the fine scale karyotypic-
evolutionary analyses of *Trillium* - east Asian as well as
western and eastern North American - of Fukuda (Fukuda 1970;
Fukuda and Grant 1980) as well as the classic studies on the
evolution of wheat by Kihara (1959) continued by the elegant work
of Tsunewaki (Tsunewaki and Ogihara 1983) using cytoplasmic
constituents particularly restriction enzyme studies of
chloroplast DNA, to the same end.

The last of the selected examples comes from Israel in the work
of Zohary (1975) on phytogeography, biosystematics, etc. and that
of Nevo (Kahler *et al.* 1980) on the association of allozymes
and environment in races of *Avena barbata.*

What do these examples tell us of the development of
Biosystematics and its status in 1983? They paint a broad
and varied picture of Biosystematics. First, there are numerous
detailed cytological studies often of chromosome numbers, but also
of karyotypes, genome analyses, chromosome pairing, DNA content,
cytotypes, etc. We see extensive genetic studies of not only the
inheritance of clusters of characteristics, i.e., coherence and of
particular characteristics including allozymes and pigments, but
of intercompatibility, barriers to gene exchange and even the
reconstruction and hybridization of polyploid species. We see a
variety of ecologically oriented studies from the analysis of
clines, to transplant studies of ecotypes, to inter-continental
comparisons, to the study of pollination systems. Also, new
approaches and methods, such as allozyme comparisons are being
used increasingly. Biosystematics up to and including 1983 is
evolutionarily, often strongly evolutionarily, oriented. It is
based primarily on three main streams of experimental studies -
cytologic, genetic and ecologic - with numerous variations and
combinations. Much of the rest of this symposium will bring us to
the actual "state-of-the-art" of biosystematics in a broad
spectrum of approaches and plant groups.

Clearly biosystematists have stressed the experimental cytogenetic approach championed by Camp and Gilly (1943), the ecologic dimension characteristic of the Carnegie studies (Clausen *et al.* 1940) as well as the underlying philosophical framework of evolutionary change (Darwin 1859). However, the terminology of biosystematics except for some use of ecotype, semi-species, or cline is rarely used. Occasionally, we see "evolutionary species" suggesting that a taxon has achieved that level of differentiation. Often, we see "biological species" usually implying some experimental basis for its delimitation, or, "classical species," typically implying only a descriptive basis. For example, in the monograph of *Crepis* approximately 1/3 of the species are "biological" and 2/3 are "classical" (Babcock 1947a,b). We do not see the wide acceptance of any of the many, many specialized terms proposed over the years by so many biosystematists. The old term *species* is remarkably durable.

THE CHALLENGE OF NEW APPROACHES TO SYSTEMATICS

Earlier, I mentioned challenges to Biosystematics and we can see them emerging in the form of new approaches to systematics that go beyond the original cytologic, genetic and ecologic ones. The new approaches arose over the last several decades in the laudable, continuing search for greater objectivity. One of the first such efforts – made feasible by computer technology – is numerical taxonomy developed by Sneath and Sokal (1973). If one uses the 100+ characters they suggest, it is a reasonably objective and very useful approach (Ornduff and Crovello 1968). However, the method is not particularly sensitive to convergent evolution, sibling species or to isolating mechanisms, in general. It needs to be used with care in biosystematic studies as, for example, in the studies of W. Grant (Grant and Zandstra 1968; Koshy *et al.* 1972).

Chemotaxonomy was pioneered here at McGill by Gibbs (Iltis and Grant, personal commun.; see Gibbs 1974) and at Texas by Alston and Turner (1962, 1963a) in the further pursuit of objective measures of similarity, differences and relationships. Chemicals such as pigments which can be assayed readily as by conventional paper, thin layer, or more recently, high proficiency liquid chromatography (HPLC), are used to compare the populations, varieties, species or hybrids under study. For example, the species of *Capsicum* (Ballard *et al.* 1970), the species of *Lotus* (Harney and Grant 1965; Grant and Zalite 1966), the species of *Baptisia* and their naturally occurring interspecific hybrids (Turner and Alston 1959; Alston and Turner 1963b) and the species of *Lasthenia* and their artificial hybrids (Ornduff *et al.* 1973) could be accurately separated. Chemotaxonomy, while labor intensive, does add significantly to our knowledge of the biology of plants and their taxa.

Allozymic comparisons of relatedness as a taxonomic method grew

out of the early studies of Hubby and Lewontin (1966) and Harris
(1966). Their electrophoretic separation of enzymes revealed an
unsuspectedly high genetic variability. This is true in all
species studied (Selander *et al.* 1970). However, all this
variability enhances the usefulness of allozyme comparisons which
provide a relatively unbiased, objective measure of genetic
similarity of, and distance between, the populations, varieties or
species under study (Nei 1972). For example, Levin (1977) used
this method to distinguish between the entities comprising the
Phlox drummondii complex. Crawford and Smith (1982) used
allozyme comparisons to elucidate a progenitor-derivative species
pair in *Coreopsis*. Allard employed the method to make fine
distinctions between ecotypes (Hamrick and Allard 1972; Kahler
et al. 1980; Shumaker *et al.* 1982). Mitton and coworkers
(Mitton *et al.* 1979) were able to demonstrate altitudinal
clines and seasonal responses in Ponderosa Pines. Gottlieb (1982)
has shown the presence and emphasized the importance to
biosystematics of isozymes in several subcellular compartments
particularly the chloroplasts - recall also Tsunewaki's work.
McLeod, Eshbaugh and Guttman (1979) have been able to distinguish
the species of *Capsicum* electrophoretically. Kelley and
Adams (1977, 1978) have shown seasonal, inter- and intrapopulational variations in isozymes in *Juniperus*. Nevo and
Zohary (Nevo *et al.* 1982) have investigated the interaction
of environment and allozyme variation in wild wheat. All in all,
allozyme comparisons have become a richly rewarding method of
working out biological relationships.

DNA/DNA hybridization (McCarthy and Bolton 1963) is a means of
assessing the similarity of species at the molecular level of the
genetic material itself. Similarity is assessed by ascertaining
the percentage reannealing of the DNA (single-stranded) of one
species with the DNA (also, single-stranded) of another species.
The rapidly reannealing fraction comprises the short, highly
repetitive DNA generally thought to be concerned with gene
regulation (Britten and Davidson 1969, 1976; Rose and Doolittle
1983) whereas the slowly reannealing fraction represents the
unique, structural genes. In a study using the fast fraction
Bendich and McCarthy (1970) were able to show the close similarity
of *Secale* to *Triticum* and the more distant relationship
of *Hordeum*. Using the same methods, Goldberg *et al.* (1972)
compared the repetitive DNA's of a number of *Cucurbita*
species and the degrees of relatedness ascertained were in good
agreement with previous taxonomic studies. The proportion of
repetitive DNA varies greatly from plant group to plant group
(Flavel *et al.* 1974) and correlates positively with the
quantity of nuclear DNA present (Mikschi and Hotta 1973). In a
study of slow-annealing, unique DNA in *Atriplex*, Belford and
Thompson (1981a,b) were able to trace elegantly the relationships
of the species of *Atriplex*. Studies involving either
repetitive or unique DNA sequences assess relationships on the
basis of the similarity of classes of genes - a refinement over

cytogenetic analyses of chromsomal homologies. However, DNA/DNA hybridization studies are slow and time consuming to perform.

Comparative protein sequencing (Ingram 1963) provides a tool for the direct comparison of gene products at the molecular level. Cytochrome-C has been sequenced for a number of organisms and the number of differences in amino acids used to construct a dendrogram of evolutionary relationships (Dobzhansky *et al.* 1977; Demoulin 1981; Baba *et al.* 1981). Since genes evolve at different rates, it soon became apparent that 6 or 8 suitable proteins would need to be sequenced and compared and the results combined to construct a more probable phylogenetic tree of the groups under study. Protein sequencing, while fundamentally informative, is too expensive and slow to be widely used.

The new approaches to classification of *numerical taxonomy, chemotaxonomy, comparative allozymes, DNA/DNA hybridization and comparative protein structure* are viewed by many workers as extensions of Biosystematics, but by others as separate, distinct approaches. So, here is a very real problem. Are the new approaches - one, or all - competing with biosystematics? Or, are they complementary to biosystematics? Or, are they actually part of biosystematics? Presently, various authors (Potvin *et al.* 1983; Soltis *et al.* 1983) are assessing these problems and possibilities. However, let me summarize my own more comprehensive study comparing four of the new approaches with the standard, cytogenetic-biosystematic approach.

CRITICAL COMPARISONS OF NEW APPROACHES USING SECTION *ERYTHRANTHE* OF THE GENUS *MIMULUS*

Wullstein (1980) and I (Vickery 1978; Vickery and Wullstein 1981) analyzed nine populations that well represent all the six species and two varieties of section *Erythranthe* of the genus *Mimulus* (Table I). The two varieties of *M. lewisii* Pursh. are pink-flowered whereas all the other species are bright red-flowered. The group ranges from Alaska to central Mexico and from the Pacific to the Rocky Mountains. We studied the same set of nine populations using (1) numerical taxonomic, (2) chemotaxonomic, (3) comparative allozyme, (4) DNA/DNA hybridization and (5) standard cytogenetic approaches. The numerical analysis distinguished four groups (Fig. 1). The chemotaxonomic investigation based on floral pigments delimited two groups. The comparative allozyme study also segregated two groups, but different ones. The DNA/DNA hybridization experiments based on fast annealing, repetitive DNA's grouped most of the section together except for three of the red-flowered taxa. lastly, the cytogenetic study revealed $n = 8$ in all forms but a spectrum of F_1 hybrid fertilities that revealed an interfertile group of five taxa. Overall, the five approaches yield results that are only partially congruent (Fig. 2). The two pink-flowered *M. lewisii* varieties form a distinctive group. The rare endemic, *M. rupestris*, is only loosely tied in to the complex.

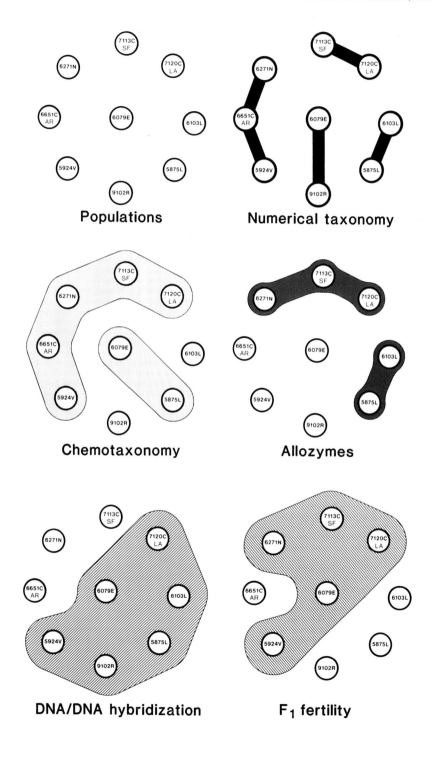

Populations

Numerical taxonomy

Chemotaxonomy

Allozymes

DNA/DNA hybridization

F₁ fertility

TABLE I. *Populations of section* Erythranthe *of the genus*
Mimulus *used in the comparative study of the* Numerical
taxonomic, Chemotaxonomic, Comparative allozyme, DNA/DNA
hybridization *and* F_1 fertility (cytogenetic) *approaches
to classification.*

Mimulus cardinalis Douglas
     6651C   Santa Catalina Mtns. Pima Co., Arizona, USA
     7113C   Los Trancos Creek, San Mateo Co., California, USA
     7170C   San Antonio Peak, Los Angeles Co., California

M. eastwoodiae Rydberg
     6079E   Bluff, San Juan Co., Utah, USA

M. lewisii Pursh.
     5875L   Albion Basin, Salt Lake Co., Utah
     6103L   Ice Lake, Placer Co., California

M. nelsonii Grant
     6271N   Devil's Backbone, Durango, Mexico

M. rupestris Greene
     9102R   Sierra de Tepotzlan, Morelos, Mexico

M. verbenaceus Greene
     5924V   Grand Canyon, Coconico Co., Arizona

In a sense each approach views the complex from its own
perspective. The numerical approach "sees" phenetic similarity,
possibly the result of a common evolutionary history and of
similar environmental pressures. The chemotaxonomic approach
"sees" the floral pigment differences and probably reflects
differing pollination systems as well. The comparative allozymes
may reflect common descent and similar environmental pressures
also. The DNA/DNA hybridization results "see" the similarity of
fast-annealing DNA, i.e. presumably of regulatory genes and

FIG. 1. Upper left: *Populations of section* Erythranthe
under study (see Table I). Upper right: *Groupings of most
closely related populations according to Numerical Taxonomy.*
Center left: *Two groups of most similar populations based
on Chemotaxonomy.* Center right: *Two groups of most similar
populations according to their Allozymes.* Lower left: *Six
populations indistinguishable on the basis of their repetitive
DNA/DNA hybridization results.* Lower right: *Five fully
inter-fertile populations based on the fertilities of the
interpopulation* F_1 *hybrids. Similarities are based on dendrograms
of relationships for each approach (Vickery and Wullstein 1981).*

suggest close, basic genetic similarity. The cytogenetic results "see" the location of barriers to gene exchange. *Each approach adds to our information about the biology of the group - often permitting distinctions to be drawn between entities that other approaches did not resolve. Clearly, they all are complementary. Therefore, it seems fair to conclude that they all can be thought of as part of biosystematics in the broad sense.* I think this study of ours nicely illustrates how the concept of Biosystematics has broadened over the years to include, legitimately, any and all new approaches that add to our understanding of the biological and evolutionary relationship of the populations and species under study.

THE NEWEST APPROACH - COMPARATIVE, MOLECULAR EVOLUTION OF GENES

What are the new opportunities I alluded to? They arise from the rapidly advancing field of molecular genetics (Watson and Tooze 1981). For example, the sequences of bases of the DNA of a gene can be worked out (Maxam and Gilbert 1980) more readily now than the sequence of amino acids in a protein (Watson and Tooze 1981). Techniques using reverse transcriptase permit the construction of a complementary DNA (cDNA) molecule from a messenger RNA molecule. Comparisons of such cDNA molecules to their genes reveal that much of the DNA of a gene is not translated into protein in higher organisms. For example, the ovalbumin gene is five times the length of its messenger RNA (Chambon 1981). Soon, it was found that the genes of higher organisms typically consist of a promoter region followed by a series of expressed sequences (exons) separated by nonexpressed intervening sequences (introns) that are spliced out in the process of maturing the active messenger RNA. This complex structure of eukaryotic genes means that mutation of a base pair may have many more effects than was previously thought possible. If the mutation changed a splicing recognition site the mRNA and gene product might be greatly lengthened or shortened. A single mutation can lead to a large, sudden - possibly functional - change in a gene product. We need to keep this in mind in assessing biosystematic relationships. For example, Rabinow (oral commun.) has found that the alcohol dehydrogenase (ADH) genes of two species of *Drosophila* differ only by one or two such small mutations. The genes have two promoter regions. The effect of the small mutations is to put the first promoter in an intron in one species and express the rest of the gene and to put the second promoter in an intron instead, in the second species and express the somewhat different residue of the gene thus leading to two significantly different forms of the important ADH enzyme being produced from very slightly modified genes. And, such differences in the structure of the ADH gene are common in *Drosophila* (Langley *et al.* 1982).

The molecular geneticist's discovery of restriction enzymes (Linn and Arber 1968) has made possible an important biosystematic approach to the study of relationships. Treating the DNA of

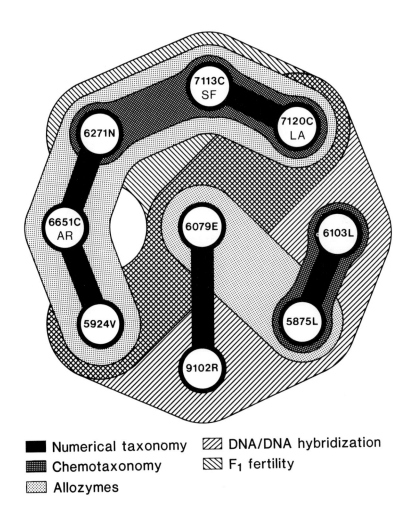

FIG. 2. *Comparative groupings of the nine taxa of section* Erythranthe *by the 5 different approaches to classification. Note the close relationships of 5875L and 6103L, the two varieties of* M. Lewisii *and of 7113C and 7120C, the two California populations of* M. cardinalis. *Note also the relative distinctness of the Arizona variety of* M. cardinalis, *6651C and of the rare, endemic* M. rupestris, *9102R.*

species A with a particular restriction enzyme, e.g. *Eco R1*,
cleaves the DNA wherever a specific sequence of bases occurs (G-C,
A-T, A-T, T-A, T-A, C-G) thus producing a characteristic set of
DNA fragments. The same may be done for species B and the
resulting sets carefully compared electrophoretically for
differences. This was the basis for Langley's (Langley *et al.*
1982) identification of different ADH genes in *Drosophila*.
Restriction maps are becoming increasingly common (Lansman *et al.*
1981). Consider, for example, the beautiful study of Palmer and
Zamir (1982) using 25 restriction enzymes to cut chloroplast DNA
and compare the various entities of *Lycopersicon* and work out
their phylogeny. Incidentally, the study showed that *Solanum
pennellii* is, in fact, a *Lycopersicon* and not a *Solanum*.

The restriction enzyme technology has made possible the cloning
of eukaryotic genes in plasmid or phage lambda vectors (Watson and
Tooze 1981). Cloned genes may be grown in quantity and used as
radioactive probes to locate the actual locus of the gene in the
genome. The results are surprising! Often there are several
loci, that is, there is a "family" of genes consisting of three
classes of genes: active ones, silent ones and "shipwrecked" or
"fossil" genes (pseudogenes). Silent genes may be turned on as
Roth and coworkers (Roth and Schmid 1981; Ciampi *et al.* 1982)
have shown when a transposon brings a promotor region near a
silent gene and actually turns it on. So, silent genes can be
reactivated leading to the reappearance, amplification or
modification of a trait. And, the transposons themselves evolve
(Biel and Hartl 1983). Thompson and coworkers at Carnegie
(Stanford) have carefully analyzed slow annealing genes
(structural genes) in Peas and Mung Beans and find families of
genes consisting of both active genes and ones that have mutated
significantly and are good examples of fossil genes (Murray *et
al.* 1981; Preisler and Thompson 1981a,b). They find evidence
in them of ancient repeats and substitutions. Gojobori *et al.*
(1982) find preferential directions for substitutions in active
genes compared to fossil ones leading to significant differences
in the A-T/G-C ratios in the two groups. Since DNA structure is
persistent and long-lived (Dickinson oral commun.) each organism,
each plant carries its own "fossil record!"

What are the opportunities molecular genetics holds for
biosystematics? Clearly, we are on the threshold of a whole new
era of understanding of the molecular events that underlie
character and species differences. Our better understanding of
the structure of genes and the nature of mutations will help
clarify our understanding of relationships. Restriction enzyme
analysis of DNA sequence similarities and differences is a
powerful new approach to the study of the relationships of
moderate length DNA species, e.g., those of genes, chlorpolasts
and mitochondria. The finding of families of genes including the
ancient "shipwrecked" or fossil genes means that we have *a
built-in molecular fossil record that is of vast potential use in
elucidating the phylogenies of groups of plants*. These methods

are hardly the tools of working systematists, but they can be
used, and are being used, in critical and important cases. They
are becoming fundamental to a real understanding of the biology of
speciation.

In conclusion, Biosystematics 1983 has assimilated the
challenges of new approaches to form a much more broadly based
experimental, evolutionary approach to classification. The
opportunities provided by the methods of molecular genetics are
opening up exciting theoretical vistas for the years ahead as well
as providing the possibilities and capabilities for solving
intricate problems of relationships and phylogenies. Ours is a
healthy, growing science, "an unending synthesis," as Constance
(1964) so aptly described systematics.

REFERENCES

Alam, M.T. and Grant, W.F. 1972. Interspecific hybridization in
    birch (*Betula*). *Naturaliste Can.* 99: 33-40.
Alston, J.L. and Bradshaw, A.D. 1966. Evolution in closely
    adjacent plant populations. II. *Agrostis stolonifera* in
    maritime habitats. *Heredity* 21: 649-664.
Alston, R.E. and Turner, B.L. 1962. New techniques in analysis of
    complex natural hybridization. *Proc. Natl. Acad. Sci. U.S.A.*
    48: 130-137.
Alston, R.E. and Turner, B.L. 1963a. *Biochemical Systematics.*
    Prentice-Hall, Englewood Cliffs, N.J. 494 pp.
Alston, R.E. and Turner, B.L. 1963b. Natural hybridization among
    four species of *Baptisia* (Leguminosae). *Am. J. Bot.* 50:
    159-173.
Anderson, E. 1949. *Introgressive Hybridization.* Wiley, New
    York. 109 pp.
Baba, M.L., Darga, L.L., Goodman, M. and Czelusniak, J. 1981.
    Evolution of cytochrome C investigated by the maximum parsimony
    method. *J. Mol. Evol.* 17: 197-213.
Babcock, E.B. 1920. *Crepis* - a promising genus for genetic
    investigations. *Am. Nat.* 54: 270-276.
Babcock, E.B. 1931. Cytogenetics and the species concept. *Am.
    Nat.* 65: 5-18.
Babcock, E.B. 1947a. The genus *Crepis.* Part I. The taxonomy,
    phylogeny, distribution and evolution of *Crepis.* Univ.
    Calif. Publ. Bot. No. 21. 197 pp.
Babcock, E.B. 1947b. The genus *Crepis.* Part II. Systematic
    treatment. Univ. Calif. Publ. Bot. No. 22 1030 pp.
Baker, H.G. 1954. The experimental taxonomy of *Armeria
    maritima* (Mill.) Willd. and its close relatives. Rapp. Comm.
    8eme Congr. Int. Bot., Paris, sect. 10: 190-191.
Baker, H.G. and Baker, I. 1977. Intraspecific constancy of floral
    nectar amino acid compliments. *Bot. Gaz.* 183: 183-191.
Baker, H.G. and Stebbins, G.L. 1965. *The Genetics of Colonizing
    Species.* Academic Press, New York, 588 pp.
Ballard, R.E., McClure, J.W., Eshbaugh, W.H. and Wilson, K.G.

1970. A chemosystematic study of selected taxa of *Capsicum*. *Am. J. Bot.* 57: 225–233.

Barlow, B.A. and Wiens, D. 1971. The cytogeography of the Loranthaceous mistletoes. *Taxon* 20: 291–312.

Bartholomew, B., Eaton, L.C. and Raven, P.H. 1973. *Clarkia rubicunda:* A model of plant evolution in semiarid regions. *Evolution* 27: 505–517.

Belford, H.S. and Thompson, W.F. 1981a. Single copy DNA homologies in *Atriplex*. I. Cross reactivity estimates and the role of deletions in genome evolution. *Heredity* 46: 91–108.

Belford, H.S. and Thompson, W.F. 1981b. Single copy DNA homologies in *Atriplex*. II. Hybrid thermal stabilities and molecular phylogeny. *Heredity* 46: 109–122.

Bell, C.R. and Constance, L. 1957. Chromosome numbers in Umbelliferae. *Am. J. Bot.* 44: 565–572.

Bendich, A. and McCarthy, B.J. 1970. DNA comparisons among barley, oats, rye, and wheat. *Genetics* 65: 545–565.

Bentham, G. and Hooker, S.D. 1862–1883. *Genera Plantarium ad Imprimis in Herbarus Kewensibus Serva Difinita*. 3 vols., London.

Bessey, C.E. 1915. Phytogenetic taxonomy of flowering plants. *Ann. Mo. Bot. Gard.* 2: 1–155.

Biel, S. and Hartl, D.L. 1983. Evolution of transposons: Natural selection for Tn5 in *Escherichia coli*. *Genetics* 103: 581–592.

Böcher, T.W. 1949. Racial divergences in *Prunella vulgaris* in relation to habitat and climate. *New Phytol*. 48: 285–314.

Böcher, T.W. 1966. Experimental and cytological studies on plant species. XI. North Atlantic tetraploids of the *Campanula rotundifolia* complex. *Ann. Bot. Fenn.* 3: 287–298.

Böcher, T.W. 1972. Variational pattern in *Clinopodium vulgare* L. *Symp. Biol. Hung.* 12: 23–29.

Briggs, F.N. and Knowles, P.F. 1967. *Introduction to Plant Breeding*. Reinhold, New York. 426 pp.

Britten, R.J. and Davidson, E.H. 1969. Gene regulation for higher cells: A theory. *Science* 165: 349–357.

Britten, R.J. and Davidson, E.H. 1976. DNA sequence arrangement and preliminary evidence on its evolution. *Fed. Proc.* 35: 2151–2157.

Camp, W.H. 1942. A survey of the American species of *Vaccinium,* subgenus *Euvaccinium*. *Brittonia* 4: 205–247.

Camp, W.H. and Gilly, C.L. 1943. The structure and origin of species. *Brittonia* 4: 323–385.

Carr, R.L. and Carr, G.D. 1983. Chromosome races and structural heterozygosity in *Calycadenia ciliosa* Greene (Asteraceae). *Am. J. Bot.* 70: 744–755.

Chambon, P. 1981. Split genes. *Sci. Am.* 244: 60–71.

Chetverikov, S.S. 1927. On the genetic constitution of wild populations. In *Evolutionary Genetics* (D.L. Jameson, ed.), Dowden, Hutchinson & Ross, Stroudsburg, Pennsylvania. 332 pp.

Cheng, R.I–J. and Grant, W.F. 1973. Species relationships in the

Lotus corniculatus group as determined by karyotype and
cytophotometric analyses. *Can. J. Genet. Cytol.* 15: 101–115.

Ciampi, M.S., Schmid, M.B. and Roth, J.R. 1982. Transposon Tn10
provides a promoter for transcription of adjacent sequences.
Proc. Natl. Acad. Sci. U.S.A. 79: 5016–5020.

Clausen, J. 1926. Genetical and cytological investigations on
Viola tricolor L. and *V. arvensis* Murr. *Hereditas* 8: 1–156.

Clausen, J. and Hiesey, W.M. 1958. Experimental studies on the
nature of species. IV. Genetic structure of ecologic races.
Carnegie Inst. Wash. Publ. 615, Washington, D.C. 312 pp.

Clausen, J., Keck, D.D. and Hiesey, W.M. 1934. Experimental
Taxonomy. Carnegie Inst. Wash. Year Book No. 33, 173–177.

Clausen, J., Keck, D.D. and Hiesey, W.M. 1939. The concept of
species based on experiment. *Am. J. Bot.* 26: 103–106.

Clausen, J., Keck, D.D. and Hiesey, W.M. 1940. Experimental
studies on the nature of species. I. Effect of varied
environments on western North American plants. Carnegie Inst.
Wash. Publ. 520, Washington, D.C. 452 pp.

Clausen, J., Keck, D.D. and Hiesey, W.M. 1945. Experimental
studies on the nature of species. II. Plant evolution through
amphiploidy and autoploidy, with examples from the Madiinae.
Carnegie Inst. Wash. Publ. 564, Washington, D.C. 174 pp.

Clausen, J., Keck, D.D. and Hiesey, W.M. 1948. Experimental
studies on the nature of species. III. Environmental responses
of climatic races of *Achillea*. Carnegie Inst. Wash. Publ.
581, Washington, D.C. 129 pp.

Cleland, R.E. 1954. Evolution of the North American
Euoenotheras: the *Strigosas*. *Proc. Am. Phil. Soc.* 99: 189–203.

Constance, L. 1964. Systematic botany – an unending synthesis.
Taxon 13: 257–273.

Crawford, D.J. and Smith, E.B. 1982. Allozyme variation in
Coreopsis nuecensoides and *C. nuecensis* (Compositae), A
progenitor – derivative species pair. *Evolution* 36: 379–386.

Darwin, C. 1859. *The Origin of Species by Means of Natural
Selection.* Murray, London (facsimile ed. 1964, Harvard).

De Candolle, A. 1824–1873. *Prodromus Systematis Naturalis
Regni Vegetablis.* 17 vols. Masson et cie., Paris.

Demoulin, V. 1981. Cytochrome C sequence data and the origin of
higher fungi. XIII Bot. Congr. Sydney, Australia. p. 142.

Dobzhansky, Th., Ayala, F.J., Stebbins, G.L. and Valentine, J.W.
1977. *Evolution.* Freeman, San Francisco. 572 pp.

Ehrendorfer, F. 1953. Okologisch-geographische mikro-
differenzierung einer population von *Galium pumilum* Murr.
s. str. *Osterr. Bot. Z.* 100: 616–638.

Ehrendorfer, F. 1961. Akzeessorische chromosomen bei *Achillea:*
Struktor, cytologisches verhalten, zahlenmassige instabilitat
und entstehung. *Chromosoma* 11: 523–552.

Ehrlich, P.R. and Raven, P.H. 1965. Butterflies and Plants: a
study in coevolution. *Evolution* 18: 586–608.

Eisley, L. 1961. *Darwin's Century*. Anchor, Doubleday, N.Y.

Ellerström, S. and Sjödin, J. 1966. Frequency and vitality of

aneuploids in a population of tetraploid Red Clover.
Hereditas 55: 166–182.

Engler, A. and Prantl, K. 1897–1915. *Die naturlichen Pflanzenfamilien.* Leipzig.

Flavell, R.B., Bennett, M.D., Smith, J.B. and Smith, D.B. 1974. Genome size and the proportion of repeated nucleotide sequence DNA in plants. *Biochem. Genet.* 12: 257–269.

Fukuda, I. 1970. Chromosome compositions of the southern populations in *Trillium kamtschaticum.* Sci. Reports nos. 12–14, Tokyo Woman's Christian College, Tokyo.

Fukuda, I. and Grant, W.F. 1980. Chromosome variation and evolution in *Trillium grandiflorum. Can. J. Genet. Cytol.* 22: 81–91.

Gadella, Th. W.J. 1964. Cytotaxonomic studies in the genus *Campanula. Wentia* 11: 1–104.

Gadella, Th. W.J. and Kliphuis, E. 1969. Cytotaxonomic studies in the genus *Symphytum.* II. Crossing experiments between *Symphytum officinale* L. and *Symphytum asperum* Lepech. *Acta. Bot. Neerl.* 18: 544–549.

Gerard, J. 1597. *The herball or General Historie of Plants.* Norton, London. 1392 pp.

Gibbs, R.D. 1974. *Chemotaxonomy of Flowering Plants.* McGill-Queen's Univ. Press, Montreal.

Gojobori, T., Li, W.H. and Graur, D. 1982. Patterns of nucleotide substitution in pseudogenes and functional genes. *J. Mol. Evol.* 18: 360–369.

Goldberg, R.B., Bemis, W.P. and Siegel, A. 1972. Nucleic acid hybridization studies within the genus *Cucurbita. Genetics* 72: 253–266.

Goodspeed, T.H. 1923. A preliminary note on the cytology of *Nicotiana* species and hybrids. *Sv. Bot. Tidskr.* 17: 472–478.

Goodspeed, T.H. 1938. Three new species of *Nicotiana* from Peru. *Univ. Calif. Publ. Bot.* 18: 137–151.

Goodspeed, T.H. 1954. The genus *Nicotiana.* Chronica Botanica, Waltham, Mass. 536 pp.

Gottlieb, L.D. 1982. Conservation and duplication of isozymes in plants. *Science* 216: 373–380.

Gottlieb, L.D. and Bennett, J.P. 1983. Interference between individuals in pure and mixed cultures of *Stephanomeria malheurensis* and its progenitor. *Am. J. Bot.* 70: 276–284.

Grant, V. 1950. Genetic and taxonomic studies in *Gilia* I. *Gilia capitata. El Aliso* 2: 239–316.

Grant, V. 1957. The plant species in theory and practice. In *The Species Problems,* Am. Assoc. Advan. Sci., Washington, D.C. pp. 39–80.

Grant, V. and Grant, A. 1960. Genetic and taxonomic studies in *Gilia* XI. Fertility relationships of the diploid cobwebby Gilias. *El Aliso* 4: 435–481.

Grant V. and Grant, K.A. 1980. Clonal microspecies of hybrid origin in the *Opuntia lindheimeri* group. *Bot. Gaz.* 141: 101–106.

Grant, V. and Grant, K.A. 1982. Natural pentaploids in the *Opuntia lindheimeri-phaeacantha* group in Texas. *Bot. Gaz.* 143: 117-120.

Grant, W.F. and Zalite, I.I. 1966. The cytogenetics of *Lotus*. XII. Thin layer chromatography in the separation of secondary phenolic compounds in *Lotus* (Leguminosae). *J. Chromatog.* 24: 243-244.

Grant, W.F. and Zandstra, I.I. 1968. The biosystematics of the genus *Lotus* (Leguminosae) in Canada. II. Numerical chemotaxonomy. *Can. J. Bot.* 46: 585-589.

Gregor, J.W. 1939. Experimental taxonomy. IV. Population differentiation in North American and European Sea Plantains allied to *Plantago maritima* L. *New Phytol.* 38: 293-322.

Gunther, R.T. 1959. *The Greek Herbal of Dioscorides*. Halfner, New York. 701 pp.

Hall, H.M. 1926. The taxonomic treatment of units smaller than species. *Proc. Int. Congr. Plant Sci.* 2: 1461-1468. Ithaca, N.Y.

Hall, H.M. and Clements, F.E. 1923. The phylogenetic method in taxonomy. Carnegie Inst. Wash. Pub. 326, 355 pp. Washington.

Hamrick, J.L. and Allard, R.W. 1972. Microgeographic variation in allozyme frequencies in *Avena barbata*. *Proc. Natl. Acad. Sci. U.S.A.* 69: 2100-2104.

Harberd, D.J. and McArthur, E.D. 1980. Meiotic analysis of some species and genus hybrids in the Brassicaceae. In Brassica *Crops and Wild Allies: Biology and Breeding*. (S. Tsuneda, K. Hinata and C. Gomez-Campo, eds.). Japan Sci. Soc. Press, Tokyo. 354 pp.

Harney, P.M. and Grant, W.F. 1965. A polygonal presentation of chromatographic investigations on the phenolic content of certain species of *Lotus*. *Can. J. Genet. Cytol.* 7: 40-51.

Harris, H. 1966. Enzyme polymorphisms in man. *Proc. R. Soc.* London, *Ser. B.* 164: 298-310.

Hiesey, W.M. Nobs, M.A. and Bjorkman, O. 1971. Experimental studies in the nature of species. V. Biosystematics, genetics, and physiological ecology of the *Erythranthe* section of *Mimulus*. Carnegie Inst. Wash. Publ. 628, Washington, D.C.

Hiesey, W.M. and Nobs, M.A. 1982. Experimental studies in the nature of species. VI. Interspecific hybrid derivatives between facultatively apomictic species of Bluegrasses and their responses to contrasting environments. Carnegie Inst. Wash. Publ. 636, Washington, D.C. 119 pp.

Hubby, J.L. and Lewontin, R.C. 1966. A molecular approach to the study of genetic heterozygosity in natural populations. I. The number of alleles at different loci in *Drosophila pseudoobscura*. *Genetics* 54: 577-594.

Hunziker, J.H., Naranjo, C.A., y Zeiger, E. 1973. Las relaciones evolutivas entre *Hordeum compressum* y otras especies diploides Americanas afines. *Kurtziana* 7: 7-12.

Ingram, V.M. 1963. The hemoglobins in genetics and evolution. Columbia Univ. Press, New York. 165 pp.

Kahler, A.L., Allard, R.W., Krzakowa, M., Wehrhahn, C.F. and Nevo, E. 1980. Associations between isozyme phenotypes and envornment in the Slender Wild Oat (*Avena barbata*) in Israel. *Theor. Appl. Genet.* 56: 31–47.

Keck, D.D. 1936. Studies in pensteman. II. The section *Hesperothamnus*. *Madrono* 3: 200–219.

Kelley, W.A. and Adams, R.P. 1977. Seasonal variation of isozymes in *Juniperus scopulorum*: Systematic significance. *Am. J. Bot.* 64: 1092–1096.

Kelley, W.A. and Adams, R.P. 1978. Analysis of isozymes variation in natural populations of *Juniperus ashei*. *Rhodora* 80: 107–134.

Kihara, H. 1959. Fertility and morphological variation in the substitution and restoration backcrosses of the hybrids *Triticum vulgare* X *Aegilops caudata*. *Proc. Xth Int. Cong. Genet.* 1: 142–171.

Koshy, T.K., Grant, W.F. and Brittain, W.H. 1972. Numerical chemotaxonomy of the *Betula caerulea* compex. *Symp. Biol. Hung.* 12: 201–211.

Langley, C.H., Montgomery, E. and Quottlebaum, W.F. 1982. Restrtiction map variants in the *Adh* region of *Drosophila*. *Proc. Natl. Acad. Sci. U.S.A.* 79: 5631–5635.

Lansman, R.A., Shade, R.O., Shapira, J.F. and Avise, J.C. 1981. The use of restriction endonucleases to measure mitochondrial DNA sequence relatedness in natural populations III. Techniques and potential applications. *J. Mol. Evol.* 17: 214–226.

Lawrence, G.H.M. 1951. *Taxonomy of Vascular Plants*. MacMillan, New York. 823 pp.

Levin, D.A. 1977. The organization of genetic variability in *Phlox drummondii*. *Evolution* 31: 477–494.

Lewis, H. 1962. Catastrophic selection as a factor in speciation. *Evolution* 16: 257–271.

Lewis, H. 1966. Speciation in flowering plants. *Science* 152: 167–172.

Lewis, H. and Raven, P.H. 1958. Rapid evolution in *Clarkia*. *Evolution* 12: 319–336.

Lewis, H. and Roberts, M.R. 1956. The origin of *Clarkia lingulata*. *Evolution* 10: 126–138.

Lewis, W.H. 1959. A monograph of the genus *Rosa* in North America. I. *R. acicularis*. *Brittonia* 10: 1–24.

Lewis, W.H. ed. 1980. *Polyploidy, Biological Relevance*. Plenum Press, New York. 583 pp.

Linhart, Y.B. 1973. Ecological and behavioral determinants of pollen dispersal in Hummingbird-pollinated *Heliconia*. *Am. Nat.* 107: 511–523.

Linn, S. and Arber, W. 1968. Host specificity of DNA produced by *Escherichia coli* X. *In vitro* restriction of phage fd replication form. *Proc. Natl. Acad. Sci. U.S.A.* 59: 1300–1306.

Linné, C.A. ca. 1737. *Systema Natura*. Gmelin, Holland, Vol. 2, 1661 pp.

Linné, C.A. 1778. *Genera Plantarum*. 9th ed. Francofurti ad Moenum. 571 pp. and index.

Löve, A. 1983. IOPB Chromosome number reports LXXIX. *Taxon* 32: 320-324.

Löve, D. 1975. The genus *Acanthoxanthium* (DC.) Fourr. revived. *Lagascalia* 5: 55-71.

Löve, A. and Kjellquist, E. 1964. Chromosome numbers of some Iberian Cistaceae. *Port. Acta Biolog.* 8: 69-80.

Löve, A. and Löve, D. 1956. Cytotaxonomical conspictus of the Icelandic flora. *Acta Horti Gotob.* 20: 65-290.

Löve, A. and Löve, D. 1957. Arctic polyploidy. *Proc. Genet. Soc. Can.* 2: 23-27.

Löve, A. and Löve, D. 1973. Cytotaxonomy of the boreal taxa of *Phyllitis*. *Acta. Bot. Acad. Sci. Hung.* 19: 201-206.

Löve, A. and Solbrig, O.T. 1964. 1OPB Chromosome number reports I. *Taxon* 13: 99-110.

Lovis, J.D. and Vida, G. 1969. The resynthesis and cytogenetic investigation of *Asplenophyllitis microdon* and *A. jacksonii*. *Brit. Fern Gaz.* 10: 53-67.

Manton, I. 1950. *Problems of Cytology and Evolution in the Pteridophyta*. Cambridge Univ. Press, Cambridge. 316 pp.

Maxam, A.M. and Gilbert, W. 1980. Sequencing end-labeling DNA with base-specific chemical cleavages. *Methods Enzymol.* 65: 499-560.

McCarthy, B.J. and Bolton, E.T. 1963. An approach to the measurement of genetic relatedness among organisms. *Proc. Natl. Acad. Sci. U.S.A.* 50: 156-164.

McLeod, M.J., Eshbaugh, W.H. and Guttman, S.I. 1979. An electrophoretic study of *Capsicum* (Solanaceae): The purple flowered taxa. *Bull. Torrey Bot. Club* 106: 326-333.

Mez, C. 1926. Die bedeutung der serodiagnostik fur die stammesgeschichtlich forchung. *Bot. Arch.* 16: 1-23.

Miksche, J.P. and Hotta, Y. 1973. DNA base composition and repetitious DNA in several conifers. *Chromosoma* 41: 29-36.

Millar, C.I. 1983. A steep cline in *Pinus muricata*. *Evolution* 37: 311-319.

Mitton, J.B., Linhart, Y.B., Sturgeon, K.B. and Hamrick, J.L. 1979. Allozyme polymorphisms detected in mature needle tissue of Ponderosa Pine. *J. Hered.* 70: 856-889.

Moore, D.M. and Raven, P.H. 1970. Cytogenetics, distribution and amphitropical affinities of South American *Camissonia* (Onagraceae). *Evolution* 24: 816-823.

Morishima, H. and Oka, H.I. 1970. A survey of genetic variations in the populations of wild *Oryza* species and their cultivated relatives. *Jpn. J. Genet.* 45: 371-385.

Müntzing, A. 1932. Cyto-genetic investigations on synthetic *Galeopsis tetrahit*. *Hereditas* 16: 105-154.

Murray, M.G., Peters, D.L. and Thompson, W.F. 1981. Ancient repeated sequences in the Pea and Mung Bean genomes and implications for genome evolution. *J. Mol. Evol.* 17: 31-42.

Negrul, A.M. 1968. Origin and evolution of cultivated grape vine.

Proc. XII Int. Cong. Genet. Tokyo 1: 322

Nei, M. 1972. Genetic distance between populations. *Am. Nat.* 106: 283–292.

Nevo, E., Golenberg, E., Berles, A., Brown, A.H.D. and Zohary, D. 1982. Genetic diversity and environmental associations of wild wheat, *Triticum dicoccoides* in Israel. *Theor. Appl. Genet.* 62: 241–254.

Nordenskiöld, H. 1951. Cyto-taxonomic studies in the genus *Luzula* L. *Hereditas* 37: 325–355.

Nygren, A. 1962. Artificial and natural hybridization in European *Calamagrostis*. *Symb. Bot. Ups.* 17: 1–105.

Oka, H.I. 1978. An observation of wild rice species in tropical Australia. Contr. Natl. Inst. Genet. Japan. No. 1216. 26 pp.

Oka, H.I., Morishima, H., Sano, Y. and Koizumi, T. 1977. Observations of rice species accompanying savanna plants on the southern fringe of Sahara desert. Contr. Natl. Inst. Genet. Japan. No. 1215. 94 pp.

Ornduff, R. 1969. Reproductive biology in relation to systematics. *Taxon* 18: 121–133.

Ornduff, R. 1975. Heterostyly and pollen flow in *Hypericum aegypticum* (Guttiferae). *Bot. J. Linn. Soc.* 71: 51–57.

Ornduff, R. 1976. Speciation and oligogenic differentiation in *Lasthenia*. *Syst. Bot.* 1: 91–96.

Ornduff, R., Bohm, B.A. and Saleh, N.A.M. 1973. Flavonoids of artificial interspecific hybrids in *Lasthenia*. *Biochem. Syst.* 1: 147–151.

Ornduff, R. and Crovello, T.J. 1968. Numerical taxonomy of Limnanthaceae. *Am. J. Bot.* 55: 173–182.

Ornduff, R. and Weller, S.G. 1975. Pattern diversity of incompatibility groups in *Jepsonia heterandra* (Saxifragaceae). *Evolution* 29: 373–375.

Ownbey, M. 1950. Natural hybridization and amphiploidy in the genus *Tragopogon*. *Am. J. Bot.* 37: 487–499.

Palmer, J.D. and Zamir, D. 1982. Chloroplast DNA evolution and phylogenetic relationships in *Lycopersicon*. *Proc. Natl. Acad. Sci. U.S.A.* 79: 5006–5010.

Peterson, F.P. and Fairbrothers, D.E. 1983. A serotaxonomic appraisal of *Amphipterygium* and *Leitneria* – two amentiferous taxa of Rutiflorae (Rosidae). *Syst. Bot.* 8: 134–148.

Potvin, C., Bergeron, Y. and Simon, J.P. 1983. A numerical taxonomic study of selected *Citrus* species (Rutaceae) based on biochemical characters. *Syst. Bot.* 8: 127–133.

Preseisler, R.S. and Thompson, W.F. 1981a. Evolutionary sequence divergence within repeated DNA families of higher plant genomes. I. Analysis of reassociation kinetics. *J. Mol. Evol.* 17: 78–84.

Preseisler, R.S. and Thompson, W.F. 1981b. Evolutionary sequence divergence within repeated DNA families of higher plant genomes. II. Analysis of thermal denaturation. *J. Mol. Evol.* 17: 85–93.

Raven, P.H. 1975. The bases of Angiosperm phylogeny: Cytology.

Ann. Mo. Bot. Gard. 62: 724-764.

Raven, P.H. 1977. The systematics and evolution of higher plants. The Changing Scenes in Natural Sciences, 1776-1976. *Acad. Nat. Sci., Philadelphia Publ.* 12: 59-83.

Raven, P.H., Raven, T.E. and West, K.R. 1976. The genus *Epilobium* (Onagraceae) in Australasia; a systematic and evolutionary study. N.Z. Dept. Sci. Ind. Res. Bull. 216, 321 pp. Christchurch.

Raven, P.H., Kyhos, D.W. and Cave, M.S. 1971. Chromosome numbers and relationships in Annoniflorae. *Taxon* 20: 479-483.

Rollins, R.C. 1957. Interspecific hybridization in *Lesquerrella* (Cruciferae). *Contrib. Gray Herb. Harv. Univ.* 181: 1-40.

Rollins, R.C. 1963. The evolution and systematics of *Leavenworthia* (Cruciferae). *Contrib. Gray Herb. Harv. Univ.* 192: 3-98.

Rose, M.R. and Doolittle, W.F. 1983. Molecular biological mechanisms of speciation. *Science* 220: 157-162.

Roth, J.R. and Schmid, M.B. 1981. Arrangement and rearrangement of the bacterial chromosome. *Stadler Symp.* 13: 53-70.

Schweizer, D. and Ehrendorfer, F. 1976. Giemsa banded karyotypes, systematics, and evolution in *Anacyclus* (Asteraceae-Anthemideae). *Plant Syst. Evol.* 126: 107-148.

Seavey, S.R. and Raven, P.H. 1977. Chromosomal differentiation and the sources of the South American species of *Epilobium* (Onagraceae). *J. Biogeogr.* 4: 55-59.

Selander, R.K., Yang, S.Y., Lewontin, R.C. and Johnson, W.E. 1970. Genetic variation in the Horseshoe Crab (*Limulus polyphemus*), a phylogenetic "relic". *Evolution* 24: 402-414.

Shumaker, K.M., Allard, R.W. and Kahler, A.L. 1982. Cryptic variability at enzyme loci in three plant species, *Avena barbata, Hordeum vulgare* and *Zea mays*. *J. Hered.* 73: 86-90.

Smith, E.B. 1983. Phyletic trends in sections *Eublepharis* and *Calliopsis* in the genus *Coreopsis* (Compositae). *Am. J. Bot.* 70: 549-554.

Sneath, P.H.A. and Sokal, R.R. 1973. *Numerical Taxonomy.* Freeman, San Francisco. 573 pp.

Solbrig, O.T. 1964. Infraspecific variation in the *Guttierrezia sarothrae* complex (Compositae - Asteraceae). *Contr. Gray Herb.* 193: 67-115.

Solbrig, O.T. 1971. *Principles and Methods of Plant Biosystematics.* Collier-Macmillan, Toronto. 226 pp.

Soltis, D.E., Bohm, B.A. and Nesom, G.L. 1983. Flavonoid chemistry of cytotypes in *Galax* (Diapensiaceae). *Syst. Bot.* 8: 15-23.

Somaroo, B.H. and Grant, W.F. 1972. Chromosome differentiation in diploid species of *Lotus* (Leguminosae). *Theor. Appl. Genet.* 42: 34-40.

Somaroo, B.H. and Grant, W.F. 1972. Crossing relationships between synthetic *Lotus* amphidiploids and *L. corniculatus*. *Crop Sci.* 12: 103-105.

Stebbins, G.L. Jr. 1950. *Variation and Evolution in Plants.*
  Columbia Univ. Press, New York. 643 pp.
Stebbins, G.L. Jr. and Babcock, E.B. 1939. The effect of
  polyploidy and apomixis on the evolution of species in
  Crepis. *J. Hered.* 30: 519–530.
Taper, L.J. and Grant, W.F. 1973. The relationship between
  chromosome size and DNA content in birch (*Betula*) species.
  Caryologia 26: 263–273.
Tsunewaki, K. and Ogihara, Y. 1983. The molecular basis of
  genetic diversity among cytoplasms of *Triticum* and *Aegilops*
  species. II. On the origin of polyploid wheat cytoplasms as
  suggested by chloroplast DNA restriction fragment patterns.
  Genetics 104: 155–171.
Turner, B.L. and Alston, R. 1959. Segregation and recombination
  of chemical constitutents in a hybrid swarm of *Baptisia
  laevicaulis* X *B. viridis* and their taxonomic implications.
  Am. J. Bot. 46: 678–686.
Turreson, G. 1922. The genotypical response of the plant species
  to the habitat. *Hereditas* 3: 211–350.
Uhl, C.H. 1982. The problem of ploidy in *Escheveria*
  (Crassulaceae). I. Diploidy in *E. ciliata.* *Am. J. Bot.*
  69: 843–854.
Valentine, D.H. 1950. Interspecific compatibility and hybrid
  fertility as taxonomic criteria. Proc. 7th Int. Bot. Congr.
  Stockholm. Almquist and Wiksell, Stockholm. pp. 285–286.
Vickery, R.K., Jr. 1978. Case studies in the evolution of species
  complexes in *Mimulus.* In *Evolutionary Biology* (M.K. Hecht
  et al., eds.). Plenum, New York. pp. 404–506.
Vickery, R.K., Jr. and Wullstein, B.C. 1981. A comparison of six
  taxonomic methods using section *Erythranthe* of the genus
  Mimulus. XIII Int. Bot. Congr. Sydney, Australia. p. 142.
  Abstr.
Wagner, W.H., Smith, A.P. and Pray, T.R. 1983. A Cliff Brake
  hybrid, *Pallaea bridgesii* X *mucronata,* and its systematic
  significance. *Madrono* 30: 69–83.
Watson, J.D. and Tooze, J. 1981. *The DNA Story.* Freeman,
  San Francisco. 605 pp.
Weller, S.G. and Ornduff, R. 1977. Cryptic self-incompatibility
  in *Amsinkia grandiflora.* 31: 47–53.
Wells, P.V. and Hunziker, J.H. 1976. Origin of the Creosote bush
  (*Larrea*) deserts of Southwestern North America. *Ann. Mo. Bot.
  Gard.* 63: 843–861.
Westergaard, M. 1946. Aberrant Y chromosomes and sex expression
  in *Melandrium album.* *Hereditas* 32: 419–443.
Wullstein, B.C.M. 1980. Comparison of deoxyribonucleic acid
  hybridization with experimental hybridization and numerical
  taxonomy as a method for determining taxonomic relationships in
  Erythranthe. Ph.D. Diss., Univ. Utah, Salt Lake City.
  149 pp.
Zohary, M. 1975. Principles of phytogeographical subdivision of
  the earth. *XII Int. Bot. Congr., Leningrad.* 1: 176. Abstr.

Cytology and Biosystematics: 1983

Keith Jones
The Royal Botanic Gardens
Richmond, Surrey, England

INTRODUCTION

As an introduction to our printed program we have been given a
definition of "Biosystematy" as proposed by Camp and Gilly and its
derivative "Biosystematics" is included in the title of almost
every lecture. It might seem that there is today some universal
agreement as to its meaning, but I doubt it. I should make it
clear at the commencement of my contribution that I take it to
mean, in the context of the chromosome, the exposure and analysis
of cytological variation in natural, and some artificial,
populations as a means of detecting the ways in which chromosomes
control recombination and evolution. I do not see it as being
equated with experimental taxonomy neither do I regard its aims as
being similar. Both Camp and Böcher have protested at the
perversion of the original definition of "Biosystematy" (Camp
1961; Böcher 1970) and I find myself very much in agreement with
them. Böcher also reminded his audience that Biosystematics is a
synthetic study which depends on the contributions from
specialists in the various contributing fields. These may prefer
to call themselves cytologists, biochemists, palynologists etc.
whose studies will influence the subject though not directly. I
wonder therefore how many biosystematists there are in the world –
though this week we all gather under the same terminological
umbrella.

These thoughts have been passing through my mind in preparing
this talk in an attempt to decide whether I should attempt to be a
synthesizer or a specialist. Having thought of myself as a
population cytologist I have concluded that it would be best to
stick to my last and allow the symposium to develop its own
synthesis as the meeting progresses. But in addition since I
believe as I did in Seattle (Jones 1970) that we have much to
learn about the activities and potential of the chromosome itself

PLANT BIOSYSTEMATICS

I would prefer to bring to your attention some of the advances
which change our concepts and our interpretations of the
chromosome and its evolution. In doing this I am mindful that
later speakers will be dealing with some of these new views of the
chromosome in some detail. I will therefore let them expand on
points which I will only touch upon.

Although the organisers of the meeting have chosen the
publication of Camp and Gilly's paper as a reference point for us
all it did not represent a watershed. It has been
preceded by "The Evolution of Genetic Systems" (Darlington 1939),
"The New Systematics" (Huxley 1940), "Genetics and the Origin of
Species (Dobzhansky 1941) and by other numerous publications which
focussed attention on the central role of the chromosome, and the
genetic system as a whole, in the evolution and categorisation of
populations. The principles of chromosome change and behavior
which were established at that time have stood us in good stead
and the main framework added to later by such works as "Variation
and Evolution in Plants" (Stebbins 1950) remain with us today.
But of course the intense investigations of chromosomes both in
and out of natural populations in the 50's and 60's provided new
insight into the chromosome, new examples of chromosome control
and situations of evolutionary complexity. The discovery of the
structure of the DNA molecule however drew many cytologists away
from the population and into the chemical laboratory to examine
gene structure and gene action leaving it to the medical men who
had only recently discovered the chromosome to undertake the most
intensive population study of all time – that of *Homo sapiens*
himself (herself? personself?). It is from these "post DNA"
researches that we have derived the information and techniques
which now influence both our view of the chromosome and the way we
deal with it.

We must address ourselves to chromosome organisation as far as
we understand it as a basis for our thoughts on the inception and
meaning of chromosome change.

The chromosome turns out to be both more simple and also more
complex than previously imagined. It is more simple in being a
single molecule of DNA consisting only of a series of nucleotide
sequences whose coding and reproduction we seem to understand.
Its complexities lie in the nature and repetition of sequences
which seem to have no genetic effect and their ability for
unsuspected reproduction and movement. Much has been written and
is being written about this but if I can attempt to distil the
present information as it appears to those of us who still have to
see the chromosome through lenses, the view would be this. A
chromosome though still regarded as a linear array of genes may
consist of only a small proportion of nucleotide sequences which
code – the rest of the molecule is made up of sequences repeated
several hundred times (moderate repeats) or millions of times
(highly repeated sequences) and which with the rare exception of
moderate repeats coding for ribosomal RNA, are not transcribed and
which therefore have no known genetic effect. These repeat

sequences have a capacity for changing their amount on a single chromosome by tandem duplication and some (transposable elements) can move from one chromosome to another spreading themselves across a whole genome. The chromosome and the genome is then capable of changes which had not been imagined particularly considerable enlargement in length by replication of regions which in former days was though possible only by relocation of existing chromatin. But although such changes of dimensions may be remarkable in extent being the result of addition of genetically inert DNA, the content of structural genes may remain the same. As in most other types of chromosome changes these quantitative ones can proceed in either direction so we see diminution as well as enlargement. Armed with this information we can take a new look at the matter of the remarkable differences in chromsome size both within and between complements which have intrigued cytologists ever since they began their surveys of species. Today we know this problem as the C-value paradox —which if we are unable to fully understand we can at least partly explain. I shall refer to this question later.

Associated with these molecular findings have been methods for estimating the total quantity of DNA per nucleus and for detecting both the amount, position and quality of heterochromatin. The cytophotometric methods and the banding techniques are easy to use and should now form part of the normal procedures for the analysis of the karyotype. Since banding can also discriminate different genomes its use with meiotic chromosomes can also identify pairing chromosomes with obvious advantage in hybrid analysis. For the population cytologist these methods are sufficient for most purposes, but if further detail of the organisation of the chromosome is needed, molecular probes and other devices can locate specific gene sequences. Hutchinson's paper is a convenient introduction to the methods and literature on this subject (Hutchinson 1983).

In addition to these researches into the organisation of the chromosome other discoveries have made possible insights into chromosome arrangement and chromosome pairing and indeed their interrelationship. Dr. Bennett will be speaking on this topic (Bennett 1984) and I will not elaborate further. I will however make brief mention of the synaptonemal complex. This has helped to explain the basis of chromosome pairing but has also enabled details of pachytene pairing to be observed in close detail in ultrastructural studies. Initially this type of study was very laborious involving the examination of many serial sections, but now techniques are available for spreading SC in such a way as to allow them all to be viewed at once in the electron microscope. Developed first in animals the technique has now been modified for use in higher plants and has been beautifully applied to *Zea mays* (Gillies 1981, 1983). Although not yet used widely in plants it has a great deal of promise for the future investigation of pairing patterns. The serial section technique has already been used with considerable success for observing the details of

pairing in plant hybrids as exemplified by those investigations
which seek to understand the distribution of extra DNA in
chromosomes of differing size (Jenkins and Rees 1983).

So much for the new view of the chromosome and some methods
available for their analysis. I want now to briefly consider some
general topics which can be of fundamental importance in various
levels of analysis.

THE KARYOTYPE

The number and form of the chromosome complement constituting this
karyotype still represents the only cytological information for
most plants and for many of these it is reduced to number only.
We all fully realise too that the sampling of species is so meagre
that often we have no idea of the occurrence or extent of
variation. Nevertheless what can be deduced from the karyotype
morphology when we have it and have it in reasonable population
samples?

There are no rules of procedure but there is one general dogma
which sits upon what may be an intuitive feeling which declares
that symmetry of chromosomes and karyotypes in some way represents
the simple, perhaps primitive condition, which gives way to
asymmetry with evolutionary advancement. Levitsky's limited
observations on a small group of plants first gave voice to this
view (Levitsky 1931) which was taken up and developed by Stebbins
(1950, 1971) and others. It has become a conscious and
subconscious view but one which lies in stark contrast to that of
zoologists who have always held asymmetry as primitive and
symmetry as advanced and have ample evidence for their views.

We can now look at the karyotype in its Feulgen outline and
with the benefit of banding to which we can add the measurement of
DNA amount. These features interact in different ways and can
have distinct implications. Let us take the karyotype outline
first.

The establishment and maintenance of any particular karyotype
pattern is considered generally to have meaning for heredity
rather than development. It represents an arrangement of genetic
and nongenetic chromatin advantageously disposed for the purpose
of controlling recombination within and between chromosomes. We
may assume that the pattern remains reasonably constant in
interbreeding plants particularly diploids as any marked variation
can disturb pairing relations, chiasma location and the
maintenance of coadapted sequences. We have therefore the concept
of karyotype stability within actually or potentially
interbreeding groups which may or may not be recognised as
taxonomic species. But the influence of pattern may spread
further than that. It may characterize a cluster of species
within a genus, an entire genus or a group of genera implying
either that there has been no requirement for change, that there
has been no opportunity for change or that pattern is so important
that change cannot be tolerated. Of course, we use the pattern

similarity to indicate relatedness, but we should attempt to
determine its fundamental significance. The Aloes offer an
opportunity to do this.

The strongly bimodal, $2n = 14$, karyotype is the constant
feature of *Aloe, Gasteria* and *Haworthia* and a few smaller genera
consisting in all of several hundred species. It is maintained in
the face of an ample supply of structural mutation which could
modify its shape (Brandham 1976). Occasional mutants heterozygous
for a visibly altered chromosome do occur, but despite careful
study no homozygotes can be found in the wild. Hybridisation both
within and between genera give no indication of any involvement of
symmetrical changes in the evolution of genomes although they
frequently show that they are differentiated to a high degree by
unidentified modifications. Karyotypes can on the other hand show
substantial differences in overall size due to an increase in
total DNA. It is from hybridisation of these that it is concluded
that genomes differentiate by the insertion of small segments of
chromatin throughout the length of each chromosome (Brandham 1983)
and it is likely that this even distribution of changes is the way
in which karyotypes of similar DNA amount have evolved. This
extreme sensitivity to change of pattern - this high degree of
karyotype conservatism is a very puzzling phenomenon and it leads
us to consider that pattern in this group is important to
development as well as heredity and we may think that the very
definition of these plants in terms of their physiological,
structural and morphological attributes is dependent on the
maintenance of their gene characters within the particular
chromosome shapes which have been adopted. But then in many other
groups of plants with similarly strong evolutionary alliance
karyotypes can be extremely diverse as in *Crocus* where almost
everyone of the 70 or so taxonomic species has a karyotype
distinctive in form and/or number of chromosomes (Brighton 1976,
1977), and in some cases like the remarkable *Crocus speciosus*
subspecies *speciosus* seven karyotypes with $2n$ ranging from 8 to
18 and differing in chromosome morphology exist without producing
phenotypic effects sufficient to allow even the sharp-eyed
taxonomist to separate them (Brighton *et al.* 1983). The number
of examples of chromosome variation with little phenotypic effect
increases as more intensive sampling of populations takes place
and *Scilla autumnalis* is another emphatic case involving some
substantial differences in DNA amount (Ainsworth *et al.* 1983).
Although not put to the test it is unlikely that these different
karyotypes are capable of free interbreeding and we must recognize
that they are resulting in evolutionary independence whatever the
consequences of that may be. For a very thoroughly studied and
most striking example of karyotype divergence within a taxonomic
species I want to mention the case of the Australian grasshopper
Caledia captiva. This species occurs as two principal
chromosome races, i.e. the Moreton race with most chromosomes
metacentric and the Torresian where they are all acrocentric, the
difference being attributable to eight asymmetric pericentric

inversions (Shaw and Coates 1983). F_1 hybrids are fertile but heterozygosity for the inversions, which pair straight and not with inversion loops, causes chiasmata to shift to positions which are novel for both paired chromosomes. The F_2 generation is inviable not passing beyond the early embryonic stage, and on the basis of this and other evidence the authors conclude that this failure can be attributed to the disruption of coadapted gene sequences by the shift in crossing over. They see this as suggesting that karyotype organisation is important to development as well as heredity. The intensive study of the pathology of *Homo sapiens* has for long pointed in the same direction.

The Feulgen karyotype is clearly highly significant although it cannot be said that its meaning is fully understood. Dr. Bennett will be saying more about its features in connection with the spatial distribution of the haploid genomes and direct your attention to a totally new approach to the study of karyotype evolution initiated by himself and his coworkers. I think it will become apparent that there is more to be discovered about their basic chromosome features than was thought in their earlier days when cytological information was so confidently employed in unravelling the problems of evolution.

Dr. Bennett will also be dealing with C-band features of karyotypes and this allows me to be brief. It has become evident that C-band polymorphism occurs within a fairly rigid Feulgen karyotype. Some of this variation within an interbreeding group probably reflects segmental realignment or tandem duplication. But where populations are markedly different in C-band patterns we have reasons to suspect that they point to substantial changes in homology which can interfere with interbreeding potential. Like the Feulgen karyotype the C-band detail has been found to be a useful marker of phylogenetic affinity as in the case of *Anemone* (Marks and Schweizer 1974), *Allium* (Vosa 1976) and *Scilla* (Greilhuber and Speta 1977) nevertheless it too can show considerable divergence even within some taxonomic species as in *Gibasis karwinskyana* (Kenton 1978). In this last case hybrids between C-band distinctive populations show reduced meiotic stability and a lowering of fertility. C-band patterns as an additional chromosome marker must therefore be used cautiously and in combination with other types of information.

Although Professor Rees will be dealing with changes in DNA amount in the differentiation of species (Rees 1984), I need to make brief mention of it here. It is clear that DNA amount per genome can increase or decrease in evolution, it can be spread amongst all members of the chromosome complement in equal amounts or in quantities proportional to chromosome length, or it may be confined to one or more chromosomes. There are species where additional DNA has left the Feulgen karyotype pattern unchanged as for example the two $x = 7$ genomes in *Scilla autumnalis* (Ainsworth *et al. l.c.*) and in *Gibasis* (Kenton 1983a,b). The DNA changes may be an interspecific character or will be present within the taxonomic species. Populations of *Microseris douglasii* show

several different levels of DNA amount (Price *et al.* 1981)
as do those of *Gibasis venustula* (Kenton 1983a,b). Some
correlations between DNA amount and ecological conditions were
suggested in the former case and a very positive correlation with
altitude in the latter. As with other significant variations of
the karyotype DNA differences can interfere with interbreeding and
there are preliminary indications of this in the *Gibasis*
study. DNA amount is clearly a character of importance which
needs to be taken into account. The fact that additions or
deletions may be of genetically inert material should not blind us
to the influence of the total DNA quantity which is known to
affect cell cycle time and the speed of development of tissues,
organs and individuals (Van't Hof and Sparrow 1963; Evans and Rees
1971). Signs of its adaptive qualities are present in such cases
as *Gibasis* and in the distribution of C-values in some crop
plants (Bennett 1976).

   If in summation we consider the total investigation of the
karyotype by assessing the quantity, quality and disposition of
DNA as a means of determining relatedness and direction of
evolution I can do no better than to recommend to you the papers
of Dr. Greilhuber and his colleagues which are excellent
illustrations of the use of these modern methods (Greilhuber 1977,
1979; Greilhuber and Speta 1977, 1978).

   I hope I have made some points about the karyotype which
illustrate how the deeper understanding of its form and function
can improve the employment of it in population studies. I am
certain that when Dr. Bennett has concluded his lecture you will
be more than impressed that it is not a mere silhouette to be
attached to a species description.

   Now I wish to turn to one type of karyotype change the
appreciation of which can fundamentally influence the
interpretation of some complex situations. We have seen over the
years how common interchanges are in evolution. They have been
observed in the heterozygous condition as a frequent and sporadic
mutant, in high frequency as floating interchanges and as
permanent complex heterozygotes in special cases. It is a simple
type of mutation but one of substantial effect. I don't want to
dwell on this subject but I think it is useful to point out that
we have to add some interesting new cases to the classical
literature amongst which the frequency of heterozygotes in
Viscum and the multiple interchanges involving a sex
chromosome are of particular note (Wiens and Barlow 1973, 1979;
Barlow and Wiens 1975). New cases of complex interchange
heterozygotes have also appeared including a rich array in the
Onagraceae of S. America as described by Dietrich (Dietrich 1977).
But there are a number of other very interesting occurrences
which can be added to the good old originals of the N. American
Oenotheras and *Rhoeo*, all of which are reviewed in a recent
publication (Holsinger and Ellstrand 1983). To these we can add
one other example from the Commelinaceae – *Gibasis pulchella*
($2n$ = 10) first noticed by Handlos (1970) and now under

study at Kew. I look forward to hearing more about *Isotoma*
and its evolution from Dr. James shortly (James 1984).

The above types of interchange can be recognized because of
their effects on meiotic pairing and in the same way the
involvement of interchanges in speciation as homozygous
differences can be detected in hybrids. I want now to say
something about a special type of interchange which is less easily
seen and whose presence by causing change of chromosome number
produces varying degrees of confusion. I refer to the uniting of
the long arms of two acrocentrics to a common centromere which is
known as Robertsonian fusion. This mutation has long been
recognized as a common and important mechanism of chromosome
evolution in animals but has not been regarded as being of the
same significance in plants.

I have discussed Robertsonian fusion and its reciprocal,
centric fission, at some length elsewhere (Jones 1978) and have
described the evolution of a single homozygous fusion in *Gibasis*
(Jones 1974) and a series of multiple fusions in *Cymbispatha*
(Jones 1977; Jones *et al.* 1981). I will not repeat myself
here and will confine my remarks to some general points.

A Robertsonian fusion like other structural changes arises
first as a heterozygote and such sporadic mutants occur in natural
populations. When in a diploid it becomes homozygous the basic
chromosome number is reduced by one and the number of possible
interchromosomal recombinations in gametes reduced by 50%. A new
linkage group is formed and usually redistribution of chiasmata in
the new chromosome can produce further modification of
recombination. It should be noted also that we have an enlarged
chromosome which can be twice the size of the former acrocentrics
when they had long arms of similar size. The chromosomal and
potential genetic changes caused by this one simple event are
profound.

The progression of Robertsonian fusion in a diploid lineage can
be readily traced where this is a preferred change and not
obscured by later overlapping changes in structure which shift
centromere position. Where polyploidy intervenes either imposing
itself upon a Robertsonian change or when fusion occurs in a
polyploid the matter becomes rather more confusing since the
fusion can develop towards complete (quadriplex) homozygosity in
the case of tetraploidy, via a duplex condition. The impact of
such developments on basic number can be seen in Fig. 1. The
number of major chromosome arms will remain the same and in this
example, based on the *Cymbispatha* situation, it will always
be a multiple of 7 -the basic number of the 14A plant which starts
the series.

This type of evolution generates dysploidy and disguises
polyploidy including the eventual production, as in *Cymbispatha*
of a $2n = 14$ metacentric condition which though seemingly
diploid with the commonest diploid number in the angiosperms, is
in fact constitutionally tetraploid. All this can be demonstrated
unequivocally in favorable material which retains the clues of

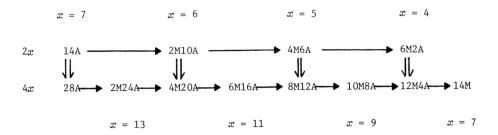

FIG. 1. *The consequences of Robertsonian fusion at diploid and tetraploid levels.*

ancestry including supportive meiotic evidence, but the disappearance of intermediate conditions, or strictly allopolyploid ancestry of the polyploids could cause the situation to be interpreted in ways which would be far from the truth.

In my 1978 paper I listed examples of probable Robertsonian change in the evolution of plants in groups as diverse as the Cycads and the Iridaceae and since then other examples are coming to light. I would mention in particular the orchid genus *Paphiopedilum* where numbers of $2n$ = 26,28,30,32,34,36,38,40,41,42 have arm numbers which are always a multiple of 13. Karasawa's paper beautifully displays this variation and leads to other relevant studies in the genus (Karasawa 1979). In this and other cases centric fission has to be considered as well as fusion and it may be difficult or impossible to distinguish between the two.

Finally, there are two points that should be noted. Firstly, chromosome evolution can move from asymmetry to symmetry and secondly, in this development chromosomes can increase in size. Here then is another mechanism for enlarging the dimension of chromosomes. If for example we compare the $2n$ = 14A and $2n$ = 14M extremes, the extra DNA here has been contributed not by the tandem duplication of repeat sequences, but by the replication of chromosome sets. In the long term of evolution it may not be possible to distinguish between these possibilities. Polyploidy as we have seen can convert by fusion to apparent diploidy and we have to consider that it may have been involved in successive cycles of karyotype change since the emergence of higher plants. This leads me on to my final subject polyploidy.

Polyploidy has probably been written about more often than any other cytological topic. This understandable because it is such a common type of change. Arguments will continue on the relative frequency of auto- and allopolyploids and their evolutionary (and taxonomic) importance. One thing we can say is that the frequency of polyploids showing very high levels of multivalent formation continues to increase - the study of the new world Commelinaceae alone adding a substantial number.

Although it has been demonstrated that genetic control can reduce moderate levels to complete bivalent pairing, there is no evidence that selection for improved fertility reduces multivalent formation in autopolyploids, and it may be the case that for all intents and purposes the frequency of autopolyploidy is reflected in the extent of multivalent pairing plants today. Segmental allopolyploidy is as we know an intermediate condition which may be resolved by the application of methods for comparing expectation with observation as developed by our colleagues in Texas (Jackson and Huber 1982; Jackson and Casey 1980, 1982) from whom we shall be hearing later in this meeting (Jackson 1984).

The consequences of polyploidy depend on the case and on the observer. We cannot doubt its dampening effect on recombination between it and its diploid ancestors despite the possible channel of gene flow via triploids (Zohary and Nur 1959) and through tetraploids produced directly from tetraploid × diploid crosses (Carrol and Borrill 1965). There is no question either that polyploidy whether auto- or allo produces a whole series of changes affecting the mucleus, development, meiosis, recombination and the phenotype and it must be regarded as one of the most important types of chromosome evolution through which most, if not all higher plants, have passed through. However polyploidy tends to be disregarded as a significant change when it occurs within the taxonomic species even when the polyploids are in high frequency and have attained a distinctive distribution. Some, for reasons I cannot comprehend, even see them as evolutionary noise.

FINALE

Although it is quite beyond the scope of a review of this length one must acknowledge the enormous increase in chromosome information and understanding over these last 40 years. It has been a period which saw not only the deciphering of the DNA molecule and the exposure of the complexities and evolutionary potentialities of the chromosome but also the discovery of new ways in which genes control chromosomes, chromosomes control genes, new patterns and pathways of evolution. Much more has been revealed about the B chromosomes and their influence on chiasma frequency, choice of partner and a variety of quantitative characters outside the genetic system - so much so that they now have a book to themselves (Jones and Rees 1982). Polyploidy has continued to attract attention and has had a conference devoted entirely to it (Lewis 1980). And we are all aware of the theories and syntheses which have emerged in the many writings of Darlington, Grant, Stebbins, Lewis and John and others. From the earlier endeavors of Manton came the exposure of fern cytology which stimulated so much research in the pteridophytes (Manton 1950) whilst the bryophytes too had their share of attention (cf. Smith 1978) including the application of the most modern methods

(Newton 1977, 1981).

Despite all this accumulated information there are some basic aspects of chromosome structure and influence which are not understood and we must be prepared to modify our approaches to the plant population as new knowledge comes to hand.

You may think that in this review I have concentrated too much on the chromosome and not on the application of chromosome knowledge. Well that may be but I put the understanding of the chromosome in the first position since that is the foundation upon which all our conclusions are based. If we fail to appreciate its full potential then we cannot possibly erect meaningful hypothesis of evolution.

CYTOLOGY AND BIOSYSTEMATICS

The emphasis which has been placed on chromosome organization and the karyotype in this talk is intended to demonstrate uncertainties of knowledge and interpretation of the basic organelle which we employ as our guide in tracing significant evolutionary events. If we do not become aware of new findings or if we fail to appreciate the capabilities and influences of the chromosome we will misuse it to our detriment. For there are no rules of chromosome evolution — we cannot make predictions in any general sense but we have to examine each situation as it arises. We know that chromosomal DNA can increase or decrease as can indeed chromosome number; most structural changes can proceed in reverse directions and karyotype patterns may or may not disclose major degrees of internal differentiation. I could not therefore be facile and suggest that chromosomes can be used in any general way for they are unpredictable.

What has become clear over the years is that only close, intensive and often prolonged investigations are capable of revealing what is going on in natural populations. Let us remember the painstaking work that went into the study of *Crepis* (Tobgy 1943), *Oenothera* (Cleland 1972), *Isotoma* (James 1965, 1970), *Haplopappus* (Jackson 1962, 1965, 1973), *Brachycome* (Carter *et al.* 1974; Carter 1978; Kyhos *et al.* 1977 etc.), Aloineae (Brandham 1983, 1976, etc.), to name but a few examples in plants, and the extremely informative studies of whelks, mice, gerbils, grasshoppers and man in the animal world. Each investigation tells us something about the complexities of change in the group examined and although often broadening our understanding in ways which assist all our studies do not offer a general panacea for all our problems. Indeed the more we look the more complex nature is in its diversity and more difficult to understand — too complex perhaps for the simple questions we ask.

There is no doubt that chromosome studies have influenced many taxonomic treatments and will continue to do so, but as Camp and Gilly and others have realised, taxonomy is incapable of accommodating all the situations that are uncovered — so be it. We are more concerned with explaining the events which bring about

discontinuities and genetic independence in evolution and leave it
to others to classify whatever of our results that they can.

REFERENCES

Ainsworth, C.C., Parker, J.S. and Horton, D.M. 1983. Chromosome
  variation and evolution in *Scilla autumnalis*. In *Kew Chromosome
  Conference II* (P.E. Brandham and M.D. Bennett eds.). Allen &
  Unwin, London. pp. 261–268.
Barlow, B.A. and Wiens, D. 1975. Permanent translocation
  heterozygosity in *Viscum hildebrandtii* Engl. and *V. engleri*
  Tregl. (Viscaceae) in East Africa. *Chromosoma* 53: 265–272.
Bennett, M.D. 1976. DNA amount, latitude, and crop plant
  distribution. *Environ. Exp. Bot.* 16: 93–108.
Bennett, M.D. 1984. The genome, the natural karyotype, and
  biosystematics. In *Plant Biosystematics* (W.F. Grant,
  ed.). Academic Press, Toronto.
Böcher, T.W. 1970. The present status of biosystematics. *Taxon*
  19: 3–5.
Brandham, P.E. 1976. The frequency of spontaneous structural
  change. In *Current Chromosome Research* (K. Jones and P.E.
  Brandham eds.). Amsterdam, Elsevier. pp. 77–87.
Brandham, P.E. 1983. Evolution in a stable chromosome system.
  In *Kew Chromosome Conference II* (P.E. Brandham and M.D.
  Bennett eds.). Allen & Unwin, London. pp. 251–260.
Brighton, C.A. 1976. Cytological problems in the genus *Crocus*
  (Iridaceae). I *Crocus vernus* aggregate. *Kew Bull.* 31: 33–46.
Brighton, C.A. 1977. Cytological problems in the genus *Crocus*
  (Iridaceae). II *Crocus cancellatus* aggregate. *Kew Bull.*
  32: 33–45.
Brighton, C.A., Mathew, B., and Rudall, P. 1983. A detailed study
  of *Crocus speciosus* and its ally *C. pulchellus*
  (Iridaceae). *Pl. Syst. Evol.* 142: 187–206.
Camp, W.H. 1961. The pattern of variability and evolution in
  plants. In *A Darwin Centenary* (P.J. Wanstall, ed.).
  B.S.B.I. London. pp. 44–70.
Carrol, C.P. and Borrill, M. 1965. Tetraploid hybrids from
  crosses between diploid and tetraploid *Dactylis* and their
  significance. *Genetica* 36: 65–82.
Carter, C.R. 1978. The cytology of *Brachycome* and the
  inheritance, frequency and distribution of B chromosomes in *B.
  dichromosomatica (n = 2)*, formerly included in *B. lineariloba*.
  Chromosoma 67: 109–121.
Carter, C.R., Smith-White, S. and Kyhos, D.W. 1974. The cytology
  of *Brachycome lineariloba*. 4. The 10 chromosome Quasi-
  diploid. *Chromosoma* 44: 439–456.
Cleland, R.E. 1972. Oenothera *Cytogenetics and Evolution*.
  Academic Press, N.Y. pp. 370.
Darlington, C.D. 1939. *Evolution of Genetic Systems*. Cambridge
  Univ. Press.
Dietrich, W. 1977. The South American species of *Oenothera*

sect. *Oenothera (Raimannia, Renneria;* Onagraceae).
Ann. Mo. Bot. Gard. 64: 425–626.

Dobzhansky, Th. 1941. *Genetics and the Origin of Species.*
Revised ed. Columbia Univ. Press, N.Y. 446 pp.

Evans, G.M. and Rees, H. 1971. Mitotic cycles in dicotyledons and
monocotyledons. *Nature* 233: 350–351.

Gillies, C.B. 1981. Electron microscopy of spread maize pachytene
synaptonemal complexes. *Chromosoma* 83: 575–591.

Gillies, C.B. 1983. Spreading plant synaptonemal complexes for
electron microscopy. In *Kew Chromosome Conference II*
(P.E. Brandham and M.D. Bennett eds.). Allen & Unwin, London.
pp. 115–122.

Greilhuber, J. 1977. Nuclear DNA and heterochromatin contents in
the *Scilla hohenackeri* group, *S. persica* and *Puschkinia
scilloides* (Liliaceae). *Pl. Syst. Evol.* 128: 243–257.

Greilhuber, J. 1979. Evolutionary changes of DNA and
heterochromatin amounts in the *Scilla bifolia* group
(Liliaceae). *Pl. Syst. Evol. Suppl.* 2: 263–280.

Greilhuber, J. and Speta, F. 1977. Giemsa karyotypes and their
evolutionary significance in *Scilla bifolia, S. drunensis*
and *S. vindobonensis* (Liliaceae). *Plant Syst. Evol.*
127: 171–190.

Greilhuber, J. and Speta, F. 1978. Quantitative analyses of
C-banded karyotypes and systematics in the cultivated species of
the *Scilla siberica* group (Liliaceae). *Pl. Syst. Evol.*
129: 63–109.

Handlos, W.L. 1970. Cytological investigations of some
Commelinaceae from Mexico. *Baileya* 17: 6–33.

Holsinger, K.E. and Ellstrand, N.C. 1983. The evolution and
ecology of permanent translocation heterozygotes. *Am. Nat.*
In press.

Hutchinson, J. 1983. *In situ* hybridisation mapping of plant
chromosomes. In *Kew Chromosome Conference II* (P.E. Brandham
and M.D. Bennett, eds.). Allen & Unwin, London, pp. 27–34.

Huxley, J.W. 1940. *The New Systematics.* Oxford.

Jackson, R.C. 1962. Interspecific hybridization in *Haplopappus*
and its bearing on chromosome evolution in the *Blepharodon*
section. *Am. J. Bot.* 49: 199–132.

Jackson, R.C. 1965. A cytogenetic study of a three-paired race of
Haplopappus gracilis. Am. J. Bot. 52: 946–953.

Jackson, R.C. 1973. Chromosomal evolution in *Haplopappus
gracilis:* a centric transposition race. *Evolution* 27: 243–256.

Jackson, R.C. 1984. Chromosome pairing in species and hybrids.
In *Plant Biosystematics* (W.F. Grant, ed.) Academic Press,
Toronto.

Jackson, R.C. and Casey, J. 1980. Cytogenetics of polyploids. In
Polyploidy: Biological Relevance (W.H. Lewis, ed.).
Plenum, N.Y. pp. 17–44.

Jackson, R.C. and Casey, J. 1982. Cytogenetic analyses of
autopolyploids: models and methods for triploids to octoploids.
Am. J. Bot. 69: 489–503.

Jackson, R.C. and Hauber, D.P. 1982. Autotriploid and autotetraploid cytogenetic analyses: correction coefficients for proposed binomial models. *Am. J. Bot.* 69: 646-648.

James, S.H. 1965. Complex hybridity in *Isotoma petraea*. I. The occurrence of interchange heterozygosity, autogamy and a balanced lethal system. *Heredity* 20: 341-353.

James, S.H. 1970. Complex hybridity in *Isotoma petraea*. II. Components and operation of a possible evolutionary mechanism. *Heredity* 25: 53-77.

James, S.H. 1984. The pursuit of hybridity and population divergence in *Isotoma petraea*. In *Plant Biosystematics* (W.F. Grant, ed.). Academic Press, Toronto.

Jenkins, G. and Rees, H. 1983. Synaptonemal complex formation in a *Festuca* hybrid. In *Kew Chromosome Conference II* (P.E. Brandham and M.D. Bennett, eds.). Allen & Unwin, London. pp. 233-242.

Jones, K. 1970. Chromosome changes: reliable indicators of the direction of evolution. *Taxon* 19: 172-179.

Jones, K. 1974. Chromosome evolution by Robertsonian translocation in *Gibasis* (Commelinaceae). *Chromosoma* 45: 353-368.

Jones, K. 1977. The role of Robertsonian fusion in karyotype evolution in higher plants. *Chromosomes Today* 6: 121-129.

Jones, K. 1978. Aspects of chromosome evolution in higher plants. *Adv. Bot. Res.* 6: 119-194.

Jones, K., Kenton, A., and Hunt, D. 1981. Contributions to the cytotaxonomy of the Commelinaceae IV. Chromosome evolution in *Tradescantia* sect. *Cymbispatha*. *Bot. J. Linn. Soc.* 83: 157-188.

Jones, R.N. and Rees, H. 1982. *B Chromosomes*. Academic Press, London. pp. 266.

Karasawa, K. 1979. Karyomorphological studies in *Paphiopedilum*, Orchidaceae. *Bull. Hiroshima Bot. Gard.* 2: 1-149.

Kenton, A. 1978. Giemsa C-Banding in *Gibasis* (Commelinaceae). *Chromosoma* 65: 309-324.

Kenton, A. 1983a. Qualitative and quantitative chromosome changes in the evolution of *Gibasis*. In *Kew Chromosome Conference II*. (P.E. Brandham and M.D. Bennett eds.). Allen & Unwin, London. pp. 273-282.

Kenton, A. 1983b. Chromosome evolution in Mexican species of *Gibasis* (Commelinaceae). Ph.D. Thesis. Univ. Reading.

Kyhos, D.W., Carter, C.R. and Smith-White, S. 1977. The cytology of *Brachycome lineariloba*. 7. Meiosis in natural hybrids and race relationships. *Chromosoma* 65: 81-101.

Lewis, W.H. 1980. *Polyploidy: Biological Relevance*. Plenum Press, N.Y. pp. 583.

Levitsky, G.A. 1931. The karyotype in systematics. *Bull. Appl. Bot. Genet. Pl. Breed.* 27: 200-240.

Manton, I. 1950. *Problems of Cytology and Evolution in the Pteridophyta*. Cambridge Univ. Press.

Marks, G.E. and Schweizer, D. 1974. Giemsa banding differences in *Anemone* and *Hepatica*. *Chromosoma* 44: 405–416.

Newton, M.E. 1977. Heterochromatin as a cyto-taxonomic character in liverworts: *Pellia*, *Riccardia* and *Cryptothallus*. *J. Bryol.* 9: 327–342.

Newton, M.E. 1981. Evolution and speciation in *Pellia* with special reference to the *Pellia megaspora-endiviifolia* complex (Metzgeriales). II Cytology. *J. Bryol.* 11: 433–440.

Price, H.J., Chambers, K.L., and Bachmann, K. 1981. Geographic and ecological distribution of genomic DNA content variation in *Microseris douglasii* (Asteraceae). *Bot. Gaz.* 142: 415–426.

Rees, H. 1984. Nuclear DNA variation and the homology of chromosomes. In *Plant Biosystematics* (W.F. Grant, ed.). Academic Press, Toronto.

Shaw, D.D. and Coates, D.J. 1983. Chromosomal variation and the concept of the coadapted genome in a direct cytological assessment. In *Kew Chromosome Conference II* (P.E. Brandham and M.D. Bennett, eds.). Allen & Unwin, London. pp. 207–216.

Smith, A.J.E. 1978. Cytogenetics, biosystematics and evolution in the Bryophyta. Adv. Bot. Res. 6: 195–276.

Stebbins, G.L. 1950. *Variation and Evolution in Plants*. Columbia Univ. Press, N.Y.

Stebbins, G.L. 1971. *Chromosomal Evolution in Higher Plants*. Edward Arnold, London. 216 pp.

Tobgy, H.A. 1943. A cytological study of *Crepis fuliginosa, C. neglecta* and their F_1 hybrid and its bearing on the mechanism of phylogenetic reduction in chromosome number. *J. Genet.* 45: 67–111.

Van't Hof, J. and Sparrow, A.H. 1963. A relationship between DNA content, nuclear volume and minimum cell cycle time. *Proc. Natl. Acad. Sci. U.S.A.* 49: 897–902.

Wiens, D. and Barlow, B.A. 1973. Unusual translocation heterozygosity in an East African mistletoe *Viscum fischeri*. *Nature* 243: 93–94.

Wiens, D. and Barlow, B.A. 1979. Translocation heterozygosity and the origin of dioecy in *Viscum*. *Heredity* 42: 201–222.

Vosa, C. 1976. Heterochromatic banding patterns in *Allium*. II. Heterochromatin variation in species of the paniculatum group. *Chromosoma* 57: 119–133.

Zohary, D. and Nur, U. 1959. Natural triploids in the orchard grass, *Dactylis glomerata* L., polyploid complex and their significance for gene flow from diploid to tetraploid levels. *Evolution* 13: 311–317.

Added in Proof:

Hutchinson, J. 1983. In situ hybridization mapping of plant chromosomes. In *Kew Chromosome Conference II* (P.E. Brandham and M.D. Bennett, eds.). Allen & Unwin, London. pp. 27–34.

The Genome, the Natural Karyotype, and Biosystematics

Michael D. Bennett

Plant Breeding Institute
Trumpington, Cambridge, England

THE POSSIBILITY OF STUDYING THE GENOME

For centuries plant taxonomists identified and classified taxa
using phenotypic characters alone. The existence and importance
of the genome was sometimes suspected, but its physical and
chemical characters were unresolved or unrecognised. However,
within one human lifetime it has become possible to study the
genome directly, and at several levels (Table I).

Not until early in this century, when the chromosome theory of
inheritance was accepted, were the first direct and meaningful
studies of plant genomic characters possible. Fortunately the
chromosomes of many plants are large (often 1-10 nm in length)
compared with the resolving power of the light microscope (about
0.1 µm), so that observations of their numbers, shapes and
behavior in many species could be made.

The title of this symposium, "Plant Biosystematics: 40 Years
Later" invites us to look back. Cytology and genetics have
histories spanning the whole of this period, but it was only 30
years ago that Watson and Crick (1953a, b) published their benchmark papers on the structure and function of DNA. Since then
molecular biology has revolutionised our understanding of the
genome, making it possible (within the last 10-20 years) to
characterize and compare plant genomes at the most fundamental
level - the base sequences of their DNAs.

THE GENOME - PROGRESS AND PUZZLES

While some developments have produced a profoundly increased
understanding of the genome (e.g. the elucidation of the genetic
code, and the establishment of the central dogma that "DNA makes
RNA makes protein"), others have presented aspects of bewildering
complexity. For example, cytologists have long recognised some

TABLE I. *Some genomic characters of potential biosystematic interest**

CYTOLOGICAL [chromosome(s)]	GENETICAL [gene(s)]	MOLECULAR & CYTOCHEMICAL [DNA and DNA sequence(s)]
a. Chromosome number (gametic number – n, basic number – x, ploidy level)	a. Physical gene map positions (i.e. locations obtained by *in situ* hybridization	a. DNA amount per genome or per chromosome
b. Chromosome shape(s) (karyogram)	b. Gene linkage map (in centimorgans showing separations in proportion to recombination frequency	b. Base composition and AT:GC ratio
c. Detailed chromosome morphology (i.e. distribution of: constrictions, puffs, heterochromatic knobs and segments, chromomeres, etc.		c. DNA density
d. Chromosome behavior during meiosis (pairing and recombination patterns		d. Amounts of unique and repeated sequences in the genome, and their ratio
e. Intranuclear spatial order of chromosomes		e. Amounts of coding and noncoding sequences in the genome, and their ratio
		f. Repeated sequence copy number
		g. Repeated sequence length (i.e. number of nucleotides)
		h. Interspersion patterns of repeated sequences
		i. Order of bases in a coding or noncoding DNA sequence

*Although these characters are grouped under three headings (cytological, genetical and molecular), this classification is somewhat arbitrary, and cannot be rigidly enforced. For example, the physical mapping of a gene *in situ* hybridization (listed as a genetical character) usually involves light microscope cytology to observe its chromosomal position, and depends on a probe developed using molecular techniques.

differentiation in chromosome morphology due to "knobs" or heterochromatic segments, but without perceiving any rules which govern their occurrence or distribution. In the last 10 years the linear differentiation seen in plant chromosomes has dramatically increased following the development of banding techniques (Schweizer 1973, 1980). For example, 280 bands or interband segments can now be recognised in the C-banded karyotype of *Triticum aestivum* (Dr. A.G. Seal - pers. comm.), but the function (if any) of most of these structures is unknown.

While cytologists puzzle over the meaning of rococo chromosomes, molecular biologists are trying to discover the significance of baroque DNA! The simplistic view of the plant genome as comprising only different types of coding sequences was short-lived, having been replaced by the knowledge that, in many angiosperms, coding sequences comprise less than 1% of the nuclear genome (Flavell 1982), while even coding genes may contain several intervening sequences (introns) with no known function (Sun *et al.* 1981).

GENOMIC CHARACTERS AND CLASSIFICATION

Genomic characters may be viewed in two distinct ways when utilized for the purpose of classification. In one sense they are strictly anatomical information, for example: the size of the genome, the number and shapes of the chromosomes, the number and spatial arrangement of genes on a chromosome, and the number and sequence of bases in a given piece of DNA. In theory, all these characters could be handled for taxonomic (phenetic) purposes just like the number, shapes and spatial ordering of the parts of a flower. However, in practice only two genomic characters have been widely used in this way, namely: chromosome number and crossability.

Perhaps the most noticeable difference between the "Handbook of the British Flora" (Bentham 1930) and the "Flora of the British Isles" (Clapham *et al.* 1952) is that while the former nowhere mentions chromosomes, the latter gives the chromosome number(s) for most species. Information concerning hybrids is also mentioned in some Floras, and is available in a collected form (e.g. Stace 1975). Thus, in practice, the influence of genomic characters on plant phenetics has been minimal.

In 1960, Heslop-Harrison (while clearly recognising the importance of cytological evidence for plant taxonomy), wrote with regard to its practical contribution that, "In the majority of cases, it has actually turned out that the data from this source have simply served to justify or reinforce existing classifications based upon morphological criteria. At times cytological evidence has helped to decide between competing morphological classifications, or to suggest improvements or amendments. Only rarely has evidence from this source resulted in major taxonomic revisions." In 1983 the same words accurately summarize the practical contribution of genomic evidence, despite the consider-

able increase in cytological, genetical and molecular knowledge
which has occurred. Thus, the conclusions of recent systematic
studies involving three different genomic characters cited below
are typical. With regard to chromosome banding in *Scilla* and
related genera, Greilhuber (1977) wrote, ".....C-banded karyotypes
are consistent with systematic affinities based on morphological
similarities," while of genome size (i.e. DNA amount) he wrote,
"DNA contents confirm taxonomic relationships supported by
morphology and chromosome number.." (Greilhuber 1979). Similarly,
after constructing a phylogeny for 15 taxa from the Solanaceae
based on parsimony analysis of shared mutation in chloroplast DNA
(detected by restriction endonucleases) Palmer and Zamir (1982)
concluded, "This phylogeny is generally consistent with
relationships based on morphology and crossability but provides
more detailed resolution at several places."

THE SPECIAL IMPORTANCE OF GENOMIC CHARACTERS FOR BIOSYSTEMATICS

In a second sense, genomic characters constitute a uniquely
special type of information, and so they cannot be treated like
other characters. The structure of DNA is unique – it alone
contains the encoded genetic information which is the prime cause
of all other plant anatomy (e.g. floral anatomy is its effect).
The spatial arrangement of genes on the chromosomes is of special
importance because it can affect their expression (e.g. Nicoloff
et al. 1979). The chromosomes too are uniquely important
because their number, and their behavior at fertilization and
meiosis, affects the combination and recombination of genetic
information, and can regulate the amount of variation available to
a population. It is this second view which assumes greater
importance in biosystematic (phylogenetic) studies.

There has been disagreement as to which genomic characters
should be given special emphasis, and to what extent [e.g.
Darlington (1956, 1981) would emphasize the special importance of
the chromosomes much more than Stace (1980)]. Similarly there is
disagreement as to the relative value of cytological *versus*
(I use the term advisedly) molecular genomic studies when trying
to understand developmental and evolutionary processes. For some
forthright and entertaining comments on this issue compare Crick
(1977) with John (1983). However, it is widely agreed that, as
the genome is the quintessential matter of both ontogeny and
phylogeny, then ultimately genomic characters must be of greater
fundamental importance than phenotypic characters, both for
defining taxa, and for understanding the processes which give rise
to them. In other words, it is the implicit or explicit
assumption of much current genomic research that, in order to
answer two of the great questions in plant taxonomy and
biosystematics (notably, "What is a species?" and, "What are the
essential processes in speciation?"), it is essential to study and
compare genomic characters.

The potential of genomic studies to contribute to answers is

undisputed, but, so far, no causal correlation has been
established between variation in any genomic character
(cytological or molecular) and morphological change for evolution.
Some lines from T.S. Elliot's "The Hollow Men" seem apt:

> *Between the idea*
> *And the reality..*
> *Falls the shadow*
>
> *Between the potency*
> *And the existence*
> *Between the essence*
> *And the descent*
> *Falls the shadow*

Certainly, speciation has been shown to involve variation in all
of the genomic characters listed in Table I in different
particular instances. However, the overall picture is still a
mass of special cases (and perhaps it always will be, since the
basis of speciation is likely to differ for different species).
For example, although speciation may involve a change in
chromosome number, there is no predictable correlation between
variation in this caracter and morphological change, as shown by a
species like *Cardamine pratensis* L. which has $2n$ = 16, 24, 28,
30, 32-38, 40-46, 48, 52-64, 67-96 (Clapham *et al*., 2nd Ed., 1962).
Similarly, speciation may involve detailed karyotypic rearrangement, or contrariwise may apparently occur without any significant
change in even the most detailed aspect of karyotype morphology,
since the more than 100 species of Hawaiian picture wing
Drosophila show no significant alterations in the detailed
quantitative appearance of their polytene chromosome banding
(Carson 1981).

Partly because of such findings concerning cytological
characters, interest in recent years has focused increasingl on
molecular genomic characters. A lot of exciting new information
has emerged from such molecular studies on how specific genes have
evolved and diverged. These results may be of considerable value
for understanding the molecular processes responsible for
evolutionary change (Dover 1982). However, the impact of these
and other genomic studies on biosystematics has been minimal, so
far, because (1) such analyses are time consuming, and therefore,
relatively few have been published; and (2) as noted above, no
causal correlation has been established so far, relating the
origin of species to the evolutionary behavior of particular
genome characters (Rose and Doolittle, 1983).

THE NEED FOR GENOMIC DATA FOR BIOSYSTEMATICS

Undoubtedly genomic studies can provide kinds of information about
organisms which are unobtainable by any other means, and are of
prime biosystematic interest. However, their practical
contribution to biosystematics is often minimized by the amounts,
and types, of information which are available. The history of the

genomic studies shows salutary examples of failures to use
established techniques to obtain various kinds of genomic data
either in sufficient quantity *or* with balanced
emphasis. For example, it has been possible to estimate total
genome size (i.e. DNA C-value) for at least 25 years. However,
DNA C-values have been published for less than 2,000 (out of the
estimated 240,000) angiosperm species (Bennett and Smith 1976;
Bennett *et al.* 1982), and even this sample is grossly
unrepresentative of the world flora; yet on average the number of
new estimates published each year is already falling.

Similarly, it has been possible to map genes for more than 40
years. The total number of different genes in one angiosperm
species has been estimated to be about 30,000 (Goldberg *et
al.* 1978). However, the total number of genes mapped by
genetical analysis for all 250,000 angiosperms is about 4,000, of
which the overwhelming majority are in less than 10 species (e.g.
see O'Brien 1982). Thus, if as seems likely, gene maps could make
important practical contributions to biosystematics (not to
mention plant breeding), then clearly those contributions are
almost all still to come! (While the thirst for genomic data of
biosystematic interest is unlikely to change this situation, man's
dependence on angiosperms for food and life may soon provide the
motivation for a concerted program to map and identify more genes
of agronomic importance in a wide range of crop species.
Meanwhile it is interesting to note that in a list of genetic
variation in the important world crop *Vicia faba* published
by ICARDA (Chapman 1981) only 4 genes were assigned to a
chromosome, while in 1983 an international workshop of *faba*
bean scientists set as a target for the next five years the
mapping of 30 new genes in the species!)

It has been possible to count chromosomes for nearly a century,
yet counts exist for only about 6% of bryophytes, 20% of
pteridophytes, and 15-20% of angiosperms (Stace 1980). If Peter
Raven is correct in his assessment (given at the last Botanical
Congress in Sydney in 1981) that as much as 20-25% of all the
diversity of all the living things on earth may become extinct by
early in the next century, then it is a sad reflection on our
species that the rate of reduction in the unfinished task is very
much faster because of the extinction of uncounted species, than
because of the activities of cytologists.

Much of the available data concerning plant genomic characters,
which is of biosystematic interest, arose as a by-product of
experiments made for other reasons. This situation seems unlikely
to change, and so the acquisition of all types of genomic data
will probably continue to depend mainly on the direction and
extent of other plant research.

The molecular approach to studying plant genomes is very
powerful, but it is also technically demanding and expensive.
Thus, it will probably not be possible to apply its sophisticated
methods to enough genomes and genes for it to have any widely
based general influence on plant classification on a 40-year

timescale. Instead, its biosystematic influence will probably be limited to detailed specialist studies comparing mainly the few groups of plants which are of greatest interest to the farmer or the scientist.

THE NEED FOR AN AGREED SYSTEM OF KARYOTYPE NOMENCLATURE

Another factor preventing the wider use of genomic studies for biosystematics is the failure of cytologists to agree to rules for the systematic description or portrayal of karyotypes in a way that geneticists have adopted general rules for gene nomenclature, legalised the binomial. As Stace noted (1980), "Some workers have devised formulae to summarize karyological data concisely, but none of these methods has been generally adopted. Nevertheless, the acceptance of a shorthand formula to designate the karyotype is surely a prerequisite for the more widespread use of cytogenetic data, and would probably lead to its routine inclusion in Floras, etc." Indeed, not only is there no widely accepted shorthand system for describing the shapes of the chromosomes in a karyotype, but often there is also little agreement as to how to refer to a particular chromosome, or chromosome arm in a single species. The chromosomes in the karyotypes of well-known angiosperms have been numbered variously according to: (i) their relative sizes, (ii) their increasing arm ratios, (iii) the order in which they were located as monosomics, and (iv) their homoeology with chromosomes of another species. Thus, the same chromosome in *Secale cereale* (rye) $2n = 2x = 14$ has been numbered 7 (e.g. Heneen and Casperson 1973) because it is the nucleolar organizer, or 1R (e.g. by Bennett and Smith 1975) because it is homoeologous with chromosomes of homoeologous group 1 in bread wheat. Similarly, the two arms of a biarmed chromosome may be variously distinguished as L and S (for long and short), α and β, or p and q in different species, and even in the same species (Sears and Sears 1979).

Many plant cytogeneticists are envious when they compare the state of plant karyotype nomenclature with, for example, the more general agreement on the nomenclature of the standard human karyotype with its precisely defined arms, bands and interband regions (Shows and McAlpine 1979; McKusick 1982), and the ease with which interspecific comparisons of chromosome morphology have been made for different mammals (e.g. Yunis and Prakash 1982). Thus, a trend in plant genomic studies is for the adoption of much more detailed agreed standard karyotypes (especially for species with many clear bands, or knobs in their karyotypes). For example a workshop on rye chromosome nomenclature held in 1982 agreed to establish such a generalized stylized karyotype for *Secale cereale* (Fig. 1).

The objection might be raised that intraspecific variation in chromosome morphology and banding patterns is commonplace, so that a standard karyotype is of limited or no value. This may seem to be a problem at first sight since, for example, the number of

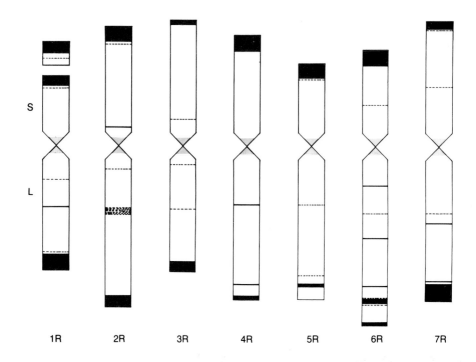

S

L

1R   2R   3R   4R   5R   6R   7R

Secale cereale

FIG. 1 *A generalized karyogram for* Secale cereale *(reproduced
by kind permission of Dr. J.P. Gustafson), established by an
international workshop on rye nomenclature and homoeology
relationships at Wageningen, The Netherlands, in March 1982.
(N.B. The chromosome numbers indicate homoeology with* Triticum
aestivum; *L and S signifies long and short arms of rye;
commonly and less frequently observed bands are shown solid and
dashed, respectively).*

knobs in the karyotype of *Zea mays* can vary from zero to
over 20, and it is correlated with environmental characters such
as latitude and altitude (Anderson and Brown 1952; Bennett 1976).
However, provided there are a limited number of relatively fixed
sites where knobs may be present or absent, then a standard
karyotype showing all their positions is of considerable value.
Detailed studies have already shown this to be so in *Zea mays*
(e.g. see McClintock *et al.* 1981). Moreover, considerable
variation in details of the banding pattern, and in chromosome
morphology in the human population have not rendered the standard
human karyotype of limited value. Far from it: comparisons with
the standard karyotype have allowed such variations to be
routinely identified and accurately described (e.g. Stoll 1980).

THE NATURAL KARYOTYPE

As plant cytogeneticists increasingly mind their p's and q's in establishing detailed standard karyotypes in particular species, the need for a single system of plant karyotype nomenclature to facilitate interspecific comparisons will intensify. The workshop on rye chromosome nomenclature mentioned above adopted a system giving (as its main criterion) the closest possible parallel with wheat chromosome nomenclature. While this system will undoubtedly facilitate useful comparisons of biosystematic interest on a limited scale, nevertheless, it is most unlikely that homoeology with wheat chromosomes (which are numbered according to the order in which they were first isolated as monosomics) would provide an acceptable basis for a widely applicable general scheme. To have maximal significance a system of karyotype nomenclature should be based on biological rather than artificial principles, and moreover, on biological principles of the widest fundamental importance and general application. But what biological principles are applicable? Numbering chromosomes from 1 to n in order of decreasing size seems unlikely to be a sound biological principle, and while giving genetically related chromosomes (homoeologues) in different species the same number does have a biological basis, there is no sound rationale for numbering particular linkage groups 1 or n.

The problem would be solved were it to be shown that nature herself constructs karyotypes of functional significance based on common principles of widespread application, for then the naturally occurring karyotype would be adopted as the correct order of chromosome for any species. Recent work at Cambridge has led to the suggestion that there may indeed be such naturally occurring karyotypes, and that chromosomes may be ordered according to simple principles of widespread biological application and functional significance (Bennett 1982, 1983; Heslop-Harrison and Bennett 1983a, b). This work was undertaken to investigate the spatial arrangement of mitotic metaphase chromosomes in several grass species and hybrids (all with biarmed chromosomes) using electron microscope, serial-section, 3-D reconstruction techniques to study physically undistorted cells. These studies have shown that:

(1) There is a highly significant tendency for haploid genomes to be spatially separated in both interspecific diploid hybrids, and in diploid species. In hybrids the tendency was for a concentric separation of parental genomes (Fig. 2a) (Finch *et al.* 1981), while in the species there was a tendency for side-by-side separation of two haploid genomes (Fig. 2b) (Bennett 1982, 1983). Such results were interpreted as showing that the haploid genome is a basic structural unit in nuclear architecture. In other words, the basic set of chromosomes portrayed together on paper in a karyogram, does tend to be spatially organised as a real unit *in vivo*.

(2) The spatial arrangement of different chromosome types

a.   b.

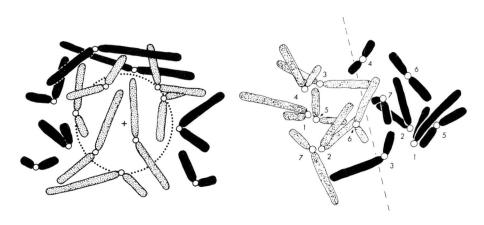

FIG. 2 *Polar views of reconstructed metaphase plates of (a) a root tip cell of* Hordeum vulgare *cv.* Sultan *(stippled)* X Secale africanum *(solid) showing concentric parental genome separation, and (b) a male archesporial cell of* H. vulgare *cv.* 'Tuleen 346' *at premeiotic mitosis showing side-by-side separation of centromeres of two haploid sets of chromosomes (one stippled and one solid). (N.B. The numbers in (b) refer to the linkage groups of the barley centromeres.)*

(heterologues) within a separate haploid set is nonrandom. Indeed, the mean spatial order of heterologues is predictable using a simple model which orders a complete simple haploid genome so that each chromosome is associated with two constant neighbors. Details of the model have been published (Bennett 1982, 1983; Heslop-Harrison and Bennett 1983a). In essence it is based on associating in pairs the most similar long, and the most similar short, chromosome arms throughout the karyotype, except at one point (termed the "discontinuity", or, the "ends"). Fig. 3 shows the chromosomes of *Hordeum bulbosum* and their mean order predicted using the Bennett model. The predicted mean order of chromosomes was significantly expressed (indeed, it was demonstrably the best expressed out of all 360 possible orders) in five out of five tests involving the haploid complements of four grass species: *Aegilops umbellulata*, *Hordeum vulgare*, *H. bulbosum* and *Secale africanum*. All four species were tested in somatic (root tip) cells (Heslop-Harrison and Bennett 1983a) and *H. vulgare* was also tested in germ line cells at premeiotic mitosis (Bennett, unpublished).

(3) The order predicted by the model was shown to be significantly expressed using pooled data for only 10 or fewer reconstructed cells for each genome in each test. In other words, the predicted mean order of heterologues was strongly expressed in undistorted somatic and germ line cells.

Given these results it seems reasonable to conclude that there is a naturally occurring mean order of chromosome types for each of the basic haploid genomes tested (and so hereafter such an order will be referred to as the "natural karyotype"). However, it is important to ask whether the natural karyotype is merely a packing phenomenon, or whether it has other more fundamental implications. The available evidence indicates that the natural karyotype is an important structure whose existence may provide the basis for understanding and interrelating variation in genomic characters as diverse as the gene map, the distribution of C-bands, aspects of chromosome mechanics, karyotype architecture and genome evolution.

THE NATURAL KARYOTYPE AND CHROMOSOME MECHANICS

Although the natural karyotype has been portrayed and analysed as a continuous sequence of chromosomes (Heslop-Harrison and Bennett 1983b) nevertheless its structure is apparently differentiated to contain two chromosomes at what has been conveniently called its "ends" (Bennett 1982; see Ashley 1979). Thus, a "discontinuity" has been noted in the natural karyotype (e.g. Fig. 3), which seems to be much more than a mathematical quirk of the model used to predict chromosome order. The end arms of chromosomes at this discontinuity appear to be functional sites, rich in genes with major effects on the mechanical behavior of chromosomes.

For example, such end arms (Fig. 4) are the sites of the *min* gene (which appears to suppress anaphase in somatic cells of *Hordeum vulgare*); and the *am* (ameiotic) gene and K (abnormal 10) (which inhibit meiosis and cause all knobs to respond precociously as neocentromeres at the two meiotic divisions, in *Zea mays*). Perhaps significantly these genes all map to positions near to the telomeres of the two end arms of the chromosomes at a discontinuity (Fig. 4). Perhaps similar genes controlling similar mechanical functions occur at corresponding sites near the telomeres of end arms in widely unrelated species. Unfortunately, it is not yet possible to test this hypothesis, given the current state of the gene maps for higher plants (as noted earlier).

THE NATURAL KARYOTYPE AND KARYOTYPE EVOLUTION

The karyotypes of many species are nonrandomly organized. For example, arranging chromosomes in order of relative length reveals a strikingly regular rectilinear distribution in many species (Shchapova 1971). Similarly, in karyotypes ordered in this way, acrocentric and telocentric chromosomes occur more frequently next

a.

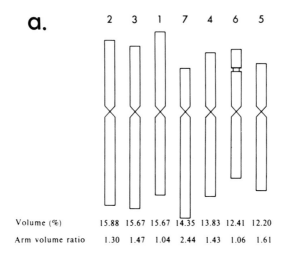

	2	3	1	7	4	6	5
Volume (%)	15.88	15.67	15.67	14.35	13.83	12.41	12.20
Arm volume ratio	1.30	1.47	1.04	2.44	1.43	1.06	1.61

c.

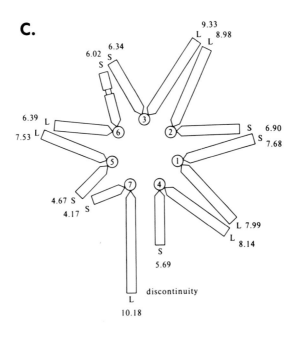

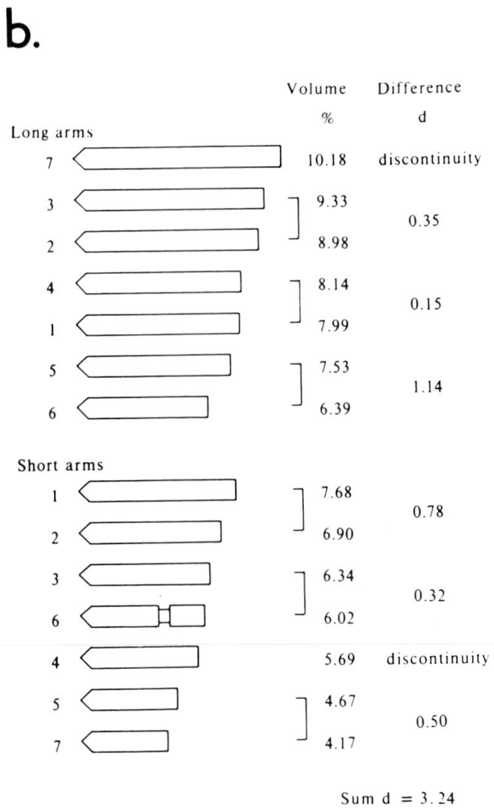

FIG. 3 *The prediction of the natural karyotype of* Hordeum
bulbosum - *clone L6 (2n = 2x = 14). (a) The mean relative
volume (as a percentage) and arm volume ratio of the chromosomes
of* H. bulbosum *clone L6 identified in five reconstructed
root-tip cells in a hybrid with* H. vulgare cv. *'Tuleen 346'.
Numbers above the idiogram refer to presumed homoeologues in the
natural karyotype of* H. vulgare cv. Sultan *(see Fig. 6).
Volumes do not add to 100% because of rounding. (b) The
chromosome arms ranked by decreasing volumes (as a percentage of
the haploid complement) for long and short arms. Those with most
similar volumes, predicted to be adjacent by the model, are linked
and the differences in percentage volume between members of each
pair is shown and differences are summed. This sum of differences
(sum = 3.24) is the minimum possible for all order of arms. (c)
The haploid complement ordered as predicted by Bennett's model.
The numbers beside each long arm (L) and each short arm (S) refer
to their volume as a percentage of the total haploid set. (N.B.
There is one discontinuity where a long and a short arm are
associated by the model.)*

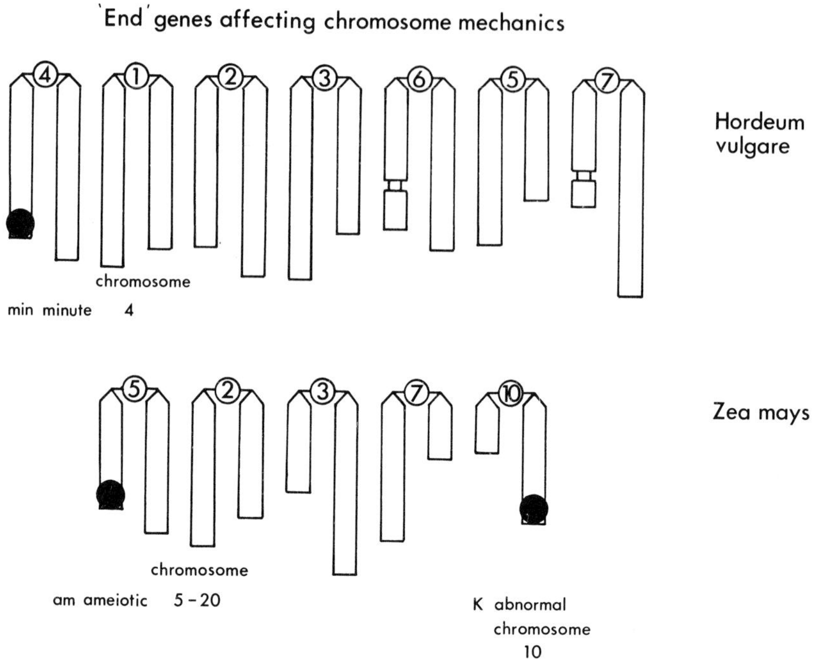

FIG. 4 *Genes, mapped to "end" arms (at the discontinuity) of the natural karyotypes of* Hordeum vulgare *and* Zea mays, *which have profound mechanical effects on the behavior of the whole genome at mitosis or meiosis.*

to other acro- and telocentric chromosomes, than would be the case if there were no correspondence between chromosome length and shape within the karyotype (Bengtsson 1975). Knowledge of the rules for predicting chromosome order may help to explain such phenomena. Thus, nature often appears to favor a karyotype which (for a given range of arm sizes) tends to maximize the differences between pairs of nonadjacent corresponding arms, while perhaps minimizing the differences within pairs of adjacent corresponding arms (Bennett 1983). Many different karyotypic architectures are possible which obey this formula (e.g Fig. 5), but once a particular design is adopted, the rules governing a chromosome's position in the natural karyotype according to its arm length ranking, will often place constraints on further karyotype evolution. The existence of these constraints may help to explain why the relative shapes of the chromosomes are often retained during evolution, even when this involves large changes in DNA C-values (Rees *et al.* 1978; Seal and Rees 1982). One option for changing the DNA C-value while retaining the original ranking of arm lengths, is to vary the amount of DNA in each arm by a constant proportion. Another option is to vary the DNA content of each arm by a constant absolute amount. Both of these options have been taken by different organisms (compare Seal and Rees 1982 with Rees *et al.* 1978; and see the chapter by Rees in this volume). However, the rules of designing stable natural karyotypes may provide the common denominator explaining both strategies in karyotype evolution.

The rules governing a chromosome's position in the natural karyotype according to its relative arm lengths will often place narrow constraints on its ability to undergo unilateral changes in size without also changing its position. Nevertheless, considerable variation in the relative sizes of some chromosome arms is possible while still retaining the order of chromosomes in an adopted natural karyotype design (Fig. 5). For example, an association between the two longest arms would remain intact even if (i) the longest arm became longer, or (ii) the second longest arm became the longest arm. While the length of most arms associated in pairs is highly constrained, arms at the discontinuity may vary in size independently. Thus, when the relative sizes of arms at all corresponding positions in the natural karyotypes of closely-related species are compared, arms at the discontinuity are expected to show the greatest interspecific differences. So far only one test of this hypothesis is possible (comparing the known natural karyotypes of *Hordeum vulgare* and *H. bulbosum*) and this fits the expectation (Fig. 6).

The natural karyotype is formed by associations in pairs of corresponding arms from heterologues, and so the two arms within such pairs are expected to vary nonindependently in genome evolution. The distribution of major C-bands [especially those at the telomeres, where C-bands can be rapidly gained or lost (Gustafson *et al.* 1983)] may provide important evidence relating to this aspect of genome evolution (Bennett 1982).

THE NATURAL KARYOTYPE AND C-BAND DISTRIBUTION

The distribution of the major C-bands on the chromosomes of *Secale africanum* arranged in the classical karyogram reveals no obvious pattern (Fig. 7a). However, superimposing this C-banding pattern on the known natural karyotype (Bennett 1982) shows that most of the variation (except significantly that at the discontinuity) is accounted for either by the concurrence of C-bands of matching size, or by the absence of C-bands at both telomeres of adjacent corresponding arms (Fig. 7b). Similar distributions of C-bands have been suggested for other plant species (Greilhuber and Loidl 1983).

The distribution of major C-bands in *Secale* species corresponds with the quantitative distribution of several noncoding, highly repeated DNA sequences (Bedbrook *et al.* 1980; Jones and Flavell 1982). Thus, the karyotypic distribution of at least some noncoding DNA sequences is nonrandom, insofar as

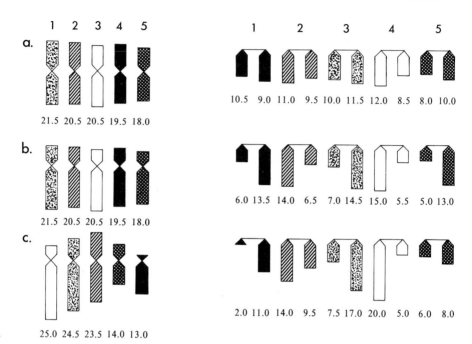

FIG. 5. *Comparisons of the classical (left) and the natural (right) karyotypes of three hypothetical, related species (all 2n = 2 x = 10) showing considerable interspecific variation in chromosome shape and relative total size (left) while retaining a basic design with a constant order of heterologues in the natural karyotype with the same pairs of long, and pairs of short, arms associated (right). Note the likely greater biological significance for understanding karyotype evolution of comparing chromosomes with corresponding numbers and positions in the natural, rather than in classical, karyotypes. (N.B. Chromosomes with the same shading are homoeologues. Numbers under chromosomes and arms are volumes (as a percentage of the total haploid complement).*

they tend to occur most frequently at corresponding sites on adjacent heterologues. If this is true for noncoding sequences in heterochromatin, it is obviously important to discover whether the distribution of related coding sequences (genes) is similarly nonrandom.

THE NATURAL KARYOTYPE AND NONRANDOM GENE DISTRIBUTION

The first test of whether the distribution of related genes between chromosomes is random, or whether they tend to be grouped in super-domains extending over heterologues which are adjacent in

the natural karyotype, used Zea mays ($2n$ = 20) because it has
probably the most extensive and reliable gene map of any higher
plant species. The Bennett model strongly suggests that the 10
chromosome types are divided into two natural karyotypes, each
with 5 chromosomes (Fig. 8a). A Procrustes analysis (Heslop-
Harrison 1983) of chromosome dispositions in reconstructed cells
of *Z. mays* cv. 'Seneca 60' showed a significant expression of the
predicted orders. Genes were taken to be related if they were
paralogous (i.e. had the same name). Analysis of the distribution
of paralogous genes which mapped to 2 or 3 different chromosomes
showed a significant excess of paralogous genes on (i) pairs of
chromosomes at corresponding positions in the two predicted
natural karyotypes (e.g. Fig. 8B), and on (ii) adjacent
heterologues within the two predicted natural karyotypes (e.g.
Fig. 8c) compared with random expectation (Bennett 1983). These
results were interpreted as showing (i) that *Z. mays* is a
tetraploid with considerable homoeology between chromosomes at
corresponding positions in its two natural karyotypes, and (ii)
that the distribution of coding DNA sequences is nonrandom - since
related genes show a strong tendency to be grouped on adjacent
heterologues.

If, as seems likely, the results for *Zea* are typical, then
it will be important to discover whether such nonrandom gene
distributions are merely a consequence of the mean relative
proximities of chromosomes in the natural karyotype *in vivo*, or
whether they have a functional significance, perhaps for gene
action.

PERCEPTION AND PORTRAYAL OF THE GENOME FOR BIOSYSTEMATICS

Taken together, the above results provide a novel way of
perceiving the genome. It starts by reassembling the classical
karyotype as a natural karyotype, then superimposes on this the
physical locations of DNA sequences detected as coding genes or as
repeated noncoding sequences (e.g. many heterochromatin bands),
converts the resulting 2-dimensional structure into dynamic 3-
dimensional nuclear architecture, and finally asks whether
modulation of this structure has a role in controlling gene
expression including some of the major steps in ontogeny. Thus,
this new view takes the somewhat disconnected perspectives of the
cytologist, the geneticist and (to some extent) that of the
molecular biologist, and integrates them into a unified concept in
which genome function modifies genome form (and perhaps *vice
versa*) in both ontogeny and phylogeny.

If the model described above correctly predicts the mean
spatial order of chromosomes in the basic karyotypes of a wide
range of plant species with bi-armed chromosomes, then the natural
karyotype may provide an acceptable new system of karyotype
nomenclature with a sound biological basis. If so, the
chromosomes in all such taxa would be numbered according to their
order in the natural karyotype - those at the discontinuity being

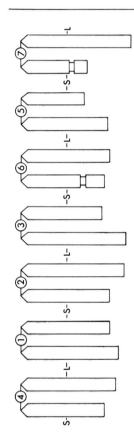

HORDEUM VULGARE CV. SULTAN (V)

VOLUME (%)   6.84 7.69   7.93 7.21   7.09 8.30   8.35 6.53   6.66 7.15   6.99 5.05   5.26 8.98

HORDEUM BULBOSUM CLONE L6 (B)

VOLUME (%)   5.69 8.14   7.99 7.68   6.90 8.98   9.33 6.34   6.02 6.39   7.53 4.67   4.17 10.18

DIFFERENCE   −1.15+0.44 +0.06+0.47 −0.19+0.68 +0.98 −0.19 −0.64−0.76 +0.54−0.38 −1.09 +1.20

RANKED DIFFERENCE
BETWEEN ARMS (B−V)

1 +1.20 7L discontinuity
2 −1.15 4S discontinuity
3 −1.09 7S
4 +0.98 3L
5 −0.76 6L
6 +0.68 2L
7 −0.64 6S
8 +0.54 5L
9 +0.47 1S
10 +0.44 4L
11 −0.38 5S
12 −0.19 2S/3L
13 −0.19 2S/3L
14 +0.06 1L

Figure 6

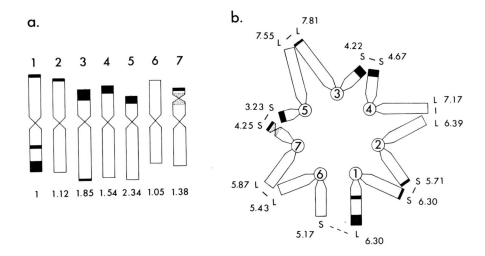

FIG. 7 *The haploid complement of* Secale africanum
showing the major C-bands (a) the classical karyotype numbered in
order of decreasing total chromosome size (from left to right; arm
volume ratios are given below each chromosome); (b) the natural
karyotype ordered according to Bennett's model (numbered by arms
are their mean volumes in μm^3 *in root tip cells) which was*
significantly expressed in reconstructed cells (Bennett 1982).

numbered 1 and n (e.g. Fig. 10). If different taxa exhibit
either some form of constant polarity throughout their
natural karyotypes, *or* some regular differentiation in form
or function between the two chromosomes at the discontinuity, this
would provide a natural basis for deciding which chromosome in a
natural karyotype should always be called number 1.
   The genome is 3-dimensional *in vivo*, but it will probably be more
convenient to portray the natural karyotype in 2 dimensions,
perhaps with centromeres arranged either in a linear array (as in

---

FIG. 6 *A comparison of the natural karyotypes of* Hordeum
vulgare *and* H. bulbosum. *The numbers in circles are*
the linkage groups in H. vulgare, *and their presumed*
homoeologues in H. bulbosum. *Numbers beneath each short*
(S) and long (L) arm refer to their volumes as a percentage of
the total haploid set in each species. The differences between
arms at all 14 corresponding positions in the two natural
karyotypes (bulbosum – vulgare) are shown underlined.
Ranking these differences (right) shows that the greatest
interspecific variation involved the two "end" arms, at the
discontinuity (see also Fig. 3).

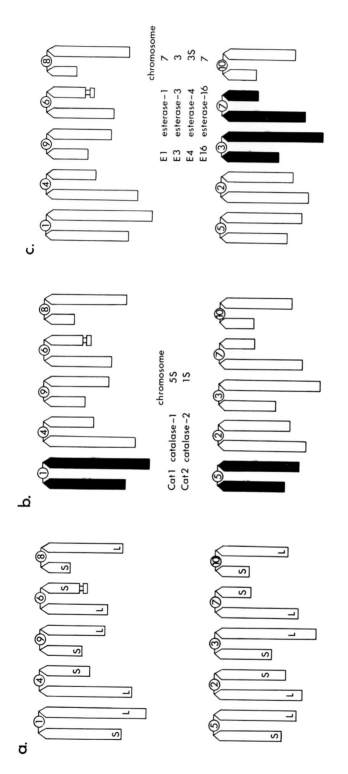

FIG. 8 The probable natural karyotypes of Zea mays with two sub-sets (genomes?) of chromosomes; (b) two paralogous genes mapped to a pair of putative adjacent heterologues. (N.B. L and S designate Long and Short arms; numbers in circles are linkage groups; other information is from the 1983 Maize Genet. Coop. News Lett.)

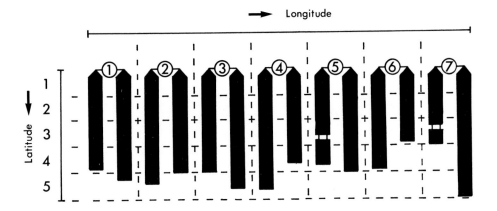

FIG. 9 *The natural karyotype of* Hordeum vulgare *L. cv.
'Sultan' with a superimposed grid allowing features to be
compared, in terms of their genomic 'latitude' and 'longitude'.
[N.B. Chromosomes are numbered, not according to their linkage
groups (as in Fig. 6), but according to their positions in the
natural karyotype starting at the discontinuity (as in Fig.10)].*

Fig. 4), or in a polygon (as in Fig. 3). Any final decision must
depend on understanding which type of configuration has the most
biological significance.

There are good reasons to expect a tendency for related
chromosome segments and DNA sequences to occur at physically
similar locations within the natural karyotype (Bennett 1982;
Greilhuber & Loidl 1983). Thus, it might be useful to superimpose
a grid on the natural karyotype and to refer to the position of
each segment in terms of what might be called its latitude and
longitude (Fig. 9). Thus, segments occurring at similar physical
distances from the centromere on different chromosomes would have
a similar karyotypical latitude but different longitudes, while
related genes occurring on adjacent heterologues would have
similar karyotypic longitudes. This way of portraying the genome
would have further significance if the natural karyotype were
shown to exhibit a relatively constant tertiary structure during
interphase, so that particular segments in a constant cell type
have relatively fixed mean latitudes and longitudes *in vivo*.

There are also good reasons to expect a tendency for related
DNA sequences to tend to occur in similar locations in the natural
karyotypes of related or even unrelated species. For example, the
probable order of chromosomes in the natural karyotype of
Aegilops umbellulata (expressed in terms of their homoeology with
bread wheat) —4-3-7-2-6-5-1 is almost identical with the supposed
order (similarly expressed) in *Secale cereale*, 1-3-7-2-4-6-5-
(Heslop-Harrison & Bennett 1983b). Apparently only one

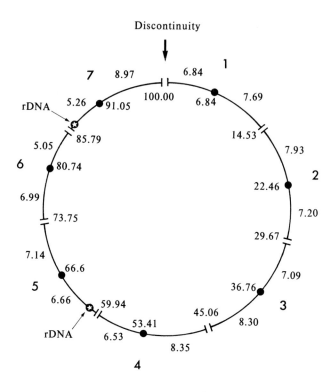

FIG. 10 *The natural karyotype of* Hordeum vulgare *cv 'Sultan'*
(from Heslop-Harrison and Bennett 1983a) portrayed diagrammati-
cally as a circle whose circumference equals 100 percentage units
(with clockwise polarity). Chromosomes, numbered 1 to 7 (large
numbers outside the circle) in a clockwise direction starting at
the discontinuity, are represented by arcs. Small numbers outside
the circle are the relative volumes of the 14 arms (as a
percentage of the total haploid set). Telomeres of arms at the
discontinuity meet at 100% and telomeres of pairs of adjacent
corresponding arms meet at intervals proportional to chromosome
volume. The positions of centromeres (solid circles) and
telomeres (radial lines at the ends of chromosome arcs) within the
whole genome are given (numbers inside the circle) as the
cumulative percentage of arm volumes moving clockwise from the
discontinuity.

chromosome (4R) may be moved in its position with respect to the
order in *Ae. umbellulata.* Thus, homoeologues may tend to occur
in corresponding positions in the natural karyotypes of related
species. If so, there is a good reason to expect corresponding
genes to tend to occur in corresponding locations in closely
related natural karyotypes. Interestingly, the location of the

rRNA genes (the only coding genes whose physical locations are accurately known in both *Hordeum bulbosum* and *H. vulgare*) in the natural karyotype of *Hordeum bulbosum* corresponds closely with one of their two locations in the natural karyotype of *H. vulgare* (Fig. 6).

In order to facilitate biosystematic comparisons of widely unrelated taxa, it may also be meaningful to compare genomes portrayed as either a straight line or as a circle (Fig. 10) of constant relative length (i.e. 100%). The latter might be most useful as it would facilitate comparisons with the circular DNA molecules of bacterial genomes. For eukaryotes the circle would be subdivided into arcs representing chromosome arms according to their relative physical sizes or DNA contents, and according to their order in the natural karyotype, with the discontinuity in a constant position (e.g. Fig. 10). The physical locations of a feature could then be given not only within a chromosome, but also (for the first time) in terms of a relative position in the genome as a whole, independent of the number of chromosomes in the natural karyotype. The positions of all telomeres and centromeres as cumulative percentages within the genome (e.g. Fig. 10) may provide a simple shorthand system precisely describing the number, shapes and order of chromosomes in the natural karyotype. Moreover, interspecific comparisons of genomes portrayed in this way would show whether or not some features of genome architecture are relatively independent of variation in characters such as genome size or basic chromosome number (x).

REFERENCES

Anderson, E. and Brown, W.L. 1952. Origin of corn belt maize and its genetic significance. In *Heterosis* (J.W. Gowen, ed.). pp. 124–148.

Ashley, T. 1979. Specific end-to-end attachment of chromosomes in *Ornithogalum virens*. *J. Cell Sci.* 38: 357–367.

Bedbrook, J.R., Jones, J., O'Dell, M., Thompson, R.D. and Flavell, R.B. 1980. Molecular characterisation of telomeric heterochromatin in *Secale* species. *Cell* 19:545–560.

Bengtsson, B.O. 1975. Mammalian chromosomes similar in length are also similar in shape. *Hereditas* 79: 287–292.

Bennett, M.D. 1976. DNA amount, latitude and crop plant distribution. In *Current Chromosome Research* (K. Jones and P.E. Brandham, eds.), North Holland, Amsterdam. pp. 151–158.

Bennett, M.D. 1982. The nucleotypic basis of the spatial ordering of chromosomes in eukaryotes and the implications of the order for genome evolution and phenotypic variation. In *Genome Evolution* (G.A. Dover and R.B Flavell, eds.), Academic Press, London. pp. 239–261.

Bennett, M.D. 1983. The spatial distribution of chromosomes. In *Kew Chromosome Conference II*, (P.E. Brandham and M.D. Bennett, eds.). Allen & Unwin, London. pp. 71–79.

Bennett, M.D. and Smith, J.B. 1975. Confirmation of the

identification of the rye chromosome in 1B/1R wheat-rye
chromosome substitution and translocation lines. *Can. J.
Genet. Cytol.* 17: 117-120.

Bennett, M.D. and Smith, J.B. 1976. Nuclear DNA amounts in
angiosperms. *Philos. Trans. R. Soc. Lond. Ser.B* 274: 227-274.

Bennett, M.D., Smith, J.B. and Heslop-Harrison, J.S. 1982.
Nuclear DNA amounts in angiosperms. *Proc. R. Soc. Lond. Ser.
B* 216: 179-199.

Bentham, G. 1930. *Handbook of the British Flora* (5th ed.
revised by Sir J.D. Hooker; 7th ed. revised by A.B. Rendle).
Reeve, Ashford, Kent.

Carson, H.L. 1981. Homosequential species of Hawaiian
Drosophila. In *Chromosomes Today 7* (M.D. Bennett
et al. eds.). Allen & Unwin, London. pp. 150-164.

Chapman, G.P. 1981. Genetic variation within *Vicia faba*. Faba
Bean Inf. Serv.

Clapham, A.R. Tutin, T.G. and Warburg, E.F. 1952. *Flora of the
British Isles*. Cambridge Univ. Press.

Crick, C.D. 1977. Postscript, *Chromosomes Today 6* (A. de la
Chapelle and M. Sorsa, eds.), Elsvier/North Holland, Amsterdam.
pp. 403-406.

Darlington, C.D. 1956. *Chromosome Botany* (and the origins
of cultivated plants). Allen & Unwin, London.

Darlington, C.D. 1981. Chromosomes and organisms: the
evolutionary paradoxes. *Chromosomes Today 7* (M.D. Bennett,
et al. eds.). Allen & Unwin, London. pp. 1-6.

Dover, G.A. 1982. Molecular drive; a cohesive mode of species
evolution. *Nature (London)* 299: 111-117.

Finch, R.A., Smith, J.B. and Bennett, M.D. 1981. *Hordeum* and
Secale mitotic genomes lie apart in a hybrid. J. Cell Sci.
52, 391-403.

Flavell, R.B. 1982. Chromosome DNA sequences and their
organisation. In *Encyclopedia of Plant Physiology New
Series, 14B. Nucleic aicds and protein in plants II*. (B.
Parthier and D. Boulter, eds.). Springer-Verlag, Berlin. pp.
46-74.

Goldberg, R.B., Hoschek, G. and Kamalay, J.C. 1978. Sequence
complexity of nuclear and polysomal RBA in leaves of the tobacco
plant. *Cell* 14: 123-131.

Greilhuber, J. 1977. Nuclear DNA and heterochromatin contents in
the *Scilla hohenackeri* Group, *S. persica,* and
Puschkinia scilloides (Lilliaceae). *Plant Syst.
Evol.* 128: 243-257.

Greilhuber, J. 1979. Evolutionary changes of DNA and
heterochromatin amounts in the *Scilla bifolia* group
(Liliaceae). *Plant Syst. Evol.* Suppl. 2: 263-280.

Gustafson, J.P., Lukaszewski, A.J. and Bennett, M.D. 1983.
Somatic deletion and/or redistribution of telomeric
heterochromatin in the genus *Secale* and in Triticale.
Chromosoma. In press.

Heneen, W.K. and Caspersson, T. 1973. Identification of the

chromosomes of rye by distribution patterns of DNA. *Hereditas* 74: 259-272.

Heslop-Harrison, J. 1960. *New concepts in flowering-plant taxonomy*. *2nd reprint*. Heinemann, London.

Heslop-Harrison, J.S. 1983. Chromosome disposition in *Aegilops umbellulata*. In *Kew Chromosome Conference II* (P.E. Brandham and M.D. Bennett, eds.). Allen & Unwin, London. pp. 73.

Heslop-Harrison, J.S. and Bennett, M.D. 1983a: Prediction and analysis of spatial order in haploid genome complements. *Proc. R. Soc. Lond. Ser. B* 218: 211-213.

Heslop-Harrison, J.S. and Bennett, M.D. 1983b. The spatial order of chromosomes in root-tip metaphases of *Aegilops umbellulata*. *Proc. R. Soc. Lond. Ser. B* 218: 225-239.

John, B. 1983. Through the looking glass: a sceptical cytologist in a molecular wonderland. In *Kew Chromosome Conference II* (P.E. Brandham and M.D. Bennett, eds.). Allen & Unwin, London. pp. 305-310.

Jones, J.D.H. and Flavell, R.B. 1982. The structure, amount and chromosomal localisation of defined repeated DNA sequences in species of the genus *Secale*. *Chromosoma* 86: 613-641.

McClintock, B., Takeo, A.K.Y. and Blumenschein, A. 1981. Chromosome constitution of races of maize. *Col. Postgraduados*, Chapingo, Mexico.

McKusick, V.A. 1982. The human genome through the eyes of a clinical geneticist. *Cytogenet. Cell Genet.* 32: 7-23.

Nicoloff, H., Anastassova-Kristeva, M., Rieger, R. and Kunzel, G. 1979. "Nuclear Dominance" as observed in barley translocation lies with specifically reconstructed SAT chromosomes. *Theor. Appl. Genet.* 55: 247-251.

O'Brien, J. 1982. *Genetic Maps 2*. pp. 406, (Natl. Cancer Inst., Frederick, Maryland, U.S.A.)

Palmer, J.D. and Zamir, D. 1982. Chloroplast DNA evolution and phylogenetic relationships in *Lycopersicon*. *Proc. Natl. Acad. Sci. U.S.A.* 79: 5006-5010.

Rees, H. 1984. Nuclear DNA variation and the homology of chromosomes. In *Plant Biosystematics* (W.F. Grant, ed.). Academic Press, Toronto.

Rees, H., Shaw, D.D. and Wilkinson, P. 1978. Nuclear DNA variation among acridid grasshoppers. *Proc. R. Soc. Lond. Ser. B* 202: 517-525.

Rose, M.R. and Doolittle, W.F. 1983. Molecular biological mechanisms of speciation. Science 220: 157-162.

Schweizer, D. 1973. Differential staining of plant chromosomes with Giesma. *Chromosoma* 40: 307-320.

Schweizer, D. 1980. Fluorescent chromosome banding in plants: applications, mechanisms, and implications for chromosome structure. In *The Plant Genome*. 4th John Innes Sympos. and 2nd Int. Haploid Conf. (D.R. Davies and D.A. Hopwood, eds.). Norwich. pp. 61-72.

Seal, A.B. and Rees, H. 1982. The distribution of quantitative DNA changes associated with the evolution of diploid

Festuceae. *Heredity* 49: 179–190.

Sears, E.R. and Sears, L.M.S. 1979. The telocentric chromosomes of common wheat. 5th Int. Wheat Genet. Symp. pp. 389–407.

Shchapova, A.I. 1971. On the karyotype pattern and the chromosome arrangement in the interphase nucleus. (Russian, English summary). *Tsitologia* 13: 1157–1164.

Shows, T.B. and McAlpine, P.J. 1979. The 1979 catalog of human genes and chromosome assignments. *Cytogenet. Cell Genet.* 25: 117–127.

Stace, C.A. 1975. *Hybridisation and the flora of the British Isles*. Academic Press, London.

Stace, C.A. 1980. *Plant Taxonomy and Biosystematics*. Edward Arnold, Pitman Press, Bath.

Stoll, C. 1980. Nonrandom distribution of exchange points in patients with reciprocal translocations. *Hum. Genet.* 56: 89–93.

Sun, S.M., Slightom, J.L. and Hall, T.C. 1981. Intervening sequences in a plant genome–comparison of the partial sequence of cDNA and genomic DNA of French bean phaseolin. *Nature (London)* 289: 37–41.

Watson, J.D. and Crick, F.H.C. 1953a. Molecular structure of nucleic acids. A structure for deoxyribose nucleic acid. *Nature* 171: 737–738.

Watson, J.D. and Crick, F.H.C. 1953b. Genetical implications of the structure of deoxyribonucleic acid. *Nature (London)* 171: 964–967.

Yunis, J.J. and Prakash, O. 1982. The origin of man: A chromosomal pictorial legacy. *Science* 215: 1525–1529.

Chromosome Pairing in Species and Hybrids

R. C. Jackson
Department of Biology
Texas Technical University
Lubbock, Texas, U.S.A.

INTRODUCTION

Many biosystematic studies over the past 40 years have included cytological observations, but all too frequently the authors have not derived the maximum amount of information from their preparations. The majority of papers could be classified as alpha cytology in the sense of White (1978). This is regrettable because a considerable amount of labor is involved in collecting, fixing, staining, and examining cytological preparations, and one should obtain as much information as possible from the efforts expended. I suspect that the shortcomings in the observations are due mainly to the lack of training of the particular investigator. Many biosystematists have learned to count chromosomes from someone who knows the techniques, but neither person may have had formal training in cytogenetics or have taken the time to read a suitable textbook. Hopefully, this problem will be alleviated in the next few years, and biosystematists as a group will take advantage of current techniques and methods of data analyses. A number of papers in this symposium should serve as exemplars.

A major shortcoming in the cytological analyses of species and hybrids has been the inadequate quantification of data and the misuse of terms. In reading biosystematic papers that discuss cytological findings, one often finds a statement like "pairing at meiosis was good" or some other not very quantitative term. The use of the adjectives good or poor may be highly subjective, depending on the meiotic stage analyzed and the sophistication of the observer. Further examination of many articles brings out the fact that pairing was not studied at all, only stages following synapsis were observed. Pairing or synapsis is by definition initiated at zygotene and is completed at pachytene insofar as a particular genetic system dictates. Therefore, the general statement that "pairing was good" may indicate that only

diakinesis or metaphase I stages were examined. An example of the
disparity between what is observed at pachytene and metaphase I
was provided many years ago by Levan (1940). He found that the
autotetraploid *Allium porrum* regularly had one to as many as
eight quadrivalents per cell at pachytene, but quadrivalents at
metaphase I were so rare that an intense study was needed to find
one. A search of 250 meiocytes on one slide yielded only one
quadrivalent, and a study of 130 meiocytes on a second slide
produced only four. These are larger cell sample sizes than
reported for most studies, so in similar situations and with
smaller sample sizes the probability of seeing a quadrivalent
would be very small indeed. A sample size of about 750 meiocytes
would be needed to have a 95% chance of observing a quadrivalent
in the first slide, and a sample size of 95 would be required in
the second slide.

Not only is the quantification of data important but the kind
of information collected should be considered carefully. Over the
past four years, my colleagues and I often have been frustrated
during our literature search for cytological data on polyploids.
Frequently some missing vital data items could have been supplied
by the author if it had been thought necessary. The lesson to be
learned from this is that we should quantify and present at least
the following kinds of data that can be collected from quality
light microscope preparations: (1) basic and somatic chromosome
numbers; (2) relative or actual mitotic metaphase chromosome
lengths, arm ratios, and differentially staining regions if
present; (3) number of meiocytes analyzed at each stage; (4)
synapsis as observed at early pachytene; (5) number or frequency
of chiasmata per cell, and any differences among meiotic
configurations; (6) univalent frequency; (7) diakinesis or
metaphase I (MI) configurations and their frequencies; (8) early
anaphase I (AI) disjunction patterns, especially for multivalents
or heteromorphic bivalents; (9) chromosome behavior at AI and AII;
and (10) pollen stainability. Other data may be taken as the
occasion demands.

With these introductory admonitions and pleas for more
effective data taking, I would like to briefly summarize current
findings on chromosome pairing, develop some models for chiasma
distribution and synapsis in diploids, discuss chromosome pairing
in species and hybrids, and suggest some new perspectives on the
evolutionary implication of chromosome pairing.

CHROMOSOME SYNAPSIS

By definition, the beginning of synapsis is zygotene. When
pairing has been completed as far as dictated by a particular
genetic system, the pachytene stage has been attained. Not all
chromosomes complete synapsis, and examples are known among
both plants and animals (Darlington 1937; cf. Moens and Short
1983; Oakley 1983). Interest in the early pachytene stage stems
from a desire to determine if the paired chromosomes are

structurally the same because small differences can be detected at
this stage that may not be observed later, viz., duplications,
deletions, inversions, translocations, transpositions, and
insertions.

Beginning with the seminal article by Moses (1956), there has
been an increasing number of papers describing the ultrastructure
of leptotene, zygotene, pachytene, and diplotene stages of
meiosis. A summary of the general findings are the following:
(1) The leptotene chromosomes are attached to the nuclear membrane
at least by their ends, and the lateral components (LCs) of the
synaptonemal complex (SC) are present at this stage. If they are
not so positioned initially, the leptotene chromosomes must be
brought into close proximity (ca. 200–300 nm) before the central
component (CC) of the tripartite SC is formed. (2) The length of
the SC is proportional among bivalents in a nucleus, but
internuclear variation may occur. There may be relatively small
changes in SC length from zygotene to late pachytene, but those
are proportional within a nucleus. (3) The amount of DNA
"trapped" in the LCs is less than 1% of the total of the genomes.
(4) Normal crossing over does not occur in the absence of SCs, but
the present of normal-appearing SCs does not insure crossing over,
thus indicating the complexity of the process. (5) In species
lacking chiasmata in one sex, the SC may persist in a normal or a
modified form until MI so as to assure normal AI disjunction.
(6) Species that may differ by as much as 40% in total DNA may
form normal and uninterrupted SCs and have regular crossing over.
(7) Excess amounts of DNA in one chromosome of a bivalent or other
differences in structure that cause buckles may undergo secondary
adjustment during mid-to-late pachytene in a process originally
described by Darlington (1935) as torsion pairing. The
elimination of buckles does not cause an interruption of the SC.
(8) In normal systems, the breakdown of SCs initiate the diplotene
stage, and SC fragments are usually absent by late diakinesis
(Menzel and Price 1966; Moses 1968; Westergaard and von Wettstein
1972; Ting 1973; Gillies 1974, 1975, 1981; Stern et al. 1975;
Holm 1977; Moens 1978; Moses et al. 1979; Rees and Jenkins 1982).

MODELS OF CHROMOSOME PAIRING IN DIPLOIDS

Normal Pairing and Nonrandom Distribution of Chiasmata:
Normal synapsis is expected in diploids, and deviations from
normality are due to mutations that affect the entire chromosome
complement in the same way or in a chromosome specific fashion.
If synapsis is normal, pairing is complete in both arms of
homologous chromosomes. Synapsis is complete also in telocentric
chromosomes. The remaining factors that will affect meiotic
configurations at diakinesis or MI is the number and distribution
of chiasmata among synapsed homologous chromosomes. Chiasma
frequency may vary among genotypes of a population due both to
genotypic (Rees 1961; Jones 1967) and environmental effects
(Grant 1952; Wilson 1959; cf. Jackson 1976; 1982; Jackson and

Casey 1980). The distribution of chiasmata among bivalents of
normal diploids is not random (Jones 1967; Jackson 1982; Jackson
and Hauber 1982) because each bivalent will have at least one
chiasma; univalents are rare and unpredictable. However, models
can be derived that will predict such nonrandom distribution, and
these can be used to detect deviations from normality in species
and hybrids.

In the equations derived and used in the following models, I
assume that n = 7 and that all of the bivalents have the
same chiasma frequency and are of about equal size. Symbols   used
consistenly throughout are the following: Cx = mean chiasma
frequency per cell; n = number of bivalents; II = bivalent
at diakinesis or metaphase; II_1, II_2, II_3, II_4 are, respectively,
bivalents with one, two, three, and four chiasmata;
I = univalent. P is the value obtained by dividing the observed
mean chiasma number per cell by the maximum expected. In some
case n is subtracted from the mean and maximum chiasma
number prior to division.

In a model with a maximum of two chiasmata per bivalent, only
II_1 and II_2 are expected at diakinesis. Assume that in analyzing
a species with n = 7, the average chiasma frequency per cell
is 12. If there is a genetic requirement of one chiasma per
bivalent, then each of the seven bivalents will *a priori*
have one. Thus 12 − 7 = 5 remaining chiasmata that can be
allocated at random. Since each bivalent already has one chiasma,
each of the remaining five will be allocated randomly so that
there are five II_2 per cell; it follows that $7II$ _ $5II_2$ = $2II_1$ per
cell. Another way to look at this mathematically is the
following:

$$II_2 = (CX - n) \qquad\qquad\qquad \text{Eq. 1}$$
$$\therefore\ n - (Cx - n) = II_1 \qquad\qquad \text{Eq. 2}$$

The methodology is somewhat more complex for a system with a
maximum of three chiasmata per bivalent. As in the preceding
example, the mean chiasma frequency per cell is first determined.
Then the bivalent number of chiasmata is subtracted from the mean
because one chiasma per bivalent is allocated *a priori*.
The remaining number is then divided by $3n - n$ because one
of the theoretical maximum of three chiasmata per bivalent
already has been allocated. The quotient from this is the P
value and represents the probability of chiasma formation; Q or
1-P is the probability of not forming a chiasma. The equations
are as follow:

$$\frac{Cx - n}{3n - n} = \frac{Cx - n}{2n} = P \qquad \text{Eq. 3}$$

$$\therefore 1 - \frac{CX - n}{2n} = Q \qquad \text{Eq. 4}$$

Values of P and Q can then be used to determine the distribution of chiasmata in excess of the one per bivalent already allocated; this cannot exceed two. Thus random distribution by the binomial $(P + Q)^2$ gives

$$P^2 + 2P^1Q^1 + Q^2 = \Sigma \text{ of excess chiasmata} \qquad \text{Eq. 5}$$
$$\text{distribution}$$

which represents the probabilities of two, one, and zero chiasmata reading from left to right. Note that one chiasma per bivalent was allocated *a priori*, but this chiasma is added only to the subscript chiasma number for each kind of bivalent after calculating the frequencies for equation 5. The coefficients and terms for the average diakinesis configuration per cell then are as follow: $II_3 = (P^2) \cdot n$; $II_2 = (2P^1Q^1) \cdot n$; $II_1 = (Q^2) \cdot n$. Note that (P^2) indicates a two chiasmata bivalent, but remember that one chiasma was already present so this gives a three chiasmata bivalent, II_3, and so on for the other bivalent types. An example will demonstrate the methodology for use. Assume $n = 7$, a maximum of three chiasmata per bivalent, and $Cx = 18$. The P value is equal to 18 − 7 divided by 21 − 7 = 0.7857; 1 − 0.7857 = 0.2143 = Q (Equations 3,4). Therefore, $II_3 = (0.7857)^2 \cdot 7 = 4.3214$; $II_2 = 2(0.7857 \times 0.2143) \cdot 7 = 2.3572$; $II_1 = (0.2143)^2 \cdot 7 = 0.3214$. Note that the total bivalents sum to seven and the total chiasmata utilized is 18 so unity is achieved.

The methodology for bivalents with a maximum of four chiasmata is much the same as that for three. One chiasma per bivalent is allocated *a priori* and is used only to indicate the final chiasma number per bivalent. It does not enter into the binomial calculations. The P value is derived as before with modifications only for the change in the maximum number of chiasmata as:

$$\frac{Cx - n}{4n - n} = \frac{Cx - n}{3n} = P \qquad \text{Eq. 6}$$

$$\therefore 1 - \frac{CX - n}{3n} = Q \qquad \text{Eq. 7}$$

The binomial for the random distribution of chiasmata remaining after subtracting one for each bivalent is:

$$P^3 + 3P^2Q^1 + 3P^1Q^2 + Q^3 = \Sigma \text{ of excess chiasmata} \qquad \text{Eq. 8}$$
$$\text{distribution}$$

Each coefficient and term of the expansion from left to right
represents bivalents with 4, 3, 2, and 1 chiasmata because after
the final calculation each of the terms has the *a priori*
chiasma added. Thus the coefficients and terms for cell means
become: $II_4 = (P^3) \cdot n$; $II_3 = 3(P^2Q^1) \cdot n$;
$II_2 = 3(P^1Q^2) \cdot n$; $II_1 = (Q^3) \cdot n$. Proof of unity for bivalent
and chiasma number can be shown by the following example. Assume
$n = 7$ and $Cx = 20$. The P value (Eq. 6) is 0.6190;
$Q = 0.3810$. Expected diakinesis configurations and their
frequencies are the following; $II_4 = (0.6190)^3 \cdot 7 = 1.6606$;
$II_3 = 3(0.6190^2 \times 0.3810^1) \cdot 7 = 3.0658$; $II_2 = 3(0.6190^1 \times 0.3810^2) \cdot 7 = 1.8866$; $II_1 = (Q^3) \cdot 7 = 0.3870$.
      It is obvious that deriving models for bivalents with greater
numbers of chiasmata requires only that the maximum number of
chiasmata in the numerator of equation 6 be changed and the
binomial expansion use one less than the maximum chiasmata number
expected.
      In the models and methods presented, all bivalents were of the
same size and behaved alike in that each was capable of forming
the same maximum number of chiasmata. However, not all genomes
have such symmetry, and probably a more typical situation is when
there are significant differences among bivalents in their chiasma
forming capabilities. In examples where asymmetry exists, each
bivalent may require a different analysis, using the appropriate
two, three, or four etc. chiasmata per bivalent models. The
results are then summed. In some organisms several bivalents may
be placed in the same subclass if they are not significantly
different *inter se,* and each subclass is then multiplied by
the appropriate n before summing.
      In considering the models for chiasmata distribution, the basic
assumption is that there has been natural selection for genetic
systems with a minimum of one chiasma per bivalent. However, it
is quite possible that some bivalents with a maximum of three,
four, or more chiasmata may have a higher minimum number. If this
occurs, corrections in the equations are easily made as in
Chorthippus discussed later.

Normal Pairing but Random Distribution of Chiasmata: This
model assumes that synapsis has occurred but crossovers are
allocated randomly whatever their number.
      In a bivalent with zero to a maximum of two chiasmata per
bivalent, the P value is obtained by:

$$\frac{Cx}{2n} = P \qquad\qquad\qquad\qquad \text{Eq. 9}$$

$$\therefore\ 1 - \frac{Cx}{2n} = Q \qquad\qquad\qquad \text{Eq. 10}$$

With the P value, the random distribution of chiasmata can be obtained by $(P + Q)^2$ where P equals probability of chiasmata and Q equals failure of chiasma formation. Thus the expansion of the binomial

$$P^2 + 2P^1Q^1 + Q^2 = \Sigma \text{ of bivalent types.} \qquad \text{Eq. 11}$$

The bivalent types can be read directly from the exponents of P. Thus, $II_2 = (P^2) \cdot n$; $II_1 = 2(P^1Q^1) \cdot n$; $I = 2(Q^2) \cdot n$. Calculations of expected numbers of the different configurations is straightforward once the P value is known.

For a model with zero to three chiasmata per bivalent, the equation for the P value determination is

$$\frac{Cx}{3n} = P \qquad\qquad\qquad\qquad \text{Eq. 12}$$

$$\therefore\ 1 - \frac{CX}{3n} = Q \qquad\qquad\qquad \text{Eq. 13}$$

and the binomial is $(P + Q)^3$. On expansion this yields the terms and coefficients for the different kinds of bivalents as follow: $II_3 = (P^3) \cdot n$; $II_2 = 3(P^2Q^1) \cdot n$; $II_1 = 3(P^1Q^2) \cdot n$; $I = 2(Q^3) \cdot n$.

Equations and solutions to models with four or more chiasmata are easily obtained. The only change is to substitute the maximum chiasmata number in the denominators of equations 12 and 13.

Random Pairing: I am not aware of any rigorous study in which there has been a test of the relationship of pairing efficacy and chiasma formation. However, it appears that a one-to-one relationship is usually assumed between an observed chiasma in a bivalent and normally paired arms at pachytene. But there is evidence that synapsis of only a short segment may produce a chiasma regularly (Darlington 1937; Oakley 1983).

A model for bivalents with two synaptic initiation sites is much the same as that for the two chiasmata per bivalent model. Each set of synaptic arms behaves independently of the other so the synaptic value, S, can be calculated as

$$\frac{\text{observed number of paired arms}}{\text{maximum number of paired arms}} = S \qquad Eq. \ 14$$

The S value can be used for the binomial $(S + U)^2$ where S is the probability of pairing and U $(= 1 - S)$ is the lack of pairing. Thus $(S^2) \cdot n$ would yield the expected mean for completely paired arms per cell, $2(SU) \cdot n$ the frequency of one paired and one unpaired arm, and $2(U^2) \cdot n$ the univalent frequency at pachytene.

These terms and coefficients cab be used in two ways. First, they allow calculation of expected values for the three classes based on a given sample so that the sample can be tested for randomness. Second, the expected classes can be tested against the observed bivalent classes in systems that have the two chiasmata per bivalent model.

Three or more synaptic sites can be modeled and tested in a way similar to the above method. However, the process may be so tedious that few will attempt it.

CYTOLOGICAL ANALYSES

Synapsis: The section that summarized current findings on chromosome pairing suggests some potential strictures in using such data to determine chromosome homology in species and hybrids. First, less than 1% of the DNA is bound by the lateral components and is thus available for crossing over. Therefore, only a very small part of the total potential homology can be viewed in any one cell, and it is likely that the same bivalent may have different segments in the synaptonemal complex in different meiocytes. Second, structural differences that are expressed as buckles or other anomalies may be observed only in early pachytene and not be evident at late pachytene because of torsion pairing (McClintock 1933; Tobgy 1943; Jackson 1962; Menzel and Price 1966; Rees and Jenkins 1982). Moses (1977) has shown by EM that duplication buckles and inversion loops formed by homologous pairing at early pachytene subsequently undergo secondary adjustment so that the segments are nonhomologously paired at late pachytene. I believe this is the same phenomenon that was described by Darlington (1937) as torsion pairing, and this is the term that should be used. Darlington (1935) pointed out that if pairing has occurred at two separate points, the coiling of the chromatin between the two contact areas would be sufficient to bring the intervening region into contact. In molecular terms, we might assume that the strength of the DNA coiling process is sufficient to adjust the proteinaceous lateral and central components of the synaptonemal complex. The reason such adjustment might not occur is if a long segment of duplicated or nonhomologous DNA is anchored or synapsed only on one side. Differences in appearance of early and late pachytene bivalents

clearly demonstrate the necessity of analyzing the early stage.
Following the development of techniques for studying whole
mounts of synaptonemal complexes (Counce and Meyer 1973), there
has been an increasing use of the methods in both plants and
animals to analyze structural differences per se and to examine
synapsis more closely in hybrids (Moses *et al.* 1979).
Techniques were developed recently for plants that are more
difficult to work with than maize (Stack 1982). What kind of
pairing occurs between homoeologous chromosomes that differ
radically in DNA amounts? Striking examples have been
demonstrated in species of *Lolium* and *Festuca* (Rees and Jenkins
1982; Jenkins and Rees 1983). *Lolium temulentum* has about
40% more DNA than *L. perenne,* but their hybrids produce viable
gametes, and the chiasma frequency is not significantly different
from *L. perenne.* Backcrosses to *L. perenne* showed a normal curve
of DNA amounts between the F_1 and recurrent parent, and the DNA
amounts were not correlated with chiasma frequency, indicating
that the amount of extra DNA does not affect crossover capability.
Festuca drymeja has about 50% more DNA than *F. scariosus* yet
there is a considerable amount of chromosome pairing and formation
of a normal synaptonemal complex in some bivalents. It is not yet
clear whether the occurrence of univalents is due to disparity in
DNA amounts of homoeologous chromosomes or to a distance
relationship (Jackson 1982) discussed later.
Nonhomologous synapsis recognized by pairing partner change has
been reported in the *Festuca* hybrid described above, and
Jenkins and Rees (1983) have suggested this was caused by repeated
sequences found throughout the genome. Hobolth (1981) has
described extensive multivalent SC complexes at zygotene in
hexaploid *Triticum aestivum*, but only bivalents were observed
at late pachytene and MI. This example and the one from *Festuca*
show no evidence of crossing over from this noneffective pairing
and may indicate that such DNA is excluded from the final
synaptonemal complex.

Diakinesis and MI: The number and distribution of chiasmata
at these stages indicate the efficacy of synapsis and subsequent
events involved in crossingover. In the analysis of hybrids, a
fundamental consideration is whether chiasma frequency is reduced
from that of the parental taxa and whether there is a change in
position of the crossovers. This is important for several
reasons. The extent of recombination in the F_2 or backcross
progeny is determined by numbers and positions of chiasmata.
Crossingover in certain positions of structural heterozygotes can
lead to various levels of genetically unbalanced gametes and in
some cases to gene duplication in the progeny. A chiasma
frequency reduced too low causes univalent formation and aneuploid
gametes. For these and other reasons, chiasmata data are
important in making meaningful comparisons of genome
relationships.
Using the models developed earlier, it is possible to determine

whether distribution of chiasmata is nonrandom and normal or random and anomalous. Comparison of parental taxa and hybrids will show whether or not there has been a reduction in frequencies of different kinds of bivalents and chiasma frequency in the hybrids.

Unfortunately, adequately quantified data are usually not available for testing of bivalents with four or more chiasmata. However, John and Lewis (1965) have given chiasma frequencies for each of the eight autosomal bivalents of *Chorthippus brunneus* in which the data are unequivocal. Three of the bivalents (L1, L2, L3) have a maximum of three and a minimum of two chiasmata with rare exception, one (M4) has a maximum of two and a minimum of one, and four (M5, M6, M7, S8) have only one chiasma, and only one of the latter type is used in the analysis.

As mentioned earlier, it is likely that bivalent systems with a minimum of two chiasmata may occur, and this may prove to be the rule as it is in the L1, L2, and L3 bivalents of *Chorthippus brunneus* that normally have a maximum of three chiasmata. Data for the bivalents are presented in Table I where the observed numbers are compared with those expected according to the nonrandom and random models of chiasmata distribution. The equations for the bivalent model with a minimum of two and a maximum of three chiasmata are as follow: $II_3 = Cx - 2n$ (Eq. 15); $II_2 = n - (Cx-2n)$ (Eq. 16). Results of equations 1, 2, 15, and 16 are a fixed part of the total sample much in the way as Mendelian ratios and do not lose a degree of freedom in their calculation.

In each of the bivalent types tested (Table I), the nonrandom models give a better fit than the random ones, and the difference between them widens as the chiasma frequency decreases. This is due mainly to expectation of increasing number of univalents and a decrease in bivalent numbers in the random model. These results and those presented by Jackson (1982) and Jackson and Hauber (1982) are clearly in disagreement with the model presented by Driscoll *et al.* (1979). I do not know of any normal diploid organism that fits the latter model when chiasma frequencies approach one per bivalent. Therefore major deviations from the normal nonrandom distribution of chiasmata must be considered anomalous behavior, and this conclusion will be used in the interpretations of further examples in this paper.

Data are more abundant for species and hybrids that have a minimum of one and a maximum of two chiasmata per bivalent. Table II gives data drawn from several families, and they are biased only in the requirements that the data are quantifiable and lack translocation heterozygosity in the hybrids. Additional examples are given by Jackson (1982).

Perusal of Table II shows that all hybrids listed have lower mean chiasma frequencies than parental species. When two parental taxa differ in chiasma frequency and the F_1's have none or a low number of univalents, the hybrids are usually intermediate. The distribution of chiasmata among parental bivalents clearly fits

the normal nonrandom model, and this is true also for the hybrids
with but two exceptions. One of the two *Larrea* and the *Hordeum*
hybrids did not fit either model. The reason for the lack of fit
of the *Larrea* hybrid is unknown, but the aberrant distribution
in *Hordeum vulgare* X *H.bulbosum* probably is related to somatic
chromosome instability. Only 40% of the latter hybrid's cells
had 14 chromosomes, and elimination of *H. bulbosum* chromosomes
is known to occur in the F_1 (Kasha 1974).

Hybrids of *Lolium temulentum* and *L. perenne* are treated
separately for two reasons. First, the two species are known to
differ in DNA amount by about 40%, and the species and hybrids
have been studied extensively for many years. Second, regular F_1
hybrids have been compared to those with B chromosomes. Table III
gives data for nine diploid hybrids, and Table IV also provides
information on nine hybrids, but these each have two B chromosomes
in addition to their normal diploid number.

Examination of the data in Table III shows that all but one of

TABLE I. *Observed and expected bivalent (II) types in 20
primary spermatocytes of* Chorthippus brunneus. *NR =
nonrandom expected;* R = *random expected, and L1, L2,
L3, M4, and M5 refer to five of the eight autosomal
bivalents.*

Observed and Expected	II_4	II_3	II_2	II_1	I	Chi square probability
L1 Ob	1	15	3	1	0	
NR	0	16	4	0	0	>0.70
R	0	16.26	3.50	0.25	0.01	>0.50
L2 Ob	1	6	13	0	0	
NR	0	8	12	0	0	>0.30
R	0	10.24	7.68	3.00	0.32	<0.001
L3 Ob	0	2	17	1	0	
NR	0	1	19	0	0	>0.20
R	0	6.38	8.87	4.11	1.27	<0.001
M4 Ob	0	0	12	18	0	
NR	0	0	12	18	0	>0.95
R	0	0	12.80	6.40	1.60	<0.001
M5 Ob	0	0	0	20	0	
NR	0	0	0	20	0	>0.95
R	0	0	5	10	10	<0.001

Data from John and Lewis (1965).

TABLE II. *Diakinesis or metaphase I bivalent (II) types in species and hybrids and their observed and expected numbers according to Nonrandom (NR) and Random (R) models.*

Taxa	Cx	No. cells	II_2	II_1	$2I$	Chi square probability
Prosopsis (x = 14)[1]						
alba	22.98	47	422	236	0	
NR			422	236	0	>0.90
R			443.21	193.64	42.30	<0.001
ruscifolia	22.54	46	393	251	0	
NR			393	251	0	>0.95
R			417.46	202.08	48.91	<0.001
P. a × *r*	21.58	59	447	379	0	
NR			447	379	0	>0.95
R			490.47	292.05	86.95	<0.001
Larrea (x = 13)[2]						
divaricata	22.47	92	871	325	0	
NR			871	325	0	>0.95
R			893.08	280.84	44.16	<0.001
tridentata	25.40	100	1240	60	0	
NR			1240	60	0	>0.95
R			1240.68	58.62	1.38	>0.30
L. d × *t* (102)	24.41	34	389	52	2	
102 NR			388	54	0	>0.90
R			389.65	50.70	3.30	<0.05
L. d × *t* (110)	23.06	69	722	147	28	
110 NR			694	203	0	<0.001
R			706.21	179.81	11.48	<0.001
Helianthus (x = 17)[3]						
mollis	23.90	50	345	505	0	
NR			345	505	0	>0.95
R			420	354.99	75	<0.001
giganteus	23.58	50	329	521	0	
NR			329	521	0	>0.95
R			408.84	360.77	75.59	<0.001
H. m × *g*	22.90	50	295	555	0	
NR			295	550	0	>0.95
R			385.60	373.81	90.60	<0.001

TABLE II (Con't)

Taxa	Cx	No. cells	II_2	II_1	2I	Chi square probability
Hordeum (x = 7)[4]						
vulgare	13.88	60	413	7	0	
NR			413	7	0	>0.95
R			413.02	6.95	0.03	
bulbosum	13.78	50	339	11	0	
NR			339	11	0	>0.95
R			339.09	10.82	0.09	
H. v X b	7.24	49	110	137	96	
NR			97.96	245	0	<0.001
R			92.89	171.21	7839	<0.001

Data from [1]Hunziker *et al.* (1975) [2]Yang *et al.* (1977),

[3]Jackson (unpublished), [4]Kasha and Sadasivaiah (1971).

the nine plants fit the nonrandom chiasma distribution model.
Conversely, all but two of these plants have an acceptable fit to
the random model but usually with lower probability. Since the
populations were grown under uniform conditions, it appears that
some segregation is occurring through the *Lolium perenne*
genomes because the *L. temulentum* samples were based on
selfed single heads through several generations.
    The data in Table IV are remarkable when compared to those in
Table III. All nine plants with two B chromosomes have a much
reduced chiasma frequency, as pointed out in the original study by
Evans and Macefield (1973). The highest mean chiasma frequency
per cell is 5.2, well below a one per bivalent average, and the
nonrandom chiasmata distribution model cannot be used in this
situation. Thus all comparisons in Table IV are with the random
model. All nine plants with $2n = 14 + 2$ Bs have a random
distribution of chiasmata. This is more extreme than the
distribution pattern in the hybrids without B chromosomes.
Therefore, these data from the two kinds of progeny of the same
interspecific cross provide clear evidence that the B chromosomes
are the major causative agent for both a reduced chiasma frequency
and a distinctive change in the way chiasmata are allocated among
the bivalents. A similar situation has been found in hybrids
between *Lolium multiflorum* and *L. perenne* (Evans and Macefield
1974), and a somewhat reduced effect caused by autosomal genes was
later discovered in both *L. perenne* and *L. temulentum* genomes
(Taylor and Evans 1977).

TABLE III. *Metaphase I bivalent (II) types in F_1 hybrids of* Lolium temulentum *and* L. perenne *and their expected numbers based on the non-random (NR) and random (R) models of chiasma distribution. Twenty meiocytes were analyzed for each plant*

Plant no.	Observed and Expected	II_2	II_1	$2I$	Chi square probability
1	Ob	101	39	0	
	NR	101	39	0	>0.95
	R	103.72	33.57	2.72	>0.05
2	Ob	71	64	5	
	NR	66	74	0	>0.10
	R	75.78	54.44	9.78	<0.05
3	Ob	98	40	2	
	NR	96	44	0	>0.50
	R	99.46	37.08	3.46	>0.30
4	Ob	95	43	2	
	NR	96	44	0	>0.70
	R	96.94	39.11	3.94	>0.30
5	Ob	59	68	13	
	NR	46	94	0	<0.001
	R	61.78	62.44	15.78	>0.20
6	Ob	71	62	7	
	NR	64	76	0	>0.05
	R	74.31	55.37	10.31	>0.10
7	Ob	80	57	3	
	NR	76	64	0	>0.30
	R	83.31	49.37	7.31	>0.50
8	Ob	100	40	1	
	NR	96	44	0	>0.70
	R	99.46	37.08	3.46	>0.20
9	Ob	56	80	4	
	NR	50	90	0	>0.10
	R	64.46	61.07	14.46	<0.001

SYNTHESIS

Unfortunately, many studies of species and their hybrids have not
coupled a careful and detailed analysis of pachytene chromosomes
with the following meiotic events. One is then left with later
stages to draw conclusions on synaptic success and chromosome
homology. Bivalent numbers and chiasma frequencies at diakinesis
or MI are used to assess pairing efficacy, and in many cases this
is adequate if the same kind of data are available for parental
taxa and their hybrids. The major reason for carrying out such
studies is to obtain information on the homology of the parental
taxa. In general usage, homology is usually considered complete
if at diakinesis or MI there are no significant differences in
bivalent numbers and chiasma frequencies between the parental taxa
and their hybrids. A much stricter definition of homology would
require that the chromosome segments be identical with respect to
their loci, but this requirement is not followed in most studies
because suitable markers usually are lacking.

A recurrent theme in systematic and evolutionary literature is
that lack of pairing in intertaxon hybrids is due to extensive
chromosomal repatterning caused by inversions, translocations,
deletions, duplications, transpositions and other possible changes
below the light microscope level of resolution. Furthermore, such
changes at the diploid level are believed to result in
preferential pairing or differential affinity in polyploids
derived from hybrids heterozygous for such changes. Species and
hybrids without overt structural rearrangements characterize the
examples in Tables II, III and IV, and they were used to make a
point. This is that pairing efficacy may be reduced in
interspecific hybrids in the absence of detectable structural
rearrangements. Chiasma frequency was reduced in each of the
hybrids in Table II below that of the highest parent and in most
instances below the level of both parents. The data in Tables III
and IV do not allow comparison of bivalent types with parental
Lolium species because their frequencies were not included
in the original paper. Most of the *Lolium* hybrids in Table
III do have univalents at MI, and there are rather large
differences in the DNA amounts between the two genomes. But Rees
and Jenkins (1982) have indicated that the hybrids do not differ
significantly from the *Lolium perenne* parent in chiasma
frequency. The presence of B chromosomes drastically affect
chiasma frequency by reducing it to a point where univalents are
numerous (Table IV). This is a highly significant fact because
the great decrease in chiasma frequency and bivalent number occur
in the complete absence of any additional structural change in the
genomes. This demonstrates unequivocally that bivalent and
chiasma frequencies are affected by genes on the B chromosomes.
These genes and others on the autosomes have been likened to the
Ph gene in hexaploid bread wheat that is largely responsible
for homologous pairing (Evans and Macefield 1973; Taylor and
Evans 1977). Others have suggested a similar explanation for

TABLE IV. *Metaphase I bivalent (II) types in F_1 hybrids of*
Lolium temulentum × L. perenne *with 2n = 14 + 2 chromosomes*
in each plant. Data analyzed by the random (R) model of
chiasma distribution. Twenty meiocytes were analyzed for
each plant

Plant no.	Observed and Expected	II_2	II_1	$2I$	Chi square probability
1	Ob	23	53	53	
	R	17.50	64	58.50	>0.05
2	Ob	5	53	82	
	R	7.08	48.83	83.09	>0.30
3	Ob	18	65	57	
	R	11.22	64.64	57.22	>0.90
4	Ob	7	50	83	
	R	7.31	49.37	83.32	>0.90
5	Ob	13	58	69	
	R	12.60	58.8	68.60	>0.90
6	Ob	20	69	51	
	R	19.31	65.37	55.31	>0.30
7	Ob	24	62	54	
	R	21.61	66.79	51.61	>0.30
8	Ob	9	41	90	
	R	6.22	46.57	87.22	>0.20
9	Ob	18	65	57	
	R	18.22	64.57	57.22	>0.90

Data from Evans and Macefield (1973).

examples from different plant species, but in many cases the data
are not sufficiently quantified to test for such an effect.
    It should be clear from these limited examples of a much more
extensive literature that the occurrence of univalents at MI can
be caused by one or relatively few gene differences among or
within species. The appropriate question then is, do structural
rearrangements cause univalent formation in the same frequencies?
Inversion heterozygotes are known to have reduced crossingover in
the loop, but this has been found to increase recombination

outside this region and also to increase crossingover in
nonhomologous chromosomes. There is no doubt that a univalent can
result from a translocation heterozygote if the chiasma frequency
is two per quadrivalent, but the resulting trivalent-univalent
configuration is evidence for prior synapsis and random
crossingover among the four homologously-paired segments.
Further, we need only to look at numerous representatives of the
Onagraceae and several other families to find permanent
translocation heterozygotes in which many chromosomes are
involved, yet such species regularly produce orderly arrays of
multivalents, and univalents are notably lacking.

I have proposed a very simple model that considers both pairing
and chiasmata failure as due to the same cause (Jackson 1982).
This model assumes that a genome has a specific attachment site in
the prezygotene nucleus. The logical anchor site is the nuclear
membrane, and site specificity is under genetic control. In order
to have normal synapsis, the homologous chromosomes must either be
attached close enough at prezygotene or undergo some movement on
or via the nuclear membrane to bring them close enough for
synaptonemal complex formation.

Mutations in a population that change genome attachment sites
could have various effects on chromosome pairing. At one extreme,
the distance between homologous genomes may be so great that
synapsis is impossible. The normal distance relationship could be
considered the converse of this extreme mutational effect. But
the hiatus between these two situations is mostly what is
encountered in intertaxon hybrids and is reflected in different
frequencies of univalents. If the genomes of the parental taxa
are essentially equal in size and ability to form chiasmata, then
crossovers should be distributed randomly among bivalents, and
sampled univalents should represent random sets of homologous or
homoeologous chromosomes. The occurrence of univalents that fit
the random distributions models are, I believe, an indication of
heterozygosity for nuclear attachment site (NAS) genes (Jackson
and Hauber 1983). However, I hasten to add that asynaptic and
desynaptic mutants may fit these nonrandom models; the later could
fit due to mutations that impose a limitation on some diffusable
enzymes needed for recombination.

Theoretically, it should be possible to have univalents despite
synapsis if we assume that early prophase consists of highly
synchronized processes. Thus, if two homologous chromosomes are
slightly beyond the normal pairing distance, synapsis may
ultimately be successful, but recombination events may be
desynchronized and result in few successful crossovers.

Without further analysis, it is difficult to distinguish
between the two possible causes of chiasma failure. Is
failure due to genes preventing synapsis, or is it caused by genes
involved only in crossingover? If chiasma failure is due to
recombination defective genes, then a tetraploid of the hybrid
should show the same defect and have about twice as many
univalents. Conversely, if chiasma failure is caused by too great

a distance between homologues for normal synapsis (a NAS mutant), then doubling the genomes should restore normal pairing, and univalents should be absent or much reduced in frequency. Quadrivalents should occur less frequently than expected in normal autotetraploids (Jackson and Hauber 1982) due to what has been called preferential pairing or differential affinity, but the terms lose their original meaning if such pairing is due solely to distance relationships.

The logical conclusion from the above deductions leads to the proposition that what we have in the past referred to as allopolyploids need not have been derived from hybrids between taxa with distantly related and structurally different genomes. Rather, alloploid-like behavior can be expected from intra- and interpopulation hybrids as well as those derived from intertaxon crosses. If alloploids are derived from taxa with appropriate NAS mutations, then there is not an underlying structural difference that causes preferential pairing; synaptic partners are determined by distance relationships.

If structural differences determine preferential synapsis, then one should not be able to induce so-called homoeologous pairing in strict alloploids. However, colchicine can bring about such intergenomal pairing and crossingover in strict alloploids if it is applied at the premeiotic stage (Driscoll *et al.* 1967; cf. Jackson and Murray 1983). Because it has been demonstrated that this alkaloid does not affect synapsis or chiasma formation (Driscoll and Darvey 1970), its effect has to be one that disrupts the genetically controlled positioning in the nucleus. Therefore, the logical conclusion must be that strict alloploids may have homologous chromosomes, and that these will synapse and form chiasmata normally if given the opportunity.

This takes us full circle back to the initial paragraph of this section, and one must conclude that lack of chromosome homology in hybrids cannot be deduced from the lack of bivalents.

REFERENCES

Counce, S. and Myer, G. 1973. Differentiation of the synaptonemal complex and the kinetochore in *Locusta* spermatocytes studied by whole mount electron microscopy. *Chromosoma* 44: 231–253.

Darlington, C.D. 1935. The internal mechanics of chromosomes. III. Relational coiling and crossing-over in *Fritillaria*. *Proc. R. Soc. London, Ser. B.* 117: 79–114.

Darlington, C.D. 1937. *Recent Advances in Cytology*. 2nd edit. Churchill, London.

Driscoll, C.J., Bielig, L.M. and Darvey, N.L. 1979. An analysis of frequencies of chromosome configurations in wheat and wheat hybrids. *Genetics* 91: 755–767.

Driscoll, C.J., Darvey, N.L. and Barber, H.N. 1967. *Nature* 216: 687–688.

Driscoll, C.J. and Darvey, N.L. 1970. *Science* 169: 290–291.

Evans, G.M. and Macefield, A.J. 1973. *Chromosoma* 41: 63–73.

Evans, G.M. and Macefield, A.J. 1974. The effect of B chromosomes on homoeologous pairing in species hybrids II. *Lolium multiflorum* × *Lolium perenne*. *Chromosoma* 45: 369–378.

Gillies, C.B. 1974. *Chromosoma* 48: 441–453.

Gillies, C.B. 1975. Synaptonemal complex and chromosome structure. *Annu. Rev. Genet.* 9: 91–109.

Gillies, C.B. 1981. *Chromosoma* 83: 575–591.

Grant, V. 1952. *Chromosoma* 5: 372–390.

Hobolth, P. 1981. Chromosome pairing in allohexaploid wheat var. Chinese Spring. Transformation of multivalents into bivalents, a mechanism for exclusive bivalent formation. *Carlsberg Res. Commun.* 46: 129–173.

Holm, P.B. 1977. Three dimensional reconstruction of chromosome pairing in the zygotene stage of meiosis in *Lilium*. *Carlsberg Res. Commun.* 42: 103–151.

Hunziker, J.H., Poggio, L., Naranjo, C.A. and Palacios, R.A. 1975. Cytogenetics of some species and natural hybrids in *Prosopis* (Leguminosae). *Can. J. Genet. Cytol.* 17: 253–262.

Jackson, R.C. 1962. Interspecific hybridization in *Haplopappus* and its bearing on chromosome evolution in the *Blepharodon* section. *Am. J. Bot.* 49: 119–132.

Jackson, R.C. 1976. Evolution and systematic significance of polyploidy. *Annu. Rev. Ecol. Syst.* 7: 209–234.

Jackson, R.C. 1982. Polyploidy and diploidy: New perspectives on chromosome pairing and its evolutionary implications. *Am. J. Bot.* 69: 1512–1523.

Jackson, R.C. and Casey, J. 1980. Cytogenetic of polyploids. In *Polyploidy: Biological Relevance* (W.H. Lewis, ed.), Plenum Press, New York. pp. 17–44.

Jackson, R.C. and Hauber, D.P. 1982. Autotriploid and autotetraploid cytogenetic analyses: correction coefficient for proposed binomial models. *Am. J. Bot.* 69: 644–646.

Jackson, R.C. and Hauber, D.P. 1983. Epilogue. In *Benchmark Papers in Genetics*, Vol. 12 (R.C. Jackson and D.P. Hauber, eds.), Hutchinson Ross, Stroudsburg. pp. 363–367.

Jackson, R.C. and Murray, B.G. 1983. Colchicine induced quadrivalent formation in *Helianthus*: Evidence of ancient polyploidy. *Theor. Appl. Genet.* 64: 219–222.

Jenkins, G. and Rees, H. 1983. Synaptonemal complex formation in a *Festuca* hybrid. In *Kew Chromosome Conference II* (P.E. Brandham and M.D. Bennett, eds.), George Allen and Unwin, London. pp. 233–242.

John, B. and Lewis, K.R. 1965. The meiotic system. *Protoplasmatologica* 6: 1–331.

Jones, G.H. 1967. *Chromosoma* 22: 69–90.

Kasha, K. 1974. Haploids from somatic cells. In *Haploids in Higher Plants: Advances and Potential* (K. Kasha, ed.), Univ. Guelph, Ontario. pp. 67–86.

Kasha, KI. and Sadasivaiah, R. 1971. *Chromosoma* 35: 264–287.

Levan, A. 1940. *Hereditas* 26: 454–462.

McClintock, B. 1933. The association of non-homologous parts of
    chromosomes in mid-prophase of meiosis in *Zea mays*. *Z.*
    Zellforsch. Mikrosk. Anat. 19: 191-237.
Menzel, M. and Price, J.M. 1966. Fine structure of synapsed
    chromosomes in F_1 *Lycopersicon esculentum-Solanum*
    lycopersicoides and its parents. *Am. J. Bot.* 53: 1079-1086.
Moens, P.B. 1978. *Annu. Rev. Genet.* 12: 443-450.
Moens, P. and Short, S. 1983. Synaptonemal complexes of bivalents
    with localized chiasmata in *Chloealtis conspersa*
    (Orthoptera). In *Kew Chromosome Conference II* (P.E.
    Brandham and M.D. Bennett, eds.), George Allen and Unwin,
    London. pp. 99-106.
Moses, M.J. 1956. *J. Biophys. Biochem. Cytol.* 2: 215-218.
Moses, M.J. 1968. *Annu. Rev. Genet.* 2: 363-412.
Moses, M.J. 1977. *Chromosomes Today* 6: 71-82.
Moses, J., Karatsis, P.A. and Hamilton, A.E. 1979. Synaptonemal
    complex analysis of heteromorphic trivalents in *Lemur*
    hybrids. *Chromosoma* 70, 141-160.
Oakley, H.A. 1983. Male meiosis in *Mesostoma ehrenbergii*.
    In *Kew Chromosome Conference II* (P.E. Brandham and M.D.
    Bennett, eds.), George Allen and Unwin, London. pp. 195-199.
Rees, H. 1961. *Bot. Rev.* 27: 288-312.
Rees, H. and Jenkins, G., 1982. Assays of the phenotypic effects
    of changes in DNA amounts. In *Genome Evolution* (G.A. Dover
    and R.B. Flavell, eds.), Academic Press, pp. 287-297.
Stack, S. 1982. Two-dimensional spreads of synaptonemal complexes
    from solanaceous plants. I. The technique. *Stain Technol.*
    57: 265-272.
Stern, H., Westergaard, M. and von Wettstein, D. 1975.
    Presynaptic events in meiocytes of *Lilium longiflorum* and
    their relation to crossing over: a preselection hypothesis.
    Proc. Natl. Acad. Sci. U.S.A. 72: 961-965.
Taylor, I.B. and Evans, G.M. 1977. The genotypic control of
    homoeologous chromosome association in *Lolium temulentum* X
    Lolium perenne hybrids. *Chromosoma* 62: 57-67.
Ting, Y.C. 1973. *Cytologia* 38: 497-500.
Tobgy, H.A. 1943. A cytological study of *Crepis fuliginosa, C.*
    neglecta, and their F_1 hybrid, and its bearing on the
    mechanisms of phylogenetic reduction in chromosome number.
    J. Genet. 45: 67-111.
Westergaard, M. and von Wettstein, D. 1972. The synaptinemal
    complex. *Annu. Rev. Genet.* 6: 71-110.
White, M.J.D. 1978. *Modes of Speciation*. Freeman, San
    Francisco.
Wilson, J.Y. 1959. *Genetica* 29: 290-303.
Yang, T.W., Hunziker, J.H., Poggio, L. and Naranjp, C.A. 1977.
    Hybridization between South American Jarilla and North American
    diploid creosotebush (*Larrea*, Zygophyllaceae). *Plant Syst.*
    Evol. 126: 331-346.

Nuclear DNA Variation and the Homology of Chromosomes

H. Rees
Department of Agricultural Botany
University College of Wales
Aberystwyth, Wales, U.K.

INTRODUCTION

Surveys within many genera of both plants and animals tell us that
the evolution and divergence of species are often accompanied by
massive changes in nuclear DNA amount. A spectacular example is
provided by the dicotyledonous genus *Vicia*. Among diploid
species of this genus the DNA amounts range from 27.07 pg in *V.
faba* to 3.85 pg in *V. monantha*. Indeed one pair of chromosomes
in *V. faba* carries more DNA than the whole chromosome
complement of *V. monantha* (Raina and Rees 1983). We have
yet to explain satisfactorily, in genetical and physiological
terms, the causes and consequences of such an astonishing
paradox.

In most of the surveys to date the emphasis has been upon the
distribution of DNA changes *between* the chromosome
complements of different species. The present work dwells upon
the way in which nuclear DNA changes are distributed *within*
the chromosome complements. The aims were to determine how
individual chromosomes within complements evolve and change
relative to one another and to consider the consequences of such
change upon the divergence and evolution of species.

PATTERNS OF CHANGE

Consider, by way of a theoretical example, a diploid species with
a basic chromosome number of 5, with DNA amounts per metaphase
chromosome of 1.0, 1.5, 2.0, 2.5 and 3.0 pg, a total of 10 pg per
2C haploid complement. Suppose now that the nuclear DNA amount is
doubled. There are two obvious alternatives by which the doubling
could be achieved. (I) is by doubling of the DNA amount in each
chromosome, i.e. an increase per chromosome that is proportional

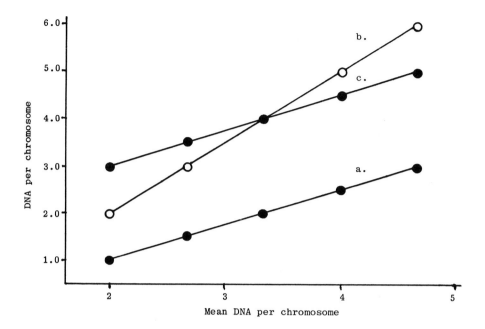

FIG 1. *DNA amount per chromosome plotted against the mean for each of the five chromosomes. The complements of* b *and* c *have twice as much DNA as* a. *In* b *increase per chromosome was proportional to initial DNA amount. In* c *each chromosome has the same DNA increment.*

to its initial DNA content. The new complement would contain chromosomes with 2.0, 3.0, 4.0, 5.0 and 6 pg. (II) is by distributing the extra 10 pg of DNA equally throughout the complement, namely 2.0 pg to each chromosome, independently of its initial DNA content. The new complement would contain chromosomes with 3.0, 3.5, 4.0, 4.5 and 5.0 pg. (I) and (II) are readily distinguished graphically as shown in Fig. 1, where the DNA amounts for individual chromosomes in each species are plotted against the mean DNA amount per chromosome. With (1) the regression slopes diverge (Fig. 1b); with (2) they are parallel (1c).

Before considering the results of surveys within a range of genera it is worth considering the implications of the different patterns of change referred to above. If we assume, as is reasonable, that amplification, or perhaps deletion, of base sequences which generate the changes in DNA amount within chromosomes occur at random along the DNA molecules the expectation is that DNA change within each chromosome of the complement would be proportional to its DNA content, i.e. as in (I) above. Departures from this pattern, including (II) above,

would argue for constraints, either upon the initiation of change
or upon its preservation.

The Question of Homology

It is an easy matter to ascertain the DNA distribution among
chromosomes within complements, because the DNA amount per
chromosome is proportional to chromosome size, measured as area or
volume at metaphase (e.g. Raina and Rees 1983). In the
theoretical example presented rank in order of DNA content
corresponded with rank in order of homology or, more strictly,
with order of homoeology, and, therefore, DNA changes in
individual chromosomes were readily established. In practice a
problem arises in determining which chromosomes from the
complements of different species are homoeologous with one
another. Where it is possible to make hybrids between different
species the problem is solved by measurements at first metaphase
of meiosis when the homoeologous chromosomes are identifiable by
virtue of their capacity to pair and to form chiasmata. Measurements at meiosis in species hybrids in *Lolium, Festuca* (Seal
and Rees 1982) and in *Allium* (unpubl.) justify the
assumption that rank in order of chromosome size and DNA content
in related species does in general correspond closely with rank in
order of homoeology.

SURVEYS IN ANGIOSPERM GENERA

Figure 2 shows the distributions of DNA amounts within diploid
complements of species in *Lolium, Festuca, Lathyrus, Vicia*
and in *Allium*. It will be observed that the slopes of the
regressions within all genera are virtually parallel. As
explained earlier, this means that increase in nuclear DNA amount
within each genus is achieved by equal DNA increments to each
chromosome within the complement. DNA change associated with
speciation was clearly, therefore, not randomly distributed within
the chromosome complement. On the contrary there was in all cases
a constraint upon the distribution of DNA change, a constraint
that determines an equal DNA change for all chromosomes, large and
small alike (as in II above).

CONSEQUENCES TO THE EVOLUTION OF CHROMOSOME COMPLEMENTS

Symmetry

Since, with increasing nuclear DNA, the small chromosomes acquire
as much extra DNA as the large chromosomes, it follows inevitably
that the relative size and DNA content of chromosomes within a
complement become more and more alike. In other words the
chromosome complements become more symmetrical as the nuclear DNA
amount increases. The phenomenon is made manifestly clear from a
comparison of the chromosome complements of *Lathyrus* species

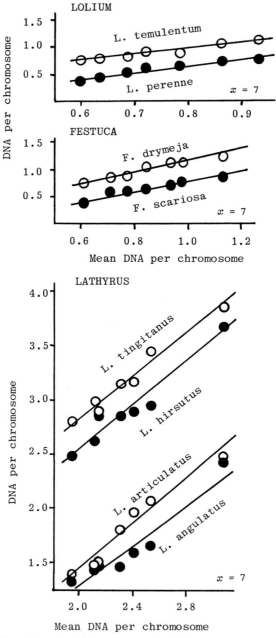

FIG 2. *The distribution of DNA amounts in chromosomes of species in five genera of flowering plants. DNA amounts in Lolium and Festuca are from measurements at meiosis in species hybrids (from Seal and Rees 1982). In the other genera the DNA*

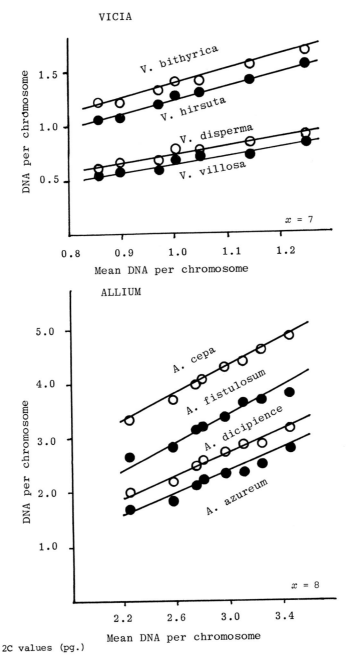

amounts were estimated from measurements at mitosis. Data on
Vicia are from Raina and Rees 1983; on Lathyrus from Narayan
1983; on Allium from Jones 1967.

in Fig. 3.

The question of symmetry within chromosome complements in the context of species divergence and evolution has of course been considered extensively by Stebbins (1950). The present results are in no way at variance with his arguments. They serve, rather, to explain how such changes in symmetry are generated.

The Consequences of Chromosome Pairing

Failure of effective chromosome pairing at meiosis is a characteristic feature of interspecific hybrids. Such failure is often attributable to a diminution of homology between chromosomes and is both a cause and a consequence of the divergence of species. The loss of homology, in turn, is a reflection of structural changes in the chromosomes. The massive changes in chromosomal DNA amount to which we have referred must, in themselves, of course generate large differences between the chromosomes of different species. In the genera described it will be recalled that with increase in nuclear DNA the small chromosomes acquire as much "extra" DNA as the large chromosomes. It follows that between species with high and low DNA amounts the difference in size and structure between the small chromosomes must be greater, in relative terms, than between the larger chromosomes of the complements. One may predict, therefore, that failure of chromosome pairing in hybrids resulting from loss of homology will be greater between the smaller than the larger chromosome "pairs". The prediction begs the question of course as to what extent the large scale quantitative DNA changes affect homology, the capacity of chromosome pairing. An investigation of a *Festuca* species hybrid provides the answer to the question and a test of the prediction.

Festuca drymeja X *F. scariosa*. Table I shows the DNA amounts of homologous, or rather homoeologous, chromosomes of *Festuca drymeja* and *F. scariosa*. The greater asymmetry within the four smaller pairs is reflected clearly by the percentage differences.

Figure 4 shows a pachytene from a pollen mother cell in the species hybrid. Buckles and loops in the four smaller bivalents

TABLE I. *The DNA amounts, in picograms, in metaphase chromosomes of the haploid complements of* Festuca drymeja *and* F. scariosa

Chromosome	1	2	3	4	5	6	7
F. drymeja	1.33	1.11	1.12	1.03	0.94	0.89	0.76
F. scariosa	0.93	0.83	0.74	0.64	0.59	0.53	0.46
Difference	0.40	0.28	0.38	0.39	0.35	0.36	0.30
% difference	22	34	51	61	59	68	65

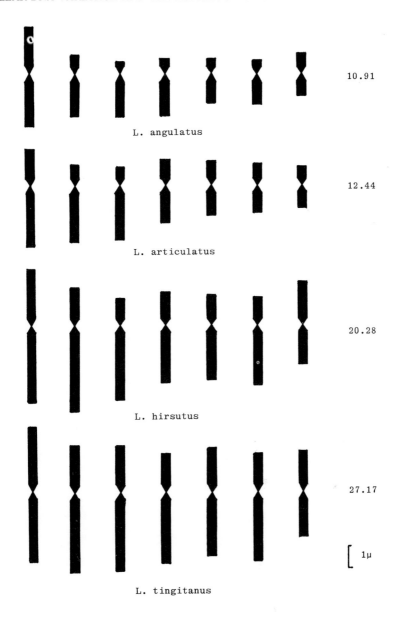

FIG 3. *The haploid mitotic complements of four* Lathyrus *species. 2C DNA amounts (in picograms) on the right. Note the increasing symmetry of complements with increase in DNA amounts.*

(4 to 7) indicate abnormal pairing which is characteristic of
these bivalents. In the three larger bivalents (1 to 3) pairing
is complete and end-to-end. E.M. serial sections confirm that
synaptonemal complex formation is normal and complete in the large
bivalents, abnormal and incomplete in the smaller (Jenkins and
Rees 1983).

At first metaphase of meiosis the average chiasma frequency is
about three per cell. The chiasmata are formed almost exclusively
by the larger chromosomes (Seal and Rees 1982).

The prediction of greater failure of pairing in the smaller
chromosome pairs is confirmed. This is not in the least
surprising in view of their marked asymmetry in size and DNA
content. What is surprising is the remarkable effectiveness of
pairing and chiasma formation in the three larger bivalents,
bearing in mind that they differ in DNA content, on average, by
over 30%.

DISCUSSION

The consistent nature of the pattern of quantitative DNA changes

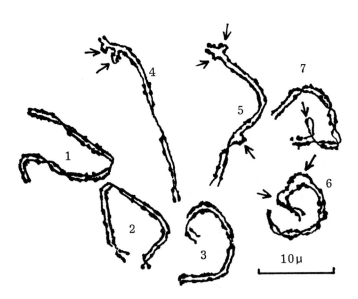

FIG 4. *Pachytene in a pollen mother cell of the haploid*
Festuca drymeja X F. scariosa. *Bivalents numbered in
decreasing order of size. Arrows point to loops and buckles in
the smaller bivalents (with the permission The Systematics
Association)*

within complements of species in all five genera investigated
suggests that the pattern is widespread among flowering plant
genera. This is not to say that it is exclusive (cf. Brandham
1983). Nor does it imply, of course, that quantitative DNA change
within chromosomes of a complement is the result only of
amplification or deletion of DNA base sequences. Other kinds of
structural change, particularly Robertsonian change, may
drastically alter the relative size and DNA content of individual
chromosomes (cf. Raina and Rees 1983).

Of particular interest is that the surveys show that the DNA
changes within a complement are not at random, that is to say
not proportional to chromosome size and DNA content. Each
chromosome, large and small alike, is equally affected. This
constraint argues for selection and, in turn, that the pattern of
change is of adaptive importance. It is tempting to suggest that
relative chromosome size and shape are in themselves of adaptive
importance. There is indeed good evidence to this effect (e.g.
Bennett 1982). If this be the case it is of interest to emphasise
that while the changes in nuclear DNA may be of importance in
speciation the nature of the nuclear changes may in themselves
bear no direct relation to the morphological and physiological
consequences of the divergence and evolution of species. In other
words these cytological changes may not be of diagnostic benefit
to the taxonomist.

Finally, the pattern of change described has predictable
consequences upon the homology of large and small chromosome pairs
in hybrids. At the same time, however, the observations in the
Festuca hybrid are instructive in demonstrating tl : large
differences in DNA amount between homoeologous chromosomes, of the
order of 30% or more, have surprisingly little effect upon the
capacity for effective chromosome pairing at pachytene of meiosis
(see also Rees *et al*. 1982). This, of course, raises the
question of what precisely it is that determines homology.

Leaving that aside, it would appear most unlikely that the
large scale quantitative DNA changes often associated with
speciation are to be explained simply on grounds of generating
barriers to gene exchange by alterations in the homology of
chromosomes.

SUMMARY

1. Large scale nuclear DNA variation is often associated with
   speciation in the Flowering Plants. Surveys in a number of
   genera show that the distribution of DNA change among
   chromosomes within complements is not at random. Each
   chromosome irrespective of its initial size and DNA content is
   affected to the same degree. For example, with increase in
   nuclear DNA amount the small chromosomes acquire as much
   "extra" DNA as the large chromosomes.
2. Two consequences of this pattern of change are,
   (a) the chromosome complements of species with high DNA amount

are more "symmetrical" than related species with low DNA amount.

(b)   In related species with different nuclear DNA amount the difference in size and DNA content between the smaller homoeologous chromosomes is greater than for the larger homoeologues. It would be expected that in hybrids between such species that chromosome pairing would be less effective between small than large homoeologous chromosomes. The prediction is confirmed.

3. Despite large differences in DNA content between homoeologous chromosomes pairing and synaptonemal complex formation is often normal at pachytene and effective in forming chiasmata. In other words the effect of change in chromosomal DNA upon homology is surprisingly small.

4. From a taxonomic standpoint, variation in nuclear DNA amount between species is not a good index of their morphological divergence.

REFERENCES

Bennett, M.D. 1982. Nucleotypic basis of spatial ordering of chromosomes in eukaryotes and the implications of the order for genome evolution and phenotypic variation. In *Genome Evolution* (G.A. Dover and R.B. Flavell, eds.), Academic Press, London, pp. 239–261.

Brandham, P.E. 1983. Evolution in a stable chromosome system. Kew Chromosome Conference II. Allen and Unwin, London.

Jenkins, G. and Rees, H. 1983. Synaptonemal complex formation in a *Festuca* hybrid. Kew Chromosome Conference II. Allen and Unwin, London.

Jones, R.N. 1967. Ph.D. thesis, University of Wales, Aberystwyth.

Narayan, R.K.J. 1983. Chromosome changes in the evolution of *Lathyrus* species. Kew Chromosome Conference II. Allen and Unwin.

Raina, S.N. and Rees, H. 1983. DNA variation between and within chromosome complements of *Vicia* species. *Heredity,* in press.

Rees, H., Jenkins, G., Seal, A.G. and Hutchinson, J. 1982. Assays of the phenotypic effects of changes in DNA amounts. In *Genome Evolution* (G.A. Dover and R.B. Flavell, eds.), Academic Press, London, pp. 287–297.

Seal, A.G. and Rees, H. 1982. The distribution of quantitative DNA changes associated with the evolution of diploid *Festuceae*. *Heredity,* 49: 179–190.

Stebbins, G.L. 1950. *Variation and Evolution in Plants.* Columbia Univ. Press, New York.

Chromosome Banding and Biosystematics

Ichiro Fukuda
Division of Biology
Tokyo Woman's Christian University
Tokyo, Japan

INTRODUCTION

Chromosome banding methods for linear differentiation have been greatly improved in the cytogenetical field, and now have come to be used widely as effective techniques in systematics and evolutionary studies for undertaking biosystematic approaches.

Analyses of these chromosome banding techniques are carried out with the following aims:

1. To obtain genetic information at a more micro-level for chromosome identification.
2. To obtain information on genetic variation at a population level with regard to speciation.
3. To determine the relationships between species with regard to phylogeny.

In biosystematics the approach of chromosome banding is an extremely useful tool. I will summarize the research on chromsome banding including some of my original *Trillium* data. I will consider the following items:

1. The history of chromosome banding research.
2. Plants which have been used in chromosome banding techniques.
3. Chromosome banding and systematics.
4. Chromosome banding and evolution.
5. Future problems of chromosome banding in biosystematics.

THE HISTORY OF CHROMOSOME BANDING RESEARCH

The pioneers of chromosome banding investigations were Darlington and LaCour (1938, 1940) for *Paris* and *Trillium* species in the Monocotyledons; they discovered heterochromatic segments on chromsomes induced by cold treatment. In the Dicotyledons, Geitler (1940) carried out the first chromosome banding for

Adoxa moschatellina by cold treatment.

This method has since been tried for many plants by a number of individuals. However, it has been found that cold-induced banding techniques have so far been adopted for only species of the following plants: *Vicia faba* (Leguminoseae), *Adoxa* (Adoxaceae), *Cestrum* (Solanaceae), *Trillium, Paris* and *Kinagusa* (Trilliaceae), *Fritillaria* and *Tulbaghia* (Liliaceae), *Secale* and *Hordeum* (Graminae) (Fig. 1).

Yamasaki (1956) was successful in inducing chromosome banding (Hy-banding) by subjecting unfixed tissues to a mixture of HCl and acetic acid at temperatures between 60° and 80°C for the orchid *Cypripedium debile*. Takehisa (1968) also applied this method to *Vicia faba* and found that Hy and cold-induction showed similar banding patterns.

The use of acridine derivatives such as quinacrine, as pioneered by Caspersson *et al.* (1969a, b) has allowed the visualization by means of ultraviolet light the characteristic sequences of light and dark banding (Q-banding).

Giemsa C-banding (C means constitutive heterochromatin) was first applied to animal chromosomes (Pardue and Gall 1970). It involves the denaturation-reassociation of DNA, with the highly repetitive DNA reassociating faster, and appearing as dark bands. This C-banding method has now been adopted for many plants as the discussion below will show.

G-bands are produced by pretreatment with trypsin, pronase,

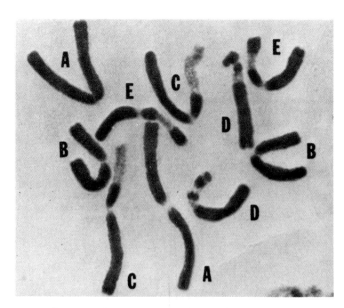

FIG. 1. *Somatic metaphase chromosomes of* Trillium grandiflorum *showing allocyclic regions (Fukuda and Grant 1980).*

warm saline or an alkali solution and by the staining of chromosomes with Giemsa. The staining pattern is essentially the same as that which appears with quinacrine fluorescence. This method does not seem well adapted to plants (see below).

N-bands are induced in the nucleolar organizing regions of chromosomes, that is, the groups of genes coding for ribosomal RNA. The N-banding method of Funaki *et al.* (1975) is useful for plant material.

In *Trillium*, the distribution of highly fluorescent bands has given a good agreement with the heterochromatic regions as induced by the cold treatment method of Darlington and La Cour (Caspersson *et al.* 1969a). The cold-induced banding battern is partially related to the Giemsa staining techniques (Takehisa and Utsumi 1973).

Caspersson and colleagues (1972) have shown that the Quinacrine Mustard stained fluorescent patterns for different tissues, blood, skin and testis of human chromosomes are similar; comparisons between the banding patterns from root tip cells and endosperm cells in *Vicia faba* showed no observable difference. In *Scilla,* identical banding patterns were found in cells from root tips, endosperm and also pollen tubes (Caspersson *et al.* 1972).

With regard to the inheritance of the Giemsa C-banding pattern, Hadlaczky and Kalman (1975) showed that hybrids contained the parent banding patterns (Fig. 2).

PLANTS WHICH HAVE BEEN USED IN CHROMOSOME BANDING TECHNIQUES

During the more than 40 years since the first investigation of chromosome banding, we must ask what are the different kinds of plants with which researchers have worked and what are the kinds of banding techniques they have used.

Table I lists representative plants (genera) that have been used successfully and the banding techniques utilized during the history of banding research. From these data, I make the following points:

1. Many Liliaceae, *Allium, Scilla* and *Trillium* plants have been subjected to banding investigations as well as many of the Graminae such as *Hordeum, Secale* and *Triticum*. The comparatively large chromosomes of these plants can easily be analyzed. However, we will need to analyze plants with smaller chromosomes in the future. Although such analyses will require greater patience, some workers have already been successful. For example, Gostev and Asker (1979) have studied species of *Salvia* (Labiatae), *Plantago* (Plantaginaceae), *Rosa* (Rosaceae), *Coriandrum, Foeniculum* and *Pimpinella* (Umbelliferae) using Feulgen-Giemsa banding. Ladizinsky *et al.* (1979) have also studied the many small chromosomes of *Glycine max* (soybean) by means of Giemsa staining.

2. The banding technique most often used successfully has been the Giemsa C-banding method. Even this now successful method has

a research history of only a decade. It is very useful for many
plants. On the other hand, the G-banding method cannot be adopted
for plants. Grielhuber (1977) wrote an enlightening paper, "Why
plant chromosomes do not show G-bands?" Although no clear
understanding of the G-banding mechanism has yet emerged, it seems
to be related to the tightness of the condensation of the
chromatin. I have found only one paper on G-banding chromosomes;
a study of *Pinus resinosa* (Drewry 1982). An improved,

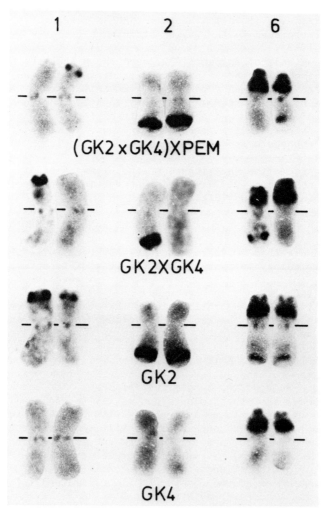

FIG. 2. *Differential staining patterns in chromosomes 1, 2 and
6 of* Zea mays *in hybrids and parental stocks after Giemsa
staining (Hadlaczky and Kalman 1975).*

TABLE I. *Representative genera and banding techniques*

Family/Genera	Authors	Banding technique*
Pinaceae		
Pinus	Drewry 1982	G
Chenopodiaceae		
Beta	de Jong and Oud 1979	C
Ranunculaceae		
Anemone	Marks 1976	C
Hepatica	Marks and Schweizer 1974	C
Nigella	Marks 1975	C
Saxifragaceae		
Boykinia, Heuchera, Mitella, Sullivania Tiarella and Tolmiea	Soltis 1982	Hy
Leguminosae		
Glycine	Ladizinsky 1979	C
Lotus	Shankland and Grant 1976	C
Phaseolus	Schweizer 1976	C,Q
	Schweizer and Ambrose 1979	C,Ag
Pisum	Lamm 1981	C
Vicia	McLeish 1953	L
	Dobel *et al.* 1973	C
	Funaki *et al.* 1975	N
	Greilhuber 1975	Hy,C
	Takehisa *et al.* 1976	L,Hy
	Hizume *et al.* 1980	C
	Pignone and Attolico 1980	C
	Rowland 1981	L,C,F
Labiatae		
Salvia	Gostev and Asker 1978	C
Solanaceae		
Nicotania	Narayan and Rees 1974	C
	Merritt and Burns 1974	C
Petunia	Dietrich *et al.* 1981	C
Adoxaceae		
Adoxa	Geitler 1940	L
	Greilhuber 1979a	C,L
Compositae		
Anacyclus	Ehrendorfer 1977	C

Family/Genera	Authors	Banding technique*
Crepis	Tanaka and Komatsu 1977	C
	Siljak-Yakovlev and Cartier	
	1982	C
Graminae		
Aegilops	Gill 1981	C
Agropyron	Gill 1981	C
Avena	Yen and Filion 1977	C
Elymus	Hadlaczky and Kalman 1975	C
	Gill 1981	C
Haynaldia	Gill 1981	C
Hordeum	Vosa 1976c	C
	Noda and Kasha 1978	C
	Linde-Laurson *et al.* 1982	C
	Singh and Tsuchiya 1982	C
Lolium	Thomas 1981c	C
Secale	Sarma and Natarajan 1973	C
	Vosa 1974	C,F
	Gill and Kimber 1974b	C
	Verma and Rees 1974	C
	Funaki *et al.* 1975	N
	Singh and Lelley 1975	C
	Singh and Röbbelen 1975, 1977	F
	Weinmark 1975	F
	Gustafson *et al.* 1976	F
	Lelley *et al.*1976	C
	Appels *et al.* 1978	C
	Iordansky *et al.* 1978b	F
	Jones 1978	C
	Naranjo and Lacadena 1982	C
	Ziauddin and Kasha 1982	C
Triticum	Natarajan and Sarma 1974	C
	Gill and Kimber 1974a	C
	Hadlczky and Belea 1975	C
	Gerlach 1977	N
	Iordansky *et al.* 1978a	C
	Zurabishvili *et al.* 1978	C
	Armstrong 1982	N
	Seal and Bennett 1982	C
Zea	Sartori and Ting 1974	C
	Funaki *et al.* 1975	N
	Hadlaczky and Kalman 1975	C
	Ward 1980	C
	Chow and Larter 1981	C
Commelinaceae		
Gibasis	Kenton 1978	C
Liliaceae		
Allium	Kurita 1958	L
	Stack and Clarke 1973	C

Family/Genera	Authors	Banding technique*
	Greilhuber 1974	Hy
	Fiskesjö 1975	C
	Elkington *et al.* 1976	C,Q
	Vosa 1976a,b 1977	C
	Badr and Elkington 1977	C
	Loidl 1979	C
Fritillaria	LaCour 1951, 1978	L,C
Leapoldia	Bentzer and Landström 1975	C
Lilium	Holm 1976	C,Q
	Kongsuwan and Smyth 1977	C,Q
	Son 1977	C
Paris	Darlington and LaCour 1938	L
	Shaw 1959	L
	Filion and Vosa 1980	Q
Scilla	Greilhuber 1974, 1978, 1979b	Hy,C
	Greilhuber and Speta 1976, 1977, 1978	C
	Vosa 1979	C
	Greilhuber *et al.* 1981	C
	Greilhuber and Deumling 1982	C
Trillium	Darlington and LaCour 1940	L
American	Wilson and Boothroyd 1944	L
Trillium	Bailey 1958	L
	Boothroyd and Lima-de-Faria 1964	L
	Fukuda and Channell 1975	L
	Chinnappa and Morton 1978	C,L
	Fukuda and Grant 1980	L
Japanese	Kurabayashi 1952, 1960	L
Trillium	Haga and Kurabayashi 1953	L
	Fukuda *et al.* 1960	L
	Yamasaki 1971	Hy
	Takehisa and Utsumi 1973	L,C
	Utsumi and Takehisa 1974	Ag
American and	Darlington and Shaw 1959	L
Japanese	Dyer 1963, 1964	L
Trillium	Fukuda 1973	L
Tulbaghia	Vosa 1973	Q
Tulipa	Blakey and Vosa 1981, 1982	C
Iridaceae		
Iris	Greilhuber 1974	Hy
Orchidaceae		
Cyprepedium	Yamasaki 1956, 1971	Hy
Cephalanthera	Schwarzacher and Schweizer 1982	C,F

*G = Giemsa; C = C-banding; Hy = Hy-banding; Ag = silver-banding;
L = cold-induced; N = N-banding; F or Q = Fluorescent or
Q-banding.

shortened method was adopted using Hy-banding by Greilhuber
(1974). The cold-induced banding technique is still useful
because it enables squashing and dispersal of the chromosomes and
the bands are easily made visible.

3. Chromosome banding in *Anacyclus*, *Secale*, *Allium* and
Scilla has been used to aid in determining systematic
reltionships. Certainly this approach is effective as a tool in
taxonomical research, as will be shown later.

4. In studying agricultural problems we use many hybrid lines,
such as wheat-rye and wheat-barley. In this case, we can identify
each chromosome in a hybrid line in terms of its parents from an
analysis of the banded chromosomes (Darvey and Gustafson 1975;
Hadlaczky and Koczka 1974; Iordansky *et al.* 1978b; Islam
1980; Schlegel and Weyszko 1979; Ziauddin and Kasha 1982; Seal
and Bennett 1982).

5. To distinguish chromosomes of a complement. In *Lotus
pedunculatus*, the two smallest chromosomes of the complement
are almost identical morphologically. Banding, however, clearly
distinguished them (Shankland and Grant 1976).

6. To show variations between plants from an evolutionary
viewpoint where population analyses are required, we need many
preparations for analyses. *Trillium* is an extremely useful
plant for this purpose. It is an exceptionally good example to
use for chromosome banding in biosystematics.

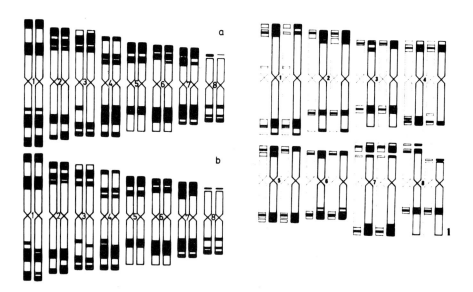

FIG. 3. *C-banded karyotype of two plants (a and b) of*
Allium flavum. *Karyotype of a plant of* A. stamineum *showing
Q-bands (left) and C-bands (right) (Vosa 1976b).*

CHROMOSOME BANDING AND SYSTEMATICS

With regard to chromosome banding in the field of botany, Vosa at
Oxford, and Schweizer and Greilhuber in Vienna, have carried out
many studies and have contributed significantly to this field.

One of the many investigations by Vosa was a survey of
Allium (1976a, b, 1977). He analyzed the *Allium cepa* and
paniculatum groups. He found characteristic banding patterns for
each species in their groups, with some intraspecific variation.
For example, *A. flaxum* and *A. stamineum* showed different C-
banded karyotypes. Likewise, there was some variation in the
banding pattern between two plants from Hungary (Fig. 3).

Blakey and Vosa (1981, 1982) have shown heterochromatin and
chromosome variation in *Tulipa*. The C-banding patterns
observed in different species in the two subgenera aided in the
systematics of the genus. Schweizer and Ehrendorfer (1976) have
attempted to establish the systematics and evolution of the genus
Anacyclus (Compositae) by using the "Banding style"
approach. They define "Banding style" as the entirety of band
location and equilocality, band number, heterochromatin quantity,
band width and width differences, and heterochromatin
heterogeneity. From a detailed comparison using "Banding style",
the authors presented a line of evolutionary differentiation for
the species of *Anacyclus*.

Greilhuber and Speta (1976, 1977, 1978) have carried out
considerable work on the quantitative analyses of C-banded
karyotypes of *Scilla* species. They have analyzed the
species by means of karyology and C-banding, in addition to
morphology. They have indicated that the banding data support the
systematic grouping proposed on a morphological basis and have
provided additional evolutionary evidence. They have established
the remarkable karyotypic stability of *Scilla vindobonensis*
as opposed to the variability of *S. bifola* and *S. drunensis*.
They explain their results on the labile and relatively stable
environmental conditions for these taxa during the Pleistocene
glacial periods.

CHROMOSOME BANDING AND EVOLUTION

The evolutionary problems of variation in chromsome banding have
been analyzed in detail by Fukuda using species of *Trillium*
(Liliaceae) both in Japan and North America.

Trillium ovatum occurs in the Pacific coast region of
North America and also in the Rocky Mountain region. Collections
were made from 14 localities throughout the distributional area of
the species. Sufficient numbers of plants were obtained from each
locality and examined for chromosome banding and statistically
analyzed. For chromosome studies, cold-induced banding methods
were used including hydrolysis in 1N HCl and Feulgen staining
procedure after cold treatment. The results of analyses showed
that the six populations from the Pacific coast region exhibited

little variation and considerable homogeneity in chromosome
composition of the banding patterns. In contrast, the eight
populations from the Rocky Mountain region showed considerable
variation and heterogeneity of chromosome composition within and
between populations. The difference is clearly evident in

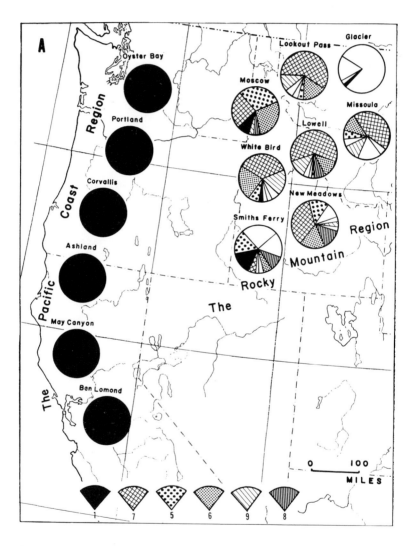

FIG. 4. *Distribution and frequency of chromosome A patterns in
natural populations of* Trillium ovatum. *Polygons show the
frequencies of the chromosome patterns indicated at the bottom of
the map (Fukuda and Channell 1975).*

chromosome A; all populations in the Pacific coast region have only a single banding pattern, whereas in the Rocky Mountain region each population shows considerable variation (Fig. 4). Such variation in chromosome banding pattern is also correlated with morphological characters; data are shown for the leaf index in both the Pacific coastal and the Rocky Mountain populations.

Fukuda and Channell (1975) have discussed how these evolutionary differences in *Trillium ovatum* may have originated in response to differences in habitat. In the pacific coast region the plants are associated with the humid redwood and Douglas fir forests. On the other hand, in the Rocky Mountain region the plants occur under more arid conditions in a Ponderosa and Lodgepole pine habitat. The populational heterogeneity in the Rocky Mountains may reflect various and fluctuational, climatological and paleoecological habitat conditions, whereas the populational homogeneity in the Pacific coast may reflect the comparatively more stable habitat conditions during the initial movement of the Pleistocine glaciers, their subsequent advances and their final retreat.

In eastern North America, *Trillium grandiflorum* grows around the Great Lake region. Fukuda and Grant (1980) examined cold-induced banded chromosomes from 25 populations throughout the natural range of the species. The banding patterns of *T. grandiflorum* were fairly uniform within and between populations. There was a greater homogeneity of chromosome composition from populations in the northern range of the species in contrast to those in the more southern populations (Fig. 5) located in the Appalachian Mountain region. It seems reasonable to assume that these populations of *T. grandiflorum* have been stabilized by natural selection presumably since the Pleistocene age. This stabilization could perhaps be due to the comparatively uniform environmental conditions of the lowlands around the Great Lakes. In the southern populations plants grow under considerably more heterogeneous conditions in terms of climate, elevation and latitude.

At present, in progress is a population analysis of *Trillium erectum,* which is distributed in the Appalachian Mountain region. Recent observations are interesting in that the variation in chromosome banding patterns of *T. erectum* were found to be only moderate within each population, but highly heterogeneous between populations. As an explanation, it is known that the southernmost part of the Appalachian Mountain range was never glaciated and thus a habitat has been provided for *Trillium* for a long period. As a result, each population of the Appalachian Mountains has adapted under different environmental conditions; different elevation, different latitudes and different climates, as reflected in the chromosome banding patterns.

The breeding system is also a very important factor in determining chromosome composition (Fukuda 1967). *T. grandiflorum* and *T. erectum* are self-compatible and undergo both self- and cross-pollination. A homogamous system is

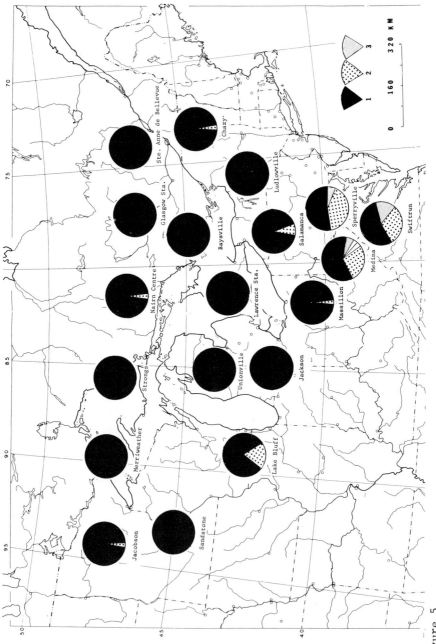

Figure 5

developing in both species. In *T. grandiflorum,* the pistil
and stamens are in close contact at the time of anthesis, just
touching each other. Flowers of *T. erectum* are of the
nodding type (bent over) and crossing can take place easily within
the flower by wind pollination. As a result, self-pollination is
the dominant mode of pollination in both *T. grandiflorum* and
T. erectum.

FUTURE PROBLEMS OF CHROMOSOME BANDING IN BIOSYSTEMATICS

In conclusion, I would like to point out the following problems
for the future.
  1. Considering the early knowledge on cold-induced banding, we
may say that the history of banding goes back many years.
However, the major developments have occurred only in the past
decade. We need to extend the application of plant-banding
methods to other species, for example as previously discussed, to
species with small chromosomes. We also need improved banding
techniques for those species which have shown that they can be
banded, but for which an increase in the number of bands or
greater differentiation is desirable.
  2. For evolutionary and biosystematic analyses, a considerable
number of preparations and observations are needed. We are now
just beginning quantitative, populational analyses using
chromosome banding. I hope there will be more development in this
field using chromosome banding in the future.
  3. From the brief results which I have given, it is clear that
chromosome banding methods are an effective tool for systematic
studies. Moreover, this approach along with other genetical,
ecological, chemical and physiological methods will become an even
more powerful tool for use in biosystematic analyses in the
future.

REFERENCES

Appels, R., Driscolle, C. and Peacock, W.J. 1978. Heterochromatin
    and highly repeated DNA sequences in rye *(Secale cereale).*
    Chromosoma 70: 67-89.
Armstrong, K.C. 1982. N-banding in *Triticum aestivum*
    following Feulgen hydrolysis. *Theor. Appl. Genet.* 61: 337-339.
Badr, A. and Elkington, T.T. 1977. Variation of Giemsa C-band and
    fluorochrome banded karyotype, and relationships in *Allium*
    subgen. *Molium. Pl. Syst. Evol.* 128: 23-35.
Bailey, P.C. 1958. Differential chromosome segments in eight
    species of *Trillium. Bull. Torrey Bot. Club.* 85: 201-214.

FIG. 5. *Polygons for chromosome A banding in* Trillium
grandiflorum *for 25 populations showing the frequency of the
standard pattern (A1) and variants (A2, A3) (Fukuda and Grant
1980).*

Bentzer, B. and Landström, T. 1975. Polymorphism in chromosomes of *Leopoldia comosa* (Liliaceae) revealed by Giemsa staining. *Hereditas* 80: 219–232.

Blakey, D.H. and Vosa, C.G. 1981. Heterochromatin and chromosome variation in cultivated species of *Tulipa* subg. *Eriostemones* (Liliaceae). *Pl. Syst. Evol.* 139: 47–55.

Blakey, D.H. and Vosa, C.G. 1982. Heterochromatin and chromosome variation in cultivated species of *Tulipa* subg. *Leiostemones* (Liliaceae). *Pl. Syst. Evol.* 139: 163–178.

Boothroyd, E.R. and Lima-de-Faria, A. 1964. DNA synthesis and differential reactivity in the chromosomes of *Trillium* at low temperature. *Hereditas* 52: 122–126.

Caspersson, T., Zech, L., Modest, E.J., Foley, G.E., Wagh, U. and Simonsson, E. 1969a. Chemical differentiation with fluorescent alkylating agents in *Vicia faba* metaphase chromosomes. *Exp. Cell Res.* 58: 128–140.

Caspersson, T., Zech, L., Modest, E.J., Foley, G.E. Wagh, U. and Simonsson, E. 1969b. DNA-binding fluorochromes for the study of the organization of the metaphase nucleus. *Exp. Cell Res.* 58: 141–152.

Caspersson, T., de la Chapelle, A., Schröder, J. and Zech, L. 1972. Quinacrine fluorescence of metaphase chromosomes. Identical patterns in different tissues. *Exp. Cell Res.* 72: 56–59.

Chinnappa, C.G. and Morton, J.K. 1978. Heterochromatin banding patterns in two species of *Trillium*. *Can. J. Genet. Cytol.* 20: 475–481.

Chow, C. and larter, E.N. 1981. Centromeric banding in maize. *Can. J. Genet. Cytol.* 23: 255–258.

Darlington, C.D. and LaCour, L.F. 1938. Differential reactivity of the chromosomes. *Ann. Bot.* 2: 615–625.

Darlington, C.D. and LaCour, L.F. 1940. Nucleic acid starvation of chromosomes in *Trillium*. *J. Genet.* 40: 185–213.

Darlington, C.D. and Shaw, G.W. 1959. Parallel polymorphism in the heterochromatin of *Trillium* species. *Heredity* 13: 89–121.

Darvey, N.L. and Gustafson, J.P. 1975. Identification of rye chromosomes in wheat-rye addition lines and triticale by heterochromatin bands. *Crop Sci.* 15: 239–243.

Dietrich, A.J.J., de Jong, J.H. and Mulder, R.J.P. 1981. Location and variation of the constitutive heterochromatin in *Petunia hybrida*. *Genetica* 55: 85–91.

Döbel, P., Rieger, R. and Michaelis, A. 1973. The Giemsa banding patterns of the standard and four reconstructed karyotypes of *Vicia faba*. *Chromosoma* 43: 409–422.

Dreury, A. 1982. G-banded chromosomes in *Pinus resinosa*. *J. Hered.* 73: 305–306.

Dyer, A.F. 1963. Allocyclic segments of chromosomes and the structural heterozygosity that they reveal. *Chromosoma* 13: 545–576.

Dyer, A.F. 1964. Heterochromatin in American and Japanese species of *Trillium*. 1. Fusion of chromocentres and the distribution of H-segments. *Cytologia* 29: 155–170.

Ehrendorfer, F., Schweizer, D., Greger, H. and Humphries, C. 1977. Chromosome banding and synthetic systematics in *Anacyclus* (Asteraceae – Anthemidae). *Taxon* 26: 387–394.

Elkington, T.T., Badr, A., El-Gadi, A., Hussain, L. and White, S. 1976. Giemsa C-band and quinacrine banded karyotypes and systematic relationship in *Allium*. In *Current Chromosome Research* (K. Jones and P.E. Brandham, eds.) pp. 214–215.

Filion, W.G. and Vosa, C.G. 1980. Quinacrine fluorescence studies in *Paris polyphylla*. *Can. J. Genet. Cytol.* 22: 417–420.

Fiskesjö, G. 1975. Chromosomal relationships between three species of *Allium* as revealed by C-banding. *Hereditas* 81: 23–32.

Fukuda, I. 1967. The formation of subgroups by the development of inbreeding systems in a *Trillium* population. *Evolution* 21: 141–147.

Fukuda, I. 1973. Comparative study of chromosome variation in the Japanese and American *Trillium* species. *Sci. Rep. Tokyo Woman's Christian Coll.* 29–31: 361–367.

Fukuda, I. and Channell, R.B. 1975. Distribution and evolutionary significance of chromosome variation in *Trillium ovatum*. *Evolution.* 29: 257–266.

Fukuda, I. and Grant, W.F. 1980. Chromosome variation and evolution in *Trillium grandiflorum*. *Can. J. Genet. Cytol.* 22: 81–91.

Fukuda, I., Hiraizumi, Y., Narise, T. and Kurabayashi, M. 1960. Evolution and variation in *Trillium*. VI. Migrations among natural populations of *T. kamtschaticum* across the Ishikari depression. *Evolution* 14: 224–231.

Fukuda, I., Matsui, S., and Sasaki, M. 1975. Location of nucleolar organizers in animal and plant chromosomes by means of an improved N-banding technique. *Chromosoma* 49: 357–370.

Geitler, L. 1940 Temperaturbedingte Ausbildung von Spezialsegmenten an Chromosomenenden. *Chromosoma* 1: 554–561.

Gerlach, W.L. 1977. N-banded karyotypes of wheat species. *Chromosoma* 62: 49–56.

Gill, B.S. 1981. Evolutionary relationships based on heterochromatin bands in six species of the Triticinae. *J. Hered.* 72: 391–394.

Gill, B.S. and Kimber, G. 1974a. Giemsa C-banding and the evolution of wheat. *Proc. Natl. Acad. Sci. U.S.A.* 71: 4086–4090.

Gill, B.S. and Kimber, G. 1974b. The Giemsa C-banded karyotype of rye. *Proc. Natl. Acad. Sci. U.S.A.* 71: 1247–1249.

Gostev, A. and Asker, S. 1978. Polytene chromosomes in glandular hairs of *Salvia horminum*. *Hereditas* 88: 133–135.

Gostev, A. and Asker, S. 1979. A C-banding technique for small plant chromosomes. *Hereditas* 91: 140–143.

Greilhuber, J. 1974. Hy-banding: A new quick technique for heterochromatin staining in plant chromosomes.

Naturwissenschaften 61: 170–171.

Greilhuber, J. 1975. Heterogeneity of heterochromatin in plants: Comparison of Hy- and C-bands in *Vicia faba*. *Pl. Syst. Evol.* 124: 139–156.

Greilhuber, J. 1977. Why plant chromosomes do not show G-bands. *Theor. Appl. Genet.* 50: 121–124.

Greilhuber, J. 1978. DNA contents, Geimsa banding, and systematics in *Scilla bifolia, S. drunensis* and *S. vindobonensis* (Liliaceae). *Pl. Syst. Evol.* 130: 223–233.

Greilhuber, J. 1979a. C-band distribution, DNA content and base composition in *Adoxa moschatellina* (Adoxaceae), a plant with cold-sensitive chromosome segments. *Pl. Syst. Evol.* 131: 243–259.

Greilhuber, J. 1979b. Evolutionary changes of DNA and heterochromatin amounts in the *Scilla bifolia* group (Liliaceae). *Pl. Syst. Evol.*, Suppl. 2: 263–280.

Greilhuber, J. and Deumling, B. 1982. Characterization of heterochromatin in different species of the *Scilla siberica* group (Liliaceae) by in situ hybridization of satellite DNAs and fluorochrome banding. *Chromosoma* 84: 535–555.

Grieilhuber, J. and Speta, F. 1976. C-banded karyotypes in the *Scilla hohenackeri* group, *S. persica* and *Puschkinia* (Liliaceae). *Pl. Syst. Evol.* 126: 149–188.

Greilhuber, J. and Speta, F. 1977. Giemsa karyotypes and their evolutionary significance in *Scilla bifolia, S. drunensis* and *S. vindobonensis* (Liliaceae). *Pl. Syst. Evol.* 127: 171–190.

Greilhuber, J. and Speta, F. 1978. Quantitative analyses of C-banded karyotypes and systematics in the cultivated species of the *Scilla siberica* group (Liliaceae). *Pl. Syst. Evol.* 129: 63–109.

Greilhuber, J., Deumling, B. and Speta, F. 1981. Evolutionary aspects of chromosome banding, heterochromatin, satellite DNA and genome size in *Scilla* (Liliaceae). *Ber. Deutsch. Bot. Ges.* 94: 249–266.

Gustafson, J.P., Evans, L.E. and Josifek, K. 1976. Identification of chromosomes in *Secale montanum* and individual *S. montanum* chromosome additions to 'Kharkov' wheat by heterochromatin bands and chromosome morphology. *Can. J. Genet. Cytol.* 18: 339–343.

Hadlaczky, G. and Belea, A. 1975. C-banding in wheat evolutionary cytogenetics. *Plant Sci. Lett.* 4: 85–88.

Hadlaczky, G. and Kalman, L. 1975. Discrimination of homologous chromosomes of maize with Geimsa staining. *Heredity* 35: 371–374.

Hadlaczky, G. and Koczka, K. 1974. C-banding karyotype of rye from hexaploid *Triticale*. *Cereal Res. Commun.* 2: 193–200.

Haga, T. and Kurabayashi, M. 1953. Genome and polyploidy in the genus *Trillium*. IV. Genome analysis by means of differential reaction of chromosome segments to low temperature. *Cytologia* 18: 13–28.

Hizume, M., Tanaka, A. Yonezawa, Y. and Tanaka, R. 1980. A
technique for C-banding in *Vicia faba* chromosomes.
Jpn. J. Genet. 55: 301-305.

Holm, P.B. 1976. The C- and Q-banding patterns of chromosomes of
Lilium longiflorum. *Carlsberg Res. Commun.* 41: 217-224.

Iordansky, A.B., Zurabishvili, T.G., Badaev, N.S. 1978a. Linear
differentiation of cereal chromosomes. I. Common wheat and its
supposed ancestors. *Theor. Appl. Genet.* 51: 145-152.

Iordansky, A.B., Zurabishvili, T.G., Badaev, N.S. 1978b. Linear
differentiation of cereal chromosomes. III. Rye, triticale and
'Aurora' variety. *Theor. Appl. Genet.* 51: 281-288.

Islam, A.K.M.R. 1980. Identification of wheat-barley addition
lines with N-banding of chromosomes. *Chromosoma* 76: 365-373.

Jones, G.H. 1978. Giemsa C-banding of rye meiotic chromosomes and
the nature of "terminal" chiasmata. *Chromosoma* 66: 45-57.

De Jong, J.H. and Oud, J.L. 1979. Location and behaviour of
constitutive heterochromatin during meiotic prophase in *Beta
vulgaris* L. *Genetica* 51: 125-133.

Kenton, A. 1978. Giemsa C-banding in *Gibasis* (Commelinaceae).
Chromosoma 65: 309-324.

Kongsuwan, K., and Smyth, D.R. 1977. Q-bands in *Lilium* and
their relationships to C-banded heterochromatin. WBChromosoma
60: 169-178.

Kurabayashi, M. 1952. Differential reactivity of chromosomes in
Trillium. *J. Fac. Sci. Hokkaido Imp. Univ., Ser. V* 6:
233-248.

Kurabayashi, M. 1960. Evolution and variation in Japanese species
of *Trillium*. *Evolution*. 12: 286-310.

Kurita, J. 1958. Heterochromaty in the *Allium* chromosomes.
Mem. Ehime Univ. Sect. II. Sci. 3: 23-28.

LaCour, L.F. 1951. Heterochromatin and the organization of
nucleoli in plants. *Heredity* 5: 37-50.

LaCour, L.F. 1978. The constitutive heterochromatin in
chromosomes of *Fritillaria* sp., as revealed by Giemsa
banding. *Philos. Trans. R. Soc. Lond. Ser. B* 285: 61-71.

Ladizinsky, G., Newell, C.A. and Hymowitz, T. 1979. Giemsa
staining of soybean chromosomes. *J. Hered.* 70: 415-416.

Lamm, R. 1981. Giemsa C-banding and silver-staining for
cytological studies in *Pisum*. *Hereditas* 94: 45-52.

Lelley, T., Josifek, K. and Kaltsikes, P.J. 1978. Polymorphism in
the Giemsa C-banding pattern of rye chromosomes. *Can. J.
Genet. Cytol.* 20: 307-312.

Linde-Laursen, I., Doll, H. and Nielsen, G. 1982. Giemsa C-
banding patterns and some biochemical markers in a pedigree of
European barley. *Z. Pflanzenzüchtg.* 88: 191-219.

Loidl, J. 1979. C-band proximity of chiasmata and absence of
terminalisation in *Alium flavum* (Liliaceae). *Chromosoma*
73: 45-51.

Marks, G.E. 1975. The Giemsa-staining centromeres of *Nigella
damascena*. *J. Cell Sci.* 18: 19-25.

Marks, G.E. 1976. Variation of Giemsa banding patterns in the

chromosomes of *Anemone blanda* L. *Chromosomes Today* 5: 179–184.
Marks, G.E. and Schweizer, D. 1974. Giemsa banding: Karyotype
   differences in some species of *Anemone* and in *Hepatica nobilis*.
   Chromosoma 44: 405–416.
McLeish, J. 1953. Heterochromatin segments in *Vicia faba*.
   Annu. Rep. John Innes Hort. Inst. 47: 31.
Merritt, J.F. and Burns, J.A. 1974. Chromosome banding in
   Nicotiana otophora without denaturation and renaturation.
   J. Hered. 65: 101–103.
Naranjo, T. and Lacadena, J.R. 1982. C-banding pattern and
   meiotic pairing in five rye chromosomes of hexaploid triticale.
   Theoret. Appl. Genet. 61: 233–237.
Narayan, R.K.J. and Rees, H. 1974. Nuclear DNA, heterochromatin
   and phylogeny of *Nicotiana* amphidiploids. *Chromosoma*
   47: 75–83.
Natarajan, A.T. and Sarma, N.P. 1974. Chromosome banding
   patterns and the origin of the B genome in wheat. *Genet.
   Res.* 24: 103–108.
Noda, K. and Kasha, K.J. 1978. A modified Giemsa C-banding
   technique for *Hordeum* species. *Stain Technol.* 53: 155–162.
Pardue, M.L. and Gall, J.G. 1970. Chromosomal localization of
   mouse satellite DNA. Science 168: 1356–1358.
Pignone, D. and Attolico, M. 1980. Chromosome banding in four
   groups of *Vicia faba* L. *Caryologia* 33: 283–288.
Rowland, R.E. 1981. Chromosome banding and heterochromatin in
   Vicia faba. *Theor. Appl. Genet.* 60: 275–280.
Sarma, N.P. and Natarajan, A.T. 1973. Identification of
   heterochromatic regions in the chromosomes of rye. *Hereditas*
   74: 233–238.
Sartori, L. and Ting, Y.C. 1974. Giemsa banding in the
   chromosomes of haploid maize. *Am. J. Bot. (Suppl.)* 61
   (5): 63.
Schlegel, von R. and Weryszko, E. 1979. Intergeneric chromosome
   pairing in different wheat-rye hybrids revealed by the Giemsa
   banding technique and some implications on karyotype evolution
   in the genus *Secale*. *Biol. Zbl.* 98: 399–407.
Schwarzacher, T. and Schweizer, D. 1982. Karyotype analysis and
   heterochromatin differentiation with Giemsa C-banding and
   fluorescent counterstaining in *Cephalanthera* (Orchidaceae).
   Pl. syst. Evol. 141: 91–113.
Schweizer, D. 1976. Giemsa and fluorochrome banding of polytene
   chromosomes in *Phaseolus vulgaris* and *P. coccineus*. In
   Current Chromosome Research (K. Jones and P.E. Brandham, eds.).
   Elsevier/North Holland, Amsterdam. pp. 51–56
Schweizer, D. and Ambrose, P. 1979. Analysis of nucleolus
   organizer regions (NORs) in mitotic and polytene chromosomes of
   Phaseolus coccineus by silver staining and Giemsa C-banding.
   Pl. Syst. Evol. 132: 27–51.
Schweizer, D. and Ehrendorfer, F. 1976. Giemsa banded karyotypes,
   systematics and evolution in *Anacyclus* (Asteraceae –
   Anthemidae). *Pl. Syst. Evol.* 126: 107–148.

Seal, A.G. and Bennett, M.D. 1982. Preferential C-banding of wheat or rye chromosomes. *Theor. Appl. Genet.* 63: 227-233.

Shankland, N.E. and Grant, W.F. 1976. Localization of Giemsa bands in *Lotus pedunculatus* chromosomes. *Can. J. Genet. Cytol.* 18: 239-244.

Shaw, G.W. 1959. The nature of differential reactivity in the heterochromatin of *Trillium* and *Paris* species. *Cytologia* 24: 50-61.

Siljak-Yakovlev, S. and Cartier, D. 1982. Comparative analysis of C-band karyotypes in *Crepis praemorsa* subsp. *praemorsa* and subsp. *dinarica*. *Pl. Syst. Evol.* 141: 85-90.

Singh, R.J. and Lelley, T. 1975. Giemsa banding in meiotic chromosomes of rye, *Secale cereale* L. Z. *Pflanzenzüchtg.* 75: 85-89.

Singh, R.J. and Röbbelen, G. 1975. Comparison of somatic Giemsa banding pattern in several species of rye. Z. *Pflanzenzüchtg.* 74: 270-285.

Singh, R.J. and Röbbelen, G. 1977. Identification by Giemsa technique of the translocation separating cultivated rye from three wild species of *Secale. Chromosoma* 59: 217-225.

Singh, R.J. and Tsuchiya, T. 1982. Identification and designation of telocentric chromosomes in barley by means of Giemsa N-banding technique. *Theor. Appl. Genet.* 64: 13-24.

Soltis, D.E. 1982. Heterochromatin banding in *Boykinia, Heuchera, Mitella, Sullivania, Tiarella* and *Tolmiea* (Saxifragaceae). *Am. J. Bot.* 69: 108-115.

Son, J.-H. 1977. Karyotype analysis of *Lilium lancifolium* Thunberg by means of C-banding method. *Jpn. J. Genet.* 52: 217-221.

Stack, S.M. and Clarke, C.R. 1973. Differential Giemsa staining of the telomeres of *Allium cepa* chromosomes; observations related to chromosome pairing. *Can. J. Genet. Cytol.* 15: 619-624.

Takehisa, S. 1968. Heterochromatic segments in *Vicia* revealed by treatment with HCl-acetic acid. *Nature* 217: 567-568.

Takehisa, S. and Utsumi, S. 1973. Heterochromatin and Giemsa banding of metaphase chromosomes in *Trillium kamtschaticum* Pallas. *Nature (New Biol.)* 244: 286-287.

Takehisa, S., Döbel, P., Rieger, R. and Michaelis, A. 1976. Differential response to cold and HCl-acetic acid treatment of heterochromatin in reconstructed *Vicia faba* karyotypes. *Chromosoma* 54: 165-173.

Tanaka, R. and Komatsu, H. 1977. C-banding patterns in somatic and meiotic chromosomes of *Crepis capillaris. Proc. Jpn. Acad., Ser. B.* 53: 6-9.

Thomas, H.M. 1981. The Giemsa C-band karyotypes of six *Lolium* species. *Heredity* 46: 263-267.

Utsumi, S. and Takehisa, S. 1974. Heterochromatin differentiation in *Trillium kamtschaticum* by ammoniacal silver reation. *Exptl. Cell Res.* 86: 398-401.

Verma, S.C. and Rees, H. 1974. Giemsa staining and distribution
of heterochromatin in rye chromosomes. *Heredity* 32: 118-120.

Vosa, C.G. 1973. Quinacrine fluorescence analysis of chromosome
viariation in the plant *Tulbaghia leucantha*. *Chromosomes
Today* 4: 345-349.

Vosa, C.G. 1974. The basic karyotype of rye (*Secale cereale*)
analyzed with Giemsa and fluorescence methods. *Heredity*
33: 403-408.

Vosa, C.G. 1976a. Heterochromatic patterns in *Allium*. I.
The relationship between the species of the *cepa* group and
its allies. *Heredity* 36: 383-392.

Vosa, C.G. 1976b. Heterochromatic banding patterns in *Allium*.
II. Heterochromatin variation in species of the *paniculatum*
group. *Chromosoma* 57: 119-133. Vosa, C.G. 1976c. Chromosome
banding patterns in cultivated and wild barleys (*Hordeum*
spp.). *Heredity* 37: 395-403.

Vosa, C.G. 1977. Heterochromatic patterns and species
relationship. *Nucleus* 20: 33-41.

Vosa, C.G. 1979. Chromosome banding patterns in the triploid
Scilla sibirica Harv. var. "Spring Beauty". *Nucleus* 22: 32-33.

Ward, E.J. 1980. Banding patterns in maize mitotic chromosomes.
Can. J. Genet. Cytol. 22: 61-67.

Weimark, A. 1975. Heterochromatin polymorphism in the rye
karyotypes as detected by the Giemsa C-banding technique.
Hereditas 79: 293-300.

Wilson, G.B. and Boothroyd, E.R. 1944. Temperature-induced
differential contraction in the somatic chromosomes of
Trillium erectum L. *Can. J. Res. C.* 22: 105-119.

Yamasaki, N. 1956. Differentielle Färbung der somatischen
Metaphasechromosomen von *Cypripedium debile*. *Chromosoma*
7: 620-626.

Yamasaki, N. 1971. Karyotypanalyse an Hand des Farbungsmusters
der Metaphase chromosomen von *Cypripedium debel* und
Trillium kamtschaticum. *Chromosoma* 33: 372-381.

Yen, S.T. and Filion, W.G. 1977. Differential Giemsa staining in
plants. V. Two types of constitutive heterochromatin in species
of *Avena*. *Can. J. Genet. Cytol.* 19: 739-743.

Ziauddin, A. and Kasha, K.J. 1982. Giemsa C-band identification
of rye chromosomes in some advanced lines of winter triticale.
Can. J. Genet. Cytol. 24: 721-727.

Chromosome Evolution and Adaptation in Mistletoes

B. A. Barlow
Division of Plant Industry
C.S.I.R.O.
Canberra, Australia

Nicole J. Martin
Cytogenetics Unit
Royal Brisbane Hospital
Brisbane, Australia

INTRODUCTION

Mistletoes can be broadly defined as shrubby evergreen parasites which grow attached to woody tree branches. When so defined, they form a diverse group, although essentially Santalalean. Currently most mistletoe species are assigned to one of two families, Loranthaceae (900 species in 70 genera) and Viscaceae (400 species in 7 genera).

The contribution of biosystematic studies to an understanding of evolution and adaptation in mistletoes can be easily appreciated if we take a starting point consistent with the theme of this symposium. Forty years ago the accumulated biosystematic literature on mistletoes was very modest, at least in terms of the overall size of the group and its unusual biological features. The only species for which an extensive and intensive literature existed was the traditional European mistletoe, *Viscum album,* and even here some significant features such as the female-biased sex ratio and the sex-associated interchange complexes still awaited discovery.

Two decades ago almost all the species commonly described as mistletoes were placed together in a single family Loranthaceae, variously reported as having 500 to 1,000 species in 20 to 60 genera. Even though the earlier taxonomic works of van Tieghem and Danser favored the acceptance of many genera, the bulk of the species were usually assigned to one large genus *Loranthus,* with most of the remainder placed in *Viscum* or *Phoradendron.* Because the highest proportion of the species occurs in the tropics, there was a general presumption that the family was of tropical origin, and that the mistletoe habit represented an

adaptive strategy similar to epiphytism, allowing growth under
favorable light conditions in closed forests. Even Danser (1931)
constructed a phylogeny based on a presumed center of origin in
the Malesian tropics.

Karyological knowledge of mistletoes 40 years ago consisted of
a small number of casual chromosome counts, mostly derived from
studies of life history, and some of them erroneous. The numbers
were insufficient for determination of basic numbers, and made
very little contribution to knowledge of natural relationships
within the mistletoes.

Morphological and biosystematic knowledge of mistletoes has
been extended on a number of fronts in the last two decades.
Studies of reproductive anatomy, led by the Maheshwari School, and
reviewed by Johri and Bhatnagar (1961), have highlighted the
unusual ovary structure and fertilization patterns. Branching
patterns, inflorescence structures and pollen structures have been
investigated by Kuijt (especially 1959, 1970, 1981; Feuer and
Kuijt 1978). Biogeographical analyses have been made by Barlow
and Wiens (1971), Wiens and Barlow (1971) and Barlow (1983b).
Phytochemical studies of high molecular weight proteins have been
made by Samuelsson (especially 1973). Karyological studies have
been reviewed by Barlow and Wiens (1971), Wiens and Barlow (1971),
Wiens (1975) and Martin (1983). Host/parasite interactions have
been studied by Barlow and Wiens (1977) and are being extended by
Atsatt (1983 and unpublished data). Comparative studies of
haustorial structure have been made by Thoday (especially 1961),
Hamilton and Barlow (1963), Kuijt (1964) and Weber (1980 and
unpublished data). Comparative studies of pollination biology and
incompatibility levels have been reviewed by Bernhardt et al.
(1980) and Bernhardt (1982). From a synthesis of all these
investigations, the mistletoes now emerge as a group in which
knowledge of biological adaptation and coadaptation has made a
significant contribution to an understanding of their internal
relationships.

It is beyond the scope of this paper to integrate the data from
all of these sources. Emphasis is given here to the utilization
of karyological data in interpreting natural relationships. In
fact, patterns of chromosomal structure and behavior have made an
important contribution to biosystematic knowledge of mistletoes,
and clearly reflect the overall phylogeny derived from an
evolutionary synthesis of all data.

LORANTHACEAE AND VISCACEAE – ONE FAMILY OR TWO?

The classification of mistletoes into higher taxonomic categories
has been principally a question of assessment of rank. Two
natural groups of mistletoes, the loranthoids and viscoids, have
long been recognized, and very few genera have been found
difficult to place in one group or the other. The first
suggestion that the two groups should be treated as distinct
families was made by Miers (1851), but until 1960 they were almost

universally regarded as subfamilies Loranthoideae and Viscoideae of the Loranthaceae.

Much of the impetus for the acceptance of Loranthaceae *sens. str.* and Viscaceae as distinct families came from embryological studies. It was already well known that loranthoids and viscoids differed in several significant aspects of perianth structure, pollen structure and sexual expression. Maheshwari and co-workers showed additional differences in embryo sac development, embryo development and fruit vasculature (Johri and Bhatnager 1961; Dixit 1962). These differences, when taken together with those in high molecular weight protein composition (Samuelsson 1973) and in chromosome number, suggest that the two groups have had independent origins within the Santalales, perhaps from different ancestral nonmistletoe stocks (Kuijt 1968). This basis for their treatment as distinct families (Barlow 1964) is now generally accepted.

Karyological knowledge of the two families is now extensive. In Loranthaceae chromosome numbers have been accumulated for 57 genera and 184 species, representing 81% of the genera and 21% of the species (Barlow and Wiens 1971; Wiens 1975). In Viscaceae chromosome numbers are known for 6 genera and 125 species, representing 86% of the genera and 31% of the species (Wiens and Barlow 1971; Wiens 1975). Chromosome numbers are generally conservative, often remaining constant within large suites of related genera. Even polyploidy is of rare occurrence, being reported in 5 (possibly 6) species of Loranthaceae (2.7%), and in 9 species of Viscaceae (7.2%), usually as an infraspecific phenomenon. The significant lack of polyploidy in these two families is still not adequately explained (Wiens 1975).

In Loranthaceae the primary basic chromosome number is $x = 12$ (Barlow 1963), and is characteristic of several apparently primitive genera. The other basic numbers are $x = 11, 10, 9$ and 8, and they indicate progressive dysploid reduction (see below). There is a general inverse relationship between chromosome number and size in the family. In the Viscaceae the primary basic chromosome number is $x = 14$ (Wiens 1975), prevailing in 5 of the 6 genera studied. The other basic numbers are $x = 17, 15, 13, 12, 11$ and 10, indicating limited dysploid increase as well as reduction. The basic number $x = 14$ has not been recorded in *Notothixos* ($x = 13, 12$), but only 2 of its 8 species have been studied. Chromosome size is generally large except for *Notothixos*, in which they are relatively small (Barlow 1983a).

There can be little doubt that the basic genomes of the two families are of independent derivation. There is a range of basic numbers within the Santalales, and the two families have apparently been derived from cytologically different stocks. Shared chromosome numbers in the two families ($x = 12, 11, 10$) occur in a few distinctive, primitive, often relictual genera of Loranthaceae, but only in a few specialized, derived species of *Viscum* and *Notothixos* in Viscaceae; they can be no indicator of common origin.

Three somewhat anomalous genera, *Eremolepis* (x = 10),
Eubrachion, and *Antidaphne* (x = 13) differ from Viscaceae
sens. str. in a number of important vegetative and floral
features, and Kuijt (1968) reestablished the family Eremolepi-
daceae to accommodate them. More recently, Feuer and Kuijt (1978)
have proposed transfer of *Lepidoceras* and *Tupeia* to this family.
With this exclusion, the Loranthaceae *sens. str.* and
Viscaceae *sens. str.* are homogeneous groups with significant
morphological and karyological differences sufficient to justify
their treatment as distinct families. This view is strongly
supported by biogeographical analysis of the two groups,
summarized below.

CHROMOSOMAL EVOLUTION AND DIFFERENTIATION IN LORANTHACEAE

Loranthaceae as an Old Southern Family

Karyological features have been a key to an understanding of
the biogeography of the Loranthaceae (Barlow 1983b). The primary
basic number x = 12 is represented in both Old and New World
loranths (Barlow and Wiens 1971, 1973), occurring in several
apparently relictual, mostly monotypic, primitive genera in south
temperate habitats. This number otherwise occurs only in a few
larger and more specialized but still fundamentally rather
primitive genera of the Malesian/Australasian region. The basic
number x = 11 occurs only in two or three relictual southern
genera, and x = 10 is found only in the monotypic *Ligaria*
of temperate South America. The latter two basic numbers appear
themselves to be south temperate and relictual in their
distribution.
The number x = 9 is characteristic of the derived and
advanced genera of the Old World, from Australia to Asia and
Africa. Similarly, x = 8 is characteristic of the large,
advanced tropical genera of the New World. The cytogeographic
pattern therefore suggests that an ancestral stock with x = 12
was widespread in the southern lands, and that dysploid reduction
to x = 9 and 8 preceded massive evolutionary radiation of
the family in the tropics. That this radiation has been
independent in the Old and New Worlds is indicated by the fact
that the derived stocks, with x = 9 and 8 respectively, have
closer affinities with old southern groups than with each other.
A limited independent radiation in the Malesian region has
apparently also occurred in a stock having x = 12.
Although now predominantly pantropical in distribution, the
Loranthaceae are thus Gondwanan rather than equatorial in origin.
The likely history of the loranths in relation to the
paleogeography of the southern lands has been summarized by Barlow
(1983b). Although specialized parasites, the Loranthaceae must
have attained a wide distribution in Gondwanaland by late
Cretaceous times at the latest. Considerable differentiation had
already occurred by that time, including the establishment of

different dysploid stocks. Following the separation of Africa, Australia, New Zealand and South America these differentiated stocks were isolated, and underwent independent secondary radiations. Radiation in a stock with $x = 8$ was confined entirely to South America, while radiation in a stock with $x = 12$ was confined to the Papuasian/Malesian region. The early separation of Africa/Madagascar/India from South America/ Antarctica/Australia/New Zealand apparently also separated two floristic stocks with $x = 9$. All African loranths (ca. 300 spp.) have $x = 9$, and they differ from Australian/Papuasian 9-paired stocks in several floral and inflorescence characters. There are indications that the ancestral 9-paired genomes were themselves rather different with respect to chromosome size and therefore perhaps DNA organization.

From our knowledge of the climates of the Cretaceous and early Tertiary in Gondwanaland, it is apparent that the Loranthaceae have evolved in closed forests under mesic conditions. Therefore their parasitism did not evolve in response to environmental water shortage. The aerial hemi-parasitic mistletoes are derived from terrestrial root-parasitic ancestors, and it seems more likely that parasitism gave the mistletoes easy access to mineral nutrients in forests growing on highly leached soils (Barlow 1983b).

There are few families in which cytogeographic data have been such useful indicators of change at higher taxonomic levels as they have in Loranthaceae. The rather striking correlation of chromosome evolution with the biogeography of the family, and with southern hemisphere paleogeography, is due largely to the genomic stability in the family. This stability seems to represent a kind of conservative strategy maintained during the extensive adaptation and divergence which has occurred in the family. As mentioned above, polyploidy occurs, but has hardly been utilized as a cytogenetic mechanism. The basic machinery for dysploid change is certainly present, and structural aberrations such as interchanges and inversions have been encountered in routine karyological surveys (Barlow 1963, 1974). Dysploidy, however, appears to have occurred only in the very early stages of diversification in the family; the correlation with paleogeographic events suggests that there may have been no dysploid derivatives established throughout the Tertiary period.

In an attempt to quantify the genomic changes which have occurred, Martin (1980, 1983) undertook karyotypic analysis of 55 species from 15 genera of Loranthaceae in the Australasian region. Conventional karyotypes were derived from mitotic preparations, with raw data for each chromosome arm in the haploid genome expressed as mean percent length of the entire genome. For purposes of comparison, chromosome sizes were represented by multiplying all mean percent arm lengths by the relative nuclear DNA content of the species, determined by Feulgen microdensitometry. On the assumptions that chromosomes in individual mitotic cells are uniformly condensed, and that total nuclear DNA is evenly distributed along the chromosomes of the set, the karyotypes produced should allow comparisons of shapes and sizes of individual chromosomes within and between species. Heterochromatin C-banding resulted in only the telomeres being affected, although the acid hydrolysis necessary for maceration may have suppressed other bands. Heterochromatin is apparently terminal on all chromosomes, and C-banding was therefore not useful for karyotyping. Similarly, N-banding contributed no additional information, but did confirm that visible secondary constructions were nucleolar organizing regions. Relative DNA values for the genera studied are summarized in Table I, and some representative karyotypes are illustrated in Fig. 1.

Although all karyotypes are relatively symmetrical, the 12-paired genomes are less uniform, and contain some acrocentric and subacrocentric chromosomes. This situation is consistent with an evolutionary sequence of dysploid reduction from $x = 12$ to $x = 9$ and 8 through Robertsonian translocations. Chromosome sizes in putatively primitive genera such as *Nuytsia, Alepis, Muellerina* and *Ileostylus* are obviously smaller than in more specialized, derived genera such as *Macrosolen* and *Lysiana*, clearly illustrating the evolutionary increase in DNA content in the family. Among the 9-paired stocks, the chromosomes in genera such as *Dendrophthoe* and *Benthamina* are somewhat smaller, on average, than in *Amyema* and *Diplatia*. The former genera are possibly satellites of the *Taxillus* alliance of Africa and Asia, while the latter are representative of the *Amyema* alliance of the Papuasian region.

From Table I it can be seen that the inverse correlation between basic chromosome number and chromosome size is only slight. Increase in chromosome size has occurred apparently independently in different basic karyotypic lineages. It is likely that increase in chromosome size has resulted from amplification of middle repetitive DNA, while single copy DNA has remained relatively constant (Martin 1983). Martin has further suggested that range expansion and diversification in loranths may have been facilitated by such amplification, especially through an increase in recombination potential. This cytogenetic mechanism appears to have operated continuously in the differentiation of genera, of species and of local biotypes within species, and is discussed in more detail below.

Evolution Within a Basically Conservative Genome

Within the framework of overall genomic stability within the main karyotypic lineages there is now clear evidence of variation at a finer level, arising from studies of a number of Australian, Papuasian and New Zealand species. Australia and New Guinea, each with about 70 species of Loranthaceae, represent significant centres of secondary radiation within genomically stable stocks. Adaptation and speciation in Australia have involved extensive colonization of arid and semi-arid habitats which have arisen and

TABLE I. *Relative DNA contents of Australasian genera of Loranthaceae*

Genus	x	Average relative DNA content*			
		Lowest		Highest**	
		Species	DNA Value	Species	DNA Value
Nuytsia	12			floribunda	21.90 ± 1.40
Alepis	12			flavida	29.20 ± 2.15
Peraxilla	12	tetrapetala	40.01 ± 2.59	colensoi	47.54 ± 1.89
Tupeia	12			antarctica	30.81 ± 2.59
Decaisnina	12	brittenii	60.26 ± 4.54	hollrungii	112.90 ± 5.27
Amylotheca	12			dictyophleba	103.16 ± 7.14
Macrosolen	12			cochinchinensis	131.58 ± 8.32
Lysiana	12	casuarinae	82.73 ± 2.68	exocarpi	114.65 ± 5.25
Muellerina	11	bidwillii	37.22 ± 2.85	eucalyptoides	48.48 ± 1.55
Ileostylus	11			micranthus	28.86 ± 3.63
Dendrophthoe	9	glabrescens	20.59 ± 1.07	falcata	46.50 ± 3.11
Benthamina	9			alyxifolia	75.95 ± 3.22
Sogerianthe	9			sessiliflora	51.82 ± 2.30
Amyema	9	artense	62.16 ± 4.20	pendulum	130.30 ± 5.87
Diplatia	9	grandibractea	117.55 ± 3.80	furcata	120.13 ± 6.57

*100.0 = Nuclear DNA content of *Vicia faba* (53.31 picograms).
**Data derived from a single species only are placed in this column.

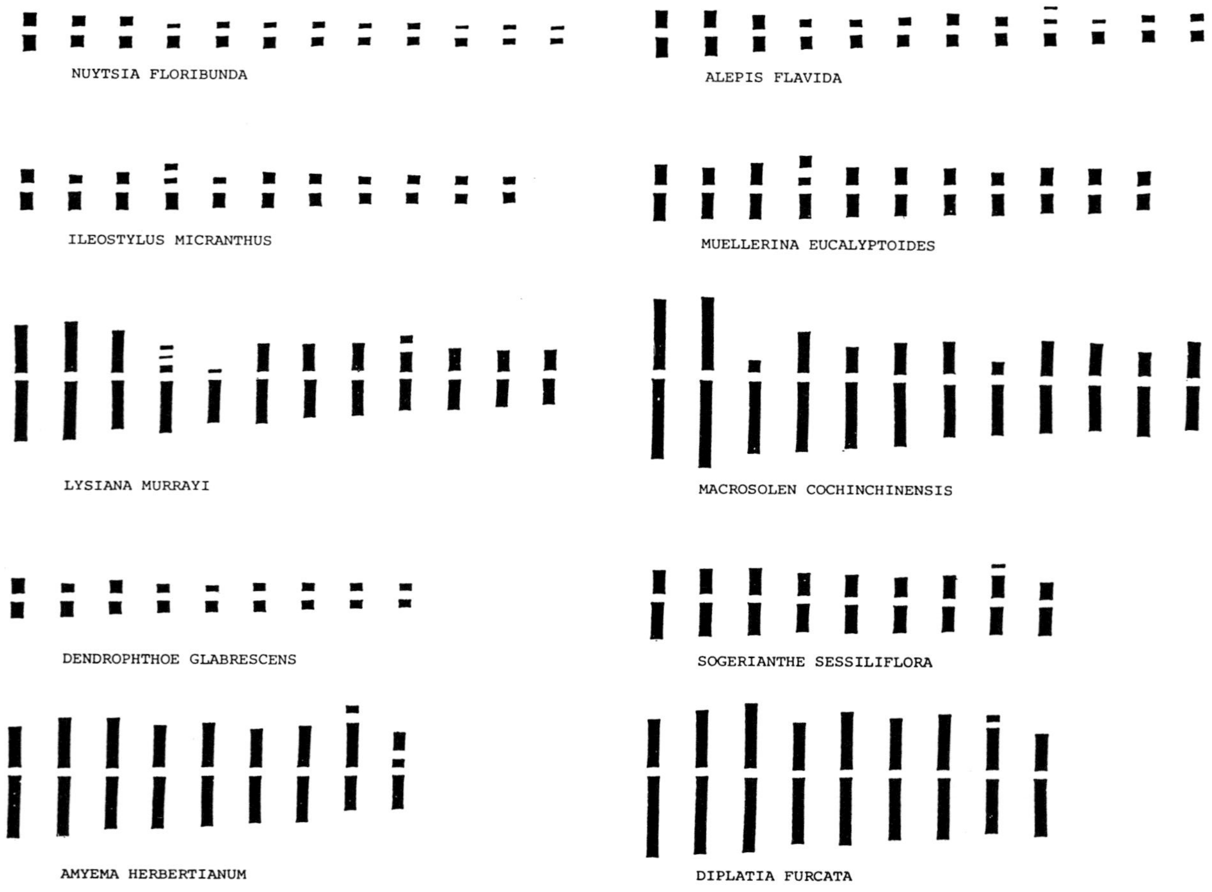

ALEPIS FLAVIDA

MUELLERINA EUCALYPTOIDES

MACROSOLEN COCHINCHINENSIS

SOGERIANTHE SESSILIFLORA

DIPLATIA FURCATA

NUYTSIA FLORIBUNDA

ILEOSTYLUS MICRANTHUS

LYSIANA MURRAYI

DENDROPHTHOE GLABRESCENS

AMYEMA HERBERTIANUM

spread since mid-Tertiary times, and correspondingly significant changes in mistletoe ecology and coadaptation. In some cases, for example, these changes have involved attainment of highly specialized relationships between parasite species and host species. There can be little doubt that an available host represents an important habitat parameter in speciation in the family. Differentiation and adaptation to preferred hosts appears to be linked with shifts in levels of self incompatibility, even though floral morphology is rather uniform (Bernhardt *et al.* 1980). In Australia the adaptation of the parasite to the host is often associated with striking cryptic mimicry, so that the mistletoe is not easily distinguished against the background of the host. This mimicry has been explained in terms of escape from predation (Barlow and Wiens 1977), coevolution with avian dispersal agents (Barlow 1983c), and selection for growth hormone compatibility (Atsatt 1983). The existence of adaptation and diversification on this scale suggests that some variation in genomic organization in the mistletoes might be expected.

An example of genomic variation within and between species is provided by the work of Martin (1980, 1983), who investigated nuclear DNA content in 27 species of *Amyema* from Australia and Papua New Guinea. Average DNA values for the species are listed in Table II, and some representative karyotypes are illustrated in Fig. 2. Even though all species have relatively similar symmetrical karyotypes, there is considerable variation between species, with the highest average DNA value being more than twice the lowest. Minor relative variations between genomes in chromosome length are apparent. The best indicators of karyotypic change, however, are variations in the number and location of the nucleolar organizing regions. These indicate that some structural rearrangements may have been incorporated into different genomes, although the changes could have resulted from differential amplification of chromosome segments (see below), or more likely from activation and inactivation of different nucleolar organizing sites.

There are no clear latitudinal or habitat-related gradients in average DNA content in *Amyema* which are comparable, for example, with those found in *Picea* by Miksche (1968, 1971). There is a general tendency for tropical species of *Amyema* to have lower DNA values than the species of temperate or arid habitats, but the overlap is considerable. Similarly, no strong correlations exist with host specificity, host mimicry, habitat seasonality or size of geographic area. In this respect the variation between species is similar in nature to the recorded variation within the species studied (Martin 1980), which is summarized in Fig. 3.

FIG. 1. *Karyotypes of selected species of Loranthaceae, represented as mean percent arm lengths multiplied by DNA content. (Modified from Martin 1983).*

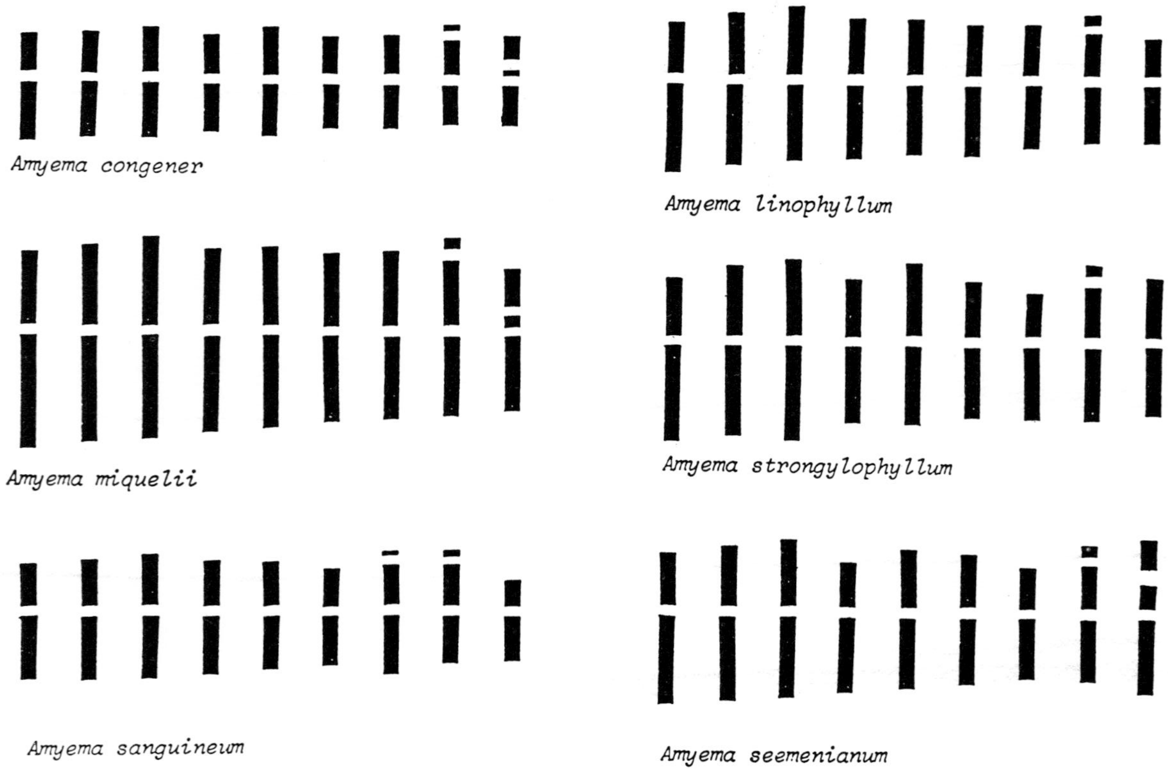

Amyema congener

Amyema linophyllum

Amyema strongylophyllum

Amyema miquelii

Amyema seemenianum

Amyema sanguineum

TABLE II. *Average relative DNA contents of Australasian species of* Amyema (x = 9)

Species	DNA content	Species	DNA content
artense	62.16 + 4.20	*melaleucae*	86.12 + 4.26
villiflorum	65.62 + 3.84	*herbertianum*	91.45 + 6.24
lucasii	68.73 + 4.33	*bifurcatum*	92.57 + 4.13
conspicuum	70.56 + 4.47	*cambagei*	93.78 + 2.64
quandang	72.30 + 4.59	*nestor*	96.16 + 5.35
maidenii	72.36 + 3.78	*linophyllum*	98.10 + 4.93
congener	72.51 + 4.85	*seemenianum*	99.95 + 4.91
queenslandicum	73.88 + 3.27	*friesianum*	103.31 + 6.26
hillianum	74.13 + 4.67	*finisterrae*	104.07 + 6.34
miraculosum	75.01 + 4.05	*strongylophyllum*	112.99 + 6.52
gibberulum	75.60 + 4.64	*preissii*	115.12 + 4.74
mackayense	77.16 + 5.95	*miquelii*	127.41 + 5.76
sanguineum	78.86 + 4.22	*pendulum*	130.30 + 5.87
gaudichaudii	79.00 + 4.90		

Most species showed significant variation between populations, even after a few extreme values were eliminated because of suspected or observed polyploidy or endoreduplication. In general, the widest ranges of variation were found in the most widespread species. In *A. miquelii,* for example, there was a relatively continuous variation in which the highest value recorded was 46% greater than the lowest. More extensive sampling of some species may therefore have resulted in a wider range of DNA values. However in some widespread species, such as *A. maidenii* and *A. quandang,* relatively uniform values were obtained from widely separated localities. It is probable that some species exhibit greater ranges of variation in nuclear DNA content than others.

The widest range of variation in DNA values in diploid nuclei was recorded in *Amyema miquelii,* and their geographic distribution is shown in Fig. 4 as an example. Although no clear

FIG. 2 *Karyotypes of selected species of* Amyema, *showing variation between species in chromosome morphology and size, and in positions of secondary constrictions.*

DISTRIBUTION OF DNA VALUES IN SPECIES OF

Amyema.

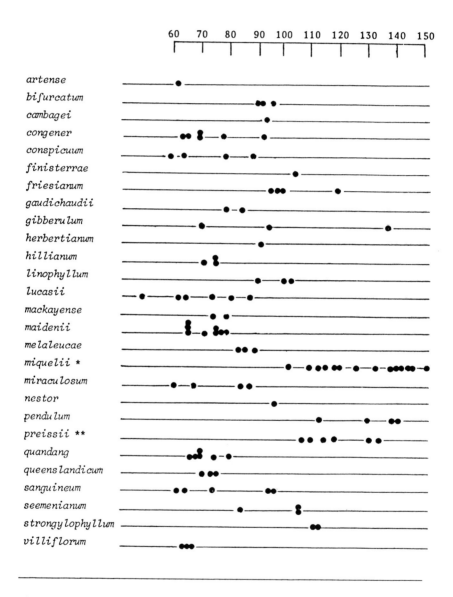

---

* One suspected tetraploid individual (239.34 ± 20.69) excluded

** One value from established tetraploid population (195.44 ± 4.49) excluded

geographic pattern exists, it is likely that local biotypes within this species differ significantly in nuclear DNA content, and that adaptation and differentiation in local populations has involved adaptive changes in DNA content. The karyotype of *A. miquelii* is constant throughout this range of variation (Fig. 5), showing how DNA gain or loss is apparently distributed uniformly over all chromosomes of the genome. Studies by Martin (1980) on smaller numbers of species in *Decaisnina, Dendrophthoe* and *Lysiana* have shown patterns of chromosome variation similar to that in *Amyema*.

Chromosomal Evoloution and Adaptation in Loranthaceae - A Summary

DNA content can vary in three general ways (Rees and Jones 1972): (i) supernumerary chromosomes, (ii) polyploidy and aneuploidy, and (iii) amplification or deletion of segments within chromosomes. In the Loranthaceae it is clear that not all of these strategies have been utilized, and that they have been utilized sequentially rather than simultaneously. Supernumerary chromosomes, for example, are unknown in Loranthaceae, although they occur in Viscaceae (see below), and in about 15% of angiosperm species generally. Polyploidy occurs at a much lower frequency than in angiosperms generally, and being mostly infraspecific, the existing polyploidy must be of very recent establishment. It is presumably involved occasionally in the genetic conservation of locally adaptive biotypes.

Aneuploidy (dysploidy), on the other hand, is the original mechanism of genetic change in the family, and appears to have been restricted to the early differentiation of the family in Gondwanaland. It appears to have generated a limited number of karyotypes which have subsequently remained unusually constant. Dysploidy presumably accelerated the isolation and conservation of diverging adaptive lines through the restructuring of linkage groups. This sequential pattern of dysploidy succeeded by polyploidy appears to have occurred in a number of other old southern families, including Rutaceae (Smith-White 1954), Casuarinaceae (Barlow 1959) and Epacridaceae (Smith-White 1959), and may represent a common response to the environmental sequences which prevailed in the southern lands.

Amplification or deletion of chromosome segments has been the major genetic and cytoevolutionary strategy in the family through a long time period, embracing the differentiation of many genera, speciation within the genera, and local biotype differentiation within the species. The general trend has been towards amplification, and has led to chromosome sizes amongst the largest in the plant kingdom. Since amplification has apparently not involved

FIG. 3. *Variation in relative DNA content within and between species of* Amyema. *Each dot represents the mean value of a population sample. (After Martin 1983)*

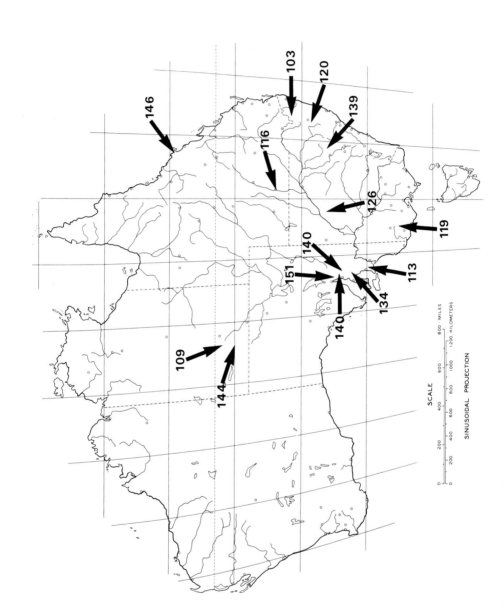

SCALE

SINUSOIDAL PROJECTION

large scale changes in satellite DNA of heterochromatin or the repetitive DNA of nucleolar organizing regions, it has probably involved the middle repetitive variety. This situation has been documented in salamanders (Macgregor 1978), where increase in chromosome size similarly occurs without change in karyotype. Amplification of middle repetitive DNA may occur as spasmodic change, under genetic control, involving the entire genome (Narayan and Rees 1976).

Several physiological and genetic effects of amplification of DNA can be predicted, including changes in cell size, duration of mitotic cycle and recombination rate. They can occur without disruption of existing adaptive sequences of functional genes. Changes such as these apparently have been of significance in the Loranthaceae as they have specialized in different habitats and on different hosts. In fact, general amplification of DNA in stabilized karotypes appears to have been the major basis of adaptive change in the family over perhaps 50 million years of evolutionary radiation.

CHROMOSOMAL EVOLUTION AND DIFFERENTIATION IN VISCACEAE

Viscaceae as a Laurasian Family

Chromosome numbers are known in 6 of the 7 genera of Viscaceae, and in all except *Notothixos* the basic number is clearly $x = 14$. In *Notothixos* the two records of $n = 13$ and $n = 12$ indicate dysploid reduction from the basic number of the family. Dysploid variation also occurs in *Dendrophthora*, and is very extensive in *Viscum*. Chromsome size is generally large, and the largest genomes in *Viscum* probably exceed those of the Loranthaceae in DNA content. In comparison, the chromosomes of *Notothixos* are relatively small.

Levels of advancement or specialization in the family are difficult to identify, and there are no clearly relictual genera. The history of the family has therefore been deduced from conventional analysis of geographic distribution, areas of species richness and host specializations in the seven genera (Wiens and Barlow 1971; Barlow 1983b). The family appears to be Laurasian, with a centre of origin in tropical or subtropical Asia. Progenital stocks were probably widespread in Laurasia by the Paleocene, and have possibly reached North America via the Beringian land connection in two independent migrations at different times in the Tertiary period.

The Viscaceae therefore appear to have evolved completely independently of the Loranthaceae. Many of the similarities between the two groups of mistletoes are probably the result of parallel or convergent evolution. In particular, it is likely

---

FIG. 4. *Geographic distribution of relative DNA values in* Amyema miquelii.

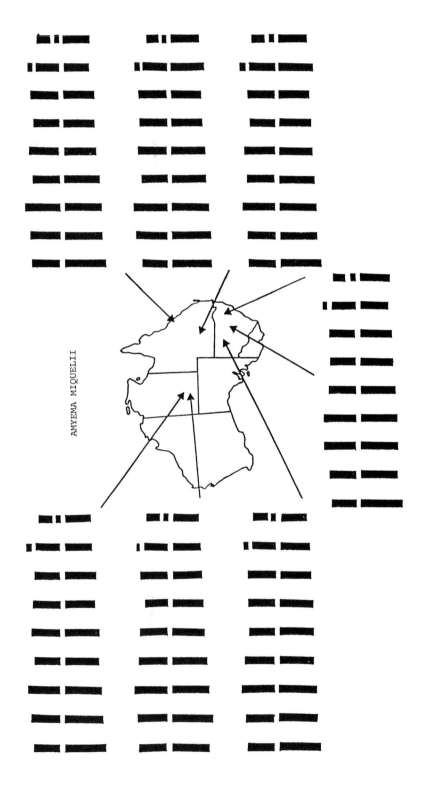

AMYEMA MIQUELII

that massive increases in DNA content have occurred in Viscaceae much as they occurred in Loranthaceae. It is tempting to postulate that some aspect of the hemiparasitic habit is directly or indirectly linked with the role of amplification of DNA in evolution and adaptation in the two families.

Speciation in Viscum - Dioecy and Interchange Heterozygosity

While the genera *Phoradendron* and *Arceuthobium* are well known in terms of chromosome numbers, the genus in which the most intensive biosystematic analysis of cytogenetic data has been made is *Viscum*. This analysis relates primarily to the origin and maintenance of dioecy in the genus. *Viscum* comprises ca. 100 species, with its main centres of species richness in Madagascar, Africa and southern Asia. About 60% of the species are dioecious; the inflorescences have the same basic structure as those of the monecious species but all flowers are of one sex instead of having the central flower of the cyme of opposite sex to the lateral ones. A very generalized indication of the relative frequencies of monoecious and dioecious species through the range of the genus is given in Fig. 6. In Africa monoecious and dioecious species have the same general distribution, with dioecious species being slightly more numerous. In Madagascar there is a strong predominance of dioecious species. Outside Africa-Madagascar there is little overlap in the distributions of monoecious and dioecious species, and the number of dioecious species is much lower; the more numerous monoecious species have an obvious centre of species richness in India/China, and extend with diminishing species density through Malesia to eastern Australia. There can be no doubt that dioecy in *Viscum* is derived from monoecy, which is fixed or predominant in all other genera of the family.

The cytogenetic basis of dioecy in *Viscum* has been the subject of detailed karyological analysis (Wiens and Barlow 1973, 1975, 1979; Barlow and Wiens 1975, 1976; Mechelke 1976; Barlow *et al*. 1978; Barlow 1981). Permanent sex-associated interchange heterozygosity is present in the majority of the dioecious species, in which male plants consistently show a multivalent at meiosis ranging from a ring-of-four ($\Theta 4$) to $\Theta 6$, $\Theta 8$, $\Theta 10$ and even $\Theta 12$ (Fig. 7). In addition there are high levels of floating interchange heterozygosity, that is, of interchanges which are not sex-associated, and not maintained by balanced lethals, so that both homozygotes and heterozygotes may occur in either sex. In a number of cases the frequencies of such heterozygosity are so high that their maintenance by positive selection for the interchange heterozygotes seems likely.

FIG. 5. *Geographic distribution of karyotypes in* Amyema miquelii. *The karyotypes are based on mean percent arm lengths only, and are not comparable as to DNA content.*

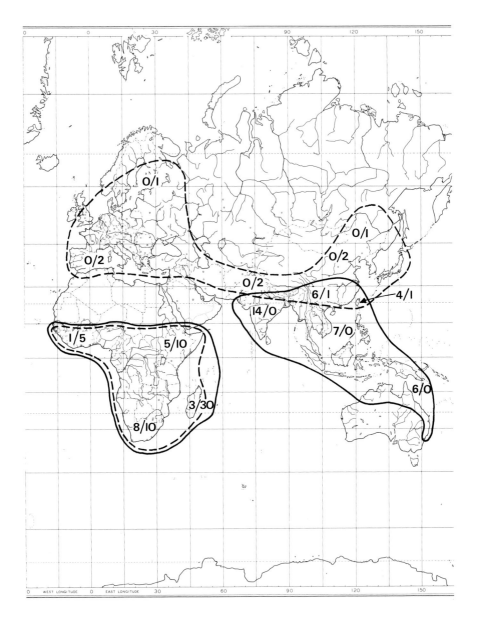

FIG. 6. *World distribution of* Viscum, *showing areas of monoecious species (solid outline) and dioecious species (broken outline). Numerals indicate approximate numbers of monoecious species (first number) and dioecious species (second number) in various regions of the total area. (After Barlow 1983b).*

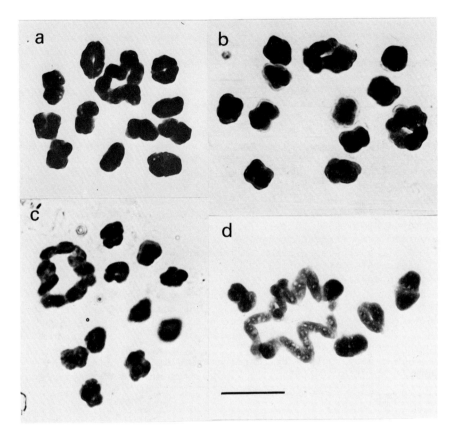

FIG. 7. *Meiotic configurations in male plants of dioecious species of* Viscum. *Bar represents 25 µm. a*, V. combretiocola, *11II + ⊖6. b*, V. hildebrandtii, *10II + 2 ⊖4. c*, V. hildebrandtii, *10II + ⊖8. d*, V. album, *4II + ⊖12. (a from Wiens & Barlow 1979; b and c from Barlow and Wiens 1975; d from Barlow 1981).*

Surprisingly, floating interchanges are almost absent in monoecious species of *Viscum,* even though the probability of inbreeding should increase selection pressure for the maintenance of heterozygosity.

Many related dioecious *Viscum* species seem to have the same sex-associated multivalents in the male plants (Wiens and Barlow 1979), and it is likely that some of the interchanges were established before or during the evolution of the dioecious species. They may therefore have had an essential role in the evolution of dioecy in the genus. Since it is likely that genes for maleness and femaleness in dioecious angiosperms are non-allelic (Ross and Weir 1976; Charlesworth and Charlesworth 1978),

interchanges would have been important in bringing factors for
dioecy located on different chromosomes into genetic linkage (Fig.
8). On stabilization of dioecy, the heterogamic (male) sex would
thus be permanently heterozygous for the interchange complexes
involved.

The smallest and simplest sex-associated interchange complexes
occur in the dioecious African species with x = 14. In these
species, male plants mostly have a sex-associated Θ4 or Θ6, while
Θ8 is rare. In the Asian-European species with x = 10, a
sex-associated Θ8 is most common, but Θ10 occurs frequently and
Θ12 is rare. The process of enlargement of the interchange
complexes through incorporation of additional interchanges has
thus occurred along with the dysploid reduction in chromosome
number from x = 14 to x = 10. This strongly suggests that
dioecy in *Viscum* has originated in Africa, and that the
dioecious species of Europe/ Asia are derived ones which have
colonized from African progenitors. Similarly it appears that
almost all of the speciation in Madagascar has been from dioecious
colonizing progenitors, and that the few monoecious species there
represent separate introductions.

Because dioecy now predominates in *Viscum,* and because dioecy
is derived from monoecy, it appears that more rapid and extensive
speciation has occurred in dioecious stocks than in monoecious
ones. This could be a consequence of dioecy *per se,* or of the
interchange heterozygosity which characterizes dioecious species.
Although dioecy confers obligate outbreeding, the population
dynamics and floral biology in *Viscum* suggest that homozygosity

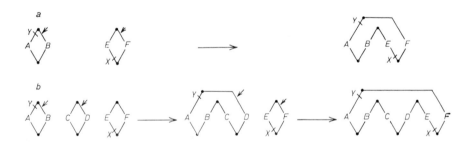

FIG. 8. *Models for the origin of dioecy in* Viscum
*through interchanges. Hypothetical sites of nonallelic factors
for maleness (Y) and femaleness (X) are indicated. Hypothetical
breakage points for interchanges are indicated by arrows. a,
formation of a sex-associated Θ4 by an interchange involving
chromosomes carrying the sex factors. b, formation of a sex-
associated Θ6 by interchanges linking X and Y indirectly through
two different interchanges. (After Wiens and Barlow 1979; Barlow
1981).*

may remain relatively high in natural populations (Wiens and Barlow 1979; Barlow 1981). Dioecy *per se* may therefore have little effect on genetic structure of populations or on speciation rate. Interchange heterozygosity, on the other hand, provides a mechanism for the accumulation in large linkage groups of adaptive gene complexes, and is associated with biotype differentiation and speciation, especially in Onagraceae. In *Viscum,* therefore, dioecy may have promoted a secondary phase of speciation indirectly through its association with interchange heterozygosity.

In addition to the sex-associated fixed interchange complexes, the floating interchange complexes seem to be characteristic of the dioecious state, occurring in almost all dioecious species studied. The floating interchanges appear to be maintained by selection. They usually occur as one or two Θ4s and/or Θ6s, but rarely may be as large as Θ8, Θ10 or Θ12 (Wiens and Barlow 1979). In *V. album* there are also floating Θ4s and Θ6s which occur only in female plants, apparently because they are the result of interchanges within the large sex-associated complexes (Barlow 1981). A possible explanation for their general absence in monoecious species and abundance in dioecious species is that they may have been derived from fixed sex-associated rings in dioecious species. Adaptive gene complexes may have accumulated while the complexes were fixed, and when later structural rearrangements led to their release as floating complexes (Fig. 9), they may have been conserved by selection. In comparison, interchanges in monoecious species cannot be fixed, and are immediately exposed to selection as raw floating interchanges. Prior sex association may therefore facilitate pre-adaptation of floating interchange complexes, which might eventually contribute to differentiation and speciation in the genus through generation and conservation of local adaptive biotypes.

Supernumerary chromosomes

Supernumerary chromosomes are present in several species of *Phoradendron* (Wiens 1964) and *Viscum* (Wiens and Barlow 1979). In

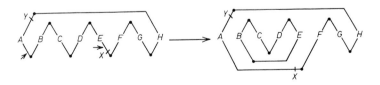

FIG. 9. *Model showing rearrangement in a sex-associated Θ8 to produce a floating Θ4 and a sex-associated Θ4, bringing the sex factors X and Y into closer linkage. (After Barlow 1981).*

V. album in Japan up to 8 supernumeraries of at least three different types may occur in a single plant; 21% of plants had supernumeraries and the average frequency was 0.41 per genome (Barlow 1981). The apparent association of supernumeraries with the dioecious state in *Viscum* suggests that they have originated as centric fragments through the breakage and reunion events involved in interchanges. They probably have the function of further boosting genetic variability in the genus.

Chromosome Evolution and Adaptation in Viscaceae - Summary and Comparison

In general, patterns of cytogenetic change in Viscaceae are similar to those in Loranthaceae. These two groups are natural, irrespective of the rank given them, and the similarities therefore represent parallelisms. Common strategies include dysploid reduction and amplification of DNA, with polyploidy having a minor role.

Dysploid change in Viscaceae has been more restricted than in Loranthaceae, and occurs primarily as an infrageneric phenomenon. It has probably persisted to a later stage of differentiation than in Loranthaceae. However a strict comparison at the generic level may be misleading, because the Viscaceae comprise many fewer genera which are on average much larger, and the difference may reflect different generic concepts in the two families. Increase in chromosome size has not been studied experimentally, but probably has resulted from general amplification of DNA over entire genomes. These two types of change may have had the same evolutionary roles as in the loranths. Polyploidy is somewhat more frequent in Viscaceae than in Loranthaceae, and appears to occur sometimes at the specific level. However, its incidence is still comparatively low, and no evolutionary radiation appears to have occurred at a polyploid level.

The additional strategies which have been discovered in *Viscum* are all linked to the origin of dioecy. It appears that dioecy *per se* may not have had a dramatic effect on genetic variability in the genus, but its origin has been associated with cytogenetic phenomena which have been important. Dioecy has been the means of achieving permanent interchange heterozygosity and also of adaptive floating interchange heterozygosity. Linked with this is the accumulation of supernumerary chromosomes. These features have been associated with a secondary phase of speciation in the genus.

REFERENCES

Atsatt, P.R. 1983. Mistletoe leaf shape: a host morphogen hypothesis. In *The Biology of Mistletoes,* D.M. Calder and P. Bernhardt (eds.). Academic Press, New York (in press).
Barlow, B.A. 1959. Chromosome numbers in the Casuarinaceae. *Aust. J. Bot.* 7: 230-237.

Barlow, B.A. 1963. Studies in Australian Loranthaceae. IV. Chromosome numbers and their relationships. *Proc. Linn. Soc. N.S.W.* 88: 151-160.
Barlow, B.A. 1964. Classification of the Loranthaceae and Viscaceae. *Proc. Linn. Soc. N.S.W.* 89: 268-272.
Barlow, B.A. 1974. A revision of the Loranthaceae of New Guinea and the south-western Pacific. *Aust. J. Bot.* 22: 531-621.
Barlow, B.A. 1981. *Viscum album* in Japan: chromosomal translocations, maintenance of heterozygosity and the evolution of dioecy. *Bot. Mag.* 94: 21-34.
Barlow, B.A. 1983a. A revision of the genus *Notothixos* (Viscaceae). *Brunonia* 6: (in press).
Barlow, B.A. 1983b. Biogeography of Loranthaceae and Viscaceae. In *The Biology of Mistletoes,* D.M. Calder and P. Bernhardt (eds.). Academic Press, New York (in press).
Barlow, B.A. 1983c. Mistletoes in South Australia's forests and woodlands. In *Ecology of South Australia's Forests and Woodlands,* H.R. Wallace (ed.). Govt. Printer, Adelaide (in press).
Barlow, B.A. and Wiens, D. 1971. The cytogeography of the loranthaceous mistletoes. *Taxon* 20: 291-312.
Barlow, B.A. and Wiens, D. 1973. The classification of the generic segregates of *Phrygilanthus (= Notanthera)* of the Loranthaceae. *Brittonia* 25: 26-39.
Barlow, B.A. and Wiens, D. 1975. Permanent translocation heterozygosity in *Viscum hildebrandtii* Engl. and *V. engleri* Tiegh. (Viscaceae). *Chromosoma* 53: 265-272.
Barlow, B.A. and Wiens, D. 1976. Translocation heterozygosity and sex ratio in *Viscum fischeri. Heredity* 37: 27-40.
Barlow, B.A. and Wiens, D. 1977. Host-parasite resemblance in Australian mistletoes: the case for cryptic mimicry. *Evolution* 31: 69-84.
Barlow, B.A., Wiens, D., Wiens, C., Busby, W.H. and Brighton, C. 1978. Permanent translocation heterozygosity in *Viscum album* and *V. cruciatum:* sex association, balanced lethals, sex ratios. *Heredity* 40: 33-38.
Bernhardt, P. 1982. Interspecific incompatibility amongst Victorian species of *Amyema* (Loranthaceae). *Aust. J. Bot.* 30: 175-184.
Bernhardt, P., Knox, R.B. and Calder, D.M. 1980. Floral biology and self-incompatibility in some Australian mistletoes of the genus *Amyema* (Loranthaceae). *Aust. J. Bot.* 28: 437-451.
Charlesworth, B. and Charlesworth, D. 1978. A model for the evolution of dioecy and gynodioecy. *Am. Nat.* 112: 975-997.
Danser, B.H. 1931. *Bull. Jard. Bot. Buitenz.* 11: 232-519.
Dixit, S.N. 1962. *Bull. Bot. Sur. India* 4: 49-55.
Feuer, S. and Kuijt, J. 1978. Can. J. Bot. 56: 2853-2864.
Hamilton, S.G. and Barlow, B.A. 1963. Studies in Australian Loranthaceae. II. Attachment structures and their interrelationships. *Proc. Linn. Soc. N.S.W.* 88: 74-90.
Johri, B.M. and Bhatnagar, S.P. 1961. Embryology and taxonomy of

the Santalales. I. *Proc. Natl. Inst. Sci. India* 26, Part B (suppl.): 199–220.

Kuijt, J. 1959. *Acta Bot. Neerl.* 8: 506–546.

Kuijt, J. 1964. *Can. J. Bot.* 42: 1243–1278.

Kuijt, J. 1968. *Brittonia* 20: 136–147.

Kuijt, J. 1970. Mem. Torrey Bot. Club 22: 1–38.

Kuijt, J. 1981. *Blumea* 27: 1–73.

Macgregor, H.C. 1978. Some trends in the evolution of very large chromosomes. *Philos. Trans. R. Soc. London Ser.* B 283: 309–318.

Martin, N.J. 1980. Karyotype and nuclear DNA variation in the Australasian Loranthaceae. Ph.D. thesis, Flinders Univ., Adelaide, South Australia.

Martin, N.J. 1983. Nuclear DNA variation in the Australasian Loranthaceae. In *The Biology of Mistletoes,* D.M. Calder and P. Bernhardt (eds.). Academic Press, New York (in press).

Mechelke, F. 1976. Naturwissenchaften 8: 390.

Miers, J. 1851. *Ann. Mag. Nat. Hist.* 8: 161–184.

Miksche, J.P. 1968. Can. J. Genet. Cytol. 10: 590–600.

Miksche, J.P. 1971. *Chromosoma* 32: 343–352.

Narayan, R.K.J. and Rees, H. 1976. Chromosoma 54: 141–154.

Rees, H. and Jones, R.N. 1972. *Int. Rev. Cytol.* 32: 53–92.

Ross, M.D. and Weir, B.S. 1976. *Evolution* 30: 425–441.

Samuelsson, G. 1973. Mistletoe toxins. *Syst. Zool.* 22: 566–569.

Smith-White, S. 1954. *Aust. J. Bot.* 2: 281–303.

Smith-White, S. 1959. Cytological evolution in the Australian flora. *Cold Spring Harbor Symp. Quant. Biol.* 24: 273–289.

Thoday, D. 1961. Modes of union and interaction between parasite and host in the Loranthaceae. VI. A general survey of the Loranthoideae. *Proc. R. Soc. London Ser.* B 155: 1–25.

Weber, H.C. 1980. Untersuchungen an australischen und neuseeländischen Loranthaceae/Viscaceae. 1. Zur Morphologie und Anatomie der unterirdischen Organe von *Nuytsia floribunda* (Labill.) R. Br. *Beitr. Biol. Pflanz.* 55: 77–99.

Wiens, D. 1964. Chromosome numbers in North American Loranthaceae: *(Arceuthobium, Phoradendron, Psittacanthus, Struthanthus).* Am. J. Bot. 51: 1–6.

Wiens, D. 1975. Chromosome numbers in African and Madagascan Loranthaceae and Viscaceae. *Bot. J. Linn. Soc.* 71: 295–310.

Wiens, D. and Barlow, B.A. 1971. The cytogeography and relationships of the viscaceous and eremolepidaceous mistletoes. *Taxon* 20: 313–332.

Wiens, D. and Barlow, B.A. 1973. Unusual translocation heterozygosity in an East African mistletoe. *Nature (New Biol.)* 243: 93–94.

Wiens, D. and Barlow, B.A. 1975. *Science* 187: 1208–1209.

Wiens, D. and Barlow, B.A. 1979. Heredity 42: 201–222.

Differentiation and Evolution of the Genus *Campanula* in the Mediterranean Region

J. Contandriopoulos
Laboratoire de Cytotaxinomie végétale
Université de Provence
Centre de Saint-Charles
Marseille, France

INTRODUCTION

Cytotaxonomic studies thus far completed on the genus
Campanula L. have confirmed the amazing complexity of this
genus which consists of more than 500 species. Moreover, at the
subgeneric level, the simple, but at times equivocal,
morphological characters are far from allowing any suppositions to
be made as to specific divisions on a morphological, or likewise,
a karyological basis.

1. BIOGEOGRAPHY

If two-thirds of all campanula grow in the Mediterranean region,
the Pontic region and west and central Asia, the remaining one-
third are spread over an immense territory. It extends as far
north as the Eurasiatic-Arctic regions and Japan, as well as in
North America, where 22 indigenous taxa have been recorded. To
the south we find some species in eastern Africa and in the
southern Sahara mountains as far as Morocco. The perennial
species occupy either regions of very extensive geographical
distribution, or are confined in restricted ranges and constitute
an important endemic element. The annuals are found almost
exclusively in the mesogenous region as far as the Himalayas.
They are present at times in the continuous circum-Mediterranean
ranges (*C. erinus* L.), are scattered on the perimeter of the
Mediterranean (*C. fastigiata* Duf.), or in the eastern
Mediterranean (many endemics). Nevertheless, here and there, in
areas occupied by perennials there exist some annuals. For
example, in Kenya, the perennial *C. edulis* Fors. ($2n = 56$)
has given rise to the annual *C. keniensis* Thul. ($2n = 54$)

(Thulin 1975). In the Himalayas, the perennial *C. cashmeriana*
Royle ($2n$ = 28) is related to the annual *C. colorata* Wall.
($2n$ = 28). Lastly, it is interesting to point out the
presence of annuals in California. Perhaps they originated, as in
the Mediterranean, at the time of an analogous warming up of the
climate, but from North American perennial species with n = 17.

2. CLASSIFICATION OF THE CAMPANULAS

Depending on the authors, the genus *Campanula* has been
subjected to different treatments according to which morphological
criteria were considered as most important. While de Candolle
(1830) distinguished two major sections on the character "the
presence or absence of the appendices between the lobes of the
calyx", Boissier (1875) placed more importance on the mode of
dehiscence of the capsule. It is from the results of Boissier
that subsequent classifications were established and were refined
further by Fedorov (1957) and Damboldt and Phitos (1978).

For the campanulas of the USSR, Fedorov recognized the two
major sections of Boissier, *Medium* (DC.) Boiss. and *Rapunculus*
(Fourr.) Boiss., but he subdivided them into many subsections and
series, regrouping the related species.

For those of Turkey, Damboldt and Phitos elaborated a new
classification which better interprets Boissier's sections by
distinguishing six subgenera:
- the subgenus *Rapunculus* (Fourr.) Charadze and the subgenus
Brachycodonia (Fed.) Damb. = the monotypic genus *Brachycodon*
(C. fastigiata Duf.) which formed the section *Rapunculus* (Fourr.)
Boiss.
- the subgenera *Campanula* = sect. *Medium* (DC.) Boiss
(pro parte max.); Rouce=la (Feer) Damb. = series *Annuae* Boiss.
(pro parte incl. type); *Megalocalyx* Damb. = series *Annuae*
(pro parte excl. type) and the monotypic subgenus *Sicyodon*
(C. macrostyla Boiss. et Heldr.). These four subgenera
constituted the section *Medium*.

An interesting endeavor was that made by Gadella (1964) who
used the criterion "chromosome number" to establish relationships
likely to exist between species. This author is of the opinion
that many subsections recognized by Fedorov are not natural
because he has been able to cross species belonging to different
subsections: *C. trachelium* L. (subsect. *Eucodon*) X *C. glomerata*
(subsect. *Involucratae)* or with *C. alliariaefolium* Willd.
(subsect. *Latilimbus*). In our opinion, this experiment reveals
preferably the relationship existing between the different
subspecies.

Gadella proposed, thus, a new classification which takes into
account the chromosome numbers, chromosome size, morphological
characters (presence or absence of calyx appendages, styles –
glabrous or not –, number of cells of the capsule, its position –
upright or inclined –, apical or basal dehiscence) and the life
cycle of the plant.

He distinguishes seven groups:

I: $x = 8$, $2n = 16$, 32 V: $x = 12(?)$, $2n = 24$

II: $x = 10$, $2n = 18$, 20,   VI: $x = 14(?)$, $2n = 28$
    40, 80

III: $x = 13$, $2n = 26$ VII: $x = 15$, 17, $2n = 30$, 60,
                                        90, 34, 68, 102, 32

IV: $x = 18(?)$, $2n = 36$

Now, since the works of Gadella, cytotaxonomic knowledge of the genus *Campanula* has increased considerably (cf. References). The unpublished basic numbers ($x = 7$ and $x' = 11$) were discovered; new numbers (often accompanied by polyploidy, aneuploidy or dysploidy) were found. Taking into consideration these facts, Gadella's classification, based on a small number of species, belonging mainly to the subgenus *Rapunculus*, cannot be used to generalize, since chromosome numbers which do not necessarily have the same significance, nor the same origin, have been found in other subgenera. The number $2n = 26$, depending on the species, derives from $x = 14$ or from $x = 12$. The number $2n = 80$ which, for Gadella, belongs to group II with $x = 10$ (in this case we are speaking of an octoploid of *C. patula* (subgenus *Rapunculus*)), characterises also an endemic of Hoggar (Sahara) of subgenus *Campanula*, *C. bordesiana* Maire, amphiploid of *C. edulis* Fors. ($2n = 56$) X*C. filicaulis* Dur. ($2n = 24$).

3. PHYLOGENY OF THE CAMPANULAS

Although Gadella did not establish a link between the seven groups, his efforts seemed to us to be well-founded. We have thus classified the species by subgenera and sections (morphology) and in relation to cytotaxonomy, life cycle, and geographic distribution. We have selected 317 taxa whose chromosome numbers are known (around three-fifths of the campanulas). We have prepared Table I which illustrates the extreme complexity of the cytotaxonomy of the genus *Campanula* taking into account the chromosomal races whose geographic distribution is, at times, very useful for disclosing the centers of origin and differentiation.

3.1 CHROMOSOME NUMBER DIVERSITY

There is in the Campanulaceae, at the cytotaxonomic level, a large range of chromosome numbers from $x = 6$ (and more likely 5) to $x' = 17$. Now it would seem, at first glance, that originally, the basic number was not fixed and that a certain period of instability could have intervened. It is undoubtedly for this reason that, in polybasic families, complex genera possessing a more or less complete series of basic numbers exist in the family (*Campanula, Phyteuma, Asyneuma*), alongside other usually smaller, monobasic genera (*Jasione,* $x = 6$).

To describe this phenomenon Stebbins (1966) speaks of aneuploid

TABLE 1.   *Nombres chromosomiques chez* Campanula. *% annuelles/viv-*
aces - bisannuelles = 10,09; seulement 6 polyploides chez les
annuelles.($\underline{x}$ = 7).   Chromosome numbers of Campanula. *% annuals/*
perennials - biennials = 10.09; only 6 polyploid annuals.($\underline{x}$ = 7).

$2\underline{n}$	Espèces Species	Annuelles Annuals	$\underline{x}$	$\underline{x}'$	%
34	175	3	–	17	
68	28	–	–	17	69.08
102	$\frac{16}{219}$	–	–	17	
16	15	4	8	–	
32	13	–	8	–	9.14
48	$\frac{1}{29}$	–	8	–	
30	6	1	–	15	
60	2	–	–	15	3.47
90	$\frac{3}{11}$	–	–	15	
14	1	1	7	–	
28	11	3	7	–	
56	3	1	7	–	5.36
54	1	1	7	–	
58	$\frac{1}{17}$	–	7	–	
26	$\frac{4}{4}$	–	–	13	1.26
24	5	1	6	–	
48	2	–	6	–	2.52
72	$\frac{1}{8}$	–	6	–	
22	$\frac{1}{1}$	–	–	11	+
20	21	15	5	–	
40	1	–	5	–	7.57
80	$\frac{2}{24}$	–	5	–	
18	2	2	9	–	
36	$\frac{2}{4}$	–	9	–	1.25
80'	$\frac{1}{1}$	–	–	40 (28 + 12)	+

changes of the primary number and Favarger (*in litteris*) of
"oscillating dysploidy" or "fluctuating dysploidy". Favarger and
Huynh (1980) explain the formation and evolution of polybasic
genera in the following way: "The primary expansion which gave
rise to the genera is rapidly followed by a secondary expansion
which is responsible for the principal sections and series.
Finally, a tertiary variation of the basic number can occur in
some species".

In the case of the campanulas, the complete series of
chromosome numbers known in this genus is found in the
Mediterranean region. Among some of these can also be found
examples of oscillating chromosome numbers. Fernandes (1962) has
counted two chromosome numbers in an annual of subgenus
Rapunculus; C. lusitanica L. endemic to Portugal: $n = 9$
for the subspecies *lusitanica* and $n = 10$ for the subspecies
transtagana R. Fern. Diploids ($n = 12, 13$) and tetraploids
($n = 24, 25, 26$) exist in two perennial species of the
subspecies *Campanula (C. filicaulis* Dur. endemic to North
Africa and *C. mollis* L., endemic to the Ibero-Riffain region);
a hexaploid also exists in *C. filicaulis*. It was the
frequency of one or the other of these numbers which led us to
choose $x' = 12$ as the basic number, the chromosomal races
are not yet well fixed and are not easily distinguishable at a
morphological level (Contandriopoulos *et al.* 1983). In the
subgenus *Megalocalyx*, we have established $n = 10, 11$ for the
annual *C. reuterana* Boiss. & Bal., endemic to Iran and Turkey
(but $n = 10$ is most frequent).

In these examples, we discover the dysploidy phenomenon in its
primitive phase which provides a better understanding of what may
have happened during the formation of a genus.

The basic numbers of *Campanula* are, according to us,
$x = 5, 6, 7, 8$ and 9. This last exceptional number could
have been formed, perhaps directly, during the course of the
initial period of chromosomal instability, and would have evolved
independently of 8. It is only known in one annual: *C.
fastigiata* Duf. (Geslot *in litteris*). As Fedorov (1957) has
suggested, the morphological, karyological and ecological
specialization of this annual will support it belonging to the
monotypic genus *Brachycodon*.

The major problem in the genus *Campanula* is the origin of
$x' = 17$, the secondary basic number which is found in 69% of
the taxa studied and which also gave rise to the polyploid races
(cf. Table I). $x' = 17$ is represented in the subgenus
Campanula by 83.5% of the taxa (perennials, biennials, di-
tetra- and hexaploids) and in the subgenus *Rapunculus* by
21.3% (perennials and three diploid annuals). In the subgenera
formed solely from annuals the chromosome numbers are lower.
$x' = 17$, very common in the Campanulaceae, appears in many
monobasic genera (*Adenophora, Symphyandra,* etc.). The
origin of this number is very old, given the diversity of the
genera which it characterizes and, the numerical importance of the

species which possess it. In the campanulas, it is found mainly
in the relict species of Taurus, in those sections localized only
in western Asia. It is associted with the endemics of the eastern
Mediterranean Basin, of the Balkans, the southern Alps, the
Pyrenees, etc. But this number is also present in species of wide
geographical distribution, such as *C. rotundifolia* L. which
has differentiated, in different parts of its range, small diploid
or polyploid races in the Alps (Podlech 1965), the Pyrenees
(Geslot 1982), in Central (Bielawska 1968) or Arctic Europe, as
well as in North America (Schetler 1982).

3.2   ORIGIN OF THE SECONDARY BASIC NUMBER x' = 17

To explain the origin of a secondary basic number, authors have
resorted to amphiploidy occurring between two primary numbers as,
for example, Mangenot (1977) for *Phyllanthus* or again
Celebioglu and Favarger (1982) for *Minuartia*. Let us see
what this would mean in the case of *Campanula*.

3.2.1   AMPHIPLOIDY FROM x = 10 and x = 7

The present taxa of x = 10 are not very numerous and belong
in part to complexes of Mediterranean distribution which spread,
in the case of some perennials, to central Europe. Including the
minor races, we have counted 15 annuals out of 24 taxa. In the
subgenus *Megalocalyx*, composed of annuals, the species with
n = 10 together with n = 8 and n = 12 form a narrowly
related group of species. In the subgenus *Rapunculus*, the
perennial species with n = 10 are part of complexes which
are widely distributed throughout Southern and Central Europe
(*C. rapunculus, C. patula* which had differentiated polyploid
races), or are more localized (*C. olympica*, common in its
Irano-Turkish range). The annuals, more Mediterranean species,
are either very diversified (groups of *C. sparsa* or of *C.
spathulata*) or are very localized endemics. Finally,
n = 10 characterizes the monotypic subgenus *Sicyodon* with *C.
macrostyla* Boiss. and Heldr., endemic to Taurus of Cilicie.
    The group of species with n = 10 do not appear to have
strict ecological requirements. They are found in forests as well
as in meadows. We believe that we are dealing with young dynamic
elements which are giving rise to large, diversified complexes.
They, themselves, are perhaps polyploids descended from an
ancestor with x = 5 and not x = 10. This hypothesis would
better explain their diversification and their expansion.
Finally, x = 10 is totally absent in the most important
subgenus: *Campanula*. Quite rare in the family, this number
is found mainly in the genus *Legousia*.
    x = 7 is known, in the diploid state, only in *C.
scutellata* Griseb, an annual endemic to the South Balkans.
More numerous are the tetraploids, both endemic (perennials) and
those with a wider range (annuals). The hyper- or hypoploid

octoploids can most often be likened to neopolyploids. As a consequence of the early origin of the number $x = 17$, an amphiploidy 10 + 7 seems impossible.

3.2.2 AMPHIPLOIDY FROM $x = 9$ AND $x = 8$

This is the hypothesis supported by Tischler (1950) and Fernandes (1962). Here too, there are some difficulties in accepting this process of origin since one would have to imagine an ancestor with $x = 9$ whose descendants have almost completely disappeared. This number exists only in the monotypic subgenus *Brachycodonia* and in *C. lusitanica* but associated with 10 in the latter.

$2n = 36$ is found in two endemic perennials, *C. lactiflora* Bieb. (Caucasia and Asia Minor) and *C. primulifolia* Brot. (Portugal). The study of Gadella (1964) shows that the chromosome number of these two vicariants derives instead from $x' = 17$ and that it is not a polyploid with a base of 9. The form and shape of the chromosomes separates these tetraploids from the diploid annuals and brings them closer to those species with $n = 17$ which they differ from in the length of their chromosomes (less than 1 µm, an evolved character).

Finally, $x = 9$, quite rare in the family, is found only in the genus *Wahlenbergia*. In the campanulas, there does not seem to be any direct relationships between $x = 8$ and $x = 9$ and an amphiploidy based on these two numbers, in our opinion, should be rejected.

3.2.3 TRISOMY FROM $x = 8$

This is the hypothesis of Böcher (1960). The number $x' = 17$ results from a trisomy based on $x = 8$ according to this diagram:

$$n = 8 \quad \rightarrow \quad 2n = 16 \quad \rightarrow \quad 2n = 17 \quad \rightarrow \quad 2n = 34$$

His studies and those of Geslot (1982) on *C. rotundifolia* demonstrate certain phenomena (presence of multivalents, chains, multiple translocations and "stickiness" during meiosis) which could not explain the existence of amphiploidies with 9 + 8 or 10 + 7. Species with $n = 17$ are perhaps comparable to the permanent heterozygote complexes of *Oenothera* (Stebbins 1971). Moreover, Böcher (1964) discovered in *C. persicaefolia* $(n = 8)$ a similar meiotic behavior. He established a clear analogy between the behavior of these two species which reinforces this highly plausible hypothesis. However, Böcher has most of the other chromosome numbers, especially $2n = 56$ (heptaploid) derived from 16. This seems, to us, not possible. We have studied *C. creutzburgii* Greut. $(2n = 56)$, closely related to *C. erinus* $(2n = 28)$, which would be either an autopolyploid of the latter (Contandriopoulos 1970), or perhaps an amphiploid of *C. erinus* and *C. drabifolia* $(2n = 28)$.

3.3 KARYOLOGICAL EVOLUTION OF *CAMPANULA*

We have seen that the genus *Campanula* is polybasic. With
the exception of $x = 9$ which must consist of an only slightly
evolved, independent phylum (subgenus monotypic, very isolated,
quite long, asymmetric chromosomes similar to those of species
with $n = 8$), we believe that there is a direct link between
the basic numbers $x = 8$, 7, 6 and 5 (Table V).

3.3.1: $x = 8$

$x = 8$ corresponds, in our opinion, to the ancestral number
of the campanulas. From this number evolved two separate lines:
one as depicted by Böcher, the other, by descending dysploidy, to
give rise to $x = 7$, 6 and 5. Based on these numbers, chromosome
races were formed. Morphological, karyological, biological and
geographical results confirm this hypothesis.

3.3.1.1: $x' = 17$

In the subgenus *Campanula*, 17 is the most frequent chromosome
number and this is true for the entire range of its distribution.

It is the only number known in 14 sections (+ polyploidy). In
three sections, it is associated with 16 and two with 16 and 14.
In section *Involucratae*, it is allied to 15 (one species to
13). Based on the morphological affinities existing between the
taxa with $n = 17$ and 15 it seems possible that 15 is derived
from 17. Polyploid races evolved from 15 are mainly found in
Caucasian and Irano-Turkish regions. The number $n = 13$,
counted for *C. involucrata* Auch. is more difficult to
explain. Perhaps it is a question of the morphological
convergence of a taxon towards the end of a phylum instead of a
true parentage. In the subgenus *Rapunculus*, the species
with $n = 17$ are disseminated over the entire range. Note in
California, that the annual *C. angustifolia* East. is the
only species with $n = 15$ and it cohabits with annuals and
perennials with $n = 17$ (Morin 1980).

3.3.1.2: $x = 8$

The basic number $x = 8$ appears in the diploid state in four
polytypic subgenera in the Mediterranean region and in the west
asiatic. In the subgenus *Campanula*, it characterizes a
monocarpic, endemic, *C. psylostachya* Boiss. & *al.* which is
the only diploid known in its subgenus. In the subgenus
Rapunculus, it is found in relic complexes such as that of
C. aizoon Boiss. & *Heldr. with C. aizodes* Zaff. and
C. columnaris Contand. & *al.* The range of their distribution
is of the "fossil type" according to Greuter (1971), that is to

say that there exists a hiatus between the actual localization of the endemics and the paleogeographical history of the region (Greece and Crete; Contandriopoulos *et al.* 1973). The two annual subgenera, characterize the endemics of the eastern Mediterranean. Lastly, it is a question of diploids or polyploids; this number is associated with endemism (with the exception, however, of *C. persicaefolia* whose European-Mediterranean range of distribution is very vast).

3.3.2: $x = 7$

We have already spoken of those species with $x = 7$ which are found in the Mediterranean basin, West and Central Asia and Eastern Africa. To the North, their limit of distribution does not go beyond the Alps. There are some very localized polyploid endemics (only one annual diploid) and species of wide polyploid distribution which have given rise to annual endemics. The presence, in certain sections, of species with $2n = 28$, 32 and 34 indicates the affinities related to $x = 8$ and $x = 7$.

3.3.3: $x = 6$; $x' = 12$

This number is known only in the polyploid state and only appears in Mediterranean species. In the subgenus *Campanula*, in the group of Moroccan campanulas, it is sometimes associated with 13. In the subgenus *Rapunculus*, it is represented by a hypotetraploid form ($n = 11$) in *C. hawkinsiana*, endemic to Pinde (Balkans) restricted to serpentine. The subgenus *Megalocalyx*, $n = 12$, is characterized by *C. dichotoma* whose distribution is in the Central and Western Mediterranean.

3.3.4: $x = 5$; $x' = 10$

The species with $n = 10$ have a distribution which covers the Mediterranean and the Near East. Of those which have invaded Europe, only the ones found in the southern part of their range have differentiated endemics. They appear as young elements compared to the other campanulas.

CONCLUSIONS

In Tables II, III and IV are sketched the phylogenetic relationships of the subgenera. In the two large subgenera,

we observe a continuous series of basic numbers, namely $x = 8$, 7, 6 and 5 (only subgenus *Rapunculus*) and the series of secondary basic numbers (especially $x' = 17$). In the annual subgenera we have $x = 8$ and 7 (*Roucella*) and $x = 8$, 6 and 5 (*Megalocalyx*). If we return to the two large sections of Boissier, we find, in each, the complete series of basic numbers.

TABLE II. PHYLOGENETIC OUTLINE (Esquisse phylogénétique) OF THE SUBGENUS *CAMPANULA*

Sous-genre *CAMPANULA*

Plantes vivaces, bisannuelles ou monocarpiques (2 annuelles).
Capsules à déhiscence généralement basilaire
Appendices calicinaux, présents, absents, peu visibles

FORMES ANCESTRALES

TAXONS ACTUELS		2n
Campanula	Sections	34
Dasystigma Fedorov		34
Dictyocalyx (Fed.) Damboldt		34
Elatae (Boiss.) Damboldt		34,68
Heterophylla (Nym.)Fedorov		34,68,102
Platysperma Damboldt		34
Quinqueloculares (Boiss.) Phitos		34
Scapiflorae (Fomin) Damboldt		34
Sibiricae (Fomin) Charadze		34,(26?)
Spinulosae (Fomin) Fedorov		34
Symphyandriformes (Fomin) Charadze		34
Tracheliopsis (Buser) Damboldt		34
Trigonophyllon Fedorov		34
Tulipella Fedorov		34
Megalocodon Damboldt		32
C. herminii Hoffm. parfois rattaché à la sect. *Heterophylla*		32

x'=17

x=8 ⟶ x'=16 ⟶ x'=16 ⟶ x'=16

Latilimbus Fedorov 48,34
Spicatae (Fomin) Damboldt 32,34
Isophyllae Damboldt 32,34

Involucratae (Fomin) Charadze 34,26 / 30,60,90

Rupestres (Boiss.) Charadze 32,34,28
Saxicolae (Boiss.) Charadze 16,34,28
C. cashmeriana Royle 28
C. colorata Wall.⊙ 28
C. edulis Forsk. 56
C. keniensis Thul⊙ 54

C. velata Pomel 26

Groupe des campanules nord-africaines:
C. atlantis Gatt&Maire 24
C. numidica Durr. 24
C. mollis L. 24,48
C. filicaulis Durr. 24,48,72

x'=16,17
x'=17,15,13
x'=16,17,14
x'=14 -
x'=13
x=7
x'=12
x=6

Nombres chromosomiques connus : 255.

Taxons % des taxons possédant chacun de ces nombres.

$x=8$	1	+
$x'=12$	7	2,8
$x'=13$	3	1,2
$x'=14$	7 (2 ⊙)	2,8
$x'=15$	10	4,0
$x'=16$	12	4,8
$x'=17$	213	83,6

C. bordesiana Maire = C. edulis Forsk. x C. filicaulis Durr.
2n=80   2n=56   2n=24

TABLE III: Esquisse phylogénétique des ssg. *RAPUNCULUS* et *BRACHYCODONIA*.

Sous-genre *BRACHYCODONIA*

monotypique, annuel
capsule à déhiscence apicale
appendices calicinaux absents
ramifications fastigiées
fleurs peu visibles
C. fastigiata

n=9 →

Sous-genre *RAPUNCULUS*

plantes vivaces, bisannuelles, annuelles
capsule à déhiscence apicale à médiane
appendices calicinaux absents

TAXONS ACTUELS. → *FORMES ANCESTRALES* → *TAXONS ACTUELS vivaces*
annuels

n=17 → x'=16,17   x'=16,17 → n=17 → n=18
↓   n=8   x=8 → n=8 → n=16
n=15

n=14 → x'=14 →   x=7 → x'=14 → n=14 → n=29
      x=6 → x'=12 → n=13   n=11

n=10 → x'=10   x=5 → x'=10 → n=10,20,40
↓
n=9

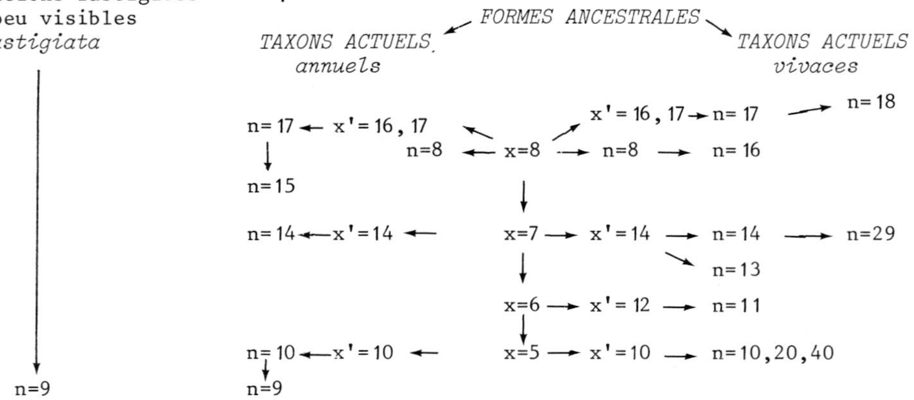

Nombres chromosomiques connus : 47

Taxons Vivaces Annuels % des taxons possédant
chacun de ces nombres.

n=8	9	8	1	18,95
n=9	1	1	1	+
n=10	17	8	9	35,42
n=11	1	1		+

n=13	1	1		+
n=14	3	2		6,25
n=15	1	1		+
n=16	2	2		4,16
n=17	10	7	3	20,83
n=18	2	2		4,16
n=29	1	1		+
Totaux 48	32	16		ca 100%

TABLE IV : Esquisse phylogénétique des ssg. *SICYODON, MEGALOCALYX, ROUCELLA*.

Sous-genre *SICYODON*
monospécifique, annuel
capsule à déhiscence basilaire
appendices calicinaux présents
style longuement exert
C. macrostyla

Sous-genre *MEGALOCALYX*
annuel
capsule à déhiscence basilaire
appendices calicinaux présents

Sous-genre *ROUCELLA*
annuel
capsule à déhiscence basilaire
appendices calicinaux absents.

TAXONS ACTUELS — *FORMES ANCESTRALES* → *TAXONS ACTUELS*

n=10

n=8 ← x=8 → n=8
x=7 → n=7 → n=14 → n=28
x=6
x=5

n=12 → x'=12
n=10 → x'=10
n=8:3 taxons
n=12:2 taxons
n=10:4 taxons

n=8: 1 taxon
n=7: 1 taxon
n=14:2 taxons
n=28:1 taxon

TABLE V

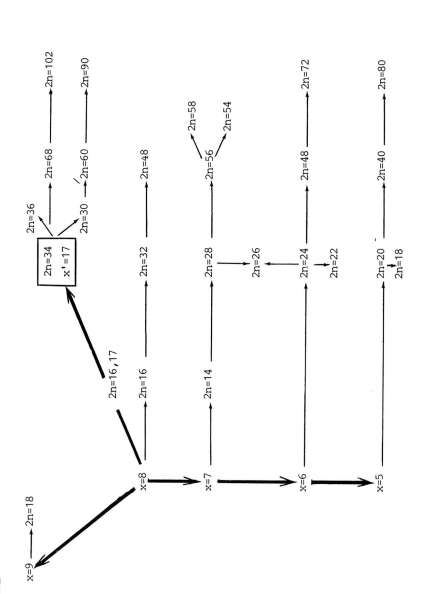

The studies of Schulkina (1980) on the different types of life
forms also demonstrate parallelism between the sections of
Boissier.

However, these diagrams do not take into account all the
realities. The genus *Campanula*, of Asiatic origin, belongs
to the arctictertiary flora which has distributed itself widely in
the Mediterranean Basin, Eurasia as far as East Asia, and North
America, as well as East Africa. The presence in Western Asia and
in the Mediterranean of an important paleoendemism (diploids and
polyploids) demonstrates the age of this flora. But the abundance
of annuals and the frequency of neopolyploids (perennials and
annuals) is an indication of an expanding flora. Near the relic
centers (Near and Middle East), the Eastern Mediterranean, the
Balkans, the Southern Alps, the Western Mediterranean Basin, also
exist sources of formation and differentiation. For example, the
annuals are polyphyletic. Some of them evolved in different
places, from perennials already polyploid and having different
chromosome numbers (cf. Sec. 1 and 3.2.3).

Reflecting on the theme of speciation and the evolution of the
campanulas, this study is not an end in itself, but rather reminds
us of the complexity of a genus which has yet to reveal all its
mysteries, in view of its numerical importance, its variety and
its immense geographic distribution.

ACKNOWLEDGMENTS

We thank those who helped us during the preparation of this paper:
Prof. C. Favarger (Botanical Institut, Neuchâtel), Mrs. R.
Verlaque, research associate of C.N.R.S. (Marseille), M. Charpin,
Librarian (Botanical Conservatory of Geneva).

RÉSUMÉ

Le genre *Campanula* est extrêmement complexe en raison de son
importance numérique (plus de 500 espèces), sa très large
distribution géographique (Eurasie, Amérique du Nord, région
méditerranéenne, Afrique orientale et centrale), l'hétérogénéité
de certains sous-genres ou sections à caractères morphologiques
convergents, la variété des nombres chromosomiques (de $n = 7$
à $n = 17$), séries polyploïdes à partir de différents nombres
de base, dysploidie, aneuploidie. La confrontation des données
biogéographiques, morphologiques et caryologiques met en évidence:
des espèces à large distribution géographique et des endémiques
étroitement localisées, une instabilité des nombres primaires et
un foisonnement de nombres de base secondaires ($x = 10$, 11,
12, 13, 15 et 17); l'importance de $x' = 17$ (chez les 3/4
environ des espèces étudiées) présent dans l'aire globale du

TABLE V. *Karyological evolution of* Campanula/*Evolution
caryologique des Campanules*.

genre, les autres nombres appartenant à des espèces
méditerranéennes ou localisées dans le proche et le moyen Orient
et l'Afrique, souvent chez des annuelles. Un schéma
phylogénétique est proposé à partir du nombre de base x = 8
qui a donné naissance à 3 phylums. Le premier à x = 9 est
représenté par le sous-genre annuel, monotypique *Brachycodonia*.
Le deuxième, par trisomie, a engendré le nombre de base secondaire
x' = 17 qui a formé des séries polyploides à n = 17, 34 et 51.
Le troisième, par dysploidie descendante, a donné naissance aux
nombres de base x = 7, 6 et 5. À partir de ces nombres se
sont formées des séries polyploides hyper- et hypoploides.

REFERENCES

Bielkawska, H. 1968. Cytogenetic relationships between lowland
    and montane species of *Campanula rotundifolia* L. group. I.
    C. cochleariifolia Lam. and *C. rotundifolia* L.*Acta Soc.
    Bot. Pol.* 33: 15-44.
Böcher, T.W. 1960. Experimental and cytological studies on plant
    species. V. The *Campanula rotundifolia* complex. *Biol. Skr.
    Dan. Vid. Selsk.* 11: 1-69.
Böcher, T.W. 1964. Chromosome connections and aberrations in the
    Campanula persicaefolia group. *Sv. Bot. Tidskr.* 58: 1-17.
Boissier, E. 1875. *Flora Orientalis*, III. Genève et Bâle.
Bolkhovskikh, Z. et al. 1969. *Chromosome Numbers of Flowering
    Plants*. Fedorov ed.
Candolle, A. de. 1830. *Monographie des Campanulées*. Paris.
Celebioglu, T. et Favarger, C. 1982. Contribution à la
    cytotaxonomie du genre *Minuartia* L. en Turquie et dans
    quelques régions voisines. *Biol. Ecol. Médit.* 9: 139-160.
Contandriopoulos, J. 1964. Contribution à l'étude cytotaxonomique
    des Campanulacées de Grèce. *Bull. Soc. Bot. Fr.* 111: 225-235.
Contandriopoulos, J. 1966. ____. II. *Bull. Soc. Bot. Fr.*
    113: 453-474.
Contandriopoulos, J. 1970. Contribution à l'étude cytotaxonomique
    des Campanulacées du Proche Orient. *Bull. Soc. Bot. Fr.*
    117: 55-70.
Contandriopoulos, J. 1972. ____. III. *Bull. Soc. Bot. Fr.*
    119: 75-95.
Contandriopoulos, J. 1976. ____. IV. *Bull. Soc. Bot. Fr.*
    123: 33-46.
Contandriopoulos, J. 1980a. Contribution à l'étude cytotaxonomique
    des Campanulacées de l'Iran. *Biol. Ecol. Médit.* 7: 27-36.
Contandriopoulos, J. 1980b. Contribution à l'étude cytotaxonomique
    du genre *Campanula* L. en Afrique du nord et centrale.
    Bol. Soc. Brot. 53: 887-906.
Contandriopoulos, J. et al. 1972. Campanulacées nouvelles du
    pourtour méditerranéen oriental. *Ann. Univ. Provence, Sci.*
    46: 53-61.
Contandriopoulos, J. et Quezel, P. 1973. À propos des campanules du
    groupe *aizoon* en Grèce méridionale et en Crète. *Bull. Soc. Bot.*

Fr. 120: 331-340.

Contandriopoulos, J., Favarger, C. et Galland, N. 1983. Contribution à l'étude cytotaxonomique des Campanulaceae du Maroc. In press.

Damboldt, J. 1965. Zytotaxonomische Revision der Isophyllen *Campanulae* in Europaea. *Bot. Jb.* 84: 302-358.

Damboldt, J. and Phitos, D. 1978. *Campanulaceae, Flora of Turkey and the East Aegean Islands.* Edinburgh Press. pp. 2-89.

Favarger, C. et Huynh, K.L. 1980. Contribution à la cytotaxonomie des Caryophyllacées méditerranéennes. *Bol. Soc. Brot.* 53: 493-514.

Fedorov, A.A. 1957. *Flora of the U.S.S.R.*, vol. 24 Campanulaceae. Translated from Russian, Jerusalem, 1972.

Fedorov, A.A. and Kovanda, M. 1976. *Flora Europaea. Campanula* 4: 74-93.

Fernandes, A. 1962. Sobre a cariologia de *Campanula lusitanica* L. ex Loefl. e *C. transtagana* R. Fernandes. *Bol. Soc. Brot.* 36: 129-137.

Gadella, Th. W.J. 1964. Cytotaxonomic studies in the genus *Campanula. Wentia.* 11: 1-104.

Geslot, A. 1982. Les campanules de la sous-section *Heterophylla* (Wit.) Fed. dans les Pyrénées. Une étude biosystématique. Thèse d'état Marseille.

Greuter, W. 1971. Bertrachtungen zur Pflanzengeographie der Südägäis. *Opera Bot.* 30: 49-64.

Mangenot, G. et al. 1977. Caryologie du genre *Phyllanthus* (Euphorbiaceae, Phyllanthoideae). *Ann. Sci. Nat. Bot. Paris.* 18: 71-116.

Moore, R.J. 1967-1977. Index to plant chromosome numbers. *Regnum Veg.* 90, 91, 96.

Morin, N. 1980. Systematics of the annual California Campanulas (Campanulaceae). *Madrono* 27: 149-163.

Podlech, D. 1965. Revision der subsect. *Heterophylla* (Vit.) Fed. der gattung *Campanula. Feddes Repert.* 71: 50-187.

Phitos, D. 1964. Trilokuläre *Campanula*-Arten. *Osterr. Bot. Zeitschrift.* 111: 208-230.

Phitos, D. 1965. Die quinquelokulären *Campanula*-Arten. *Osterr. Bot. Zeitschr.* 112: 449-498.

Quezel, P. 1953. Les Campanulacées d'Afrique du nord. *Feddes Repert.* 56: 1-56.

Rechinger, K. and Schimann-Czeika, H. 1965. *Flora of Iranischen Hochlandes und der Umrahmenden Gebirge. Campanulaceae.* Graz Austria.

Schetler, S.G. 1982. *Variation and Evolution of the Nearctic Harebells (Campanula subsect. Heterophylla).* Cramer.

Shulkina, T.V. 1980. The significance of life-form characters for systematics, with special references to the family Campanulaceae. *Pl. Syst. Evol.* 136: 233-246.

Stebbins, G.L. 1966. Chromosomal variation and evolution. *Science* 152: 1463-1469.

Stebbins, G.L. 1971. *Chromosome Evolution in Higher Plants*.
   Arnold, London.   pp. 216.
Thulin, M. 1975. *Campanula keniensis* Thulin *sp. nov.*
   and notes on allied species. *Bot. Not.* 128: 350–356.

The Biological Species Concept Reexamined

Bengt Jonsell
Bergius Botanic Garden
Stockholm University
Stockholm, Sweden

INTRODUCTION

The titles of my lecture in the English and French versions of the
program, "The Biological Species Concept Reexamined" and "Le
Concept Biologique de l'Espèce: Une Revue," do not, as I
understand them, express exactly the same meaning. A
reexamination critical and bold enough to lead to the proposal of
a new definition, or new ideas as to how the species as a category
should be viewed, will not be found here, nor will this be an
entirely historical review, but perhaps something in between. The
subject has constantly received much attention in the last 10 or
15 years. Most commentators have, from various different
starting-points, been critical of the concept; only a few have
defended it. Among the latter is, in particular, . Grant (1981);
the former include Ehlrich and Holm (1962), Bennett (1964),
Ehrlich and Raven (1969), Sokal and Crovello (1970), Sokal (1973),
Raven (1976, 1980), Van Valen (1976), Cronquist (1978) and Wiley
(1978, 1981). The opinions of these authors will be considered in
the following discussion. I am myself so far totally uncommitted
on this delicate subject, and this is my only excuse for having
taken on the topic. It will be clear, however, that I have
considered in some depth the critical aspects of the subject and
have reached at least an ad hoc view-point on the matter.

The biological species concept (BSC) is the most rigorously
defined species concept so far proposed. However, most of its
advocates have regarded it as being of very general validity for
the organisms with breeding systems that are considered of
particular importance in evolution and speciation. Its failure to
explain the various evolutionary "dead ends" has perhaps not been
taken seriously enough.

The BSC may be seen as a hypothesis open to testing. How have
such tests stood up? Firstly, has this hypothesis been carefully

and unconditionally scrutinized? Or has this model been taken for granted to such an extent that there has been more of a trend to seek its confirmation, to make it fit the scheme, and to regard nonconforming observations as exceptions or anomalies? It is only too easy to see what you wish to see, as many have said before.

One must remember that the empirical basis for the BSC mainly comes from studies with mammals and birds; the plant studies which have contributed are all based on vascular plants and, for us, maybe even more significant is the fact that nearly all the plants studied are from temperate or Mediterranean climates with hardly any results coming from tropical plants. Under such circumstances it is perhaps surprising that the BSC has become so widely accepted, that it is extended to organisms about which virtually nothing of relevance is known and that it is regarded as, if not the only existing model for these entities, at least the one which is evolutionarily important.

The rather unreserved adoption of the BSC is surprising also because botanists, to a much higher degree than vertebrate taxonomists, have for a long time been aware of the complications of defining species. Since the start of taxonomy botanists have realized that there were "good", "less good" and "bad" species, even disregarding agamic or polyploid species, etc. So why has the BSC been so easily accepted? It is tempting to explain away many of the anomalies by saying that the entities in question are incomplete stages in the formation of biological species, but this may well be to oversimplify the matter.

There are, however, some obvious explanations for this state of affairs: (1) the BSC is beautifully simple to comprehend; (2) it conforms elegantly with current speciation theories, stressing the primary importance of the origin of reproductive barriers. This is, of course, only another facet of the BSC itself. One reason for its success is at the same time one of its major drawbacks: it is so extremely laborious to test the model by experiments, and so difficult to interpret patterns of variation and differentiation, that the results are not easily open to disproof. The success of the concept then lies in the very inconclusiveness of the limited data obtained, which are tempting to interpret in terms of a BSC.

SOME VIEWS ON THE BIOLOGICAL SPECIES CONCEPT

Pheneticists (Sokal and Crovello 1970) have claimed that it is *a priori* an impossibility to define the BSC experimentally. The sampling of the material to be tested necessarily makes use of other criteria before the one of real significance, the degree of interbreeding, can be approached (cf. Sokal and Crovello 1970). The first criterion in choosing the samples to be studied is that they should be in some geographical contact with each other, giving them the possibility of interbreeding in nature. The next criterion is that there should be some degree of phenetic similarity, which must unavoidably underlie the choice of

experimental material. It has been argued that this very
operation excludes the possibility of defining the BSC. A number
of phenetic assumptions, which are unlikely to be reconcilable
with the biological species definition, have to be made before a
biological species can be circumscribed. In this sense the BSC is
nonoperational; it cannot, however much research is undertaken, be
defined without a phenetic element. And because of the labor
involved in most biological experiments, the phenetic component is
likely to be considerable in relation to the amount of fertility
tests.

It seems to me, however, that these difficulties are of an
operational and not of a theoretical nature. They do not preclude
the biological species definition from being valid for a
substantial number, maybe even for a majority of biparental
vascular plants, though our tools to prove it may be too crude
and, for many critics, unconvincing.

It is possible to plan experiments and relevant population
studies in such a way that a BSC can be refuted or confirmed?
Some situations are of course quite impossible to handle in this
way, most obviously the interfertile allopatric entities. This
situation is, *mutatis mutandis,* similar to that confronted
by physicists studying certain elementary particles – as soon as
you interfere with the object of your experiments you disturb the
prerequisites and make the experiment invalid. For plants it is
on the whole not possible, as has been suggested for allopatric
populations of birds and mammals, to take morphological difference
as a measure of reproductive isolation. A redefining of the BSC
to make laboratory tests definitive would make the concept still
less "biological" – the situation in nature is the one that
matters and greenhouse results must be interpreted against that
background.

This is relevant for the situation met in contiguous entities,
too. The crossability *in se* is not of real interest, but
the degree of miscibility is important, and this can only be
studied in the field in relation to the ecological situation.
However carefully selected and performed, experiments result in
only a minute fraction of the material in space and time being
studied and one must be careful not to draw general conclusions
from what are perhaps limiting local and/or temporal situations.
Moreover, this is valid only as far as the biological species
definition really is in accordance with our theories about the
evolutionary role of hybridization, a matter I will touch upon
below.

Whether using computerized statistical methods on population
samples, or the conventional working methods necessary for
haphazardly gathered herbarium specimens, the phenetic element
will usually be the dominant one, both quantitatively and as the
source from which to draw conclusions. Some people seem inclined
to go to the extreme and throw biological considerations largely
overboard (Cronquist 1978), except for indirectly considering
reproductive isolation as a necessary corollary of phenetic

discontinuity, which may be far from true. Traditions and conventions as to the species criteria within the various plant groups will take a dominant role, with Protagoras: "Man is everything's measure." I find it unsatisfactory to take the species concept to the extreme, where it is void of "biological" content, a view which is also unhistorical.

From the early days – Ray's true-breeding criterion, the crossings by Koelreuter and even by Linnaeus himself (Jonsell 1978; Raven 1980) – ideas of the species role in nature were incorporated into taxonomy, culminating for the period with Gaertner's summarizing work about the rise of hybrids among plants. The early 19th century saw a fairly penetrating discussion of the theme (cf. Raven 1980). To avoid making this paper a course of history, we shall move forward: with the "evolutionary synthesis" the BSC, essentially reproductional, was considered to reflect the biological role of the species category. There was a time when the BSC was considered to be universal and applicable to all biota (Löve 1964), an overoptimistic extrapolation from observations in certain well-studied animals. That difficulties were early encountered by botanists is reflected by the growing discrepancies in botany between taxonomic and biological species concepts, particular terminologies, and so forth. With the growing insights that there may not be any "normal" biological species – any that can be called more normal than others – and an understanding of the complexities and variation among species-as-taxa, these have to be phenetically defined, but preferably not by crude rule of thumb. Where reproductive, ecological and other information is available and applicable it should be included. Implicitly the context of the species in nature would be in the mind of all except the most ardent nominalists.

THE INTERNAL COHESION

I will now turn from the operational difficulties connected with a BSC, whether this model is valid or not, to the matter of whether, according to our present state of knowledge, the reproductional and evolutionary prerequisites upon which the BSC is built are in a higher, generalizing degree valid for vascular plants.

There are, as we all know, two elements of the biological species definition, the first one, with Mayr's formation (1963), dealing with "groups of actually or potentially interbreeding populations." The importance of gene recombination and heterozygosity for the evolutionary process was one of the novelties which appeared with the "synthesis" and it is not surprising that gene-exchange became considered a process of wide-ranging importance. Gene-flow was generally considered the mechanism behind the homogeneity within a biological species, and it is striking that this seems to have been taken for granted. Little can be found in the way of attempts to explain how this gene-flow really operates and how it can work effectively among scattered populations. A difficult aspect of the definition is of course the word "potentially", which is very vague and may be interpreted in various ways (cf. Sokal 1973, p. 132, 134). It is surprising that it survived for many years though it was finally dropped even by Mayr (1969).

The word "potentially" as opposed to "actually" meant that the entities would interbreed under experimental conditions and were maybe connected in nature via ancestors and descendants. An exchange significant enough to keep the species together, to be the cohesive force, is thus difficult to imagine. As pointed out by Heslop-Harrison (1955) potentiality for interbreeding already exists between numerous spatially isolated taxa with more or less distinct habitat preferences, i.e. ecologically isolated groups of populations. How much weight should be given to the potentiality of interbreeding versus ecological specialisation is undoubtedly a recurring problem when vascular plant species are to be circumscribed – in sharp contrast to the situation among higher vertebrates.

In recent years doubts as to the existence of the cohesive gene-flow have become all the more clearly expressed (Ehrlich and Raven 1969, Sokal 1973, Levin 1979). In such a strong advocate for the BSC as V. Grant, we see a shift in attitude between the first and the second editions of his classical *Plant Speciation* (Grant 1971, 1981). In the first, gene-flow is hardly discussed, in the second arguments for and against are presented and the author largely avoids coming to a conclusion: "extensive interbreeding within the population system is not an essential property of biological species, non interbreeding with other population systems is" (Grant 1981, p. 91). He underlines the important difference between dispersal and gene-flow through time – the latter falls off precipitously compared with the former. Moreover, effective dispersal seems mostly to be surprisingly short-reaching (Ehrlich and Raven 1969).

As in the case for gene-flow, the more homogeneous structure of continuous populations is cited in contrast to the greater heterogeneity of those with disjunct distribution. It is possible, however, that a complex of factors, including selective forces, founder effects etc., may be of no less importance for such situations.

The obvious circumstance that species are homogeneous over large areas in spite of having been long split up into spatially isolated populations calls for some other cohesive factor in explanation. Raven (1976, 1980) and others have strongly argued that the clue is to be found in selection, in line with the current trend to give more weight to such factors. Under fairly constant conditions it may be imagined that stabilizing selection, consistently adapting the populations to the proper ecological niche of the species, is functioning. The degree of homogeneity will probably depend on where the species is placed on a generalist-specialist scale for various environmental factors.

There is also the problem of explaining how a homogeneous

system of disjunct populations, or population groups, originated. Some stepwise dispersal, without founder effects being too much involved, is a hypothetical explanation for which evidence seems lacking.

Whatever the cause behind the homogeneity among populations of a species - gene-flow or selectional forces - the pattern cannot be said to be at variance with the biological species as a model, although the balance may shift from a genetic to a more ecological viewpoint.

THE EXTERNAL BARRIER

The other element of the biological species definition states that the groups in question should be reproductively isolated from other such groups (e.g. Mayr 1963). The reproductive barriers have been regarded, often by emphatic advocates, as making the species a closed genetic system, the members of which share a common gene-pool. The discovery of barriers across a morphologically/ecologically homogeneous or at least continuous entity has led to the concept of sibling species. I think that most students have regarded such species as more or less rare exceptions for which morphological and genetical differentiation does not keep step. But if it becomes apparent that barriers within traditional species are a recurring feature another attitude may be needed. If it is true that gene-flow is not the dominant cohesive force, it seems only logical to presume that barriers to gene exchange may often be discovered without being reflected in the phenetic variation.

As pointed out by Raven (1976), barriers of this kind have been found especially in annual herbs in a variety of families, e.g. *Clarkia, Holocarpha, Gilia.* Intersterility between populations may be largely caused by chromosomal rearrangements, more or less autoploidal situations would give similar results. Biosystematists should consider it an urgent task to look for such patterns, see how frequent they are and estimate their evolutionary role. Perhaps the sibling species concept will have to be abandoned for vascular plants, although it has been useful at a certain stage in our understanding of the BSC.

The reproductive isolation between groups of populations has, as hinted above, been regarded as the fundamental basis of the BSC. The students' demands upon isolating mechanisms have varied from the most rigorous, approving only supposedly internal chromosomal barriers (Löve 1964), to the current set of pre- and post zygotic isolating mechanisms found in every textbook on the subject. Obviously valid for higher vertebrates, these criteria have to a large extent been considered appropriate among vascular plants too. Most examples studied have, both for practical and geographical reasons, been groups with short-lived individuals, dependent on regular seed production for survival; that is, they are probably more or less so-called r-strategists. For such plants hybrid progeny would be effectively selected against in sympatric or parapatric situations. For a long time this has been considered the "normal" model, though it has always been clear, not least to floristically orientated people, that numerous groups, and indeed many trees and shrubs - less apt to experimental work - do not conform to that model, they are "critical" and often form fertile hybrids. Their barriers are not internal but seasonal, ecological, spatial, etc., which easily permits the formation of hybrids, especially in a changing environment.

The significance of these circumstances has become apparent in recent years, particularly as a result of Raven's papers (1976, 1980), which include examples of long-lived perennials of various life forms and geographical origin. It has been emphasized that taxa of these groups are largely K-strategists, for which comparatively little resources have to be channelled into reproduction. The perennial habit, often combined with vegetative dispersal, guarantees survival for long periods.

The external and internal barriers have mostly been regarded as alternative or combined ways of securing the closed genetic system, the gene-pool, gradually changing with time and occasionally splitting to produce the initials of potentially new species. Hybrid situations such as swarms, introgression, etc., have been mostly considered as local break-downs of the barriers. In other cases, hybrids may have passed as separate species. In quite a few groups the process has resulted in "syngamia" or "multispecies" (*Salix, Quercus*, etc.), numerous species have produced numerous hybrid populations among themselves, where they are or have been in contact.

An alternative way of looking at such situations is to regard hybrid formation as a step in the speciation process, which Raven (1976, 1980) called "interspecific recombination." This process would yield numerous genotypes from among which some, by dispersal and strong selection - maybe reinforced by autogamy, would exploit new ecological conditions. This possibility has been indicated in textbooks, but usually with strong reservations about the probability of the process. The differentiation patterns of many genera seem, however, possible to reconcile with fairly rapid speciation along these lines (e.g. *Epilobium* in New Zealand according to Raven and Raven 1976). If such models really turn out to be characteristic for perennials, which are by far the majority of vascular plants, the idea of a biological species as a closed community within reproductional barriers must be fundamentally revised.

ALTERNATIVES TO THE BSC

On the whole, it would be highly desirable to confront the BSC with the new ideas of possible pathways to speciation. There is a growing realization that there are for vascular plants alternatives to the classical, so-called "normal", allopatric, gradual model. They would, however, not exclude each other and there must be all sorts of intermediary models.

With a critical attitude to gene-flow it must be difficult to uphold the gene-pool of the biological species as an evolutionary unit except for entities with very restricted distribution.   If stabilizing selection is considered important for the homogeneity of the species, evolutionary stasis over long periods of time may often be the case rather than gradual evolution of the species as a whole.   This brings us to consider as a probable unit for speciation a population, or close groups of populations (cf. Raven 1976, 1980), perhaps marginal and with ecological changes as an important driving force.   The ecological shifts may be drastic and may correspond to the model of the punctuated equilibrium (Gould and Eldredge 1977), affecting a whole set of populations of various species. After rapid remodelling under intraspecific, and, according to the punctualists, particularly interspecific competition, presumably coupled with hybridization between related species, new species adapted to new conditions may arise.   There are distribution patterns among vascular plants that might be interpreted that way, but the interpretations are conjectural and biosystematic and other approaches are needed to confirm them.

In this context, as well as in several others mentioned above, the ecological aspect of processes and results seem to be of outstanding importance.   "A common ecological role" is a phrase now often used in connection with groups of populations which constitute a species, and the opinion was clearly expressed by Van Valen (1976) that "species are maintained for the most part ecologically, not reproductively."   In accordance with this view other species concepts, no more operational than the BSC have been suggested.   Van Valen (1976) spoke contrastingly of an ecological and a reproductional one, the latter a synonym of the BSC.   With the more flexible view we now are able to take towards species structures and speciation possibilities among vascular plants, I feel too, that the term "biological" is a bit pretentious.

I shall not go more closely into Van Valen's interesting ecological species concept, but instead observe that the ecological aspect also largely underlies the older and, with cladistics, increasingly popular evolutionary species concept (ESC).   Both view the species as an ancestral-descendant lineage. One finds this in Simpson's (1961) original definition of the ESC, "a unitary evolutionary role" and "own evolutionary tendencies," the latter maintained also in Wiley's (1978, 1981) recent formulations.   Behind these expressions are ideas of distinctive ecological roles, of the fitting of a species into a particular ecological niche.

The biological species may be considered as a special case of the evolutionary species (cf. e.g. Grant 1971).   These lines, which are the evolutionary species, must in a broad sense be reproductively isolated from each other, and for outbreeders the well-known mechanisms may operate.   The ESC is indeed less rigid and also less informative than the BSC, but this gives it some advantages.   It is not tied to a particular model of speciation, and this gives room for more open-minded approaches; it may include allopatric entities without considering their reproductive behavior with each other, and perhaps most importantly, the reproductive system, whether biparental, uniparental, clonal, etc., is irrelevant. The ESC versus cladistics will not be taken up here. From a non-cladistic viewpoint the ESC is at least as little operational as the BSC, though it may be attractive as a more inclusive model when we now realize that we know less than we might have thought during the heyday of the BSC.

Consideration of the phenetic element when defining species-as-taxa seems to be unavoidable. Phenetic variation and discontinuities will continue to be the dominant way of defining taxa and the patterns obtained must not be regarded as reflecting particular situations, e.g. concerning reproductive isolation, in the population systems studied. The species may be morphologically defined, but at the same time we may also be aware of deeper considerations and   hypotheses corresponding to one or the other of possible   models. it will be a task for biosystematists, together with ecologists, geneticists and others to test the hypotheses. This work will not primarily be part of taxonomy, although it will give us a much deeper understanding of our taxa.

I will end by saying that it would indeed have been alarming if we still saw the BSC as we did several decades ago. That would have meant a standstill. Today, the discussion is overwhelmed by new ideas - or old ones in new dresses. The scene will probably continue to change and open-minded biosystematists will have more than enough to deal with for decades to come.

REFERENCES

Bennett, E. 1964. Historical perspectives in genecology. *Scott. Plant Breed. Stn. Rec.* 1964: 49-115.

Cronquist, A. 1978. Once again, What is a species? *Beltsville Symp. Agric. Res.* 2: 3-20.

Ehrlich, P.R. and Holm, R.W. 1962. Patterns and populations. *Science* 137: 652-657.

Ehrlich, P.R. and Raven, P.H. 1969. Differentiation of populations. *Science* 165: 1228-1232.

Gärtner, C. 1849. Versuche und Beobachtungen über die Bastarderzeugung im Pflanzenreich. Stuttgart.

Gould, S.J. and Eldredge, N. 1977. Punctuated quilibria: the tempo and mode of evolution reconsidered. *Palaeobiology* 3: 115-151.

Grant, V. 1971. *Plant Speciation.* Columbia Univ. Press, New York.

Grant, V. 1981. *Plant Speciation.* 2nd ed., Columbia Univ. Press, New York.

Heslop-Harrison, J. 1955. The conflict of categories. In *Species Studies of the British Flora* (J.E. Lousley, ed.). pp. 160-172.

Jonsell, B. 1978. Linnaeus's views on plant classification and evolution. *Bot. Not.* 131: 523–530.

Levin, D.A. 1979. The nature of plant species. *Science* 204: 381–384.

Löve, A. 1964. The biological species concept and its evolutionary structure. *Taxon* 13: 33–45.

Mayr, E. 1963. *Animal Species and Evolution*. Harvard Univ. Press, Cambridge, Mass.

Mayr, E. 1969. *Principles of Systematic Zoology*. McGraw-Hill, New York.

Raven, P.H. 1976. Systematics and plant population biology. *Syst. Bot.* 1: 284–316.

Raven, P.H. 1980. Hybridization and the nature of species in higher plants. *Can. Bot. Assoc. Bull.* 13, Suppl. 1: 3–10.

Raven, P.H. and Raven, T.E. 1976. The genus *Epilobium* (Onagraceae) in Australasia; a systematic and evolutionary study. *New Z. D.S.I.R. Bull.* 216. Christchurch.

Simpson, G.G. 1961. *Principles of Animal Taxonomy*. Columbia Univ. Press, New York.

Sokal, R.R. 1973. The species problem reconsidered. *Syst. Zool.* 22: 360–374.

Sokal, R.R. and Crovello, T.J. 1970. The biological species concept: a critical evaluation. *Am. Nat.* 104: 127–153.

Van Valen, L. 1976. Ecological species, multispecies and oaks. *Taxon* 25: 233–239.

Wiley, E.O. 1978. The evolutionary species concept reconsidered. *Syst. Zool.* 27: 17–26.

Wiley, E.O. 1981. *Phylogenetics. The theory and practice of phylogenetic systematics*. Wiley, New York.

The Pursuit of Hybridity and Population Divergence in *Isotoma petraea*

S. H. James

Department of Botany
University of Western Australia
Nedlands, Western Australia

INTRODUCTION

There are several events of signal importance in evolutionary
biology which occurred about 50 years ago. The introduction by
Camp and Gilly of the word biosystematy into our lexicon was one.
Another was the publication in 1932 of the first edition of C.D.
Darlington's *Recent Advances in Cytology*. This work
contained a germinal last chapter which was deleted from the 2nd
edition but which was expanded to become *The Evolution of
Genetic Systems* in 1939. Here, Darlington developed the idea
that the genetic material determines its own manipulative
machinery and capabilities and is itself subject to genetic
variation and evolution. Cytological studies, and especially the
cytological studies of hybrids, are thus an integral part of
biosystematy, providing arrowheads in branching and reticulated
phylogenies and defining boundaries of taxonomic significance.
   But cytology does more than this; it focuses attention on
special events in speciation and it probes the machinery
modulating the generation of selectively malleable variations. In
this way, it allows an assessment of the evolutionary potentials
of lineages and a rational and independent basis for interpreting
evolutionary sequences. Darlington recognized this, and told the
world. However, in 1932 his book "was such a mixture of fact and
fancy that the inductively operating cytologists of the day were
highly suspicious of it" and in some, perhaps most laboratories,
the book was received "with stiff attitudes of outrage, anger and
ridicule" (Carson 1981). But now, 50 years later, this antipathy
is dwindling, and the impact and relevance of Darlington's
"prolegomenon to every future theory of evolution" (Haldane 1937)
is being more widely appreciated if not understood (see Darlington
1981). Indeed, Bell (1982) in the concluding chapter of his book

The Masterpiece of Nature, now proposes the term
"metagenetics" to name a science epitomized by the title (if not
the content?) of Darlington's book, *The Evolution of Genetic
Systems.*

Darlington's incisive enterprise allowed him to formulate
several guiding metagenetic principles. These include the
idea that genetic system evolution in diploid sexuals and their
derivatives, at least, may be interpreted in terms of the "pursuit
of hybridity" and the "control of recombination." He also
developed the metagenetic concepts of orthogenetic evolution down
evolutionary blind alleys, the escape from sterility, the ideal
ancestor and two-track heredity. These ideas were built on
fundamental contributions to biological science, including the
verification of parasynapsis, the nature of chiasmata and the
cytological basis of complex hybridity and its de Vriesian
mutability. In order to test his hypothesis of the evolution of
complex hybridity in *Oenothera* as a pursuit of hybridity
mediated by natural selection favoring genetic heterozygotes in
chromosomally preadapted inbreeders. Darlington simulated the
process experimentally in *Campanula persicifolia* with
considerable success. When he heard that complex hybridity had
been discovered in the granite rock inhabiting *Isotoma petraea*,
Darlington wrote to my then supervisor, Spencer Smith-White,
expressing his confidence that the evolutionary mechanism could be
unequivocally determined in this species. This essay aims to show
our current position in understanding the patterns of population
divergence in *Isotoma* and to emphasize the strength and
validity of Darlingtonian interpretations.

ISOTOMA PETRAEA AND ITS PURSUIT OF HYBRIDITY

Isotoma petraea is an herbaceous perennial member of the
Lobeliaceae occurring on granite rocks and other rocky areas
throughout the Eremaean Province of Australia. Because of this
habitat preference, it exists in relatively small populations
which are isolated from each other so that genetic communication
between populations is very rare and consequently of profound
importance.

In addition to the effects of population structure, high levels
of inbreeding in the species are promoted by its substantially
autogamic breeding system. When the plants are vigorously growing
and in flower, the stigmas emerge from the syngenesious anther
tubes in the normal Lobeliaceous manner in high frequency. Cross
pollination may be affected in such flowers by native bees (e.g.
Lasioglossum sp.) which feed from the stigmatic surfaces.
However, as the season progresses, the frequency of protruding
stigmas declines so that autogamous self-pollination occurs within
the anther tube. The vast majority of seeds produced in natural
populations is the result of self-pollination.

Two types of populations exist in *Isotoma petraea*. Most
populations are composed of structurally homozygous plants

exhibiting 7 bivalents at meiosis (7II) but include a small
percentage (about 10%) of plants exhibiting small interchange
rings (Θ 4, 2 Θ 4, Θ 6), usually rings-of-4 (Θ 4). These have
been referred to as "floating interchange populations". In the
second type of population, every plant in the population is hybrid
for multiple interchanges. These complex hybrid populations are
strikingly uniform cytologically, although some variability does
exist within them. Thus Θ 12 and 2 Θ 6 plants both occur at
Elachbutting Rock; (Θ 8 + Θ 6) and (Θ 10 + Θ 4) plants occur at
Warrachuppin Hill. The complex hybrid populations are confined to
the south-west tip of the species distributional range and show a
more or less progressive increase in ring size from Θ 6 at Pigeon
Rocks to Θ 14 in the most south westerly populations, as at Mt.
Stirling.

Only two populations have been shown to combine complex hybrid
and floating interchange characteristics. These are the Pigeon
Rock Θ 6 population which is considered to be the source of
complex hybridity, and the 3-mile Rock Θ 10 population which is
considered to be in *statu nascendi* towards a stabilized
complex hybrid condition.

Allozyme analyses of structural homozygote populations shows
that there is a deficiency of heterozygotes at enzyme coding loci
in these populations. Rare alleles at these loci in structurally
homozygous populations occur more often as homozygotes than as
heterozygotes. This situation is most readily interpreted as
the consequence of inbreeding on neutral alleleic variation,
but we prefer to recognize the various allozymic homozygotes
as "natives" and "immigrants" according to their relative
frequencies, and the heterozygotes as interpopulational
hybrids and their derivatives. Preliminary analyses indicate
that the floating interchanges in otherwise structurally
homozygous populations are confined to the interpopulational
hybrids and their derivatives. This interpretation implies
that interpopulational hybrids are able to generate
interchanges and these are maintained in the population if
they are associated with heterosis. Heterotic segments may
be fortuitously associated with allozymic markers which may
themselves be neutral. The generation of interchanges, in
turn, implies that the native and immigrant components of the
structurally homozygous populations may be differentiated for
transpositions of internal chromosome segments.

In structurally homozygous *Isotoma petraea*, there
are seven bivalents at meiosis, each with two terminally
localized chiasmata. The effective recombination index is 7,
for the chiasmata are generally insignificant $_{-7}$
recombinationally. In selfed progeny, only 2 , i.e. one in
128 of the offspring will have hybridity levels equivalent to
that of the parent. However, with 700 to 1500 ovules in each
ovary, and with hundreds of flowers produced per plant each year,
the seed fecundity of *Isotoma petraea* is such that there is
a multiplicity, if not an abundance, of fully heterozygous seed

produced amongst the selfed progeny each season. In Θ 4 plants,
this yield is doubled (except for the slight reduction associated
with meiotic irregularity of the Θ 4), and this is proposed to be
the basis for the selective advantage of rings in these
cytologically polymorphic populations.

At Pigeon Rock, we believe, there was, effectively, an
association of two transpositions involving three chromosomes in a
hybrid enabling it to assemble ring-of-six (Θ 6) complex hybrids
in its selfed progeny. This hybrid is a complex hybrid generator.
The initial complex hybrid so generated carried duplications and
deletions in its complexes. The deletions acted as recessive
lethal genes and so assembled the zygotically acting balanced
lethal system. The duplications enhanced the synaptic
possibilities of the complexes when associated with other genomes
in interpopulational hybrids, and made them into complex hybrid
generators. In association with novel transpositions and
interchanges derived from the recipient genome, these subsequent
interpopulational hybrids were able to generate larger ringed
complex hybrids which trapped the new levels of segmental
hybridity associated with hybridization. Thus, the genetic
system, initiated on Pigeon Rock, migrated through populations to
the south west, pillaging the genetic resources of those
populations and creating stabilized hybrids with progressively
larger complexes.

Most notably, the distribution of genetic variation in complex
hybrid populations differs from that in structural homozygote
populations when assessed by electrophoretic analysis. Generally,
all the allozyme variability in the complex hybrid populations is
present as fixed hybridity, and the individual heterozygosity is
some 12.5 times that of the structurally homozygous populations.
Within the complex hybrid populations, however, there is some
genotypic variation at allozyme determining loci; heterozygotes
and homozygotes may be represented at particular loci in the one
complex hybrid population. This variation may well define native
and immigrant conditions amongst the components within such
populations, and/or different derivatives of the generative
interpopulation hybrids.

It appears that once allozyme hybridity has been established at
a particular locus, in the complex hybrid population system, it
remains there, but not necessarily forever. Thus, an immigrant
complex hybrid may establish, through interpopulational
hybridization, improved levels of heterotic interaction at
particular chromosomal segments which may be incorporated into the
new complex hybrid. While that chromosome segment may have been
flagged by allozyme heterozygosity in the immigrant, it may be
homozygous in the new (James *et al.* 1983; see Levin 1975;
Levy and Levin 1975), or *vice versa*.

Since there is a finite number of complex hybrid populations in
Isotoma petraea, it is conceivable that continued allozymic
analysis, combined with other "fingerprinting" procedures, may
eventually provide an exact description of the evolutionary

distribution of complex hybridity throughout this population
system.
     The interpretation of the evolution of complex hybridity in
Isotoma petraea given above clearly supports the
Darlingtonian contentions that the pursuit of hybridity is an
important directive influence in cytoevolution, and that the
control of recombination (in chiasma localization and in the
suppression of independent assortment through interchange
hybridity) is an important device in facilitating that pursuit.
Nevertheless, the model proposed is in fact, not the one favored
by Darlington. Darlington (1931) described three models for the
evolution of complex hybridity in *Oenothera* (and elsewhere).
Of these, he favored the first (Fig. 1a) which envisaged an
endogenous origin of a sequence of interchanges in a single
inbreeding chromosomally preadapted lineage. The other two models
were rejected on the grounds that they required novel
heterozygotes to be adaptively superior to genotypes honed by
natural selection. While Darlington's favored model may be true
for the evolution of Θ 6 complex hybrids in *Isotoma petraea*
on Pigeon Rock, the expansion of those complexes was achieved with
sequential interpopulational crosses as indicated in the second
model (Fig. 1b). This second model was favored by Catcheside
(1932) who also proposed the construction of the balanced lethal
systems in *Oenothera* via deficiencies and duplications. The
third model (Fig. 1c) is that favored by Cleland and American
cytogeneticists generally. It is rejected, for *Isotoma*, on
several grounds. Firstly, no fixation of new chromosome end
sequences in *Isotoma* structural homozygote populations has
been detected, while the primitive "*axillaris*" chromosome
end sequence has been shown to occur right across Australia and it
predates the divergence of *I. petraea, axillaris* and
anaethifolia from each other. The primitive
"*axillaris*" sequence occurs as one of the complexes in Θ 6
Pigeon Rock, and the total numbers of interchanges involved in the
construction of *Isotoma* complex hybrids tends to be little
more than that required for a minimal divergence from the
primitive sequence (James 1965, 1970). Secondly, complex
hybridity arose in *Isotoma* subsequent to the establishment
of high levels of autogamy, and indeed, in that population known
to be the most inbreeding (James 1970, and unpublished data),
rather than in outbreeding taxa and subsequently stabilized by the
adoption of autogamy (see Cleland 1960). Thirdly, meiotic
irregularity in the large ringed *Isotoma* complex hybrids is
such as to deny adaptive utility in relation to the conservation
of adapted haploid genomes, whereas it is sufficient for adaptive
utility in relation to the pursuit of hybridity (James 1970).
Fourthly, the sequential enlargement of rings from the Θ 6 Pigeon
Rock condition through intermediate sized rings to full rings-of-
14 is paralleled by progressive changes in other aspects of the
genetic system. In particular, the lethal systems in the complex
hybrids exhibit progressive improvement from relatively late

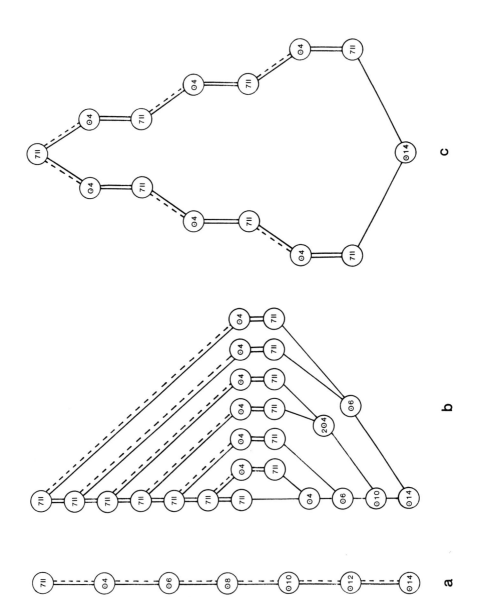

a

b

c

acting conditions in the Pigeon Rock Θ 6 to earlier acting situations in the large rings such that it becomes difficult to differentiate between postzygotic abortion and unfertilized meiotically sterilized ovules in large rings. Furthermore, the lethal systems began as balanced zygotic lethals (Beltran 1971, Beltran and James 1970) but during the course of the evolution of larger complexes, gametic lethals maintaining genetic interdependence between meiotically independent rings were assembled (James 1970) and we have recently obtained evidence that in some populations, e.g. Mt. Stevens Θ 14, gametic lethals have been assembled into balanced heterogamy as characterizing the α - β complexes in *Oenothera*. These observations, indeed, offer an alternative to the proposal by Steiner (1956, 1961) that in *Oenothera* the gametic lethal system was derived from the incompatibility system of outbreeding progenitors. This alternative becomes more acceptable when the zygotic lethal system of *Oenothera lamarkiana* is taken into account.

The model of the evolution of complex hybridity in *I. petraea* outlined above is based upon the assumption of transposition heterozygosity in key hybrids. Evidence supporting the assumption of transposition hybridity within the complexes of Θ 12 Bencubbin has been adduced from the curious epistatic ratios associated with the numbers of early and late death inclusions amongst the selfed seeds of the twin hybrids derived by crossing Θ 12 Bencubbin to an alethal structural homozygotes (Beltran and James 1970). The assumption is also supported by the occurrence of "sticky connections" associating otherwise nonhomologous chromosome segments at meiosis, especially in synthetic hybrids. But firm and unequivocal evidence of transposition heterozygosity, as may be possible through banding or *in situ* hybridization studies remains to be gathered. However, mechanisms to generate transpositions, including overlapping interchange mutations and, presumably, insertion sequence mediated transposition, would not appear to be wanting in *Isotoma*.

The model of the evolution of complex hybridity in *Isotoma petraea* outlined above also assumes the generally universal superiority of heterozygous genotypes. Evidence supporting that assumption is available. Firstly, it is clear that complex hybrids have displaced their structurally homozygous progenitors, and complex hybrids are 12.5 times as heterozygous at enzyme

---

FIG. 1. *Three possible methods of Θ 14 origin (after Darlington 1931). Solid connecting lines represent unchanged gametes; broken connecting lines represent new interchanges. Methods b and c depend upon outbreeding and hybridization between segmentally differentiated lines to produce progressively larger rings (in b) or a large ring at one stroke (in c). Method a depends upon the sequential fixation of interchanges in the heterozygous condition as a response to inbreeding within a single lineage.*

determining loci than structural homozygotes. Secondly,
interpopulational hybrids amongst structural homozygote
populations exhibit heterosis, while interpopulational hybrids
between complex hybrid populations, in which hybridity is already
conserved, exhibit negative heterosis (Beltran and James 1974).
Thirdly, we have recently shown that the level and quality of
hybridity in seed lots determines the outcome of competition in
mixed seed-bed competition experiments (Cohen and James 1983).
Thus, intrapopulational crosses in structurally homozygous
Isotoma populations provide the most competitive seed;
complex hybrid selfs are the next best, then come
interpopulational crosses from structurally homozygous
populations, and structural homozygote selfed seed is the least
competitive. The results of these competition experiments were
remarkably determinate, except that the interpopulational crosses
provided the least predictable seed and they could sometimes
outcompete the most competitive seed types. The interpopulational
hybrids have also been shown to survive longer in situations of
imposed lethal stress (drought).
     Whilst the detailed reasons for heterozygote superiority are
still unclear; it is, however, abundantly clear that heterozygotes
are superior in important biological attributes, and that the
pursuit of hybridity through a control of recombination provides
an overall summary of the processes accounting for the evolution
of complex hybridity in *Isotoma petraea*. This Darlingtonian
concept is a very powerful one, and it can be applied with
confidence to the interpretation of other cytoevolutionary
situations. See, for example, James (1981, 1982, 1984).

ACKNOWLEDGMENTS

I thank the Organizing Committee of this Conference for inviting
me to present this paper, the Australian Research Grants Scheme
and the University of Western Australia for supporting the
associated research, my students and colleagues, and the lovely
Fiona Webb for processing the words.

REFERENCES

Bell, G. 1982. *The Masterpiece of Nature*. Crom Helan,
     London.
Beltran, I.C. 1971. Embryology, balanced lethal systems and
     heterosis in *Isotoma petraea*. Ph.D. Thesis, Univ.
     Western Australia. 145 pp.
Beltran, I.C. and James, S.H. 1970. Complex hybridity in *Isotoma
     petraea* III. Lethal system in θ 12 Bencubbin. *Aust. J.
     Bot.* 18: 223-232.
Beltran, I.C. and James, S.H. 1974. Complex hybridity in *Isotoma
     petraea* IV. Heterosis in interpopulational hybrids. *Aust.
     J. Bot.* 22: 251-264.
Carson, H.L. 1981. Cytogenetics and the neo-Darwinian synthesis.

In *The Evolutionary Synthesis* (E. Magr and W.B. Provine eds.).
Harvard Univ. Press, Cambridge, Mass.
Catcheside, D.G. 1932. The chromosomes of a new haploid
Oenothera. *Cytologia* 4: 68–113.
Cleland, R.E. 1960. A case history of evolution. *Proc. Indiana
Acad. Sci.* 69: 51–64.
Cohen, N. and James, S.H. 1983. Complex hybridity in *Isotoma
petraea* VI. The effects of hybridity on competitive ability.
Aust. J. Bot. submitted.
Darlington, C.D. 1931. The cytological theory of inheritance in
Oenothera. *J. Genet.* 24: 405–474.
Darlington, C.D. 1932. *Recent Advances in Cytology*.
Churchill, London.
Darlington, C.D. 1939. *The Evolution of Genetic Systems*
Cambridge Univ. Press, Cambridge.
Darlington, C.D. 1981. The evolution of genetic systems:
Contributions of cytology to evolutionary theory. In *The
Evolutionary Synthesis* (E. Mayr and W.B. Provine, eds.).
Harvard Univ. Press, Cambridge, Mass.
Haldane, J.B.S. 1938. Forward. In *Recent Advances in
Cytology* by C.D. Darlington. 2nd Ed. Churchill, London.
James, S.H. 1965. Complex hybridity in *Isotoma petraea* I.
The occurrence of interchange heterozygosity, autogamy and a
balanced lethal system. *Heredity* 20: 341–353.
James, S.H. 1970. Complex hybridity in *Isotoma petraea* II.
Components and operation of a possible evolutionary mechanism.
Heredity 25: 53–77.
James, S.H. 1981. Cytoevolutionary patterns, genetic systems and
the phytogeography of Australia. In *Ecological Biogeography
of Australia* (A. Keast, ed.). pp. 763–782. Junk, The Hague.
James, S.H. 1982. Coadaptation of the genetic system and the
evolution of isolation among populations of Western Australian
native plants. In *Mechanisms of Speciation*. (C. Barigozgi,
ed.). Liss, New York.
James, S.H. 1984. Genetic systems in components of the Western
Australian sand plain flora. In *Biology of the Sand Plains*.
(J.S. Pate and J.S. Beard, eds.). In Press.
James, S.H., Wylie, A.P., Johnson, M.S., Carstairs, S.A. and
Simpson, G.A. 1983. Complex hybridity in *Isotoma petraea*
V. Allozyme variation and the pursuit of hybridity. *Heredity*
(submitted).
Levin, D.A. 1975. Genic heterozygosity and protein polymorphism
among local populations of *Oenothera biennis*. *Genetics*
79: 477–491.
Levy, M. and Levin, D.A. 1975. Genic heterozygosity and variation
in permanent translocation heterozygotes of the *Oenothera
biennis* complex. *Genetics* 79: 493–512.
Steiner, E. 1956. New aspects of the balanced lethal mechanism in
Oenothera. *Genetics* 41: 486–500.
Steiner, E. 1961. Incompatibility in the complex heterozygotes of
Oenothera. *Genetics* 46: 301–315.

The Role of Hybridization in the Evolution of *Bidens* on the Hawaiian Islands

Fred R. Ganders
Department of Botany
University of British Columbia
Vancouver, B.C., Canada

Kenneth M. Nagata
Harold L. Lyon Arboretum
Honolulu, Hawaii, U.S.A.

INTRODUCTION

Adaptive radiation in the animals and plants of the Galapagos Islands inspired Charles Darwin's thoughts on the evolution of species. Adaptive radiation on oceanic archipelagos still provides a model system for the experimental investigation of evolutionary divergence at the junction between microevolution and macroevolution. Rather few examples of adaptive radiation in plants have been studied biosystematically. In several cases, including *Bidens* (Gillett and Lim 1970; Gillet 1975) and *Scaevola* (Gillet 1972) in Hawaii and *Epilobium* (Raven and Raven 1976) in New Zealand, interspecific hybridization has been postulated as an important evolutionary mechanism in the adaptive radiation of flowering plants on oceanic islands.

We have been investigating the biosystematics and genetics of adaptive radiation in Hawaiian species of *Bidens*. Our results suggest a reinterpretation of the role of hybridization in the evolutionary divergence of *Bidens* on the Hawaiian islands.

ADAPTIVE RADIATION IN HAWAIIAN *BIDENS*

The Hawaiian islands are usually considered the most isolated archipelago on earth (Carlquist 1970). All are giant submarine volcanoes that arose from a hot spot beneath the Pacific Plate. As the Pacific Plate moved northwestward new volcanoes formed new islands and the old islands, severed from their volcanic roots, became extinct and were eroded to reefs and shoals. Of the present major islands, Kauai is the oldest with an age of 5.7 million years, while Hawaii is youngest, being less than 700,000 years old. The ancestors of all the indigenous terresterial species of plants and animals arrived by accidental long-distance dispersal. Many groups of organisms have undergone spectacular examples of adaptive radiation in the Hawaiian islands (Carlquist 1970). *Bidens* in the Hawaiian islands has evolved into a morphologically and ecologically diverse group of 19 species and 8 subspecies as we currently classify them (Ganders and Nagata 1983a). Hawaiian *Bidens* occur from sea level to over 2,200 m in elevation; in semi-desert with rainfall less than 0.3 m annually to rain forests and montane bogs with rainfall exceeding 7.0 m annually. All the evidence indicates that they evolved from a single ancestral immigrant species (Gillet and Lim 1970; Ganders and Nagata 1983b; Helenurm and Ganders 1983; Marchant *et al.* 1983).

Some of the range of morphological diversity in Hawaiian *Bidens* can be illustrated by brief descriptions of four of the species.

Bidens hillebrandiana is a species of coastal bluffs and rocky shorelines on Molokai, east Maui and northwest Hawaii in areas of moderately heavy rainfall. It is a low, spreading plant, less than 0.3 m tall with somewhat succulent, pinnately or bipinnately compound leaves with crenately-lobed leaflets. Relatively inconspicuous inflorescenses with several small flower heads terminate the main stem and lateral branches (Fig. 1). The achenes are straight, wingless, setose on the margins and faces, with barbed awns (Fig. 5).

Bidens mauiensis occurs from sea level to about 700 m elevation in dry to mesic habitats on Maui and Lanai. It also is low and spreading, less than 0.3 m tall. Leaf shape is variable, from simple and serrate to pinnately or bipinnately compound. Leaves tend to be dull grey-green in color. Flower heads are medium-large and solitary on terminal peduncles 40–150 mm long (Fig. 2). Achenes are straight, brown, glabrous, with broad, flat marginal wings. The awns are reduced to smooth teeth, decurrent into the wings (Fig. 5).

Bidens cosmoides is a shrub 1–3 m tall, endemic to very wet rain forests from 350 to over 1200 m in elevation on Kauai. Leaves are pinnately compound with 5–9 large leaflets. One to three flower heads terminate lateral branches only. The main stem of the plant is indeterminant in growth and never flowers. The heads are large and pendant. Disk flowers are orange with enormously-elongated styles that project 20 mm or more beyond the corollas (Fig. 3). Achenes are irregularly curved, or slightly twisted, and each is enveloped by its subtending receptacular chaffy bract. Some bristles and reduced awns may be present apically (Fig. 5).

Bidens menziesii occurs in dry habitats on Molokai, west Maui and Hawaii. On Hawaii it is especially common on cinder cones from 700–2200 m in elevation. The plants are shrubs 1–3 m tall with leaves bipinnately divided into long linear divisions less than 3 mm wide. Dense inflorescences of small flower heads terminate the main stem and branches (Fig. 4). Achenes are

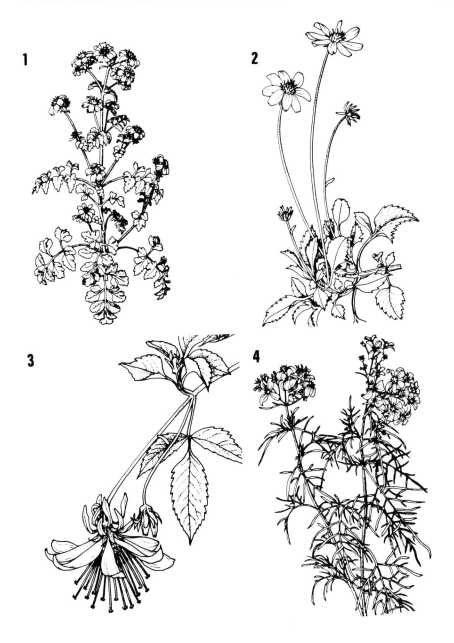

FIG. 1. Bidens hillebrandiana ssp. polycephala *from Maui.*
FIG. 2. Bidens mauiensis *from Maui.* FIG. 3. Bidens
cosmoides *from Kauai.* FIG. 4. Bidens menziesii ssp.
filiformis *from Hawaii.*

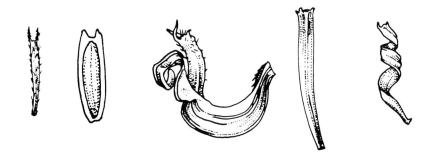

FIG. 5. *Achenes of some Hawaiian species of* Bidens.
From left to right: B. hillebrandiana ssp. polycephala, B.
mauiensis, B. cosmoides, B. menziesii ssp. filiformis, B. torta.

straight, glabrous, wingless and usually awnless (Fig. 5).

Additional novelties that have evolved in other species of
Hawaiian *Bidens* include completely fused inner involucular
bracts that split irregularly at anthesis in *B. amplectens,*
and spirally coiled achenes in *B. torta* (Fig. 5) and some
other species.

The Hawaiian species of *Bidens* exhibit a greater range of
morphological and ecological diversity than the rest of the genus
on five continents. This adaptive radiation in morphology and
ecological tolerance does not involve differentiation at isozyme
loci. The different species of Hawaiian *Bidens* are as
similar genetically as are populations of a single species in most
plants (Helenurm and Ganders 1983). This suggests that the
species are a monophylletic group and that evolutionary divergence
has proceeded much more rapidly at loci controlling morphological
characters than at loci controlling isozymes. Adaptive radiation
in morphological and ecological preferences without isozyme
differentiation is an evolutionary pattern also exhibited by the
land-snail genus *Partula* on Tahiti and Moorea (Johnson *et al.*
1977), and may be typical of adaptive radiation on oceanic
islands.

CROSSABILITY OF HAWAIIAN *BIDENS*

Gillett (1975) obtained fertile hybrids among 11 of the 19 species
and 2 of the subspecies of Hawaiian *Bidens*. We have extended
this crossing program so that hybrids have now been obtained
involving all Hawaiian species and all subspecies except *B.
hillebrandiana* ssp. *hillebrandiana, B. micrantha* ssp. *kalealaha*
and *B. campylotheca* ssp. *waihoiensis* (Table I). All hybrids
that have flowered so far are completely pollen and seed fertile.
All taxa that have been counted have the same chromosome number,
$2n = 72$ (Gillett and Lim 1970; Gillett 1975; Gerald Straley,
unpublished).

Gillett and Lim (1970) were unable to cross *B. cosmoides* with other Hawaiian species and Gillet (1975) concluded that there had been two separate introductions of *Bidens* into the Hawaiian islands, one that gave rise to *B. cosmoides,* and one that evolved into all of the other species. We found that *B. cosmoides* would cross freely with the other species. This is strong evidence that all evolved from a single ancestral species. And it is only necessary to assume one successful immigration of the genus into the Hawaiian islands (Ganders and Nagata 1983b).

TAXONOMY OF HAWAIIAN *BIDENS*

Sherff (1937) recognized 43 species and more than 20 varieties and forms of *Bidens* endemic to the Hawaiian islands. Sherff had little concept of the amount of variation that occurs within individual populations of Hawaiian *Bidens*, nor even of the amount of intra-plant variation. He erected many species on the basis of leaf characters which are extremely variable within and among populations, and are rather unreliable as taxonomic characters in this group. At the other extreme, Gillett (1975) suggested that all Hawaiian *Bidens* should be included in only two species, one with perhaps seven subspecies, because of their genetic compatability and natural hybridization.

All Hawaiian *Bidens* appear to be interfertile so that interfertility cannot be used as a criterion for species delimitation, unless one is willing to consider them all a single species which exhibits greater morphological diversity than the rest of the genus. Biosystematics in the last forty years has clearly shown that the taxonomic category "species" consists of different types of populations in plants with different life forms or mating systems (Grant 1971). Different criteria must be used to define species in different groups, but the criteria should be consistent within higher taxa with similar kinds of species.

Evolutionary divergence of taxa with unique genetically controlled morphological characters, differing habitat requirements, and distinctive geographical ranges has occurred in *Bidens* in the Hawaiian islands without the evolution of gametic or postzygotic isolating mechanisms. We have therefore used a species concept based on morphological characters and ecogeographical considerations. To identify lineages on these criteria, several correlated characters must be used. Any single character or pair of characters might be correlated fortuitously in some populations, and such correlation may change from generation to generation. The taxa must also have distinctive geographical ranges. If two similar taxa are sympatric over most of their ranges or have both been collected in most known localities, they are likely to represent extreme variants of a species variable in a few characters. For example, five species, one variety and one form recognized by Sherff (1937) represent leaf-shape variants of *B. sandvicensis* ssp. *sandvicensis.*

TABLE I. *Experimental hybrids produced in Hawaiian Bidens*

	B	C	D	E	F	G	H	I	J	K	L	M	N	O	P	Q	R	S	T	U	V	W	X	Y*
A*	X	X		X		X	X	X	X	X	X	X	X	X	X	X	X	X	X	X	X			X
B				X		X	X	X	X	X	X	X	X	X	X	X	X	X	X	X	X	X		
C									X					X	X					X	X	X		
D				X			X		X			X	X	X	X		X	X	X					
E						X	X	X	X	X	X	X	X	X	X	X	X	X	X	X	X			X
F																X	X	X						
G							X	X	X	X	X	X	X	X	X	X	X	X	X	X	X			X
H								X	B	X	X	B	X	B	X	B	B	B	B	G	X	X		
I										X	X	X	X	X	X	X	X	X	X	X	X			X
J										X	B	B	X	B	X	X	X	B	B	B	B			X
K											G	G	X	B	X	G	B	X	G	G	X		G	X
L												X	X	B	X	X	X	X	X	X	X		G	X
M													G	G	B	G	B	X	X	X	X		G	X
N*														B	X	G	X	X	X	X	X		G	X

	B	C	D	E	F	G	H	I	J	K	L	M	N	O	P	Q	R	S	T	U	V	W	X	Y*
O														X		G		B		B	B	X	G	X
P															X	X	X	X	G	X	X	X	X	X
Q																					G		G	
R																	X	G	B	X	X	X	X	X
S																		X	X	X	X	X	X	X
T																			X	X		X	X	
U																				X		X	X	
V																						X	X	

*Parental taxa: A) *B. amplectens*; B) *B. asymmetrica*; C) *B. campylotheca* ssp. *campylotheca*; D) *B. campylotheca* ssp. *pentamera*; E) *B. cervicata*; F) *B. conjuncta*; G) *B. cosmoides*; H) *B. forbesii* ssp. *forbesii*; I) *B. forbesii* ssp. *kahiliensis*; J) *B. hawaiensis*; K) *B. hillebrandiana* ssp. *polycephala*; L) *B. macrocarpa*; M) *B. mauiensis*; N) *B. menziesii* ssp. *menziesii*; O) *B. menziesii* ssp. *filiformis*; P) *B. micrantha* ssp. *micrantha*; Q) *B. micrantha* ssp. *ctenophylla*; R) *B. molokaiensis*; S) *B. populifolia*; T) *B. sandvicensis* ssp. *sandvicensis*; U) *B. sandvicensis* ssp. *confusa*; V) *B. torta*; W) *B. valida* X *B. wiebkei*; Y) an F$_1$ hybrid *B. valida* X *B. molokaiensis*.
Hybrids produced by: X = Ganders; B = Ganders and Gillett; G = Gillett.

The polyacetylenes accumulated in the roots and leaves have
been identified in all Hawaiian taxa except *B. hillebrandiana*
ssp. *hillebrandiana* and *B. campylotheca* ssp. *waihoiensis*
(Marchant *et al.* 1983). All except two pairs of taxa can be
distinguished by their arrays of root and leaf polyacetylenes, but
only one compound is unique to a single taxon. Even
morphologically similar subspecies can be distinguished by
polyacetylene chemistry. *Bidens sandvicensis* ssp.
sandvicensis does not accumulate polyacetylenes in the leaves,
but ssp. *confusa* does. No variation occurred within taxa except
in *B. torta* in which each of four populations sampled had a
unique array of polyacetylenes. Polyacetylene chemistry supports
the taxa recognized on the basis of morphology and ecogeography.
The polyacetylenes also support the monophyletic origin of
Hawaiian *Bidens*, because the rather limited array of
polyacetylenes can be derived biosynthetically from a single
precursor compound, oleic acid. And with one exception, the
individual compounds are shared widely among the taxa (Marchant
et al. 1983).

NATURAL HYBRIDIZATION IN HAWAIIAN *BIDENS*

Because all indigenous Hawaiian species of *Bidens* are
capable of hybridization it is not surprising that natural
hybridization has been reported. Sherff (1937) and Degener (1946)
reported several cases of putative interspecific hybrids. Gillett
and Lim (1970) refuted three of the putative hybrids reported by
Sherff (1937), and we agree with the interpretation of Gillett and
Lim in these cases. Gillett and Lim (1970) accepted 16 putative
natural interspecific hybrids, and concluded that ten of the
species recognized by Sherff were actually interspecific hybrids.
Gillett (1975) added another putative natural hybrid to the list
and reinterpreted the parentage of one of his previously accepted
hybrids.
    Six of the natural interspecific hybrids postulated by Gillett
and Lim (1970) involved parental species endemic to different
islands. In four of these cases the supposed hybrid occurred on a
different island than either putative parent species. Gillett
(1975) assumed that the parental species had previously occurred
sympatrically on the same island as the hybrid derivative, but one
or both had subsequently become extinct. Although not impossible,
this seems highly unlikely without strong evidence.
    Sherff (1937) relied heavily on unreliable leaf characters in
describing species. Gillett and Lim (1970) were able to produce
experimental interpsecific hybrids with leaves resembling species
described by Sherff, and they seemed to have relied heavily on
leaf characters in equating Sherff's species to putative
interspecific hybrid combinations. In many cases they were
unwarranted in assigning hybrid status to Hawaiian taxa of *Bidens*,
ignoring both the intrapopulational variation in leaf morphology
characteristic of many species of *Bidens* and the floral and fruit

characters distinguishing the species. Their criteria for
deciding which populations represented species and which
represented interspecific hybrids also seem to have been somewhat
arbitrary. It appears that those populations used in their
crossing program were usually regarded as parental species and
populations and taxa that were not used were regarded as
interspecific hybrids.

We have resynthesized several of the hybrids Gillett produced
or reported from nature, including *B. hawaiensis* X *B.
sandvicensis* ssp. *sandvicensis* (Gillett used the synonyms *B.
skottsbergii* and *B. coarctata,* respectively). The hybrid is
intermediate between the parental species in outer involucral
bract length, peduncle length, number of ray flowers and size of
heads. However, the simple leaves and glabrous achenes of *B.
hawaiensis* are dominant in the F_1 over the pinnately compound
leaves and setose achenes of *B. sandvicensis*. Gillett and Lim
(1970) cited specimens of this putative hybrid combination that
are in fact three different taxa, *B. campylotheca* ssp.
campylotheca, *B. micrantha* ssp. *kalealaha* (which Gillett
called *B. distans*), and *B. micrantha* ssp. *micrantha*.
Bidens campylotheca is not intermediate in any character
(having glabrous twisted achenes, distinctive heads and
inflorescences and very long, lax peduncles and large outer
involucral bracts). The two subspecies of *B. micrantha* have
smaller outer involucral bracts than either putative parent and
their heads are as small or smaller than either parent and have
five or fewer ray flowers. Achenes are glabrous but the gene or
genes for glabrous achenes are not homologous with those of *B.
hawaiensis* since they are recessive rather than dominant in
interspecific crosses with setose-fruited species. Leaf shape is
intermediate but variable in both taxa. Gillett, however,
obviously based his determination of hybrids primarily on leaf
characters.

Of the 17 natural interspecific hybrid combinations recognized
by Gillett, we accept only three as genuine hybrids. Six are leaf
form variants of variable species that show no convincing evidence
of interspecific hybridization. Eight are taxa that Gillett did
not recognize as well as some misidentifications. However, we
have found evidence of six additional genuine or highly probable
natural interspecific hybrids not reported by Gillett. Of the
total of nine interspecific natural hybrids, we regard six as
certain and three as highly probable. Our determinations of
Gillett's putative hybrids and documentation of genuine hybrids
are detailed below. All specimens are at BISH unless indicated
otherwise.

1. *B. fecunda* X *B. torta*: *B. fecunda* is a synonym of the Oahu
   population of *B. cervicata* and the specimens cited by Gillett
   and Lim are *B. cervicata*.

2. *B. amplectens* X *B. torta*: Hybrid swarms between these two
   species occupy a relatively large area of the northwest tip of
   the Waianae Range on Oahu from the head of the Makua Valley to

Kaena Point. The two parental species are morphologically very different and hybrids can be easily recognized.

3. *B. amplectens* × *B. waianensis*: *B. waianensis* is a synonym of *B. torta* so this reputed hybrid combination is the same as the above.

4. *B. mauiensis* var. *cuneatoides* × *B. menziesii*: *B. mauiensis* var. *cuneatoides* is the simple-leaved form of *B. mauiensis*. The putative hybrids cited by Gillett and Lim (1970) are the pinnately and bipinnately compound leaved form of *B. mauiensis*. Synthetic hybrids have heads intermediate in size and ray flowers number on branched inflorescences, achenes with slightly developed wings, and an intermediate growth habit. The specimens cited by Gillett show no evidence of these intermediate characters. However, herbarium specimens, such as *Degener 12405 b&c* (UC) do seem to be genuine interspecific hybrids of this parentage. These specimens are from west Maui, but hybrids appear to be rare.

5. *B. torta* × *B. waianensis*: *B. waianensis* is a synonym of *B. torta* so that these plants are *B. torta*.

6. *B. forbesii* × *B. menziesii*: Gillett regarded *B. waianensis*, the bipinnately compound leafed form of *B. torta*, as a hybrid of this combination. These plants as well as *B. fulvescens* are just part of the range of leaf shape found in *B. torta*. They are typical of *B. torta* in all of their characters.

7. *B. hillebrandiana* × *B. mauiensis* var. *cuneatoides*: This hybrid combination was reported by Sherff (1973) but the specimens were determined as typical *B. mauiensis* by Gillett. Other specimens considered to be this hybrid by Gillett and Lim (1970) and also *B. mauiensis*. *B. hillebrandiana* and *B. mauiensis* both occur on Maui but are not sympatric.

8. *B. hillebrandiana* × *B. wiebkei*: The putative hybrids of this combination cited by Gillett and Lim (1970) are *B. sandvicensis* ssp. *sandvicensis*. *B. sandvicensis* does not occur on the same islands as either putative parent.

9. *B. coartata* × *B. molokaiensis*: *B. coartata* is a synonym for *B. sandvicensis*. Specimens cited as hybrids of this parentage by Gillett and Lim (1970) include *B. campylotheca* ssp. *campylotheca* (*Cowen 889*), *B. asymmetrica* (*Caum s.n.*), *B. pilosa* (*Stone 2795*), *B. populifolia* (*Degener et al. 2514, 4114, 20875*), and the rest are *B. macrocarpa*. Gillett considered hybrids of this combination to be the species described by Sherff as *B. populifolia*. We regard *B. populifolia* as a distinct species.

10. *B. coartata* × *B. skottsbergii*: *B. coartata* is a synonym of *B. sandvicensis*. *B. skottsbergii* is a synonym of *B. hawaiensis*. The hybrids postulated by Gillett and Lim (1970) occur on different islands from either parent. *Munro 505* is *B. campylotheca* ssp. *campylotheca*. The other specimens from Lanai are *B. micrantha* ssp. *kalealaha*. The specimens from west Maui are *B. micrantha* ssp. *micrantha*.

11. *B. forbesii* × *B. menziesii*: There is no evidence for

hybrids of this parentage. The specimens cited by Gillet and
Lim (1970) are all *B. sandvicensis* ssp. *sandvicensis* except
Gillett 1888 and *Degenger 23868, 27189* which are *B.
sandvicensis* ssp. *confusa, Degenger 27183* which is *B.
cervicata,* and *Degener 21488* which is probably *B.
cervicata* but could be *B. sandvicensis* (the specimen was too
young to determine).

12. *B. mauiensis* var. *cuneatoides* X *B. molokaiensis: The small
population of B. molokaiensis* that occurs near the summit
of Diamond Head in Honolulu has previously been called *B.
cuneata.* Gillett (1975) postulated that *B. cuneata* was a
hybrid between *B. mauiensis* and *B. molokaiensis* despite
the fact that one parent occurs on Molokai, the other on Maui
and the presumed hybrid on Oahu. He stated that achene wings
were inherited maternally in this cross and since *B. cuneata*
has wingless achenes, the female parent must have been *B.
molokaiensis.* In other interspecific crosses involving
B. mauiensis achene wings were not inherited maternally
and hybrids were intermediate with slightly developed wings
(Gillett and Lim 1970). We have examined Gillett's vouchers
of his hybrids and conclude that all except one were
accidental self-fertilizations. One voucher at BISH has
intermediate achenes with narrow wings and is a genuine
hybrid; its achenes do not resemble those of the Diamond Head
population and there is no evidence that this population of
B. molokaiensis involves hybridization with *B.
mauiensis.*

13. *B. fulvescens* X *B. macrocarpa: B. fulvescens* is a synonym of
B. torta. Specimens cited as hybrids of this parentage by
Gillett and Lim (1970) include *B. sandvicensis* ssp.
*sandvicensis (Degener et al. 3529), B. asymmetrica (St.
John 13024, Degener et al. 4133, 10066)* and the rest are
B. campylotheca ssp. *campylotheca.*

14. *B. macrocarpa* X *B. wiebkei:* The specimens cited by
Gillett and Lim (1970) as hybrids of this parentage occur on
west Maui while one putative parent is on Oahu and the other
on Molokai. The specimens cited are actually *B. conjuncta
(Forbes 468M, Degener & Weibke 2164, 2178)* or *B. micrantha*
ssp. *micrantha (Forbes 2363M, Rock 8144, Degener 22032).*

15. *B. menziesii* ssp. *menziesii* X *B. wiebkei:* The hybrids cited
by Gillett and Lim (1970) from Molokai are just forms of *B.
menziesii* ssp. *menziesii. Munro 122* from Lanai is *B.
micrantha* ssp. *kalealaha.*

16. *B. ctenophylla* X *B. menziesii* var. *filiformis: B. ctenophylla*
is a synonym of *B. micrantha* ssp. *ctenophylla.* This
hybrid was reported by Gillett and Lim (1970) and investigated
by Mensch and Gillett (1972), who documented natural
hybridization and hybrid swarms between the two species near
Puu Waa Waa, Hawaii. Later, Gillett (1975) revised his
opinion of this situation concluding that *B. ctenophylla*
itself was a hybrid or hybrid segregate from hybridization

between *B. menziesii* ssp. *filiformis* and *B. skottsbergii* (a synonym of *B. hawaiensis*). He also considered *B. micrantha* and *B. campylotheca* to be segregates from this hybridization. We consider the original interpretation of Mensch and Gillett (1972) to be correct. There is no good evidence that *B. hawaiensis* has been involved in these hybrids.

There are some anomalous specimens of *B. micrantha* ssp. *micrantha* from west Maui that may be hybrids with *B. menziesii* ssp. *menziesii,* including *Forbes 2269M* from Olawalu Valley and *Hobdy 811, 812* from Lihau Peak (HLA). *B. menziesii* ssp. *menziesii* does occur at these localities.

17. *B. hillebrandiana* X *B. menziesii:* Plants from Waihoi Valley, east Maui, considered to be hybrids of this parentage by Gillet (1975) are *B. campylotheca* ssp. *waihoiensis.*

Of the preceding 17 putative interspecific hybrids there is evidence that only three are genuine: *B. amplectens* X *B. torta, B. mauiensis* X *B. menziesii,* and *B. micrantha* X *B. menziesii.* In addition, we have seen two additional certain natural hybrids, two probable hybrids and two possible cases of interpsecific hybridization.

1. *B. asymmetrica* X *B. sandvicensis* ssp. *sandvicensis:* Hybridization between these two species is highly probable where their ranges abut in the southeastern Koolau Range, Oahu, on the ridges north of Honolulu. Evidence includes twisted achenes on plants in populations we regard as *B. sandivicensis* from Waahila Ridge and Nuuanu Pali Overlook and sparsely setose achenes in populations we regard as *B. asymmetrica* on the Manoa Cliffs trial on Tantalus. The type specimens of *B. gracilis* Nutt., which we regard as a synonym of *B. asymmetrica* following Sherff (1937), shows evidence of hybridization with *B. sandvicensis.*

2. *B. cervicata* X *B. forbesii* ssp. *forbesii:* These two taxa appear to intergrade along the Kalalua Valley Trail on the Napali Coast of Kauai and hybridization seems possible. However, no definite hybrid plants have yet been documented.

3. *B. conjuncta* X *B. micrantha* ssp. *micrantha:* Some specimens that possibly represent hybrids of this combination have been seen from high elevations on west Maui, including *Hobdy 916* (HLA) from 1000 m elevation in the Iao Valley.

4. *B. mauiensis* X *B. micrantha* ssp. *micrantha:* Hybrids of this parentage have been collected from west Maui and Lanai. They have intermediate achenes which are slightly winged apically, long pedunculate heads on inflorescences with few branches, and the heads and outer involucular bracts are intermediate in size between the two parental species. The Maui plants have simple leaves and are from the region where *B. micrantha* and

sometimes *B. mauiensis* have simple leaves. The Lanai
specimen has bipinnately compound leaves as do most plants of
B. mauiensis on Lanai. *B. micrantha* has pinnately compound
leaves. The following specimens are hybrids of this parentage:
Munro 536 (Lanai), *Davis & Silva 21, 49, Degener 21943,
Nagata, Kimura, & Hobdy 1905,* Forbes *2318N* (all from
West Maui).
5. *B. hillebrandiana* ssp. *polycephala* X *B. molokaiensis:*
This hybrid was reported to be growing with both of its parents
by Degener (1946) and we have seen recent collections from the
north shore of Molokai that undoubtedly represent this hybrid
or a recombinant of it. It was not discussed by Gillett and
Lim (1970).
6. *B. cervicata* X *B. sandvicensis* ssp. *sandvicensis:* Plants
that are probably hybrids of this parentage have been seen on
Ohikilolo Ridge in the northwest Waianae Range, Oahu.

INTERSPECIFIC ISOLATING MECHANISMS IN HAWAIIAN *BIDENS*

With 19 interfertile species of *Bidens* in the Hawaiian islands
there are 171 possible interspecific hybrid combinations. We have
found evidence for possibly nine different interspecific hybrids
and conclude that natural hybridization is a relatively rare
phenomenon in Hawaiian *Bidens.* An analysis of the distribution
of the species reveals the reason for this. One hundred and two,
or 60%, of the possible parental combinations involve species that
are allopatric on different islands. Forty-three, or 25%, involve
parents that occur on the same island, but are allopatric on
different mountain ranges or occur at different ends of the
island. Thus, 85% of the possible interspecific hybrid
combinations are prevented by geographical isolation. The
remaining 26 possible parental combinations involve species that
have parapatric or partly sympatric distributions. Of these,
seven combinations involve species that occur at different
elevations and/or in habitats that differ in precipitation, and no
evidence of hybridization between these parents has been found.
Five combinations (3%) involve species that differ in flowering
time or pollinators and no evidence of hybridization has been
found. No evidence has been found for five other combinations
(3%), although no reason for the isolation is presently known and
only nine of the interspecific hybrid combinations (5%) are
presently known or suspected.
    Allopatry is the most important isolating mechanism in the
Hawaiian species of *Bidens,* but ecological isolation, mainly
by altitude, is a mechanism of secondary importance. Differences
in pollintors probably prevent interspecific hybridization between
the bird pollinated species, *B. cosmoides,* and sympatric or
parapatric insect pollinated species on Kauai. Hybridization
between *B. valida* and *B. forbesii* ssp. *kahiliensis*which
are sympatric on Mt. Kahili, Kauai, seems to be prevented by
seasonal isolation. *Bidens valida* flowers in the fall and

B. forbesii ssp. *kahiliensis* flowers in winter.

SPECIATION IN HAWAIIAN *BIDENS*

Clues to the speciation process in Hawaiian *Bidens* are provided
by an analysis of the distribution of subspecies within a species
and of closely related species pairs. There are four cases where
different subspecies of the same species occur on the same island,
two on Kauai and two on Maui. In one case the subspecies are
allopatric and occupy different mountain ranges: *B. micrantha*
ssp. *micrantha* occurs on the west Maui mountains and *B. menziesii*
ssp. *kalealaha* occurs on Haleakala on east Maui. *Bidens forbesii*
ssp. *forbesii* is relatively widespread in northern and central
Kauai from sea level to 460 m in elevation. *Bidens forbesii*
ssp. *kahiliensis* is presently known from above 600 m
elevation on two nearby mountain peaks, Kahili and Kapalaoa,
at the southern edge of the range of ssp. *forbesii*. *Bidens*
sandvicensis ssp. *sandvicensis* occurs at elevations
below 600 m from Waimea Canyon across the southern part of Kauai and
northward on the eastern side of the island. Subspecies
confusa is confined to the west side of Waimea Canyon from
670 to 1050 m in elevation. On Maui, *B. campylotheca* ssp.
pentamera occurs from 1000 to 2100 m elevation in foggy rain
forests on Haleakala. Subspecies *waihoiensis* is restricted
to the Waihoi Valley between 850 and 1000 m elevation. Subspecies
pentamera also occurs in the upper reaches of Waihoi Valley
at higher elevations than ssp. *waihoiensis*. In all three of
these cases one of the subspecies has a restricted range at the
margin of the range of the more widespread subspecies, and the
rarer subspecies occurs at a different elevation. Subspeciation
has involved marginal populations that have invaded different
altitudinal zones.

The same pattern is exhibited by pairs of closely related
species. *Bidens conjuncta* is most similar to *B. micrantha*
ssp. *micrantha* and both occur on the mountains of west Maui,
but *B. conjuncta* is restricted to high elevation bogs and
rain forests near the summit of Mt. Eke. *Bidens micrantha*
usually occurs at lower elevations. Here speciation has probably
involved marginal populations invading new habitats at a higher
elevation. On Kauai *B. cervicata* is closely related to
B. forbesii and is parapatric in distribution, occurring
west of the distribution of *B. forbesii* in drier habitats.

In four cases different subspecies of a species occur on
different islands. *Bidens menziesii* ssp. *menziesii* occurs
on Molokai and Maui and ssp. *filiformis* occurs on Hawaii,
B. hillebrandiana ssp. *polycephala* occurs on Molokai and Maui,
while ssp. *hillebrandiana* occurs on Hawaii, *B. micrantha* ssp.
kalealaha occur on Maui while ssp. *ctenophylla* occurs on Hawaii,
B. campylotheca ssp. *campylotheca* occurs on Oahu, Lanai and
Hawaii while ssp. *pentamera* and *waihoiensis* occur on Maui.
In these species, subspeciation has followed colonization of a

different, usually younger island.

Speciation has also occurred following colonization of younger islands. *Bidens molokaiensis* and *B. mauiensis* are closely related species that share the derived characters of low growing, spreading habit and solitary heads on long peduncles. *Bidens mauiensis* is most advanced with brown, glabrous, winged achenes and occurs on the younger islands of Maui and Lanai while *B. molokaiensis* occurs on Oahu and Molokai.

Successful inter-island dispersal and colonization by Hawaiian *Bidens* has been rare. Of the 27 taxa, 19 are endemic to a single island, six are found on two adjacent islands, and only two taxa occur on three islands. Of the eight taxa occurring on more than one island, more than half are very local on one of the islands they inhabit. *Bidens cervicata* is found on Oahu only on the sides of one valley, *B. mauiensis* occurs in only one gorge on Lanai, *B. menziesii* ssp. *menziesii* is found in only one canyon on Maui, and *B. molokaiensis* occurs only on Diamond Head on Oahu. In most cases colonization of neighboring islands has resulted in evolutionary and taxonomic differentiation. However, endemic taxa occur only on the five largest islands and the percentage of endemic species is correlated with the age and size of the island. Kauai, the oldest island, has the highest percentage of endemic *Bidens* species, but Oahu, which is larger and has two mountain ranges, has the greatest number of endemic species.

Four factors may be responsible for the remarkable divergence in *Bidens* on the Hawaiian islands. One factor is the chronologically and geographically linear sequence of islands in the archipelago. A second is the diversity of habitats, particularly differences in elevation and rainfall, on each island. Third, *Bidens* populations are usually relatively isolated from each other on different ridges or mountain peaks separated by deep valleys in which *Bidens* rarely occurs. Fourth is the rarity of dispersal. Without effective native animal vectors, dispersal appears to have been very local.

In Hawaiian *Bidens* speciation and the beginnings of macroevolutionary change largely involve morphological and ecological characters, some of which are obviously adaptive. Phytochemical differentiation in polyacetylenes has also occurred but to a lesser extent. Genetic differentiation at loci controlling isozymes has not been significant. The pattern of speciation is consistent with natural selection and genetic drift in founder populations as the mechanism for evolutionary change. Initial geographical or ecological isolation, often involving single marginal or immigrant populations, has been necessary to initiate cladogenesis. But the volution of intrinsic, gametic or chromosomal isolating mechanisms has been unnecessary. Hybridization does occur but appears to be of little significance in the evolutionary divergence of this group. Gradual genetic divergence in allopatric populations has been the most important evolutionary phenomenon.

In summary, the 19 species and eight subspecies of *Bidens*
endemic to the Hawaiian Islands evolved from a single ancestral
species. Morphologically they are more diverse than the
Bidens on any continent, but all Hawaiian species are
interfertile. Contrary to Gillett's conclusion, however, natural
hybridization is relatively rare. Good evidence exists for only
5% of the possible hybrid combinations. Eighty-five percent of
the possible hybrid combinations are prevented by geographical
isolation and 8% by ecological or seasonal isolation. Rather than
hybridization, the most important evolutionary phenomenon has been
genetic divergence in marginal populations that have invaded
different habitats or colonized adjacent, younger islands.
Factors promoting adaptive radiation are the linear chain of
islands, habitat diversity on each island, isolation of
appropriate habitats, and the rarity of dispersal.

REFERENCES

Carlquist, S. 1970. *Hawaii, A Natural History*. *Natural History
Press, N.Y.*
Degener, O. 1946. *Flora Hawaiiensis*. Published by the author,
Honolulu.
Ganders, F.R. and Nagata, K.M. 1983a. New taxa and new
combinations in Hawaiian *Bidens* (Asteraceae). *Lyonia,* in pr.
Ganders, F.R. and Nagata, K.M. 1983b. Relationships and floral
biology of *Bidens cosmoides* (Asteraceae). *Lyonia,* in press.
Gillett, G.W. 1972. The role of hybridization in the evolution of
the Hawaiian flora. In *Taxonomy, Phytogeography and Evolution*
(D.H. Valentine, ed.), Academic Press, London.
Gillett, G.W. 1975. The diversity and history of Polynesian
Bidens section Campylotheca. *Univ. Hawaii, H.L. Lyon Abor.
Lect.* 6: 1-32.
Gillett, G.W. and Lim, E.K.S. 1970. An experimental study of the
genus *Bidens* in the Hawaiian Islands. *Univ. Calif. Publ. Bot.*
56: 1-63.
Grant, V. 1971. *Plant Speciation*. Columbia Univ. Press, N.Y.
Helenurm, K. and Ganders, F.R. 1983. Adaptive radiation and
genetic differentiation in Hawaiian *Bidens*. *Evolution,* in press.
Johnson, M.S. Clarke, B. and Murray, J. 1977. Genetic variation
and reproductive isolation in *Partula*. *Evolution* 31: 116-126.
Marchant, Y.Y., Ganders, F.R., Wat, C.-K. and Towers, G.H.N.
1983. Polyacetylenes in Hawaiian *Bidens* (Asteraceae).
Biochem. Syst. Ecol. in press.
Mensch, J.A. and Gillett, G.W. 1972. The experimental
verification of natural hybridization between two taxa of
Hawaiian *Bidens* (Asteraceae). *Brittonia* 24: 57-70.
Raven, P.H. and Raven, T.E. 1976. The genus *Epilobium*
(Onagraceae) in Australasia: a systematic and evolutionary
study. *N.Z. Dept. Sci. Ind. Res. Bull.* 216: 1-321.
Sherff, E.E. 1937. The genus *Bidens*. *Field Mus. Nat. Hist.
Bot. Ser.* 16: 1-721.

Hybridization in the Domesticated-Weed-Wild Complex

Ernest Small
Biosystematics Research Institute
Agriculture Canada
Ottawa, Ontario, Canada

INTRODUCTION

In the perhaps 11,000 years of agriculture (Harlan, 1969), plants have undergone dramatic coevolution with man as domesticates and as weeds. Domesticates have been modified to serve man, a complete intergradation existing between plants barely altered from their wild progenitors, to forms no longer capable of independent survival. The degree of weediness also spans a broad spectrum. There are many weedy plants without close domesticated relatives, although the majority of major crops are accompanied by conspecific weeds. Just what a weed is has been controversial (Baker, 1974; Holzner, 1982) but Harlan's (1970) definition, "a race or species adapted to habitats disturbed by man or his activities," is useful for this paper. While some may have existed in their present form prior to agriculture, it has been suggested that most modern weeds evolved under landscape disturbance caused by civilizing man (Harlan and De Wet, 1965). The third component of interest is also designated by a somewhat ambiguous label, "wild." Weeds would seem to be wild, but the word wild will be employed in its more restrictive sense to designate plants unaltered by man's activities. Unfortunately, as noted by De Candolle (1886), the first great student of the domesticated-weed-wild complex, it is often very difficult to distinguish genuinely wild ancestors of domesticates from their feral descendants.

The evolutionary association of weeds and domesticates has received much study since the pioneering work of Vavilov (e.g. 1951). There are now available excellent reviews of the nature and evolutionary dynamics of the relationships between weeds, domesticates, and their wild progenitors (e.g. Harlan, 1969, 1975; Baker, 1972; Harlan *et al.* 1973; De Wet and Harlan, 1975; De Wet, 1981; Pickersgill, 1977, 1981), and the popularity

of the subject will be evident in the contributions to this
symposium of Drs. Brandenburg and Zohary. Hybridization of course
represents only one of the major evolutionary forces; mutation,
selection and genetic drift are also usually of critical
relevance. In the limited space available, however, it will be
difficult to more than sketch the importance of hybridization. I
will include under this term the combination or transfer of
germplasm, irrespective of its quantity. Except for the review of
evolutionary pathways of hybridization, examples chosen are from
personal research.

GOUTWEED, EXEMPLIFYING IMPORTANCE OF MUTATION AND SELECTION

A very simple but salutary illustration of evolution in the
domesticated and weedy man-made habitat is provided by the
ornamental *Aegopodium podagraria,* a European perennial
frequently cultivated in a variegated form having pale leaves with
white margins. The whiteness is due to a periclinal chimera,
which results in an absence of chlorophyll in the pallisade spongy
layer of cells of the leaves, and in all of the cells of the leaf
margins. The normal form has about a 50% higher net
photosynthetic rate, and is much more vigorous (Small, 1973).
Normal branches often arise, and as the plant propagates
vegetatively, comparatively aggressive all-green "back-mutations"
frequently come to dominate a planting. In this example, a
domesticate arose by a mutation that was artificially selected by
man, and a "weedy mimetic" was similarly generated spontaneously,
but was selected naturally. Although the genetic basis of this
domesticated-weed-wild complex is exceptionally simple, it does
demonstrate that such complexes can evolve and function without
hybridization. Indeed, it is well to keep in mind the opinion of
most students of the crop-weed complex, that mutation and
selection are more important than hybridization. However,
hybridization does play an extensive role in many groups, as will
be demonstrated.

EVOLUTIONARY PATHWAYS OF HYBRIDIZATION

The frequency of amphidiploids bears witness to the importance of
this result of hybridization, although it is much less clear how
significant is transfer of smaller quantities of germplasm than
whole genomes, in part because of difficulty in distinguishing
hybridization products from the results of other processes
(Heiser, 1953). Much of our present understanding is derived from
research on the Gramineae, particularly the cereals (e.g. Harlan
et al., 1973; De Wet, 1975), and there is need for study of
other groups. In the following, a brief review of the variety of
hybrid pathways by which germplasm may flow between, and indeed
within, the phases of the domesticated-weed-wild complex, is
presented.

Possible pathways of sexual transfer of germplasm from wild ancestors into either weeds or domesticates are shown in Fig. 1. The germplasm may vary in quantity from an allele to a genome; and the transfer may result from a single or of multiple hybridizations; and could occur once or repetitively. At a given stage one can define a set of plants that constitute the original wild ancestors, and the domesticates and/or weeds derived from them (termed "unsupplemented" for this review) by pathways D1 and W1. Further addition/substitution of germplasm may then occur from three different sources, to produce "supplemented" domesticates and/or weeds. By pathways D2 and W2 the sources are other domesticates or weeds, respectively; by pathways D3 and W3 the sources are "supplementary wild ancestors;" and by D4 and W4, as a result of germplasm combination or transfer between domesticates and weeds, a new form may arise suited to a cultivated or weedy niche, or the domesticate or weed may be altered by introgression. A few examples from the literature follow.

Pathway D1 is of venerable age, Linnaeus proposing that crop plants originated by hybridization (Roberts, 1929). In cases such as the cultivated tobaccos *Nicotiana tabacum* and *N. rustica,* allopolyploids unknown in the wild except as escapes (Gurstel, 1976), it is possible that the domesticate came into existence by hybridization *per se.* By contrast, there is good evidence that the cultivated tetraploid New World cottons *Gossypium hirsutum* and *G. barbadense* combined wild genomes of Old and New World origin prior to domestication (Pickersgill and Heiser, 1977). The parallel pathway W1, producing weeds rather than domesticates, is illustrated by the well known derivation of *Spartina anglica,* a European mudbinder that is weedy in recreational shorelands. It originated from two basically non weedy species, the American *S. maritima* and the European *S. alterniflora* (Marchant, 1968).

Pathways D2 and W2 respectively denote the evolution by hybridization of domesticates from preexisting domesticates, and of weeds from preexisting weeds. At least two of the cultivated potatoes seem to have arisen as hybrids between other cultivated potatoes (Hawkes, 1981): *Solanum chaupa (3x)* from *S. stenotomum (2x)* × *S. tuberosum* subsp. *andigena (4x);* and *S. curtilobum (5x)* from *S. tuberosum* subsp. *andigena* × *S. juzepczukii (3x).* Ownbey (1950) described two tetraploid weeds, *Tragopogon miscellus* and *T. mirus,* that arose by hybridization between three weedy diploids, *T. dubius, T. porrifolius,* and *T. pratensis.* The new species went on to establish ranges in the Pacific Northwest of the United States (Hitchcock and Cronquist, 1973).

D3 and W3 respectively denote the evolution by hybridization of domesticates from the combination of a preexisting domesticate and a wild plant; and of a weed from a preexisting weed and a wild plant. Hexaploid wheat, *Triticum aestivum* (genome ABD) probably originated by hybridization between cultivated tetraploid wheat (*T. turgidum,* genome AB) and wild diploid *T. tauschi* (genome D) (Feldman, 1976). Pathway W3 is illustrated by Heiser's

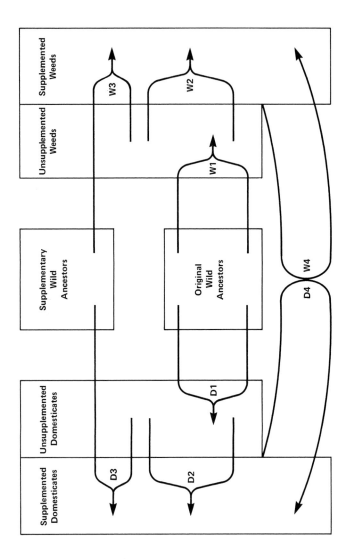

Fig. 1. *Hybrid pathways for origin of domesticates and weeds from wild plants.*

(e.g. 1965) studies of *Helianthus annuus,* demonstrating that
weedy sunflower considerably expanded its geographical and
adaptive ranges by absorbing germplasm from a variety of wild
species of *Helianthus.*

D4 and W4 denote the products of hybridization between weeds
and domesticates, pathways that are often of comparatively recent
occurrence. If there is appreciable divergence between the
domesticate and weedy plants involved, the hybrid is
characteristically inferior in adaptation to the weedy habitat,
and cannot survive there. However, it often finds a niche in
cultivated fields, being maintained deliberately by man (D4) or
surviving by mimicing the parental crop (W4). Once again,
domesticated potatoes are exemplary (Hawkes, 1981): *Solanum
stenotomum (2x,* cult.) × *S. sparsipilum (2x,* wild) produced *S.
tuberosum* subsp. *andigena* (cult., *4x); S. stenotomum* × *S. acaule
(4x,* wild) produced *S. juzepczukii (3x,* cult.); and *S.
stenotomum* × *S. megistacrolobum (2x,* wild) produced *S. ajanhuiri
(2x,* cult.). Illustrating W4, the Bolivian weed potato *S.
sucrense* seems to have arisen from *S. tuberosum* subsp. *andigena
(4x,* cult.) × *S. oplocense (4x,* wild) (Astley and Hawkes, 1979).
Similar examples of genome amalgamation producing weeds may be
found in *Sorghum* (De Wet, 1978), *Beta* (O'Connor and Grogan,
1974), and *Raphanus* (Panetsos and Baker, 1967).

Introgression between domesticated and weed forms, a variant of
pathway W4, may be quite extensive and indeed may result in
considerable transfer of germplasm, although introgression is
typically understood as implying restricted gene movement. Well
documented examples of introgression are between maize and
teosinte (Wilkes, 1967, 1970; cf. Iltis and Doebley 1984), weedy
and domesticated sorghums in Africa (De Wet, 1978), and
domesticated tomatoes and wild *Lycopersicon
pimpinnellifolium* in South America (Rick, 1958).

CARROT: SUBSTANTIAL ADAPTIVE AND SELECTIVE BARIERS TO GERMPLASM
TRANSFER

The *Daucus carota* complex comprises several intergrading
wild, weedy, and domesticated forms (Small, 1978a), all of which
are perfectly interfertile. The domesticated carrot variant with
orange storage organs, familiar in Western culture, provides an
example of extreme differentiation between a domesticate and its
companion weed, despite an absence of intrinsic breeding barriers.
In contrast with the weed, the domesticate possesses relatively
brittle, palatable, pigmented, larger, unbranched storage organs,
fewer but larger relatively erect leaves, strict biennialness,
larger fruits, comparatively rapid growth, relatively foliose
involucral bracts, fewer central purple flowers in the umbels,
smaller petals, fewer fruit spines, brittle mericarp spine hooks,
basally branched primary mericarp bristles, and relatively
efficient water use. The first 11 differences may be interpreted
as the results of directional selection by man, and the last seven

as changes from lack of stabilizing selection for characters of value in the wild (Small, 1978a; Small and Desjardins, 1978). Where wild carrot is common in Canada, it has been found to be very difficult to produce uncontaminated commercial carrot seed because of considerable spontaneous natural crossing. Wild-type carrots regularly make an appearance in carrot fields, but are rogued away prior to seed production. In North America, there is no evidence of introgression of domesticated characters into weedy plants, which are very homogeneous.

The extreme differentiation of the familiar domesticated form, its inability to perpetuate as an escape, and the absence of its traits in the weed suggest that domesticated features are simply too inadaptive to wild existence for transferred genes to survive. Natural selection in the wild coupled with strong artificial selection in cultivation seem to effectively block gene exchange.

On the other hand, this situation may not hold for a less familiar form of domesticated carrot, once prevalent in Asia, but now rarely encountered outside of Afghanistan (Small, 1978a). The "Eastern Carrot" differs from the familiar orange-rooted kind in a number of respects, most noticeably in the purple color of the storage organ. Exceptionally primitive, or perhaps more accurately degenerate variants of this have been collected from cultivation on the Indian subcontinent. Local wild/weedy variants share some of the morphological traits of the Eastern Carrot. Whether hybridization has played a role in obscuring the separation of domesticated and wild carrot in southern Asia is presently unclear. The situtation is provocative, however. The semi-selected or degenerate forms maintained in cultivation by primitive agriculturists could act as a bridge facilitating gene transfer between domesticated and wild/weedy forms.

ALFALFA: MODERATE DOMESTICATED-WILD DIVERGENCE, EPISODIC HYBRIDIZATION

Hybridization has played an especially extensive role in alfalfa, the world's oldest and most important forage crop. Alfalfa is the domesticated form of *Medicago sativa,* a highly polymorphic Old World perennial. Diploid wild forms are native to northern Eurasia. Tetraploids are sympatric with these, but occur over a much wider area of Eurasia. The domesticate is tetraploid, except for some experimentally produced lines. It seems to have arisen in Turkey, the Caucasus, and Iran during prehistoric times dating back to the 5th millenium B.C. when agriculture was evolving and the horse was proving to be a major implement of warfare (Lesins and Lesins, 1979). Tetraploids may have been selected by the early agriculturists, or may simply have arisen earlier spontaneously.

Much of the complexity of alfalfa is due to hybridization between two entities, which will be denoted here as the "species" *M. sativa s. str.* and *M. falcata.* These are readily identified, the former possessing purple flowers and highly coiled pods, the

latter with yellow flowers and straight to curved pods. The two
are completely interfertile, and are regarded by many taxonomists
as simply components of *M. sativa s.l.* (Small, 1981c; Small and
Brookes, 1983). Both occur as diploids and tetraploids, and both
have been domesticated. They are quite differnt ecologically,
physiologically and agronomically (Small and Brookes, 1982; Small
and Lefkovitch, 1982), but have become widely intermingled
genetically and geographically, both as wild and as cultivated
plants. In the wild, hybrid swarms occur at both levels of ploidy
(Small and Bauchan, 1983). But the matter does not end with
transient hybrid swarms, as it seems to so often in other
hybridizing groups (Heiser, 1953). Stabilized hybrid derivatives
are pervasive, at both the diploid and tetraploid levels. Not
surprisingly, this has led to preposterous taxonomic inflation,
dozens of stabilized hybrid forms being labelled as species.

The *Medicago sativa* complex is an example of a polyploid
complex, but not one that conforms to the stereotyped conception
of such. As noted by Zohary (1970), in such well-analyzed
polyploid complexes as *Dactylis, Triticum-Aegilops,* and
blue-stem grasses *(Andropogon),* all Gramineae, there are
marked differences in the genetic systems of the diploids and
polyploids. The diploids or so-called "diploid pillars" are
ecologically and morphologically well defined and are separated by
very strong breeding barriers. But the superimposed "polyploid
superstructure" exhibits substantially continuous variation, the
result of comparatively free combining of the separate diploid
gene pools. The relatively free tolerance to gene exchange at the
polyploid level has been explained as a consequence of genetic
buffering. In the *M. sativa* complex, by contrast, there appears
to be not only considerable hybridization and introgression
between the polyploid variants, but equally free exchange of
germplasm between the diploids. Pollen sterility is about 30% in
both diploids and tetraploids, with no significant difference
between the pure species and the hybrid derivatives (Small, 1983),
suggesting that wide-ranging hybridization is reasonably well
tolerated. This is to be expected in a perennial that is adapted
to outcrossing, and exhibits rapid deterioration when inbred.

During recent field studies in northeastern Turkey (Small,
1981b, 1982) I made a particular effort to check for possible
introgression between wild and domesticated plants. In the wild,
diploid and tetraploid hybrid swarms between *M. falcata* and
M. sativa were frequent, as well as a variety of stabilized
recombinants at both chromosome levels, and even some ploidally
mixed populations (Small and Bauchan, 1983). It seemed probable,
therefore, that evidence of hybridization between alfalfa
plantations and roadside weeds would be equally forthcoming.
There is only one cultivar grown in quantity in Turkey, an ancient
form known as 'Kayseri', and occasionally its characteristics
could be discerned in roadside weeds, suggesting that at least a
small amount of gene transfer was taking place from the cultivated
to weedy plants. As 'Kayseri' and indeed all alfalfa in Turkey is

purple-flowered, introgression into alfalfa plantings from yellow-flowered or hybrid *sativa-falcata* weeds growing nearby would have been easily detected. In Turkey, much of the roadside weed comprises hybrid forms, which seem to be more successful in this niche than the parental species. In hundreds of situations with yellow- or yellowish-flowered weeds growing beside the purple-flowered cultigen, however, there was no hint of introgression into the cultigen. It may be, however, that the modern monocultural plantation is a poor model for evaluating past exchange between wild and cultivated plants, and germplasm transfer might be facilitated by small, scattered plantings, as seem to have occurred in past times. Indeed, Jones *et al.* (1971) observed that crossing between alfalfa plots increases notably as plot size diminishes.

The observation that apparently no transfer of wild *falcata* germplasm has occurred into the cultivated *M. sativa* of Turkey stands in sharp distinction to the accepted historical record concerning alfalfa in other parts of the world. Infiltration of wild *falcata* genes into cultivated alfalfa in Germany and northern France during the 16th century is commonly thought to have provided the winter hardiness that allowed expansion of the crop into northern areas of the world, and the tolerance to relatively acidic soils that led to considerable expansion of the range of the crop in North America (Bolten *et al.*, 1972).

Adaptive divergence between the cultigen and wild/weed form in alfalfa is much less than in carrot, so it is not surprising that there has been movement of germplasm between the phases. Alfalfa domesticates have larger leaf blades and a more erect habit than wild plants, and often differ physiologically, with a concentration of the root system in the cultivatable layer of soil, rapid growth, and in some forms adaptation to irrigation. However, divergence is only moderate, and the domesticate is perfectly able to perpetuate in the wild, and often escapes from cultivation.

HOP: MINIMAL DOMESTICATED–WILD DIVERGENCE, EXTENSIVE GERMPLASM TRANSFER

Hops are the fruit of *Humulus lupulus*, employed to flavor beer. Wild hops are widely distributed in the North Temperate World (Small, 1978b). Four quite distinctive allopatric taxa occur, respectively in Europe (var. *lupulus*), Japan (var. *cordifolius*), the Cordillera of Western North America (var. *neomexicanus*) and the Midwest of the U.S. (var. *pubescens*). The European variety has been widely introduced in eastern North America, and a fifth, comparatively poorly delimited taxon (var. *lupuloides*) in eastern North America may be the result of hybridization between the European introduction and plants of the indigenous American taxa.

Of the various locations where wild hops are native, the use of

hops in brewing was established in pre-Columbian times only in
Europe, reliable evidence for cultivation dating to 9th century
Bavaria (Wilson, 1975). European cultivars are essentially
indistinguishable morphologically from wild European hop (Small,
1980, 1981a).

In the New World, hop growing began in New Netherlands (New
York) in 1629, but did not become important until about 1800
(Schwartz, 1973). The industry first thrived in New England and
New York (where var. *lupuloides* occurs), and after the Civil
War, in Wisconsin (where var. *pubescens* is found). Hop
cultivation began in the Pacific Coast states (where var.
neomexicanus is distributed) in the 1860's, this region soon
becoming the leading hop-producing area of North America. In the
east, the hop louse, mildew, and prohibition led to the demise of
hop growing after about 1920. The historical record reveals that
imported European cultivars were grown in North America. American
beer differs in aroma and flavor from European beer, in large part
due to the hops used, indicating that the original imported hops
cultivated in the early years have been altered. Indeed, the
morphology of present-day American cultivars is about midway
between that of the European variety, and the Cordilleran and
Eastern indigenous North American varieties, showing that there
has been a combining of domesticated European and wild American
germplasm. There is no evidence that the Midwestern U.S. var.
pubescens, the most geographically restricted taxon, has
played a role in the evolution of extant American beer cultivars,
but ornamental hop vines, like the beer cultivars, were also
originally imported from England, and the ornamental vines of the
American Midwest often exhibit morphology intermediate between the
European var. *lupulus* and the local var. *pubescens.* Thus the
cultivated European import in North America has managed to absorb
germplasm of all the New World's principal variants.

In Japan, a parallel situation prevails. Hops were not used in
any appreciable way in Japan before the present century. A
governmental hop garden was established as early as 1877 (Ehara,
1955), but it was not until after the Second World War that
interest in hops for brewing became substantial. Most of the
imported hops appear to have been of European origin, some of
American. Not unexpectedly, present Japanese cultivars are
intermediate in morphology between the distinctive native Japanese
variety and the European variety, once again a substantial
transfer of germplasm from the wild into the domesticated having
occurred.

Wild and domesticated hops (a separate weedy phase is not
recognizable) are quite similar, and it is not surprising that
germplasm transfer has taken place readily. Hops are propagated
vegetatively by rhizome cuttings, and so particular cultivars are
often identifiable by minor morphological peculiarities maintained
by vegetative propagation. However, no consistent morphological
differences have been found between the wild and cultivated phases
(at least in Europe). Prominent epidermal glands on the fruit

bracts secrete aliphatic acids and other resinous constituents of
brewing interest, and there has been disruptive selection for
resin content, the domesticate being much more productive.
Because the European cultigen is indistinguishable from the
European wild plant, genic infiltration has not been detected.
Undoubtedly such has occurred, since outside of Europe the
European cultigen has been infiltrated by quite divergent
germplasm pools. Hop cultivars suffer from a variety of pests and
physiological diseases, and of course as vegetatively propagated
crops have limited capacity to evolve adaptations. Hop growers
have been quite unaware of the likely beneficial influx of wild
genes. Indeed, just how beneficial are the effects of crop-weed
gene transfer is a topic deserving of considerable future
research.

Using a figure of five years for the life expectancy of hop
plantings, there have been perhaps less than 40 generations in
North America and less than ten in Japan since hop cultivation
began in earnest. Since hops are vegetatively propagated, it
might seem surprising that so much genetic infiltration has
occurred. Hops are dioecious, and the staminate plants are
removed as soon as they can be recognized in hop yards, since they
contribute no fruits, and indeed seedless hops are often desired.
But in fact even in "seedless" hop yards, one frequently
encounters seeds and seedlings, male plants managing to make an
appearance in or near to the yards. Much of the genetic
"contamination" by wild plants seems to have been due to
unscrupulous suppliers of the rhizome cuttings used by hop
planters in the past, the indistinguishable and more easily
obtained wild plants often being sold (Davis, 1957). The
"contaminated" product obviously has proven acceptable in North
America and Japan. One wonders whether other food plants and
other cultures have acquired distinctive regional tastes as a
result of local hybridization.

HEMP: CONTINUOUS GERMPLASM TRANSFER BETWEEN DOMESTICATE AND WEED

Cannabis sativa exemplifies a class of domesticate-weed
complex in which despite notable divergence, there is no
significant adaptive discontinuity between weed and domesticate
(for this and references for the following information on
Cannabis, see Small, 1979). Such plants often seem to have
developed a curious love-hate relationship over the years with
man, sometimes existing as a weed, and at other times as a
cultigen. As is well known, *Cannabis* was once a respected
crop in North America, although presently efforts are being made
to eradicate it, both as a crop and a weed. Hemp is the most
ancient of the fibre crops, dating to at least 5,000 B.C., and is
the classical example of a camp-follower, whose attributes have
been selected for close coexistence with man. Other crops that
have managed to alternate existence as a weed during periods of
unpopularity include comfrey *(Symphytum* spp.), African rice *(Oryza*

glaberrima), Bermuda grass *(Cynodon dactylon),* and sweetclover *(Melilotus* spp.) (cf. Harlan, 1982).

Cannabis has been disruptively selected for at least three products, and the three forms of domesticate are modally distinguishable. Hemp, grown for the textile fibre in the phloem, is generally tall, hollow-stemmed, with long internodes and relatively few, large leaves. All of these characteristics can be interpreted as maximizing yield or obtainability of the fibre. Normally dioecious, many modern fibre cultivars have been selected for monoecy (a compromise between the desirable fibre characters of the males and the greater vigor of the females). Plants domesticated for the edible, drying oil in the achenes generally are shorter, highly branched, with small internodes, relatively solid stems, and numerous achenes. Although oil cultivars do not have larger achenes, or achenes with more oil, they do produce a larger quantity of achenes than fibre and drug forms of *Cannabis.* "Marihuana" strains, selected for their narcotic properties, are similar in habit to oil cultivars, but differ in the qualitative composition of the resin secreted by the epidermal glands covering the shoots, possessing much more intoxicating material than do oil and fibre strains.

As with many plants with wild and domesticated phases, there are notable morphological and physiological differences in the propagules. Wild achenes have a well-developed basal abscission area; an elongated base (often serving as an "eliasome", i.e. an oil-containing source attracting insects which serve as a dispersal vector); a perianth-derived, marbled outer covering (serving as camouflage); smaller size; a harder, more protective wall; and dormancy and comparative longevity. These features are due to disruptive selection pressures between cultivated and uncultivated plants, and similar variation occurs in many domesticate-weed complexes. However, in *Cannabis* there is no "separation", but rather an unbroken continuum of variation between the weed and the domesticates. One frequently encounters cultivated plants with wild features, and vice versa. Domesticated plants easily escape cultivation and take up wild existence. Study of plants in the Ottawa area, derived from escaped fibre plantings, has indicated that the propagule adaptations of wild plants are reacquired in less than half a century. But even without these adaptatations, escaped cultigens thrive as weeds.

Occasionally one encounters "marker characters" in a crop or weed that show up in the opposite phase, and so allow the inference that genetic transfer has occurred. For example, the release of purple-leaved varieties of rice in India led to the appearance of purple-leaved weed rice, and the release of a durum variety of wheat with unique characters for Israel led to the appearance of these characters in local wild emmer (Harlan, 1965, 1969). In *Cannabis,* there is a striking geographical distribution of chemical races that is suggestive of domesticate-weed germplasm exchange. At least three races have been noted.

In Europe and north-temperate North America, hemp cultivars
historically grown, and weedy populations, are low in intoxicant
ability by virtue of having little of the intoxicant chemical
delta-9-THC, but considerable nonintoxicating CBD. Conversely,
"narcotic" cultivars of southern Asia and weedy races there and in
Central and South America where narcotic cultivars have been
grown, exhibit the reverse chemical profile. And in Japan,
strains grown for fibre, and weedy strains, tend to have trace
amounts of another chemical, CBGM, unlike strains from other
regions.

Evaluating the contribution of hybridization to germplasm
transfer between weed and domesticate in a plant such as
Cannabis is confounded by an apparent lability to selection.
As noted earlier, atavistic recovery of wild achene characters
seems to occur rapidly in escapes. And despite disruptive
selection for fibre, oil, and drug, most plants (including the
weed) yield appreciable quantities of each of these components,
and so have often been employed and selected for each. What
appear to be cultivated characters in domesticated plants, and
vice versa, could simply be a stage in artificial selection or in
readaptation to the wild, respectively. However, as a wind-
pollinated plant producing prodigious quantities of pollen (a
single plant can yield more than a million grains), and lacking
any breeding barriers, there would seem little doubt that there is
widespread cross-pollination between domesticate and weed, and
subsequent gene exchange. Such may be both advantageous and
disadvantageous. In personal studies of wide-ranging hybrids
between diverse kinds of *Cannabis*, including crosses between
weeds and cultivars, I was impressed by the invariable hybrid
vigor of the F_1's. However, hemp farmers at the zenith of the
cultivation of fibre-type *Cannabis* in North America during
the late 19th century seem to have suffered from influx of weedy
pollen. Hemp quality decreased each year seed was generated in
areas where weed hemp was common, and the practice developed of
importing hemp seed from weed-free areas every several years.

EXTINCTION BY HYBRIDIZATION

There has been considerable debate concerning the possibility of
the existence of wild as opposed to weed hemp, and its possible
indigenous range. I suspect that dispersal of the weed and crop
by man and resulting hybridization during the last five millenia
has almost certainly obliterated *C. sativa* in its form prior
to the dawn of agriculture. The possibility of extinction of the
wild progenitors of crops and weeds by hybridization has been
raised before. Mangelsdorf *et al.* (1964) hypothesized that
"wild maize" may have been extirpated by contamination from the
cultigen (others, such as De Wet and Harlan, 1972, believe that
teosinte is wild maize). Eshbaugh (1976) postulated that the
pepper *Capsicum frutescens* represents a wild ancestral
species being hybridized out of existence by domesticated

C. chinense. Purseglove (1965) suggested that the progenitors of date palm (*Phoenix dactylifera*) and sweet pea (*Lathyrus odoratus*) may also have been obliterated by hybridization.

CONCLUSIONS

In this brief review based on a selected sample of outcrossing domesticate-weed-wild complexes, it does appear that at least several considerations determine the extent of germplasm transfer. When there is radical divergence between crop and weed, artificial and natural selection may be expected to effectively nullify transfer. At the other extreme, with limited divergence and a genetic system favoring wide crossing, it is possible for considerable genetic continuity to be maintained between crop and weed through hybridization. In widespread, ancient, widely-crossing groups, it would seem unlikely that wild progenitors would not be contaminated genetically, if indeed not exterminated. In the origin of many crops, hybridization has clearly played a critical role in transferring germplasm to the cultigens from wild and weedy relatives; however, evidence of transfer of <u>adaptive</u> germplasm from domesticates to weeds and wild plants is much harder to demonstrate, and does not seem to be of general occurrence. Finally, it may be noted that the cultural practices of man have often been responsible for the evolutionary history of hybridization in weeds and domesticates.

REFERENCES

Astley, D. and Hawkes, J.G. 1979. The nature of the Bolivian weed potato species *Solanum sucrense* Hawkes. *Euphytica* 28: 685-696.

Baker, H.G. 1972. Human influences on plant evolution. *Econ. Bot.* 26: 32-43.

Baker, H.G. 1974. The evolution of weeds. *Ann. Rev. Ecol. Syst.* 5: 1-24.

Bolton, J.L., Goplen, B.P. and Baenziger, H. 1972. World distribution and historical developments. In Alfalfa Science and Technology (C.H. Hanson, ed.), pp. 1-34, Am. Soc. Agron., Madison, Wisc.

Davis, E.L. 1957. Morphological complexes in hops (*Humulus lupulus* L.) with special reference to the American race. *Ann. Mo. Bot. Gard.* 44: 271-294.

De Candolle, A. 1886. *Origin of Cultivated Plants* (reprint, 2nd ed., 1959), Hafner, New York.

De Wet, J.M.J. 1975. Evolutionary dynamics of cereal domestication. *Bull. Torrey Bot. Club* 102: 307-312.

De Wet, J.M.J. 1978. Systematics and evolution of *Sorghum* sect. *Sorghum* (Gramineae). *Am. J. Bot.* 65: 477-484.

De Wet, J.M.J. 1981. Species concepts and systematics of domesticated cereals. *Kulturpflanze* 29: 177-198.

De Wet, J.M.J. and Harlan, J.R. 1972. Origin of maize: the

tripartite hypothesis. *Euphytica* 21: 271-279.
De Wet, J.M.J. and Harlan, J.R. 1975. Weeds and domesticates:
  evolution in the man-made habitat. *Econ. Bot.* 29: 99-107.
Ehara K. 1955. Comparative morphological studies on the hop
  (*Humulus lupulus* L.) and the Japanese hop (*H. japonicus*
  Sieb. et Zucc.). I. *J. Fac. Agric. Kyushu Univ.* 10: 209-
  232 + 5 pl.
Eshbaugh, W.H. 1976. Genetic and biochemical systematic studies
  of chili peppers (*Capsicum* - Solanaceae). *Bull. Torrey Bot. Club*
  102: 396-403.
Feldman, M. 1976. Wheats. In *Evolution of Crop Plants*
  (N.W. Simmonds, ed.), pp. 120-128, Longman, London.
Gurstel, D.U. 1976. Tobacco. In *Evolution of Crop Plants*
  (N.W. Simmonds, ed.), pp. 273-277, Longman, London.
Harlan, J.R. 1965. The possible role of weedy races in the
  evolution of cultivated plants. *Euphytica* 14: 173-176.
Harlan, J.R. 1969. Evolutionary dynamics of plant domestication.
  Jpn. J. Genet. 44(Suppl. 1): 337-343.
Harlan, J.R. 1970. Evolution of cultivated plants. In *Genetic
  Resources in Plants - their exploration and conservation*
  (O.H. Frankel and E. Bennett, eds.), pp. 7-32, Blackwell,
  Oxford.
Harlan, J.R. 1975. *Crops and Man*, Am. Soc. Agron., Madison, Wisc.
Harlan, J.R. 1982. Relationships between weeds and crops. In
  Biology and Ecology of Weeds (W. Holzner and N. Numata,
  eds.), pp. 91-96, W. Junk, The Hague.
Harlan, J.R. and De Wet, J.M.J. 1965. Some thoughts about weeds.
  Econ. Bot. 19: 16-24.
Harlan, J.R., De Wet, J.M.J. and Price, E.G. 1973. Comparative
  evolution of cereals. *Evolution* 27: 311-325.
Hawkes, J.G. 1981. Biosystematic studies of cultivated plants as
  an aid to breeding research and plant breeding. *Kulturpflanze*
  29: 327-335.
Heiser, C.B. Jr. 1953. Introgression re-examined. *Bot. Rev.*
  39: 347-366.
Heiser, C.B. Jr. 1965. Sunflowers, weeds and cultivated plants.
  In *The Genetics of Colonizing Species* (H.G. Baker and
  G.L. Stebbins, Jr., eds.), pp. 391-401, Academic Press, New
  York.
Hitchcock, G.L. and Cronquist, A. 1973. *Flora of the Pacific
  Northwest,* Univ. Wash. Press, Seattle.
Holzner, W. 1982. Concepts, categories and characteristics of
  weeds. In *Biology and Ecology of Weeds* (W. Holzner and
  N. Numata, eds.), pp. 3-20, W. Junk, The Hague.
Iltis, H.H. and Doebley, J.F. 1984. *Zea* - A biosystematic
  odyssey. In *Plant Biosystematics* (W.F. Grant, ed.).
  Academic Press, Toronto.
Jones, L.G., Feather, J.T., Marble, V.L. and Ball, R.B. 1971.
  Barriers to outcrossing in alfalfa seed production. *Calif.
  Agric.* 23: 10-12.
Lesins, K.A. and Lesins, I. 1979. *Genus* Medicago

(Leguminosae). *A taxogenetic study,* W. Junk, The Hague.

Manglesdorf, P.C., MacNeish, R.S. and Willey, G.C. 1964. Origins of agriculture in Middle America. In *Handbook of Middle American Indians* (R. Wauchope, ed.), Vol. 1, pp. 427–445, Univ. Texas Press, Austin.

Marchant, C.J. 1968. Evolution in *Spartina* (Gramineae). II. Chromosomes, basic relationships and the problem of *S.* X *townsendii* agg. *J. Linn. Soc.* (Bot.) 60: 381–409.

O'Conner, L.J. and Grogan, V. 1974. 'Annual' bolters or 'weed beet': a new weed problem? *Biotas* 28: 171–173.

Ownbey, M. 1950. Natural hybridization and amphiploidy in the genus *Tragopogon. Am. J. Bot.* 37: 385–499.

Panetsos, C.A. and Baker, H.G. 1967. The origin of variation in "wild" *Raphanus sativus* (Cruciferae) in California. *Genetica* 38: 243–274.

Pickersgill, B. 1977. Taxonomy and the origin and evolution of cultivated plants in the New World. *Nature* 268: 591–595.

Pickersgill, B. 1981. Biosystematics of crop–weed complexes. *Kulturpflanze* 29: 377–388.

Pickersgill, B. and Heiser, C.B., Jr. 1977. Origins and distribution of plants domesticated in the New World Tropics. In *Origins of Agriculture* (C.A. Reed, ed.), pp. 803–835, Mouton, The Hague.

Purseglove, J.W. 1965. The spread of tropical crops. In *The Genetics of Colonizing Species* (H.G. Baker and G.L. Stebbins, Jr., eds.), pp. 375–386, Academic Press, New York.

Rick, C.M. 1958. The role of natural hybridization in the derivation of cultivated tomatoes of Western South America. *Econ. Bot.* 12: 346–367.

Roberts, H.F. 1929. *Plant Hybridization before Mendel*. Princeton Univ. Press, Princeton.

Schwartz, B.W. 1973. A history of hops in America. In *Steiner's Guide to American Hops,* pp. 37–71, Steiner, U.S.A.

Small, E. 1973. Photosynthetic ecology of normal and variegated *Aegopodium podagraria. Can. J. Bot.* 51: 1589–1592.

Small, E., 1978a. A numerical taxonomic analysis of the *Daucus corota* complex. *Can. J. Bot.* 56: 248–276.

Small, E. 1978b. A numerical and nomenclatural analysis of morphogeographic taxa of *Humulus. Syst. Bot.* 3: 37–76.

Small, E. 1979. *The species Problem in* Cannabis: *Science & Semantics.* 2 vols., Corpus, Toronto.

Small, E. 1980. The relationships of hop cultivars and wild variants of *Humulus lupulus. Can. J. Bot.* 58: 676–686.

Small, E. 1981a. A numerical analysis of morpho-geographical groups of cultivars of *Humulus lupulus* based on sample of cones. *Can. J. Bot.* 59: 311–324.

Small, E. 1981b. An alfalfa germplasm expedition in Turkey. *Forage Notes* 25(2): 56–66.

Small, E. 1981c. A numerical analysis of major groupings in *Medicago* employing traditionally used characters. *Can. J. Bot.* 59: 1553–1577.

Small, E. 1982. *Medicago* collecting in Turkey. *Plant Genet. Res. Newsl.* 49: 11-12.

Small, E. 1983. Pollen ploidy-prediction in the *Medicago sativa* complex. *Pollen Spores*, in press.

Small, E. and Bauchan, G.R. 1983. Chromosome numbers of the *Medicago sativa-falcata* complex in Turkey. *Can. J. Bot.*, in press.

Small, E. and Brookes, B.S. 1982. Coiling of alfalfa pods in relation to resistance against seed chalcids. *Can. J. Plant Sci.* 62: 131-135.

Small, E. and Brookes, B.S. 1983. Taxonomic circumscription and identification in the *Medicago sativa-falcata* (alfalfa) continuum. *Econ. Bot.*, in press.

Small, E. and Desjardins, R.L. 1978. Comparative gas exchange physiology in the *Daucus carota* complex. *Can. J. Bot.* 56: 1739-1743.

Small, E. and Lefkovitch, L.P. 1982. Agrochemotaxometry of alfalfa. *Can. J. Plant Sci.* 62: 919-928.

Vavilov, N.I. 1951. *The Origin, Variation, Immunity, and Breeding of Cultivated Plants* (K.S. Starr, ed.). Chronica Botanica, Waltham, Mass.

Wilkes, H.G. 1967. *Teosinte: the Closest Relative of Maize.* Bussey Inst. Harvard, Cambridge.

Wilkes, H.G. 1970. Teosinte introgression in the maize of the Nobogame Valley. Bot. Mus. Leafl. Harvard Univ. 22: 297-311.

Wilson, D.G. 1975. Plant remains from the Graveney boat and the early history of *Humulus lupulus* in W. Europe. *New Phytol.* 75: 627-648.

Zohary, D. 1970. Centers of diversity and centers of origin. In *Genetic Resources in Plants - their Exploration and Conservation* (O.H. Frankel, and E. Bennett, eds.), pp. 33-42, Blackwell, Oxford.

Plant Reproductive Strategies

Krystyna M. Urbanska
Geobotanisches Institut ETH
Stiftung Rübel
Zürich, Switzerland

INTRODUCTION

Considering the phenomenon of plant reproduction, a student in
ecological genetics is faced with such questions as: Should
reproduction assure a certain rate of population increase or
should it rather reinforce the population ability to persist
unaltered in its given environment? Is reproduction oriented
towards an increase of, or maintaining, genetic diversity in a
population or should it rather assure its maximum genetic
stability? Is either goal achieved by always the same, consistent
reproductive behavior, or is a range of behavioral patterns,
codified *a priori* in the genotypical program, available for
choice from the following current environmental conditions? These
and other questions refer to reproductive strategies involving not
only the actual production of propagules, but also their
functioning in subsequent phases of plant life as well as their
eventual success or failure.

Plant reproductive strategies have recently been more and more
discussed and various approaches have been suggested. One
approach has involved considerations on energy budget and focused
on resource allocation patterns (e.g. Harper and Ogden 1970;
Putwain and Harper 1972; Ogden 1974; Sarukhan 1976; Pitelka 1977;
Gross and Soule 1981). A quantitative assessment of resource
allocation expressed in biomass or calorific values undoubtedly
represents, from an ecological point of view, an interesting
approach. However, an accurate evaluation of energy expenditure
on various structures is exceedingly difficult in plants and
virtually impossible in taxa that combine sexual and asexual
reproduction. Particular organs in perennial plants frequently
have a multiple function (Hickman 1975; Kawano and Nagai 1975;
Bostock and Benton 1979; Urbanska 1981). There also seems to be a

PLANT BIOSYSTEMATICS

211

considerable translocation of materials from vegetative to
reproductive structures (Levin 1976; Pitelka 1977; Urbanska
unpubl.). Last but not least, calories and nutrients may be
allocated differently than biomass to structures of vegetative and
sexual reproduction (Abrahamson 1980).

Descriptions of plant reproductive strategies based on life-
history components obviously lack the information on energy costs;
this approach comprises, on the other hand, an important evolutionary element relating reproduction and survival to genetic and
ecological factors. This importance has long been recognized but
our knowledge in the subject is still rather limited and inadequate, especially as far as various natural populations are
concerned. The results of studies using artificial populations
and environments are not always conclusive in understanding
reproductive strategies observable in natural conditions.

The strategy governing the entire reproduction process in
iteroparous plants can be considered on three levels:
(1) population systems viz. species, semi-species or races show
basic patterns of reproduction selected for in the past;
(2) genetic polymorphism for reproductive traits observable in
local populations reflects their adaptive variability;
(3) particular individuals show environmentally controlled phenotypical plasticity in reproductive behavior. Detailed studies
dealing with two latter aspects accumulate evidence that plants
are frequently equipped with alternative or complementary tactics
for realizing a given reproductive strategy; some of these tactics
can be habitually combined, others are only exceptionally shown.
Despite a general similarity in reproductive characteristics,
disparate reproductive strategies are observable even in closely
related taxa living in the same area.

To illustrate the very reality of differences in reproductive
strategies of iteroparous plants, several examples have been
chosen. The material presented is intended to be merely exemplary
and no attempts have been made to discuss the subject in an
exhaustive way.

1. *ANTENNARIA* GAERTN.

Three principal reproductive traits occurring within the genus
viz. (1) sexual reproduction, (2) agamospermy i.e. asexual
reproduction by seed and (3) vegetative propagation are combined
in various ways by particular taxa. The five *Antennariae*
discussed below represent respectively the sect. *Carpaticae,
Dioicae* and *Alpinae*. The sect. *Carpaticae* is
characterized by virtually exclusive sexual reproduction, agamospermy being absent and vegetative propagation exceptionally rare.
Taxa of the stoloniferous sect. *Dioicae* combine reproduction
by seed with vegetative propagation, but in tetraploids seeds are
produced sexually whereas higher polyploids are agamospermous.
The sect. *Alpinae* is characterized by predominant asexual
reproduction both by seed and leafy rosette-bearing stolons; taxa

of this section are most frequently high polyploid.

1.1 SECT. CARPATICAE: *ANTENNARIA ANAPHALOIDES, A. PULCHERRIMA*
A. anaphaloides occurs in the Western Pacific Range of North
America. It shows a distinct preference for open sunny sites with
dry, loose and permeable soils (Urbanska 1983a,b). The mostly
small or medium-large populations comprise virtually no seedlings;
nonreproducing and reproducing male and female individuals usually
are scattered within a given site. *A. anaphaloides* reproduces
exclusively by sexual means. Tall-growing flowering shoots bear
very numerous capitula both in male as well as female plants;
however, achene output is low (Fig. 1).

   Seed dormancy in *A. anaphaloides* is very pronounced: in
a laboratory trial, only 22.6% of seeds germinated within 70 days
(Fig. 1) and the germination proceeded very slowly. Interestingly
enough, germinability of seeds does not seem, on average, to
diminish with increasing seed age; in a trial carried out three
years after harvest in the wild, germination rate in first three
weeks remained virtually the same as in the harvest year (6.0 vs.
6.5%) and was even slightly higher in five-year-old seeds (8.9%).
It should be noted that the achenes of *A. anaphaloides* used
in our experiments were air-dry stored i.e. endured conditions

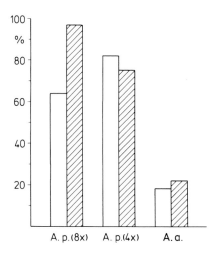

FIG. 1. *Seed output and germination in* Atennaria   pulcherrima
(A.p. 8x = octoploids, A.p. 4x = tetraploids) and
A. anaphaloides *(A.a.)*.

generally considered as greatly affecting seed viability (e.g.
Wareing 1966; Roberts 1972; Harrington 1972; Viliers 1973; Cook
1980).

The reproductive strategy of *A. anaphaloides* is apparently
oriented towards persistence of populations in their harsh and
unpredictable environments. The apparently slow release and
recruitment of new genotypes reflects an adaptation to prolonged
periods of aridity characteristic of the biotopes inhabited by the
taxon; it may also be influenced by other factors limiting the
carrying capacity. The large seed bank in the soil, probably
comprising various generations, not only assures the population
turn-over but also may carry enough latent genetical variability
to cope with more radical changes in ecological conditions.

Antennaria pulcherrima is closely related to *A. anaphaloides*
but its habitat preferences are quite different. The North
American taxon occurs in moist alluvial soils frequently subject
to intermittent flooding; its representative sites correspond to
river flats or stream banks. The best developed populations of
A. pulcherrima, observable in northern parts of its total
distribution area viz. Alaska, Yukon, northern British Columbia
and Alberta are octoploid ($2n = 56$); tetraploids ($2n = 28$)
occurring farther to the south viz. Wyoming, Colorado, form small
isolated colonies (Urbanska 1983a,b). Notwithstanding these
differences, the population structure of *A. pulcherrima* is
basically similar in various areas of occurrence of the taxon:
numerous seedlings often occur in groups and alternate with
reproducing rosettes. Male and female individuals usually grow
side by side. Population density varies from one sector to
another and may be pronounced in some sites.

Sexual reproduction in *A. pulcherrima* is very strong. The
taxon produces tall-growing shoots with numerous capitula. Achene
output is good and so is rapidly progressing germination (Fig. 1).
Seedlings grow fast and are vigorous. Seeds of *A. pulcherrima*
do not seem to remain viable for a long time in dry conditions;
three years after harvest, germination onset was retarded and
germination rate was greatly reduced (2.4 vs. 98.1%). No seeds
germinated five years after harvest.

A. pulcherrima is practically the only taxon of the sect.
Carpaticae that not only reproduces by sexual means but also
propagates vegetatively; it forms slender creeping rhizomes that
surface at some distance from "mother" rosettes and develop
ramets. Clones are often fragmented. Reproductive strategy of
A. pulcherrima is tuned to labile, difficult conditions of a
river bar, especially in northern latitudes. A massive production
of achenes, rapid germination and good growth of seedlings may
compensate, on the one hand, for losses suffered by populations
during inundation periods; on the other hand, they may be advantageous when an opportunity for further colonizing appears.
Vegetative propagation not only enhances the survival but also the
reproductive potential of particularly fit genets and increases
the population density.

1.2 SECT. DIOICAE: *ANTENNARIA DIOICA*

A. dioica has a circumpolar distribution; it occurs in various
sites, adaptive radiation on the tetraploid level (2*n* = 28) being
very pronounced. Alpine populations of *A. dioica* in Switzerland
grow on various substrata, most frequently in sunny, dry and wind-
exposed sites that may remain snow-free in winter. The stoloniferous taxon develops rather small clones that are often fragmented. Populations on the whole are small or medium large; they
usually comprise both male and female individuals as well as
nonreproducing rosettes. Seedlings are very scarce.

The achene output in *A. dioica* varies from one alpine
site to another. For instance, it comported 46.2% in a population
from an acidic siliceous area near Davos (about 2350 m a.s.l.) but
did not exceed 31% in a neighboring population from serpentine
(Urbanska unpubl.). In laboratory conditions, germination rate in
the material from acidic silicate comported 48% (Fig. 2), but only
3% of seeds germinated in a trial that was carried out in parallel
with siliceous soil from the study area (Fossati 1980). In the
material harvested and sown in serpentine soil at about 2300 m
a.s.l., no germination has been observed during two consecutive
seasons (Zuur-Isler 1981; Gasser unpubl.). On the other hand,
seeds of *A. dioica* remain viable for several years (Urbanska,
unpubl.).

Reproductive strategy of *A. dioica* from ecologically extreme

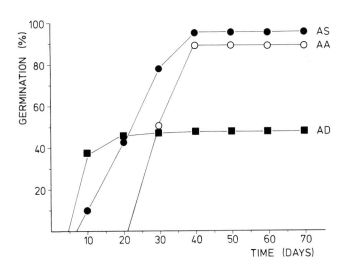

FIG. 2. *Germinating behavior in* Antennaria dioica *(AD),*
A. alpina *(AA) and* A. stolonifera *(AS).*

alpine sites clearly shows low-risk elements (e.g. clonal growth,
vegetative propagation, pronounced seed dormancy). On the whole,
persistence of populations seems to be well assured, the obligatory allogamy acting apparently in a certain balance with clonal
growth. The fecundity in *A. dioica* should be further investigated. For the time being, it is impossible to say whether a
rather low achene output reflects an adaptation to hazardous life
conditions of the alpine tundra or merely represents annual
fluctuations.

1.3 SECT. ALPINAE: *ANTENNARIA ALPINA, A. STOLONIFERA*

A. alpina has a holarctic distribution; it frequently occurs
in subarctic and arctic tundra. Populations of *A. alpina*
presented below were observed in solifluxion-influenced mountain
soils of subarctic Scandinavia. The colonies were rather large,
semi-continuous and characterized by a variable density. Seedlings and very young plants were often observed; they mostly
occurred among single nonreproducing rosettes whereas reproducing
clones locally formed a rather dense cover. Male individuals are
exceedingly rare in *A. alpina*; in the study area, only two
clones were noted. They manifested a diminished vegetative vigor,
produced very few capitula and the pollen was highly sterile.
A. alpina is autonomously agamospermous, both the embryo and the
the endosperm developing without fertilization (Juel 1900; Bergman
1935; Urbanska 1974; Urbanska unpubl.).

Female clones of *A. alpina* produced numerous flowering shoots
and the achene output per ramet was high (on average 84%).
Germination was very good (91%, Fig. 2) and the development of
individuals proceeded fast, formation of stolons beginning sometimes in three-week-old rosettes. Clones of *A. alpina* are often
fragmented, probably as a result of soil movement; the ramets can
be transported over a short distance when the soil is slipping or
being washed out. *Antennaria stolonifera* occurs in alpine
slopes of central Alaska and SE Yukon Territory as well as on
east slope of the MacKenzie Mts. (Porsild and Cody 1980). The
data presented below refer to a population found in Kigluaik Mts.,
Seward Peninsula, NW Alaska. About 17 small clones and a few
nonreproducing rosettes were scattered in a barren, wind-exposed
soil over a rather limited area. Some of the rosettes could have
been identified as fragmented clones. Only female individuals
were present; in *A. stolonifera*, staminate plants are unknown.
The plants flowered rather abundantly and the achene output per
ramet was of 71.2%; germination was very good (97.8%, Fig. 2).
Individual development proceeded rapidly and rosettes in two-week-
old plants consisted most frequently of 8 leaves, one or two
lateral buds at the base of the rosette being already observable.
Seeds of *A. stolonifera* apparently remain viable for several
years: in a laboratory trial run three years after harvest,
germination rate on 20th day comported 32.8% whereas in the

harvest year it was 42.1% (Fig. 2).
   Reproductive strategies of *A. alpina* and *A. stolonifera*
not only enable both taxa to persist and function in their very
harsh, unpredictable environments but also have a high colonizing
potential. It is very interesting that two different biological
processes viz. seed production and clonal growth resulting in
vegetative propagation represent two facets of asexual reproduction in the *Alpinae*.

2. LEMNACEAE

The duckweed family is characterized by an exceedingly strong
asexual reproduction. The vegetative propagation resulting from
clonal growth accompanied by a spontaneous clone fragmentation is
representative of the *Lemnaceae*; their populations may
accumulate considerable biomass, but their genetical stability is
pronounced (see e.g. Urbanska 1980a). Another interesting aspect
of asexual reproduction observed in some taxa of the family is the
formation of turions that essentially serve as survival units.
Turions are specialized structures: they are smaller, usually
more pigmented and store compounds of higher molecular weight than
normal fronds. Sexual reproduction occurs as well within the
Lemnaceae, but particular taxa greatly differ form one
another in this respect. For a comprehensive study on the
duckweed family and their behavioral patterns, the reader is
referred to the excellent monograph by Landolt (soon to appear).
In the present paper, only a few aspects are briefly mentioned.

2.1 *SPIRODELA POLYRRHIZA*

S. polyrrhiza is very widely distributed and apparently has
spectrum of tolerances both as to climatic conditions as well as
the actual composition of water in which it is growing (Landolt
1957; 1982; Landolt and Wildi 1977; Lüönd 1980). Its response
towards particular nutrients are characterized by a strong
plasticity in frond size (Zimmermann 1981) and/or multiplication
rate (e.g. Landolt 1957; Hillmann 1961; Lüönd 1980).
   In unfavourable conditions, *S. polyrrhiza* forms turions that
sink to the bottom of the water basin and stay dormant until the
life conditions improve. Factors stimulating germination of
turions are very complex and apparently depend on conditions under
which the turions were developed and stored. For instance,
Sihasaki and Oda (1979) observed in *S. polyrrhiza* at least two
different types of turion dormancy in respect to light and nitrate
requirements. Turions produced under nitrogen deficiency
comprised so-called "young" (Y) and "old" (O) types, whereas
turions induced by low temperatures were mainly of the "young"
type; the Y- and O-types differed from each other in their
germinating behaviour. Perry (1968) and Lacor (1969) found that
gibbelleric acid in light enhanced the germination in freshly
harvested, unchilled turions.

Sexual reproduction in *S. polyrrhiza* seems to be exceedingly rare. Very few records are known from the wild (Hicks 1932; Daubs 1965) and mature fruits were reported only twice, from USA and India, respectively (Hegelmaier 1871; Maheshwari and Maheshwari 1963). Germinating behavior of seeds has not been investigated to date. Experimental results of Wolek (1974) and Krajncic (1980) suggest that sexual reproduction in *S. polyrrhiza* remains under control of rather unusual ecological conditions, difficult to stimulate.

Reproductive strategy of *S. polyrrhiza*, apparently well adapted to large fluctuations that may rapidly occur in aquatic habitats, is very valuable both for population persistence as well as its increase. The recurrent rejuvenation of clones enhances the phenotypical plasticity; populations are thus able to respond more flexibly towards their changing environment. Should the life conditions become too extreme for survival of normal fronds, *S. polyrrhiza* is often able to produce turions. Varied types of turion dormancy suggest self-regulatory mechanisms that provide a safeguard against a total propagule loss. Sexual reproduction in *S. polyrrhiza* usually remains not expressed but still seems to form part of a genetic makeup of the taxon and might exceptionally result in release of new genets.

2.2 LEMNA TURIONIFERA AND L. GIBBA

L. turionifera and *L. gibba* are very closely related but patterns of their geographical distribution suggest different adaptations. *L. turionifera* occurs in continental temperate areas with cold winters (to -40°C), whereas *L. gibba* appears in regions with often dry summers and mild winters (Figs. 3-4, see also Landolt 1982).

Reproduction in *L. turionifera* is exclusively asexual. Depending on the ecological conditions, the taxon may propagate by fronds or form turions. The behavior of turions has not been investigated so far. *L. gibba* propagates vegetatively but does not form turions; on the other hand, it frequently reproduces sexually. In some sites, the seed production is apparently a prerequisite for the regeneration of the population (Witztum 1977). Seeds of *L. gibba* are drought-resistant (Landolt 1957; Rejmankova 1976) and can also withstand temperatures about 0°C (Rejmankova 1976). No after-ripening period is required for germination (Kandeler and Hügel 1974). The seed viability largely depends on storage conditions, temperature and humidity being of particular importance (Wilson 1830; Rejmankova 1976). The interesting study of Rejmankova (1976) on germinating behavior of seeds in *L. gibba* is, to the best of our knowledge, the only work done with materials from the wild. Seed samples originating from two climatically different sites in Tschechoslovakia consistently showed some differences in germinating behavior (Fig. 5). This pattern seems to be genetically influenced and suggests a racial differentiation in *L. gibba*.

Reproductive strategy of *L. turionifera* and that of *L. gibba* are both oriented towards persistence and increase of populations; however, phenotypical plasticity and a maximum genetic stability represent the principal features of strategy in *L. turionifera*, whereas genetic diversity apparently plays an important role in *L. gibba*. These differences show clearly when the population survival is taken into consideration, the same biological function being assured in *L. turionifera* by vegetative structures and in *L. gibba* by sexually produced seeds.

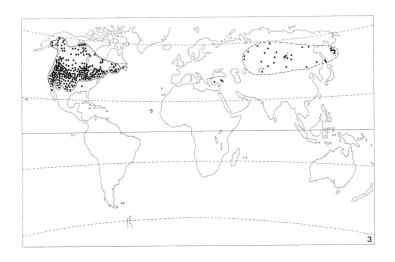

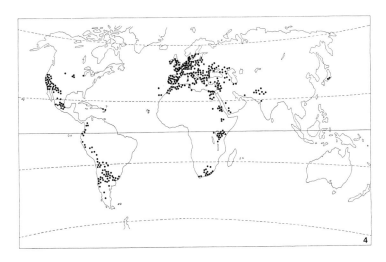

FIG. 3-4. *Geographical distribution of* Lemna turionifera *(3) and* L. gibba *(4). Courtesy of Dr. E. Landolt.*

3. *CARDAMINE RIVULARIS, C. INSUETA* AND *C. SHULZII*

C. rivularis, C. insueta and *C. Schulzii* are phylogenetically
linked: *C. insueta* is a triploid hybrid between *C. rivularis*
and *C. amara* originating with a part of an unreduced gamete of
of *C. rivularis. C. Schulzii*, on the other hand, represents an
autoallopolyploid species formed as a result of doubling of the
chromosome set in *C. insueta*. The three taxa occur together
at Urnerboden, Central Switzerland. Their reproduction was
described in detail elsewhere (Urbanska 1977a,b, 1980); in the
present paper, brief characteristics are given together with some
unpublished data. The three taxa of *Cardamine* show that
even closely related plants occurring in the same limited area may
have distinct reproductive strategies and differ from one another
not so much in kind but in the actual functioning of their
reproductive structures, sexual and asexual alike. *C. rivularis*
occurs at Urnerboden in rather small and semi-isolated colonies.
It does not stand competition for light and grows mostly in open

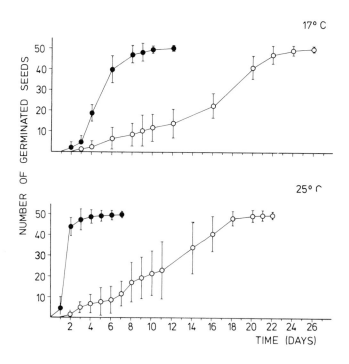

FIG. 5. Lemna gibba: *seed germination in samples from two climatically different sites in Tschechoslovakia. (Redrawn from Rejmankova, 1976).*

sites with meagre, swampy and slightly acidic soil (Urbanska and Landolt 1978). Nonreproducing and reproducing individuals most frequently are scattered within a given population sector; seedlings are very scarce.

Reproduction in *C. rivularis* is nearly exclusively sexual, an average seed output being 63.9% (Fig. 6); the germination rate is medium-high (Fig. 6). Seeds of *C. rivularis* retain their viability for at least two years (Urbanska unpubl.). Vegetative propagation in form of a specialized foliar vivipary (Urbanska 1980a) occurs only occasionally. Reproductive strategy of *C. rivularis* corresponds to the nonaggressive behavior of the taxon and is mostly oriented towards its persistence in the harsh environment. The predominant allogamy and partial seed dormancy represent important adaptive features, the former mechanism generating genetic variability, the latter – controlling its release.

C. insueta forms a rather large population that is exceedingly dense in some sectors (Fig. 9). The tall-growing flowering shoots bear numerous flowers but *C. insueta* is most frequently male-sterile (Fig. 7). No seedlings appear within fenced, manured hay meadows corresponding to the representative biotope of *C. insueta*; very numerous juvenile forms occurring there were identified as

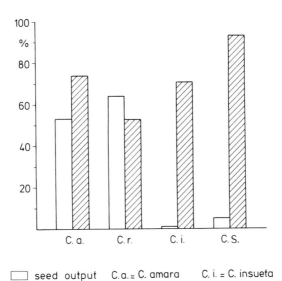

FIG. 6. *Seed output in taxa of* Cardamine L. *from Urnerboden. For comparison, data on* C. amara *are included.*

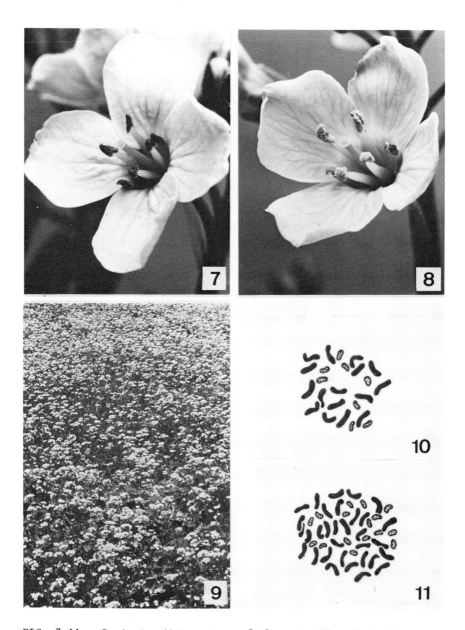

FIG. 7-11.   Cardamine X insueta *and the autoallopolyploid* C.
Schulzii. *7. Flower of* C. insueta *with sterile, nondehiscent*
anthers. 8. Flower of C. Shulzii *with fertile, abundant*
pollen. 9. Luxuriant but male-sterile C. insueta *in manured*
hay meadows at Urnerboden.   10. Somatic metaphase of C.

propagules (Urbanska 1980a, 1981). In open pastures, where *C. insueta* occurs much less abundantly, rare seedlings occur intermingled with other age variants.

Reproduction in *C. insueta* is predominantly asexual, its most prominent feature being the formation of very numerous plantlets on the surface of rosette leaves as well as in axils of the flowering shoots. The propagule-bearing fragments of the shoots fall to the ground when the hay meadows are being cut and the propagules root in easily. In addition to this peculiar, very effective mode of propagation, *C. insueta* forms short runners with supplementary rosettes. Sexual reproduction in *C. insueta* is also rather unusual. The taxon is allogamous and yet breeds true for its genomic constitution RRA (i.e. two genomes of *C. rivularis* and a single one of *C. amara*, Fig. 10). The seeds are produced by means of polarized segregation controlled by complex selective mechanisms (Urbanska 1977a). The seed output is accordingly very limited (on average 1%, Fig. 6). Germination onset is rather delayed (Urbanska 1980a) but germination rate is high (70.7%, Fig. 6) and seedlings grow fast.

Reproductive strategy in *C. insueta*, oriented towards a maximum stabilization of the triploid RRA genome, shows features that are directly adaptive to the conditions occurring in the prevailing biotope of the hybrid. As a result, not only the population density may locally increase but also neighboring surfaces may be colonized. Interestingly enough, not only vegetative propagation but also sexual reproduction contribute to the preservation of the intact hybrid structure.

C. Shulzii (2n = 48, Fig. 11) occurs in a rather limited sector; it is actually spreading within the station, apparently at the expense of both *C. insueta* as well as *C. rivularis*. The population of the hexaploid consists of seedlings, juvenile forms of various age, as well as adult individuals. Reproducing plants mostly are very luxuriant and produce abundant flowers with large amounts of pollen (Fig. 8). Reproduction in *C. Schulzii* comprises both sexual and asexual mechanisms, but the newly arisen taxon reproduces mainly by seed. Seed output is limited (Fig. 6); on the other hand, germination rates are very high (Fig. 6) and the seedling development proceeds rapidly. *C. Schulzii* propagates sometimes vegetatively in the same way as *C. insueta* but the production of propagules occurs much less frequently.

Reproductive strategy in *C. Schulzii* seems to be oriented towards expansion. The low seed output may reflect selective mechanism, seed number being sacrificed in favor of their fitness; the spreading population may support better loss of some zygotes than a high mortality of young individuals. A balanced

---

insueta: 2n = 24. 11. *Somatic metaphase of* C. Schulzii: 2n = 48. c. X2000. *Figs. 7-9 from Urbanska, 1981; reproduced by permission from Vierteljahrschrift der Naturforschenden Gesellschaft in Zürich.*

combination of genetic variability and stabilization of successful
genets is apparently advantageous for the polyploid born into a
crowded community; it provides an opportunity for finer adaptive
adjustments to arise and become established in later generations.

DISCUSSION AND CONCLUSIONS

Two principal themes repeatedly emerging in the present paper are
(1) the type of environment to which a given reproductive strategy
is tuned and (2) different tactics involved in the realization of
a given strategy. They bring us to the problem of habitat
characterization as well as that of the biological significance of
various reproductive mechanisms.

Habitats are described in a number of terms that neither are
always synonymous nor mutually exclusive. For instance, a
distinction is sometimes made between a harsh environment and an
unpredictable one; however, a habitat can be both harsh and
unpredictable as exemplified by arctic regions (Lloyd 1980). The
same opinion applies largely to other sites e.g. those occurring
in the alpine vegetation belt. On the other hand, a habitat that
is subject to large fluctuations can accordingly be considered as
unstable or unpredictable even if overall life conditions do not
qualify as harsh (e.g. some sites inhabited by the *Lemnaceae*).
Harsh environments are characterized by a high stress that may be
caused by various factors operating simultaneously or in seasonal
succession (Grime 1979). For that reason, generalizations based
on only one principal type of environmental hazard may sometimes
be rather misleading. A certain coordination in use of particular
terms for the characterization of sites should be most helpful.

The unpredictability or predictability of environment results
obviously from both physical as well as biotic factors that may
influence in various way different phases of plant life.
Reproductive plant strategies may accordingly be quite distinct
from those involved in the established nonreproductive phase (see
e.g. Grubb 1977). Moreover, features that increase e.g. seed
survival are not necessarily advantageous for survival of
seedlings or very young plants (Grime 1978; Fossati 1980; Sarukhan
1980). It follows that only rarely can the environmental
unpredictability be assessed by data referring to a single life
phase (e.g. Wilbur 1976). As well formulated by Abrahamson
(1980), the function of reproduction is to perpetuate parental
genes in time and space; they are carried by descendants forming a
new generation in the demographic sense. Survival and dispersal
represent thus important aims in plant reproductive strategies.
Seeds serve these purposes very well; their dormancy and increased
longevity may represent successful adaptations to the unpredictability of environment (Levins 1969; Harper 1976; Cook 1980). It
should be emphasized, however, that seed production is not to be
automatically associated with sexual mechanisms and recombination.
We stress this point for there still is a great deal of confusion
on the subject, even in some recent reviews (e.g. Williams 1975,

in particular the comparison between sexual and asexual offspring;
see also Abrahamson 1980). Plants in which seed production
results from polarized segregation admittedly represent a small
and exceptional group (Cleland 1962; Smith-White 1948, 1955;
Täckholm 1922; Urbanska 1977a, 1980a). On the contrary, there is
nothing so exceptional about the numerous agamospermous taxa; they
frequently have high fecundity rates, often display some other
forms of asexual reproduction and apparently are able to cope well
with environmental hazards. Biological functions viz. dispersal
and survival are naturally the same in sexually and asexually
produced seeds; their behavior may accordingly comprise identical
aspects (e.g. dormancy or density-dependent mortality of seedlings). Important demographic parameters viz. size, density or
age structure are accordingly not different in sexual and agamospermous populations; on the other hand, their genetical structure
is diametrically different and long-term evolutionary consequences
are obvious. Another intriguing point in plant reproductive
strategies is that some specialized vegetative structures may
rather effectively substitute for seeds. A very good example to
represent this aspect is the turions produced in some *Lemnaceae*;
their behavior, in particular dormancy and germinability (Malek
and Oda 1980) offer an exciting field for future research.
Another interesting case is the bulbils produced in the head of
scapes in *Allium Grayi*; they are playing a function almost
comparable to seeds as far as production, dispersability and
germinability are concerned (Kawano and Nagai 1975).

Abrahmson (1980) argued recently that the general adaptive
significance of asexual reproduction was essentially the same
regardless of the method. We propose that a distinction be made
at least between the agamospermy and other forms of asexual
reproduction. It seems that the adaptive significance of given
reproductive structures may be primarily influenced by their
biological function and not their sexual or asexual origin. In
this perspective, an optimal fitness might sometimes be ascribed
not so much to an advantageous combination of sexual and asexual
reproduction but to the reproduction *by seed* (sexual or asexual)
combined with vegetative propagation. Agamospermous plants with
clonal growth and duckweeds combining reproduction by fronds with
environmentally cued turion formation should be very instructive
in this respect; unfortunately, only fragmentary field data are so
far available.

An understanding of the reproductive strategy of a population
is essential not only for a complete demographic accounting but
also for a better comprehension of the genetic population structure. It may be very important in interpreting the forces governing population dynamics and its evolution (Meagher and Antonovics
1982). Results of further, urgently needed studies are awaited
with interest.

ACKNOWLEDGMENTS

Elias Landolt permitted me to use some of his not yet published
data on the *Lemnaceae* and critically read the manuscript; his
help is greatly appreciated. Ms. E. Wohlmann made the drawings;
Ms. S. Dreyer typed the manuscript. Sincere thanks of the author
are addressed to these persons.

REFERENCES

Abrahamson, W.G. 1980. Demography and vegetative reproduction.
In *Demography and Evolution in Plant Populations* (O.T. Solbrig,
ed.). Blackwell, Oxford, pp. 89-106.
Bergman, B. 1935. Zur Kenntnis der skandinavischen *Antennaria*-
Arten. *Sven. Bot. Tidskr.* 26: 99-106.
Bostock, S.J. and Benton, R.A. 1979. The reproductive strategies
of five perennial *Compositae*. *J. Ecol.* 67: 91-107.
Cleland, R.E. 1962. The cytogenetics of *Oenothera*. *Adv. Genet.*
11: 147-237.
Cook, R. 1980. The biology of seeds in soil. In *Demography and
Evolution in Plant Populations* (O.T. Solbrig, ed.). Blackwell,
Oxford, pp. 107-129.
Daubs, E.H. 1965. A Monograph of *Lemnaceae*. III. *Biol. Monogr.*
34, Univ. Illinois Press, Urbana.
Fossati, A. 1980. Keimverhalten und frühe Entwicklungsphasen
einiger Alpenpflanzen. *Veröff. Geobot. Inst. ETH, Stift. Rübel
Zürich* 73.
Grime, J.P. 1978. Interpretation of small-scale patterns in the
distribution of plants in space and time. In *Structure and
Functioning of Plant Populations*, (A.H.J. Freyson and J.W.
Woldendorp, eds.). North-Holland, Amsterdam, pp. 101-121.
Grime, J.P. 1979. *Plant Strategies and Vegetation Responses*.
Wiley, Chichester.
Gross, K.L. and Soule, J.D. 1981. *Am. J. Bot.* 68: 801-807.
Grubb, P.J. 1977. The maintenance of the species richness in plant
communities: the importance of the regeneration niche. *Biol. Rev.*
52: 107-145.
Harper, J.L. 1976. *Population Biology of Plants*. Academic Press.
Harper, J.L. and Ogden, J. 1970. The reproductive strategy of
higher plants. I. The concept of strategy with special
reference to *Senecio vulgaris* L. *J. Ecol.* 58: 681-698.
Harrington, J.F. 1972. Seed storage and longevity. In *Seed
Biology* (T.T. Kozlowski, ed.). Academic Press, 3: 145-245.
Hegelmaier, F. 1871. Ueber die Fructifikationstheile von
Spirodela. *Bot. Zeit.* 29: 621-629, 645-666.
Hickman, J.C. 1975. Environmental unpredictability and plastic
energy allocation strategies in the annual *Polygonum
cascadense (Polygonaceae)*. *J. Ecol.* 63: 689-701.
Hicks, L.E. 1932. Flower production in the *Lemnaceae*. *Ohio.
J. Sci.* 32: 115-131.
Hillman, W.S. 1961. *Bot. Rev.* 27: 221-287.

Kandeler, R. and Hügel, B. 1974. Development in vitro of flower primordia of *Lemnaceae*. In 3rd Int. Congr. Plant Cell Culture. Univ. Leicester, p. 60. Abstr.

Kawano, S. and Nagai, Y. 1975. The productive and reproductive biology of flowering plants. I. Life history strategies in three *Allium* species in Japan. *Bot. Mag.* 88: 281-318.

Krajncic, B. 1980. Report on photoperiodic responses in *Lemnaceae* from Slovenia. *Ber. Geobot. Inst. ETH, Stift. Rübel Zürich* 47: 75-86.

Lacor, M.A. 1969. On the influence of gibberellic acid and kinetin on the germination of turions of *Spirodela polyrrhiza* (L.) Schleiden. *Acta Bot. Neerl.* 18: 550-557.

Landolt, E. 1957. Physiologische und ökologische Untersuchungen an Lemnaceen. *Ber. Schweiz. Bot. Ges.* 67: 271-410.

Landolt, E. 1982. Distribution pattern and ecophysiological characteristics of the European species of the *Lemnaceae*. *Ber. Geobot. Inst. ETH, Stift. Rübel Zürich* 49: 127-145.

Landolt, E. 1983. The Family of *Lemnaceae* – a Monographic Study. *Veröff. Geobot. Inst. ETH, Stift. Rübel Zürich* 71, in preparation.

Landolt, E. and Wildi, O. 1977. Oekologische Felduntersuchungen bei Wasserlinsen (*Lemnaceae*) in den südwestlichen Staaten der USA. *Ber. Geobot. Inst. ETH, Stift. Rübel Zürich* 44: 104-146.

Levin, D.A. 1976. *Annu. Rev. Ecol. Syst.* 7: 121-159.

Levins, R. 1969. *Soc. Exp. Biol. Symp.* 23: 1-10.

Lloyd, D.G. 1980. Demographic factors and mating patterns in Angiosperms. In *Demography and Evolution in Plant Populations* (O.T. Solbrig, ed.). Blackwell, Oxford, 67-88.

Lüönd, A. 1980. Effects of nitrogen and phosphorus upon the growth of some *Lemnaceae*. *Veröff. Geobot. Inst. ETH, Stift. Rübel Zürich* 70: 118-141.

Malek, L. and Oda, Y. 1980. *Plant Cell Physiol.* 21: 357-362.

Maheshwari, S.C. and Maheswari, N. 1963. The female gametophyte, endosperm and embryo of *Spirodela polyrrhiza*. *Beitr. Biol. Pflan.* 39: 179-188.

Meagher, T.R. and Antonovics, J. 1982. *Ecology* 63: 1690-1700.

Ogden, J. 1974. *J. Ecol.* 62: 291-324.

Perry, T.O. 1968. Plant Physiol. 43: 1866-1869.

Pitelka, L.F. 1977. Energy allocation in annual and perennial lupines (*Lupinus: Leguminosae*) L. *Ecology* 58: 1055-1065.

Porsild, A.E. and Cody, W.J. 1980. *Vascular Plants of Continental Northwest Territories, Canada*. Nat. Mus. Nat. Sci. Ottawa.

Putwain, P.D. and Harper, J.L. 1972. *Studies in the dynamics of plant populations. V. Mechanisms governing the sex ratio in Rumex acetosa and R. acetosella*. J. Ecol. 60: 113-129.

Rejmankova, E. 1976. Germination of seeds of *Lemna gibba*. *Folia Geobot. Phytotaxon.* 11:261-267.

Roberts, H.N. 1972. *Viability of Seeds*. Syracuse Univ. Press.

Sarukhan, J. 1976. On selective pressures and energy allocation in populations of *Ranunculus repens* L., *R. bulbosus* L. and *R. acris* L. *Ann. Mo. Bot. Gard.* 63: 290-308.

Sarukhan, J. 1980. Demographic problems in tropical systems. In
  Demography and Evolution in Plant Populations (O.T. Solbrig,
  ed.). Blackwell, Oxford. pp. 161–188.
Sibasaki, T. and Oda, Y. 1979. Heterogeneity of dormancy in the
  turions of *Spirodela polyrrhiza*. *Plant Cell Physiol.* 20: 563–571.
Smith-White, S. 1948. *Heredity* 2: 119–129.
Smith-White, S. 1955. Heredity 9: 79–91.
Täckholm, G. 1922. Zytologische Studien über die Gattung *Rosa*.
  Acta Horti Berg. 7: 91–381.
Urbanska, K. 1974. L'agamospermie, système de reproduction
  important dans la spéciation des Angiospermes. *Bull. Soc.
  Bot. Fr.* 121: 329–346.
Urbanska, K. 1977a. Reproduction in natural triploid hybrids
  ($2n$ = 24) between *Cardamine rivularis* Schur and *C. amara*
  L. *Ber. Geobot. Inst. ETH, Stift. Rübel Zürich* 44: 42–85.
Urbanska, K. 1977b. An autoallohexaploid in *Cardamine* L.,
  new to the Swiss flora. *Ber. Geobot. Inst. ETH, Stift. Rübel
  Zürich* 44: 86–103.
Urbanska, K. 1980a. Reproductive strategies in a hybridogenous
  population of *Cardamine* L. *Acta Oecol. Plant.* 1: 137–150.
Urbanska, K. 1980b. Cytological variation within the family of
  duckweeds (*Lemnaceae*). *Veröff. Geobot. Inst. ETH,
  Stift. Rübel Zürich* 70: 30–101.
Urbanska, K. 1983a. *Antennaria carpatica* (Wahlb.) Bl. et
  Fing. s.l. in North America. I. Chromosome numbers,
  geographical distribution and ecology. *Ber. Geobot. Inst.
  ETH, Stift. Rübel Zürich* 50: 33–66.
Urbanska, K. 1983b. Cyto-geographical differentiation in
  Antennaria carpatica s.l. *Bot. Helv.* 93.
Urbanska, K. and Landolt, E. 1978. Recherches démographiques et
  écologiques sur une population hybridogène de *Cardamine* L.
  Ber. Geobot. Inst. ETH, Stift. Rübel Zürich 45: 30–53.
Viliers, T.A. 1973. Ageing and the longevity of seeds in field
  conditions. In *Seed Ecology* (W. Heydecker, ed.), Pa.
  State Univ. Press, University Park, pp. 265–388.
Wareing, P.F. 1966. The ecological aspects of seed dormancy and
  germination. In *Reproductive Biology and Taxonomy of Vascular
  Plants* (J. Hawkes, ed.). Pergamon, New York, pp. 103–121.
Williams, G.C. 1975. *Sex and Evolution*. Princeton Univ.
Wilson, W. 1830. *Lemna gibba*. *Hooker Bot. Misc.* 1: 145–149.
Witztum, A. 1977. *Israel J. Bot.* 26: 36–38.
Wolek, J. 1974. Experimental control of flowering in *Spirodela
  polyrrhiza* (L.) Schleid., strain 7401 – a preliminary report.
  Ber. Geobot. Inst. ETH, Stift. Rübel Zürich 42: 140–162.
Zimmerman, A. 1981. Einfluss von Calcium und Magnesium auf das
  Wachstum von mitteleuropäischen Lemnaceen-Arten. *Ber.
  Geobot. Inst. ETH, Stift. Rübel Zürich* 48: 120–160.
Zuur-Isler, D. 1981. Germinating behaviour and early life phase
  of some species from alpine serpentine soils. *Ber. Geobot.
  Inst. ETH, Stift. Rübel Zürich* 49: 76–107.

The Relationships between Self-incompatibility, Pseudo-compatibility, and Self-compatibility

David L. Mulcahy

Department of Botany
University of Massachusetts
Amherst, Massachusetts, U.S.A.

INTRODUCTION

Stebbins (1957) presented the following indications that self-compatible taxa were obviously derived from self-compatible relatives:
1. Self-fertilizing species appear to be clearly more specialized in morphological characteristics than many of their cross-fertilizing relatives.
2. Many self-fertilizing species possess structures which could have a high selective value only in connection with self-fertilization.
3. In some groups, self-fertilization has originated in historical times from cross-fertilizing species.
4. The same system of cross-fertilization is present in large taxonomic groups, indicating a common and ancient origin.
    The logic of these arguments is clear, the supporting examples convincing, and the significance of this conclusion for biosystematics is great. Thus I suggest that we might speak of "Stebbins's Rule" as the generalization that inbreeding species are derived from outbreeding species.
    In the present paper, I would like to extend this generalization somewhat, considering the genetic bases for a three stage progression: self-incompatibility, pseudo-self-compatibility, and self-compatibility. Also, I would like to suggest why the transition from self-incompatibility to pseudo-self-compatibility is a spontaneously reversible step while the next step, pseudo-self-compatibility to self-compatibility, is not.

SELF-INCOMPATIBILITY AND PSEUDO-SELF-COMPATIBILITY

East (1927) was perhaps the first to describe as, "pseudo-self-

PLANT BIOSYSTEMATICS

229

compatibile", individuals within a self-incompatible population
which set seeds upon self-pollination. Such individuals exhibit a
relatively low degree of self-incompatibility, and their offspring
may be self-incompatible. In the following paragraphs, I shall
suggest a possible genetic model for pseudo-self-compatibility and
its relation to self-incompatibility.

In order to introduce this model, it is first necessary to
explain that self-incompatibility, specifically, gametophytic
self-incompatibility, is subject to two quite different interpretations. The first of these is known as the "oppositional" model,
because it assumes that incompatible pollen is actively inhibited
(opposed) by specific pollen tube inhibiting molecules (see Lewis
1954). The second model assumes that self-incompatibility is a
passive phenomenon, with incompatible pollen tubes failing to
reach the ovary for lack of appropriate stimulation or conditions.
This latter interpretation is termed the "heterosis" model because
it assumes that there are heterosis-like interactions between
pollen tubes and styles. Compatible pollen tubes are those which
obtain sufficient stimulation to allow them to reach the ovules in
time for normal fertilization to occur.

The heterosis model is well illustrated by a recent study of
Beta vulgaris (Larsen 1978). In fact, this particular study
is one of the cornerstones for the heterosis model. Unlike the
classic opositional model which invokes only one multiallelic
locus (Lewis 1954), or two, reported in the Gramineae (Lundquist
1975), the heterosis model involves several loci. In the case of
B. vulgaris, at least four loci are involved in the control
of self-incompatibility. With *Beta*, pollen-tube growth is
positively correlated with a number of allelic differences between
the pollen and the stylar genotypes. To illustrate this point,
consider pollen of genotype A1B1C2D2. (A1 indicates that the
incompatibility locus "A" is present as allele #1, etc.) In a
style of genotype A1A1B1B1C1C1D1D1 (homozygous at all four
incompatibility loci), pollen of genotype A1B1C2D2 will grow more
rapidly than will A1B1C1D2 pollen which, in turn, will grow more
rapidly than will A1B1C1D1 pollen in the same style. So far, the
system resembles the classical model in that the presence of
common alleles in pollen and styles retards pollen-tube growth.
In other features, the results with *Beta,* and the predictions
of the heterosis model, differ from what would be expected
according to the classical model. For example, Larsen found that
pollen of genotype A1B1C1D1 will grow faster in the style
A1A2B1B2C1C2D1D2 (heterozygous at all four loci) than it will in
the style A1A1B1B1C1C1D1D1 (homozygous at all four). This
dissimilarity between the *Beta* observations and the classical
expectations points to the radical differences between the two
interpretations of gametophytic self-incompatibility. The
classical model suggests that incompatibility is a qualitative
character, a pollen genotype being either incompatible or
compatible. Thus A1B1C1D1 pollen tubes should be inhibited in any
style which includes the alleles A1B1C1D1 within its diploid

genotype. Consequently, the degree of pollen-tube inhibition
should not vary according to the degree of heterozygosity
exhibited in the stylar genotype. The heterosis model, in
contrast, assumes that there are heterosis-like interactions
between pollen and stylar genotypes. Thus, while the presence of
an A1 allele in the style may retard, and certainly not stimulate,
the growth of A1 pollen tubes, the simultaneous presence of A2 in
the heterozygous style (A1A2) *will* stimulate it. Growth of
A1 pollen tubes should thus be faster in A1A2 styles than in A1A1
styles. The crucial and, for biosystematic considerations, the
relevant, differences between the two interpretations is that,
with the classical model, all plants within a self-incompatible
population should be strictly self-incompatible whereas with the
heterosis model, the degree of self-incompatibility could vary
from 0% to 100% within a single population. The heterosis model
thus predicts and explains a free and reversible interconverti-
bility between self-incompatibility and pseudo-self-compatibility.

In deference to both the theme of this meeting and its
orientation toward historic and changing perspectives, I would ask
the reader to consider the following question. If the hererosis
model of self-incompatibility is even approximately correct, how
did the radically different oppositional model come to occupy its
present position of eminence? I believe that the answer is to be
found in specific wording of East's classic 1927 discussion of
self-incompatibility. I was first attracted to this paper by its
running title, "Peculiar Genetic Results." The point is perhaps
best made by quoting directly from East. Italics and material
within square brackets are mine.

"The reason why these studies ought to prove helpful to others
is because the results are comparatively uncomplicated by
extraneous variables, due to the fact that in strains of Nicotiana
used, the differences between compatibile matings is complete,
i.e., under ordinary circumstances incompatible matings produce no
seeds. *Such material is unusual. Ordinarily the block to
fertilization which produces the difference between an
incompatible and a compatible union is incomplete...* [With
compatible matings]...fertilization occurs in about 3 days.
Clearly any conditions which extend the 'life' of the flower or
accelerate the rate of pollen-tube growth in an incompatible
mating, tend to promote fruitfulness in such unions; and I have
called such fruitfulness 'pseudo-fertility'. In practice,
'pseudo-fertile' unions can be obtained with most of the self-
sterile strains of Nicotiana used by pollinating the young bud.
But unfortunately, material which is serviceable for self-
sterility studies is often so 'pseudo-fertile' under ordinary
conditions that an analysis of the results is very difficult
without a *good working hypothesis as a guide*. This is
because the average difference in time required for
incompatible pollen-tubes and for compatible pollen-tubes to
reach the micropyle is so small that the frequency

distributions of the two types overlap each other.
  By selecting the material advantageously and by
growing it under such favorable conditions that pseudo-
fertility is eliminated, it has been possible to isolate
seven multiple allelomorphs which control the behavior of
self-sterile plants in crosses with each other. These
allelomorphs have been termed S1, S2...etc. No second locus
for self-sterility has been found. *One cannot say, of
course, that no other loci exist; but if such is the case,
the plants with which we have dealt are homozygous for these
loci.*" [According to the heterosis model, individuals
heterozygous at other loci would probably exhibit "pseudo-
fertility" and thus be excluded from the "advantageously
selected material."]

  In sum, East explained that, in order to develop a good
working hypothesis *as a guide*, it was necessary to
simplify the system, selecting plants that produced
peculiarly convenient genetic results. However, we, his
followers, have forgotten that this was intended as a *guide*
and, instead, have since used it as an *answer*. According
to East, strict self-incompatibility is, "peculiar" and pseudo-
self-compatibility is extremely common. [Williams *et al.* (1983)
recently reported that in *Lycopersicon peruvianum*, another
classical example of gametophytic self-incompatibility, 1 out of
the 5 plants they studied appeared to be pseudo-self-compatibile.]
  It is instructive to consider East's statements on the number
of loci involved in self-incompatibility. He specifically leaves
open the possibility that more than one locus may be involved,
although this was not the case in his, "peculiar", material.
Subsequent workers apparently, and understandably, recalled East's
words as "Only one incompatibility locus exists." Over the next
half century, this misunderstanding would, we have suggested,
(Mulcahy and Mulcahy 1983) cause students of self-incompatibility
some difficulties in trying to reconcile their data with the
single locus model. The heterosis model was suggested in order to
resolve some of these difficulties and as such, it might seem to
be a radical departure from a concept which has long been widely
accepted. In reality, however, it seems that the heterosis
interpretation is functionally quite close to what East originally
had in mind.
  Whatever its origin, if the heterosis model of gametophytic
self-incompatibility is correct, then biosystematists must not
equate occasional self-fertility with inbreeding. Pseudo-self-
comptibility, as demonstrated by East, is very likely a common
phenomenon. In the statement of Stebbins's Rule, inbreeding
species include only those which exhibit true self-compatibility.
  Those individuals which are pseudo-self-compatible should be
included with outbreeding species.

WHY IS SELF-COMPATIBILITY DIFFICULT TO REVERSE?

If pseudo-self-compatibility is merely a quantitative (and reversible) variation in self-incompatibility, it may seem reasonable to assume that true self-compatibility is a further extension of this quantitative shift. However, there are two possible means whereby the change to self-compatibility could become a qualitative modification, one from which it might be difficult to reattain true self-incompatibility.

The first condition that could block the return to self-incompatibility from self-compatibility is suggested by the heterosis interpretation of gametophytic self-incompatibility. According to that interpretation, self-incompatibility indicates the presence of deleterious recessive alleles which are homozygous in the style and present also in the pollen. When pseudo-self-compatible plants are inbred, selection will eliminate many of these deleterious recessives, and, after a period of inbreeding depression, a line may be largely free of them. From there, the return to self-incompatibility will be difficult since the raw materials of self-incompatibility have been purged from the system.

A second condition which could make the return from self-compatibility to self-incompatibility difficult is illustrated by studies of at least three crop species. Each of these three show that, in highly inbred lines, nonself-pollen tubes can only infrequently outcompete self-pollen tubes. For example, with *Zea mays,* Jones (1928) applied mixtures of self and nonself pollen to the styles of highly inbred lines. He found that, in 20 out of 21 pollen mixtures, the pollen tubes from self-pollen grew more rapidly than did those from nonself-pollen. In a later study, Pfahler (1967) made the same type of mixed pollinations in *Zea mays* although he used only F_1 hybrids. Instead of a preponderance of self fertilizations, Pfahler found that neither self nor nonself exhibited a clear advantage over the other. In some mixtures, self-fertilizations were the more common. The apparent disparity between the Jones and Pfahler studies should probably be explained by the single obvious difference between the two: namely, that Jones used only highly inbred lines whereas Pfahler used F_1 hybrids. Indeed, it now seems likely that this difference should be sufficient to cause the observed results. Most recently and clearly, Ottaviano *et al.* (1983) have demonstrated the selection for rapid pollen-tube growth can induce significant changes in pollen-tube growth rates. This is so even when the pollen being subjected to selection is derived from a single sporophyte. This latter fact indicates that there are genes which are expressed, and thus subjected to selection, in the gametophytic portion of the life cycle. When a plant is self pollinated, the number of pollen grains applied to the stigma may greatly exceed the number of ovules available for fertilization, and this will create selection for pollen-tubes which penetrate

that stylar environmet very rapidly. If selfing continues for a
number of generations, one of the results will be a gametophytic
genotype which is highly selected for rapid growth in that style.
If, at that point, a mixture of self- and nonself-pollen types is
applied to the stigma, one would expect that the self pollen,
specialized for that environment, would be the better competitor.
In fact, that is what Jones (1928) found. To test this
hypothesis, Johnson and Mulcahy (1978) made a series of mixed
pollinations using, F_1, F_2, F_7, and F_{20} generations of *Zea mays*.
In each case, self pollen was mixed with a standard tester pollen.
According to the pollen selection hypothesis, self pollen should
improve its competitive ability relative to the standard tester as
the generations become more highly inbred, and this is what was
observed (see Fig. 1).

Other demonstrations of selfed pollen outcompeting nonself
pollen have been given by Hornby and Shin-Chai (1975), working
with *Lycopersicon esculentum* and Currah (1983), with *Allium
cepa*. Very likely, Pfahler's data reflect the fact that, in an
F_1 hybrid, each population of gametophytes is entirely new and
unselected. Lacking the advantage of previous selection, self
pollen of F_1 hybrids is, in some pollen mixtures, the superior
competitor but, in others, it is not.

Consider what this implies for a highly inbred, weedy species.
Not only might the floral structure be profoundly modified, so

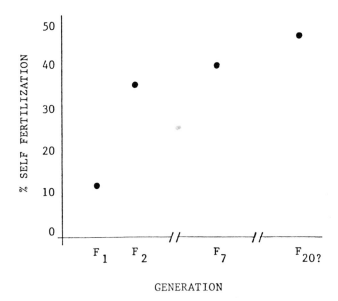

FIG. 1. *The relative competitive ability of self pollen in
different generations of* Zea mays.

that cross pollination is unlikely, but also the self gametophytic genotypes have been subjected to many generations of selection for rapid growth through that stylar environment. Thus, what little nonself pollen does reach the stigma will be unlikely to outcompete the highly selected self pollen. This may have the direct consequence of making it difficult for a highly inbred individual to return to outbreeding.

In conclusion, not only is the generalization that self-incompatibile species give rise to self-compatible species (Stebbins's Rule) well supported by numerous examples, but at least two hypotheses exist which may explain it. Whether or not these two hypotheses are correct is not yet known. The generalization, however, seems to be clearly established.

REFERENCES

Currah, L. 1983. Pollen competition and the breeding system in onion (*Allium cepa* L.). In *Pollen: Biology and Implications for Plant Breeding*. (D. Mulcahy and E. Ottaviano, eds.), Elsevier Biomedical, N.Y. pp. 375-379.

East, E.M. 1927. Peculiar genetic results due to active genetic gametophytic factors. *Hereditas* 9: 49-58.

Hornby, C.A. and Li, Shin-Chai 1975. Some effects of multiparental pollination in tomato plants. *Can. J. Plant Sci.* 55: 127-132.

Jones, D.F. 1927. *Selective Fertilization*. Univ. Chicago Press.

Johnson, C.M. and Mulcahy, D.L. 1978. Male gametophyte in maize II. Pollen vigor in inbred plants. *Theor. Appl. Genet.* 51: 211-215.

Larsen, K. 1978. Oligoallelism in the multigenic incompatibility system of *Beta vulgaris*. *Incompatibility Newslett.* 10: 23-28.

Lewis, D. 1954. Comparative incompatibility in angiosperms and fungi. *Adv. Genet.* 6: 235-285.

Lundquist, A. 1975. Complex self-incompatibility systems in angiosperms. *Proc. R. Soc. Lond. B.* 188: 235-246.

Mulcahy, D.L. and Mulcahy, G.B. 1983. Gametophytic self-incompatibility reexamined. *Science* 220: 1247-1251.

Ottaviano, E., Sari-Gorla, M. and Arenari, I. 1983. Male gametophytic competitive ability in maize. Selection and implications with regard to the breeding system. In *Pollen: Biology and Implications for Plant Breeding*. (D. Mulcahy and E. Ottaviano, eds.), Elsevier Biomedical, N.Y. pp. 367-373.

Pfahler, P. 1967. Fertilization ability of maize pollen grains II. Pollen genotype, female sporophyte, and pollen storage. *Genetics* 57: 513-521.

Stebbins, G.L. 1957. Self-fertilization and population variability in higher plants. *Am. Nat.* 91: 337-354.

Williams, E.G., O'Neill, P. and Hough, T. 1983. The pollen tube as an experimental system. In *Pollination '82* (E.G. Williams *et al.* eds.), Plant Cell Biol. Center, Univ. Melbourne, pp. 15-27.

Apomixis and Biosystematics

Sven Asker
Institute of Genetics
University of Lund
Lund, Sweden

INTRODUCTION

This review is written by a plant geneticist rather than a taxonomist. This is why, as some readers may feel, the discussion is not strictly limited to problems of biosystematics.

What is apomixis? As shown by Table I, vegetative reproduction, in my opinion, should not be included among the apomictic phenomena. Apomixis is then synonymous with *agamospermy*, or asexual seed formation. If residual sexual seed set occurs, we speak of facultative apomixis.

Some apomicts are independent of pollination, as both embryo and endosperm formation takes place autonomously. In this case, normal pollen formation often does not occur. In other cases, pollination is necessary for endosperm formation (*pseudogamy*). The central nucleus is then usually fertilized by a male sperm cell.

In *nucellar embryony*, an embryo is formed directly from a somatic cell in the ovule without the intervening formation of embryo-sacs and egg cells. In *gametophytic apomixis*, unreduced embryo-sacs are formed, whose egg cells (rarely some other cell, *apogamety*) develop parthenogenetically into an embryo. In *diplospory* and *apospory*, these unreduced embryo-sacs are formed from the primary EMCs by circumvention of meiosis and from somatic cells, respectively.

The author has mainly studied gametophytic apomixis in angiosperms. Special problems of nucellar embryony or of apospory in ferns, interesting as they may be, will not be discussed here.

Apomixis occurs in several angiosperm families. Gametophytic apomixis occurs especially in Asteraceae (Compositae), Rosaceae and Poaceae (Gramineae). Apomixis must have originated independently on several occasions, and under rather different conditions. It is tempting, but risky, to make wide generaliza-

TABLE I. *Principal modes of asexual reproduction in angiosperms*

VEGETATIVE REPRODUCTION

APOMIXIS (= agamospermy, asexual seed formation)
   Obligate or facultative.
   With or without pseudogamy (= fertilization necessary to
   initiate endosperm development).

I. GAMETOPHYTIC APOMIXIS
    Unreduced embryo-sacs formed.
    Parthenogenesis or apogamety (formation of an embryo
    without fertilization from the egg cell or another cell
    of the gametophyte, respectively).

A. *Diplospory:*   Embryo-sac develops from cells of the
                    archespore without reduction.
B. *Apospory:*   Embryo-sac develops from somatic cells of
                    the nucellus or integument.

II. ADVENTITIOUS (NUCELLAR) EMBRYONY
    Embryos arise directly from sporophytic tissue in the
    ovule.

---

tions proceeding only from material that one is familiar with.
   Most apomicts are polyploid, and many, if not all, are of
hybrid origin. Where sexual ancestors are known, they are usually
cross-breeding, often self-incompatible or dioecious.
   The consequences of apomictic reproduction are well known.
Many apomictic clones occur within the area of an agamic complex,
often separated by small differences only, and are usually related
to sexual taxa within the same area. They are often very
constant, their offspring maternal and uniform. Such clones are
often described as agamospecies, or apomictic microspecies. This
may be disputed by non-taxonomists; but the study of such
apomictic taxa has been of great importance, for instance, in
plant-geographical discussions.
   Classical studies in apomixis were reviewed by Gustafsson
(1946-1947). Later, less extensive reviews on this subject were
published, *inter alios,* by Stebbins (1950), Nygren (1954) and
Grant (1971). Among important earlier studies, which have in many
ways been models for later work, may be mentioned those of the
Crepis occidentalis complex (Babcock and Stebbins 1938), of the
Potentilla gracilis complex (Clausen *et al.* 1940) and of *Rubus*
(European blackberries; Gustafsson 1943).
   Concerning the taxonomic treatments of such groups, two extreme
opinions have been presented. Each apomictic clone could be
recognized as a species, as has partly been done in European
material of *Hieracium* and *Taraxacum*. This is practicable only

in groups of obligate apomicts that are relatively few in number
and separated by patent differences. The other extreme could be,
on the basis of a continuous series of intergradations between
various apomictic and sexual forms, to regard the entire complex
as a single species. Of course, the satisfactory treatment of
agamic complexes is one which is intermediate between these
extremes (Babcock and Stebbins 1938). Still, as stated by Grant
(1971), "no large complex has yet been thoroughly explored
taxogenetically."

Missing to some extent in the earlier investigations, and also
in most later ones, is a profound knowledge concerning the basis
of apomictic reproduction: its gene regulation and dependence on
environmental and physiological factors, and why and how it
originates and becomes fixed in natural populations. We also know
too little about the degree of genetic variation within and
between apomictic and related sexual populations. Although this
may seem a secondary theme for biosystematics, I think it is of
the utmost importance for the understanding of evolution and
differentiation of agamic complexes. Thus, such problems will be
the main topic of this paper.

APOMIXIS VERSUS SEXUALITY

Adaptations in Apomicts

As far as it is known, gametophytic apomixis always implies two
main changes as compared with normal sexual reproduction: first,
the formation of unreduced embryo-sacs and egg cells; second, the
predominant capacity of the egg cells to undergo parthenogenetic
development. These two changes seem to be of independent origin,
and to be regulated by different genes or groups of genes.

Some other changes of the reproductive cycle are also known
from apomictic taxa. Deviating endosperm formation, obviously an
adaptation in connection with pseudogamy, has been studied
especially in *Ranunculus auricomus* (Nogler 1978). Here, double
fertilization of central nuclei occurs, as well as fertilization
of single polar nuclei. Related sexual diploids have normal
endosperm formation. Precocious egg cell division, which
practically renders fertilization impossible is another modification of this kind, as well as, perhaps, the inhibition and
degeneration of meiosis in primary EMCs of certain aposporous
taxa.

Occasional Apomixis in Sexual Taxa

In sexual diploid plants unreduced egg cells - probably formed by
some kind of diplospory - are known to give rise to triploid
offspring at a low frequency. Their importance increases in
distant crosses and when meiosis is impaired, for instance because
of hybrid structure. Sometimes, in such cases, only unreduced egg
cells give rise to viable offspring. Nonfunctional aposporous

embryo-sacs have been observed especially within the Asteraceae.
Formation of haploids by occasional parthenogenetic development
of reduced egg cells is also a well-known phenomenon. The
frequency of haploid formation can be increased by delayed
pollination, or by the use of pollen inactivated by X-ray or UV
radiation, or by certain chemical compounds.

"Matromorphy," occasional parthenogenetic development of
unreduced egg cells, has been demonstrated in certain sexual
species – partly as a result of wide crosses – and has especially
been studied in *Brassica* (Eenink 1974). Because the occurrence
of matromorphy can only be proved with the aid of suitable markers
in crosses, the phenomenon is very likely to have been overlooked
and may be rather common.

Genetic Regulation of Apomixis

In sexual plants, the tendencies towards apomictic reproduction
are obviously, to a great extent, under polygenic control. In
plants where apomictic reproduction predominates, also single
genes with a strong effect upon the mode of reproduction seem to
be of importance.

Mutant genes – both spontaneous and induced – that promote
parthenogenesis or nonreduction of egg cells have been discovered
in various plant genera. In the obligately nonapomictic barley,
for instance, one recessive mutation, *tri*, in the homozygous
state leads to the formation of 50% flat kernels with triploid
embryos (Ahokas 1977), due to restitution in the 2nd meiotic
division on the female side. Male meiosis is regular. Another
barley mutant *hap* causes, when similarly in the homozygous
state, formation of 30% or more haploid offspring by parthenogenesis (Hagberg and Hagberg 1981).

Much evidence indicates that genes regulating the principal
components of apomixis – viz., as stated before, nonreduction and
capacity for parthenogenesis, respectively –are more or less
closely linked, for example, in *Ranunculus auricomus* (Nogler
1978) and in *Panicum maximum* (Savidan, personal communication).
In these two cases, a simple regulation by one pair of genes has
been demonstrated for apospory (presence of aposporous embryo-
sacs) versus sexuality, which is, however, to some extent blurred
by minor genes and environmental influence.

In these and perhaps some other cases, the genes for apomixis
may be included in some kind of a "super-gene" for apomixis. As a
super-gene has a tendency to accumulate additional units
(Darlington 1958; Ford 1964), it is possible that, for instance,
genes regulating endosperm development and earliness of egg-cell
division can also be included, but we know nothing about the
genetic regulation of these secondary phenomena.

APOMIXIS: DEPENDENCE ON ENVIRONMENTAL AND PHYSIOLOGICAL FACTORS

Various environmental factors, such as light and temperature
regimes, influence the balance between apomictic and sexual
reproduction (Knox and Heslop-Harrison 1963; Saran and de Wet
1976; Schmidt 1977). More studies in this field, using controlled
environments, are badly needed.

In facultative apomicts, the mode of reproduction is, in large
measure, decided by the choice of pollinator. Pollination from
distantly related taxa increases the rate of apomictic
reproduction.

Apomixis is strongly correlated with hybridization and
polyploidy. Paradoxically, hybridization and polyploidization may
cause a breakdown in apomictic reproduction. This may ensue in a
cross between two not too closely related facultative apomicts,
and it is reported to be the result of "incompatibility" between
different genetic systems for apomixis (Asker 1971).

As regards the effects of polyploidization and haploidization,
I will give an example from my work in *Potentilla argentea*.
Müntzing (1928) established the occurrence of apomictic seed
formation; the apomicts reproduce by apospory combined with
pseudogamy. Much cytogenetic work has been done in Eurasian
material of this and other apomictic *Potentilla* complexes (see,
e.g. Rutishauser 1948; Müntzing 1958). However, their taxonomy
remains a bit diffuse. This is understandable, however, when the
geographical area of such complexes is taken into consideration.
The Linnaean species *P. argentea,* according to the monograph
by Wolf (1908), is distributed from Spain across most of Europe
and part of western Asia as far as Mongolia and northwestern
China.

Of interest is the occurrence in northern Europe of diploid
apomicts, which are known from very few other genera with
gametophytic apomixis (Table II). Recently, I have also found
apomictic diploids (*P. argentea* L. *sensu stricto*) in material
from the Pyrenees. Morphologically deviating sexual diploids
(*P. calabra* Ten.) occur in other parts of southern Europe.
Apomictic polyploids (*P. neglecta* Baumg.), predominantly
hexaploids, occur in northern and central Europe. The diploid
apomicts are facultative as contrasted especially to the higher
polyploids, which are obligate apomicts. It must be kept in mind,
however, that the cytogenetic studies cover only a small part of
the total distributional area of *P. argentea* L. (coll.).

DIPLOID-TETRAPLOID-DIPLOID (DIHAPLOID) CYCLES IN *Potentilla*
AND IN THE PANICOIDEAE

After chromosome doubling of diploid *P. argentea* apomicts through
colchicine treatment, it turned out that the resulting autotetraploids were totally sexual (Asker 1971). The same was true also
of autotriploids and at least certain aneuploids (Table III).

TABLE II. *Variation in* Potentilla argentea L. *(coll.)*

Apomicts (apospory + pseudogamy)	$2n = 14$ facultative	spp. *argentea* L.
	$2n = 28,35,42,56$ more or less obligate	spp. *neglecta* Baumg.
Sexuals (southern Europe)	$2n = 14$	spp. *calabra* Ten.

Chromosome doubling and other changes of ploidy here seem to interfere with the processes leading to formation of aposporous embryo-sacs. However, "recombined diploids" in the offspring from triploid or trisomic plants were again facultatively apomictic, as shown by the outcome of test-crosses.

A monoploid with 7 chromosomes set a lot of seeds after pollination from diploids, but the seeds gave rise to diploid offspring only (Asker 1983). Either its 7-chromosome egg cells had no capacity for parthenogenetic development or the 7-chromosome embryos and/or seedlings were inviable. Even in polyploid *argentea* biotypes, chromosome doubling may lead to a partial breakdown of apomixis.

This alternation between apomixis and sexuality, dependent on level of ploidy, brings into mind the so-called diploid-tetraploid-dihaploid cycles in grasses belonging to the Panicoideae (de Wet 1968; de Wet and Harlan 1970; Savidan 1978). As compared with other taxa, the latter group has one great advantage for apomictic studies: they have four-nucleated aposporous embryo-sacs, that can easily be distinguished from the eight-nucleated sexual ones, even in squash preparations.

In the *Bothriochloa-Dichanthium* complex, tetraploid facultative apomicts predominate, but a small part of the populations consists of sexual diploids. Cultivated material of *Panicum maximum* consists of apomictic tetraploids, but local populations in Kenya contain sexual diploids. Sexual tetraploids, produced from them by chromosome doubling, are used in the breeding programme. Apomictic diploids do not exist. On higher levels than tetraploidy, only apomicts are known. Thus, the relationship between diploids and tetraploids here presents quite a contrast to the conditions in *Potentilla argentea*.

ORIGIN AND FIXATION OF APOMIXIS

What factors make possible the origin and fixation of apomixis in natural populations? The course of events must be different in hermaphroditic and dioecious populations, and in the case of diplospory (associated with obligate apomixis and occurrence of triploid apomicts) as compared with facultative apospory (where

TABLE III. *Effects of changed ploidy in diploid apomictic* Potentilla argentea

"Recombined" Diploids	*facultative apomict*
↑ X diploid	
Trisomics A.O. Aneuploids	(mainly)*sexual*
↑ X diploid	
Autotriploid	*sexual*
↑ X diploid	
Autotetraploid	*sexual*
↑ colchicine	
Diploid Argentea Biotype	*facultative apomict*
I X tetraploid	
Monoploid	*sexual?*

triploids do not often occur).

Although from theoretical points of view, sexuality could imply a waste of resources in comparison with asexual reproduction (Maynard—Smith 1971; Williams 1975), apomixis has hardly become established in diploid plants. It must have usually arisen in polyploids of hybrid origin derived from outbreeding, strongly heterozygous sexual species.

It has been suggested that genes necessary for functional apomixis (which, occurring alone, could sometimes reduce fitness) are brought together by hybridization to directly produce an apomict. Often, however, functional apomixis is not produced by hybridization. The vigorous, but sexually sterile, hybrid has means to maintain itself obligately by vegetative reproduction. Here, mutations favoring asexual seed formation must have an enormous selective value (the same would be true for an odd polyploid). Mass reproduction and long distance propagation of highly fit genotypes would become possible, independent of meiotic disturbances and (as regards nonpseudogamous taxa) of problems relating to pollination.

Once apomictic reproduction has become established, its genetic system is likely to be further improved by natural selection. A final step could be events leading to a close coupling between the genetic determinants. By translocations or transpositions, they may be transferred to the same chromosome. Then a super—gene is formed: the genes become linked by inversions, are brought more closely together by new translocations, or else the chiasma frequency between them is adaptively strongly reduced.

During recent years, plant breeders have discussed the possibility of fixation of heterosis in sexual crops by inducing apomictic seed formation. Different means to achieve this have been proposed or tried (Asker 1979, 1980; Matzk 1982): namely, crossing crops with related wild apomicts followed by backcrosses to obtain cultivars with genes for apomixis; intergeneric crosses and selection for apomictic reproduction among the sterile hybrids; and bringing together mutant genes from different sources to form a functional system for apomixis. So far, no new apomictic cultivars of commercial importance have been produced. But this work has already yielded important data concerning the regulation of apomixis, which will be of relevance for future biosystematic studies.

BIOTYPE FORMATION AND VARIATION WITHIN AGAMIC COMPLEXES

The evolutionary fate of the newly-formed apomict depends of course, to a great extent, on its degree of residual sexuality. But even within obligately apomictic taxa, evolution does not come to a standstill. Mutations – even somatic ones – accumulate and favorable genotypes are preserved. Unreduced egg cells are sometimes fertilized by pollen from the same or other biotypes, giving rise to taxa with higher degrees of ploidy; and pollen especially from pseudogamous apomicts may fertilize sexual forms.

The somewhat cryptic term autosegregation refers to "female changes or rearrangements of the genotype" (Gustafsson 1946-1947). Hypo- and hyperploid aberrant types are formed in *Taraxacum* as a result of irregular female meiosis. In other cases, aberrant forms with unchanged chromosome number occur, for instance following occasional bivalent formation and crossing over in diplosporous plants.

Morphological and cytological variation within an agamic complex with a low degree of outbreeding can be exemplified by *Potentilla tabernaemontani* L. (coll.) and related species on the island of Gotland. Here, close to their northern border, these taxa are very common and polymorphous, especially on limestone. Certain varieties and forms are described by Johansson (1905), known for his flora of Gotland. I had no difficulty in identifying his taxa in the field. They can be regarded as good, rather "broad" agamospecies, usually well separated (although rare intermediates occur), but each taxon does also contain some morphological variants. These taxa are morphologically, perhaps also genetically, much more separated than several *Taraxacum* and *Hieracium* microspecies.

To date, chromosome numbers have been determined for 58 of 82 collected clones. As seen by Table IV, two thirds of the clones have the hexaploid number $2n = 42$. The other numbers observed so far are, in decreasing frequency, $2n = 49$, 56, and 63. Cytological variation occurs within most morphological types, a picture similar to that in *Poa* and within *Boehmeria* (Yahara 1983).

The *tabernaemontani* populations (*croceolata* is sometimes given specific rank) are connected with the related species *P. arenaria* and *P. collina* by intermediate forms, probably of hybrid origin. No sexual forms have been revealed here, or in other Scandinavian localities, but probably the overall frequency of sexual reproduction in this group – with apospory combined with pseudogamy, like *P. argentea* – is higher than in apomictic

TABLE IV. *Chromosome number in* Potentilla tabernaemontani
apomicts from Gotland

	2n =	42	49	56	63	?	Total
"normal type"		11	6	1	–	4	22
"*concaviflora*"		8	2	–	–	3	13
"*parviflora*		7	–	–	–	5	12
"*erythrodes*"		4	2	3	1	3	13
"*obcordipetala*"		3	–	2	1	5	11
"*croceolata*"		4	–	–	–	2	6
"X *incana*"		1	1	–	1	2	5
TOTAL		38	11	6	3	24	82

Nomenclature according to Johansson (1905)

Taraxaca and Hieracia.

The populations on Gotland ought to have been isolated from
those in the Swedish mainland and other Baltic islands for perhaps
more than 5,000 years. But some of the morphologically different
types, which have partly different ecological preferences –
although sometimes three or four types occur in close proximity –
may be of much greater age, as taxa similar to some of the
varieties reappear in material from other parts of Europe, for
example in mountainous regions of Spain and Italy.

A measure of the degree of inter- and intrapopulational genetic
variation can be obtained by isozyme studies. Unfortunately, as
far as I know, no comprehensive study of this kind has been done
in apomictic plants. The polyploidy of apomictic taxa need not
render such studies impossible, as evidenced, for instance, by
work on parthenogenetic insects and their sexual relatives by
Suomalainen and Saura (1973) and Lokki *et al.* (1975). These
studies allowed important conclusions concerning the origin and
evolution of apomixis in weevils (Curculionideae) and the moth
genus *Solenobia*.

The problems to deal with in comparisons between related
apomictic taxa, where crossing experiments are impracticable, are
similar to those in comparisons between genera and higher
taxonomic units in groups with sexual reproduction. Especially
in animal taxonomy, DNA and amino acid sequencing have proved to
be good tools in such cases. Although some apomictic plant
complexes are of recent origin, and the taxonomic distance between
their parts might not have become very large, I think biochemical
techniques in this field will become very important in the future.

THE EVOLUTIONARY PROSPECTS OF APOMICTS

The short-term advantages of apomixis have already been touched
upon. Darlington coined the expressions "a stabilization of
hybridity" (1932) and "an escape from sterility" (1939). Apomixis

is one of the adaptations promoting immediate fitness in temporary
habitats (Stebbins 1958). The preservation of heterosis, and
sometimes the independence of pollination, are useful in this
connection.

But at least when obligate apomixis is concerned, most authors
agree that the future prospects are not very good. The loss of
sexuality and meiotic recombination obstructs the formation of new
biotypes capable of survival in the wake of drastic environmental
changes. The conclusions are that apomixis is "a blind alley of
evolution" (Darlington 1939) and "evolutionary opportunism carried
to its limit" (Stebbins 1950). Agamic complexes with only
obligate apomicts seem, however, to be fewer in number than was
once believed. During more recent years, traces of sexuality have
been found in taxa within *Taraxacum*, *Hieracium* and *Potentilla*
all of which were formerly thought to be obligate apomicts (see,
e.g. Asker 1970; Gadella 1972; Malecka 1973; Richards 1970;
Skalinska 1976).

Complexes with facultative apomicts should be better equipped,
since they are capable to combine the advantages of sexual and
apomictic reproduction. The opinions concerning the evolutionary
potential of such complexes are, however, somewhat discordant.
According to Grant (1971), "intermediate degrees of asexuality, as
in facultative apomixis, restrict the generation of variability to
an intermediate extent... Retention of sexuality in a plant group
is no guarantee of long-range evolutionary success." Khokhlov
(1976), on the other hand, regarded apomixis as "a regular step of
progressive evolution." An indirect proof, however, of the
restricted evolutionary potential of apomicts could be the fact
that apomictic complexes have not given rise to new families or
genera.

We know of examples of agamic complexes that obviously
represent different stages of development. There are young
complexes with a restricted area of distribution and mature
complexes with a multitude of agamospecies related to sexual
diploids (and polyploids). Finally, there are old, "dying"
complexes like *Houttuynia* (Babcock and Stebbins 1938), where
the sexual relatives are extinct. What is left are a few
agamospecies without future prospects. It is difficult to be
certain if this is the final state of facultatively, as well as
obligately, agamic complexes. But as stated by Gustafsson (1946-
1947), "is this not typical of all species and genera? Like
separate individuals, species and genera are born, bloom and
die."

REFERENCES

Ahokas, H. 1977. A mutant of barley: Triploid inducer. *Barley
Genet. Newsl.* 7: 4-6.
Asker, S. 1970. Apomixis and sexuality in the *Potentilla
argentea* complex. II. Crosses within the complex. *Hereditas*
66: 189-204.

Asker, S. 1971. Apomixis and sexuality in the *Potentilla argentea* complex. III. Euploid and aneuploid derivatives (including trisomics) of some apomictic biotypes. *Hereditas* 67: 111-142.

Asker, S. 1979. Progress in apomixis research. *Hereditas* 91: 231-240.

Asker, S. 1980. Gametophytic apomixis: elements and genetic regulation. *Hereditas* 93: 277-293.

Asker, S. 1983. A monoploid of *Potentilla argentea*. *Hereditas*, in press.

Babcock, E.B. and Stebbins, G.L. 1938. The American species of *Crepis*. Their interrelationships and distribution as affected by polyploidy and apomixis. *Carnegie Inst. Washington Publ.* 504.

Clausen, J., Keck, D. and Hiesey, W. 1940. Experimental studies on the nature of species. *Carnegie Inst. Washington Publ.* 520.

Darlington, C.D. 1932. *Recent Advances in Cytology*. 1st ed. Churchill, London.

Darlington, C.D. 1939. *The Evolution of Genetic Systems*. 1st ed. Cambridge Univ. Press. 1958, 2nd ed.

De Wet, J.M.J. 1968. Diploid-tetraploid-haploid cycles and the origin of variability in *Dichanthium* aganospecies. *Evolution* 22: 394-397.

De Wet, J.M.J. and Harlan, J.R. 1970. Apomixis, polyploidy and speciation in *Dichanthium*. *Evolution* 24: 270-277.

Eenink, A.H. 1974. Matromorphy in *Brassica oleraceae* L. I. Terminology, parthenogenesis in Cruciferae and the formation and usability of matromorphic plants. *Euphytica* 23: 429-433.

Ford, E.B. 1964. *Ecological Genetics*. Methuen, London.

Gadella, T.W.J. 1972. Biosystematic studies in *Hieracium pilosella* L. and some related species of the subgenus *Pilosella*. *Bot. Not.* 125: 361-369.

Grant, V. 1971. *Plant Speciation*. Columbia Univ. Press, N.Y.

Gustafsson, A. 1943. The genesis of the European blackberry flora. *Lunds Univ. Arsskr. Avd. 2, 39*: 3-199.

Gustafsson, A. 1946-1947. Apomixis in higher plants. *Lunds Univ. Arsskr. Avd. 2, 42-43*: 1-370.

Hagberg, G. and Hagberg, A. 1981. Haploid initiator gene in barley. *Proc. 10th Int. Barley Genet. Symp.*, Edinburgh, pp. 686-689.

Johansson, K. 1905. Kenntnis des Formenkreises der *Potentilla verna* (L. ex P.) LEHM. et auct. plur., mit besonderer Berücksichtigung der Gottländischen Formen. *Ark. Bot.* 4: 1-18.

Khokhlov, S.S. 1976. Evolutionary-genetic problems of apomixis in angiosperms. In *Apomixis and Breeding, Amerind., New Delhi*. pp. 3-17.

Knox, R.B. and Heslop-Harrison, J. 1963. Experimental control of aposporous apomixis in a grass of the Andropogoneae. *Bot. Not.* 116: 127-141.

Lokki, J., Suomalainen, E., Saura, A. and Lankinen, P. 1975.

Genetic polymorphism and evolution in parthenogenetic animals.
II. *Genetics* 79: 513-525.

Malecka, J. 1973. Problems of the mode of reproduction in
microspecies of *Taraxacum* section *Palustria* Dahlstedt.
Acta. Biol. Crac., Ser. Bot. 16: 37-84.

Matzk, F. 1982. Vorste lungen über potentielle Wege zur Apomixis
bei Getreide. *Arch. Zuchtungsforsche.* 12: 183-195.

Maynard-Smith, J. 1971. The origin and maintenance of sex. In
Group Selection (G.C. Williams ed.), *Aldone-Atherton, Chicago,*
pp. 163-175.

Müntzing, A. 1928. Pseudogamie in der Gattung *Potentilla*.
Hereditas 11: 267-283.

Müntzing, A. 1958. Heteroploidy and polymorphism in some
apomictic species of *Potentilla*. *Hereditas* 44: 280-329.

Nogler, G.A. 1978. Zur Zytogenetik der Apomixie bei *Ranunculus
auricomus*. *Habilitationsschrift ETH Zurich,* 916, 704: 218 H.

Nygren, A. 1954. Apomixis in the angiosperms II. *Bot. Rev.*
20: 577-649.

Richards, A.J. 1970. Eutriploid facultative agamospermy in
Taraxacum. *New Phytol.* 69: 761-774.

Rutishauser, A. 1948. Pseudogamie und Polymorphie in der Gattung
Potentilla. *Arch. Julius Klaus-Stift. Vererbungsforsch.*
23: 267-424.

Saran, S. and De Wet, J.M.J. 1976. Environmental control of
reproduction in *Dichanthium intermedium* (Gramineae).
Bull. Torrey Bot. Club 97: 6-13.

Savidan, Y. 1978. Genetic control of facultative apomixis and
application in breeding *Panicum maximum*. *Communication XIV
Int. Congr. Genet., Moscow,* 21-30/8 1978. *Off. Rech. Sci.
Tech. Outre-Mer, Abidjan, Ivory Coast.*

Schmidt, H. 1977. Contributions to the breeding of apple stocks.
4. On the inheritance of apomixis. *Z. Pflanzenzücht.* 78: 3-12.

Skalinska, M. 1976. Cytological diversity in the progeny of
octoploid facultative apomicts of *Hieracium aurantiacum*.
Acta. Biol. Crac., Ser. Bot. 19: 39-46.

Stebbins, G.L. 1958. Longevity, habitat, and release of genetic
variability in the higher plants. *Cold Spring Harbor Symp.
Quant. Biol.* 23: 365-378.

Suomalainen, E. and Saura, A. 1973. Genetic polymorphism and
evolution in parthenogenetic animals. I. Polyploid
Curculionideae. *Genetics* 74: 489-508.

Williams, G.C. 1975. *Sex and Evolution.* Princeton Univ. Press.

Wolf, T. 1908. Monographie der Gattung *Potentilla*. *Bibl. Bot.*
16: 71.

Yahara, T. 1983. A biosystematic study on the local populations
of some species of the genus *Boehmeria* with special
reference to apomixis. *J. Fac. Sci. Univ. Tokyo, Sect.
III,* 13: 217-261.

Constraints on the Evolution of Plant Breeding Systems and Their Relevance to Systematics

C. J. Webb

Botany Division
D.S.I.R.
Christchurch, New Zealand

INTRODUCTION

Stebbins (1973), in a discussion of the evolution of the
angiosperm inflorescence, outlined two principles which are of
importance in understanding how evolution is guided by natural
selection. The first of these, *conservation of organisation*
(Stebbins 1973), states that structures which are formed by a
complex sequence of developmental processes tend to be conserved
over long periods of evolution. The second is essentially the
converse – that *adaptive modification occurs along the lines of
least resistance* (Ganong 1901; Stebbins 1950, 1967). These two
principles underlie what might be more simply termed *phylogenetic
constraint:* the way in which the evolutionary history of a
taxon influences the ease with which particular characters can
respond to selection. The importance of phylogenetic and
developmental constraints in modifying the course of evolution as
determined by natural selection has also been reemphasised
recently in the zoological literature (e.g. Gould and Lewontin
1979; Alberch 1982).

When considering the evolution of the breeding system or
pollination system of particular species or genera, it is
important to appreciate the constraints which operate on
reproductive characters in that family. The extent to which
particular characters are constant or labile will also determine
their usefulness to biosystematics. In general, more conservative
characters will be useful in defining families and orders whereas
more labile characters may be useful at generic and specific
levels.

This paper examines four plant groups in the Umbelliferae,
Gentianaceae, Ranunculaceae, and Elaeocarpaceae which illustrate
different levels of constraint on stamen number. Stamen number is
then related to pollen production, ovule number, sexual system,
and pollination system in order to demonstrate how the particular
groups respond differently to selective pressures on pollen/ovule
ratio. Finally, the usefulness of these reproductive characters
in the biosystematics of these groups is examined.

MATERIALS AND METHODS

Data presented are based on studies of entomophilous plants, all
indigenous to New Zealand except *Muntingia calabura* which
was observed in a riparian forest in lowland Guanacaste Province,
Costa Rica. Further information on the reproductive biology of
this species has been published in detail elsewhere (Bawa and Webb
1983; Webb 1984). Habitat and distribution information for all
other species are given in Tables I, II, IV and V.

For most species, the number of pollen grains per anther was
determined separately for each of four plants from which a sample
of five anthers was removed. Pollen was placed in a viscous
medium of glycerol and lactic acid, agitated, and then four counts
made using a haemocytometer. For each species of *Ranunculus*
two anthers were used from each of ten herbarium sheets. In a few
species with very low pollen numbers, absolute counts were made on
squashed anthers stained with safranin.

In the Umbelliferae, species of all mainland indigenous genera
were included in the sample. For larger genera several species
were selected to cover a range of habits and breeding systems.
For those umbellifers with only hermaphrodite flowers,
descriptions of sexual system, dichogamy, and pollination are
based on observations of field plants and plants grown under
controlled greenhouse conditions at Lincoln, Canterbury, New
Zealand. For umbellifers with unisexual flowers, counts of flower
numbers and floral sexuality were made from samples of one
inflorescence per plant for 30 plants in each population. Sex
ratios were based on at least 100 plants, or on all plants present
in smaller field populations. Pollen/ovule ratios (henceforth
abbreviated P/Os) for sexually dimorphic populations were
calculated from the average pollen and ovule production of male
and female plants, the relative flower numbers of the two sexes,
and the population sex ratio. Breeding systems of these species
are defined following Webb (1979).

The 13 species of *Gentiana* were selected to include
species which are annual, perennial but monocarpic, and perennial
and polycarpic. Ovule counts were made for 20 or more flowers,
each flower collected from a different plant. Descriptions of
dichogamy and life history are based on field observations for
most species and detailed studies in the greenhouse for *G.
saxosa* and *G. serotina*.

All five lowland New Zealand species of *Ranunculus* sect.

Epirotes (Fisher 1965) were examined. For each species, ten
herbarium specimens (CHR) were selected to cover the range of the
species in the North and South Islands of mainland New Zealand.
The number of stamens and carpels per flower were counted for two
young flowers or flower buds from each sheet to give a sample of
20 flowers for each species.

RESULTS AND DISCUSSION

1. Umbelliferae: stamen number constant (5), ovule number
   constant (2).

   As a family the Umbelliferae are very constrained in respect of
variation in the number of floral parts. Stamen number is almost
invariably five, and ovule number, at least for hermaphrodite
flowers, is almost always two. There are rare exceptions where
ovule number is reduced to one (Magin 1980) and reduction in
stamen number of hermaphrodite flowers to four or two is even less
common (Webb 1980). Nevertheless, there are other characters
associated with the distribution of stamens, and with pollen
production, which might be of use to biosystematics.
   New Zealand Umbelliferae can be conveniently divided into two
groups considered separately here: cosexual species where all
individuals are able to produce functional pollen and seeds, and
sexually dimorphic species where individuals specialise as seed or
pollen parents.
   All of the cosexual species have only hermaphrodite flowers and
in all except one species most or all flowers produce a 2-seeded
schizocarp. The exception is *Actinotus novae-zelandiae*
which produces only one seed from each flower and is further
distinguished by having only two stamens per flower. All species
are polycarpic perennials except *Daucus glochidiatus* which
is annual. Although andromonoecism is a common sexual system in
Umbelliferae none of the New Zealand cosexual species is
andromonoecious; interestingly, the sexually dimorphic species
appear to have evolved from andromonoecious ancestors (Webb 1979).
   As might be expected P/Os can be related to outcrossing within
this group (Table I). Species with larger flowers (e.g. *H.
novae-zelandiae*) or flowers clustered in conspicuous heads
(e.g. *E. vesiculosum*) or larger umbels (e.g. *A. prostratum,
S. haastii*) have more pollen per anther and consequently higher
P/Os than species with small, fewer-flowered heads (e.g. *A.
novae-zelandiae, H. americana*).

Cosexual Species

   In this group of umbellifers a change in P/O is not achieved by
a change in stamen number per flower (except within the genus
Actinotus), nor is it effected by the proportion of
unisexual flowers or individuals. Differences in P/O are solely
accounted for by changes in the amount of pollen per anther.

TABLE I *P/Os for New Zealand species of Umbelliferae with only hermaphrodite flowers*

	Pollen/Anther (mean)	P/O	Flower diameter (mm)	Habitat
Hydrocotyloideae				
*Actinotus novae-zelandiae	310	610	1.4	South and Stewart Is, montane to alpine, boggy sites of wetter areas.
Hydrocotyle americana	110	280	1.5	Mainland N.Z., lowland, forest margins, clearings, scrubland.
H. elongata	320	800	2.6	Mainland N.Z., lowland to montane, forest margins, clearings and scrubland.
H. novae-zelandiae	620	1,550	2.4	Mainland N.Z., lowland to montane, glassland and forest margins.
Centella uniflora	810	2,030	1.9	Mainland N.Z., lowland, wet sites.
Schizeilema cockaynei	1,250	3,130	no data	South and Steward Is, lowland, damp places and salt meadows.
S. exiguum	1,800	4,500	2.5	South I, subalpine, herbfield.
S. haastii	3,250	8,130	2.7	Mainland N.Z., montane to subalpine, rocky places.

TABLE I (Cont'd.) *P/Os for New Zealand species of Umbelliferae with only hermaphrodite flowers*

	Pollen/Anther (mean)	P/O	Flower diameter (mm)	Habitat
Saniculoideae				
Eryngium vesiculosum	1,520	3,900	1.6	Mainland N.Z., coastal, sands and gravels.
Apioideae				
Daucus glochidiatus	240	600	1.5	Mainland N.Z., lowland, open dry sites.
Apium prostratum	930	2,330	1.9	Mainland N.Z., coastal, rocks and sand.
Oreomyrrhis colensoi	1,070	2,680	2.7	Mainland N.Z., montane to subalpine, grassland.
Lilaeopsis novae-zelandiae	1,300	3,250	2.4	South I, lowland to montane, lake margins and stream edges.

* 2 stamens/flower and 1 seed/fruit only in this species.

Stamen number per flower is of no use to systematics except that within *Actinotus* the reduction to two stamens per flower distinguishes *A. novae-zelandiae* from all but two Australian taxa. However, other characters associated with the pollination system as indicated by P/Os are of use in distinguishing species within genera. Within *Hydrocotyle,* for example, P/Os vary considerably and can be related to the pollination system. *H. elongata* has conspicuous umbels held well above the leaves (Fig. 1A), relatively large flowers within which anthers dehisce before the stimga becomes bulbous and transluscent (Fig. 1C), and the P/O is relatively high (Table I). All these characters indicate outcrossing. In contrast, *H. americana* has inconspicuous, more or less sessile umbels hidden among the leaves (Fig. 1B), smaller flowers within which anther dehiscence and stigma receptivity appear to be simultaneous (Fig. 1D), only a third the number of pollen grains per anther compared with *H. elongata,* and it is probably largely self-pollinated. Such consistent and distinctive characters have not previously been used in distinguishing species of *Hydrocotyle* in New Zealand (Allan 1961).

Sexually Dimorphic Species

All species of the other five New Zealand genera of Umbelliferae (all Apioideae) are sexually dimorphic although populations which are andromonoecious, presumably secondarily, occur in a few species of *Lignocarpa* and *Gingidia*. In most populations of *Gingidia, Scandia* and *Lignocarpa* two forms of plant are found: female plants with female flowers only and male plants with variable proportions of male and hermaphrodite flowers. *Aciphylla* and *Anisotome* are dioecious (Table II). Species of all five genera are polycarpic, long-lived perennials.

It is not obvious what method one should use to calculate P/Os for sexually dimorphic species. The values given in Table II take into account the porportion of male and hermaphrodite flowers in male plants and the frequency of female plants in the population, to give an overall P/O for the population. Using this method of calculation, P/Os for dioecious species tend to be higher than those for gynodioecious species – this might be expected as strict dioecism enforces outcrossing. It is also noticeable that the P/Os for sexually dimorphic species are in general an order of magnitude greater than for the cosexual species (Table I). Some of this difference is accounted for by the male-biased sex ratios of sexually dimorphic species, but they also have more pollen grains per anther on average.

Within *Lignocarpa,* an endemic genus of only two species, the pattern of variation in P/O with sexual system can be seen more clearly (Table III). *L. diversifolia* is near dioecious,

has a relatively high number of grains per anther and a higher
P/O. *L. carnosula* is andromonoecious in some populations,
but is usually gynodioecious with the fruit set of male plants and
sex ratio varying among populations. In those populations
examined so far, pollen per anther and P/O vary inversely with the
proportion of hermaphrodite flowers on male plants. Thus, as
fruit set is reduced on male plants and populations approach
dioecism, allocation to pollen production is increased in male
plants probably because success through pollen can increasingly be
achieved only by outcrossing.

Variation in number of grains per anther among species is not
great, in fact the highest number recorded, that for *G. enysii*,
is only 3.5 times that for the least in *An. filifolia*.
This is in marked constrast to the variation of from 110 to 3,250
grains per anther (29.6 times) found in the cosexual species or
even 5.6 times recorded among three species of *Hydrocotyle*.

This difference between the two groups may be accounted for in
part by the greater diversity of the cosexual sample which
includes representatives of all three subfamilies of the
Umbelliferae. However, the sexually dimorphic species are also
less diverse in their pollination systems as all have relatively
large, conspicuous, compound inflorescences and the flowers are
pollinated by a diversity of insects. The amount of pollen per
anther may also vary less in this group because there are two
additional ways in which P/Os can be altered, viz., pollen numbers
can be increased by adding more male flowers in male plants, and
ovule numbers vary with the frequency of female plants in the
population.

Within the sexually dimorphic umbellifers, stamen number per
flower is again of no biosystematic use. However, the extent to
which staminodes are reduced has been used as a generic character
(Dawson 1967a, b) and the breeding system is also useful in
defining generic limits (Dawson and Webb 1978). In female flowers
of both *Aciphylla* and *Anisotome,* staminodes are macroscopic
and both genera are dioecious. Staminodes are microscopic in
Gingidia, Lignocarpa, and *Scandia,* all of which are
basically gynodioecious. Further, the proportion of hermaphrodite
flowers in male plants of gynodioecious species may be used
in a few cases to distinguish species within genera as, for
example, in the two species of *Lignocarpa* (Table III). The
number of pollen grains per anther and P/O do not seem
particularly useful biosystematic characters here.

Although stamen number is under strong phylogenetic constraint
in the Umbelliferae, P/O ratios may be adjusted to suit the
plant's reproductive biology in several ways. In cosexual species
the amount of pollen per anther varies, whereas in sexually
dimorphic species P/Os vary with the proportion of unisexual
flowers. It should be noted that while adding male flowers to a
strictly hermaphrodite species affects the P/O in the same way as
increasing the amount of pollen in existing flowers, there are
additional consequences of adding male flowers - male flowers may

TABLE II *Breeding system and P/Os for the sexually dimorphic New Zealand umbelliferous genera* Aciphylla, Anisotome, Gingidia, Lignocarpa *and* Scandia

	% ⚲ Flowers on male plants (mean)	Sex ratio (♂/♀ plants)	Breeding system	Pollen/ Anther (mean)	P/O for population	Habitat
S. rosifolia	1.63	1.00	near-dioecious	1,970	7,300	North I, coastal to montane, cliffs and rock outcrops.
G. trifoliolata	35.86	25.11	gyno-dioecious	2,590	9,200	South I, montane to subalpine, bogs and river courses.
G. montana	34.12	1.88	"	2,380	9,600	Mainland N.Z., montane to subalpine, streamsides, banks and grassland.
L. carnosula	43.96	1.45	"	3,810	9,700	South I, alpine rock screes.
G. baxterae	11.40	2.47	"	2,000	13,600	South I, montane to subalpine, grassland, forest margin and scrub.
S. geniculata	27.27	13.79	"	1,800	14,700	Mainland N.Z., lowland, dry scrubland.

TABLE II (Cont'd.) *Breeding system and P/Os for the sexually dimorphic New Zealand umbelliferous genera* Aciphylla, Anisotome, Gingidia, Lignocarpa *and* Scandia

	% ♀ Flowers on male plants (mean)	Sex ratio (♂/♀ plants)	Breeding system	Pollen/ Anther (mean)	P/O for population	Habitat
G. flabellata	38.39	302.03	near-andro-monoecious	3,100	17,000	Stewart I, coastal to subalpine, peaty turf and rock crevices.
L. diversifolia	1.88	1.78	near-dioecious	4,160	29,000	South I, alpine, rock screes.
G. enysii	22.79	31.75	gyno-dioecious	4,840	50,100	South I, montane to subalpine, rock outcrops.
G. decipiens	13.36	5.30	"	3,190	51,100	South I, subalpine to alpine, herbfield and stony sites.
Ac. monroi	0	3.03	dioecious	2,060	67,500	South I, alpine, grassland and herbfield.
An. filifolia	0	4.0ʳ	"	1,380	70,000	South I, montane to subalpine, forest margin and grassland.

FIG. 1 *Inflorescences and flowers* of Hydrocotyle.
A: H. elongata, *inflorescence raised above leaves* (X0.5).
B: H. americana, *inflorescences subsessile (arrowed,* X1.75).
C: H. elongata, *anthers dehiscent, stigmas immature*
(arrowed, X10). *D:* H. americana, *anthers dehiscent, stigmas*
receptive (arrowed, X12.5).

be produced more cheaply than hermaphrodite flowers and so greatly
increase the floral display and floral rewards as well as
spreading the time over which pollen is presented.

2. *Gentiana*: stamen number constant (5), ovule number
   labile (3-116).

In New Zealand species of *Gentiana s.l.* nearly all
flowers are hermaphroditic; there are five stamens surrounding a

TABLE III. *P/Os for different populations of* Lignocarpa

	% ♀ Flowers on male plants	Sex ratio (♂ / ♀ plants)	Pollen/ Anther (mean)	P/O for population
L. carnosula - 1	80.14	no	2,190	6,830
L. carnosula - 2	52.30	1.89	2,250	5,980
L. carnosula - 3	43.96	1.45	3,810	9,680
L. diversifolia - 1	1.88	1.78	4,160	29,030

central superior ovary which is topped by a 2-lobed stigma. The only exception to this pattern is the presence of a low frequency of female plants in some populations of a few species (Burrows and Hobbs 1964; Simpson and Webb 1980); in these plants all five stamens are reduced to staminodes or occasional flowers have only 1-4 functional stamens. In contrast with the Umbelliferae, however, ovule number is somewhat variable within a population and also varies considerably among species (Table IV).

Given the constraint of five anthers per flower it is again not surprising that the number of pollen grains per anther varies widely among the 13 species studied, ranging from only 140 grains in *G. lineata* to almost 10,000 in *G. corymbifera* (Table IV). Variation in both pollen and ovule numbers reflects to some extent the size of flowers and overall size of plants. Nevertheless, P/Os also vary and can be related to the pollination system of the species. At one extreme are the larger outcrossed mainland species such as *G. corymbifera* and *G. divisa* with P/Os of 400 to 850. In these species, flowers are usually strongly protandrous, anthers are versatile, dehisce extrorsely, and are positioned below or more or less at the level of the stigma from which they are drawn back before the stigmatic lobes separate (Fig. 2A, B). At the other extreme, *G. lineata* has a P/O of only 20. This species occurs on often cloudy mountain tops of the southern South Island and Stewart Island; its anthers are dehiscent simultaneously with stigma receptivity, are positioned at the same level as the stigma, and dehisce introrsely (Fig. 2C). In fact, on overcast days the flowers of *G. lineata* barely open and the dehiscent anthers are more or less in contact with the stigma (Fig. 2C), on sunny days the fully open petals which are adnate to the filaments for over half their length, draw the stamens away from the stigma (Fig. 2D). Lower P/Os are also found in monocarpic species which are not strongly adapted to outcrossing. *G. antarctica* and *G. antipoda,* both species of New Zealand's subantarctic islands, have similar numbers of grains per anther but markedly different ovule numbers; P/Os are therefore very different although Godley (1982) noted no difference in selfing potential as indicated by the extent of dichogamy. P/Os in *Gentiana* are generally low; even those

TABLE IV Breeding systems and P/Os for some New Zealand species of Gentiana

	Ovules/Flower (mean)(range)		Polle. Anther (mean)	P/O	Life cycle*	Protandry**	Habitat
G. lineata	35.15	17-59	140	20	polycarpic	absent	Southern South I, Stewart Is, alpine herbfield.
G. gibbsii	66.90	52-88	1,100	80	annual	weak	Stewart I, alpine grassland.
G. antarctica	18.86	10-27	760	200	monocarpic	present	Campbell I, bogs and tussock grassland.
G. grisebachii	45.40	27-59	3,050	340	annual	present	Mainland N.Z., montane, grassland, wet sites.
G. montana	56.40	41-77	4,560	400	polycarpic	strong	South I, montane to subalpine, grassland and herbfield.
G. townsonii	55.90	40-69	4,470	400	polycarpic	present	South I, subalpine grassland.
G. amabilis	48.05	32-63	4,350	450	polycarpic	strong	South I, subalpine to alpine, grassland and herbfield.
G. corymbifera	93.25	70-116	9,880	530	monocarpic to polycarpic	strong	South I, montane to subalpine, grassland and herbfield

TABLE IV (Cont'd.) *Breeding systems and P/Os for some New Zealand species of Gentiana*

	Ovules/Flower (mean) (range)		Pollen/Anther (mean)	P/O	Life cycle*	Protandry**	Habitat
G. antipoda	4.85	3–9	630	650	monocarpic	present	Antipodes I, bogs and tussock grass-land.
G. serotina	47.37	30–63	6,460	680	polycarpic	strong	South I, montane grassland.
G. astonii	26.00	18–33	3,880	750	polycarpic	strong	South I, lowland to montane, grass-land.
G. divisa	33.50	27–44	5,160	770	monocarpic	strong	South I, subalpine to alpine, ridges and outcrops.
G. saxosa	29.30	22–34	5,000	850	polycarpic	strong	South and Stewart Is, coastal, rocks, sands and turf.

*Monocarpic and polycarpic both refer here to perennial species.
**Present = protandry recorded but extent of dichogamy not studied in detail.

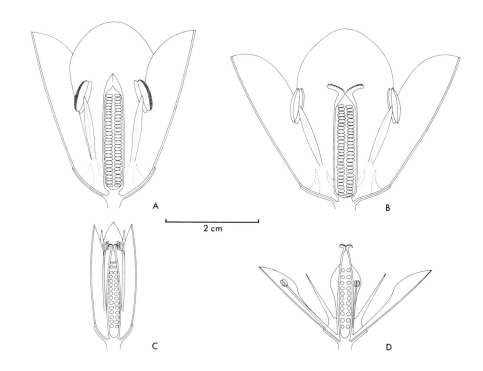

FIG. 2 *Flowers of* Gentiana. *A: newly opened flower of* G.
corymbifera, *anthers dehiscent, stigma closed.* B: 4-day old
flower of G. corymbifera, *anthers more or less empty and
drawn back, stigma open.* C: newly opened flower of G. lineata
on cloudy day, petals just parted at apex, anthers dehiscent
onto open stigma. D: newly opened flower of G. lineata on
sunny day, petals wide open, dehiscent anthers drawn back from
open stigma.

for the large, outcrossed species are much lower than is typical
for xenogamous plants (Cruden 1977).

The genus *Gentiana* in New Zealand is poorly understood
and some of the characters described here might well help in
defining species more precisely. Stamen number is clearly of no
use, but ovule number may be more useful, for example in
distinguishing *G. antarctica* and *G. antipoda* (Godley 1982).
Similarly, the number of pollen grains per anther clearly
distinguishes some species. Other characters which might be used
are the relative level of anthers and stigma, the extent of
protandry, and orientation of the anthers. Even the reaction of
the flower to receipt of pollen may vary between species (Webb and
Ulrich in prep.).

3. *Ranunculus:* stamen number labile (6–32), ovule number labile (4–54).

All five New Zealand lowland species of *Ranunculus* sect. Epirotes (Fisher 1965) are polycarpic perennials with the open dish-shaped hermaphrodite flowers typical of the genus. A low frequency of male-sterile plants has been reported for some populations of *R. limosella* (Melville 1964) but only hermaphrodite plants were present in the samples used in this study.

For *Ranunculus* there is a lack of phylogenetic constraint with respect to both stamen and ovule number and they vary considerably. The number of pollen grains per anther is surprisingly consistent among these five species (Table V). In general, both carpel and stamen number increase with flower size and one might expect species with larger flowers to be outcrossed to a greater degree – this expectation is not really reflected in the P/Os however.

R. glabrifolius and *R. rivularis* are morphologically very similar, but differ significantly in their allocation to male and female functions. *R. rivularis* has smaller flowers, fewer stamens, but more carpels when compared with *R. glabrifolius* and is probably selfed to a greater degree as suggested by the relative P/Os. Stamen and carpel number have been used as characters in taxonomic treatments of these ranunculi. Melville (1956) gives stamens 30–40, carpels 15–20 for Australian plants of *R. glabrifolius,* and stamens about 15, carpels about 10 for Australian *R. rivularis*. Mason (1973) gives the same values in a discussion of New Zealand plants. While these figures for Australian *R. rivularis* fall within the range of values presented in Table V for New Zealand plants, the figures given for *R. glabrifolius* do not agree with those reported here. This is particularly true for stamen number for which even the ranges given barely overlap. It seems that New Zealand and Australian plants of *R. glabrifolius* differ significantly in stamen number, and perhaps also in carpel number; whether this indicates a specific or subspecific difference, or just a difference in pollination system is not known. Melville (1956) noted that New Zealand plants are generally smaller and more slender.

Stamen number may vary considerably even within a species as shown for *R. insignis* by Fisher (1965); in this species stamen number varies from less than 35 in the southern part of its range to over 86 in the north.

Within the Ranunculaceae there is considerable variability in the numbers of floral parts, but within species the ratio of stamens to ovules is relatively stable (Salisbury 1973). In the five species examined here both stamen and carpel number vary somewhat from flower to flower. Both of these characters may be useful in the biosystematics of this group although neither character is likely to be diagnostic. The number of pollen grains

TABLE V Stamen and carpel number, and P/Os for New Zealand lowland species of Ranunculus sect. Epirotes. ($\bar{x}$ + SE = mean ± standard error)

	Stamens/Flower		Carpels/Flower		Pollen/Anther $\bar{x}$	P/O	Flower diameter (mm)[a]	Habitat (all from mainland N.Z.)
	$\bar{x}$ + SE	Range	$\bar{x}$ + SE	Range				
R. limosella	11.45 + 0.62	6–16	7.35 + 0.37	4–10	1,090	1,700	6–8	Lowland to montane, muddy lakeshores or aquatic.
R. rivularis	12.90 + 0.71	7–20	15.10 + 1.15***	9–30	700	600	7–12	Lowland to montane, streamsides and ponds, also aquatic.
R. acaulis	14.00 + 0.91	8–23	14.50 + 1.22**	7–28	1,900	1,050	5–9	Coastal sites, damp sands and turf, mainly coastal.
R. glabrifolius	18.55 + 1.39	6–31	12.90 + 1.06	6–21	700	1,010	12–15–(20)	Lowland, streamsides, ponds, swamps, damp coastal sands and turf.
R. macropus	24.15 + 0.89	16–32	34.00 + 2.35*	16–54	630	450	10–15	Lowland, mainly swamps.

Stamen and carpel numbers significantly correlated: *P <0.05, **P <0.01, ***P <0.001.
[a]Flower diameters are unpublished measurements by P.J. Garnock-Jones.

TABLE VI. *Stamen numbers for flowers at different positions within 3-flowered fascicles of* Muntingia calabura *(1 plant, 10 flowers for each position, $\bar{x} \pm SE$ = mean $\pm$ standard error)*

1st opening flower $\bar{x} \pm$ SE	Range	2nd opening flower $\bar{x} \pm$ SE	Range	3rd opening flower $\bar{x} \pm$ SE	Range
19.60 $\pm$ 1.57	15–32	48.10 $\pm$ 4.67	14–63	91.80 $\pm$ 4.20	66–109

per anther and P/Os would seem to be of little taxonomic use.
4. *Muntingia calabura:* stamen number labile (14–109), ovule number labile (several thousand).

In *Muntingia calabura,* stamen number is extremely variable within trees and is related to the order in which flowers open within fascicles (Table VI; see also Bawa and Webb 1983). The probability of a flower setting fruit is inversely related to stamen number. Ovule number is high and variable with fruits containing as many as 5,000 seeds. The number of pollen grains per anther is also high (22,390 $\pm$ 1,350) but does not vary with stamen number. Here variation occurs within each plant, and within an inflorescence in a more or less organised manner.

The biological significance of this sexual system is discussed in detail by Bawa and Webb (1983). Neither stamen number nor number of pollen grains per anther are useful biosystematic characters here. However, an accurate description of the variation in stamen number within plants and of the position of the stamens in relation to the petals in few- and many-stamened flowers may be useful in determining with which family this monotypic genus should be placed – Elaeocarpaceae, Tiliaceae, and Flacourtiaceae have so far been suggested (Heywood 1978; Smith 1965; Cronquist 1981).

CONCLUSIONS

Different reproductive strategies do not evolve with equal probability in particular plant families, and conversely, particular reproductive strategies are unevenly distributed among families. Within particular families one can point to the multiple evolution of a particular strategy, and to the absence of others. In Umbelliferae dioecism has evolved at least four times but monoecism and gynomonoecism are apparently absent, sexual dimorphism has evolved four times in *Cotula* sect. Leptinella (Compositae; Lloyd 1975, pers. comm.), and at least three times in *Hebe* (Scrophulariaceae; Lloyd, pers., comm.). Raven (1979) suggested that self-pollination has evolved independently at least 150 times in Onagraceae, and Rollins (1963) and Lloyd (1965) have noted that self-compatibility has arisen independently at least four times in *Laevenworthia* (Cruciferae). Repeated evolution of the same floral syndromes has been recognised in some groups, for example, in Polemoniaceae (Grant and Grant 1965), and *Pedicularis* (Scrophulariaceae; Macior 1982), and Ornduff (1979) has argued for the repeated evolution of heterostyly in Lythraceae.

If one looks at particular reproductive strategies similarly non-random patterns are found. Distyly is almost confined to families with radially symmetric, sympetalous flowers (Ganders 1979) and tristyly to families with two whorls of stamens (Yeo 1975). Gynomonoecism usually occurs in families or genera with uniovulate ovaries, the exception being Araceae where this breeding system is reported for three genera in which there are few ovules per flower (Yampolsky and Yampolsky 1922); Lloyd (1979) provides an explanation for the concentration of gynomonoecism in Compositae. Gynomonoecism is also found in Chenopodiaceae, Gramineae, and Urticaceae (Yampolsky and Yampolsky 1922), and Gunneraceae. That bird-pollinated flowers must have the ovary protected in some way from the probing bill (Faegri and van der Pijl 1979) restricts the distribution of ornithophily among plant families.

All of these examples suggest the action of phylogenetic constraint in particular families. And it is by understanding the action of phylogenetic constraint in particular groups that reproductive characters can best be assessed as to their usefulness in biosystematics.

The data summarised in this paper were collected to demonstrate phylogenetic constraint on stamen number and its consequences in relation to pollen number per anther, ovule number, P/O and breeding and pollination systems. P/Os are adjusted in varying ways in the different groups (Table VII). *Muntingia calabura* represents an extreme in stamen number lability, and in this species variability is greatest among flowers of a fascicle on individual plants rather than among flowers of different species.

In general, when the number of stamens per flower is low and usually, also, most constrained, the amount of pollen per anther is most variable. For other reproductive characters, when the number of parts is reduced the character is more likely to be constrained and variation may occur at another level. For example, Stebbins (1967) noted the extreme constancy of low ovule numbers in Compositae, Gramineae, Labiatae, and Umbelliferae and suggested that selection for increased seed production in these families is more likely to affect the number of flowers in inflorescences than the number of ovules per flower.

A difference in stamen number among species in one family may have the same biological significance as the difference in the number of pollen grains per anther in another family where stamen number is constant, yet the former is more likely to be used in systematics (e.g. Dahlgren and Clifford 1982; Johnson and Briggs 1983). In fact, characters associated with the reproductive system are given great weight in biosystematic work. There is a danger here if the biological importance of the characters used is not appreciated. In a study of the pollination and breeding

TABLE VII *Phylogenetic constraint on stamen number and other characters affecting pollen-ovule ratios*

	Stamens/Flower	Ovules/Flower	Other characters which alter P/Os
Cosexual Umbelliferae	constrained	constrained	pollen/anther
Sexually dimorphic Umbelliferae	constrained	constrained	pollen/anther unisexual flowers
Gentiana	constrained	labile	pollen/anther
Ranunculus	labile	labile	—
Muntingia calabura	labile	labile	—

systems of *Parahebe* (Scrophulariaceae) Garnock-Jones (1976a)
described two different pollination systems in *P. linifolia*
– autogamy and entomophily – which occur in different parts of the
species range. If the biological significance of the striking
differences in flower structure had not been appreciated the two
groups may have been described as distinct species, Garnock-Jones
(1976b) however treated these two pollination types as subspecies.
In a recent study of Central American species of *Solanum*,
Anderson and Levine (1982) have shown that three currently used
specific names actually apply to three morphs of a single
functionally dioecious species. The three forms differ in style
length and degree of filament fusion. This case demonstrates well
the problems which may result from distinguishing species by
reproductive characters without appreciating the biological
significance of those characters. It may then become difficult to
know what weight to give to floral and fruit characters. Should,
for example, *Malvaviscus* be treated as a distinct genus
simply because it is the only fleshy-fruited member of the
Malvaceae? Three types of information might be helpful in making
such decisions about reproductive characters. Firstly, as
outlined above, it is important to understand the functional
significance of structures associated with pollination and
dispersal. Second, the usefulness of a character may be indicated
by the number of times it has independently evolved in that
family. Thirdly, an understanding of the genetic basis of
character differences may help – character states which are the
result of single gene differences are likely to be less useful
although they may appear quite striking. It is only by
appreciating all of these factors that a realistic weighting can
be given to reproductive characters in systematics.

ACKNOWLEDGEMENTS

I am grateful to H.E. Conner, E. Edgar, P.J. Garnock-Jones, E.J.
Godley, W. Harris, P.N. Johnson, and D.G. Lloyd for useful
discussions and critical comments on the manuscript. E.J. Godley
kindly supplied material of *Gentiana antarctica* and *G. antipoda*
collected from Campbell Island and Antipodes Island, H.D. Wilson
collected material of *G. gibbsii* from Stewart Island, and
P.J. Garnock-Jones made available unpublished information on
species of *Ranunculus* and checked identification of herbarium
specimens in this group. I also thank J.B. Francis and J.E. Shand
for technical assistance, Anne Hodgins for drawing the gentian
flowers and R. Lamberts for taking the photographs.

REFERENCES

Alberch, P. 1982. Developmental constraints in evolutionary
    processes. In *Evolution and Development* (J.T. Bonner,
    ed.), Springer-Verlag, Berlin.
Allan, H.H. 1961. *Flora of New Zealand*. Vol. I. Gov. Printer,

Wellington.

Anderson, G.S. and Levine, D.A. 1982. Three taxa constitute the sexes of a single dioecious species of *Solanum*. *Taxon* 31: 667-672.

Bawa, K.S. and Webb, C.J. 1983. Floral variation and sexual differentiation in *Muntingia calabura* (Elaeocarpaceae), a species with hermaphrodite flowers. *Evolution* (in press).

Burrows, C.J. and Hobbs, J.F. 1964. Gynodioecy in New Zealand *Gentiana*. *Nature* 203: 203-204.

Cronquist, A. 1981. *An Integrated System of Classification of Flowering Plants*. Columbia Univ. Press, New York.

Cruden, R.W. 1977. Pollen-ovule ratios: A conservative indicator of breeding systems in flowering plants. *Evolution* 31: 32-46.

Dahlgren, R.M.T. and Clifford, H.T. 1982. *The Monocotyledons: A Comparative Study*. Academic Press, London.

Dawson, J.W. 1967a. New Zealand species of *Gingidium* (Umbelliferae). *N.Z. J. Bot.* 5: 84-116.

Dawson, J.W. 1967b. New Zealand Umbelliferae. *Lignocarpa* gen. nov. and *Scandia* gen. nov. *N.Z. J. Bot.* 5: 400-417.

Dawson, J.W. and Webb, C.J. 1978. Generic problems in Australasian Apioideae (Umbelliferae). In *Actes du 2ème Symp. Int. sur les Ombellifères* (Perpignan, 1977).

Faegri, K. and van der Pijl, L. 1979. *The Principles of Pollination Ecology*, 3rd ed. Pergamon Press, Oxford.

Fisher, F.J.F. 1965. *The Alpine* Ranunculi *of New Zealand*. D.S.I.R. Bulletin 165. Gov. Printer, Wellington.

Ganders, F.R. 1979. The biology of heterostyly. *N.Z. J. Bot.* 17: 607-635.

Garnock-Jones, P.J. 1976a. Breeding systems and pollination in New Zealand *Parahebe* (Scrophulariaceae). *N.Z. J. Bot.* 14: 291-298.

Garnock-Jones, P.J. 1976b. Infraspecific taxonomy of *Parahebe linifolia* (Scrophulariaceae). *N.Z. J. Bot.* 14: 285-289.

Godley, E.J. 1982. Breeding systems in New Zealand plants 6. *Gentiana antarctica* and *G. antipoda*. *N.Z.J. Bot.* 20: 405-420.

Gould, S.J. and Lewontin, R.C. 1979. The spandrels of San Marco and the Panglossian paradigm: a critique of the adaptationist programme. *Proc. Roy. Soc. Lond. Ser. B.* 205: 581-598.

Ganong, W.F. 1901. The cardinal principles of morphology. *Bot. Gaz.* 31: 426-434.

Grant, V. and Grant, K.A. 1965. *Flower Pollination in the* Phlox *Family*. Columbia Univ. Press, New York.

Heywood, V.H. 1978. *Flowering Plants of the World*. Oxford Univ. Press, Oxford.

Johnson, L.A.S. and Briggs, G.B. 1983. Myrtaceae - comments on comments. *Taxon* 32: 103-105.

Lloyd, D.G. 1965. Evolution of self-compatibility and racial differentiation in *Leavenworthia* (Cruciferae). *Contrib. Gray Herb.* 195: 3-134.

Lloyd, D.G. 1975. Breeding systems in *Cotula*. III.

Dioecious populations. *New Phytol.* 74: 109–123.

Lloyd, D.G. 1979. Parental strategies of angiosperms. *N.Z. J. Bot.* 17: 595–606.

Macior, L.W. 1982. Plant community and pollinator dynamics in the evolution of pollination mechanisms in *Pedicularis* (Scrophulariaceae). In *Pollination and Evolution.* (J.A. Armstrong, J.M. Powell and A.J. Richards eds.), Royal Botanic Gardens, Sydney.

Magin, N. 1980. Eine blütenmorphologische Analyse der *Lagoecieae* (Apiaceae). *Plant Syst. Evol.* 133: 239–259.

Mason, R. 1973. On two native *Ranunculi.* *Canterbury Bot. Soc. J.* 6: 2–4.

Melville, R. 1956. Contributions to the flora of Australia: II. Some ranunculi of Tasmania and south eastern Australia. *Kew Bull.* 1955: 193–220.

Melville, R. 1964. *Nature* 203: 204. (Untitled annotation to Burrows and Hobbs, 1964).

Ornduff, R. 1979. The morphological nature of distyly in *Lythrum* section Euhyssopifolia. *Bull. Torrey Bot. Club* 106: 4–8.

Raven, P.H. 1979. A survey of reproductive biology in Onagraceae. *N.Z. J. Bot.* 17: 575–593.

Rollins, R.C. 1963. The evolution and systematics of *Leavenworthia* (Cruciferae). *Contr. Gray Herb.* 192: 3–98.

Salisbury, E. 1973. The organization of the ranunculaceous flower with special regard to the correlated variations of its constituent members. *Proc. Roy. Soc. Lond. Ser.* B 183: 205–225.

Simpson, M.J.A. and Webb, C.J. 1980. Germination in some New Zealand species of *Gentiana:* a preliminary report. *N.Z. J. Bot.* 18: 495–501.

Smith, C.E., Jr. 1965. Flora of Panama, Part VI. Family 113. Elaeocarpaceae. *Ann. Mo. Bot. Gard.* 52: 487–495.

Stebbins, G.L., Jr. 1950. *Variation and Evolution in Plants* Columbia Univ. Press, New York.

Stebbins, G.L. Jr. 1967. Adaptive radiation and trends of evolution in higher plants. *Evol. Biol.* 1: 101–142.

Stebbins, G.L., Jr. 1973. Evolutionary trends in the inflorescence of angiosperms. *Flora* 162: 501–528.

Webb, C.J. 1979. Breeding systems and the evolution of dioecy in New Zealand apioid Umbelliferae. *Evolution* 33: 662–672.

Webb, C.J. 1980. The status of New Zealand *Actinotus* (Umbelliferae). *N.Z. J. Bot.* 18: 343–345.

Webb, C.J. 1984. Flower and fruit movements in *Muntingia calabura:* A possible mechanism for avoidance of pollinator-disperser interference. *Biotropica* (in press).

Yampolsky, C. and Yampolsky, H. 1922. Distribution of sex forms in the phanerogamic flora. *Bibliogr. Genet.* 3: 1–62.

Yeo, P.F. 1975. Some aspects of heterostyly. *New Phytol.* 75: 147–153.

Pollination by Animals and Angiosperm Biosystematics

Peter G. Kevan
Department of Environmental Biology
University of Guelph
Guelph, Ontario, Canada

INTRODUCTION

Biosystematics, although difficult to define precisely, is the
consideration of the natural relationships among taxa. It
includes the description, naming, and classification of organisms,
together with studies of their evolution and phylogeny. Evolutionary studies in biosystematics must be based on species and
speciation as it is through this taxon and this process that
phylogenetic patterns can be postulated. The species concept is
complex and connotates different things to different biosystematists or with different taxa, or both. Despite difficulties
with species concepts, be they typological, biological, or evolutionary there are over-riding themes common to most species
definitions.

1. Species are groups of similar individuals recognizably
different from other such groups (e.g. typology).

2. Species are populations which share very close genetic
relatedness (a) so that they may actually or potentially interbreed yet are reproductively isolated from other such populations
(Mayr 1969) and so continue to be very closely related or (b) as
an historical consequence of the species' origins, as in asexual
species, the population is bound by epigenetic homeostatic forces
which maintain close relatedness (see Eldredge and Gould 1972).
The degree to which organisms are related or are not related forms
the basis of taxonomic decisions and is a difficult parameter to
measure or estimate. These are instantaneous precepts which must
be modified to accommodate the forces of selection. This leads to
the evolutionary species concept (Simpson 1961) in which lineages
(ancestor-descendent populations) maintain their identities from
other such lineages.

Embodied in the various species concepts must be an understanding of breeding systems and of isolating mechanisms (see Webb,

this volume; Urbanska, this volume). It is here that pollination
biology has direct relevance to the biosystematics of animals and
plants. On an evolutionary basis, it is difficult to separate
pollinators and angiosperms as they have co-evolved since early
Cretaceous time (Kevan and Baker 1983b; Feinsinger 1983; Smart and
Hughes 1973). Nevertheless, I will constrain my remarks to
flowering plants.

In this discussion I address subjects important to biosystematics and to pollination biology. I wish to show that systematists
have much to gain through understanding pollination systems. That
is not to say that information transfer in the opposite direction
is not equally valuable. The subjects I address are as follows:

1. *Reproductive Methods*. It is through these that genetic
lineages are maintained.

2. *Floral Morphology*. It is on floral structures that many
classifications are based, that some evolutionary trends have been
analyzed, and that pollination mechanisms can be understood and
even predicted.

3. *Isolating Mechanisms*. They are represented by breeding
systems, floral anatomy, as well as by cytological characters,
flowering phenology, habitat separation, and so forth.

4. *Evolutionary trends in angiosperm flowers as they relate to
pollination*. I believe that isolating mechanisms in angiosperms
are mostly the outcome of selective forces (see Levin 1971),
rather than a by-product of isolation as suggested for animals
(cf. Mayr 1969). Thus I will address character displacement, or
more particularly the Wallace effect, in floral evolution.

However, before progressing with these subjects, a quick excur-
sion into the palaeontology of pollination will introduce some
ideas on the history of angiosperms and their relationships with
pollinators.

PALAEOECOLOGY

The origins of the angiosperms are obscure and remain the subject
for much debate. Angiosperms are characterized by flowers,
pollination, tectate pollen, double fertilization, specialized
reduction in gametophyte development, the double integument of the
ovule, the enclosed ovary, and various other features. Pollination is the transfer of pollen (a microgametophyte) from a micro-
sporangium (the anther of angiosperms) to the female receptive
structure (a micropyle in all seed plants except angiosperms, in
which it is the stigma). Pollination is not peculiar to angiosperms. In most plants, other than angiosperms, pollination is,
and probably was, by wind [see Taylor and Millay 1979; Niklas,
(1981a,b) for early seed plants; Crepet (1974) for Cycadeoidophyta
and Coniferophyta]. Certainly, in angiosperms, wind pollination
is widespread, but probably was a derived condition (Whitehead
1969; Regal 1982). Kevan *et al.* (1975) proposed that a
predisposition for arthropod vectoring of microspores has existed
since Devonian time. At that time heterospory was becoming

evident, and microspores show interesting structures which suggest adaptation to their attachment and transport to arthropods. The extinct Williamsoniaceae of Jurassic time had flowers that were mostly large and robust, but *Williamsoniella* had smaller flowers. It has been suggested that animal pollination took place in these taxa (Leppik 1960, 1963, 1971).

The Cretaceous origins of angiosperms is well demonstrated through the record of fossil pollen (see Wolfe *et al.* 1975). Leaves occur contemporaneously (Doyle and Hickey 1976). Flowers are rarely preserved in the fossil record and those of early Tertiary age, mostly Eocene and Oligocene, are small and delicate (Leppik 1971, 1963; Crepet 1979a, b). They resemble flowers of extant taxa to such an extent that Crepet (1979a,b) has been able to surmise their relationships with insects in pollination.

REPRODUCTIVE METHODS

There are a variety of ways in which higher plants reproduce. For most species, one method tends to predominate, but mixed strategies are used by many plants. Reproductive methods fall into two broad categories, sexual and asexual. Sexual, or amphimictic reproduction, may be through strict outbreeding (allogamy), or through strict inbreeding (autogamy), or a mixture of both.

A variety of methods may assure allogamous reproduction. Dioecy has been a recent topic of much debate (Bawa 1980) and well represented in the Canadian flora. Heterostyly, in which different plants have flowers which differ in the relative positions of the anthers and stigmas, has been reviewed by Ganders (1979). Incompatibility mechanisms are varied; Lewis (1979) enumerates at least 11 different self-incompatibility systems whereby self-pollen fails in fertilizing ovules. Mulcahy (this volume) discusses incompatibility systems, their evolution, and importance in biosystematics. Less well studied are the widespread phenomena of herkogamy, in which spatial separation of two anthers and stigmas make self-pollination unlikely to take place, and dichogamy in which differential timing of maturation of anthers and stigmas makes self-pollination an unlikely event. Herkogamy and dichogamy tend to go hand-in-hand. In dichogamy it is most common to find that anthers mature first (protandry) although this is less likely to prevent selfing as pollen may remain in the flowers when the stigmas become receptive. In the Asteraceae individual florets are protandrous, as it is the growing style and immature stigma which push the pollen out of the floral tube; however, the whole inflorescence does not mature simultaneously and incompatibility systems may prevent selfing (see Ganders, this volume, for examples in *Bidens*). Protandry represents sequential maturation of the floral parts from lowermost to outermost to uppermost and innermost. Thus protogyny, the maturation of gynoecium first, represents a reversal of that sequence in the last two stages.

Autogamous plants set seed by self-pollination. This can be

accomplished by the failure of outbreeding mechanisms (see
Mulcahy, this volume). Autogamy is frequently interpreted as a
method by which plants may assure their reproduction if
outcrossing fails. The anatomy of some flowers assures that self-
pollination will take place, e.g. the anthers move to touch the
stigma, or pollen falls onto the stigmas from anthers above, or in
the extreme case of cliestogamous flowers, which never open, self-
fertilization takes place in the flower bud. In general, the
flowers of autogamous plants are smaller, less showy, produce less
pollen and nectar than their allogamous counterparts (e.g.
Ornduff and Mosquin 1970; Rollins 1963) and have lower ratios of
pollen grain to ovule numbers (see Webb, this volume). Even so,
as Baker (1959) states, there is probably no known case in which
asexual species are never outcrossed. It is also important to
keep in mind that mechanisms, apparently designed to assure self-
pollination, do not necessarily do so - e.g. the filament movements which cause the anthers to touch the stigmas of *Saxifraga
oppositifolia* and to deposit pollen thereupon do not result in
self-fertilization even in the absence of outcrossed pollen (Kevan
1972).

As already suggested, autogamy is probably always associated
with some amount of outcrossing. Incompatibility mechanisms may
be less pronounced in some taxa, heterostyly may break down,
herkogamy and dichogamy may be weakened, and dioecism not fully
developed (as in gynodioecy). The majority of sexually reproducing flowering plants probably combine outcrossing and inbreeding
to various degrees.

Asexual reproduction may or may not involve the flowering
parts. Vegetative reproduction through runners, stolons, suckers,
tubers, bulbils, and so on clearly do not. Agamospermy involves
the flower because a megagametophyte and seeds are produced. In
pseudogamy pollination is required for the fertilization of the
endosperm cells and to stimulate development. Stebbins (1950) and
Grant (1981) discuss the relationships of apomixis to biosystematics of plants. I will restrict my comments to the way in which
apomixis may come about as a result of pollination and hybridization (see below). In some plants agamospermy is facultative,
e.g. *Crataegus* (see Phipps, this volume), *Poa* spp. (Baker
1959), and others and even in apparently obligate apomicts genetic
exchange between plants may occasionally take place (see Asker,
this volume).

CLASSIFICATION, CHARACTERS, AND POLLINATION

In the following discussion I have tried to arrange the characters
used in plant biosystematics in the order in which they are most
frequently used at different taxonomic levels from Division down
to Species and below.

I need not recapitulate on the importance of the closed carpel
(see Cronquist 1968; Takhtajan 1969) which may have been, in part,
selected for by insect pollination (Grant 1950; van der Pijl

1961) as a protection for the ovules, or for pollen selection and perhaps avoidance of clogging micropyles with outspecies pollen, or possibly also for protection of unripe zoochorous seeds with associated expendable tissue, arils, etc. until ready for dispersal (Corner 1949, 1953, 1954, 1964). This does not seem at loggerheads with Meeuse's ovuliferous cupules in anthocorms (1975a,b) which, as postulated, were progenitors to angiosperm ovules or ovaries (see also Meeuse 1979). One of the main bases for Meeuse's objections to the magnoliaceous ancestry of the angiosperms is the unisexual, anemophilous, apetalous nature of the flowers of the Hamamelidae (Meeuse 1975c), Salicales (Meeuse 1978), Pandanales and Palmalales (Meeuse 1971) and others in which he suggests zoophily is the derived condition. However, wind pollination and associated floral reduction has evolved independently in many groups (see Regal 1982; Whitehead 1969), sometimes with dicliny (e.g. *Thalictrum*, Chenopodiaceae, and possibly *Vitis* with mixed anemophily and entomophily). Thus, although anemophily seems common to the Hamamelidae, whether it is basic to them is unresolved (Meeuse 1975c; Forman 1964; Cronquist 1968).

The major groups of angiosperms can be divided by floral characters. The number of floral parts is diagnostic for Dicots vs. Monocots. When the parts are in definite number, in dicots they are usually in sets of 5, sometimes 4, and rarely 3, whereas in monocots they are usually in sets of 3, and seldom 4. The significance of these numerals in pollination and evolution is not clear except that they represent advanced characters in both groups; multipartite flowers are considered primitive. Leppik (1953, 1956) has shown that some insects (honeybees) can discriminate between patterns with different numbers of radiating spokes. He refers to these as form numerals and argues that they are important in flora recognition by pollinating insects. The number of floral parts, when definite in number, can function as discriminatory clues so that the pollinators remain faithful, or constant, to flowers with that number. Other floral features, size, color, odor, etc. are also important in flower recognition by pollinators. I cannot suggest that there have been, or are, strong selective forces by pollinators for the numbers of floral parts as I believe that other evolutionary and developmental constraints are more important. Nevertheless, it is useful to note most floral characters and to enumerate them and discuss what they mean in terms of pollination.

Considering further the evolutionary trends in the gynoecium of flowering plants, I disagree with Cronquist (1968) that the trend from spirally arranged carpels to cyclic carpels and to the compound pistil with a single stigma is not clearly explained through entomophily. The condensation of the gynoecium, and increasing precision in the location of the stigma or stigmas must be examined within the context of the whole flower (cf. herkogamy). Pollinators would be increasingly directed in their foraging behavior by the form of the corolla and the position of the rewards they sought. Thus, the trend would be from fusion of

the ovaries, to fusion of the styles and, subsequently, of the
stigmas. At the same time, precise positioning and reduction of
the androecium would cause pollen to be placed with increasing
accuracy on the pollinator (Kevan *et al*. 1983). These trends in
gynoecial and androecial reduction and precision, together with
increasing specialization of other floral parts (corolla,
hypantheum, nectaries, etc.) are epitomized by the Ranunculaceae
(*Caltha* vs. *Aconitum* and *Delphinium*) and their outcome
emphasized in the Asteridae and well exemplified in the
Alismatales (Kaul 1976).

In the angiosperms, the Thalamiflorae are apetalous and
choripetalous. The Corolliflorae are all sympetalous. Both have
hypogynous flowers. The Calyciflorae and Ovariflorae are
similarly separated, but have epigynous or perigynous flowers (see
Benson 1979). This classification is not natural; the characters
used have arisen separately within the six subclasses, and within
some orders, or even in one family, e.g. Rosaceae, with perigynous, epigynous, or hypogynous flowers (*Coleogyne*).
Nevertheless, the importance of ovary position and corolla
characters can be seen in the natural classifications of Takhtajan
(1969) and Cronquist (1968). The Magnoliidae fit within the
Thalamiflorae. Within the Dilleniidae and Hamamelidae most
flowers are hypogynous and chloripetalous, but in the Ericales and
Primulales the flowers are sympetalous. In the Asteridae the
flowers are hypogynous, epigynous and sympetalous. Similarly in
the Monocotyledons the Liliidae, hypogyny, perigyny, and epigyny
are all represented. It is worth noting that ovary position is a
taxonomically more stable character at the family level than at
the level of subclass and that families of angiosperms tend to be
characterized by chori- vs. sympetally and actino- vs. zygomorphy.

If one considers the thesis that the inferior ovary evolved in
response to pressures from pollinators destructive to the ovaries
in flowers (Grant 1950), then strong selective forces can be
postulated. At slightly lower taxonomic levels, corolla form is
also important (i.e. choripetally vs. sympetally) and may or may
not go hand-in-hand with trends in ovary position. Strong
selective forces can also be postulated as the accuracy of pollen
placing, on the pollinator by the plant and on the stigma by the
pollinator, requires the precise directing of the pollinator on
the flower. Thus, one can appreciate that the taxonomic usefulness of hypogyny, epigyny, and perigyny, and of sympetally and
choripetally is clearly great. However, perigyny, hypogyny and
sympetally are derived characters and have arisen many times in
different taxa from subclass to family (see Douglas 1957).
Perigyny, as an intermediate step to epigyny, would be selected
for as providing an expanded surface from which foraging insects
could feed (i.e. by placement of nectaries or nectar on the
hypanthium) drawing them away from the base of the ovaries.
Epigyny would offer further protection in open dish or bowl-shaped
flowers, but would have little importance in flowers pollinated by
gently feeding long-tongued insects, but could be important in

flowers pollinated by sharp-beaked hummingbirds. Sympetally or
synsepally (as in Silenoidea: Caryophyllaceae) restricts pollinators to those with long tongues and so protects the ovary from
chewing insects. However, these features probably developed more
often in response to greater pollination efficiency. In the
Asteridae it is difficult to reason, as above, the generality of
the sympetalous condition against the limited occurrence of
epigyny, despite the clear division of orders on this character.
Within the Mangoliidae 14 families have hypogynous flowers and 5
have perigynous to epigynous.

    As I have pointed out sympetally (or its equivalent synsepally
in Silenoidea) and zygomorphy have arisen many times at different
taxonomic levels. Sympetally is characteristic of the Asteridae,
Plumbaginales, Primulales, Ebenales, Diapensiales and most
Ericales. Within Violales, 8 of 21 families are characterized by
sympetally. Sympetally in the Rosales is found in the
Alseuosmiaceae. The polypetalous Flacourtiaceae show trends to
perigyny and epigyny, and reduction in numbers of sexual parts.
Within Ericales, sympetally is somewhat diagnostic at the family
level; Empetraceae and Pyrolaceae are polypetalous but both
polypetally and sympetally can be found in other families, e.g.
Monotropaceae, Ericaceae (e.g. *Ledum*) where sympetally is diagnostic of genera. Zygomorphy seems less basic and is characteristic
of many families, and some orders (e.g. Scrophulariales, Asterales,
Polygalales, Salicales). Zygomorphy occurs within families,
notably Ranunculaceae in which the character is diagnostic at the
generic level. Stebbins (1971) discusses floral and other character differences at various hierarchial levels.

    At the level of genus, and below, floral form is also useful as
a taxonomic criterion. Indeed, many genera are separated on the
basis of floral form, and their forms provide insights into
pollination mechanisms. This is especially true of advanced forms
as in stereomorphic and zygomorphic flowers. The importance of
pollination mechanisms cannot be underestimated and are well
exemplified by Grant and Grant (1965) for the Polemoniaceae, by
van der Pijl and Dodson (1966) for Orchidaceae, by Nur (1976) on
Musaceae, by Raven (1979) on Onagraceae. Within genera, the
importance of pollination mechanisms in the interpretation of
floral anatomy is similarly important: large numbers of papers
could be cited. Macior (1982) on *Pedicularis*, Spira (1980) on
Trichostema, and Wiebes (1979) on figs provide excellent examples.
Nevertheless, taxonomists do not always take into account the
importance of floral characters to pollination. I refer here to
the differences between flowers of *Coleus* and those of
Plectranthus, two genera considered by Morton (1962) to be one.
On the other hand *Oxytropis* and *Astragalus* are separated on
the basis of a minor apical tooth on the carinal petals. These
cases are discussed briefly by Burtt (1964) and illustrate
difficulties in taxonomy.

    Floral form conveys much information about the pollination of
the plant. The generally accepted pollination syndromes are

described by Faegri and van der Pijl (1978). The generalized
primitive flower is thought to be dish- or bowl-shaped with a
large number of pistils and stamens (e.g. *Magnolia*) and to be
pollinated by mess-and-soil insects, such as beetles. These
insects chew and consume pollen, floral parts, and secretions.
However, not all primitive flowers fit this model as Thein (1980)
and Thein *et al.* (1983) demonstrate. In *Illicium, Belliolum,
Zygogynum, Bubbia, Pseudowintera* and *Drimys*, the floral parts
are not numerous and primitive families of flies (*Diptera*) appear
to be important pollinators. Flies are important pollinators of
many plants, and visit a wide variety of open-bowl and dish-shaped
flowers. Such flowers are also visited by many other kinds of
insects as the floral form poses no restriction on their obtaining
rewards. There are specialized fly-pollinated plants, e.g.
Araceae, *Stapelia, Rafflesia, Pterostylis*, which are mimetic
in coloration and odor of dung, carrion, fungi, or rotting
vegetation and attract appropriate flies. In turn some flies,
e.g. Bombyliidae, are specialized hovering visitors to many
tubular flowers. They resemble miniature sphinx moths in
anthophilous behavior, and the flowers they visit often resemble
miniature moth pollinated flowers (e.g. *Muscari* spp.) (see below
and Kevan and Baker 1983a).

The evolutionary trend is to narrower flowers, dish- to bowl-
to bell- funnel-shaped flowers (see Leppik 1957, 1972).
Pollinators are increasingly restricted in their entry into
flowers. In larger bell-shaped flowers, insects must crawl in, at
least part way (e.g. *Vaccinium*), to feed on the floral resources,
but must do so in such a way as to touch the sporophylls (e.g.
Campanula, some *Gentiana*). These flowers are often pollinated
by bees. This trend is not always associated with sympetally. A
variety of insects visit funnel-form flowers in much the same way,
as they do bell-shaped flowers. In both campanulate and funnel-
form flowers more specialized pollinators may be involved. These
are pollinators which do not enter the flowers, e.g. hovering
moths on some *Convolvulus, Datura,* bee flies on *Muscari*, bats
on bat-pollinated flowers (see below).

Trumpet-shaped flowers, i.e. salverform, or similarly shaped
choripetalous flowers are characterized by pollination by settling
insects. They land on the platform produced by the flaring of the
corolla and feed by inserting their proboscides into the corolla
tube. Butterflies and some medium to long tongued bees and flies
feed in this way (e.g. on some *Phlox, Brassica, Erysimum,
Myosotis, Primula*). Again some are pollinated by nonlanding
pollinators, especially if the flowers are horizontally or
downwardly oriented (e.g. some *Oenothera, Aquilegia, Ipomopsis,
Silene laciniata*). Whether the pollinators land or not, pollen
is transferred from flower-to-flower exactly, mostly on the
proboscides of the pollinators which touch both anthers and
stigmas as they feed. It is through this precision that pollination is accomplished in such heterostylous plants as *Primula*.

Tubular flowers, e.g. *Nicotiana glauca, Mertensia*, are

pollinated by hovering pollinators, moths, or pollinators which
land on the plant beside the flowers, or which hang onto the
flowers.

Gullet-shaped flowers are tubular and also zygomorphic (showing
bilateral symmetry). The Labiatae, Scrophulariaceae, Lobeliaceae,
are typical. In some, bracts (as in *Acanthus*) or sepals (as in
Castilleja) are important in the gullet form. In most, the sexual
parts of the flowers are on the upper side and pollination is
accomplished nototribically (i.e. on the backs of the pollinators). Often the pollinators must force their entry to the
flowers by separating the lower lip from the upper by crawling
partway into the flower. Those pollinated by hummingbirds (e.g.
in the genera *Mimulus, Penstemon, Beloperone, Castilleja, Salvia,
Trichostema*) tend to have flowers which are open at the entrance.

Flag flowers are typified by the Leguminosae, but occur in
other families, e.g. Papaveraceae, Geraniaceae. These are
pollinated much as above, but sternotribically, i.e. the pollen is
placed on the ventor of the pollinators. In the Leguminosae, most
flowers are choripetalous, but some flag flowers are sympetalous.
The Orchidaceae are choripetalous flag flowers of fantastic
diversity in pollination, and may be sternotribically or nototribically pollinated, and have flag and gullet flowers, perhaps
in that evolutionary order vs. the reverse order in Zingiberaceae
(see van der Pijl and Dodson 1966).

Intermediate forms between trumpet-shaped and zygomorphic
flowers may be seen in *Delphinium, Verbascum, Zauschneria* and
Gilia. Nectariferous spurs are also found in trumpet-shaped and
zygomorphic flowers, and form tubes with which pollinators must
contend, e.g. in various Orchidaceae, *Linaria* (Scrophulariaceae)
Delphinium, Aconitum, Aquilegia (Ranunculaceae), *Viola*
(Violaceae), *Kenthranthus* (Valerianaceae), *Tropaeolum*
(Tropaeolaceae) and *Impatiens* (Balsaminaceae).

Leppik (1972) has discussed these sorts of evolutionary trends
in Ranunculaceae, Fabaceae, Asteraceae, and in general as
originating in mid-Tertiary time with the specialized Hymenoptera,
Lepidoptera, and Diptera (see above). Thus, as the floral form
became increasingly complex as an evolutionary progression, so the
pollinators became more and more specialized. Those complex
flowers requiring manipulation by pollinators such as bees, hawk
moths, and hummingbirds rely on the intelligence and learning
capacity of their pollinators (*cf.* Laverty 1980).

Brush blossoms are usually formed of an inflorescence,
Salicaceae, Proteaceae, and in genera in many other families,
Eupatorium, Liatris, cf. *Bidens* (Ganders, this volume)
(Asteraceae), *Phacelia, Hydrohyllum* (Hydrophyllaceae),
Thalictrum (Ranunculaceae), *Castilleja* (Scrophulariaceae),
Phyteuma, Trachelium, (Campanulaceae), Dipsacaceae, *Globularia*
(Globulariaceae), *Poskea* (Boraginaceae), *Vitex* (Verbenaceae),
Vitaceae, Umbelliferae, *Rhus* (Anacardiaceae), *Acalypha* (wind
pollinated; Euphorbiaceae), Loranthaceae, *Terminalia*
(Combretaceae), *Acacia, Mimosa* (Leguminosae), Cumoniaceae,

Piperaceae, Araceae (esp. *Anthurium*). Some taxa, such as
Myrtaceae, Lecythidaceae, have flowers which individually function
as brush blossoms. Most of these are highly specialized in their
pollination mechanisms and their pollinator requirements.
Inflorescences are discussed below.

Chemical and Physical Aspects: Nectars, Oils, Pollen,
  Scent, Color, Texture.
Chemical aspects of pollination biology are also useful in
systematic studies. Nectar sugar constituents have been examined
by Baker and Baker (1983). They point out that the Brassicaceae
and Asteraceae are generally hexose rich, whereas Ranunculaceae
and Lamiaceae are sucrose rich when the major sugars, sucrose and
the hexoses, glucose and fructose are considered. Not all
families show such consistency. In the Scrophulariaceae, some
genera are sucrose-rich or dominant, *Penstemon* spp. span the gamut
from hexose-dominant to sucrose-dominant, and *Veronica* seems to be
hexose-dominant. Table I indicates some of the trends in nectar
sugar constituents and pollinators.

Baker and Baker (1983) examined several genera, *Erythrina*,
Campsis and *Penstemon* and its relatives, and show the trends in
sugar constituents are consistent with pollinator types.
Nevertheless, there are phylogenetic constraints, i.e. similarities in nectar sugars within families together with the potential
for adaptation when new flower/pollinator relationships come
about. The nectar sugars may be those which the pollinators
prefer, but data are scanty. No generalities can be made for
other nectar sugars, some are toxic to some insects but not to
others (see Kevan and Baker 1983a).

Other nectar constituents, particularly amino acids, have also
been examined from viewpoints of taxonomy and pollination, but the
picture here does not allow for great generalization (Baker and
Baker 1975). Amino acids are in their highest concentrations in
the nectars of flowers which mimic carrion or dung and are
pollinated by flies attracted to those scents. Hawkmoth and bird-
pollinated flowers have the lowest concentrations. The amino
acids, and other constituents, must contribute to the nutrition of
the pollinators (see Kevan and Baker 1983a). Nevertheless, the
amino acid constituents of nectar are intraspecifically quite
constant (Baker and Baker 1977) and are additive in hybrid species
(Baker and Baker 1976).

Oils are produced by some flowers and collected by bees. This
phenomenon is widely distributed in the tropics. Neff and Simpson
(1981) suggest that oil collection started as a secondary association between bees visiting flowers, which had oil-secreting
patches, for a different reward. The bees involved are antho-
phorine bees with specialized tarsal brushes for collecting the
oil. Oils, and other floral rewards, e.g. sexual attractants,
resins, gums, food bodies, are highly specialized. The nature of
these rewards and the pollination mechanisms of the plants are
well exemplified in *Ophrys* (Kullenberg and Bergstrom 1976), other

orchids (Williams 1983), *Freycinetia* (Cox 1982), and *Ficus*
(Wiebes 1979) and can be used in understanding of their
biosystematics.
  Floral scents are not used much in plant systematics. Their

TABLE I. *Trends in nectar sugar constituents and pollinators*

Pollinators	Sugar	Volume	Concentration
hummingbirds	sucrose-rich to dominant	high	low
perching birds	hexose-rich	high	low
bats	hexose-rich to dominant	high	low
moths	sucrose-rich	high	low
butterflies	sucrose-rich	medium	low
bee - long tongued	sucrose-rich	medium	low-medium
- short tongued	hexose-rich	medium-low	medium
lapping flies	hexose-rich	low	high
beetles	inconclusive	-	-
wasps	inconclusive	-	-

analysis requires specialized equipment (see Williams 1983).
Nevertheless, scents can act as isolating mechanisms, as is
demonstrated in Orchidaceae (Williams and Dodson 1972; Kullenberg
1961). Scent polymorphisms within species do affect their
pollinator behavior and frequency of different types of pollinators, as has been shown by Galen and Kevan (1980, 1983) for skunk-
scented and sweet-scented *Polemonium viscosum*. Thus one might
propose incipient isolating mechanisms in this instance.
  Palynology is used by palaeontologists and systematists working
on extant plants. Nevertheless, it is difficult to relate exine
pattern to pollination (*cf*. Taylor and Levin 1975). Recently,
however, Ferguson (1983, and this volume) has been able to relate
pollen morphology in the Leguminosae (Papilionoideae) to pollinators. He notes that bird-pollinated flowers produce pollen which
is more heavily sculptured, has different exine characters, and
lacks oily substances on the surface. The insect-pollinated
species have pollen which is less ornamented, but have the
expected oily substances on the surface. Perhaps this reflects
the fact that bird feathers are already oily and thus sticky to
pollen grains, whereas insect cuticle is usually dry.
  Generalizations which can be applied to palynology are that
pollen with spiny and/or oily exines are entomophilous. Those
with dry, smooth exines are anemophilous. Woodhouse (1935) shows
a series of simplifications in exine sculpturing with the trend to
anemophily in Asteraceae. In vibratile pollination by bees pollen
grains are also smooth. This phenomenon is advanced and very
widely represented in angiosperms (ca. 6 to 8% of flowering plant
species) and in the Apoidea (bees). The anthers are characterized

by dehiscence by apical slits, spores, or valves. Some anthers,
or staminodes, may produce inviable pollen as fodder, while others
produce the pollination pollen (i.e. heteranthy and pollen
dimorphism) (see Buchmann 1983). *Cassia, Dodecatheon* and *Solanum*
seem characterized by vibratile pollination. The chemistry of
pollen does reflect the pollination mechanism of the plants.
Generally, pollen is highly nutritious, but that from anemophilous
plants is less so and less used by insects as a food (Kevan and
Baker, 1983b). Baker and Baker (1979) have discussed the relative
amounts of starch or oil in pollens of different plants. In
general, pollen of anemophilous plants with long styles and large
pollen grains have starch as the cytoplasmic energy reserve, as
well as oil, whereas pollen of entomophilous plants and smaller
pollen has only oil as the reserve. Oil is a more compact energy
reserve and probably more nutritious to insects. Starch may not
be digested by some insects (e.g. some *Diptera*) (Kevan and Baker,
unpubl.), but is more economical to accumulate in the grains.
Baker and Baker (1979) discuss the trends for advanced families to
have starchless pollen, and for plants relying on pollen-feeding
insects for pollination to have starchless pollen. There are
apparently strong phylogenetic constraints in some plant families,
but not in others.

The colors of flowers are used extensively in plant taxonomy,
but are most useful at lower levels. A general trend in flower
color is from primitive flowers which tend to be pale whitish,
creams and pinks, to more advanced open-bowl flowers, many of
which are more intensively colored white, yellow and pink, some
including ultraviolet and patterns. The more advanced floral
forms take on even more colors, with blue and red, and the mixture
thereof purple, which are mostly in combinations with other colors
to form color patterns as nectar guides (see Kugler 1963; Kevan
1978, 1983). These features are extensively used for species
discrimination by taxonomists and by pollinators. I have reviewed
the subject of floral colors and color patterns as pollinators may
see them and brought into perspective the role of ultraviolet
reflections from flowers. These reflections, as those in any
other waveband visible to pollinators, have use in taxonomy (e.g.
Eisner *et al.* 1973; McCrea 1981). They have been shown to act as
powerful isolating mechanisms (Free 1966; Levin and Schaal Kay
1982; Kevan 1983) and in stabilizing selection for flower color
(Waser and Price 1981). The flavonoids, involved in flower
coloration are used by chemotaxonomists (Levy 1978; Harborne 1975;
Scogin 1983).

Microsculptural features on the surfaces of petals are used in
taxonomy, especially in the Asteraceae (Baagoe 1977a,b, 1978; Lane
1980). These features prompted Kevan and Lane (1983) to ask about
their functional significance. Honeybees have been trained to
distinguish between microsculptural features of ligules of different species of Asteraceae when all other cues (i.e. color or
scent) are eliminated. The spacing between the microsculptural
features on the ligules is the same as the spacing between the

sensilla trichodea (microscopic mechanoreceptor hairs) on the
antennae of bees. The full implications of this discovery to
plant systematics, pollination, and insect behavior are now being
explored.

INFLORESCENCE, FLOWER SIZES, and ANTHIA

The most obvious suggestion for the function of the multi-flowered
inflorescence is in increasing the attractiveness of the flowers
to pollinators. The corporate image, be it visual or olfactory,
presumably increases the distance over which the inflorescence can
be detected by pollinators. A broader term "anthium" is used by
Faegri and van der Pijl to describe the attractive unit composed
of flowers. This term encompasses inflorescences to whole flowering plants as in alpine cushion plants and blooming trees, and
could be expanded to include dense stands of blooming plants,
particularly if rhizomatous. The effects of size are shown in the
greater attractiveness of larger anthia, or inflorescences, to
bumblebees (Kugler 1943) and of weeds and arctic flowers (Kevan
1973) to pollinators in general (Mulligan and Kevan 1973).

The formal arrangements of inflorescences is important in
outcrossing. It has been long known that bees visiting tall
inflorescences tend to start at the bottom and work up (Benham
1969; Pyke 1979; Waddington and Heinrich 1979). Inflorescences in
horizontal linear display are treated by pollinators in the same
way. The flowers of these inflorescences tend to be protandrous
so that those at the base are in female stage while those at the
tip are male. Thus, pollen is transferred to the next plant
visited.

The extent of the part of the inflorescence with receptive
female flowers is important, as is the extent of the polleniferous
part. The pollinator should gather sufficient pollen to pollinate
all the female-stage flowers visited on the next plant. Studies
of pollen carryover indicate that this is indeed the case in the
capitulate inflorescence of *Trifolium*, in which self pollinations
are generally unsuccessful, and the pollen load on the bee from
visiting one inflorescence is distributed in decreasing, but
effective doses, on the florets of the next inflorescence visited
(Plowright and Hartling 1981).

Cauliflory, the presentation of flowers on the trunks or limbs
of trees, is widespread in tropical rain forests (e.g.
Lecythidaceae, *Duria, Artocarpus, Macadamia, Theobroma*). The
pollinators fly beneath the canopy, e.g. bats, to small midges, as
that is where flight is least restricted (bats) or close to
breeding sites (midges). In association with this is the problem
of seed dispersal, often by bats, from the rather large fruits
once they are ripe (see van der Pijl 1957).

Stebbins (1974) reviews the trends in the inflorescence.
Whether the primitive inflorescence was of a single terminal
flower vs. a three or more flowered structure is unresolved.
Inflorescence morphology is complex and may be further complicated

by superposition of different patterns as in Asteraceae and
Poaceae. From a developmental viewpoint it appears that inflorescences can become separate flowers, or vice-versa, simply by
condensation or elongation of internodes (Burger 1977). Stebbins
(1974) stresses that the physical environment places the major
constraints on inflorescence types which are represented by the
flora of a particular habitat.

   The anthia of flowering bushes, trees, cushion plants, and
clones are not as readily understood. Even so, as greater understandings are acquired of the foraging movements of pollinators
(Heinrich 1979; Frankie and Haber 1983; Pyke 1978; Zimmerman 1979)
through various models and field studies, it may transpire that
growth form and flowering are related, as is growth form and
presentation of the photosynthetic apparatus (i.e. leaves) (see
Horn 1976).

CONCLUSIONS

Differences in flower structure represent the most important
taxonomic characters in angiosperms. The trends in oligomeriza-
tion and fusion of parts, the placement of the ovary, and symmetry
of floral parts are all related and must be looked at in combination (Stebbins 1951; Ehrendorfer 1973) and particularly in terms
of pollination. The variety of the forms of flowers found within
taxa must also be examined in terms of pollination. It is through
pollination that plants reproduce (derived agamosperms excepted).
Thus, selective forces for evolution must act strongly on floral
form to optimize the reproductive potential of the plant. This
can be accomplished by maximizing pollination, minimizing energy
expenditure on disposable but necessary parts (flowers except the
sporophylls, especially the gynoecium), maximizing seed production
[represented, in part, by the reduction in the ratio of the number
of pollen gains to the number of ovules (see Webb, this volume)],
and minimizing disposable but necessary dispersal structures
associated with the seed, all within the constraints of the
availability of raw materials, pollinators, and seed dispersers
and the nutritional requirements of the latter two. The evolutionary trends in floral anatomy discussed herein clearly result
in an increasing efficiency and precision of pollination and thus
strengthen reproductive isolation. Thus the zoological view of
the casual evolution of isolating mechanisms must be dismissed as
applied to plants. Isolating mechanisms in plants, be they
anatomical, chemical, phenological (see Bawa 1983) or by habitat
(see Ganders, this volume) (or combinations of these) must have
come about, to a large extent, by active selection involving
pollination (see also Levin 1971). Pollinator resource partitioning by co-occurring plant species has been documented for *Aster*
species (Graenicher 1980), *Solidago* (Gross and Werner 1983) and
subalpine plants (Inouye 1978).

   The processes of optimization lead to character divergence or
displacement in pollination systems (Kevan and Baker 1983b). This

means that plant/pollinator communities must be examined in toto
to understand the way in which both plants (flowers) and pollinators partition their respective resources between themselves, and
how selective forces may act. Grant (1966) applied the term
"Wallace effect" to shifts in population structure, the primary
advantage of which is to strengthen reproductive isolation. Levin
and Kerster (1967) apply the same idea to *Phlox*, as does
Breedlove (1969) to *Fuchsia* and McCrea (1981) to *Rudbeckia*.
However, facile arguments about wastage of male gametes are not
tenable. The argument must be made in terms of the overall
increase in the efficiency, precision, and predictability of
pollination whereby the reproductive success of the plants with
the divergent characters is enhanced. The plants involved may be
morphs of the same species or may be closely related species. In
either case they will diverge under the Wallace effect. It is
worth noting here that hybridization in zoophilous flowering
plants must come about through the action of pollinators.
Hybridization may be stabilized through amphidiploidy, introgression, or the establishment of hybrid segregates (see Stebbins
1959) which may or may not require the continued use of pollinators. Above the species level, hybridization between species
placed in different genera is rare (Stace 1975), yet the effects
of character displacement of floral features may come about in
floral evolution if competition for pollinators is invoked for
plants with similar flowers (see Waser 1983). Separating the
infraspecific Wallace effect, which implies sympatric speciation,
from interspecific and selected divergence (even of isolating
mechanisms) is not easy and will be, no doubt, an area for heated
debate.

ACKNOWLEDGMENTS

I am grateful to the grant for providing the opportunity for both
verbal and written expression of my ideas on the importance of
pollination to angiosperm biosystematics. U. Posluszny, R.
Scribailo, J. Canne, and S. Marshall have all contributed through
their thoughtful discussion. The research was supported in part
by NSERC grants No. A 8098.

REFERENCES

Asker, S. 1984. Apomixis and biosystematics. In *Plant
    Biosystematics* (W.F. Grant, ed.). Academic Press, Toronto.
Baagøe, J. 1977a. Taxonomical application of ligule
    microcharacters in Compositae. I. Anthemideae, Heliantheae,
    Tageteae. *Bot. Tidssk.* 71: 193-224.
Baagøe, J. 1977b. Microcharacters in the ligules of
    Compositae. In *Biology and Chemistry of Compositae* (V. Heywood,
    J. Harborne, and B.L. Turner, eds.). Academic Press, New York.
    Chapt. 7.
Baagøe, J. 1978. Taxonomical application of ligule micro-

characters in Compositae, Eremothamneae, Inuleae, Liabeae, Mutiseae, and Senecioneae. *Bot. Tidssk.* 72: 125–148.

Baker, H.G. 1959. Reproductive methods as factors in speciation in flowering plants. *Cold Spring Harbor Symp. Quant. Biol.* 24: 177–191.

Baker, H.G. and Baker, I. 1975. Nectar constitution and pollinator–plant coevolution. In *Animal and Plant Coevolution* (L.E. Gilbert and P.H. Raven, eds.). Univ. Texas Press, Austin. pp. 100–140.

Baker, H.G. and Baker, I. 1977. Intraspecific constancy of floral nectar amino acid complements. *Bot. Gaz.* 138: 183–191.

Baker, H.G. and Baker, I. 1979. Starch in angiosperm pollen grains and its evolutionary significance. *Am. J. Bot.* 66: 591–600.

Baker, H.G. and Baker, I. 1983. Floral nectar constituents in relation to pollinator type. In *Handbook of Experimental Pollination Biology* (C.E. Jones and R.J. Little, eds.). Scientific and Academic Editions, New York. pp. 117–141.

Baker, I. and Baker, H.G. 1976. Analysis of amino acids in flower nectars of hybrids and their parents, with phylogenetic implications. *New Phytol.* 76: 87–98.

Bawa, K.S. 1980. Evolution of dioecy in flowering plants. *Annu. Rev. Ecol. Syst.* 11: 15–39.

Bawa, K.S. 1983. Patterns of flowering in tropical plants. In *Handbook of Experimental Pollination Biology* (C.E. Jones and R.J. Little, eds.). Scientific and Academic Editions, New York. pp. 394–410.

Benham, B.R. 1969. Insect visitors to *Chamaenerion angustifolium* and their behaviour in relation to pollination. *Entomologist* 102: 221–228.

Benson, L. 1979. Plant Classification. 2nd ed. Heath, Lexington.

Breedlove, D.E. 1969. The systematics of *Fuchsia* section *Encliandra* (Onagraceae). *Univ. Calif. Publ. Bot.* 53: 1–69.

Buchmann, S.L. 1983. Buzz pollination in angiosperms. In *Handbook of Experimental Pollination Biology* (C.E. Jones and R.L. Little, eds.). Scientific and Academic Editions, New York. pp. 73–113.

Burger, W.E. 1977. The Piperales and the monocots. Alternate hypotheses for the origin of monocotyledonous flowers. *Bot. Rev.* 43: 345–393.

Burtt, B.L. 1964. Angiosperm taxonomy in practice. In *Phenetic and Phylogenetic Classification* (V.H. Heywood and J. McNeill, eds.). Systematics Association, London. No. 6, 5–16.

Corner, E.J.H. 1949. The durian theory or the origin of the modern tree. *Ann. Bot.* 13: 367–414.

Corner, E.J.H. 1953. The durian theory extended. Part I. *Phytomorphology* 3: 465–476.

Corner, E.J.H. 1954. The durian theory extended. Parts II and

III. *Phytomorphology* 4: 152-165; 263-274.

Corner, E.J.H. 1964. *The Life of Plants*. Weidenfeld and
Nicholson, London.

Cox, P.A. 1982. Vertebrate pollination and the maintenance of
dioecism in *Freycinetia*. Am. Nat. 120: 65-80.

Crepet, W.L. 1974. Investigations of North American cycadeoids:
the reproductive biology of *Cycadeoidea*. *Palaeontographica*
WA 148B: 144-159.

Crepet, W.L. 1979a. *Insect pollination: a paleontological
perspective*. BioScience 29: 102-108.

Crepet, W.L. 1979b. Some aspects of the pollination biology of
Middle Eocene angiosperms. Rev. *Palaeobot*. *Palynol*. 27: 213-238.

Cronquist, A. 1968. *The Evolution and Classification of
Flowering Plants*. Houghton Mifflin, Boston.

Douglas, G.E. 1957. The inferior ovary. *Bot. Rev*. 23: 1-46.

Doyle, J.A. and Hickey, L.S. 1976. Pollen and leaves from the
Mid-Cretaceous Potomac Group and their bearing on early
angiosperm evolution. In *Origin and Early Evolution of
Angiosperms* (C.B. Beck, ed.). Columbia Univ. Press, New York.

Ehrendorfer, F. 1973. Adaptive significance of major taxonomic
characters and morphological trends in angiosperms. In *Taxonomy
and Ecology*. Systematics Association, No. 5, Academic Press,
London. pp. 317-327.

Eisner, T., Eisner, M., Hyypio, P., Aneshansley, D. and
Silberglied, R.E. 1973. Plant taxonomy: ultraviolet patterns
of flowers visible as fluorescent patterns in pressed herbarium
specimens. *Science* 179: 486-487.

Eldredge, N. and Gould, S.J. 1972. Punctuated equilibria: An
alternative to phyletic gradualism. In *Models in Paleobiology*
(T.J.M. Schopf, ed.). Freeman, San Francisco. pp. 82-115.

Faegri, K. and van der Pijl, L. 1978. *The Principles of
Pollination Ecology*. 3rd ed. Pergamon Press, Oxford.

Feinsinger, P. 1983. Coevolution and pollination. In *Coevolution*
(D.J. Futuyma and M. Slatkin, eds.). Sinauer Associates,
Sunderland, Mass.

Ferguson, I.K. 1983. Pollen morphology in relation to pollinators
in Papilionoideae (Leguminosae). *Bot. J*. 84: 183-193.

Ferguson, I.K. 1984. Pollen morphology and biosystematics of the
subfamily Papilionoideae (Leguminosae). In *Plant Biosystematics*
(W.F. Grant, ed.). Academic Press, Toronto.

Forman, L.L. 1964. *Trigonobalanus*, a new genus of Fagaceae,
notes on the classification of the family. *Kew Bull*. 17:
381-396.

Frankie, G.W. and Haber, W.A. 1983. Why bees move among mass-
flowering neotropical trees. In *Handbook of Experimental
Pollination Biology* (C.E. Jones and R.J. Little, eds.).
Scientific and Academic Editions, New York. pp. 360-372.

Free, J.B. 1966. The foraging behaviour of bees and its effect on
the isolation and speciation of plants. In *Reproductive
Biology and Taxonomy of Vascular Plants* (J.G. Hawkes, ed.).
Pergamon Press, Oxford.

Galen, C. and Kevan, P.G. 1980. Scent and color, floral
  polymorphisms and pollination biology in *Polemonium viscosum*
  Nutt. *Am. Midl. Nat.* 104: 281–289.
Galen, C. and Kevan, P.G. 1983. Bumblebee foraging and floral
  scent dimorphisms *Bombus kirbyellis* Curtis (Hymenoptera:
  Apidae) and *Polemonium viscosum* Nutt. (Polemoniaceae).
  Can. J. Zool. 61: 1207–1213.
Ganders, F.R. 1979. The biology of heterostyly. *N. Z. J. Bot.*
  17: 607–636.
Ganders, F.R. 1984. The role of hybridization in the evolution of
  Bidens on the Hawaiian Islands. In *Plant Biosystematics*
  (W.F. Grant, ed.). Academic Press, Toronto.
Graenicher, S. 1909. Wisconsin flowers and their pollination.
  Bull. Wisconsin Nat. Hist. Soc. 7: 19–77.
Grant, V. 1950. The protection of the ovules in flowering plants.
  Evolution 4: 179–201.
Grant, V. 1966. The selective origin of incompatibility barriers
  in the plant genus *Gilia*. *Am. Nat.* 100: 99–119.
Grant, V. 1981. *Plant Speciation*. 2nd ed. Columbia Univ.
  Press. New York.
Grant, V. and Grant, K.A. 1965. *Flower Pollination in the Phlox
  Family*. Columbia Univ. Press, New York.
Gross, R.S. and Werner, P.A. 1983. Relationships among flowering
  phenology, insect visitors, and seed–set of individuals:
  experimental studies of four co–occurring species of goldenrod
  (*Solidago*: Compositae). *Ecol. Monogr.* 53: 95–117.
Harborne, J.B. 1975. The biochemical systematics of flavonoids.
  In *The Flavonoids* (J.B. Harborne, I.J. Mabry and H.
  Mabry, eds.). Academic Press, New York. pp. 1056–1095.
Heinrich, B. 1979. Resource heterogeneity and patterns of
  movement in foraging bumblebees. *Oecologia* 140: 235–245.
Horn, H.S. 1976. The Adaptive Geometry of Trees. Princeton Univ.
  Press, Princeton, N.J.
Inouye, P.W. 1978. Resource positioning in bumblebees:
  experimental studies of foraging behavior. *Ecology* 59:
  672–678,
Kaul, R.B. 1976. Conduplicate and specialized carpels in
  Alismatales. *Am. J. Bot.* 63: 175–182.
Kay, Q.O.N. 1982. Intraspecific discrimination by pollinators and
  its role in evolution. In *Pollination and Evolution* (J.A.
  Armstrong, J.M. Powell and A.J. Richards, eds.). R. Botanic
  Gardens, Sydney. pp. 9–28.
Kevan, P.G. 1972. Insect pollination of high arctic flowers.
  J. Ecol. 60: 831–847.
Kevan, P.G. 1973. Flowers, insects, and pollination ecology in
  the Canadian high arctic. *Polar Rec.* 16: 667–674.
Kevan, P.G. 1978. Floral coloration, its colorimetric analysis
  and significance in anthecology. In *The Pollination of
  Flowers by Insects* (A.J. Richards, ed.). Academic Press,
  London. pp. 51–78.
Kevan, P.G. 1983. Floral colors through the insect eye: what

they are and what they mean. In *Handbook of Experimental Pollination Biology* (C.E. Jones and R.J. Little, eds.). Scientific and Academic Editions, New York. pp. 3-30.

Kevan, P.G. and Baker, H.G. 1983a. Insects as flower visitors and pollinators. *Annu. Rev. Entomol.* 28: 407-453.

Kevan, P.G. and Baker, H.G. 1983b. Insects on flowers. In *Ecological Entomology* (C. Huffaker and R.L. Rabb, eds.), Chapt. 20. Wiley, New York.

Kevan, P.G. and Lane, M.A. 1983. Insect tactile sensitivity and floral microtexture. *Naturwissenschaften*. In press.

Kevan, P.G., Chaloner, W.G. and Savile, D.B.O. 1975. Interrelationships between early terrestrial anthropods and plants. *Palaeontology* 18(2): 391-417.

Kevan, P.G., Scribailo, R. and Posluszny, U. 1983. The functional morphology of flowers. In prep.

Kugler, H. 1943. Hummela als Blutenbesucher. *Ergeb. Biol.* 19: 143-323.

Kugler, H. 1963. UV-Musterungen auf Bluten und ihr Zustandekommen. *Planta* 59: 296-329.

Kullenberg, B. 1961. Studies in *Ophrys* L. pollination. *Zool. Bidr. Uppsala* 34: 1-340.

Kullenberg, B. and Bergstrom, G. 1976. Hymenoptera Aculeata males as pollinators of *Ophrys* orchids. *Zool. Scr.* 5: 13-23.

Lane, M.A. 1980. Systematics of *Amphiachysis, Greenella, Gutierrezia, Gymnosperma, Thurovia* and *Xanthocephalum* (Compositae: Astereae). Ph.D. thesis, Univ. Texas, Austin.

Laverty, T. 1980. The flower visiting behaviour of bumblebees: learning and flower complexity. *Can. J. Zool.* 58: 1325-1335.

Leppik, E.E. 1953. The ability of insects to distinguish number. *Am. Nat.* 87: 229-236.

Leppik, E.E. 1956. The form and function of numeral patterns in flowers. *Am. J. Bot.* 43: 445-455.

Leppik, E.E. 1957. Evolutionary relationship between entomophilous plants and anthopilous insects. *Evolution* 11: 466-481.

Leppik, E.E. 1960. Early evolution of flower types. *Lloydia* 23: 72-92.

Leppik, E.E. 1963. Fossil evidence of floral evolution. *Lloydia* 26: 91-115.

Leppik, E.E. 1971. Paleontological evidence on the morphogenic development of flower types. *Phytomorphology* 21: 164-174.

Leppik, E.E. 1972. Origin and evolution of bilateral symmetry in flowers. *Evol. Biol.* 5: 49-85.

Levin, D.A. 1971. The origin of reproductive isolating mechanisms in flowering plants. *Taxon* 20: 91-113.

Levin, D.A. and Kertser, H.W. 1967. Natural selection for reproductive isolation in *Phlox. Evolution* 21: 679-687.

Levin, D.A. and Schaal, B.A. 1970. Corolla color as an inhibitor of interspecific hybridization in *Phlox. Am. Nat.* 104: 273-283.

Levy, M. 1978. Flavonoids and pollination ecology: pigments of systematists' imagination? *Phytochem. Bull.* 2: 35-42.

Lewis, D. 1979. Genetic versatility of incompatibility in plants.

New Zealand J. Bot. 17: 637-644.

Macior, L.W. 1982. Plant community and pollinator dynamics in the evolution of pollination mechanisms in *Pedicularis* (Scrophulariaceae). In *Pollination and Evolution* (J.A. Armstrong, J.M. Powell and A.J. Richards, eds.). R. Botanic Gardens, Sydney. pp. 29-45.

Mayr, E. 1969. *Principles of Systematic Zoology*. McGraw-Hill, New York.

McCrea, K.D. 1981. Ultraviolet floral patterning, reproductive isolation and character displacement in the genus *Rudbeckia*. Ph.D. thesis, Purdue Univ. West Lafayette, Indiana.

Meeuse, A.D.J. 1971. Palm and pandan pollination. Primary anemophily or primary entomophily? *Botanique* (Nagpur) 3: 1-6.

Meeuse, A.D.J. 1975a. Floral evolution as the key to angiosperm descent. *Acta Bot. Neerl.* 3: 1-18.

Meeuse, A.D.J. 1975b. Changing floral concepts: anthocorms, flowers and anthoids. *Acta Bot. Neerl.* 24: 23-36.

Meeuse, A.D.J. 1975c. Floral evolution in the Hamamelididae. II. Interpretative floral morphology of the Amentiferae. *Acta Bot. Neerl.* 24: 165-179.

Meeuse, A.D.J. 1978. Entomophily in *Salix*: theoretical considerations. In *The Pollination of Flowers by Insects* (A.J. Richards, ed.). Academic Press, London. pp. 47-50.

Meeuse, A.D.J. 1979. Why were the early angiosperms so successful? A morphological, ecological and phylogenetic approach. Proc. K. Ned. Akad. Wet. Ser. C 82: 343-369.

Morton, J.K. 1962. Cytotaxonomic studies on West African Labiatae. *J. Linn. Soc.London, Bot.* 58: 321-383.

Mulcahy, D.L. 1984. The relationships between self-incompatibility, pseudocompatibility and self-compatibility. In *Plant Biosystematics* (W.F. Grant, ed.). Academic Press, Toronto.

Mulligan, G.A. and Kevan, P.G. 1973. Color, brightness and other floral characteristics attracting insects to the blossoms of some Canadian weeds. *Can. J. Bot.* 51: 1939-1952.

Neff, J.L. and Simpson, B.B. 1981. Oil-collecting structures in the Anthophoridae (Hymenoptera): morphology, function and use in systematics. *J. Kans. Entomol. Soc.* 54: 95-123.

Niklas, K.J. 1981a. Simulated wind pollination and airflow around ovules of some early seed plants. *Science* 211: 275-277.

Niklas, K.J. 1981b. Airflow patterns around some early seed plant ovules and cupules: implications concerning efficiency in wind pollination. *Am. J. Bot.* 68: 635-650.

Nur, N. 1976. Studies on pollination in Musaceae. *Ann. Bot.* 40: 167-177.

Ornduff, R. and Mosquin, T. 1970. Variation in the spectral qualities of flowers in the *Nymphoides indica* complex (Menganthaceae) and its possible adaptive significance. *Can. J. Bot.* 48: 603-605.

Phipps, J.B. 1984. Problems of hybridity in the cladistics of *Crataegus* (Rosaceae). In *Plant Biosystematics* (W.F.

Grant, ed.). Academic Press, Toronto.
van der Pijl, L. 1957. The dispersal of plants by bats
    (chiropterochory). *Acta Bot. Neerl.* 6: 291–315.
van der Pijl, L. 1961. Ecological aspects of flower evolution.
    II. Zoophilous flower classes. *Evolution* 15: 44–59.
van der Pijl, L. and Dodson, C.H. 1966. *Orchid Flowers, their
    Pollination and Evolution.* Fairchild Tropical Gard. and Univ.
    Miami Press, Coral Gables, Florida.
Plowright, R.C. and Hartling, L.K. 1981. Red clover pollination
    by bumblebees: a study of the dynamics of a plant–pollinator
    relationship. *J. Appl. Ecol.* 18: 639–647.
Pyke, G.J. 1978. Optimal foraging in bumblebees and coevolution
    with their plants. *Oecologia* 36: 281–293.
Pyke, G.H. 1979. Optimal foraging in bumblebees: rule of
    movement between flowers within inflorescences. *Anim. Behav.*
    27: 1167–1181.
Raven, P.H. 1979. A survey of reproductive biology in the
    Onagraceae. *N. Z. J. Bot.* 17: 575–594.
Regal, P.J. 1982. Pollination by wind and animals: ecology of
    geographic patterns. *Annu. Rev. Ecol. Syst.* 13: 497–524.
Rollins, R.C. 1963. The evolution and systematics of *Leaven-
    worthia* (Cruciferae). *Contrib. Gray Herb.* 192: 1–98.
Scogin, R. 1983. Visible pigments and pollinators. In *Handbook
    of Experimental Pollination Biology* (C.R. Jones and R.J.
    Little, eds.). Scientific and Academic Editions, New York. pp.
    160–172.
Simpson, G.G. 1961. Principles of Animal Taxonomy. Columbia
    Univ. Press, New York.
Smart, J. and Hughes, N.F. 1973. The insect and the plant:
    progressive palaeoecological integration. In *Insect/Plant
    Relationships* (H.F. van Emden, ed.). Blackwells, Oxford.
    pp. 143–156.
Spira, T. 1980. Floral parameters, breeding system and pollinator
    type in *Trichostema* (Labiatae). *Am. J. Bot.* 67: 278–284.
Stace, C.E. 1975. *Hybridization and the Flora of the British
    Isles.* Academic Press, London.
Stebbins, G.L. 1950. *Variation and Evolution in Plants.*
    Columbia Univ. Press, New York.
Stebbins, G.L. 1951. Natural selection and differentiation of
    angiosperm families. *Evolution* 5: 299–324.
Stebbins, G.L. 1959. The role of hybridization in evolution.
    Proc. Am. Philos. Soc. 103: 231–251.
Stebbins, G.L. 1971. Relationships between adaptive radiation,
    speciation and major evolutionary trends. *Taxon* 20: 3–16.
Stebbins, G.L. 1974. *Flowering Plants. Evolution Above the
    Species Level.* Belknap, Harvard Univ. Press, Cambridge, Mass.
Takhtajan, A. 1969. *Flowering Plants: Origin and
    Dispersal.* Smithsonian Inst. Press, Washington, D.C.
Taylor, T.N. and Levin, D.A. 1975. Pollen morphology of the
    Polemoniaceae in relation to systematics and pollination
    systems: scanning electron microscopy. *Grana* 15: 91–112.

Taylor, T.N. and Millay, M.A. 1979. Pollination in biology and reproduction in early seed plants. *Rev. Paleobot. Palynol.* 27: 329-355.

Thein, L.B. 1980. Patterns of pollination in primitive angiosperms. *Biotropica* 12: 1-13.

Thein, L.B., White, D.A. and Yatsu, L.Y. 1983. The reproductive biology of a relict - *Illicium floridanum* Ellis. *Am. J. Bot.* 70: 719-727.

Urbanska, K.M. 1984. Plant reproduction strategies. In *Plant Biosystematics* (W.F. Grant, ed.). Academic Press, Toronto.

Waddington, K.D. and Heinrich, B. 1979. The foraging movements of bumblebees on verticle "inflorescences": an experimental analysis. *J. Comp. Physiol. A* 134: 113-117.

Waser, N.M. 1983. Competition for pollination and floral character differences among sympatric plant species: a review of evidence. In *Handbook of Experimental Pollination Biology* (C.E. Jones and R.J. Little, eds.). Scientific and Academic Editions, New York. pp. 277-293.

Waser, N.M. and Price, D.M.V. 1981. Pollinator choice and stabilizing selection for flower color in *Delphinium nelsonii. Evolution* 35: 376-390.

Webb, C.J. 1984. Constraints on the evolution of plant breeding systems and their relevance to systematics. In *Plant Biosystematics* (W.F. Grant, ed.). Academic Press, Toronto.

Whitehead, D.R. 1969. Wind pollination in angiosperms: evolutionary and environmental considerations. *Evolution* 23: 28-35.

Wiebes, J.T. 1979. Coevolution of figs and their insect pollinators. *Annual Rev. Ecol. Syst.* 10: 1-12.

Williams, N.H. 1983. Floral fragrances as cues in animal behavior. In *Handbook of Experimental Pollination Biology* (C.E. Jones and R.J. Little, eds.). Scientific and Academic Editions, New York. pp. 50-72.

Williams, N.H. and Dodson, C.H. 1972. Selective attraction of male euglossine bees to orchid floral fragrances and its importance in long distance pollen flow. *Evolution* 26: 84-95.

Wolfe, J.A., Doyle, J.A. and Page, V.M. 1975. The bases of angiosperm phylogeny: paleobotany. *Ann. Mo. Bot. Gard.* 62: 801-824.

Woodhouse, R.P. 1935. *Pollen Grains, Their Structure, Identification and Significance in Science and Medicine.* McGraw-Hill, New York. 574 pp.

Zimmerman, M. 1979. Optimal foraging: a case for random movement. *Oecologia* 43: 261-267.

The Biosystematic Importance of Phenotypic Plasticity

Pierre Morisset and *Céline Boutin**
Département de biologie
Université Laval
Québec, Québec, Canada

INTRODUCTION

The study of phenotypic plasticity goes back to the latter part of
the last century, to Bonnier (1890, 1895) and Kerner von Marilaun
(1895), and slightly later, to Massart's (1902) classical paper on
Polygonum amphibium L. Through transplantation experiments,
Massart showed that this species will develop the characteristic
features of terrestrial or aquatic forms when it is grown in these
respective environments. Turesson (1922) commented that there
was, already in the early twenties, a considerable body of
experimental data on the nature and intensity of modifications
induced in plants by various environmental factors. Clausen *et
al.* (1940, pp. 394-407) made a brief critical review of earlier
works, and Daubenmire (1959) summarized and discussed at length
the modifications brought about by differences in soil moisture
and aeration, temperature, wind, light intensity, nutrients, etc.
The biosystematic relevance of such data was considered by, among
others, Davis and Heywood (1963, pp. 335-349) and Stace (1980, pp.
186-190). Since the publication of Turesson's (1922, 1925) and
Clausen *et al.*'s (1940, 1958) classical papers, the ubiquity
of phenotypic plasticity has been widely recognized, although its
adaptive significance was either ignored or taken for granted with
little critical evidence. Bradshaw (1965), in a seminal review,
clearly showed that plasticity of a given character with respect
to a given environmental factor varies between individuals,

---

* Present address: School of Plant Biology, University College
  of North Wales, Bangor, Gwynedd LL57 2UW, United Kingdom.

Copyright © 1984 by Academic Press Canada
All rights of reproduction in any form reserved.
ISBN 0-12-295680-X

between populations, and between species. Bradshaw (1965, 1973)
also argued convincingly that phenotypic plasticity of a character
is a response which is specific to that character, specific in
relation to particular environmental influences, specific in
direction, and under genetic control.

The importance of phenotypic plasticity in plants is clearly
recognized by population biologists and ecologists (Harper 1977;
Grime 1979). Considering that this phenomenon has a marked
adaptive value (Bradshaw 1965), it is surprising that it has been
relatively neglected by biosystematists and genecologists, except
in certain groups such as amphibious plants (Cook 1968). After
considering the various types of plasticity, we will briefly
discuss the importance of plasticity in relation to adaptation and
character evolution.

TYPES OF PLASTIC RESPONSES

It is not always realized that plants can show many kinds of
plastic responses, some adaptive and others perhaps nonadaptive.
Bradshaw (1965) and Cook (1968), while discussing mechanisms of
plasticity, distinguished between characters showing a contiuous
range of modification and characters showing only a few discrete
phenotypes. In the first case, the intensity and direction of the
plastic response are directly dependent on the intensity and
direction of the environmental stimulus. In the second case, the
environmental stimulus acts as a switch, and the response is self-
regulating after it has been initiated. A more general
classification of plastic responses is proposed here:
  1. Developmental plasticity
  2. Environmental plasticity
      2.1 Tolerance plasticity
      2.2 Compensation plasticity

Developmental Plasticity

This type of plasticity is met with when the expression of a
character is dependent on the developmental stage, such as in
heteroblastic leaf development. Bradshaw (1965) called it *"fixed
phenotypic variation"* and argued that it must not be considered
plasticity at all since the modification occurs independently of
the environment. However, the occurrence, for instance, of many
leaf shapes or else of both chasmogamous and cleistogamous flowers
on the same individual plant can be considered as a special case
of plasticity *sensu lato*. Furthermore, as pointed out by
Bradshaw (1965), the boundary between developmental plasticity and
environmentally-induced plasticity is not always clear. In
certain species heterophylly is determined and fixed by the
developmental program, whereas in other species it is controlled
by environmental stimuli (Cook 1968). It seems appropriate in
such cases to include both mechanisms under the general term of
plasticity.

Environmental Plasticity

This is plasticity in the strict sense, where the phenotypic expression of a character is modified by environmental stimuli. Bradshaw (1965) called it *"true plasticity"*. That covers a wide range of responses, which may be continuous or discrete, general or specific to certain characters of the plant, adaptive or nonadaptive. If the nature and function of environmental plasticity are to be properly understood, the various types of responses must be clearly recognized. We distinguish tolerance from compensation responses.

Tolerance Plasticity: Tolerance responses to environmental variation allow plants to grow in a wide range of nonoptimal conditions through general phenotypic modifications that are not specific to any particular environmental factor. They involve responses which, in Bradshaw's (1973) words, "are obviously due to the basic properties of living matter." They include, for instance, the slowing down of growth when the temperature is abnormally low, or when certain elements essential to growth, such as nitrogen, are lacking. In such cases, phenotypic modifications can be considered as the "passive" results if impaired growth. They allow plants to tolerate suboptimal conditions without representing specific adaptations to these precise conditions. A general lowering of size due to poor growth under abnormally low or high temperatures can scarcely be called an adaptation to temperature. Similarly, the death of terminal buds in wind-exposed trees will give rise to various growth forms in affected trees (Payette 1974); these are certainly an effect of wind action but not necessarily a specific adaptation to wind.

Compensation plasticity: Compensation plasticity involves responses which are specific to a given character in relation to a particular environmental factor, in such a way that the phenotypic modifications actually result in an improved performance of the plant under the changed conditions. Whitehead (1962) used the expression "compensating mechanisms" in the same sense. Compensation responses are therefore *prima facie* adaptative and they include most classical examples of phenotypic plasticity: heterophylly in aquatic plants (Cook 1968), phenotypic increase in stem height or leaf surface area under low light intensity, etc. Compensation plasticity can comprise both continuous and discrete responses, and it can involve morphological as well as physiological changes. In a review of photosynthetic adaptation in higher plants, Berry and Björkman (1980) grouped such responses under the expression "photosynthetic acclimatation." We prefer the term "compensation" because it overtly says that the plastic responses compensate for environmental changes by specific, directional phenotypic changes. The term "acclimatation" could be restricted to reversible compensation responses.

It is important to distinguish tolerance from compensation responses because they do not have the same biological significance. To respond to some environmental stress by a passive, general reaction such as merely slowing down growth is quite different from responding to it by a phenotypic adjustment which partly alleviates the adverse effects of this stress. Thus, the development of a hierarchical weight distribution in a high density population (White and Harper 1970) may be due to tolerance plasticity responses, whereas concomitant differences in resource allocation patterns between larger and smaller plants (Hawthorn and Cavers 1982; Trivedi and Tripathi 1982) will include compensation responses. Similarly, an increase in stem height may be a tolerance response to increased soil nutrient levels, but a compensation response to decreased light intensity.

The fact that these two types of responses are not recognized, and therefore not distinguished in published papers on phenotypic plasticity makes it almost impossible to assess their relative importance from a literature review. Attempts to measure general phenotypic plasticity in a variety of environments, and for a number of characters selected because they can be easily measured (*e.g.* Marshall and Jain 1968) cannot discriminate between tolerance and compensation responses. It would be worthwhile to devise experiments in order to separate these two types of responses in particular cases. Both are of course genetically controlled.

PLASTICITY AND ADAPTATION

The potential role of phenotypic plasticity in adaptation is one of the reasons why biosystematists should be concerned with this phenomenon. It is now well established that plastic responses may have adaptive value (Cook and Johnson 1968; Bradshaw 1973). Furthermore, it has been suggested that phenotypic plasticity may be an alternative to genetic polymorphism in adaptation to unstable or heterogeneous environments (e.g. Marshall and Jain 1968; Jain 1978). A thorough review of the adaptive function of plasticity being beyond the scope of this article, we will briefly comment on three related questions: In which conditions is plasticity adaptive? Is it always adaptive? And can plasticity be a general response of the genotype to a variety of environmental conditions?

Adaptative Plasticity

Phenotypic plasticity is the only possible adaptative strategy when the scale of environmental variation is too small to permit adaptation through genetic differentiation (Bradshaw 1965). The water-air interface is one of the most extreme examples of such small-scale environmental variation, and it is not unexpected that plants which must grow through this interface should show a strong plasticity, at least in leaf shape (Cook 1968). There are small-scale variations and fluctuations in every sort of habitat, and one can therefore hypothesize that phenotypic plasticity as an adaptive strategy will be of general occurrence in plants. One way of inquiring into this possibility is to compare plastic responses among populations or taxa whose habitats differ by their temporal and/or spatial heterogeneity. Recent data of this sort mostly come from population biology. Hickman (1975) showed that allocation of resources to reproduction in the annual species *Polygonum cascadense* Baker was proportionally much higher in harsh open habitats than in more moderate habitats, the differences being plastic rather than genetically determined. This plastic response was considered to be an adaptation to short-term environmental unpredictability. A similar behaviour was described by Cartica and Quinn (1982) for *Solidago sempervirens* L. along a coastal dune gradient. Bazzaz and Carlson (1982) showed that early successional species were more plastic in their photosynthetic response than late successional species (Fig. 1). Wilken (1977), working with the annual *Collomia linearis* Nutt., showed that plants from a disturbed site were more plastic than plants from an adjacent meadow site. This sample of examples reveals a clear tendency for pioneer, early successional species or populations to show more plasticity than later successional congeners in a number of morphological and physiological characters, as well as in growth-related features. This trend was already mentioned by Baker (1965) when he presented data showing that weedy species tend to be more plastic than non-weedy related taxa.

Non-adaptive Plasticity

The above-mentioned examples imply that all plastic responses are adaptative, but this is not necessarily so. There is no doubt that compensation responses give rise to specific adaptations, as stressed by Bradshaw (1973). There are many examples in the literature of such responses, and their adaptative value has been directly demonstrated in a few cases. For instance, Whitehead (1962) showed that exposure to wind produces in leaves phenotypic modifications that effectively decrease water loss in windy conditions. Other examples are discussed by Bradshaw (1973). However, compensation responses are adaptative only within a certain range of environmental factors. Beyond this range, a compensation response may have non-adaptative drawbacks. One example will illustrate this point. In a recent study of phenotypic plasticity in *Chrysanthemum leucanthemum* L., Boutin and Morisset (in prep.) grew plants in the field under a 24% light intensity level. These plants responded, as expected, by developing taller stems and branches, which is an example of compensation plasticity; however, it turned out that the stems were too weak to support their flowering and fruiting heads, clearly a nonadaptative outcome. *C. leucanthemum* is an open

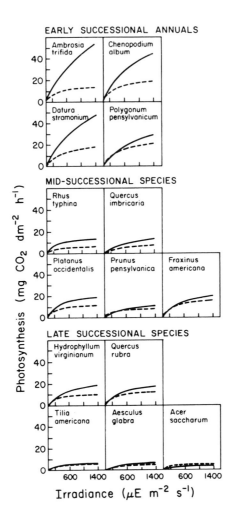

FIG. 1. *Photosynthetic response curves to light intensity for species representing early, mid-, and late successional species grown in full sunlight (solid lines) and in deep shade equal to ca 1% of full sunlight (broken lines). The distance between the solid and the broken line is a measure of the phenotypic plasticity of each species with respect to this physiological character. (Reproduced with permission from Bazzaz, 1982).*

habitat species that rarely if ever encounters a 76% cut in light intensity in its normal habitat, and it is not expected to have evolved the capacity to react to such a stress by an adaptative phenotypic response. This example also shows that adaptation through phenotypic plasticity, like genetically fixed adaptation, may involve conflicts and compromises between many characters and environmental pressures (Jones and Wilkins, 1971). The capacity for a genotype to show adaptative phenotypic plasticity can evolve only for the range of environmental variations usually encountered in the plant's environment. Exceptionally high or low levels of these factors, or other unusual stresses (late spring frost, heavy trampling, etc.) may give rise to non-adaptative responses. It is of course quite possible that the capacity of responding phenotypically in a non-adaptative fashion to an unusual stress is itself dependent on the capacity of showing adaptative responses to lower levels of stress. As far as we know this has never been clearly demonstrated. Such nonadaptative responses would belong to what we have called tolerance plasticity.

Plasticity as a General Response

The correlation between habitats and levels of plasticity leads to a consideration of the nature of plastic responses. At first sight there appears to be two main trends in the literature concerning this problem. Some authors, exemplified by Bradshaw (1973), consider that plastic responses are always quite specific in relation to certain precise environmental factors, whereas others seem to study phenotypic plasticity as if it were a general property of the plant (e.g. Wilken 1977; Wu and Jain 1978). Do these two points of view reflect the occurrence of two classes of plastic responses, or are they simply the outcome of methods used by different authors? As mentioned earlier, there do occur *specific* adaptative responses to particular environmental factors, but can phenotypic plasticity also provide a *general* adaptation to small-scale or short-term fluctuations in a complex of many environmental factors?

   A general adaptation through phenotypic plasticity is implied (although not necessarily openly) in a number of recent papers where plasticity was measured over a large number of environmental conditions. For instance, in the study already referred to, Wilken (1977) calculated the global among-environment component of variance for each of nine morphological characters, over 36 different environmental conditions representing combinations of pot size, soil type, plant density, and nutrient treatments. His analysis of the data showed that plants of the disturbed site were significantly more plastic than plants of the meadow site for six characters, but no attempt was made to relate the plasticity of any particular character to any precise factor. The inference is that plasticity must confer a general adaptation to the disturbed site conditions. Other studies using a similar approach include those of Marshall and Jain (1968) and Wu and Jain (1978). Whether such plasticity does actually reflect a general adaptative response cannot be known, since it could equally be the sum of separate specific responses to each variable aspect of the environment. On the other hand, if selective pressures have produced in one particular character (for instance leaf shape) the capacity of responding plastically to one precise environmental stimulus, this character, being less strongly canalized in its development than other features of the plant, might be sensitive to other environmental stimuli as well, even if it was not directly selected to respond to these stimuli. Such a character would then show a general plasticity in response to a number of environmental conditions. One possible example of this phenomenon may be seen in heterophyllous amphibious plants, as discussed by Cook (1968). Plastic changes in leaf shape are an adaptation to the presence or absence of water in the plant's environment, but leaf shape may also be influenced by many other factors: temperature, photoperiod, light intensity, nutritional status, etc. Some of these factors, such as photoperiod for *Ranunculus aquatilis* L., are directly involved in the adaptative response, but the reactions to other environmental stimuli and their interactions could just be secondary side effects of the morphogenetic capacity for the main purpose.

PLASTICITY AND CHARACTER EVOLUTION

A character occurring as a phenotypic modification in one population may be genetically fixed in another population of the same species. Good examples have been described by Nelson (1965) in *Prunella vulgaris* L., by Cook and Johnson (1968) in *Ranunculus flammula* L., etc. The genetic fixation (or canalization) of a phenotypic response was experimentally done thirty years ago by Waddington (1953), who named this process "genetic assimilation." If this phenomenon occurs in plants, a study of phenotypic plasticity in related species could provide insights into the evolution of characters. We would like to illustrate this point with an example drawn from the genus *Ononis* L. (Morisset 1967). *Ononis spinosa* L. ssp. *spinosa* and *O. repens* L. are two closely related species of Leguminosae occurring sympatrically in western Europe. They are both herbaceous perennials, and they grow in a variety of open habitats, meadows, roadsides, dunes, etc. In the British Isles, they differ by their chromosome number and some morphological characters, including habit, leaf shape, stem colour, and presence of spines on the stem (Table I). When *O. spinosa* is grown in low light intensity, it is modified in the direction of *O. repens* for at least four characters that differ between the two species in their normal, open habitat (Fig. 2, Table I). *O. repens* show little to no plasticity for these characters.

Beside this environmental plasticity, *O. spinosa* also shows a developmental plasticity in the same set of characters: Young, juvenile stems have the same character-states as adult stems of *O. repens*, adult characters in *O. spinosa* gradually appearing as ontogenic development proceeds towards flowering. This is illustrated in Fig. 3 for the shape of terminal leaflet. These observations suggest that *O. repens* can be considered morphologically as a neotenic variant of *O. spinosa*. The evidence

FIG. 2. *Above: Decumbent habit of* Ononis repens *grown in the greenhouse under normal light intensity; the habit remains the same in shade. Below: Ascending to erect habit of* O. spinosa *grown in the greenhouse under normal light intensity (left), and decumbent habit of a ramet from the same individual grown in shade (pot placed under the greenhouse bench) (right).*

supporting such an interpretation is as follow: (1) Juvenile
leaves of *O. spinosa* and *O. repens* have the same shape, but
leaflets become gradually narrower in *O. spinosa* (Fig. 3).
(2) Stems become geotropically negative in *O. spinosa* earlier than
in *O. repens*, hence the ascending habit of *O. spinosa* versus the
decumbent habit of *O. repens*. (3) *O. spinosa* develops spines
early in the summer, whereas *O. repens* is either unarmed or
develops spines only after the flowering period in early autumn.
(4) Young stems of *O. spinosa* are green, and they become reddish

TABLE I. *Characters distinguishing* Ononis spinosa *from* O. repens *in the British Isles**

| | O. SPINOSA | | O. REPENS |
	IN FULL LIGHT	IN SHADE	IN FULL LIGHT AND IN SHADE
Habit	Erect-ascending	Decumbent	Decumbent
Spines on stems and branches	Spiny from early summer	Unarmed	Unarmed, sometimes spiny in early autumn
Stem colour	Dark reddish	Greenish	Greenish
Shape of terminal leaflet of adult leaves	Narrow, acute	Wide, obtuse	Wide, obtuse
Chromosome number	$2n = 30$		$2n = 60$
Rhizomes	Absent		Present
Glandular hairs on stem	Absent-rare		Present
Eglandular hairs on calyx tube	Few		Many
Pod	$\geqslant$ calyx		$<$ calyx

* The first four characters show a plastic response to light
  intensity in *O. spinosa*.

early in the summer, whereas *O. repens* is either unarmed or
develops spines only after the flowering period in early autumn.
(4) Young stems of *O. spinosa* are green, and they become reddish
early in the summer; in *O. repens* stems are usually always
green, but they can occasionally become somewhat reddish in late
summer. In these four characters, juvenile stems of *O. spinosa*
are similar to adult stems of *O. repens*. When grown under
low light intensity, *O. spinosa* plastically remains in a juvenile
state. Thus, the response of *O. spinosa* to low light intensity
appears to be genetically canalized in *O. repens*. Some
observations of Jardeni (1938) are of interest in this respect.
This author described a seasonal polymorphism in *O. spinosa* ssp.
leiosperma (Boiss.) Šir. In winter (rainy, low illumination), the
shoots are unarmed and the leaves are large, but in summer (dry,
high illumination) the shoots are spiny and more erect and the
leaves are small. Anatomical differences are correlated with the
morphological ones, and according to Jardeni light is the main

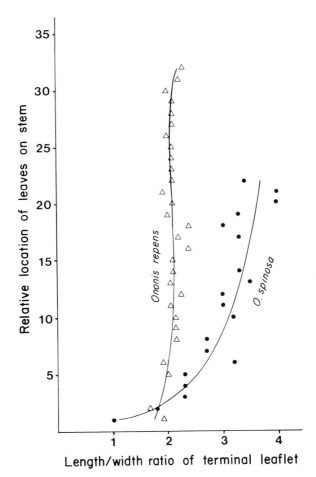

FIG. 3 *Heteroblastic leaf development in* Ononis spinosa
compared to homoblastic leaf development in O. repens. *Dots
represent successive leaves from the lower to the upper part of
the stem, and lines are hand-drawn to show the trends. Juvenile
leaves of* O. spinosa *cannot be distinguished from* O. repens
leaves on the basis of leaf shape.

*factor controlling this dimorphism. The latter is much
reminiscent of the differences both between light- and shade-grown*
O. spinosa ssp. spinosa, and between O. repens *and* O.
spinosa. This *Ononis* complex appears to be a good example of the
complex interactions that may exist between seasonal polymorphism,
phenotypic plasticity, ontogenic development, neoteny, and
character evolution, and it would certainly repay further study
along these lines.
      We do not know whether the situation in *Ononis* is exceptional

or widespread in plants. However, it is known that polyploidy generally slows down growth rate and delays flowering (Stebbins 1950). If there occur selective pressures to hasten flowering in a newly formed polyploid, one can predict that neotenic changes in vegetative characters might accompany this hastening of the flowering period. Such neoteny would be much easier to observe and study in characters showing either developmental (e.g. heteroblasty) or environmental plasticity. It would be worthwhile to look at diploid-tetraploid species pairs from this point of view. At any rate, it should be obvious that a comparative study of plastic responses in closely related taxa may provide interesting perspectives on the evolution of morphological characters in plants.

CONCLUDING REMARKS

The main function of systematics is to describe and explain the diversity of phenotypes. Since a phenotype is the outcome of interactions between a genotype and environment, its environmental component should be given as much importance as its genetical component. In retrospect, it is quite surprising that biosystematists have paid so little attention to phenotypic plasticity over the last 40 years, except in some groups such as amphibious plants (Cook 1968). Experimental designs used in the classical studies of Turesson, Gregor, Clausen and his collaborators were mainly devised to study genetic differences by levelling out environmental conditions. Much phenotypic plasticity was of course described in these studies, but it was mostly residual variation that could not be explained in genetical terms. In the stride of neodarwinism biosystematists focused their attention on genes, and they came to view the phenotype simply as a marker of the genotype. Phenotypic plasticity was then considered as mere "non-genetic" variation or, worst still, as an "often misleading source of variation" (Bell 1957, p. 50). If, as systematists, we put the genotype back in its proper place, i.e. one of the two components responsible for the making of phenotypes, the experimental study of form might become a fundamental concern of biosystematics.

Plastic responses are ontogenically very complex, as stressed by Bradshaw (1973) and as shown more recently by Steward et al. (1981). In this connection, it is quite revealing that a symposium such as this one should have included no talk on comparative morphogenesis of closely related taxa, or on character ontogeny: This absence reflects a real lack in biosystematics until now. Cladistics, population biology, and ecophysiology are now forcing us into considering the adaptation and evolution of visible phenotypes. Biosystematics needs this new direction, and genecology should give some room to a new discipline that could be called "phenecology." Considering its adaptative importance in plants, the phenomenon of phenotypic plasticity should then become a central theme in biosystematics.

ACKNOWLEDGMENTS

Luc Brouillet made useful comments on an earlier version of the
manuscript. The Natural Sciences and Engineering Research Council
of Canada provided a research grant to P.M. and a scholarship to
C.B.

REFERENCES

Baker, H.G. 1965. Characteristics and mode of origin of weeds.
In *The Genetics of Colonizing Species* (H.G. Baker and G.L.
Stebbins, eds.), Academic Press, New York. pp. 147-172
Bazzaz, F.A. and Carlson, R.W. 1982. Photosynthetic acclimitation
to variability in the light environment of early and late
successional plants. *Oecologia (Berl.)* 54: 313-316.
Bell, C.R. 1967. *Plant Variation and Classification*. Wadsworth,
Belmont, Calif., 135 pp.
Berry, J. and Björkman, O. 1980. Photosynthetic response and
adaptation to temperature in higher plants. *Annu. Rev. Plant
Physiol.* 31: 491-543.
Bonnier, G. 1890. Cultures expérimentales dans les Alpes et dans
les Pyrénées. *Rev. Gen. Bot.* 2: 513-546.
Bonnier, G. 1895. Recherches expérimentales sur l'adaptation des
plants au milieu alpin. *Ann. Sci. Nat., Bot., 7e sér.* 20: 217-
358.
Boutîn, C. and Morisset, P. Étude de la plasticité phénotypique
chez *Chrysanthemum leucanthemum* L. I. Analyse de croissance et
phénologie de la reproduction. Submitted to *Can. J. Bot.*
Bradshaw, A.D. 1965. Evolutionary significance of phenotypic
plasticity in plants. *Adv. Genet.* 13: 115-155.
Bradshaw, A.D. 1973. Environment and phenotypic plasticity. In
Basic Mechanisms in Plant Morphogenesis. Brookhaven Symp. Biol.
25: 75-94.
Cartica, R.J. and Quinn, J.A. 1982. Resource allocation and
fecundity of populations of *Solidago sempervirens* along a
coastal dune gradient. *Bull. Torrey Bot. Club* 109: 299-305.
Clausen, J., Keck, D.D. and Hiesey, W.M. 1940. *Experimental
Studies on the Nature of Species. I. Effect of varied
environments on Western North American plants.* Carnegie Inst.
Washington Publ. No. 520. 452 pp.
Clausen, J., Keck, D.D. and Hiesey, W.M. 1958. III. *Environmental
responses of climatic races of* Achillea. Carnegie Inst.
Washington Publ. No. 581. 129 pp.
Cook, C.D.K. 1968. Phenotypic plasticity with particular
reference to three amphibious species. In *Modern Methods in
Plant Taxonomy* (V.H. Heywood, ed.). Academic Press, New York.
pp. 97-111.
Cook, S.A. and Johnson, M.P. 1968. Adaptation to heterogeneous
environments. *Evolution* 22: 496-516.
Daubenmire, R.F. 1959. *Plants and Environment*. Wiley, N.Y.
Davies, P.H. and Heywood, V.H. 1963. *Principles of Angiosperm*

Taxonomy. Oliver & Boyd, Edinburgh and London. 556 pp.

Grime, J.P. 1979. *Plant Strategies and Vegetation Processes*. Wiley, New York. 222 pp.

Harper, J.L. 1977. *Population Biology of Plants*. Academic Press.

Hawthorn, W.R. and Cavers, P.B. 1982. Dry weight and resource allocation patterns among individuals in populations of *Plantago major* and *P. rugelii*. Can. J. Bot. 60: 2424–2439.

Hickman, J.C. 1975. Environmental unpredictability and plastic energy allocation strategies in the annual *Polygonum cascadense* (Polygonaceae). *J. Ecol.* 63: 689–702.

Jain, S. 1978. Adaptive strategies: Polymorphism, plasticity, and homeostasis. In *Topics in Plant Population Biology* (O.T. Solbrig, S. Jain, G.B. Johnson, and P.H. Raven, eds.). Columbia Univ. Press, New York. pp. 160–187.

Jardeni, D. 1938. The polymorphism of *Ononis leiosperma* Boiss. var. *tamarae* var. nov. *Palest. J. Bot.* 1: 235–237.

Jones, D.A. and Wilkins, D.A. 1971. *Variation and Adaptation in Plant Species*. Heinemann, London. 184 pp.

Kerner von Marilaun, A. 1895. *The Natural History of Plants*. (F.W. Oliver, ed.). Blackie, London. Vol. 2.

Marshall, D.R. and Jain, S.K. 1968. Phenotypic plasticity of *Avena fatua* and *A. barbata*. *Am. Nat.* 102: 457–467.

Massart, J. 1902. L'accomodation individuelle chez le *Polygonum amphibium*. *Bull. Jard. Bot. Etat Bruxelles* 1: 73–95.

Morisset, P. 1967. *Cytological and Taxonomic Studies in* Ononis spinosa *L.,* O. repens *L. and related species*. Ph.D. Thesis, Univ. Cambridge, Cambridge, England.

Nelson, A.P. 1965. *Brittonia* 17: 160–174.

Payette, S. 1974. Classification écologique des formes de croissance de *Picea glauca* (Moench.) Voss et de *Picea mariana* (Mill.) BSP. en milieux subarctiques et subalpins. *Naturaliste Can.* 101: 893–903.

Stace, C.A. 1980. *Plant Taxonomy and Biosystematics*. Edward Arnold.

Steward, F.C., Moreno, U. and Roca, W.M. 1981. Growth, form and composition of potato plants as affected by environment. *Ann. Bot.* 48 (Suppl. 2): 1–45.

Trivedi, S. and Tripathi, R.S. 1982. Growth and reproductive strategies of two annual weeds as affected by soil nitrogen and density levels. *New Phytol.* 91: 489–500.

Turesson, G. 1922. *Hereditas* 3: 211–350.

Turesson, G. 1925. *Hereditas* 6: 147–236.

Waddington, C.H. 1953. *Evolution.* 7: 118–126.

White, J. and Harper, J.L. 1970. *J. Ecol.* 58: 467–485.

Whitehead, F.W. 1962. *New Phytol.* 61: 59–62.

Wilken, D.H. 1977. Local differentiation for phenotypic plasticity in the annual *Collomia linearis* (Polemoniaceae). *Syst. Bot.* 2: 99–108.

Wu, K.K. and Jain, S.K. 1978. Genetic and plastic responses in geographic differentiation of *Bromus rubens* populations. *Can. J. Bot.* 56: 873–879.

A Biosystematic and Phylogenetic Study of the Dipsacaceae

R. Verlaque
Laboratoire de Cytotaxinomie végétale
Université de Provence
Centre de Saint-Charles
Marseille, France

INTRODUCTION

Within the context of those studies undertaken on the
Mediterranean flora and its history, we became interested in the
Dipsacaceae. This family presents in fact two notable
characteristics:
    (1) it seems to be associated with the Mediterranean Basin and
its climate;
    (2) it is considered by some authors (Sporne 1980) to be the
most evolved of the Dicotyledons.

TAXONOMY

The subdivision of the Dipsacaceae, which contains around 350
species, has always been quite difficult due to the apparent
morphological homogeneity of this family. The lack of "evident"
and "very conspicuous" characters led the first botanists (from
Linné 1737 to Bentham and Hooker 1873) to recognize only the two
genera *Dipsacus* and *Scabiosa*. In fact, as Coulter had revealed
by 1842, this family presents a multitude of distinctive features
which are almost all found at the level of the diaspore (= fruit
s.l.) and which later led Van Tieghem (1909) to divide the
Dipsacaceae in three tribes and 19 genera. Between these two
extremes, various intermediate positions were proposed, all of
which proved to be unreliable. Indeed, the numerous taxonomic
characters currently used do not all have the same systematic
value, especially for generic diagnoses which were based
essentially on the Mediterranean and middle-European collections,
rendering certain Asiatic and African species, particularly those
belonging to the genera *Dipsacus, Scabiosa* and *Pterocephalus*,
completely unclassifiable.
    A morphological and biogeographical examination of almost all

the representatives of the family, augmented by a biological,
anatomical, palynological and karyological study of many species,
allowed us to determine the taxonomic value of various characters
by dissociating them at times from the purely phylogenetic
criteria. Thus we now propose a new identification key which
subdivides the Dipsacaceae into 9 genera divided into 3 tribes,
slightly different from those of Van Tieghem (1909) and
Ehrendorfer (1964a).

At the tribal level, only the fruit should be taken into
consideration: the involucer-tube and the calyx. For the
identification of the genera the characteristics of the involucral
and floral (= receptacular) bracts, the calyx-setae and the
involucral-corona are taken into consideration.

```
1- Involucel-tube pedicelled, quadriveined, compressed,
   Calyx sessile, cupuliform and deciduous----------  ------KNAUTIEAE TRIBE
                                                             Knautia L. emend. Coult.
1'-Involucel-tube sessile, octoveined, quadrangular or subcylindrical

    2 -Calyx sessile, cupuliform and deciduous--------------DIPSACEAE TRIBE
       3 -Floral bracts shorter than involucral bracts------Dipsacus L.
       3'-Floral bracts longer than involucral bracts-------Cephalaria Schrader

    2'-Calyx stipitated, persistent-----------------------SCABIOSEAE TRIBE
       4 -Involucral bracts gamophyllous,
          connated in basal half--------------------------Pycnocomon Hoffm. &
       4'-Involucral bracts free, dialyphyllous                        Link
          5 -Calyx-setae absent---------------------------Succisella G. Beck
          5'-Calyx-setae present
             6 -Calyx-setae stiff, not plumose :4 or 5
                7 -Involucel-corona with 4 herbaceous lobes-Succisa Haller
                7'-Involucel-corona membranous-------------Scabiosa L.
             6'-Calyx-setae long, plumose
                8 -Setae flexible and erect : 6 to 28-------Pterocephalus Adans.
                8'-Setae rigid and spread out : 10---------Tremastelma Rafin.
```

```
1 -Tube de l'involucelle pédicellé, quadrinerve, comprimé
   Calice sessile, cupuliforme et caduc------------------TRIBU DES KNAUTIEAE
                                                          Knautia L. emend. Coult.
1'-Tube de l'involucelle sessile, octonerve, quadrangulaire à subcylindrique

    2 - Calice sessile, cupuliforme et caduc------------------TRIBU DES DIPSACEAE
        3 -Bractées florales plus courtes que les bractées involucrales---Dipsacus L.
        3'-Bractées florales plus longues que les bractées involucrales---Cephalaria Schrader

    2'- Calice pédicellé, persistant------------------------TRIBU DES SCABIOSEAE
        4 -Bractées involucrales gamophylles soudées jusqu'à mi-hauteur---Pycnocomon Hoffm. & Link
        4'-Bractées involucrales dialyphylles libres entre elles
           5 -Plateau calicinal simple et glabre------------------------Succisella G. Beck
           5'-Plateau calicinal terminé par des arêtes
              6 -Arêtes sétacées, rigides, non plumeuses : 4 ou 5
                 7 -Couronne de l'involucelle à 4 lobes herbacés---------Succisa Haller
                 7'-Couronne de l'involucelle hyaline-------------------Scabiosa L.
              6'-Arêtes plumeuses longues
                 8 -Arêtes souples dressées : 6 à 28---------------------Pterocephalus Adanson
                 8'-Arêtes rigides étalées : 10 ----------------------Tremastelma Rafin.
```

The infrageneric units (subgenera and sections) were placed in the following four genera:

```
KNAUTIA
=Perennial species --------------------------subgenus Trichera(Schrad.)Rouy
  (reduced corona, calyx 8 to 16 awned)
=Annual species(toothed corona, calyx generally ciliated)

 -Involucral bracts spread out horizontally----subgenus Tricheranthes(Schur)Szabo
 -Involucral bracts erect---------------------subgenus Knautia


DIPSACUS
= Ovoid capitula (spiny stems)-----------------section Dipsacus
= Spherical capitula (stems not spiny)---------section Sphaerodipsacus Lange


CEPHALARIA
=Immature Involucel-tube with 4 pronounced ribs
 -Involucel-corona with 4 herbaceous lobes-----subgenus Lobatocarpus Szabo
 -Involucel-corona fimbriated or scarious------subgenus Fimbriatocarpus Szabo

=Immature Involucel-tube with 8 pronounced ribs
 -Involucel-corona with 8 teeth---------------subgenus Cephalaria
 -Involucel-corona absent---------------------subgenus Phalacrocarpus(Boiss.)
                                                                      Szabo
SCABIOSA
=Trichome composed of stellated hairs---------section  Asterothrix Font Quer
=Trichome composed of simple hairs
 -Involucel-tube with 8 tips at the top--------section Trochocephalus Mertens &
 -Involucel-tube with 8 grooves                                          Koch
 .Involucel-corona completely hyaline--------section Scabiosa
 .Involucel-corona partly lignified ---------section Cyrtostemma Mertens & Koch
```

```
KNAUTIA

= Espèces pérennantes (couronne réduite, calice aristé)-------Sous-genre Trichera (Schrad.) Rouy
= Espèces annuelles (couronne dentée, calice en général cilié)
  - Bractées involucrales étalées horizontalement------------Sous-genre Tricheranthes (Schur) Szabo
  - Bractées involucrales dressées--------------------------Sous-genre Knautia


DIPSACUS   = Capitules ovoïdes (tiges épineuses)-------------Section Dipsacus
           = Capitules sphériques (tiges non épineuses)-------Section Sphaerodipsacus Lange

CEPHALARIA
= Tube de l'involucelle immature pourvu de 4 nervures marquées
  - Couronne de l'involucelle à 4 lobes herbacés-------------Sous-genre Lobatocarpus Szabo
  - Couronne de l'involucelle fimbriée à hyaline------------Sous-genre Fimbriatocarpus Szabo
= Tube de l'involucelle immature pourvu de 8 nervures marquées
  - Couronne de l'involucelle à 8 dents----------------------Sous-genre Cephalaria
  - Couronne de l'involucelle absente-----------------------Sous-genre Phalacrocarpus (Boiss.) Szabo

SCABIOSA

= Trichome composé de poils étoilés--------------------------Section Asterothrix Font Quer
= Trichome composé de poils simples
  - Tube de l'involucelle à 8 fovéoles sommitales------------Section Trochocephalus Mertens & Koch
  - Tube de l'involucelle parcouru par 8 sillons
    . Couronne de l'involucelle entièrement hyaline----------Section Scabiosa
    . Couronne de l'involucelle lignifiée puis hyaline--------Section Cyrtostemma Mertens & Koch
```

All these major characters offer the double advantage of being both unambiguous from a taxonomic point of view and of great interest at the phylogenetic level.

In this classification we have discarded numerous criteria, often used in treatments, but which are partly or entirely erroneous. Let us cite for example: – the spiny stems of *Dipsacus*, – the multi- or single-series of involucral bracts, – the coriaceous involucral and receptacular bracts of the *Dipsaceae* tribe – the shape and nature of the floral bracts, or even their absence (in *Knautia* and the majority of the *Pterocephalus*), – the spherical flower-heads (with only actinomorphic flowers) or radiant flower-heads (with peripheral zygomorphic flowers), and lastly, the tetramery or pentamery of the corolla and of the calyx (for the *Scabioseae* tribe only, since the *Knautieae* and the *Dipsaceae* always have flowers of type 4).

As we will see, even if these characters do not present a real taxonomic interest, on the other hand they all have an important evolutionary significance, their differentiation always occurring in a gradual manner.

PHYLOGENY

After this first necessary explanation, the realisation of the phylogenetic synthesis of our multidisciplinary study is proving to be a tricky task, owing to undeniable contradictions which appear when certain results are compared. In reality, these dissimilarities should not be interpreted as contradictions but more as complementary information. We have, therefore, searched for the ways in which they overlap and enhance each other. We thus propose a new phylogenetic outline of the family, quite different and much more complex than those previously elaborated (Ehrendorfer 1963; Verlaque 1976).

Biogeography

If it is true that three-quarters of the family inhabit the mesogean realm, the remaining quarter are spread out over the mountainous chains of Africa and Asia. We must emphasize the strict Mediterranean distribution of all the young or annual species, which have continuous areas, often very widespread, in the low altitude regions. On the other hand, the perennial group possesses a large geographical and altitudinal range, the older ones being more separated from each other in restricted, or split up, often relic areas. These are some of the very primitive orophyte species which are found dispersed in Africa and especially in Asia.

As for the Caprifoliaceae and Valerianaceae, central Asia seems to constitute the cradle of the family. But, due to their advanced degree of differentiation, the Dipsacaceae have colonized many territories (the Mediterranean basin, the mid-European

region, Africa), the younger ones adapting themselves very quickly
to the Mediterranean climate. The center of differentiation of
the family is thus displaced towards the West, it is presently
situated in the Middle-East.

Karyology

Many fundamental phenomena seem to govern the karyological
evolution of the Dipsacaceae, effective either on the basis of
chromosome number or on morphology (Kachidze 1929; Ehrendorfer
1963, 1965b; Verlaque 1976, 1978, 1980, 1982).

1. There is, in the first place, at the diploid level, a
progressive decrease of the primary basic number from $x = 10$
to $x = 5$. $x = 10$ characterizes the primitive groups (*Knautia,
Succisella, Succisa* and *Scabiosa* section *Asterothrix*).
$x = 9$ is the most frequent number in the perennial species,
while the numbers $x = 8$, 7 and 5 are found essentially in
the annuals. This descending dysploidy is found in young
complexes and is situated at the center of the distribution area
of the concerned group (for example, the annual complex
Scabiosa palaestina, Verlaque 1982b). Present in each
tribe, this phenomenon, primordial in evolution, is the driving
force behind some profound modifications.

2. We have observed also a second phenomenon, quite different
from the first, that is, polyploidy, which appears, in general in
those perennial taxa, situated on the border of the area or
geographically isolated. In the Dipsacaceae, the polyploidy at
times produces secondary basic numbers by the stabilization of
certain triploids (*Scabiosa*, $n' = 12$ or 13 and 14) or
aneuploids (*Knautia*, $n' = 20$, 21, 22; 30, 31, 32 and
Cephalaria, $n' = 18$, 19). This active mode of speciation
does not bring about large transformations in the family.
However, the multiplication of the genome is followed by the loss
of genetic barriers through interspecific isolation and by the
acquisition of a strong adaptive capacity due to an increased
tolerance to ecological factors. The polyploids are thus, by
virtue of their plasticity and vitality, great colonizers of new
territories. The *Knautia* constitute a significant example
(Ehrendorfer 1962; Breton-Sintes 1974a,b).

3. Finally, the comparison of idiograms and of chromosomic
formulae shows a morphological differentiation of chromosomes
which is not necessarily followed by a variation in the basic
number. With evolution, we observe, at the chromosomal level, a
general reduction in their size and a clear increase in their
complexity and their heterogeneity. At the karyotypic level there
is an evident change of symmetry towards an increasing asymmetry,
except in the perennial *Cephalaria*. Indeed, the primitive
South African subgenus *Lobatocarpus* is characterised by an
asymmetrical karyotype, but the transition to the most evolved
subgenus *Cephalaria* is accompanied by a return to a strong
chromosomal symmetry in the perennial species, which becomes again

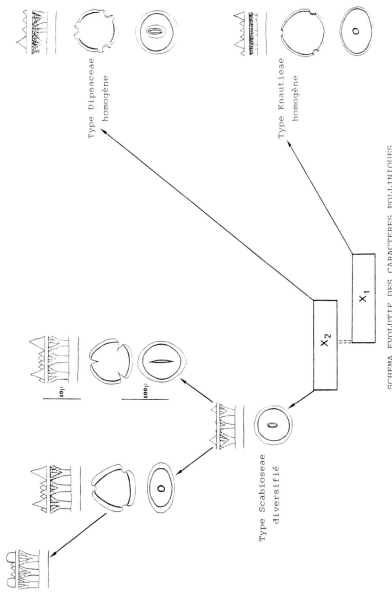

Type Dipsaceae
homogène

Type Knautieae
homogène

Type Scabioseae
diversifié

X_2

X_1

SCHEMA EVOLUTIF DES CARACTÈRES POLLINIQUES

10µ

100µ

very asymmetrical in the annual species of this genus.

Palynology

The family is subdivided into three distinct pollen types
(Verlaque 1981) which correspond exactly to the three tribes.
Evolution is accompanied by an increase of the pollen grain size
and of the exine thickness, as well as of a structural and
sculptural exine increasing complexity (see Fig. 1). The
echinulate and micro-echinulate exine becomes more and more
complex with (1) a multiplication of spine number, (2) an increase
in their height and their dimorphism due to the substitution of
the large spines by verrucae (*Scabiosa* section *Trochocephalus*
p.p.) then by bacula (*Tremastelma*).

In the *Knautieae*, the total pollini uniformity agrees
with the morphological homogeneity (subequixial pollen, triporate
with simple apertures, and the exine having the same thickness at
the poles as at the equator: 6 to 7.5 µm).

In the *Dipsaceae*, the strong palynological uniformity
contrasts significantly with the important morphological
differentiation, and suggests an undoubted discrepancy between the
evolution of these two types of characters (subequiaxial pollen,
tricolpate with complex apertures, and exine of 7 to 13 µm in
thickness).

On the other hand, in the *Scabioseae* the pollen diversity
is greater than the morphological variability which is itself
considerable (simple apertures and exine never having the same
thickness at the poles as at the equator: 7 to 32 µm). From
subequiaxial and tricolpate with three elliptic furrows having
more or less rounded extremities (*Succisella, Succisa*), in
general, the pollen becomes either breviaxial (peroblate)
triporate (*Scabiosa* section *Trochocephalus, Tremastelma,
Pycnocomon* and *Pterocephalus intermedius*), or longiaxial
(perprolate) tricolpate with three narrow furrows having pointed
extremities (*Scabiosa* sections *Scabiosa* and *Cyrtostemma,*
and the majority of the *Pterocephalus*). A particular case
occurs in four species which possess subequiaxial to longiaxial
pollen with cavities "caveae" at the poles (*Scabiosa* section
Asterothrix: tricolpate; *Pterocephalus diandrus* and *P.
centennii*:triporate).

Morphology

When we take into consideration the general appearance of the
plant, the primitive species are woody perennials with entire
leaves possessing an active vegetative reproduction. Their small,
spherical and solitary flower-heads are made up of slightly
colored actinomorphic flowers (type 4 in the *Scabioseae*).

FIG. 1. *Evolutionary scheme of the pollen characters.*

In the opposite direction the more evolved species are herbaceous,
with pinnatisect leaves, without vegetative reproduction. Their
numerous, large and radiant capitula are constituted with
zygomorphic and very vividly colored flowers (type 5 in the
Scabioseae).

More important than the appearance of the plant, we must take
into consideration in the family the anatomical structure and
morphology of the diaspores (Verlaque 1977). That is the calyx,
but especially the involucel, which is a foliar origin organ
completely enveloping the inferior ovary. Indeed, as Ehrendorfer
(1965) pointed out, the prime mover in the evolution of the
Dipsacaceae seems to be an ecological influence leading to the
specialization of the protection and the dispersion of the fruits.

PROTECTION OF THE DIASPORE: ANATOMY OF THE INVOLUCEL-TUBE

In order to reinforce seed protection, the involucel-tube becomes
progressively lignified (cf. Fig. 2).

In the *Knautieae,* only the parenchyma turns into wood
(Fig. 2-1). But, in the other two tribes, furthermore the
sclerified parenchyma, around the vascular bundles, bundle-
lignified sheaths (longitudinally fibrous), differentiate
themselves slowly and grow regularly with evolution.

In the *Dipsaceae,* there are four fibrous-sheaths in the
primitive section Sphaerodipsacus (Fig. 2-2), eight sheaths in the
section *Dipsacus* (Fig. 2-3) and in the subgenus *Lobatocarpus*
(Fig. 2-4). Finally, in the three more evolved other subgenera of
Cephalaria (Fimbriatocarpus, Cephalaria and *Phalacrocarpus),*
we observe an external continuous fibrous-ring (Fig. 2-5) which
pushes the lignified parenchyma towards the internal part.

In the *Scabioseae,* the basic process is similar but it
later becomes distinctly more complicated. There are:
- 8 sclerenchyma fibrous strands, partly surrounded over the
vascular bundles in the *Succisella* (Fig. 2-6);
- 8 fibrous sheaths of medium size in the *Succisa* and the
majority of the *Pterocephalus* (Fig. 2-7); then of
considerable size (but not joined at the grooves) in the
Scabiosa section *Scabiosa* (Fig. 2-8). This anatomical structure
characterises phylum No. 1 (cf. Fig. 3).
- an outer continuous ring of sclerified fiber in the small
phylum No. 2 which comprises the two species of the section
Asterothrix (Scabiosa saxatilis and *S. limonifolia), Pterocephalus
diandrus* and *P. centennii* (that is, the only taxa whose pollen
grains possess caveae at the poles);
- a thick, outer and continuous ring of longitudinal fiber
lined on the inside by a second ring of lignified fiber oriented
horizontally (Fig. 2-9). This anatomical structure characterises
phylum No. 3 which includes section *Cyrtostemma* of the genus
Scabiosa, the genus *Pycnocomon* and *Pterocephalus intermedius;*
- finally, in the last phylum No. 4, consisting of the
section *Trochocephalus (Scabiosa)* and the genus *Tremastelma,*

the anatomical structure is similar to the previous one, however,
horizontal sclerites supply the inner fibrous ring.

This lignification has advanced with the differentiation of
many types of involucel-tube closures, tending to form a perfectly
enclosed organ around the ovary. These different structures are
fundamental in the family. They outline, moreover, the main
phylogenetic phyla because they correspond to primordial
evolutionary stages.

ADAPTATIONS FOR DISPERSAL

In the *Knautieae,* the deciduous calyx does not play any
function; on the other hand, the quadriveined involucel is
pedicelled by a special organ called "elaiosome" which attracts
ants. In this tribe, which preferentially colonized protected
areas (forests and undergrowth) with rich soils, the seeds are
dispersed at a short distance by myrmecochory.

In the *Dipsaceae,* the deciduous calyx and the sessily
octoveined involucel seems to slowly lose importance with
evolution, because the main functions are taken over by capitulum
bracts. The latter, completely herbaceous in the Asiatic
primitive section *Sphaerodipsacus,* develop themselves,
lignify, and become spinescent, equally in the two genera

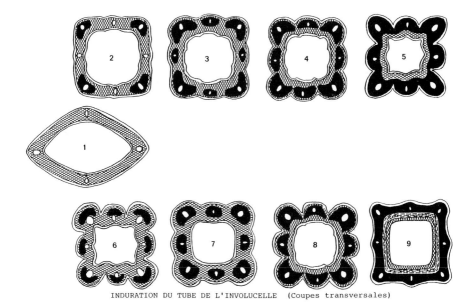

INDURATION DU TUBE DE L'INVOLUCELLE   (Coupes transversales)

FIG. 2. *Lignification of the involucel tube (transverse
section).*

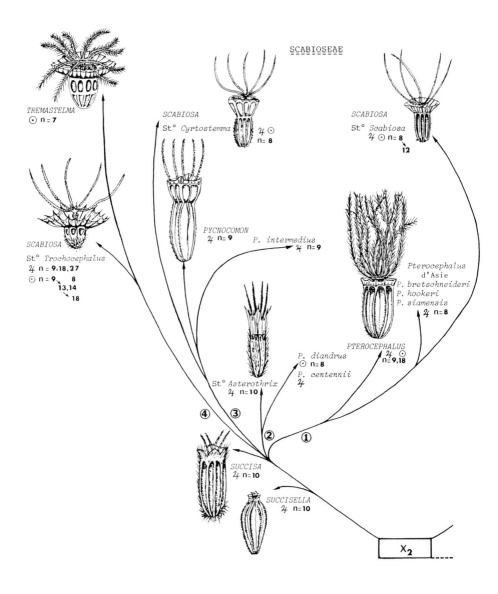

FIG. 3. *Phylogenetic outline of the Dipsacaceae.*

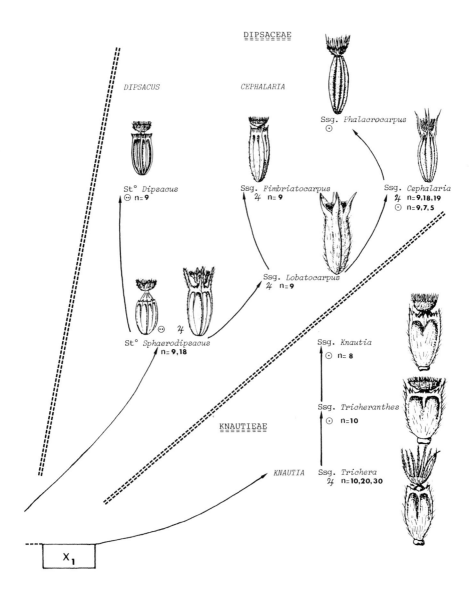

Dipsacus and *Cephalaria*. This favors the diaspores' dispersal
at a short distance by anemochory or by epizoochory due to a
system of projecting stems and a catapult mechanism. These two
types of scattering are well adapted to bushy and under-shrub
formations where the species of this tribe live.

Finally, the *Scabioseae* Tribe, which is the most
differentiated, has colonized all those open areas which are sandy
and have low vegetation. These seeds possess some remarkable and
very efficient adaptations for dispersion over great distances by
anemochory.

The simple and persistent calyx of *Succisella* develops
awns which are short in the genus *Succisa,* and then long
(numerous, flexible), plumose and erect into pappus in the genus
Pterocephalus, or rigid and more or less spread out in the
genera *Pycnocomon, Scabiosa* and *Tremastelma*.

The rudimentary involucel-corona of the *Succisella,* becomes
herbaceous in *Succisa* and in the majority of the *Pterocephalus,*
then are differentiated into a scarious membrane more or less
flared:
- completely hyaline in the genus *Tremastelma*, the genus *Scabiosa*
(sections *Trochocephalus, Asterothrix* and *Scabiosa*), as well
as the three asiatic *Pterocephalus*;
- partly lignified and "fenestrated" in *Scabiosa* (section
Cyrtostemma), *Pycnocomon* and *Pterocephalus intermedius* (phylum
No. 3: adapted to the North African desert regions).

In this tribe, the fruits are thus of the "heavy gliders type"
(= type planeurs lourds) in species with a scarious membrane, or
"roller-type" (= type rouleur) (Molinier and Muller 1938) in taxa
with lignified involucel-corona.

CONCLUSION

The synthesis of the results obtained in each discipline suggests
that the three tribes forming the Dipsacaceae differentiated very
quickly and independently of each other, with characters always
evolving in the direction of increasing complexity.

The proposed phylogenetic diagrams take into account the great
morphological, anatomical, palynological and karyological
diversity of this family (especially of the *Scabioseae* tribe)
and the more or less natural aspects of certain genera. In this
study we have attempted to show, for the first time, what a
biosystematic study could offer to taxonomy and to the
phylogenetic comprehension of a group, and secondly, where the
parallelism between the two disciplines should end. The two quite
artificial genera *Scabiosa* and *Pterocephalus*, each consisting
of four distinct groups, are good examples of this concept. It
seemed preferable to us to conserve their apparent generic unity
based on certain distinct and very conspicuous morphological
characters. According to us, these two adaptive characters
constitute the two evolutionary alternatives which arose at the
beginning of almost all the phyla deriving from *Succisa*:

either the differentiation of numerous, plumous and erect calyx-setae into pappus (in *Pterocephalus*, the involucel-corona remains rudimentary), or the development of the involucel-corona into a partial or completely hyaline membrane (in *Scabiosa* the 4 or 5 calyx-setae remain simple, rigid and often spread out).

In this family, the differentiation of all the characters did not occur simultaneously, and in the same way, causing the appearance moreover, in the different phyla, of numerous forms of convergence. The juxtaposition of these phenomena accentuates the difficulty of a "simple" approach to this group and could lead to erroneous interpretations. In fact, only one multidisciplinary study has been able to resolve the principal phylogenetic problems inherent in the Dipsacaceae. This work, therefore, allowed us to better understand certain mechanisms and directions of the fundamental processes, often complex, of evolution.

RÉSUMÉ

L'étude biosystématique des Dipsacaceae (350 espèces environ) a permis de préciser la valeur taxonomique des divers caractères et de mettre en évidence les principaux critères phylogénétiques.

Taxonomie: De nouvelle clés de détermination, basées essentiellement sur les caractères du fruit, permettent une meilleure délimitation des trois tribus, des neuf genres, de leurs sous-genres et sections.

Phylogénie: Originaires d'Asie centrale, des Dipsacaceae ont envahi le Bassin méditerranéen, la région médio-européenne et l'Afrique, leur centre de différenciation se situe actuellement au Moyen-Orient. L'évolution caryologique s réalise suivant trois processus fondamentaux: dysploïdie descendante (de $x = 10$ vers $x = 5$), polyploïdie et modifications morphologiques des chromosomes. La différenciation palynologique s'accompagne d'une complexité structurale et sculpturale croissante de l'exine. L'évolution morphologique concerne l'inflorescence (fleurs actinomorphes puis zygomorphes) et surtout le fruit: accroissement graduel de sa protection et, différenciation de plusieurs modes de dispersion adaptés aux milieux colonisés. La synthèse de ces résultats permet de tracer une nouvelle esquisse phylogénétique de la famille.

REFERENCES

Bentham, G. et Hooker, J.D. 1873. *Genera Plantarum.* 2(1), Reeve, London.

Breton-Sintes, S. 1974a. Étude biosystématique de genre *Knautia* (Dipsacaceae) dans le Massif Central. I. *Ann Sci. Nat., Bot. Biol. Vég., Sér. 12,* 15: 197-254.

Breton-Sintes, S. 1974b. Ibid. II. *Ann. Sci. Nat., Bot. Biol. Vég., Sér. 12,* 15: 277-320.

Coulter, T. 1824. Mémoire sur les Dipsacées. *Mém. Soc. Phys. Genève,* 2(2): 13-60.

Ehrendorfer, F. 1962. Beiträge zur Phylogenie der Gattung *Knautia*. I. Cytologische Grundlagen und allgemeine Hinweise. *Osterr. Bot. Z.* 109: 276–343.

Ehrendorfer, F. 1963. Cytologie, Taxonomie und Evolution bei Samenpflanzen. *Vistas Bot.* 4: 99–186.

Ehrendorfer, F. 1964a. Uber stammesgeschichtliche Differenzierungsmuster bei den Dipsacaceen. *Ber. Deutsch. Bot. Ges.* 77: 83–94.

Ehrendorfer, F. 1964b. Evolution and karyotype differentiation in a family of flowering plants: Dipsacaceae. *Genet. Today* 2: 399–407.

Ehrendorfer, F. 1965. Dispersal mechanisms, genetic systems and colonizing abilities in some flowering plant families. In *The Genetics of Colonizing Species* (H.G. Baker and G.L. Stebbins, eds.). Academic Press, New York, pp. 331–352.

Kachidze, N. 1929. Karyologische Studien uber die familie der Dipsacaceen. *Planta* 7: 482–502.

Linné, C. 1737. *Genera Plantarum I.*

Molinier, R. et Muller, P. 1938. La dissémination des espèces végétales. *Rev. Gén. Bot.* 50.

Sporne, K.R. 1980. A re-investigation of character correlations among dicotyledons. *New Phytol.* 85: 419–449.

Tieghem, P. Van. 1909. Remarques sur les Dipsacées. *Ann. Sci. Nat., Bot. Biol. Vég. Sér.* 9, 10: 148–200.

Verlaque, R. 1976. Contribution à l'étude cytotaxonomique des Dipsacaceae et des Morinaceae du Bassin méditerranéen. Thèse de spécialité, Univ. de Provence, Marseille, 250p.

Verlaque, R. 1977. Importance du fruit dans la détermination des Dipsacaceae. *Bull. Soc. Bot. Fr.* 124: 515–527.

Verlaque, R. 1978. Contribution à l'étude cytotaxonomique des Dipsacaceae et des Morinaceae du Nord de la Grèce. *Rev. Biol. Ecol. Médit.* 5: 15–30.

Verlaque, R. 1980a. Étude cytotaxonomique de quelques Dipsacaceae d'Iran. *Pl. Syst. Evol.* 134: 33–52.

Verlaque, R. 1980b. I.O.P.B. chromosome number reports. *Taxon* 29: 362–365.

Verlaque, R. 1981. Utilisation des caractères du pollen pour la réorganisation taxonomique de la famille des Dipsacaceae. *C.R. Acad. Sci. Paris* 293: 351–354.

Verlaque, R. 1982a. Étude de deux complexes vicariants d'annuelles: *Scabiosa palaestina* L. (*s.l.*) et *S. stellata* L. (*s.l.*) (section *Trochocephalus* Mertens et Koch). *Bull. Soc. Bot. Fr., Lett. Bot.* 129: 305–320.

Verlaque, R. 1982b. I.O.P.B. chromosome number reports. *Taxon* 31: 761–777.

Evolution of rDNA in *Claytonia* Polyploid Complexes

J. J. Doyle, R. N. Beachy and *W. H. Lewis*
Department of Biology
Washington University
St. Louis, Missouri, U.S.A.

INTRODUCTION

The explosive development of recombinant DNA technology has opened extensive, hitherto inaccessible areas of natural variation to the purview of the biosystematist. In this paper, we will attempt to demonstrate that simple molecular biological approaches can be used to sample and analyze variation at the DNA level. We will briefly review the techniques used, emphasizing their applicability to the unique needs of the systematist. The structure and function of ribosomal genes (rDNA) will be reviewed. Molecular analysis of rDNA variation will be illustrated with examples from our work with the *Claytonia* (Portulacaceae) polyploid complex of eastern North America.

MOLECULAR APPROACHES TO BIOSYSTEMATICS

It is (or ought to be) an axiom of proper biosystematic practice that undue reliance never be placed on a single character or method of analysis. This, along with the need for the adequate sampling of *any* character means that the systematist has very different requirements than, for example, the molecular biologist, for whom the sequencing of a particular gene from a single individual may well provide all the structural and biochemical data needed. What, then are some of the requirements that must be satisfied before methods of DNA analysis will be of use to the practicing plant systematist?

(1) DNA must be easily purified. During the last few years a variety of procedures for the isolation of high molecular weight nuclear and organellar plant DNA have been developed (Bendich *et al.* 1980; Murray and Thompson 1980; Zimmer *et al.* 1981; Palmer 1982). Of even greater promise for the systematist, who may not

have an ultracentrifuge at his disposal, are techniques in which
DNA is isolated in a matter of hours from milligram amounts of
plant material, utilizing chemicals less expensive than those used
in isozyme studies (Zimmer and Newton 1982; R. DeSalle pers.
comm.).

(2) The technique used must permit analysis of a relatively
large number of samples. Protein sequencing has not made an
appreciable impact on biosystematics, despite its great
contribution to evolutionary theory in general, because of the
time-consuming and expensive nature of the technique.
Similarly, it is unlikely that gene sequencing, though easier,
less expensive, and more informative, will in the foreseeable
future come into general use by systematists.

The analysis of specific genes with restriction endonucleases,
however, permits the efficient and rapid handling of a large
number of samples. Restriction enzymes cleave at specific
nucleotide sequences, generating reproducibly sized fragments
which can be compared among individuals or species. Organellar
DNAs, specifically that of the chloroplast, are particularly
amenable to restriction analysis, since digestion of such small
genomes yields restriction fragments that are directly visible as
bands on gels. Chloroplast DNA restriction site variation has
recently been used in several systematic studies (Palmer *et al.*
1983; Palmer and Zamir 1982; Kemble *et al.* 1983).

Nuclear DNA digested with restriction endonucleases must be
electrophoresed, hybridized to radioactively labeled DNA or RNA
probes and analyzed by autoradiography before bands representing
specific sequences are visible (Figs. 1, 2). Cloned DNA sequences
coding for a variety of genes are now available. Clearly,
hybridization probes must be chosen that will hybridize with DNA
from the taxon to be studied, and not all sequences meet this
criterion. For example, the genes that encode legume glycosylated
storage proteins show little homology even between tribes of the
Leguminosae (Doyle *et al.* unpublished). Thus, genes
encoding very conserved proteins are often chosen. For example,
Shah *et al.* (1982, 1983) found that both *Drosophilia* and
Dictyostelium actin gene sequences hybridized with higher plant
actin genes.

RIBOSOMAL GENES

Of the plant nuclear genes available for systematic studies, among
the most useful are those encoding the ribosomal genes. Two
ribosomal gene families, the genes encoding 5S RNA and those
encoding the 18S and 25S RNAs each occur as tandem repeats of
several thousand copies (Long and Dawid 1980). Figures 3 and 4
illustrate the structures of these two gene families. In both
cases the coding regions are followed by "spacer" regions which do
not encode a stable RNA molecule. The coding regions of both gene
families are highly conserved. The 18S gene of soybean, for
example, shares 75% nucleotide homology with that of yeast

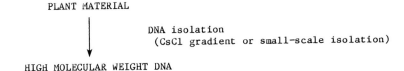

PLANT MATERIAL

| DNA isolation
| (CsCl gradient or small-scale isolation)

HIGH MOLECULAR WEIGHT DNA

1. Restriction endonuclease digestion

2. Native agarose gel electrophoresis

GEL WITH RESTRICTION FRAGMENTS
SEPARATED ACCORDING TO
MOLECULAR WEIGHT

1. Stain DNA with ethidium bromide

2. Visualize with ultraviolet light

3. Denature DNA in situ

4. Transfer DNA to nitrocellulose

5. Bind DNA to nitrocellulose by baking

NITROCELLULOSE FILTER "REPLICA"
OF GEL

1. ^{32}P-label DNA hybridization probe

2. Hybridize probe to DNA on filter

3. Wash to remove nonspecifically bound probe

4. Expose to X-ray film

5. Develop to visualize bands of hybridization

FILM WITH BANDS OF HYBRIDIZATION

FIG. 1. *Flow diagram of DNA restriction analysis methodology.*

(Eckenrode and Meagher 1983). Among flowering plants conservation
is apparent in this gene even to the level of individual
restriction enzyme recognition sequences, such as those for the
enzymes *Xba*I and *Eco*RI. Thus, the 18S:25S repeat of one plant
can be used as a hybridization probe across long evolutionary

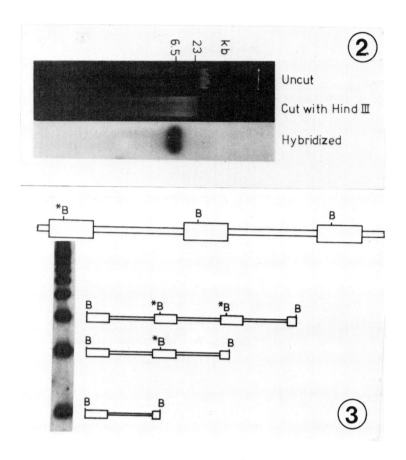

FIG. 2. *Restriction analysis of plant nuclear DNA. The first
two lanes are photographs of a 0.7% agarose gel stained with
ethidium bromide and photographed in ultraviolet light. "Uncut"
DNA was loaded prior to restriction endonuclease digestion. DNA
in the second lane was digested with the restriction endonuclease
HindIII prior to electrophoresis. Such a digestion results
in a population of fragments ranging from those uncut by this
enzyme to fragments only a few base pairs in length. The
resultant "smear" is due to the even gradation of fragment sizes
in the complex nuclear genome. The third lane is an
autoradiographic exposure of HindIII-digested DNA
transferred to nitrocellulose and hybridized to a ^{32}P-labeled rDNA
probe. The great complexity of the nuclear genome is reduced to a
single restriction fragment having homology to this probe. FIG. 3.
Structure of 5S RNA genes of plants. Open boxes represent
coding regions, connecting lines are spacers. BamHI
recognition sites (B) are indicated. Methylation of BamHI
sites (denoted by asterisks) prevents cleavage by the enzyme. The*

distances. For example, the cloned 18S:25S repeat from soybean
will cross-hybridize with DNA from monocots (Zimmer pers. comm.),
gymnosperms and ferns (Doyle unpublished) and algae (J. Harper
pers. comm.). Similarly, the 5S RNA gene from *Zea mays* will
hybridize with the 5S genes of a variety of dicots and even cycad
(Zimmer and DeSalle 1983).

The spacer regions of the ribosomal gene families are highly
variable, however, and it is these regions that account for the
great utility of these genes in systematic studies. Spacer length
can differ greatly within a single genus. In *Glycine*, for
example, the entire 18S:25S repeat varies from under 8 kilobases
(kb) to over 12 kb, with all of the variation localized in the
nontranscribed spacer (NTS) (Doyle unpublished). The restriction
site variability observed between taxa is in most cases attributable to changes in these spacer regions. In the 18S-25S genes,
variability is not evenly distributed throughout the spacer. In
wheat, the most variable region is localized in the two-thirds of
the spacer furthest from the 18S gene, an area composed of short
repeated sequences (Appels and Dvorak 1982; Dvorak and Appels
1982). Length variation within and between species of *Triticum*
appear to be due to changes in the number of these short repeats,
whose sequences also appear to diverge at a relatively rapid rate.
Less variability is observed just 5' of the 18S gene, a region
included in the primary transcript of the 18S and 25S genes
(Appels and Dvorak 1982).

The repetitive nature of the NTS is common to other higher
organisms (Long and Dawid 1980). That this variable region has
some function also appears likely. In *Drosophila melanogaster*,
the spacer is not truly "nontranscribed"--sequences capable of
binding RNA polymerase I are found throughout the region, and
transcription appears to be initiated at several sites in the
spacer, perhaps playing a role in the efficient transcription of
the rDNA genes (Coen and Dover 1982; Kohorn and Rae 1982a).
Sequence variation between promoter sites in different *Drosophila*
species appears to be correlated with the presence of species-
specific polymerases, since ability of the polymerase of one
species to transcribe the rDNA of another in *in vitro*
transcription assays is greatly reduced (Kohorn and Rae 1982b).
These data point to the potential evolutionary significance of
changes in this highly variable gene region.

Ribosomal genes, despite their high numbers, generally do not
show extreme restriction site or length variability within an
individual or population. Often only one or two predominant
repeat types are observed. This genic homogenization, termed
concerted evolution (Zimmer *et al.* 1980; Dover *et al.*

---

*random distribution of methylated sites yields, after BamHI
digestion and hybridization, a "ladder" pattern, with individual
bands representing one, two, three, etc. repeats. The smallest
band shown here is 350 base pairs in length.*

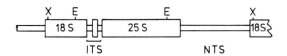

FIG. 4. *Structure of the 18S-25S RNA genes of plants. In this
tandemly repeated gene family, coding regions for the 18S and 25S
genes are separated by an internal transcribed spacer (ITS)
containing the 5.8S rRNA gene. Individual repeats are separated
by a nontranscribed spacer (NTS). Recognition sites for the
enzymes XbaI (X) and EcoRI (E) in the coding regions are found
in many higher plants.*

1982), is, for the evolutionist, a significant and useful feature
of rDNA, since the rapid fixation of a predominant repeat form can
serve as a marker for a particular evolutionary lineage. From a
technical standpoint, the presence of thousands of identical
copies of ribosomal gene repeats confers several advantages.
Digestion with a restriction enzyme having a six base recognition
sequence will generally only a few times within the 18S:25S repeat
(and even fewer times in the much shorter 5S gene), yielding a
small number of fragments each present in thousands of copies.
This is advantageous since, in contrast to the case of a low copy
number gene (such as most genes which code for proteins), a far
smaller amount of genomic DNA must be used (with a correspondingly
smaller amount of restriction enzyme), and results are attainable
in a shorter period of time. High copy number is also a valuable
feature if molecular cloning of the gene is desired.
    The salient features of restriction analysis of ribosomal genes
will be presented by discussing examples from our work with
Claytonia.

CLAYTONIA IN EASTERN NORTH AMERICA

Claytonia (Portulacaceae) is a predominantly arctic and
alpine genus composed of about 24 species, most of which are
native to western North America and eastern Asia (McNeill 1975;
Davis 1966). Two species of section *Claytonia*, *C. virginica*
and the more boreal *C. caroliniana*, are geophytic spring
ephemerals commonly occurring in mesic woodlands of eastern North
America. The two species are allopatric throughout most of their
ranges, coexisting without evidence of hybridization in the area
of most extensive sympatry, central Michigan (Voss 1968; Reznicek
and Britton 1971). Extensive cytological differentiation has long
been known to occur in these taxa, particularly in *C.
virginica*, where over 50 cytotypes having chromosome numbers
ranging from $2n = 12$ to $2n =$ ca. 191 have been described
(Rothwell and Kump 1965; Lewis *et al.* 1967; Lewis and Semple 1977).

TABLE I. *Summary of racial variation in C. virginica*

Flavo-noid race	Ploidy level	Chromosome numbers (2n)	Karyotype morphology	Average leaf L/W	Rust host	Habitats	Geographic range
I	diploid	12, 14	acrocentric	19.0	yes	rich deciduous woods	Mountains of Tennessee, North Carolina
	poly-ploid	16, 22, 28-32, 56-86	mixed meta- and acro-centric	19.0	yes	deciduous and pine woods; fields; swamps	Atlantic Coastal Plain; Kentucky to Indiana
II	diploid	14	acrocentric	24.1	yes	deciduous and pine woods	Arkansas, Louisiana and east Texas
	poly-ploid	16, 18 -42, 48	acrocentric	24.1	yes	woods, prairies, lawns, swamps	Southern U.S.; Ohio to central Kansas
III	diploid	16	metacentric	11.2	yes	rich deciduous woods	Northern U.S. and southern Canada; ?
	poly-ploid	18, 20, 36, 50	?	?	?	rich deciduous woods	Northern U.S. and southern Canada
IV	poly-ploid	24-36, 44, 46, 72; prob-ably up to ca. 191	mixed meta- and acrocentric	8.1	no	moist to dry woods; wet areas, lawns, fields	Midwestern U.S.; coastal Connecticut to Pennsylvania

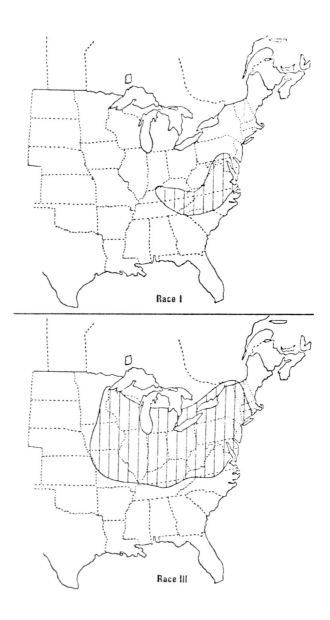

FIG. 5. *Geographical distribution of the four races of* C. virginica.

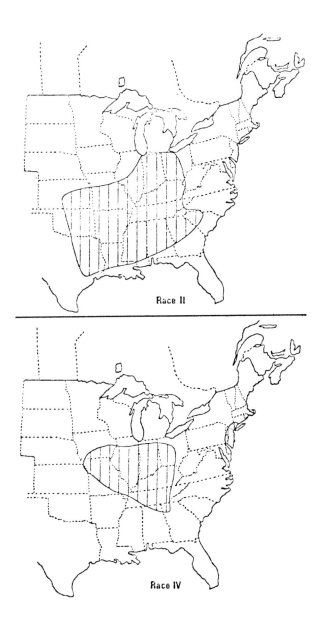

Race II

Race IV

FIG. 6. *Distribution of* C. carolini ana. *Hatched area is the range of polyploid cytotypes having diverged flavonoid profiles. Note the single disjunct diploid population in central Arkansas.*

More recently, evidence has accumulated documenting the existence of discrete infraspecific taxa within both species, identifiable on the basis of flavonoid chemical profile, chromosome number, karyotype morphology, ecology and geographical distribution (Table I; Figs. 5, 6) (Doyle 1981, 1983). Races I, II and III of *C. virginica* contain diploid cytotypes. Of these, two (I and II) are also represented by successful polyploid populations which, at least in the case of Race II, appear to be almost totally undiverged from the diploid members of the race (Doyle 1981, 1983). Race IV consists entirely of polyploids, and includes the populations in the New York metropolitan area having extremely high chromosome numbers ($2n > 100$) reported by

Rothwell and Kump (1965). The diploid progenitor(s) of this race
are unknown.
    Less is known about *Claytonia caroliniana*. Only one
major diploid cytological race ($2n$ = 16) is known from this
species, and this appears to be the major cytotype in the species.
The range of this race overlaps with *C. virginica* Race III
$2n$ = 16 diploids in Michigan; in several instances the two
grow in mixed populations. In such populations, *C. caroliniana*
is readily distinguished by its broader leves; the two groups are,
however, inseparable on the basis of such cryptic characters as
flavonoid chemistry or karyotype morphology, characters which
differentiate Race III from the other races of *C. virginica*.
A disjunct $2n$ = 16 population of *C. caroliniana* has recently
been discovered in the Ozark Mountains of Arkansas, 800 km from
the nearest *C. caroliniana* population (Lewis unpublished).
Chemically, these plants are identical to typical diploid *C.
caroliniana* (Doyle unpublished). Polyploids occur as a
predominant cytotype in this species only in the Smoky Mountains
of North Carolina and Tennessee. Some of these populations are
distinguishable from diploid *C. caroliniana* in having a
flavonoid profile like that of Race IV *C. virginica* (Doyle 1983).
    Thus, the eastern *Claytonia* complex provides an
opportunity to reexamine, with molecular approaches, a group of
wild taxa for which data from several different lines of more
conventional biosystematic data already exist. Our goals in
studying ribosomal gene variation in this complex were therefore
several: (1) To determine whether variation exists within the
eastern *Claytonia* complex. (2) To test whether patterns of
such variation, if present, are congruent with data from previous
studies involving standard biosystematic approaches, and (3) To
further elucidate evolutionary relationships within the complex.

RIBOSOMAL GENE VARIATION IN *CLAYTONIA*

5S RNA Genes

    Several populations representing three of the four races of
C. virginica and both diploid and polyploid *C. caroliniana*
were sampled for 5S gene variation. Nuclear DNA was isolated from
these plants, digested with the restriction endonuclease *Bam*HI,
electrophoresed on 2% agarose gels, transferred to nitrocellulose
(Southern 1975) and hybridized to an *in vitro* ^{32}P-labeled 5S
gene clone from *Zea mays* (Zimmer and DeSalle 1983). The results of
such an experiment are shown in Fig. 7 and Table II. As is
illustrated in Fig. 3, plant 5S genes are polymorphic for
methylation of the single *Bam*HI site in the coding region
(Goldsbrough *et al*. 1982). Since methylation of recognition
sites prevents cleavage by this enzyme, digestion of 5S sequences
with *Bam*HI yeilds a "ladder" pattern, in which the smallest
band represents a single repeat.
    It is apparent that each *C. virginica* race has a distinctive

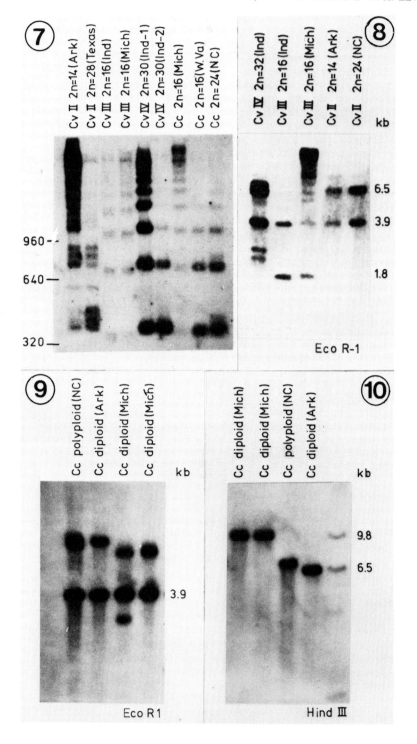

5S gene profile. In each race both populations appear to have similar patterns. Of particular interest are the two Race II populations, representing both diploid and polyploid cytotypes of this group. Both populations show a distinctive pattern in which there is an apparent size polymorphism for 5S genes, presumably resulting from heterogeneity in spacer length. Since each DNA sample is, in this case, composed of more than one individual, it is impossible to say whether this heterogeneity exists within individual plants or whether it represents a populational polymorphism. That this distinctive pattern is found in both diploids and polyploids is of note, since both chemical and karyotypic data have suggested that polyploidy within Race II has occurred without the participation of other currently recognizable *Claytonia* diploid races.

C. *virginica* Race IV polyploids also exhibit some heterogeneity, with repeat lengths that fall within the range of variability exhibited by Race II. The two populations examined from Race III, however, appear quite different from other *C. virginica* races in being monomorphic for a smaller repeat size.

The *C. caroliniana* populations examined all have a repeat length intermediate between that of *C. virginica* Race III and the smallest repeat of the other *C. virginica* races. In addition, the polyploid population also has a minor, slightly larger repeat type.

Thus 5S gene variation patterns appear to delimit the same infraspecific taxa as do flavonoid chemistry and cytology within this complex. In addition, however, it is possible to distinguish *C. caroliniana* from *C. virginica* on the basis of 5S gene profiles.

18S:25S Ribosomal Genes

18S:25S ribosomal gene variation was examined in a number of *Claytonia* populations (Fig. 11). As was the case for the 5S gene studies, most samples represent more than one individual. The methodology used was similar to that employed in the 5S

FIGS. 7-10. *Ribosomal gene variation in eastern North American Claytonia. FIG. 7. 5S rRNA genes. See text for discussion; Cv = Claytonia virginica, Cc = C. caroliniana; Roman numerals refer to geographic races; locations of samples are given as state abbreviations; molecular weights of markers (in base pairs) are given to the left of the gel. FIGS. 8-10. 18S:25S rRNA genes. In all cases, digested DNA was electrophoresed on a 0.7% agarose gel, transferred to nitrocellulose and hybridized to a ³²P-labeled rDNA clone from soybean. Molecular weights of markers are given in kilobases (kb); restriction enzymes used are given below gel; other abbreviations as in Fig. 8.*

FIG. 11. *Populations sampled for rDNA variation.*

investigation, with the exception that a number of different
restriction endonucleases were used, agarose gel percentage was
varied from 0.7 - 2.0% dpeending on the fragment size analyzed,
and the hybridization probe used was a complete 18S:25S gene
repeat from *Glycine max* (Zimmer and Walbot 1983).

The restriction endonucleases tested fall into several
categories. Some enzymes, such as *Sma*I, do not cut *Claytonia*
genomic DNA efficiently, and could not be used for analysis.
Others, among them *Bg*lII, cut efficiently but lack recognition
sites within rDNA, yielding, after autoradiography, only a band of
high molecular weight uncut DNA. Among enzymes that cut within
the *Claytonia* repeat, one enzyme, *Xba*I, has a single recognition
site within rDNA, while several others, such as *Bam*HI,
*Eco*RI, *Hind*III and *Sst*I, have two or more sites. The cleavage

TABLE II. *Repeat lengths of 5S RNA genes of* Claytonia
*populations. Molecular weights were calculated by reference to 5S
genes of* Zea mays *run on the same gel.*

Species	Race	Chromosome Number (2n)	Location of Collection	5S Gene Repeat Length (Base Pairs)
C. virginica	II	14	Arkansas	360 380 400 440 560
C. virginica	II	28	Texas	360 380 400 440 560
C. virginica	III	16	Indiana	335
C. virginica	III	16	Michigan	335
C. virginica	IV	30–32	Indiana-1	360 370
C. virginica	IV	30–32	Indiana-2	360 370
C. caroliniana	–	16	Michigan	350
C. caroliniana	–	16	West Virginia	350
C. caroliniana	–	24	North Carolina	350 400

sites of such enzymes were mapped by single and double digest
experiments to compare the different races within the complex.
The data from two enzymes, *Eco*RI and *Hind*III, will be discussed
in detail.

When rDNA from various *C. virginica* populations was
mapped with respect to *Eco*RI, differences were observed
among the populations (Figs. 8, 12). Each of the different races
represented are found to have a distinctive pattern, caused by
restriction site changes and repeat length variation. As was also
the case for 5S genes, Race III is readily distinguishable from
the other races, in this case by the presence, in all populations
examined, of an *Eco*RI site in the nontranscribed spacer.

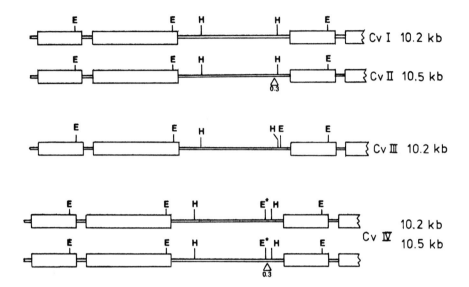

FIG. 12. *Maps of ribosomal genes of* C. virginica *races.*
Restriction site placement on map was achieved by single and
double enzyme digests. H = Hind*III recognition site;* E = EcoRI
recognition site; Triangles represent insertion in NTS.

Race II repeats have only two *Eco*RI sites; furthermore,
diploid and polyploid populations have identical restriction
profiles. Race IV plants also possess a distinctive profile.
Individual plants are polymorphic for an *Eco*RI site in the
NTS as well as for two NTS length variants, yielding a pattern of
four different repeat types. One of these repeat types (small
repeat lacking the *Eco*RI site) is also found in a polyploid
Race I population from Virginia.

Differences are also found between the two cytotypes of *C.*
caroliniana (Figs. 9, 13). In this species, however, an
*Eco*RI site polymorphism occurs within $2n = 16$ populations.
One Michigan population sample contained only repeat forms lacking
this site, while a second Michigan population and a sample from
West Virginia were polymorphic for this site. The disjunct Ozark
$2n = 16$ population lacks this *Eco*RI site and differs from the
other three diploid populations examined in having a larger repeat
length. The polyploid population from the Smoky Mountains of
North Carolina contains two repeat size classes, both of which are
larger than those of the "typical" diploid. In this respect, this
population is reminiscent of the polyploids of *C. virginica*
Race IV, to which it is chemically and karyotypically similar

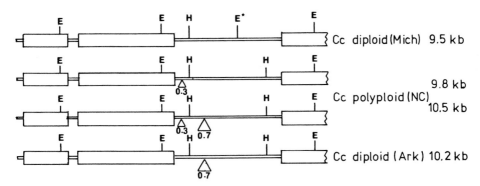

FIG. 13. *Maps of ribosomal genes of* C. caroliniana
populations. Abbreviations as in Fig. 12.

*(Doyle 1981, 1983). However, it lacks the fixed polymorphism for
the EcoRI site in the NTS.*

The digestion pattern observed with *Hind*III further
differentiates *C. caroliniana* populations (Figs. 10, 13).
There appear to be two recognition sites for this enzyme in all
C. virginica populations, bracketing the NTS (Fig. 12). Two
sites are also found in the *C. caroliniana* polyploid
population. In "typical" *C. caroliniana* diploids, however,
cleavage at one *Hind*III site does not occur, yielding a
distinctive banding pattern. In contrast to these populations,
the Ozark disjunct appears to have retained two *Hind*III sites.

Data from rDNA restriction mapping suggest several conclusions:

(1) *C. virginica* Race III diploids appear to be quite
distinct from the other eastern *Claytonia* taxa. This
conclusion is bolstered by data from other restriction enzymes as
well as from patterns of 5S gene variation already mentioned.

(2) *C. virginica* Race II diploids and polyploids are
identical in their rDNA profiles, lending further credence to the
contention that polyploidy in this group has not involved
hybridization with any other extant diploid race within the
complex.

(3) *C. virginica* Race IV polyploids are fixed for
restriction site and spacer length heterogeneity. This may
suggest an allopolyploid ancestry for this race. However, rDNA
variation patterns do little to pinpoint the diploid progenitor(s)
for this group, except to suggest that it appears closer to Race
II than to Race III or the *C. caroliniana* diploid.

(4) *C. caroliniana* 2n = 16 diploids, at least those from the
main portion of the range, are readily distinguished from all
C. virginica on the basis of *Hind*III restriction profiles. In
particular, rDNA of this group is distinct from that of the

sympatric $2n$ = 16 diploids of *C. virginica*, which are identical in flavonoid chemistry and cytology.

(5) The disjunct *C. caroliniana* diploid populations from the Ozark Mountains differs in several respects from "typical" diploids of that species, despite its identical chromosome number and flavonoid chemistry. In this respect it appears that the effect of isolation and genetic drift has operated at a higher rate on rDNA than on other biosystematic characters.

(6) A polyploid *C. caroliniana* population from the Smoky Mountains of North Carolina appears to be more similar to *C. virginica* Race IV polyploids than to *C. caroliniana* diploids. Perhaps this could represent evidence of hybridization between the two species in the Appalachians, a phenomenon reported by Uttal (1964).

All three of our goals in studying rDNA variation appear, therefore, to be attainable--(1) rDNA variation exists in the *Claytonia* complex; (2) patterns of rDNA variation are, for the most part, congruent with data from conventional systematic techniques; (3) rDNA variation appears useful in answering still-unsolved problems within the complex.

ADDITIONAL APPROACHES

The generation of restriction maps for individual enzymes, albeit of considerable utility, does not exhaust the potential of restriction enzyme analysis as a biosystematic tool. By appropriate choice of outgroups and the construction, for a large number of restriction enzymes, of trees whose branching patterns are determined by the gain or loss of restriction sites, restriction map data can be used to test phylogenetic hypotheses. Such an approach has, for example, been used recently by Templeton (1983) to infer evolutionary relationships among higher primates.

One major limitation of the 18S:25S gene system is the very feature responsible for its power as an evolutionary tool--the variable nature of the NTS. This is because, although nonhomologous probes can indeed be used to map and compare rDNA, restriction fragments located entirely within the spacer will not be detected efficiently. The transcribed region of the plant 18S-25S gene repeat covers about 6 kb (Lewin 1980). The total repeat length in *Claytonia* is around 10 kb, so the NTS covers about 4 kb. This is contrasted with the *Glycine max* rDNA repeat that was used in these experiments, whose total length is only 7.8 kb, with a NTS of only bout 2 kb. Thus there are potentially two kilobases of *Claytonia* rDNA that might not hybridize to the *Glycine* probe even were all other sequences perfectly conserved. This is probably an overestimate, since most NTS length variation (within related groups at least) occurs through the addition or deletion of repeated sequences, so that homology is retained despite length differences. However, as has been shown by Appels and Dvorak (1982) in *Triticum*, it is these repeated sequences that evolve most rapidly. Thus, as has been observed in our studies, there is virtually no homology between the NTS of *Claytonia* and *Glycine*.

The acquisition of a homologous rDNA probe would facilitate the mapping of these genes in two respects. First, it would allow us to map restriction enzyme sites directly by digestion of the clone and gel electrophoresis, without recourse to transfer to nitro-cellulose, hybridization and autoradiography. Second, NTS variation could be studied on a finer scale by using NTS-specific hybridization probes in conjunction with enzymes which produce fragments wholly within the spacer. As previously mentioned, the highly repeated nature of these genes facilitates their molecular cloning. We are currently pursuing this avenue of research.

CONCLUSIONS

We have attempted to show how a relatively simple molecular approach can be used to generate biosystematically useful data in a wild plant complex. The power of an approach which permits the investigation of actual genotypic changes is apparent, and, used in conjunction with traditional biosystematic approaches, promises to greatly enhance our understanding of the processes leading to the observed diversity of higher plants.

ACKNOWLEDGMENTS

We thank Dr. Elizabeth Zimmer and Rob DeSalle for their many helpful comments on the manuscript, and Jane Doyle for her expert technical assistance. This research was supported by a research grant to RNB from the U.S. Department of Energy; JJD was supported by a fellowship from the Monsanto Corporation.

REFERENCES

Appels, R. and Dvorak, R. 1982. The wheat ribosomal spacer DNA region: its structure and variation in populations and among species. *Theor. Appl. Genet.* 63: 337-348.

Bendich, A.J., Anderson, R.S. and Ward, B.L. 1980. Plant DNA: long, pure and simple. In *Genome Organization and Expression* (C.J. Leaver, ed.) NATO. pp. 31-33.

Coen, E.S. and Dover, G.A. 1982. Multiple Pol I initiation sequences in the rDNA spacer of *Drosophila melanogaster*. *Nucleic Acids Res.* 10: 7017-7026.

Davis, R.J. 1966. The North American perennial species of *Claytonia*. *Brittonia* 18: 285-303.

Dover, G.A., Brown, S., Coen, E., Dallas, J., Strachan, T. and Trick, M. 1982. The Dynamics of Genome Evolution and Species Differentiation. In: *Genome Evolution* (G.A. Dover and R.B. Flavell, eds.) Academic Press, N.Y. pp. 343-372.

Doyle, J.J. 1981. Biosystematic studies on the *Claytonia virginica* aneuploid complex. Ph.D. Diss. Indiana Univ., Bloomington.

Doyle, J.J. 1983. Flavonoid races of *Claytonia virginica* (Portulacaceae). *Am. J. Bot.* 70: 1085–1091.

Dvorak, R. and Appels, R. 1982. Relative rates of divergence of spacer and gene sequences within the rDNA region of species in the Triticeae: Implications for the maintenance of a repeated gene family. *Theor. Appl. Genet.* 63: 361–365.

Eckenrode, V.K. and Meagher, R.B. 1983. Primary and proposed secondary structures for 18S rRNA from soybean determined from the nucleotide sequence of cloned rDNA. In *Proc. 15th Miami Winter Sympos.* (F. Ahmad, K. Downey, J. Schultz and R. Voellmy, eds.). Abstr.

Goldsbrough, P.B., Ellis, T.H.N. and Lomonossoff, G.P. 1982. Sequence variation and methylation of the flax 5S RNA genes. *Nucleic Acids Res.* 10: 4501–4514.

Kemble, R.J., Gunn, R.E. and Flavell, R.B. 1983. Mitochondrial DNA variation in races of maize indigenous to Mexico. *Theor. Appl. Genet.* 65: 129–145.

Kohorn, B.D. and Rae, P.M.M. 1982a. Nontranscribed spacer sequences promote *in vitro* transcription of *Drosophila* ribosomal DNA. *Nucleic Acids Res.* 10: 6879–6886.

Kohorn, B.D. and Rae, P.M.M. 1982b. Accurate transcription of truncated ribosomal DNA templates in a *Drosophila* cell-free system. *Proc. Natl. Acad. Sci. U.S.A.* 79: 1501–1504.

Lewin, B. 1980. *Gene Expression 2: Eukaryotic Chromosomes.* Wiley, N.Y.

Lewis, W.H., Oliver, R.L. and Suda, Y. 1967. Cytogeography of *Claytonia virginica* and its allies. *Ann. Mo. Bot. Gard.* 54: 153–171.

Lewis, W.H. and Semple, J.C. 1977. Geography of *Claytonia virginica* cytotypes. *Am. J. Bot.* 64: 1078–1082.

Long, E.O. and Dawid, I.B. 1980. Repeated genes in eukaryotes. *Annu. Rev. Biochem.* 49: 727–767.

McNeill, J. 1975. A generic revision of Portulacaceae tribe Montioideae using techniques of numerical taxonomy. *Can. J. Bot.* 53: 789–809.

Murray, M.G. and Thompson, W.F. 1980. Rapid isolation of high molecular weight plant DNA. *Nucleic Acids Res.* 8: 4321–4325.

Palmer, J.D. 1982. Physical and gene mapping of chloroplast DNA from *Atriplex triangularis* and *Cucumis sativa*. *Nucleic Acids Res.* 10: 1593–1605.

Palmer, J.D., Singh, D.P. and Pillay, D.J.N. 1983. Structure and sequence evolution of three legume chloroplast DNAs. *Mol. Gen. Genet.* 190: 13–20.

Palmer, J.D. and Zamir, D. 1982. Chloroplast DNA evolution and phylogenetic relationships in *Lycopersicon*. *Proc. Natl. Acad. Sci. U.S.A.* 79: 5006–5010.

Reznicek, T. and Britton, D.M. 1971. Chromosome studies on the spring beauties, *Claytonia*, in Ontario. *Michigan Bot.* 10: 51–62.

Rothwell, N.V. and Kump, J.G. 1965. Chromosome numbers in populations of *Claytonia virginica* from the New York metropolitan area. *Am. J. Bot.* 52: 403-407.

Shah, D.M., Hightower, R.C. and Meagher, R.B. 1982. Complete nucleotide sequence of a soybean actin gene. *Proc. Natl. Acad. Sci. U.S.A.* 79: 1022-1026.

Shah, D.M., Hightower, R.C. and Meagher, R.B. 1983. Genes encoding acting in higher plants: intron positions are highly conserved but the coding sequences are not. *J. Molec. Appl. Biol.* 2: 111-127.

Southern, E. 1975. Detection of specific sequences among DNA fragments separated by gel electrophoresis. *J. Mol. Biol.* 98: 503-517.

Templeton, A.R. 1983. Phylogenetic inferences from restriction endonuclease cleavage site maps with particular reference to the evolution of humans and apes. *Evolution* 37: 221-244.

Uttal, L.J. 1964. A hybrid population of *Claytonia* in Virginia. *Rhodora* 66: 136-139.

Voss, E.G. 1968. The spring beauties (*Claytonia*) in Michigan. *Michigan Bot.* 7: 77-93.

Zimmer, E.A., Martin, S., Beverley, S., Kan, Y.W. and Wilson, A.C. 1980. Rapid duplication and loss of the genes coding for the chains of hemoglobin. *Proc. Natl. Acad. Sci. U.S.A.* 77: 2158-2162.

Zimmer, E.A., Riven, C. and Walbot, V. 1981. A DNA isolation procedure suitable for most higher plant species. *Plant Mol. Biol. Newslett.* 2: 93-96.

Zimmer, E.A. and Newton, K.J. 1982. A simple method for the isolation of high molecular weight DNA from individual maize seedlings and tissues. In *Maize for Biological Research* (W.F. Sheridan, ed.) Plant Mol. Biol. Assn. and U. North Dakota Press. pp. 165-168.

Zimmer, E.A. and DeSalle, R. 1983. Structure and inheritance of 5S DNA genes of maize and teosintes. *Genetics* 104: 573-574.

Zimmer, E.A. and Walbot, V. 1983. Ribosomal gene structure and evolution in maize and its ancestors. Ms. in preparation to be submitted to *J. Mol. Appl. Genet.*

Isozyme Evidence and Problem Solving in Plant Systematics

L. D. Gottlieb
Department of Genetics
University of California
Davis, California, U.S.A.

INTRODUCTION

The early biosystematists (Camp and Gilly 1943) and experimental taxonomists (Clausen, Keck, and Hiesey 1939) were well aware of the importance of genetic analysis to understand evolutionary processes and to reconstruct phylogenetic relationships. But they usually "defined" genetic analyses (Camp and Gilly 1943, p. 329) wholly in terms of cytology (to determine chromosome number, karyotype, and ploidy level), breeding tests (to assess fertility of experimentally synthesized hybrids between populations), and transplant studies (to find out whether morphological characters that distinguished plant populations in their native habitats were retained when the plants or their seed progenies were grown under uniform garden conditions). The genetic analysis of morphological differences between species or other taxa was rarely attempted although several studies (Babcock and Cave 1938; Clausen *et al.* 1947) attracted interest by demonstrating that character differences used to erect new genera were governed by one or two gene substitutions and disguised very close relationships.

In general, the genetics of interspecific differences were stymied because species are reproductively isolated making it impossible to obtain segregating progenies. Even when species are crossable, generation time can be lengthy and many plants cannot readily be grown in large numbers under uniform conditions. Another problem is that many characters that distinguish species are quantitative and yield complex segregation patterns not interpretable in terms of particular gene substitutions.

In spite of these problems, we still have to know the number and type of genetic differences between species to understand the steps by which species evolve one from another. To what extent do the novel attributes of a new species reflect the developmental consequences of simple changes in the genome of its progenitor?

Or is speciation the outcome of numerous complex changes
throughout the genome of the parent? Such questions cannot be
approached on the basis of evidence from morphology alone because
morphological similarity and difference cannot be translated
without genetic analysis into a description of gene similarities
and differences (Gottlieb 1977a). The lack of equivalence between
phenotype and genotype means that phylogenetic inferences based
primarily on morphology may often be incorrect. Much more
attention must be paid to descriptions of genetic differences
between species to specify the dynamics of speciation and to infer
accurate phylogenetic relationships.

The task is now simpler because recent advances in biochemical
genetics and molecular biology make it possible to analyze
divergence at particular genes of different species without having
to do genetic crosses. Thus electrophoretic divergence of
homologous enzymes and, thereby, divergence in their coding genes,
can be routinely described in related and unrelated species. In
addition, divergence in specific plastid and nuclear genes can be
directly assessed by restriction endonuclease analysis, and
comparisons of complete nucleotide sequences in diverse plant
species are anticipated (Doyle *et al.*, this volume). Although
the information obtained by these studies are extremely useful for
evaluating phylogenetic relationships, the genes studied to date
probably have little relevance to the mechanisms of speciation or
the evolution of adaptations. I have been asked to discuss
electrophoretic evidence in plants. Since the general topic has
been well reviewed recently (Brown 1979; Conkle 1982; Crawford
1983; Gottlieb 1977a, 1981a; Tanksley and Orton 1983), I will
emphasize three biosystematic applications.

PROGENITOR-DERIVATIVE SPECIES PAIRS

The consequences of speciation can best be revealed by study of
diploid species related as progenitor and derivative. If merely
closely related species are compared, it will not be easily
possible to determine the evolutionary steps by which a species
acquired its distinctive properties. Plant evolutionists are
fortunate because a number of pairs of species appear to have a
progenitor-derivative relationship. The examples in the plant
literature were originally identified by traditional biosystematics utilizing morphology, cytogenetic analysis of chromosome
pairing in F_1 hybrids, and evidence from ecology and geography.
Not surprisingly, species thought to derive from a particular
extant species are morphologically closely similar to their
presumed parent. Such similarity also usually suggested a
relatively recent origin for the new species. If the phylogeny is
correct, most if not all the alleles of the derivative should
still be present in the parent. This is because new alleles arise
by mutation at structural genes and are not caused by chromosomal
repatterning (which is frequently found in the species pairs
examined). Thus the fewer the generations elapsed since the

origin of a new species, the more similar it should be to its
parent.

The value of electrophoretic evidence became apparent in the
first test of a putative progenitor-derivative relationship.
Clarkia franciscana, a highly self-pollinated species native to a
single locality in San Francisco, was thought to have evolved
by rapid reorganization of chromosomes from the morphologically
similar *C. rubicunda* (Lewis and Raven 1958). But electrophoretic
analysis of eight enzymes coded by about 15 genes revealed that
the two species were totally divergent in the genes specifying
isozymes of six of the enzymes (Gottlieb 1973a). The difference
between the species translated into a very high Nei's genetic
distance (D = 1.27) which is about three times higher than that
between the average pair of congeneric species (Gottlieb 1981a).
Clarkia franciscana was later shown to have a duplicated gene
specifying alcohol dehydrogenase that was absent from *C.
rubicunda* (Gottlieb 1974a). The marked genetic divergence
between the two species indicated that their phylogenetic
separation occurred much longer ago than presumed by Lewis and
Raven, and made the proposed phylogeny quite uncertain. Lack of
concordance between the genetic evidence and the morphological and
chromosomal data suggested an important criterion for acceptance
of a progenitor-derivative phylogeny: in order to be labeled as a
recent derivative, a species must prove highly similar genetically
to its putative parent upon electrophoretic analysis (Gottlieb
1973a).

A high concordance has been obtained in six other cases of
presumed progenitor-derivative phylogenies, and the species pairs
can now be confidently used for much more intensive examination of
the consequences of speciation (e.g. Gottlieb 1979, 1983a). The
species pairs are: *Stephanomeria exigua* ssp. *coronaria* to *S.
malheurensis* - D = 0.06 (Gottlieb 1973b); *Clarkia biloba* to
C. lingulata - D = 0.13 (Gottlieb 1974b); *Gaura longiflora* to
G. demareei - D = 0.01 (Gottlieb and Pilz 1976); *Lycopersicon
chmielewskii* to *L. parviflorum* -D not available (Rick *et al.*
1976); *Coreopsis nuecensoides* to *C. nucensis* - D = 0.03
(Crawford and Smith 1982a); *Coreopsis basalis* to *C. wrightii* -
D = 0.07 (Crawford and Smith 1982b). Such low genetic distances
are similar to those observed between conspecific populations
(Gottlieb 1981a). They suggest that plant species are initially
limited genetic extractions from their parents and possess few or
no unique alleles, at least at genes coding enzymes. The
implication is that the morphological and other differences that
distinguish parent and offspring species initially result from no
more than a small number of genetic changes.

Since even very high morphological similarity and chromosomal
structural homology can mask substantial genetic divergence, each
new proposed progenitor-derivative pair must be tested. The
merits of such standard use of the electrophoretic test are
illustrated by the results of our recent study (Gottlieb, in ms)
of several populations of *Clarkia xantiana* (Onagraceae)

previously examined by Moore and Lewis (1965).

Clarkia xantiana occurs on dry slopes in the oak-digger
pine woodlands of the southern Sierra Nevada foothills of
California. Although self-compatible like all *Clarkias* (Lewis
1953), the species is generally outcrossed because the stigma is
tightly closed at anthesis and matures from two to four days after
the pollen has been shed. However, in the Kern River Canyon,
Moore and Lewis found the usual pink, large-flowered outcrossing
plants growing adjacent to two populations of small-flowered
plants, one with pink and the other with white flowers. The
small-flowered plants proved to be predominantly self-pollinated
because their stigma is open and receptive at or slightly before
anthesis and in contact with the dehiscing anthers. The
difference in stigma maturation and the height of the style
relative to that of the anthers in the outcrossing and selfing
plants was governed by one or two genes (Moore and Lewis 1965).

The two selfing populations were morphologically distinguished
only by their flower color, but could readily be separated from
the outcrossing plants in a number of floral traits. Cytogenetic
studies (Moore and Lewis 1965) showed that meiosis was regular in
the three populations, and that the pink selfer had the same
chromosome arrangement as the outcrosser, but that both differed
from the white selfer by a reciprocal translocation. The result
was that hybrids from crosses between the outcrosser and the pink
selfer were fully fertile but those between the outcrosser and the
white selfer had reduced fertility.

The evidenced led Moore and Lewis to propose that the sympatric
outcrosser was ancestral to the pink selfer which, in turn, gave
rise "perhaps very recently" to the white selfer. They rejected
independent modes of origin for the two selfing populations
because of the improbability that independent derivatives would be
found in immediately adjacent sites and in no others. The genetic
integrity of the selfers was presumed to be maintained by their
high degree of automagy, earlier flowering time, and the
preference of pollinators for the larger and more floriferous
outcrossing plants.

Electrophoretic analysis was used to test the proposed modes of
origin of the selfers. The expectation was that their divergence
would be small since the morphological divergence was small, and
changes in only a few loci accounted for the breeding system
change. Twenty enzymes were examined in the same three
populations originally investigated by Moore and Lewis. The
genetic basis of the electrophoretic variability was determined by
a combination of progeny tests, formal breeding studies (Gottlieb
1977b; Pichersky and Gottlieb 1983; Odrzykoski and Gottlieb, in
ms), and comparison of patterns of extracts from pollen and leaf
tissue to identify heterozygous genotypes (Weeden and Gottlieb
1979).

The enzymes proved to be specified by 40 loci. The outcrossing
population was polymorphic at 18 of them. In contrast, both self-
pollinating populations were completely monomorphic so that for

each of them any single individual was representative. This
remained the case even after eight of the enzymes (coded by 23
loci) were each examined under eight different electrophoretic
conditions (varying principally in gel and electrode buffer
components and in pH from 5.7 to 10.0) in a search for "hidden
variability" (Gottlieb 1981a).

Considering their conspecific status and their putative modes
of origin, the amount of genetic divergence among the three
populations was unexpectedly high. The white and pink selfers
shared the same genes at 32 loci, but were fixed for a different
allele at eight others, giving them a Nei (1972) genetic distance
of D = 0.223. Between the outcrosser and the white selfer,
D = 0.226, and between the outcrosser and the pink selfer,
D = 0.278. The difference in the latter two distance values is
perhaps more clearly revealed by noting that 36 of the 40 genes in
the white selfer were also found in the outcrosser whereas the
pink selfer and the outcrosser had 34 of the 40 genes in common.
Thus even though the white selfer and the outcrosser differed by a
reciprocal translocation, they were somewhat more similar
genetically than the pink selfer and the outcrosser which were
chromosomally alike. More striking lack of concordance has been
reported previously, for example, in *Coreopsis* (Crawford and
Smith 1982a) and *Clarkia* (Gottlieb 1974b).

The substantial genetic divergence between the selfers and the
sympatric outcrosser (conspecific populations of other species
have average distance values of about 0.04, Gottlieb 1981a,
Crawford 1983) made it reasonable to ask whether one or the other
selfer might have arisen from an allopatric population. Thus, the
sympatry of the selfing populations and the outcrosser need not be
a consequence of their mode of origin, but rather one or both of
them might have originated from an allopatric population and then
migrated to its present site in the Kern River Canyon adjacent to
the sympatric outcrosser. Therefore electrophoretic studies were
also carried out on three allopatric populations of *C. xantiana;*
the populations were geographically well-separated and from
different parts of the relatively small distribution of the
species. Only those enzymes were studied which distinguished one
or the other selfer from the sympatric outcrossing population.

The general finding was that the selfers were less similar to
any of the allopatric populations than they were to the sympatric
one. Thus only three of the six genes that were present in one or
the other selfer but not in the sympatric outcrosser turned up in
one or more allopatric populations. And, of these, only one was
in high frequency whereas the other two were detected in only one
or two individuals of one population each. The three genes were
more likely to have been lost from the sympatric outcrosser rather
than being indicative of a different phylogenetic relationship.
Therefore the electrophoretic evidence was consistent with the
basic model of Moore and Lewis that the selfers arose *in situ*.

The data also supported their proposal that the white selfer
arose from the pink selfer and not independently. The evidence

favoring the initial derivation of the pink selfer comes from
Moore and Lewis and depends on the shared flower color and the
common chromosome arrangement; the greater genetic divergence of
the pink selfer from the sympatric outcrosser presumably reflects
the accumulation of de novo mutations during the longer time
period since its origin. That the white selfer did not evolve
independently from the sympatric outcrosser is evident from the
distribution of the alleles from the polymorphic loci of the
outcrosser. If the white selfer had arisen independently from the
outcrossing sympatric population, it should have inherited
different alleles than the pink selfer at many of the polymorphic
loci. But this was the case at only three of the 15 relevant
loci. The two selfers had the same allele at 12 polymorphic loci
and at three of these loci, the allele that the selfers had in
common was a low frequency one in the outcrosser. Such loci are
particularly informative since the lower the frequency of an
allele (assuming relative stability over time), the less likely it
would be present in the hypothetical individual which was the
progenitor of each selfer, if each of them had had independent
origins. The two selfers also shared a unique GDH pattern which
they would not be expected to have in common if they had evolved
independently, since the pattern was absent from the sympatric
outcrosser.

Moore and Lewis had no qualitative characters that could have
been used to test whether one of the selfers was more similar to
an allopatric population than to the sympatric one. The
outcrossing populations of *C. xantiana* are similar
morphologically and cytologically. Thus they had to rely on an
argument that simply asserted the improbability that independent
derivatives would be found in the same or adjacent sites and no
others. In contrast, the electrophoretic analysis permitted an
explicit and direct comparison among populations because it
utilized specific genes that were either shared or not. Thus, the
phylogeny proposed by Moore and Lewis has survived a strong
challenge which could not have been undertaken with traditional
biosystematic tools. In addition, the electrophoretic evidence
provided much more information about the extent of genetic
divergence between the selfers, and confirmed that there was
little or no gene flow between them and the outcrosser (since both
are monomorphic yet grow within pollination distance of the highly
polymorphic sympatric outcrossing population).

The genetic divergence of the selfers from each other and from
the outcrossing populations also underscores differences in
taxonomic treatments in different plant genera. The apparent
reproductive isolation of the selfers and their moderate genetic
divergence might suggest that they should be regarded as species.
However, this would not be in keeping with taxonomic practise in
Clarkia since species status in the genus has been reserved
for taxa that are reproductively isolated by substantial
chromsomal repatterning in addition to a morphological distinction
(Lewis and Lewis 1955). As more information about the extent of

genetic divergence accumulates, it is likely that taxonomic
decisions will have to take into account this new evidence.

THE PHYLOGENETIC SIGNIFICANCE OF GENE DUPLICATION

Most applications of electrophoretic evidence such as on the
closely related species pairs utilize information based on the
variation of alleles (presence, number, and frequency) at large
numbers of gene loci coding enzymes. This is appropriate when the
taxa under consideration are not distantly related. But such
evidence is less convincing as phylogenetic distance increases
because electrophoretic identity is less likely to reflect
identity of coding genes and electrophoretic nonidentity is more
likely to mask a large number of mutational differences. The
exact phylogenetic distance which marks the threshold beyond which
electrophoretic evidence has limited value is not easily defined
although it probably corresponds in most cases to the genus or in
large genera to sections.

At higher taxonomic levels, another aspect of electrophoretic
evidence acquires special value: the number of isozymes of
particular enzymes. In contrast to the often substantial
variability of allozymes (allelic products), the number of
isozymes (products of different loci) of most enzymes in plants,
at least those assayed with natural (*in vivo*) substrates appears
to be highly conserved and depends on the number of subcellular
compartments in which a particular catalytic reaction is required
(Gottlieb 1982). For example, in diploid plants, all enzymes of
glycolysis and the oxidative pentose phosphate pathway are present
as two isozymes, one located in the plastids and the other soluble
in the cytosol (Gottlieb 1982). The number and subcellular
location of isozymes of other enzymes, for example, asparate
aminotransferase and malate dehydrogenase, are also highly
conserved. This makes it a simple matter to recognize plant
species with an increased number of isozymes (a reduced number
will not be observed because it would almost certainly be lethal).
In diploid species an increased isozyme number results from
duplication of a structural gene (Gottlieb 1982). Since the
occurrence and establishment of gene duplications are extremely
rare, species which possess the same duplication can be considered
to descend from the same common ancestor and therefore to belong
to a monophyletic assemblage (Gottlieb 1977b; Gottlieb and Weeden
1979; Gottlieb 1983b). About 15 duplications have been
identified during surveys of electrophoretic variation in natural
plant populations (Gottlieb 1982; Crawford and Smith 1982a;
Ellstrand *et al.* 1983). Many were recognized because related
diploid species had different numbers of isozymes.

The most thoroughly studied duplicated isozymes in plants are
the cytosolic isozymes of phosphoglucose isomerase (PGI; EC
5.3.1.9) in diploid species of *Clarkia* (Onagraceae). PGI
catalyzes the reversible isomerization of fructose-6-phosphate and
glucose-6-phosphate, a required reaction in glycolysis and

gluconeogenesis. The 43 species of *Clarkia* are mostly endemic
to California and are found primarily in open sites in oak woodlands with well-drained soils or in adjacent xeric sites. The
populations of most species are discrete, relatively small
(numbering in the few hundreds to several thousand individuals),
and generally present in the same site each year (Lewis 1953).
Flowers of outcrossed species are characteristically insect-
pollinated, though a number of them are predominantly self-
pollinated.

The detailed studies of different species groups within
Clarkia and a broad analysis of morphological and chromosomal
characters aided by numerous experimental hybridizations led to an
elegant monograph of the genus and a proposed phylogenetic
arrangement of the species (Lewis and Lewis 1955) which has come
to be regarded as a model for studies of its type. The diploid
species were placed in several well-defined sections (Rhodanthos
[formerly Primigenia], Godetia, Myxocarpa, Peripetasma,
Phaeostoma, Fibula, and Eucharidium), and three sections were
erected for polyploid members (Biortis, Connubium, and Clarkia).

On the basis of its chromosome number and generalized
morphology, Rhodanthos was regarded as most representative of the
primitive *Clarkia*, and most similar to the related genus
Oenothera. Peripetasma, Phaostoma, and Eucharidium were
recognized as highly advanced sections and were thought to derive
from different ancestral stocks within the genes (Lewis and Lewis
1955). The morphological intermediacy of Fibula suggested that it
may have evolved by hybridization without change in chromosome
number between a species of Peripetasma and one of Phaeostoma
although alternative hypotheses were not ruled out (Lewis and
Lewis 1955).

But the phylogeny of the sections was challenged (Gottlieb and
Weeden 1979) following the discovery of the PGI duplication. The
duplicated PGIs characterize all species of Phaeostoma, Fibula,
and Eucharidium, and all but one species (*C. rostrata*) of
Peripetasma. Species with the duplication have two loci
specifying cytosolic PGI isozymes and those without the
duplication have a single locus coding this function. Diploid
species of related genera of Onagraceae and diploid plants in
general have only a single cytosolic PGI isozyme.

The duplicated PGI genes assorted independently in four
species, representing two sections, that were tested (Gottlieb and
Weeden 1979). This is thought to mean that the duplication arose
by a process involving overlapping reciprocal translocations or
insertional translocations rather than by unequal crossing over
which requires only mispairing during synapsis and results in
tightly linked duplicates. Origin of duplicate genes by
translocation requires relatively rare simultaneous multiple
chromosome breaks, and the production of true-breeding duplicate
progeny which must be viable and fertile. Consequently, the
establishment of a duplication by such processes has a very low
probability of happening more than once in a given lineage. Based

on this hypothesis, species which possess the same duplication must have inherited it from the same common ancestor. The substantial chromosomal rearrangements characteristic of *Clarkia* species, and the self-compatibility of all the species which facilitates rapid achievement of structural homozygosity, provide a reasonable mechanism of origin, and permit the hypothesis that species with the duplication descended from a common ancestor.

Thus the presence of the PGI duplication in the four phylogenetically advanced sections and its absence from the three morphologically more primitive sections identified a phylogenetic dichotomy within *Clarkia,* and suggested that the former four sections are monophyletic although previously they had been thought to derive from different stocks of ancestral *Clarkia* (Lewis and Lewis 1955). This new evidence has been used by Prof. H. Lewis to modify his early phylogeny to reflect the presumed monophyletic relationship (Lewis 1980). The concordance between possession of the duplication and membership in one of the four advanced sections, and absence of the duplication in the three primitive sections, meant that it was not necessary to move species in or out of particular sections but only to realign the sectional phylogeny.

The only exception to a perfect concordance was *C. rostrata* (Peripetasma) which exhibits a single cytosolic PGI. It is not known if the species lacks an additional isozyme because the coding gene is not duplicated or because the duplicated gene is present but silenced by mutation(s). Whichever is the case, the dichotomous phylogeny within the genus remains intact.

The duplication also provided critical evidence to retain Eucharidium as a section of *Clarkia* rather than assigning it generic status because of its unique morphological characters (four rather than eight stamens and an extremely long floral tube apparently associated with a distinctive pollination system [MacSwain *et al.* 1973]). Lewis and Lewis (1955) retained Eucharidium within *Clarkia* because they thought it "almost certainly linked to section Myxocarpa" (p. 359) via the tetraploid *C. pulchella*. However, recent scanning electron microscopy of the seed surface of plants in the two species of Eucharidium revealed many unique features, unlike those found in other *Clarkia* (Raven, pers. comm.), thus weakening the argument to keep the section in the genus. The biochemical evidence supports the maintenance of Eucharidium within *Clarkia* because it possesses the PGI duplication (Gottlieb and Weeden 1979).

Gene duplications identified by electrophoretic analysis may also prove useful to group even higher taxa. We have recently shown that all diploid species of *Clarkia* possess duplicated genes specifying both the plastid and cytosolic isozymes of triose phosphate isomerase (TPI; EC 5.3.1.1) (Pichersky and Gottlieb 1983). Most diploid species have only one TPI isozyme in the plastids and a different one in the cytosol, but *Clarkias* have true-breeding multiple isozymes in both compartments. On the basis of various criteria described in Gottlieb (1983b), the

duplication of the gene specifying the cytosolic TPIs has been
identified in diploid species of six of the seven tribes of the
Onagraceae including Jussiaeeae (*Ludwigia*), Fuchsieae (*Fuchsia*),
Hauyeae (*Hauya*), Onagreae (*Clarkia, Heterogaura, Camissonia,
Gaura, Oenothera, Calyophus,* and *Gongylocarpus*), and Epilobieae
(*Boisduvalia*). Its presence in the phylogenetically primitive
Fuchsia and in *Ludwigia,* considered a distinct branch of the
family (Raven 1979), suggests the cytosolic TPI duplication may
have occurred early in the origin of the family.

The duplication has not yet been observed in the monotypic
Circaeeae (*Circaea*) or in the genus *Stenosiphon* of the Onagreae
tribe. Assuming that further study fails to identify it in these
genera, its absence is best interpreted as a loss caused by
mutation(s) either in the particular species examined or in their
progenitors. Once a duplication has been eliminated by mutation,
it has a very low probability of being "reinstated." This leads
to the hypothesis that if a particular duplication has been
established in a lineage, but lost in a derived taxon, it is
unlikely to reappear in a descendent of the derived taxon. Thus
certain conceivable phylogenetic lineages involving various genera
can be ruled out which may provide a valuable criterion to assess
relationships in the family.

In contrast to the wide distribution of the cytosolic TPI
duplication, the duplication of the gene coding plastid TPIs has
not yet been demonstrated outside of *Clarkia*, although numerous
genera have been examined (Pichersky and Gottlieb 1983). Its
absence in other genera of the family suggests it arose more
recently.

It is still premature to speculate about the ultimate value of
gene duplications in plant systematics. Their usefulness within
Clarkia has been immense, but their potential value in other
groups remains uncertain primarily because few other genera have
been sampled as extensively. The one exception is the large genus
Coreopsis (Compositae) studied by Crawford and Bayer (1981)
and Crawford and Smith (1982a, b). Crawford (1983) reported that
an apparent duplication of the gene specifying the plastid isozyme
of PGI has been discovered in several highly specialized sections
of *Coreopsis* whereas it is absent from the two most primitive
sections. This observation suggests that the duplication may be
useful to construct monophyletic groups in another genus.

ISOZYME NUMBER AND PLOIDY

Information about the number of gene loci coding isozymes of
particular enzymes can also be used to ascertain ploidy level.
Ploidy level has usually been determined primarily on the basis of
chromosome number but this may be ambiguous with gametic
chromosome numbers between $n = 9$ and $n = 12$. Also when
relatives with lower numbers are not known the species with the
lowest number in the group is often assigned diploid status.

Sharp differences have characterized the interpretation of chromosome numbers in the Astereae tribe of the Compositae. The most common numbers vary between $n = 2$ and $n = 9$, with many species having $n = 4$ or $n = 5$. Two hypotheses have been formulated: (1) $n = 9$ was the original base chromosome number of the group and lower numbers evolved by aneuploid reduction (Raven *et al.* 1960); (2) the ancestral number was $n = 4$ or $n = 5$; therefore, species with $n = 9$ are allotetraploids derived by hybridization between taxa with lower numbers (Turner *et al.* 1961; Turner and Horne 1964). Both proposals claimed support from various observations that were mostly irrelevant. Thus the association of $n = 9$ with the primitive woody habit, the high symmetry of the $n = 9$ karyotype, and the widespread phylogenetic occurrence of this gametic number were claimed as evidence favoring the first hypothesis. The rarity of species with the intermediate $n = 6$ and $n = 7$ was considered to support the second proposal.

However the essential attribute of polyploidy is not relative chromosome number, but genome multiplication and its attendant increases in gene loci (Gottlieb 1981b). Therefore, electrophoresis was used to determine if species in the Astereae with $n = 9$ had more genes coding isozymes than those with $n = 4$ or $n = 5$. The test built on two previous findings: (1) allopolyploid species display more isozymes than diploids because they inherit homeologous gene loci from their diploid parents that are frequently fixed or become so for alleles that specify enzymes having distinguishable electrophoretic mobilities (Roose and Gottlieb 1976; Hart 1979; Gottlieb 1981a); and (2) aneuploid decrease from an ancestral diploid condition does not change the number of structural genes coding isozymes (Roose and Gottlieb 1978; Crawford and Smith 1982a).

The electrophoretic test was performed using 17 enzymes extracted from five species of *Machaeranthera* with gametic chromosome numbers of $n = 4$, 5, and 9, and two species of *Aster* with $n = 5$ and $n = 9$. The number of isozymes of each of the enzymes proved the same in all of the species with no evidence of isozyme multiplicity (Gottlieb 1981b). The result made it most unlikely that the two species with $n = 9$ originated by allotetraploidy. The constancy of isozyme number in all of the species was not unexpected if the species with the lower chromosome numbers represent lineages that arose following aneuploid reduction. This process is generally thought to involve translocation of essential euchromatin and loss of only heterochromatin and centromeres. Thus, in this case, the electrophoretic test provided a direct and simple analysis which took advantage of an inherent *genetic* attribute of polyploidy.

In a different version of the test, the presence of isozyme multiplicity was used to demonstrate that species of *Cucurbita* ($n = 20$), generally considered diploid, were actually tetraploids (Weeden, pers. comm.).

CONCLUSIONS

Two of the three described applications of electrophoretic
evidence do not require high correlation between electrophoretic
similarity and genome similarity. The use of a gene duplication
as a taxonomic grouping device depends on the improbability of a
duplication of the same structural gene originating and becoming
established more than once in a single taxonomic lineage. The use
of isozyme number to identify ploidy level depends on the increase
in gene number that necessarily follows the addition of genomes
during the formation of an allopolyploid.

In contrast, the electrophoretic comparison of a species and
its putative parent does require a high degree of genetic
identity, in the range usually observed among conspecific
populations, to satisfy the proposed criterion for progenitor-
derivative relationship. Insistence on the criterion will limit
the number of species pairs with such status, and may lead to
errors in cases in which a low identity masks a recent speciation.

This possibility reveals the limitations of electrophoretic
evidence insofar as it depends on information about allelic
divergence rather than changes in the number of coding gene loci.
Electrophoretic evidence does not reveal the number of amino acid
substitutions that cause enzymes to differ in mobility. A
difference resulting from a single substitution is treated in
genetic identity statistics exactly like one resulting from
numerous substitutions. In addition, the statistics treat all
loci as equivalent items to be added up and averaged, although
some loci have a higher probability of diverging between species
(Gottlieb 1981a), and changes in different loci can have very
different biological consequences.

In practice this means that identity values by themselves are
not necessarily a useful index to phylogenetic relationships. If
the identities among three taxa are roughly similar, either high
or low, no pair can be considered more or less related. How much
divergence is required to make a claim of distant phylogenetic
relationship is also uncertain. The divergence between *Clarkia
franciscana* and *C. rubicunda* was ten times greater than that
between *C. lingulata* and *C. biloba*. The divergence values in
this case correlate well with morphology and chromosomal
structural homology (Gottlieb 1981a). The high value between the
former species pair also reflect divergence introduced by
stochastic factors associated with the predominant inbreeding and
likely history of sharp fluctuations in population size occurring
during the evolutionary history of *C. franciscana*. But often
genetic distances are more similar, and then strong conclusions
are inadvisable.

In spite of this caveat, evidence of allelic divergence has
many valuable applications at the species level. No other
procedure is so efficient in distinguishing morphologically nearly
identical "microspecies" (Jefferies and Gottlieb 1982) and just as
readily revealing high genetic similarity among morphologically

diverse species such as in *Tetramolopium* which occupy a variety
of habitats in the Hawaiian Islands (Crawford 1983). How else
validate the hypothesis that *Chenopodium incognitum* was an
artificial grouping of two distinct population systems each
properly referred to another species (Crawford and Wilson 1979)?
Many additional applications continue to be demonstrated.

Just as chromosome counts and determination of ploidy level are
routinely expected in biosystematic studies, electrophoretic
analysis must also become routine when phylogenetic relationships
among closely related species are proposed.

REFERENCES

Babcock, E.B. and Cave, M.S. 1938. A study of intra- and
    interpsecific relations of *Crepis foetida* L. *Z. Indukt.
    Abstamm. Vererbungsl.* 75: 124-160.
Brown, A.H.D. 1979. Enzyme polymorphism in plant populations.
    Theor. Popul. Biol. 15: 1-42.
Camp, W.H. and Gilly, C.L. 1943. The structure and origin of
    species. *Brittonia* 4: 324-385.
Clausen, J., Keck, D.D. and Hiesey, W.M. 1939. The concept of
    species based on experiment. *Am. J. Bot.* 26: 103-108.
Clausen, J. Keck, D.D. and Hiesey, W.M. 1947. Heredity of
    geographically and ecologically isolated races. *Am. Nat.*
    81: 114-133.
Conkle, M.T. 1981. Proc. *Isozymes of North American Forest
    Trees and Forest Insects.* USDA For. Serv. Gen. Tech. Rep.
    PSW-46, 1-64.
Crawford, D.J. 1983. Phylogenetic and systematic inferences from
    electrophoretic studies. In *Isozymes in Plant Genetics and
    Breeding* (S.O. Tanksley and T.J. Orton, eds.), Elsevier,
    Amsterdam.
Crawford, D.J. and Bayer, R.J. 1981. Allozyme divergence in
    Coreopsis cyclocarpa (Compositae). *Syst. Bot.* 6:
    373-379.
Crawford, D.J. and Smith, E.B. 1982a. Allozyme variation in
    Coreopsis nuecensoides and *C. nuecensis* (Compositae),
    a progenitor-derivative species pair. *Evolution* 36: 379-386.
Crawford, D.J. and Smith, E.B. 1982b. Allozyme divergence between
    Coreopsis basalis and *C. wrightii* (Compositae). *Syst. Bot.*
    7: 359-364.
Crawford, D.J. and Wilson, H.D. 1979. Allozyme variation in
    several closely related diploid species of *Chenopodium* of the
    western United States. *Am. J. Bot.* 66: 237-244.
Doyle, J.J., Beachy, R.N. and Lewis, W.H. 1983. Evolution of rDNA
    in Claytonia *polyploid complexes. In Plant Biosystematics*
    (W.F. Grant, ed.). Academic Press, Toronto.
Ellstrand, N.C., Lee, J.M. and Foster, K.W. 1983. Alcohol
    dehydrogenase isozymes in grain surghum (*Sorghum bicolor*):
    evidence for a gene duplication. *Biochem. Genet.* 21: 147-154.

Gottlieb, L.D. 1973a. Enzyme differentiation and phylogeny in *Clarkia franciscana, C. rubicunda* and *C. amoena. Evolution* 27: 205–214.

Gottlieb, L.D. 1973b. Genetic differentiation, sympatric speciation, and the origin of a diploid species of *Stephanomeria. Am. J. Bot.* 60: 545–553.

Gottlieb, L.D. 1974a. Gene duplication and fixed heterozygosity for alcohol dehydrogenase in the diploid plant *Clarkia franciscana. Proc. Natl. Acad. Sci. USA* 71: 1816–1818.

Gottlieb, L.D. 1974b. Genetic confirmation of the origin of *Clarkia lingulata. Evolution* 28: 244–250.

Gottlieb, L.D. 1977a. Electrophoretic evidence and plant systematics. *Ann. Mo. Bot. Gard.* 64: 161–180.

Gottlieb, L.D. 1977b. Evidence for duplication and divergence of the structural gene for phosphoglucose isomerase in diploid species of *Clarkia. Genetics* 86: 289–307.

Gottlieb, L.D. 1979. The origin of phenotype in a recently evolved species. In *Topics in Plant Population Biology* (O.T. Solbrig, S. Jain, G.B. Johnson and P. Raven, eds.), Columbia Univ. Press, N.Y.

Gottlieb, L.D. 1981a. Electrophoretic evidence and plant populations. *Progr. Phytochem.* 7: 1–46.

Gottlieb, L.D. 1981b. Gene number in species of Astereae that have different chromosome numbers. *Proc. Natl. Acad. Sci. USA* 78: 3726–3729.

Gottlieb, L.D. 1982. Conservation and duplication of isozymes in plants. *Science* 216: 373–380.

Gottlieb, L.D. 1983a. Interference between individuals in pure and mixed cultures of *Stephanomeria malheurensis* and its progenitor. *Am. J. Bot.* 70: 276–284.

Gottlieb, L.D. 1983b. Isozyme number and phylogeny. In *Proteins and Nucleic Acids in Plant Systematics* (U. Jensen and D.E. Fairbrothers, eds.), Springer-Verlag, N.Y.

Gottlieb, L.D. and Pilz, G. 1976. Genetic similarity between *Gaura longiflora* and *G. demareei. Syst. Bot.* 1: 181–187.

Gottlieb, L.D. and Weeden, N.F. 1979. Gene duplication and phylogeny in *Clarkia. Evolution* 33: 1024–1039.

Hart, G.E. 1979. Genetical and chromosomal relationships among the wheats and their relatives. *Stadler Genet. Symp.* 11: 9–29.

Jefferies, R.L. and Gottlieb, L.D. 1982. Genetic differentiation of the microspecies *Salicornia europea* (sensu strictu) and *S. ramosissima. New Phytol.* 92: 123–129.

Lewis, H. 1953. The mechanisms of evolution in the genus *Clarkia. Evolution* 7: 1–20.

Lewis, H. 1980. The mode of evolution in *Clarkia.* Symposium paper presented at International Congress Syst. and Evol. Biol. II, Vancouver, B.C.

Lewis, H. and Lewis, E.B. 1955. The genus *Clarkia. Univ. Calif. Publ. Bot.* 20: 241–392.

Lewis, H. and Raven, P. 1958. Rapid evolution in *Clarkia. Evolution* 12: 319–336.

MacSwain, J.W., Raven, P. and Thorp, R. 1973. Comparative
    behavior of bees and Onagraceae. IV. *Clarkia* bees of the
    western U.S. *Univ. Calif. Publ. Entom.* 70: 1-80.
Moore, E.M. and Lewis, H. 1965. The evolution of self-pollination
    in *Clarkia xantiana*. *Evolution* 19: 104-114.
Nei, M. 1972. Genetic distance between populations. *Am. Nat.*
    106: 283-292.
Pichersky, E. and Gottlieb, L.D. 1983. Evidence for duplication
    of the structural genes coding plastid and cytosolic isozymes of
    triose phosphate isomerase in diploid species of *Clarkia*.
    Genetics, 105: 421-436.
Raven, P. 1979. A survey of reproductive biology of Onagraceae.
    N. Z. J. Bot. 17: 575-593.
Raven, P., Solbrig, O.T., Kyhos, D.W. and Snow, R. 1960.
    Chromosome numbers in Compositae. I. Astereae. *Am. J. Bot.*
    47: 124-132.
Rick, C.M., Kesicki, E., Fobes, J.F. and Holle, M. 1976. Genetic
    and biosystematic studies on two new sibling species of
    Lycopersicon from interandean Peru. *Theor. Appl. Genet.*
    47: 55-68.
Roose, M.L. and Gottlieb, L.D. 1976. Genetic and biochemical
    consequences of polyploidy in *Tragopogon*. *Evolution* 30:
    818-830.
Roose, M.L. and Gottlieb, L.D. 1978. Stability of structural gene
    number in diploid species with different amounts of nuclear DNA
    and different chromosome numbers. *Heredity* 40: 159-163.
Tanksley, S.O. and Orton, T.J. 1983. *Isozymes in Plant
    Genetics and Breeding*. Elsevier, Amsterdam.
Turner, B.L., Ellison, W.L. and King, R.M. 1961. Chromosome
    numbers in the Compositae. IV. North American species with
    phyletic interpretations. *Am. J. Bot.* 47: 216-223.
Turner, B.L. and Horne, D. 1964. Taxonomy of *Machaeranthera*
    sect. *Psilactis* (Compositae-Astereae). *Brittonia* 16: 316-331.
Weeden, N.F. and Gottlieb, L.D. 1979. Distinguishing allozymes
    and isozymes of phosphoglucose isomerases by electrophoretic
    comparisons of pollen and somatic tissues. *Biochem. Genet.*
    17: 287-296.

Phytochemical Approaches to Biosystematics

K. E. Denford
Department of Botany
University of Alberta
Edmonton, Alberta, Canada

INTRODUCTION

The advent of Phytochemistry, or Chemosystematics, is closely
linked to the introduction of chemical analytical methods which
over the past quarter century have expanded the average
systematist's horizons (Smith 1976; Harborne 1973). Probably the
most notable and valuable technique added to the systematist's
repertoire is that of chromatography (Paper, Thin layer and HPLC
methods; Harborne *et al*. 1975; Harborne and Mabry 1982).
This single introduction is responsible for a vast accumulation of
chemosystematic data in the literature today (Harborne and Mabry
1982). At the same time there are an almost unlimited array of
chemical structures to be found in plants which can be isolated
and identified using these techniques. Some are directly
involved in day-to-day metabolism and others have functions and
distributions which appear to pose interesting and sometimes
insoluble problems (Bell and Charlwood 1980). It is often these
very compounds that attract the attention of the Biosystematist,
either as an individual or as a team member in association with a
Phytochemist (natural products chemists). Collaborative studies
of this nature are often extremely valuable, however frequently
the line of research that develops drifts off into the realms of
biosynthesis.
    Of major concern to any systematist is the utility and
availability of both techniques to be employed, and speed with
which one can generate data with meaningful value. It is of
limited value at the present time therefore, for a biosystematist
to exhaustively analyze the amino acid sequences of proteins and
other macromolecules in two varieties of a species with widely
differing distribution patterns, when a much more meaningful
analysis could be carried out on soil preferences, flowering
dates, breeding processes, and cytotaxonomic relationships. In

such a situation the line of research is usually determined by a
discriminating researcher who would appreciate that the data
obtained from such a study of just two plants would have little
benefit to both himself and other workers not possessing the
sophisticated apparatus required for such an endeavor. If one
wishes to study chemical variation at the population and
individual levels, then there are much easier methods and
approaches available to all systematists, which are within their
budgetory framework and comprehension (Harborne 1973). These
include two dimensional paper chromatography, thin layer
chromatography, column chromatography, hydrolysis and UV
spectralanalysis (Mabry et al. 1970; Wilkins and Bohm 1976).

SOURCES OF PHYTOCHEMICAL DATA

The chemical approach to biosystematics and the compounds used can
be divided into two broad categories; macromolecules and
micromolecules. For lack of a more precise definition, organic
compounds of a polymer nature, would belong to the former group
(Proteins, Nucleic acids, etc.). Many micromolecules lend
themselves easily to analysis and many have proved to be of great
utility in systematic investigation. Four of the most prominent
groups of note are the phenolics, alkaloids, terpenoids, and
nonprotein amino acids, all of which exhibit a wide variation in
chemical diversity, distribution and function (Smith 1976).

Phenolics (Polyphenols): One group of compounds in this category
are the Flavonoids which to date have been found in all groups of
plants investigated (a case of seek and you shall find!).
Flavonoids are readily extracted from both fresh and dry plant
materials and they exhibit a wide range of chemical and structural
diversity (Ribéreau-Gayon 1972; Harborne et al. 1975;
Harborne and Mabry 1982) forming the largest group of polyphenols
known to man, several thousand having been described to date.
Although there appears to be a wide array of structural variation,
all are based on the same C15 skeleton of flavone (Fig. 1), and
are derived from the same biochemical pathway involving the
condensation of malonate and phenylalanine derivatives (Hahlbrock
and Grisebach 1975). There are essentially 12 major classes of
flavonoids, each class being identified by the oxidation level of
its central pyran ring (Harborne et al. 1975). The three
major types of most widespread occurrence are the anthocyanins,
flavones and flavonols. Several hundred aglycones have been
isolated from plant tissues, however, only eight of these occur
with any great regularity in vascular plants. This of course
means that a systematist can become fairly well acquainted with
those found in his taxonomic group over a relatively short period
of time. The over-all complexity of flavonoids and a source of
almost unlimited structural variation relates to the possibility
of methylation; addition of 6- or 8-OH groups, as well as
glycosides and phenolic acids to the base structure. The

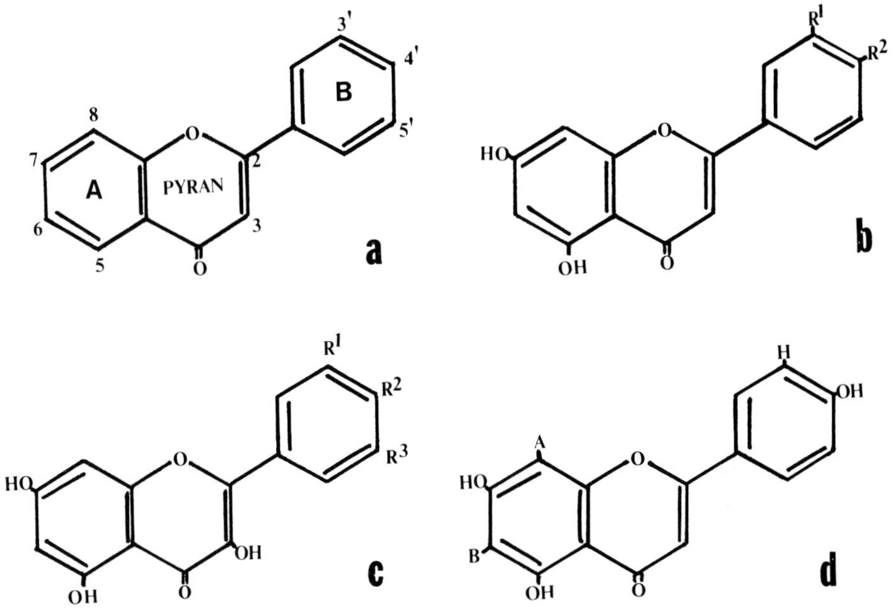

FIG. 1. *Basic flavonoid structures. i. Flavonoid ring. ii.
R1 = R2 = OH, luteolin; R1 = H, R2 = OH, apigenin. iii.
R1 = R2 = R3 = OH, myricetin; R1 = R2 = OH =, R3 = H, quercetin;
R1 = R3 = H, R2 = OH, kaempferol. iv. A, B, 8 and 6 positions for
C-glycoside attachment.*

introduction of such groups to these compounds will alter their
solubilities in both aqueous and organic solvents. Hence it is
possible to isolate widely different flavonoids during extraction
by the process of partitioning (Mues *et al.* 1979). Such an
approach enables one to isolate methylated derivatives very
efficiently using hexane against the aqueous methanol crude
extract. Similar modifications to technique can be adopted to
separate sulfated derivatives using paper electrophoresis
(Harborne and Williams 1976; Harborne 1977). Finally, differences
in distribution and location within the plant have been observed.
Certain flavonoids have been isolated from the farina of primrose
leaves as well as the waxy cuticle of *Eucalyptus*, and the
buds of *Populus* and *Alnus* indicating that not all flavonoids
are found solely in the cytoplasm or the vacuole (Harborne
1967). It is therefore possible to produce flavonoid profiles of
three major types, internal and external foliar flavonoids and
floral flavonoids. The flavonols kaempferol, quercetin and
myricetin are the most commonly found and frequently occur as co-
pigments in flowers along with their related anthocyanins.

Flavonols appear to be universally distributed in plant leaves. A
pioneering survey by Swain and Bate-Smith (1962) of over 1000
angiosperms demonstrated the presence of quercetin in 56%,
kaempferol in 48% and myricetin in 10% of species investigated,
the latter being mainly associated with woody plants. Flavones
are commonly represented by apigenin and luteolin (Fig. 1). These
compounds occur in association with flavonols and also are found
on their own in many herbaceous species. As with the flavonols
there is a great potential for structural variability, although
simple glycoside variation tends to be limited to a single 7-
position (flavones lack the 3-OH of the flavonols, hence the
distinction). There are however flavones with C-glycosides, the
sugars being attached directly to the C15 structure and not
through an -OH group (Fig. 1).

THE SOURCES OF CHEMICAL VARIATION

Hydroxylation: Hydroxylation patterns can range from the
simplest situation, that of flavone with no -OH groups to the
highly methylated Digicitrin whose parent compound, if found in
nature, must have eight or all its -OH groups present. Most of
the rare flavonoids that have been found can be derived from one
of the eight common aglycones by the addition of a -OH group in an
unoccupied site in the basic nucleus. Such substitutions occur in
both flavonols and flavones, for example, the 2' position of the
flavone luteolin when inserted with an -OH results in the
formation of isoetin, a yellow flower pigment of the Cichorieae
(Harborne 1978). In the case of kaempferol a similar substitution
results in the formation of morin, a flavonoid which appears to
control the feeding habits of silkworm leaves on *Morus* leaves
(mulberry) (Van Emden 1973).
    In many cases the addition of -OH groups results in the
transformation of an otherwise colorless flavonoid into a yellow
light visible compound. Well known examples of this occur in
Primula vulgaris, Gossypium hirsutum and *Rudbeckia hirta*
where introduction of an additional -OH at the 6 and 8 positions
results in the formation of quercetagetin and gossypetin from the
flavonol quercetin (Harborne 1967; Thompson *et al.* 1972).
Of rarer occurrence is the elimination of -OH groups as in the
case of 3-desoxyanthocyanidins produced by the removal of the 3-OH
commonly found in the anthocyanins. The resulting pigments appear
to be reds and oranges and associated with bird pollination in the
new world gesneriads (Harborne 1967).

Methylation: Although methylation was at one time thought to
be rare it has been shown to be quite widespread, with 15
methylated derivatives of quercetin described to date (Gottlieb
1975). However, the frequency of different methyl derivatives
varies, the 3-, 7- and 3'-OH are frequently methylated, whilst 4'-
and 5-methylation patterns are more infrequent. It also appears
that methylation occurs more often when a new -OH group is added

to the flavonoid, so that it is not unusual to find 6- and 8-
methylated derivatives in a taxon and no hydroxy relatives
(Harborne and Williams 1982). The addition of methylated groups
changes the solubilities of the compounds from hydrophilic to
lipophilic. The resulting derivatives are then distributed
differentially within the cell, methylated flavonoids tending to
be membrane bound whereas nonmethylated flavonoids tend to be
glycosidic and vacuolar.

Glycosides (monosaccarides): The commonly occurring
monoglycosides of the O-glycosidic type, (i.e.) linked through an
-OH group, include the glucoside, galactoside, arabinoside,
rhamnoside, xyloside and glucuronide. All occur as the stable
pyranosides, whereas the arabinoside also occurs as the unstable
furanoside (Geissman 1962). A further complication relates to the
monosaccharide structures of glucose, galactose and xylose which
are D-saccharides and -linked to the aglycone through the
appropriate hydroxyl. Rhamnose and arabinose are L-sugars and
-linked. It has been demonstrated that α and β and L- and D-forms
exist in the same plant (Geissman 1962). Recent studies have
shown that D-Apiose (Pentose), D-Allose (Hexose) and D-
galacturonic acid glycosides exist in nature also (Harborne and
Williams 1982).

Disaccharides: If there are in fact both α and β monosaccharides
in plants with the added possibility of both being associated in
the disaccharide, then over 100 combinations are possible. In
nature about 20 of these have been fully characterized, the most
widespread of which is the sophoroside (2-0-β-D-glucosyl-D-
glucoside) being found in floral parts, leaves and pollen
(Pratviel-Sosa and Percheron 1972). Certain other disaccharides
have rarely been found e.g., laminaribioside (3-0-β-D-glucosyl-D-
glucoside). Mixed disaccharides exist, the commonest of which is
rutinoside (6-0-α-L-rhamnosyl-D-glucoside) however, the 2-0 and 3-
0 isomers are extremely rare (Seshardi and Vydeeswaran 1972).
Higher numbers of saccharides can exhibit in combinations some of
which are branched (Buttery and Buzzell 1975), and in the genus
Marchantia multiple uronic acids related to cell wall structure
have been reported as being linked to flavonoids (Markham and
Porter 1975).

C-glycosides: These are a group of flavonoids that are
particularly resistant to acid hydrolysis and exhibit acid
isomerisation (Wessely-Moser rearrangements). Sugars are attached
to the A ring of the compound directly through a carbon to carbon
bond as opposed to the O-glycosides which attach through an
existing OH group. They are found rather widely in the vascular
as well as the nonvascular plants, being most often found in the
aerial parts but not restricted to them. Frequently they are
found in combination with O-glycosides, and can pose problems for
analysis as they are often present in low concentrations. Under

these conditions it is often only possible to assess their chromatographic and spectral characteristics. Often one encounters a C-glycoside with an O-glycoside attachment also and in this case it is possible to at least partially hydrolyze the compound.

There also exist in nature at least two other flavonoid derivatives, the acylated, additional phenolic acids attached (Bhutani *et al.* 1969; Collins *et al.* 1975), and the sulphated derivatives (Peryra de Santiago and Julianai 1972; Williams *et al.* 1971).

In recent years many studies have been carried out utilizing flavonoids to investigate biosystematic problems. These include the identification of autoploids and alloploid elements in populations (Packer and Denford 1974; Soltis *et al.* 1983; Murray and Williams 1973; Levy and Levin 1971), geographic variation (Levy and Fujii 1978; Wolf *et al.* 1979; Levy 1983), species limits (Elisens and Denford 1982; Whalen 1978; Denford 1981) as well as phylogenetic studies (Levy and Levin 1975; Crawford 1978; Wallace *et al.* 1983), and hybridization (Crawford 1970; Wolf and Denford 1983b).

FLAVONOID PROFILES AS BIOSYSTEMATIC AIDS

Arctostaphylos uva-ursi is a widespread common circumpolar and boreal species comprising at least five taxa (Table I). The subspecies *stipitata* (Packer and Denford 1974) is restricted in its distribution to the Yukon and high alpine regions of the Rockies (Fig. 2). An analysis of its flavonoid profile shows that it has what has become known in our lab as a depauperate profile, i.e., fewer flavonoids than its closest taxonomic relatives (Table II). The possibility that this taxon represents the remnants of a refugial entity, a survivor of pleistocene glaciation in the north west, was intimated previously by Packer and Denford (1974). Subsequent studies of about 40 taxa within the genus at large have indicated that the "*uva-ursi*" complex is a good exmple of chemical divergence within a closely knit group of taxa (Denford 1981). The pronounced reduction in flavonoid complexity in this group has led us to investigate several other genera, which have restricted elements, in order to determine if in fact the

TABLE I. *Subspecies and varieties of* Arctostaphylos uva-ursi

Taxon	*Arctostaphylos uva-ursi* Vestiture	2n
var. *coactilis*	S- L-	52, 39, 26
var. *uva-ursi*	S-	52
var. *adentotricha*	S+ L+	26
var. *stipitata*	S+	52

TABLE II. *Flavonoid Glycosides of* Arctostaphylos uva-ursi

		Adenotricha	Coactilis	uva-ursi	stipitata
Myricetin	3-O glc	+++	++	+	+
	3-O arab	++	+	+	−
Quercetin	3-O glc	+++	+++	++	+
	3-O arab	++	++	++	−
	3-O gal	++	++	++	+
	3-O rham	+	+	+	−
	3-O diglc	+	+	+	+
	3-O rhglc	+	+	+	+
	7-O glc	+	+	+	−
No. of populations		63	37	58	21

"paucity/depauperate factor" is of a common occurrence.

A genus possessing many of the chemical structures previously mentioned and exhibiting taxa of both widespread and restricted distributions, and therefore a good example of the value of a phytochemical study utilizing flavonoids, is that of *Arnica* L. a circumboreal predominantly montane genus subdivided into five subgenera (Maguire 1943) of about 32 species. Subgenus *Austromontana* includes nine species, some of which are widespread (*cordifolia, latifolia*) and others restricted in their distributions (*venosa, viscosa*). Major diversification is thought to have taken plan in the Klamath region of southwestern Oregon and northwestern California, a geologically ancient area with a high incidence of endemism (Denton 1979; Whittaker 1961; Raven and Axelrod 1978). Adaptive radiation has been demonstrated to occur in this region for several other taxa including *Crepis*, and *Sedum* (Denton 1979). Speciation within *Arnica* in northwestern North America has been accompanied by a number of ecological and morphological changes including floral morphology, flowering period, a shift from mesic to zeric habitats and substrate specificity.

A total of 22 flavonoids (11 glycosides and 11 aglycones) were found in 87 populations of the species surveyed (Table III). The glycosides quercetin 3-O-gentiobioside (87 populations) and quercetin 3-O-diglucoside (86 populations) were ubiquitous or nearly so, were not found in the other subgenera of *Arnica*, and serve to unify the subgenus *Austromontana*. Other compounds found in significant numbers include apigenin 6-methyl ether (59 populations), kaempferal 3-O-glucoside (46 populations) and quercetin 3-O-glucoside (46 populations). In general, Klamath region endemics are characterized by depauperate flavonoid profiles and/or fewer glycosides, and an increase in the number of methylated aglycones. In addition, although wide ranging species are characterized by higher flavonoid diversity and more glycosides, their Klamath region populations have fewer glycosides

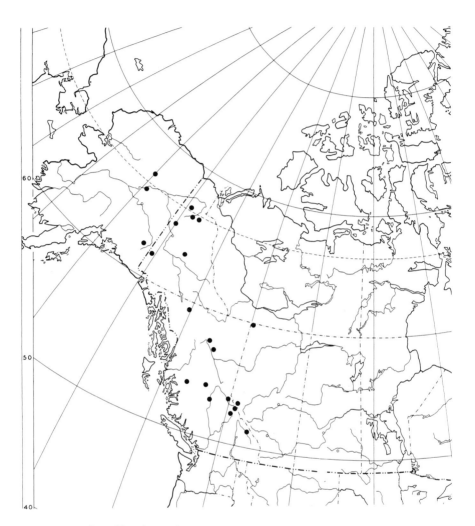

FIG. 2. *Distribution of* Arctostaphylos uva-ursi *ssp.*
stipitata *in northwestern North America.*

and more methylated aglycones. No significant systematic
differences in flavonoid content among different chromosome races
of the same species were noted.

Arnica cordifolia is a wide ranging, apomictic, polyploid
complex distributed form the Yukon Territory south to northern New
Mexico and central California, with disjunct populations in
Ontario and Michigan. Based on morphological, cytological,
ecological and geographical features it is considered the most
primitive species of *Austromontana* (Maguire 1943). Its
flavonoid profile consists of 10 compounds (7 glycosides and 3

TABLE III. *Flavonoid of* Arnica gracilis *and its putative parents.*

	cordifolia	gracilis	latifolia
Apigenin 7-OMe	+	-	-
Luteolin 4'-OMe	-	+	-
Luteolin 6-OMe	-	+	-
Luteolin 6-OMe, 7.0 glc	+	+	-
Luteolin 7.0 glc	+	+	-
Quercetin 3-OMe	+	+	-
Kaempferol 6-OMe, 30 glc	+	+	-
Quercetin 6-OMe	-	+	+
Quercetin 6-OMe, 3-0 glc	-	+	+
Kaempferol 3-0 gal	-	+	+

Ubiquitous: Apigenin, 6-OMe: Quercetin, 3-0 glc; 3-0 diglc; 3-0 gentiobioside; Kaempferol, 3-0 glc.

aglycones) and exhibits considerable inter-populational variation (Wolf and Denford 1983a). Quercetin 3-0-glucoside, quercetin 3-0-gentiobioside, quercetin 3-0-diglucoside, apigenin 6-methyl ether, and kaempferol 6-methoxy-7-0-glucoside, probably represent the ancestral flavonoid profile of *A. cordifolia*. Kaempferol 3-0-glucoside is generally lacking from most northern populations. Except for a few populations in southern Alberta, luteolin 7-0-glucoside and luteolin 6-methyl ether-7-0-glucoside are restricted to populations which occur near or north of the limits of maximum Pleistocene glaciation (Fig. 3).

Arnica latifolia is another wide ranging species, largely diploid, distributed from Alaska through Colorado and northern California. Its flavonoid profile consists of 8 compounds (2 aglycones and 6 glycosides), with the number of compounds per population ranging from a low of three to a high of seven (Wolf 1981). It differs from that of *A. cordifolia* largely by the presence of kaempferol 3-0-galactoside and the replacement of querectin 6-methoxy-3-0-glucoside by kaempferol 6-methoxy-3-0-glucoside throughout its distribution range (Table IV).

Arnica gracilis (distribution, Fig. 4). Straley (1980) suggested that this species might be a hybrid between *A. latifolia* and some other species, and indeed chemically *Arnica gracilis* appears to be a hybrid between *A. latifolia* and *A. cordifolia* (Wolf and Denford 1983a). As Harborne *et al.* (1975) noted, the flavonoid profiles of hybrids are frequently a summation of the two parental flavonoid profiles; however, novel flavonoids, i.e., those present in neither of the parents, occasionally appear in the hybrids (cf. Levy and Levin 1975). A total of 15 flavonoids have been characterized in *A. gracilis* (5 aglycones and 10 glycosides). Of the 15 compounds found in *A. gracilis,* 5 occur in both *A. latifolia* and *A. cordifolia*

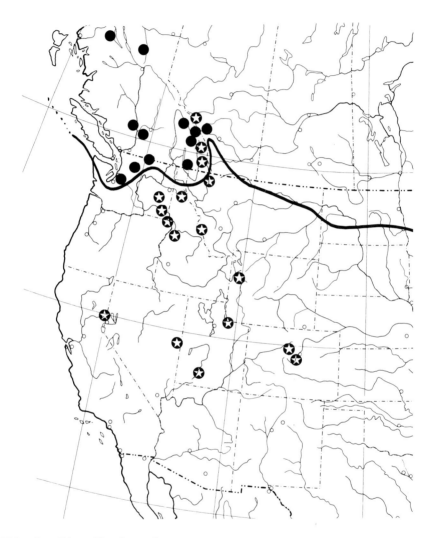

FIG. 3. *Distribution of* Arnica cordifolia *"chemotypes"*
with respect to pleistocene glacial limits. Solid line demarks
limit of glaciation. Solid circles, populations with L6Me7Glc.
Starred circles, populations lacking L6Me7Glc.

(apigenin 6-methyl ether, quercetin 6-methyl ether, kaempferol 3-0-
glucoside, quercetin 3-0-glucoside, viscosin and quercetin
gentiobioside), 3 are found in *A. latifolia* (quercetin 6-methyl
ether, quercetin 6-methyl ether-3-0-glucoside and kaempferol 3-0-
galactoside), 3 are found in *A. cordifolia* (luteolin 7-0-

TABLE IV. *Flavonoid of Arnica subgenus Austromontana.*

Compound	cord.	cern.	dis.	grac.	lati.	neva.	spath.	veno.	visc.
A 6-Me	+		+	+	+	+	+.	+	+
A7-Me	+	+	+			+	+	+	+
L 6-Me				+		+			
L 4'-Me				+		+			+
L 6OH, 4'-Me									+
L 6,-4'-di Me			+				+		+
L 3',6,7,-tri Me									+
Quercetin 3-Me	+			+					
Q 6-Me	+			+	+				
Q 3',6-di Me				+					+
A 7-0-glc	+		+	+					
L 7-0-glc	+		+	+					
L 6-Me, 7-0glc	+		+	+			+	+	+
K 3-0-glc	+	+	+	+	+		+	+	+
K 3-0-gal			+	+	+				
K 6-Me, 3-0-glc	+		+	+	+	+	+	+	+
Q 3-0-glc	+		+	+	+	+	+	+	+
Viscosin	+	+	+	+	+	+	+	+	+
Q-gentiobioside	+	+	+	+	+	+	+	+	+
Q 6-Me, 3-0-glc			+		+				+
A. viscosa #10									+
A. viscosa #11									+

FIG. 4. *Distribution of* Arnica gracilis *in northwestern North America.*

glucoside, luteolin 6-methyl ether-7-O-glucoside and quercetin 3-methyl ether) and the 3 novel compounds (luteolin 6-methyl ether, luteolin 4'-methyl ether and apigenin 7-O-glucoside) occur in neither of these two species (Table III). The compound kaempferol 3-O-galactoside, generally restricted to *A. latifolia,* and quercetin 6-methyl ether and its 3-O-glucoside establishes a clear relationship between *A. gracilis* and *A. latifolia. A. gracilis* has at times been treated as a subspecies of *A. latifolia* (Straley 1980). The presence of luteolin 7-O-glucoside, luteolin 6-methyl ether-O-glucoside and quercetin 3-methyl ether clearly

establishes a relationship between *A. gracilis* and *A. cordifolia*.
The occurrence of luteolin 6-methyl ether (an aglycone) in *A.
gracilis* is not surprising since glycosides of these compounds
also occur in *A. cordifolia*. A close relative of *A. cordifolia*
with a more southern distribution is *Arnica discoidea* Benth. which
occurs in montane habitats in the Coast-Ranges of California, the
foothills of the Sierra Nevada and less sporadically in the
foothills of the Cascades northward to southern Washington (Fig.
5). Diploid, triploid and tetraploid chromosome races exist, the
diploids being essentially restricted to the Klamath region of
California and northeastern Oregon while the polyploids occur at
the northern, eastern and southern limits of its distribuiton.
The flavonoid profile of *A. discoidea*, (4 aglycones and 7
glycosides), is strikingly similar to that of *A. cordifolia*,
with the exception of luteolin 6,4'-di-methyl ether, present only
in Klamath populations of *A. discoidea*, and apigenin 6,7-di-
methyl ether. Three compounds have significant geographical
distributions, luteolin 7-O-glucoside and its 6-methyl ether
are present only in Klamath populations of *A. discoidea* and
apigenin 7-methyl ether, present in *A. discoidea*, occurs
only in a local population of *A. cordifolia*. The flavonoid
profiles of diploid Klamath populations of *A. discoidea*
exhibit a greater diversity and number of compounds and are most
similar to the profile of *A. cordifolia*; while in contrast,
the higher ploidy levels outside this area exhibit reduced
profiles, particularly with respect to glycosides. This suggests
that *A. discoidea* has been derived from ancient diploid
Klamath populations of *A. cordifolia* and that migration has
been accompanied by polyploidization and a reduction in flavonoid
content. Also it would appear that chemically *A. discoidea*
is an important "precursor" of Klamath endemics, especially
Arnica spathulata and *A venosa*.

Arnica spathulata Greene, a serpentine endemic (Fig. 5),
is largely diploid and has probably been derived from *A.
discoidea* via saltational speciation into serpentine areas.
Its flavonoid profile (3 aglycones and 5 glycosides), is a subset
of that of *A. discoidea*. Compounds per population vary
from 3 to 7, with rare tetraploid populations all being chemically
identical containing only the same three compounds. The presence
of luteolin 6,4-di-methyl ether, found in Klamath populations of
A. discoidea, and apigenin 7-methyl ether in both *A. spathulata*
and *A. discoidea* indicate the former was derived from the
latter. Subsequently, the ancestral flavonoid profile *A.
spathulata* has been considerably reduced in more recently
derived population, i.e., the tetraploids. *Arnica venosa*
H.M. Hall is another rare, Klamath diploid endemic probably
derived from *A. discoidea* (Fig. 5). It is geographically
restricted, occurring only in the hot, dry foothills of western
Shasta and eastern Trinity Counties, California. Its flavonoid
profile (2 aglycones and 4 glycosides), is merely a subset of the
profile of *A. discoidea*. Putative hybrids between *A. venosa*

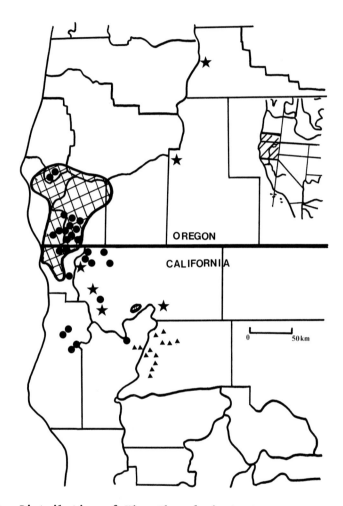

FIG. 5. *Distribution of Klamath endemic* Arnicas.
Crosshatched, A. spathulata; *Stars,* A. viscosa;
Circles, A. cernua; *Triangles,* A. venosa.

and *A. discoidea* are occasionally found in areas of sympatry,
also indicating a close relationship.

   Arnica cernua Howell, another diploid serpentine endemic
(Fig. 5), has a profile consisting of only kaempferol 3-0-
glucoside, quercetin 3-0-glucoside and quercetin 3-0-
gentiobioside, representing an extreme example of chemical
reduction in narrow edaphic endemics. It is hypothesized that
A. cernua was derived, via saltational speciation into

serpentine areas, from an ancient diploid population of *A. cordifolia* that may have already had a reduced flavonoid profile and/or its profile has subsequently been reduced during isolation on scattered serpentine areas. Such a reduced flavonoid profile, consisting of only two or three compounds, has previously been reported for a rare diploid population of *A. cordifolia* and other taxa, e.g. *Arctostaphylos* (Denford 1973), *Parthenium* (Mears 1979). *Arnica nevadensis* A. Gray is also closely related to *A. cordifolia*. It is a relatively uncommon, high montane species of the Sierra Nevada of California extending northward irregularly to the north Cascades and to the Olympic Mountains of Washington. Predominantly an apomictic tetraploid, it bears considerable morphological resemblance to *A. cordifolia*. Its profile consisting of six compounds including two quercetin glycosides characteristic of the subgenus, and the methylated aglycones, luteolin 4' methyl ether and luteolin 6-methyl ether, unique to *A. nevadensis*.

The rarest species in the subgenus *Austromontana* is *Arnica viscosa* A. Gray, a diploid, known from only seven localities, being restricted to volcanic soils at high elevations, largely in the Klamath region (Fig. 5). Its flavonoid profile, consists of 14 compounds (7 aglycones and 7 glycosides), including several unique compounds. It has an abundance of highly methylated flavones and a 6-hydroxylated flavones, both of which are considered advanced features (Mears 1979), and is probably the most recently derived species of *Austromontana* being morphologically and ecologically the most advanced species of the subgenus, occurring in volcanic habitats less than 14,000 years old. Although its derivation is uncertain, *A. viscosa* shares several morphological features with *A. latifolia* including: very narrow heads with lanceolate phyllaries, sessile leaves, the high altitude habitat and diploid chromosome number. Additionally, quercetin 6-methoxy-3-O-glucoside, which distinguishes *A. latifolia* from *A. cordifolia* and its derivatives, provides a further link between *A. viscosa* and *A. latifolia*, with which it successfully hybridizes (Straley 1980).

The above examples generally support Mears' (1979) hypothesis concerning the flavonoid chemistry of narrow endemics and their more widespread congeners. Within subgenus *Austromontana* the widespread species are characterized by a higher flavonoid diversity composed of a variety of glycosides and few methylated aglycones; while, in contrast, the derived narrow endemics are generally characterized by reduced flavonoid profiles and/or are composed of more methylated aglycones and fewer glycosides (Table IV & V). The flavonoid chemistry of *Austromontana* indicates that major chemical diversification within the subgenus has taken place within the Klamath region, as well as north of the glacial limits in the case of *A. cordifolia* which, with its primitive morphology and ecology, very diverse cytology, wide geographical distribution, and relatively primitive flavonoid profile is considered an ancestral species of the subgenus.

Furthermore, chemical evidence supports the concept that *A.*
cordifolia and *A. latifolia* (or their precursors) have hybridized
giving rise to *A. gracilis* whose distribution (Fig. 4) appears
to be dictated by post glacial boundaries. The diverse chemical
nature of this taxon, and also the "young" endemic *A. viscosa*
indicate a method for detecting ancient versus recently evolved
elements in a flora. This in combination with Mears' (1979)
observations relating to methylation patterns, indicates that
"new" elements have a much more diverse chemistry than their
mature relatives, a subsequence of recombination effects during
hybridization and early polyploidization events, as seen in
Phlox (Levy and Levin 1975).

REFERENCES

Bell, E.A. and Charlwood, B.V. 1980. *Secondary Plant*
    Products. Springer-Verlag, New York.
Bhutani, S.P., Chibber, S.S. and Seshadri, T.R. 1969. Flavonoids
    of the fruits and leaves of *Tribulus terrestris*:
    constitution of tribuloside. *Phytochemistry* 8: 299-303.
Buttery, B.R. and Buzzell, R.I. 1975. Soybean flavonol
    glycosides: identification and biochemical genetics. *Can.*
    J. Bot. 53: 219-224.
Collins, F.W., Bohm, B.A. and Wilkins, C.K. 1975. Flavonol
    glycoside gallates from *Tellima grandiflora*. *Phytochemistry*
    14: 1099-1102.
Crawford, D.J. 1970. Morphology, flavonoid chemistry, and
    chromosome number of the *Chenopodium neomexicanum* complex.
    Madrono 22: 185-194.
Crawford, D.J. 1978. Flavonoid chemistry and angiosperm
    evolution. *Bot. Rev.* 44: 431-456.
Denford, K.E. 1973. Flavonoids of *Arctostaphylos uva-ursi*
    (Ericaceae). *Experientia* 29: 939.
Denford, K.E. 1981. Chemical subdivisions within the genus
    Arctostaphylos based on flavonoid profiles. *Experientia*
    37: 1287-1288.
Denton, M.F. 1979. Factors contributing to evolutionary
    divergence and endemism in *Sedum* section *Gormania*
    (Crassulaceae). *Taxon* 28: 149-155.
Elisens, W.J. and Denford, K.E. 1982. Flavonoid studies in four
    species of the *Oxytropis campestris* complex (Fabaceae-
    Galegeae). *Can. J. Bot.* 60: 1431-1436.
Geissman, T.A. 1962. *The chemistry of flavonoid compounds*.
    Pergamon, Oxford.
Gottlieb, O.R. 1975. Flavonols. In *The Flavonoids* (J.B.
    Harborne, T.J. Mabry and H. Mabry, eds.). Chapman and Hall,
    London. pp. 296-375.
Hahlbrock, K. and Grisebach, H. 1975. Biosynthesis of Flavonoids.
    In *The Flavonoids* (J.B. Harborne, T.J. Mabry and H. Mabry,
    eds.). Chapman and Hall, London. pp. 866-915.
Harborne, J.B. 1967. Flavonoid patterns in the Bignoniaceae and

the Gesneriaceae. *Phytochemistry* 6: 1643–1651.

Harborne, J.B. 1973. *Phytochemical Methods*. Chapman and Hall, London.

Harborne, J.B. 1977. Flavonoid sulphates – a new class of natural product of ecological significance in plants. *Prog. Phytochem.* 4: 189–208.

Harborne, J.B. 1978. The rare flavone isoetin as a yellow flower pigment in *Heywoodiella oligocephala* and in other Cichorieae. *Phytochemistry* 17: 915–917.

Harborne, J.B. and Mabry, T.J. 1982. *The Flavonoids: Advances in Research*. Chapman and Hall, London.

Harborne, J.B., Mabry, T.J. and Mabry, H. 1975. *The Flavonoids*. Chapman and Hall, London.

Harborne, J.B. and Williams, C.A. 1976. Sulphated flavones and caffeic acid esters in members of the Fluviales. *Biochem. Syst. Ecol.* 4: 37–41.

Harborne, J.B. and Williams, C.A. 1982. Flavone and flavonol glycosides. In *The Flavonoids: Advances in Research* (J.B. Harborne, T.J. Mabry, eds.). Chapman and Hall, London.

Levy, M. 1983. Flavone variation and subspecific divergence in *Phlox pilosa* (Polemoniaceae). *Syst. Bot.* 8: 118–126.

Levy, M. and Fujii, K. 1978. Geographic variation of flavonoids in *Phlox carolina*. *Biochem. Syst. Ecol.* 6: 117–125.

Levy, M. and Levin, D.A. 1971. The origin of novel flavonoids in *Phlox* Allotetraploids. *Proc. Natl. Acad. Sci. U.S.A.* 68: 1627–1630.

Levy, M. and Levin, D.A. 1975. The novel flavonoid chemistry and phylogenetic origin of *Phlox floridiana*. *Evolution* 29: 487–499.

Mabry, T.J., Markham, K.R. and Thomas, M.B. 1970. *The Systematic Identification of Flavonoids*. Springer-Verlag, N.Y.

Maguire, B. 1943. A monograph of the genus *Arnica*. *Brittonia* 4: 386–510.

Markham, K.R. and Porter, L.J. 1975. Isoscutellarein and hypolaetin 8-glucuronides from the liverwort *Marchantia berteroana*. *Phytochemistry* 14: 1093–1097.

Mears, J.A. 1979. Chemistry of polyploids: A summary with comments on *Parthenium* (Asteraceae-Ambrosiinae), In *Polyploidy*, W.H. Lewis, ed., Plenum, New York.

Mues, R., Timmerman, B., Ohno, N. and Mabry, T.J. 1979. 6-methoxy flavonoids from *Brickellia californica*. *Phytochemistry* 18: 1379–1383.

Murray, B.G. and Williams, C.A. 1973. Polyploidy and flavonoid synthesis in *Briza media* L. *Nature* 243: 87–88.

Packer, J.G. and Denford, K.E. 1974. A contribution to the taxonomy of *Arctostaphylos uva-ursi*. *Can. J. Bot.* 52: 743–753.

Pereyra de Santiago, O.J. and Juliani, H.R. 1972. Isolation of quercetin 3, 7, 3', 4' – tetrasulphate from *Flaveria bidentis* L. Otto Kuntze. *Experientia* 28: 380–381.

Pratviel-Sosa, F. and Percheron, F. 1972. Les sophorosides de flavonols de quelque pollens. *Phytochemistry* 11: 1809–1813.

Raven, P.R. and Axelrod, D.I. 1978. *Origin and Relationships of the California Flora*. Univ. Calif. Press, Berkeley.

Ribéreau-Gayon, P. 1972. *Plant Phenolics*. Oliver and Boyd, Edinburgh.

Seshadri, T.R. and Vydeeswaran, S. 1972. Chrysoeriol glycosides and other flavonoids of *Rungia repens* flowers. *Phytochemistry* 11: 803-806.

Smith, P.M. 1976. *The chemotaxonomy of Plants*. Edward Arnold, London.

Soltis, D.E., Bohm, B.A. and Nesom, G.L. 1983. Flavonoid chemistry of cytotypes in *Galax* (Diapensiacea). *Syst. Bot.* 8: 15-23.

Straley, G.B. 1980. Systematics of *Arnica*, subgenus *Austromontana* and a new subgenus *Calarnica* (Asteraceae-Senecioneae). Ph. D. Diss. Univ. British Columbia, Vancouver. pp. 288.

Swain, T. and Bate-Smith, E.C. 1962. Flavonoid compounds. In *Comparative Biochemistry*. (M. Florkin, and H.S. Mason, eds.). Academic Press, New York. 3: 755-809.

Thompson, W.R., Meinwald, J., Aneshansley, D. and Eisner, T. 1972. Flavonols: Pigments responsible for ultraviolet absorption in nectar guides of flowers. *Science* 177: 528-530.

van Emden, H.F. 1973. *Insect-Plant Relationships*. Blackwell, Oxford.

Wallace, J.W., Pozner, R.S. and Gomez, L.L. 1983. A phytochemical approach to the Gleicheniaceae. *Am. J. Bot.* 70: 207-211.

Whalen, M.D. 1978. Foliar flavonoids of *Solanum* section *Androceras*: a systematic survey. *Syst. Bot.* 3: 257-276.

Whittaker, R.H. 1961. Vegetation history of the Pacific coast states and the "central" significance of the Klamath region. *Madrono* 16: 5-23.

Wilkins, C.K. and Bohm, B.A. 1976. Chemotaxonomic studies in the Saxifragaceae s.l. 4. The flavonoids of *Heuchera micrantha* var. *diversifolia*. *Can. J. Bot.* 54: 2133-2140.

Williams, C.A., Harborne, J.B. and Clifford, H.T. 1971. Flavonoid patterns in the Monocotyledons. Flavonols and flavones in some families associated with the Poaceae. *Phytochemistry* 10: 1059-1063.

Wolf, S.J. 1981. Biosystematics of *Arnica* subgenus *Austromontana*. Ph.D. Diss. Univ. of Alberta, Edmonton. pp. 284.

Wolf, S.J. and Denford, K.E. 1983a. Flavonoid variation in *Arnica cordifolia*: an apomictic polyploid complex. *Biochem. Syst. Ecol.* 11: 111-114.

Wolf, S.J. and Denford, K.E. 1983b. Flavonoids of *Arnica gracilis*, a natural hybrid. Syst. Bot. In press.

Wolf, S.J., Denford, K.E. and Packer, J.G. 1979. A study of the flavonoids in the *Minuartia rossii* complex. *Can. J. Bot.* 57: 2374-2377.

Pollen Morphology and Biosystematics of the Subfamily Papilionoideae (Leguminosae)

I. K. Ferguson
The Royal Botanic Gardens
Richmond, Surrey, England

INTRODUCTION

Recent pollen morphological studies have contributed a great deal
to the understanding of the biosystematics of the subfamily
Papilionoideae (see Polhill and Raven 1981). Preliminary results
of a large survey of the subfamily using techniques of light
microscopy (LM) together with scanning (SEM) and transmission
electron microscopy (TEM) are reported by Ferguson and Skvarla
(1981) where the main literature is reviewed. More detailed
recent investigations are reported in Ferguson (1978, 1981),
Ferguson and Skvarla (1979, 1981, 1982, 1983), Ferguson and
Strachan (1982), Graham and Tomb (1974, 1977), Horvat and Stainier
(1979, 1980), Kavanagh and Ferguson (1981), Maréchal *et al.*
(1978), Poole (1979) and Stainier and Horvat (1978, 1983).

   The purpose of this paper is to try and summarise the pollen
morphology or the subfamily discussing exine architecture in
relation to classification, to pollination and briefly to
harmomegathy (water relations and volume change).

POLLEN MORPHOLOGICAL VARIATION AT THE TRIBAL LEVEL

The basic pollen type shown by many of the tribes and widespread
throughout the Leguminosae (Guinet 1981) is a spheroidal,
tricolporate, finely reticulate pollen grain of about 30 um in
diameter with typical angiosperm exine stratification consisting,
in the mesocolpial zone of a well defined endexine layer about
equal in thickness to the foot layer with a columellate
interstitium where the columellae are well spaced and equalling or
slightly exceeding in height the combined thickness of endexine
and foot layer (nexine). There is a distinct tectum, the
thickness being less than half the height of the columellae.

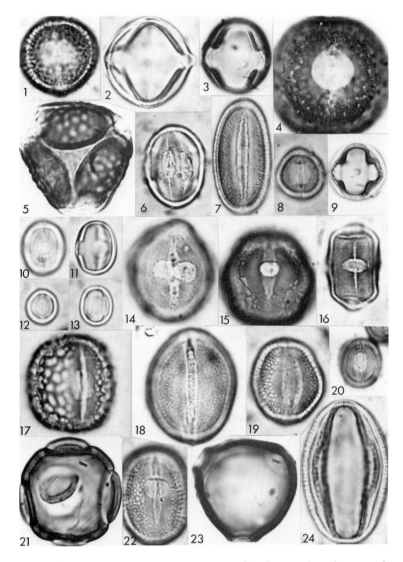

PLATE I. 1. Castanospermum australe *(Sophoreae) colpus and
endoaperture in surface view.* 2. Ateleia arsenii *(Sophoreae)
optical section showing endoapertures as interruptions in the
wall.* 3. Tipuana speciosa *(Dalbergieae) as 2.* 4. Afgekia
sericea *(Tephrosieae) large circular endoaperture in surface
view.* 5. Millettia australis *(Tephrosieae) polar view,
syncolpate (colpi fusing on pole).* 6. Baphiopsis parviflora
*(Swartzieae) 6-colporate, 2 colpi with lolongate endoapertures
visible.* 7. Dalea obovata *(Amorpheae) operculate colpus in*

However, there is considerable variation in the pollen morphology within the subfamily. Some groups have superficially very similar pollen but have remarkable differences in exine stratification seen most easily in TEM.

Pollen characters differ in significance in distinguishing taxa and in suggesting relationships from tribe to species. Apertures (Plate I) and exine stratification (Plate II) are conservative characters consistently of the greatest value in tribal classification while size, shape (Plate I) and exine ornamentation (Plates III and IV) are found to be characters of secondary importance of value at the generic and specific level. The trends in pollen characters are summarized in Table I.

The tricolporate aperture with an equatorial, usually small, circular or lalongate thinning of the endexine is the common type e.g. Plate I, 1-4, 10, 11, 16, 17, 22). There are various trends from this, towards increase in aperture number; frequently and widespread is an increase to 4 colpi while 5-7 colpi also occur (Plate I, 6). Pantoporate pollen (Plate I, 21) occurs in just two small New World genera, *Brya* and *Cranocarpus* comprising a separate subtribe of Desmodieae and reflecting a highly unusual macromorphology. The pollen of a number of groups shows a loss of lateral thinning in the endexine around the endoaperture resulting

---

surface view. 8. Amorpha canescens *(Amorpheae) large somewhat diffuse endoaperture underlying operculate colpus. 9.* Bryaspis lupulina *(Aeschynomeneae) as 3. 10-11.* Diphysa racemosa *(Robinieae). 10. surface view of colpus and circular endoaperture. 11. Endoaperture in optical section. 12-13.* Muelleranthus trifoliatus *(Bossiaeeae). 12. Surface view showing lolongate endoaperture underlying operculate colpus. 13. Endoaperture in optical section. 14.* Mecopus nidulans *(Desmodieae) surface view showing colpus and lalongate endoaperture. 15.* Psoralea macrostachya *(Psoraleeae) surface view colpus with endoaperture concentric with ectopore ("tricolpororate"). 16.* Lotus collinus *(Loteae) surface view of colus and lalongate endoaperture. 17.* Desmodium lespedesioides *(Desmodieae) surface view showing colpus and lolongate endoaperture. 18.* Chapmannia floridiana *(Aeschynomeneae) surface view of operculate colpus. 19.* Aeschynomene schimperi *(Aeschynomeneae) as 18. 20.* Adesmia boronioides *(Adesmieae) surface view of operculate colpus with underlying lolongate endoaperture. 21.* Cranocarpus mezii *(Desmodieae) pantoporate operculate. 22.* Poiretia latifolia *(Aeschynomeneae) surface view showing operculate colpus overlying circular endoaperture. 23.* Erythrina variegata *(Phaseoleae) polar view, optical section triporate. 24.* Amicia zygomeris *(Aeschynomeneae) optical section showing short operculum on colpus. All equatorial views except 5 and 23 and all X 850 .*

PLATE II. 1. Swartzia leptopetala *(Swartzieae)* *showing*
endexine, footlayer and complex granular interstitium, X10,625.
2. Psorothamnus schottii *(Amorpheae)* *colpus area with*
operculum (arrow); very thin endexine thickening and becoming
lamellated under colpus, distinct footlayer, simple columellae, X
7650. 3. Geissaspis cristata *(Aeschynomeneae)* *showing no*

in the equatorial endoaperture being concentric with an equatorial
ectopore (Plate I, 15). This type of simple endoaperture is
referred to here for convenience as "tricolpororate" or
"tripororate" as appropriate and occurs throughout the tribe
Psoraleeae and in some Phaseoleae.

A third trend in aperture structure is the loss of a defined
endoaperture with thinnings in the endexine lengthening and
becoming diffuse. Associated with this is the occurrence of an
operculum covering the colpus (Plate I, 7, 18, 19; Plate II, 2;
Plate III, 3, 4). This structure is referred to as "tricolpate",
though terminologically this is not strictly so. Various stages
of this condition occur where the operculum is present over a
well-defined endoaperture as, for example, in Adesmieae (Plate I,
8, 20).

Another trend is the shortening of the colpi towards pores.
Triporate pollen is frequent in Phaseoleae. Syncolpi, with colpi
fused on the poles of the pollen grain, occurs in various species
and genera throughout many of the tribes but is generally regarded
as a derived state (Plate I, 5).

The exine stratification seen in TEM shows two distinct trends
from the typical angiosperm condition. It is emphasised that
these, very like the situation for aperture structure, are gradual
and there are few marked discontinuities. The first trend is the
reduction or almost complete disappearance of the endexine layer
except in the aperture region, together with a thin foot layer
(Plate II, 3). The second trend is a thickening of the endexine
with reduction or loss of the foot layer. Associated with this
second change is a marked increase in complexity of the ektexine
(Plate II, 4-6). There is a change to a granular interstitium
(see Ferguson and Skvarla 1983) with almost complete loss of the
ektexine in some groups and in *Macrotyloma* (tribe Phaseoleae)
supratectal spines occur (Plate II, 5).

These trends in the aperture structure and exine stratification
correspond with each other and with other morphological
characters. Stainier and Horvat (1983) show very elegantly the
relationships between aperture structure and exine stratification
in the *Phaseolus-Vigna* complex. In the subfamily as a whole the
thinning of the endexine is associated with the diffuse
"tricolpate", operculate aperture and occurs in the predominantly
New World tribes Aeschynomeneae, Amorpheae and Adesmieae. There

---

endexine, very thin footlayer, branched columellae, X 8245. 4.
Psoralea pubescens *(Psoraleeae) thick endexine, no footlayer,
interstitium granular,* X 7480. 5. Macrotyloma axillaris
*(Phaseoleae) thick endexine, no footlayer, very short dense
columellae, thin tectum and supratectal spines,* X 3230. 6.
Erythrina coralloides *(Phaseoleae) thick endexine ± no
footlayer, somewhat granular/columellate interstitium and thick
tectum,* X 10,200. *All TEMs and all from mesocolpial area except 2.*

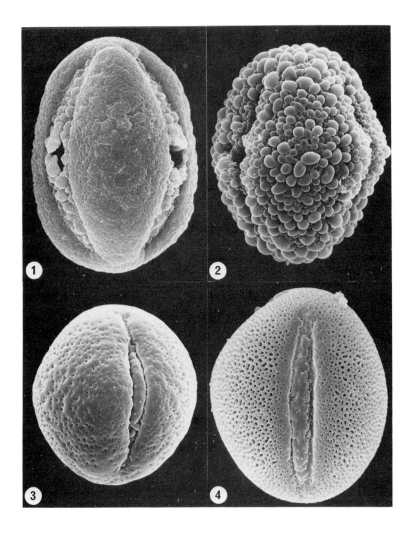

PLATE III. 1. Camoensia brevicalyx *pollen with finely
perforate ornamentation,* X 1700 *(probably insect pollinated).* 2.
Camoensia scandens *pollen with coarsely verrucate ornamentation,*
X 1700 *(probably bat pollinated).* 3. Templetonia egena *pollen
showing operculum covering colpus and finely perforate ornamentation,* X 2550 *(insect pollinated).* 4. Templetonia retusa
*pollen showing operculum covering colpus and finely reticulate
ornamentation,* X 2125 *(bird pollinated). (The differences in
pollen ornamentation shown in 3 and 4 are not correlated with
pollinator). All SEM micrographs.*

is not a very great increase in complexity in the exine stratification, the most extreme so far detected is the somewhat branched
columellae in *Geissaspis* (Plate III, 3). The thickening of the
endexine, often associated with a complexity in the stratification
of the ektexine is correlated with trends towards increase in
aperture number, porate, "pororate" and "tricolpororate"
apertures, and occurs in the more marked Old World tribes
Phaseoleae, Desmodieae, Psoraleeae and Indigofereae.

Polhill (1981) has incorporated some of this pollen data into
his diagramatic representations of the supposed relationships of
the tribes. Fig. 1 shows the results of a preliminary attempt to
analyse the pollen data for some of the tribes cladistically. The
New World tribes Aeschynomeneae, Amorpheae and Adesmieae are
compared with the generally Old World tribes Phaseoleae and
Psoraleeae as outgroups to one another. As discussed earlier it
is assumed that tricolporate pollen, a reticulate tectum, and more
or less equal representation of endexine, foot layer, columellate
interstitium and tectum are regarded as least derived. Thickening
and thinning of the endexine in relation to the foot layer is
regarded as a derived character. The operculate aperture groups
the larger part of the tribe Aeschynomeneae with Adesmieae and
Daleae genera of the tribe Amorpheae (Barneby 1977) as well as the
genus *Amorpha*. The genera *Apoplanesia, Eysenhardtia* and
Errazurizia, but excluding *E. rotundata* which from pollen
morphology at least appears to be misplaced, are clearly separable
from the rest of the tribe and this corresponds fairly closely
with Barneby's (1977) diagrammatic representations of the
relationships between genera in the Amorpheae. There are a number
of genera with relatively underived pollen in the grouping
Aeschynomeneae "A" and these are not demonstrably monophyletic and
supports the statement by Rudd (1981) that "the tribe has been
variously circumscribed but no arrangement can be fully defended
because of lack of firm, basic data." The loss of the foot layer
divides the tribe Phaseoleae and these groups can be analysed
further as with Aeschynomeneae but it is doubtful whether the
result at least with present knowledge is phyletically meaningful.
For example, the tribe Psoraleeae can be separated from
Pachyrhizus and *Calopogonium* by its granular interstitium
and usually coarsely reticulate exine ornamentation. It is noteworthy that Psoraleeae is one of the few tribes which shows a
complete discontinuity in pollen morphology separating it from all
other tribes and especially from Amorpheae where it had been
placed in earlier classifications. Likewise the genera
Erythrina and *Vigna* can be separated by exine stratification.
The pollen of *Macrotyloma* demonstrates the problem of
analysis of this type. The exine stratification is uniform and
remarkable throughout all the 24 species (Ferguson 1981) (Plate
II, 5) but 3 species have pollen with "tricolpororate" apertures
while 21 species have pollen with "tripororate" apertures. The
question arises whether the porate, the colpororate or the
increase in aperture number is the more derived or whether all the

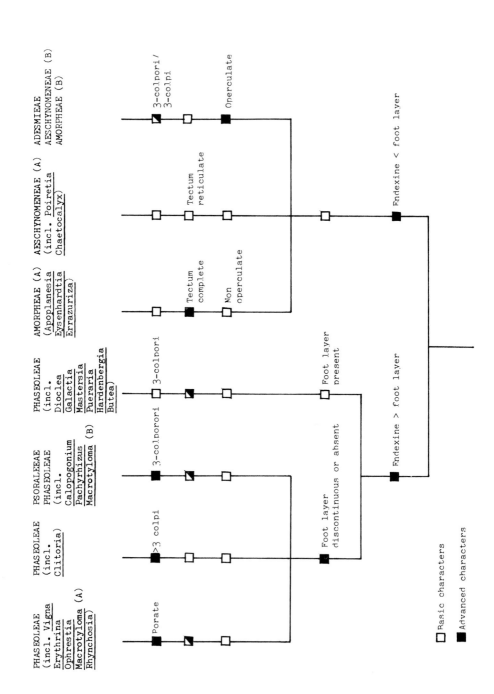

conditions have been derived independently.

These groupings in some ways amplify the pollen data more clearly than Polhill's (1981) scheme. However, the problem is posed that the tribes Indigofereae and Desmodieae do not fit clearly into this cladogram. This suggests that either the cladogram is wrong or that these tribes may not be as closely related to Phaseoleae and Psoraleeae as believed. There are a number of characters for example chromosome number and canavanine which appear to closely link Desmodieae and Indigofereae with Phaseoleae and Psoraleeae and combined cladograms incorporating other cryptic characters may alter the relationships. On the other hand Desmodieae and Indigofereae have an explosive pollination mechanism in contrast with the Phaseoleae and Psoraleeae and it can be postulated that pollen morphology has been evolved in a secondary adaptive role.

The problems of constructing a cladistic analysis such as this lies very much in the gradation of pollen characters in many of the tribes and so clearly seen in the genus *Indigofera* (Ferguson and Strachan, 1982).

POLLEN MORPHOLOGY AT THE GENERIC/SPECIFIC LEVEL

New data from examination of the pollen of further taxa since the review of Ferguson and Skvarla (1981) throws more light on the arrangement and relationships of the tribes but particularly on the position of anomolous species and genera. Considerable diversity occurs in the Swartzieae as well as in the Sophoreae (Ferguson and Skvarla 1981). But a more detailed survey of the Swartzieae (Ferguson and Skvarla in prep.) shows that there is a bigger range of exine stratification types in the genus *Swartzia* than was at first suspected, some having very complex walls (Plate II, 1) while simple walls occur less frequently but especially in the two African species *S. madagascariensis* and *S. fistuloides* which also have wide colpi without an ektexinous aperture membrane, and have a difference in chromosome number (Cowan 1981) suggesting that these species represent a distinct genus. A fuller analysis may reveal a gradation in complexity of exine stratification similar to that found by Ferguson and Strachan (1982) in *Indigofera*. *Aldina* is the only other genus in the Swartzieae where the pollen has a complex exine stratification, this supporting Cowan (1981) who proposes a close association with *Swartzia*.

In the Sophoreae complex exine stratification has been found in the Asiatic *Ormosia henryii* and a detailed study of pollen of the New World and Asiatic species could be interesting. Careful study of the pollen of *Inocarpus* shows it to be rather unspecialised, tricolporate with simple exine stratification, and this character

Fig. 1. *Cladogram of five tribes from data summarised in Table I.*

TABLE I. *Summarizing trends in pollen characters in Papilionoideae (Basic states given first)*

POLLEN CHARACTERS	SOME TRIBES WHERE BASIC AND ADVANCED STATES OCCUR
Shape	
$\pm$ equiaxy (spheroidal)/breviaxy	Desmodieae, Phaseoleae
$\pm$ " " /longiaxy	Amorpheae, Aeschynomeneae
Apertures	
3 equatorial apertures/more than 3	Swartzieae, Sophoreae, Phaseoleae, Desmodieae, Loteae.
colporate/colporoidate/colpate	Aeschynomeneae, Amorpheae, Hedysareae
colporate/shortly colporate/porate	Desmodieae, Phaseoleae
colporate/colpororate	Psoraleeae, Phaseoleae
emarginate/marginate	Aeschynomeneae, Phaseoleae, Psoraleeae
no syncolpy/syncolpy	Aeschynomeneae, Phaseoleae, Tephrosieae
operculum absent/present	Aeschynomeneae, Amorpheae, Adesmieae, Mirbelieae, Bossiaeeae, Desmodieae
" " /pseudo-operculum	Phaseoleae
Exine ornamentation	
tectal/supratectal	Phaseoleae
reticulate/verrucate	Sophoreae, Tephrosieae, Brongniartieae
reticulate/smooth	Phaseoleae, Amorpheae, Desmodieae
reticulate/areolate-rugulate	Desmodieae, Phaseoleae, Sophoreae, Tephrosieae
Exine stratification	
columellate/granular	Loteae, Swartzieae, Phaseoleae, Desmodieae, Psoraleeae, Vicieae, Indigoferae
tectum infratectum/thick tectum	Tephrosieae, Phaseoleae, Indigoferae, Desmodieae
tectum without structure/with structure	Sophoreae, Indigoferae, Swartzieae
footlayer: medium/reduced -absent	Phaseoleae, Psoraleeae, Loteae, Desmodieae, Tephrosieae
footlayer: medium/very thick	Sophoreae
endexine: medium/reduced - absent	Aeschynomeneae, Amorpheae, Genisteae, Adesmieae
endexine: medium/very thick	Phaseoleae, Psoraleeae, Desmodieae, Indigoferae, Tephrosieae

throws no further light on the position of the genus. The
operculate pollen of *Belairea* lends support to its new position
among the less derived genera of Aeschynomeneae (see Rudd 1981).
    Four species of *Vicia* (Vicieae), *V. faba*, *V. cracca*, *V.
peregrina* and *V. narbonensis*, have a granular interstitium while
Lathyrus vernus and *Lathyrus latifolius*, together with the genera
Pisum and *Lens*, have a columellate interstitium. This
potentially discriminating character between the closely related
Vicia and *Lathyrus* is worthy of further investigation. Within
the genera differences in exine ornamentation may be useful for
distinguishing species or species groups.
    There is a need for considerably more work at the specific and
generic level throughout much of the subfamily.

POLLEN AND POLLINATION

In the course of the survey of the pollen morphology of the
Papilionoideae certain pollen differences were found in what
appeared to be comparatively closely related genera and species.
The opposite situation also arises where similarities in pollen
exine ornamentation and stratification occurred in otherwise
unrelated groups. These have been assumed to be adaptive
characters associated with "bird flowers" (Ferguson and Skvarla,
1982). Flowers associated with bird or bat pollination are
usually very much larger and red or, in the case of bats, white or
greenish-white in color, while bee-pollinated flowers are smaller
and variously colored. "Bird flowers" together with the flowers
of closely related species pollinated by bees from three widely
separated tribes are shown in Fig. 2. The pollen structures found
to be associated with bird pollination fall into two types: (1)
verrucate ornamentation and (2) complex exine stratification. The
latter occurs in two unrelated genera in the tribe Sophoreae, the
New World *Alexa* and the Old World monotypic *Castanospermum*. The
complex exine stratification in *Swartzia* parallels that of
Alexa and *Castanospermum* suggesting the pollination biology of
Swartzia's remarkable flower structure requires investigation.
Until something is known of this, the proposed correlation between
this type of complex exine stratification and ornithophily is
perhaps questionable.
    The pollen of the West African genus *Camoensia* (Sophoreae)
(Plate III, 1, 2) illustrates verrucate exine ornamentation. It
is assumed to be associated with bat pollination but there is no
published data. This type of ornamentation difference occurs in
numbers of genera from widely separated tribes and also from the
subfamily Caesalpinioideae (Graham and Barker 1981) where there is
well documented pollination data. The Australian genus
Templetonia (Bossiaeeae) with some "species has two, *T. retusa*
and *T. incana* with large red "bird flowers" (Fig. 2, No. 5),
while the other nine species in the genus have small yellow or
brown and yellow flowers resembling *T. egena* (see Fig. 2, No. 6).
The pollen of these two species differs (Plate III, 3, 4) but two

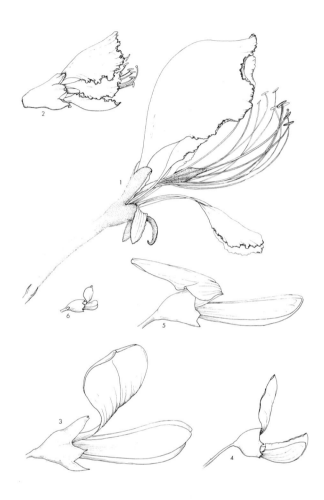

Fig. 2. *Flowers of three pairs of species from three genera.*
One species of each pair is a "bird/bat flower" and the other an
"insect flower". 1. Camoensia scandens *(Lewis, photo 82)*
(probably bat pollinated) flowers white. 2. Camoensia brevicalyx
(Keay, s.n.) (probably insect pollinated) flowers purple. 3.
Millettia theuszii *(Thoret, 91) (probably bird pollinated) flowers*
red. 4. Millettia usaramensis *(Procter, 2664) (insect*
pollinated) flowers purple. 5. Templetonia retusa *(Donner,*
2710) (bird pollinated) flowers red. 6. Templetonia egena
(Wilson, 278) (insect pollinated) flowers purple/yellow. Figs.
1-2 X.45, 3-6 X 1.35; 1-2 Tribe Sophoreae, 3-4 Tribe Tephrosieae,
5-6 Tribe Bossiaeeae.

pollen types occur in the other insect-pollinated species. For
example, *T. sulcata* has pollen with a finely perforate tectum like
T. egena, while *T. aculeata*, *T. hookeri* and *T. stenophylla*
have pollen resembling *T. retusa*. This suggests that the pollen
difference rather than having a supposedly simple functional
relationship may well have some taxonomic basis, thus supporting
Polhill's (1981) and Ross's (1982) indication that this is part of
a rather complex group. Ross (1982) in a revision of the genus,
groups *T. sulcata* and *T. egena* together and somewhat apart from
the other species.

A reevaluation of the pollen morphology of the well documented
bird pollinated genus *Erythrina* is currently being carried out
(Hemsley and Ferguson in prep.) and some preliminary results can
be summarised. New World species with long, slender, horizontally
presented, tubular flowers and vertical inflorescences are known
to be adapted to pollination by hummingbirds; these are
predominantly in subgenus *Erythrina*. The pollen is remarkably
uniform with simple, regular reticulate ornamentation (Plate IV,
1). Species with more or less gaping flowers held reflexed with
the standard petal enlarged and horizontal inflorescences are
known to allow easy access for perching birds; these are
predomonantly Old World in distribution and include Subgenera
Chirocalyx, *Micropteryx* and *Erythraster*. The pollen of these has
more varied exine ornamentation with a find reticulate pattern or
very coarse reticulate ornamentation with prominent sexinous
granules (Plate IV, 2, 3). Furthermore, during dissection of the
anthers from herbarium specimens, it was observed that generally
the pollen from species associated with hummingbird pollination
seemed to be dry and powdery in contrast with often rather sticky
pollen from species associated with passerine bird pollination
(Plate IV, 4, 5). The functional significance of these observations is not clear and they need to be tested on living material
both in cultivation and in the field if possible, but this is an
example of the type of question herbarium observation can pose.

The pollen surface coating (pollenkitt) recently investigated
extensively by Hesse (1981) in various angiosperms is now being
investigated in Papilionoideae by Hesse and Ferguson (unpublished)
to see whether pollen stickiness shows variation. Plate IV, 6, 7,
8 shows preliminary results from three different taxa in the tribe
Sophoreae. Different surface coatings are clearly present. Only
speculation is possible at this stage on the significance of these
differences but they do provide evidence of the potential for
further investigation. It is also clear that one of the primary
needs is to devise techniques to get the data on to a firm
chemical basis.

POLLEN AND HARMOMEGATHY

Another aspect of pollen function not yet discussed is water
relations and harmomegathic movement (accommodation of volume
change) which can be important. It might be postulated that much

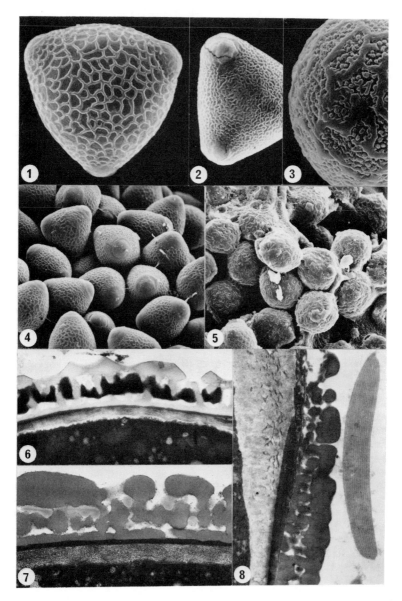

PLATE IV. 1. Erythrina costaricensis *(hummingbird pollinated)*
pollen with regular reticulate ornamentation. 2. **Erythrina**
breviflora *(passerine bird pollinated) pollen with finely*
reticulate ornamentation. 3. Erythrina variegata *(passerine*
bird pollinated) pollen with coarsely reticulate ornamentation
with coarse granules in the lumina. 1-3 all SEM micrographs X

of the similar pollen morphology found in widely separated families of angiosperms may be the result of convergent evolution in response to common functional adaptations. Some attention has recently again been focussed on functional aspects of structure and water relations. For example, in general physiological terms there is the work of Heslop-Harrison (1979a, b) and Payne (1981). In general structural terms Muller (1979) has reviewed the problems and possibilities, while in particular families, there are a number of workers who have contributed significant ideas, notably in the Compositae (Bolick, 1981; Blackmore, 1982), and in the Lythraceae and Sonneratiaceae (Muller, 1981). Little work of this type has been carried out on the pollen of the papilionoid legumes. Ferguson (1980) has discussed how harmomegathic movement might operate in the pollen of two species with markedly different exine architecture one with little endexine and very thin foot-layer and operculate apertures, the other with a thick endexine, no footlayer and inoperculate apertures. Later Misset *et al.* (1982) demonstrate the remarkable differences in shape that occur in hydrated and dehydrated pollen grains of *Ulex*. This latter work underlines the importance of fully understanding pollen shape before using it as a character for classification.

Functional significance must not be overrated. Understanding the function of pollen characters does not detract in many situations from their usefulness as taxonomic characters. This principle is true for other characters also. In our reevaluation of *Erythrina* pollen the data appear only to reaffirm the existing classification of the New World taxa (Krukoff and Barneby 1974; Barneby pers. comm.). In other bird/bat pollinated species with modified pollen the information far from being dismissed gives a broader understanding of the evolution of both characters within a group as well as of the group as a whole.

Pollen ontogeny and factors affecting exine deposition and substructure are areas which should not be overlooked although very little work has yet been done and virtually none involving Leguminosae. Results from investigations in other families (see for example Rowley *et al.* 1981; Cerceau-Larrival *et al.* 1981) suggest that eventually it may be possible to have a better understanding and assessment of the characters of the mature pollen grain used in comparative morphology.

1233. 4. Erythrina schimpffii *(hummingbird pollinated) showing "dry powdery" pollen, SEM* X *468.* 5. Erythrina variegata *(passerine bird pollinated) showing "sticky" pollen with copious pollenkitt, SEM* X *298.* 6. Sophora microphylla *thin section of pollen exine showing large quantities of black staining pollenkitt, TEM* ca.X *11,900.* 7. Sophora davidii *as 6 but showing smaller quantities of grey pollenkitt, TEM* ca. X *18,700.* 8. Castanospermum australe *as 6 and 7 but showing lamellated structures in the tapetal fluid/pollenkitt, TEM* ca. X *10,200.*

CONCLUSIONS

Pollen structure and morphology provide a significant contribution
to the systematics of the Papilionoideae both at the tribal
(apertures and exine stratification), generic and specific levels
(size, shape and exine ornamentation). Certain structural and
morphological features are now known to be under adaptive
selection in both closely related groups and in disparate groups.
Pollen structures found to be associated with bird pollination
either have verrucate ornamentation (rather than reticulate) or a
complex exine stratification (rather than simple). As in other
types of modifications for example, differential harmomegathic
movement, the apparent modifications may be the result of
convergent evolution in response to common functional
requirements. However, these do not detract from their usefulness
as taxonomic characters but rather point out that the pollen grain
is an evolutionary compromise between many external and internal
constraints. As such it should be possible to carefully evaluate
which features of the pollen grain show structural homology and
which functional homology. It is now possible using careful LM,
TEM and SEM studies coupled with studies on the living pollen
grain both in the field and in the laboratory, to clearly separate
the underlying structure and its attendant functions and so become
aware of those features which may cause taxonomic confusion. In
view of the high level of specialised technology and terminology
in the study of cryptic characters the multidisciplinary and
collaborative approach is an essential prerequisite of systematic
investigation today. Studies of the type described but also
incorporating pollen ontogenetical and fossil data where available
(very sparse or absent for subfamily Papilionoideae) are the
direction in which future pollen studies in biosystematics should
continue.

ACKNOWLEDGEMENTS

I am grateful to a large number of colleagues for help of various
kinds, but particularly to Dr. R.M. Polhill (Kew), Dr. Ph. Guinet
(Montpellier), C. Stirton (Kew), Dr. R. Barneby (New York) and
Professor J.J. Skvarla (Oklahoma) for many ideas, valuable
discussion and comments. I am especially indebted to Dr. P.
Linder (Pretoria) for introducing me to cladistics and to P.
Linder and C. Stirton for critically reading the manuscript and
making many valuable improvements. Mrs. Christine Grey-Wilson
drew Figure 2; Milan Svanderlik printed the micrographs and Mrs.
Madeline Harley made up the plates. Dr. M. Hesse (Vienna)
provided the micrograph Plate IV, 8. Living material of
Castanospermum australe was kindly collected and fixed by
Mr. B. Schrire (Durban).

REFERENCES

Barneby, R.C. 1977. Daleae Imagines. *Mem. N.Y. Bot. Gard.* 27: 1-891.

Blackmore, S. 1982. A functional interpretation of Lactuceae (Compositae) pollen. *Pl. Syst. Evol.* 141: 153-168.

Bolick, M.R. 1981. Mechanics as an aid to interpreting pollen structure and function. *Rev. Palaeobot. Palynol.* 35: 61-79.

Cerceau-Larrival, M.-Th., Abadie, M., Albertini, L., Audran, J.-C., Cornu, A., Cousin, M.-Th., Dan Dicko-Zafimahova, L., Duc, G., Ferguson, I.K., Hideux, M., Nilsson, S., Roland-Heydacker, F. and Souvre, A. 1981. Relations sporophyte-gamétophyte: assise tapétale-pollen. Résultats preliminaires. *Ann. Sci. Nat. Bot. Paris* 2 and 3, 69-92.

Cowan, R.S. 1981. Tribe 1. Swartzieae DC. 1825. In *Advances in Legume Systematics* (R.M. Polhill and P.H. Raven, eds.), pp. 209-212. R. Bot. Gardens, Kew.

Ferguson, I.K. 1978. A note on the pollen morphology of the genus *Cranocarpus* Bentham (Leguminosae). *Bradea* 2: 269-272.

Feguson, I.K. 1980. Quelques remarques sur l'ultrastructure du pollen de deux espèces de la sous-famille des Papilionoideae (Leguminosae) en relation avec le microenvironment (Hygrométrie et hydrodynamique). *Mém. Mus. Nat. Hist. Nat.* ser. B 27: 45-50.

Ferguson, I.K. 1981. The pollen morphology of *Macrotyloma* (Leguminosae: Papilionoideae: Phaseoleae). *Kew Bull.* 36: 455-461.

Ferguson, I.K. and Skvarla, J.J. 1979. The pollen morphology of *Cranocarpus martii* Bentham (Leguminosae: Papilionoideae). *Grana* 18: 15-20.

Ferguson, I.K. and Skvarla, J.J. 1981. The pollen morphology of the subfamily Papilionoideae (Leguminosae). In *Advances in Legume Systematics* (R.M. Polhill and P.H. Raven, eds.), pp. 859-896, R. Bot. Gardens, Kew.

Ferguson, I.K. and Skvarla, J.J. 1982. Pollen morphology in relation to pollinators in Papilionoideae (Leguminosae). *Bot. J. Linn. Soc.* 83: 183-193.

Ferguson, I.K. and Skvarla, J.J. 1983. The granular interstitium in the pollen of subfamily Papilionoideae (Leguminosae). *Am. J. Bot.*, 70: 1401-1408.

Ferguson, I.K. and Strachan R. 1982. Pollen morphology and taxonomy of the tribe Indigoferae (Leguminosae: Papilionoideae). *Pollen Spores* 24: 171-210.

Graham, A. and Barker, G. 1981. Palynology and tribal classification in the Caesalpinioideae. In *Advances in Legume Systematics* (R.M. Pohill and P.H. Raven, eds.), pp. 801-834, R. Bot. Gardens, Kew.

Graham, A. and Tomb, A.S. 1974. *Lloydia* 37: 465-481.

Graham, A. and Tomb, A.S. 1977. *Lloydia* 40: 413-435.

Guinet, Ph. 1981. Comparative account of pollen characters in the Leguminosae. In *Advances in Legume Systematics* (R.M. Polhill and P.H. Raven, eds.), pp. 789-799, R. Bot. Gardens, Kew.

Heslop-Harrison, J. 1979a. *Am. J. Bot.* 66: 737–743.

Heslop-Harrison, J. 1979b. *Ann. Mo. Bot. Gard.* 66: 813–829.

Hesse, M. 1981. *Rev. Palaeobot. Palynol.* 35: 81–92.

Horvat, F. and Stainier, F. 1979. L'étude de l'exine dans le complexe *Phaseolus-Vigna* et dans des genres apparentés. III. *Pollen Spores* 21: 17–30.

Horvat, F. and Stainier, F. 1980. L'étude de l'exine dans le complexe *Phaseolus-Vigna* et dans des genres apparentés. IV. *Pollen Spores* 22: 139–173.

Kavanagh, T.A. and Ferguson, I.K. 1981. Pollen morphology and taxonomy of the subtribe Diocleinae (Leguminosae: Papilionoideae: Phaseoleae). *Rev. Paleobot. Palynol.* 32: 317–367.

Krukoff, B.A. and Barneby, R.C. 1974. Conspectus of species of the genus *Erythrina*. *Lloydia* 37: 332–459.

Maréchal, R., Mascherpa, J.-M. and Stainier, F. 1978. *Boissiera* 28: 1–273.

Misset, M.-Th., Gourret, J.P. and Huon, A. 1982. Le pollen d'*Ulex* L. (Papilionoideae): morphologie des grains et structure de l'exine. *Pollen Spores* 24: 369–395.

Muller, J. 1979. Form and function in Angiosperm pollen. *Ann. Mo. Bot. Gard.* 66: 593–632.

Muller, J. 1981. *Rev. Paleobot. Palynol.* 35: 93–123.

Payne, W.W. 1981. *Rev. Paleobot. Palynol.* 35: 39–59.

Polhill, R.M. 1981a. Papilionoideae. In *Advances in Legume Systematics* (R.M. Polhill and P.H. Raven, eds.), pp. 191–208, R. Bot. Gardens, Kew.

Polhill, R.M. 1981b. Tribe 26. Bossiaeeae (Benth.) Hutch. 1964. In *Advances in Legume Systematics* (R.M. Polhill and P.H. Raven, eds.), pp. 393–395. R. Bot. Gardens, Kew.

Polhill, R.M. and Raven, P.H. 1981. *Advances in Legume Systematics*. R. Bot. Gardens, Kew.

Poole, M.M. 1979. Pollen morphology of the genus *Psophocarpus* (Leguminosae) in relation to its general morphology. *Kew Bull.* 34: 211–220.

Ross, J.H. 1982. *Muelleria* 5: 1–29.

Rowley, J.R. Dahl, A.O., Sengupta, S. and Rowley, J.S. 1981. A model of exine substructure based on dissection of pollen and spore exines. *Palynology* 5: 107–152.

Rudd, V.E. 1981. Tribe 14. Aeschynomeneae (Benth.) Hutch. (1964). In *Advances in Legume Systematics* (R.M. Polhill and P.H. Raven, eds.), pp. 347–354. R. Bot. Gardens, Kew.

Stainier, F. and Horvat, F. 1978. L'étude de l'exine dans le complexe *Phaseolus-Vigna* et dans des genres apparentés, I and II. *Pollen Spores* 20: 195–214 and 341–349.

Stainier, F. and Horvat, F. 1983. L'étude de l'exine dans le complexe *Phaseolus-Vigna* et dans des genres apparentés. V. Le sous-genre *Sigmoidotropis* (Piper) Verdcourt et *Ramirezella stroboliphora* (Robinson) Rose. *Pollen Spores* 25: 5–40.

Numerical Taxonomy and Biosystematics

J. McNeill
Department of Biology
University of Ottawa
Ottawa, Ontario, Canada

INTRODUCTION

It is both an honor and a great pleasure to be asked to contribute
to this Symposium on the topic of Numerical Taxonomy and
Biosystematics. This is because I regard these two fields as
being very closely linked. They have been linked in my own
career, but also are, I believe, necessarily so by the very nature
of biosystematics. Camp, in an address delivered on April 16,
1942, at the dedication of the M. A. Chrysler Herbarium at Rutgers
University, said "Because of its broader scope, this new phase
into which systematics is adventuring may be called *Biosystematy*,
for it no longer looks upon species as piles of named specimens
but as populations of living organisms."
    This address was published (Camp 1943) later in the same year
as the Camp and Gilly (1943) paper in which the term
"Biosystematy" first appeared in print, and I refer to it here not
from any misplaced sense of priority, but because I believe that
it presents very succinctly what I see as the fundamental nature
of biosystematics, namely that it is populationally based.
    I do not quite go back to 1943, but I do owe my own interests
in plant systematics in no small measure to a little book first
published ten years later and one of the first to embody the ideas
of biosystematics. I refer to Heslop-Harrison's (1953) *New
Concepts in Flowering Plant Taxonomy*. His strong contrast
between *Experimental* and *Classical* taxonomy, perhaps necessary
at the time, seems a false distinction today, and our techniques
have advanced enormously since 1953, but otherwise the excitement
that the book communicates about the systematics of living plants
and plant populations is still relevant.
    Any young taxonomist faced with mountains of herbarium
specimens, each slightly different from the next, could not but
feel liberated by the possibilities provided by field studies and

the collection of population samples. Population samples provide
a basis for objective judgment as to whether particular
differences are useful or not as taxonomic characters. But nearly
always analysis of population samples involves measurement of
particular features and subsequently some form of numerical or
statistical analysis.

Measurements on plant populations go back long before
biosystematics. Briggs and Walters (1969), in their historically
oriented *Plant Variation and Evolution,* describe the work of
Burkill (1985), Ludwig (1901) and Lee (1902) on variation in
stamen and carpel number in populations of *Ranunculus ficaria*.
This was done in that period of enthusiasm for biometry into which
the rediscovery of Mendel's work was launched. There was one big
difference, however, between these early biometric studies and
those that appeared as biosystematics grew and matured. This
difference can best be described by an example and as I have been
discussing the historical links between biosystematics and
numerical taxonomy, it seems appropriate to take this from the
earliest work on these topics in which I, personally, was
involved: an undergraduate study on variation in *Solidago
virgaurea* L., the common Eurasian goldenrod, in Scotland.

Figure 1 shows data derived from 210 plants from populations of
S. virgaurea from 12 localities in Scotland. For each
character, the left hand diagram shows the range, mean and
standard deviation for plants from montane populations and those
on the right the same statistics for plants from lowland
populations. It is clear that there is considerable overlap in
each and separation of montane from lowland plants would not be
possible using any of these characters on their own. What is not
being considered, however, are the associations between the
character states in each plant. Once systematists started to do
this, the close link between biosystematics and numerical taxonomy
was forged. Univariate statistical analysis, as in Fig. 1, gave
way to multivariate analysis. Figure 2 shows the results of a
discriminant analysis on a set of lowland and montane plants of
Solidago, similar to the one that I carried out at that
time, and using the same characters as illustrated in Fig. 1. The
data set comprises 81 montane and 36 lowland plants and
incorporates a subset of the plants used in Fig. 1. I will
consider discriminant analysis in more detail later in this paper,
but for the moment it is suffice to say that the first step is the
construction of correlation or covariance matrices for the groups
involved. Thus characters are not being considered singly, but
instead the co-occurrence of their states in each individual is
being taken into account.

We see, then, that one of the characteristic features of
biosystematics is its emphasis on populations rather than on
individuals and that this requires numerical studies. What,
however, is the particular contribution of what we call numerical
taxonomy?

WHAT IS NUMERICAL TAXONOMY?

So far as I am concerned, numerical taxonomy covers all aspects of quantitative taxonomy, but is primarily concerned with those that use multivariate methods. Numerical taxonomy had its origins in a phenetic approach to classification (see e.g. Sokal and Sneath 1963). This is not the place to try to characterize phenetic classification. This is well described by Jardine and Sibson (1971: 136-138) and is discussed in relation to alternative approaches by McNeill (1980). Briefly, a phenetic classification (also known as a natural or general-purpose classification) is one in which a large number of general statements can be made about the classes recognized. It is one in which "the constituent groups describe the distribution among organisms of as many features as possible" (Farris 1977), including features that were not used in the original classification, i.e. it seeks to incorporate a predictive component (McNeill 1980).

Although there has been this strong association between the growth of numerical taxonomy and a phenetic approach to classification, it is important to emphasize that the two are not synonymous. Phenetic classification, obviously, need not be numerically based, since most traditional botanical classification, at least since the latter part of the 18th century, has utilized a phenetic approach. Likewise, numerical taxonomy, although it developed in the context of phenetic philosophy, is not by any means restricted to the construction of phenetic classifications. Quantitative biosystematic data can be used in phenetic classification, but this is not the main area in which numerical taxonomy and biosystematics interact, and it is not the area that I wish to emphasize in this review of the past role of numerical taxonomy in biosystematics and of its future potential.

I have already indicated that numerical taxonomy covers essentially all quantitative studies in systematics that utilize multivariate techniques. Such a broad field cannot be covered comprehensively in the space that I have available to me in this symposium volume. *Numerical Taxonomy* (Sneath and Sokal 1973), the new and expanded version of Sokal and Sneath's (1963) *Principles of Numerical Taxonomy,* has nearly 600 pages. There has been a steady output of books and symposium volumes on various aspects of the topic. I have enumerated some of these elsewhere (McNeill 1983) and mention here only some of the more recent books that provide useful introductions to current taximetric techniques such as those by Neff and Marcus (1980), Gordon (1981), Dunn and Everitt (1982), Felsenstein (1983b) and Legendre and Legendre (1983). There have also been many reviews of aspects of numerical taxonomy, of which that on phenetics by Duncan and Baum (1981) has a particularly extensive bibliography, and that on cladistics by Felsenstein (1982) gives very valuable insights on that topic.

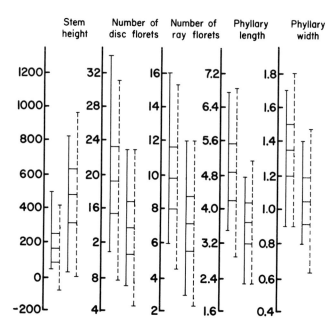

FIG. 1. *Means, ranges and standard deviations for five characters derived from 210 plants from populations of* Solidago virgaurea *from 12 localities in Scotland. For each character, the left hand diagram is based on 87 plants from 4 montane populations and those on the right on 123 plants from 8 lowland populations. The cross-bars represent the means and one standard deviation on either side of the mean; the dashed line extends for three standard deviations on either side of the mean. Measurements are in millimeters.*

So, instead of trying to provide an encyclopedic view of numerical taxonomy, I will concentrate instead on a few aspects in which I believe numerical methods can and are making important contributions to biosystematics. I will describe these briefly and point to some current trends.

FIG. 2. *Histogram showing the values of a discriminant function based on a set of 81 montane and 36 lowland plants of* Solidago virgaurea, *including a random sample of the specimens used in Fig. 1. The same five characters were used. The different hatching distinguishes the source of the plants; the lowland are to the left and the montane to the right.*

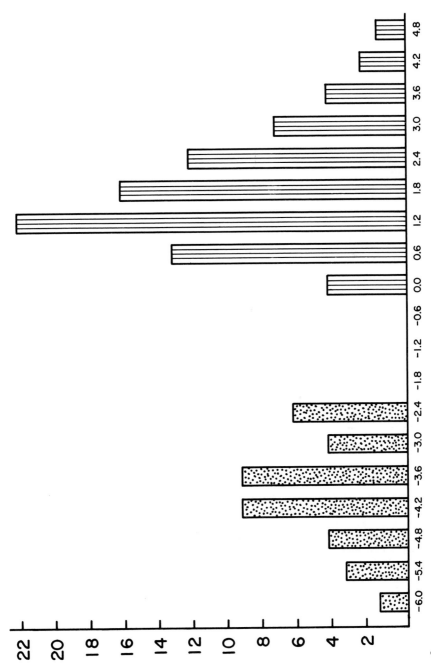

Figure 2

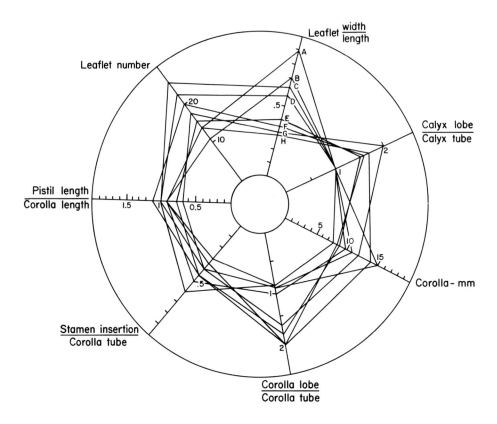

FIG. 3. *A polygonal graph of seven characters plotted from eight specimens of* Polemonium pulcherrimum. *The diagram is reproduced from Davidson,* Madrono *9: 107, 1947.*

FIG. 4. *Numerical analysis of the data on eight specimens of* Polemonium pulcherrimum *scored for seven characters, and derived from the polygonal graph in Fig. 3. The letters represent the specimens; the characters are numbered as follows: 1. Leaflet number; 2. Leaflet width/length; 3. Calyx lobe/calyx tube; 4. Corolla length; 5. Corolla tube/corolla tube; 6. Stamen insertion/corolla tube; 7. Pistil length/corolla length. The upper diagram is a phenogram derived by UPGMA clustering on pairwise euclidean distances obtained after standardization by range of the original character values. The lower diagram shows the first two axes of a principal components analysis (PCA) of the same data. The character vectors on the first two principal components show the relative contribution of each of the original characters. For clarity, the vectors are all shown at twice their actual length.*

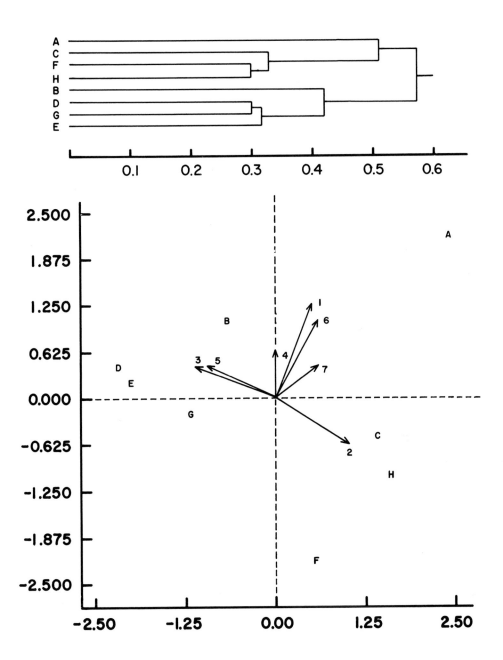

Figure 4

MULTIVARIATE ANALYSES OF BIOSYSTEMATIC DATA

To illustrate the relationship with biosystematics, let us go back
a little in time. Most multivariate analyses require
computational facilities that were not available in 1943.
Biosystematists were ingenious in inventing alternatives. One of
the first was the Hutchinsonian *polygonal graph* (Fig. 3),
introduced to plant systematics by Davidson (1947). As this
Symposium was presented in Montreal, it is perhaps of interest to
note that Hutchinson himself, was a Canadian who published the
method with quite different applications (Hutchinson 1936).
Anyone who has ever tried to construct one of these graphs will be
aware of the phenomenal labor involved, even for a small
population. Moreover, the choice of the axes (spokes) can have a
marked effect on the clarity and utility of the diagram. Yet,
until a diagram is well underway, the optimal scale and position
of the axes are not evident.

Figure 4 shows an analysis of data taken from Fig. 3 using
conventional numerical taxonomic techniques. The groupings of the
individuals are vastly more evident. The upper dendrogram,
derived by UPGMA clustering on euclidean distances standardized by
range, shows the three clusters that are influenced strongly by
the leaflet width/length and the calyx and corolla lobe/tube
characters of Fig. 3. These character associations are used in
the cluster analysis but their details are not preserved. In the
ordination, however, the information is available in terms
of the relative contributions of the original characters to each
of the principal component axes, of which the first two are shown
in the lower part of Fig. 4. These are illustrated by means of
the character vectors drawn in Fig. 4. In both cases, there is
the enormous advantage of there being no limitation on the number
of characters, nor often any need for the preliminary calculation
of ratios as was done by Davidson.

Another classical biosystematic technique for examining
multivariate patterns in populations is that of the *biometric
index* (Davis and Heywood 1963: 325), originally developed by
Anderson (1936) as the *hybrid index* method for analyzing
populations that might be showing effects of hybridization. This
technique is used by Riley (1938) to analyze four *Iris*
populations from southern Louisiana; the two most remote
populations were assigned to *I. fulva* and *I. hexagona* var.
giganticaerulea, respectively; the other two were situated
geographically between them, and were suspected of including
hybrids. Twenty-three plants were scored from each population for
seven morphological characters, all of which, save sepal length
and color of sepal blade, were qualitative assessments of
whether the individual plant had the characters of *I. fulva*
or of the variety of *I. hexagona.* Riley calculated an index
value for each plant giving a value of 0 for each qualitative
assessment of resemblance to *I. fulva* and 2 for each to *I. I.
hexagona* with 1 representing any intermediate condition. For

sepal length and blade color, he arbitrarily divided the range of
values into classes that were scored from 0 to 3 and 0 to 4
respectively; intensity of color was ignored. Histograms of the
index values obtained for each population are shown in Fig. 5.

Riley recognized the arbitrariness of his index, but numerical
taxonomic methods allow a similar analysis without the need to
impose taxonomic prejudgement or to make prior decisions about the
categorization of continuous data. Figure 6 shows a re-
examination of Riley's data, using the ordination technique of
Principal Components Analysis (PCA). Only the 0, 1, 2, scores
developed by Riley were available for the qualitative assessment
data, but the actual sepal lengths, the range of color intensities
and a complete scale for hue from 1 for orange-red to 7 for
violet-blue were used. In view of the fact that five of the eight
characters were scored identically, it is not surprising that the
overall picture is similar to that discerned by Riley. The
numerical taxonomic method makes clear, however, the degree to
which variation in each of the more continuous characters is
associated with the interspecific differences. The first axis
accommodates 83% of the variation and the second a further 11%.
The character vectors, showing that 7 of the 8 characters are
closely associated with the first axis, and that the second axis
is largely one of sepal color intensity, provide objective
evidence for Riley's claim that "depth of hue . . . seems to have
little significance" in species delimitation.

Another method developed at this time, also for analyzing
putatively hybrid populations, was Anderson's (1949, 1953)
pictorialized scatter diagram. This involved a choice of
two characters as axes and the incorporation of usually three
states of the other characters as long or short (or no) rays,
extending from the plotted points in different directions for each
character. Anderson was particularly enthusiastic about this
technique as a means of carrying out his "method of extrapolated
correlates," by which he meant that a diagram like this could
indicate the characteristics of the, perhaps as yet unknown,
species that was the source of the introgression. He illustrates
this (Anderson 1949) using Riley's *Iris* data and in particular
the population that shows the greatest influence of I. fulva.
The method, although capable of producing very elegant diagrams
(cf. that on *Oxytropis albiflora* Anderson (1953), Davis and
Heywood (1963): 477), has the same problem of arbitrariness or
subjectivity that applies to the hybrid index. A different choice
of axes or a different choice of characters, or even just a
different choice of cut-off values for character-state
delimitation (cf. Gillett 1960 for a cautionary tale in this
regard!), could make a major difference in the extrapolation.
Numerical techniques avoid these difficulties, while providing,
through ordination for example, the same type of solution.

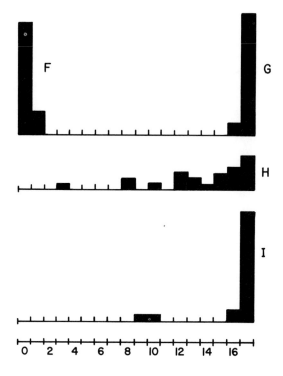

FIG. 5. *Histograms of "hybrid index" values for* Iris *populations derived from Riley (1938). Populations F and G are of* I. fulva *and* I. hexagona *var.* giganticoerulea, *respectively, whereas H and I were interpreted as showing introgressive hybridization.*

FIG. 6. *The first two axes of a principal components analysis (PCA) of the data on* Iris, *used by Riley for the construction of his hybrid index (Fig. 5). The character vectors on the first two principal components show the relative contribution of each of the original characters, providing an objective demonstration of the fact that character 3 (intensity of hue of sepal blade) does not contribute to the interspecific differences, whereas the other characters do. For clarity, the vectors are all shown at twice their actual length. Note that the positions of 38 observations overlap and are hidden.*

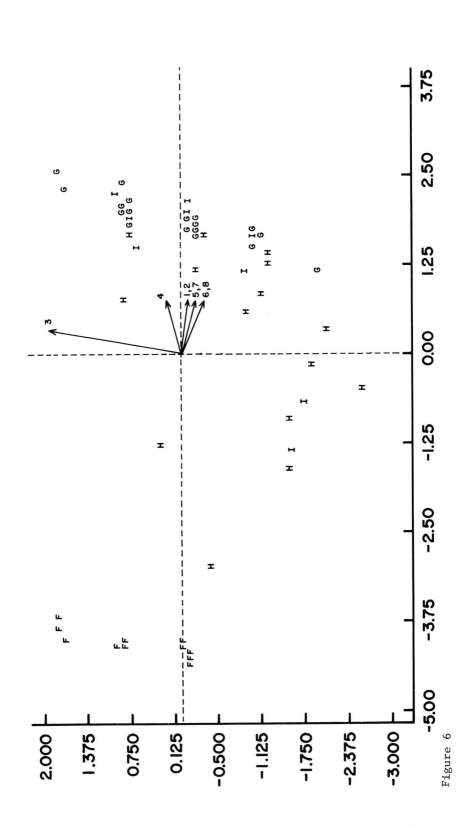

Figure 6

PATTERN RECOGNITION

Numerical taxonomic techniques are not, of course, restricted to
doing a better job at the types of multivariate analyses that grew
up with biosystematics. More importantly, they provide the means
for breaking new ground today.

In considering the current role of numerical methods in
biosystematics, I would place the greatest emphasis on *purpose*.
What is the purpose of our biosystematic research? Is there a
model to explore or a hypothesis that can be tested? There may
well not be and this should not, of itself, be a concern, so long
as the structure of the problem can be defined. Hypothesis-
testing, whatever some may claim, is not the *sine qua non* of
science. Much of biosystematics, like taxonomy, involves *pattern
recognition*. We are seeking to discern pattern and trying then
to summarize it in a manner that will be useful to other students
of biology.

Although I have illustrated their use in the polygonal graph
data on *Polemonium,* clustering methods are not usually the
most appropriate for pattern recognition. This is illustrated by
work on *Capsicum,* in which I have been engaged in cooperation with Drs. B. Pickersgill (Reading) and C. B. Heiser
(Indiana). In the wild and cultivated members of this group there
was good reason to doubt the existence of the discontinuities that
are a prerequisite for clustering methods to give meaningful
results. In such situations, other types of multivariate analysis
are often very informative. In the genus *Capsicum,* current
morphological and cytogenetic research suggests that the very
diverse assemblage of cultivated chili and bell peppers is
referable to four or perhaps five species (Heiser and Pickersgill
1969). Of these, *C. pubescens* and *C. baccata* are relatively
distinct, but cultigens or near cultigens identified with the
remaining three species are hard to disentangle morphologically.
A study of morphological features of wild and cultivated members
of this group (and of *C. baccata*) revealed an interesting
pattern of variation (Pickersgill *et al.* 1979).

A principal coordinates analysis (PCO) showed that there did
indeed appear to be character intergradation, but that there is
not a continuum of variation. Instead, the wild representatives
are relatively distinct and seem to fall into three clusters
corresponding to *C. annuum, C. frutescens* and *C. chinense*
(Pickersgill *et al.* 1979, Figs. 54.4 and 54.7). The PCO reveals
that it is only in the cultivated representatives that the
confusion exists. Moreover, a clear wild-cultivated component can
be discerned running at an angle to the first and second principal
coordinates. These ordinations reveal much of the pattern of
variation in this group. Further numerical studies are clearly
indicated: for example, the rotation of the first two axes to
allow study of the components of variation that remain after the
wild-cultivated dimension has been removed.

HYPOTHESIS TESTING

Sometimes numerical techniques reveal a pattern that in turn
suggests a hypothesis that can be tested by other numerical
analyses. An example of this is seen in investigations on the
genus *Spiranthes*.

Ordination and Canonical Variates Analysis

In 1976, R. C. Simpson, an observant naturalist, noticed unusual
plants of the genus *Spiranthes* (lady's tresses) growing in
Killbear Provincial Park in the Parry Sound District of northern
Ontario. This prompted Dr. P. M. Catling, who was then studying
the genus as part of his Ph.D. program, to sample the *Spiranthes*
in the Park and to make detailed measurements on 33 specimens,
representing these plants and the four species that are found in
that area, *S. casei, S. cernua, S. lacera* and *S. romanzoffiana*.
These are well distinguished by flower color and time of
flowering (cf. Catling and Cruise 1974; Simpson and Catling 1978),
and so it was possible, prior to making the measurements, to
assign to species all of the specimens, except for the sample of
the unusual plants originally found by Simpson. An initial
investigation of the overall pattern by projection onto the first
two principal components of a PCA, reveals three, or possibly
four, clusters with a parallel trend in each (Fig. 7). As this
trend runs in the same direction within each of the known species,
it is evident that it is not of taxonomic significance and,
indeed, turns out to be in part a size component, that may reflect
the degree of moisture and nutrient in the immediate environment
of the plants concerned. From the ordination it is equally
possible that the unusual plants are aberrantly small individuals
of *S. casei* or, as field observations suggested,
intermediates, and perhaps hybrids, between *S. romanzoffiana* and
S. lacera.

To test these alternative hypotheses, a canonical variates
analysis was carried out, using the four species as groups and
then projecting the unusual specimen into the canonical space so
produced. This has the effect of minimizing the within-species
variation so that the resultant three canonical axes will
maximally separate the species. If the unidentified specimen is
an extreme of *S. casei* it would be expected to fall
reasonably close to that species; if, on the other hand, it is a
hybrid that combines the parental characteristics, it should fall
in an intermediate position between them. Figure 8, a projection
onto the first two canonical axes, shows the result of this
investigation. It is clear that, as Simpson suspected in the
field, the unusual plants are intermediates, and possibly hybrids,
between *S. lacera* and *S. romanzoffiana*.

McNeill (1978) demonstrated the use of canonical variates
analysis in testing a hypothesis that all the plants introduced
into North America and supposedly referable to the Eurasian weed

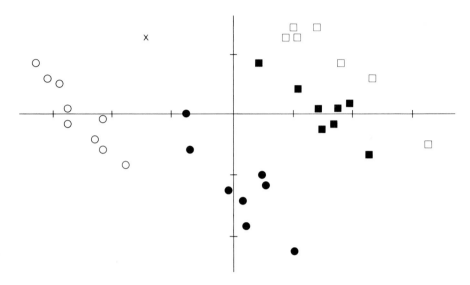

FIG. 7. *Plot of the projection of 33* Spiranthes *specimens sampled from Killbear Provincial Park, Parry Sound District, Ontario, Canada, onto the first two principal axes of a PCA. The symbols represent the species and putative hybrid distinguishable on floral characters not used in the analysis (□ S.* romanzoffiana, ■ S. cernua, ● S. casei, ○ S. lacera, *and* × *the putative hybrid).*

species, Silene latifolia Poiret (not to be confused with
S. latifolia (Miller) Britten & Rendle, a synonym of *S. vulgaris*
(= *S. cucubalus*), bladder campion.) (= *S. pratensis* = *S. alba* =
Lychnis alba) (white cockle), were, in fact, products of
hybridization with *S. dioica*. This claim had been made by
Boivin (1967, 1968), who coined the new interspecific hybrid name
Lychnis ×*loveae* for *these supposed hybrids*. McNeill *carried out
two canonical variates analyses. One was on a sample of 58
herbarium specimens, of which 17 were of S. latifolia* from
Europe, 14 of *S. dioica* also from Europe, and 27 introduced
North American "white cockle" weeds. This last group included
the type of *L.*×*loveae*. Four characters were used: leaf
breadth, calyx tube and teeth length and calyx teeth width, all of
which discriminate *S. latifolia* and *S. dioica*. The results of
this canonical variates analysis are shown in Fig. 9. This shows

-----

FIG. 8. *Plot of the projection of 33* Spiranthes *specimens onto
the first two canonical axes obtained from a canonical variates
analysis using as groups the species to which 32 of the specimens
had been independently assigned. Symbols and source of specimens
as in Fig. 7.*

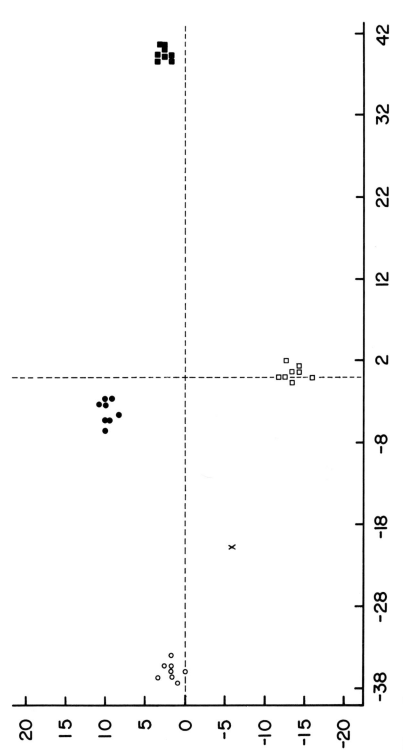

Figure 8

very clearly that, whereas *S. dioica* is readily separable,
there is no clear distinction between the European and North
American white cockle samples, even although the analysis is
designed to maximize the separation of all three groups.

   McNeill's (1978) second analysis was essentially the same but
utilized 161 plants grown in a greenhouse from seed collected in
both Europe and North America. In this case 15 characters were
used and an almost identical result was obtained, with, as in Fig.
9, the first axis separating *S. dioica* from the other two
groups, and the second axis attempting to discriminate between the
two white cockle samples. Whereas the difference between the
canonical means of *S. dioica* and each of the other two
groups was significant at $p=0.0001$, that between the two white
cockle groups was scarcely significant ($p=0.05$). Moreover, even
this difference was in other features than those discriminating
the species because, along the "interspecific" first axis, the
canonical mean of the North American white cockles was more
distant from that of *S. dioica* than was that of their European

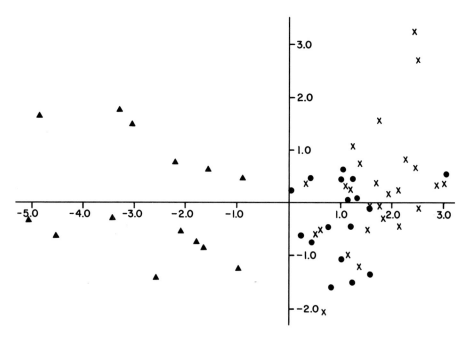

FIG. 9. *Plot of the projection of 58* Silene *specimens onto the
first two canonical axes obtained from a canonical variates
analysis, using as groups specimens of* S. dioica *and European and
North American samples of* S. latifolia (= S. pratensis = L. ×
loveae). (▲ S. dioica, ● *European* S. latifolia, × *North American
plants of "white cockle").*

counterparts. McNeill concluded from this study that the
introduced North American weeds were not distinguishable
taxonomically from European *S. latifolia,* and that the
hypothesis of a hybrid origin for the North American plants was
not tenable. This was supported by observations on flowering
time, that did not reveal any differences between plants of
European and North American origin.

Discriminant Analysis

One of the earliest and best known multivariate techniques is that
of discriminant analysis. Because it requires that groups be
known *a priori,* it has a limited role in numerical taxonomy,
where much of the emphasis is on group recognition. It does
contribute very valuably, however, in one area of biosystematics.
This is in situations in which groups are recognized on the basis
of a particular biosystematically interesting feature, and the
aim is to establish whether or not these groups are also
recognizable by other means. The most familiar example of this
situation is when two or more cytodemes exist within a taxon, and
one wishes to discover if these can be distinguished on
morphological features alone, and, if so, to identify to
chromosome race, specimens, perhaps in a herbarium, for which no
chromosomal data are available.

In *Polygonum* sect. *Polygonum,* a number of ploidy levels have
been reported on a base number of $x = 10$ (Löve and Löve 1956;
Mertens and Raven 1964; McNeill 1981; Wolf and McNeill 1984).
The widespread native eastern North American weed, *P. achoreum,*
is a member of this section and most counts made on plants of this
species have shown it to be tetraploid with $2n = 40$. A few
hexaploid plants have recently been discovered, however, in one
location in Québec (Wolf and McNeill 1984). No evident
morphological differences could be discerned but measurements were
made on voucher specimens of a number of features that are often
of taxonomic significance in this group, such as the length and
degree of division of the perianth and the size and shape of the
nutlet.

The results of a discriminant analysis on these data are shown
in Fig. 10, in which a separation of the two ploidy levels on
morphological grounds is shown. The size of the available sample,
15 flowers from three tetraploid plants and 10 from two
hexaploids, is too small to permit generalization of these
results, but they do indicate that further investigation of
possible morphological differences is justified, despite the
original failure to discern any consistent differences.

OPPORTUNITIES UNNNUMBERED

The examples quoted above do little more than scratch the surface
of the ways in which numerical taxonomy contributes to
biosystematic research. I do not have space to do more than

mention some of the many other possibilities.

Common garden experiments have generally confined themselves to comparisons between populations as grown under the relatively uniform garden conditions. Integration of data from both wild and cultivated material would often be helpful and utilization of residuals after an appropriate regression analysis, can permit study of variation after the removal of a phenotypic plasticity component.

Isozyme studies of allele frequencies benefit greatly from numerical analysis. A number of efficient programs for analysing allele frequencies have been developed in recent years (cf. Felsenstein 1983a; Swofford and Selander 1981) and applications of numerical taxonomic techniques to this type of data and to sequence data – both amino acid sequences of proteins and base sequences of DNA – are rapidly expanding fields.

CONCLUSION

The development and availability of electronic computers allowed the growth of numerical taxonomy in the late 1950's and in the 1960's. Without the techniques of numerical taxonomy, biosystematics could not have utilized effectively the advantages

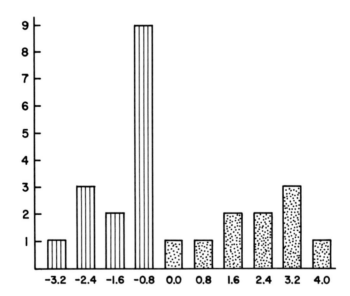

FIG. 10. *Histogram showing the values of a discriminant function based on measurements on 25 flowers from voucher specimens of plants of* Polygonum achoreum *determined to be tetraploid (2n=40) and hexaploid (2n=60), respectively. The different hatching distinguishes the ploidy level; the tetraploids are on the left, the hexaploids on the right.*

that its populational basis implies. This is evident from the
inadequacies of early attempts to represent multivariate data
through polygonal graphs, pictorialized scatter diagrams and the
like.

But the number of standard techniques in numerical taxonomy –
even just clustering techniques – is almost infinite, so that it
is important in biosystematics to be clear of the purpose of the
investigation. I believe that, initially, the purpose is often
one of pattern-seeking. Only later are there specific hypotheses
to be tested. Numerical taxonomic techniques appropriate to both
situations exist and are being increasingly utilized.

Biosystematics also contributes to numerical taxonomy in that
biosystematic studies in which there is a conscious and proper
effort to discard environmental influences make it more evident to
the numerical taxonomist that goodness of fit to the original data
is rarely a good measure of the quality of a numerical technique.

Although growth of biosystematics, at least so far as
population studies are concerned, has been largely dependent on
numerical taxonomy, there is still need for wider use of
multivariate techniques in biosystematics. For the future, I see
continuing essential collaboration between numerical taxonomists
and biosystematists in the analysis of populationally based data,
including data from rapidly developing techniques, such as amino
acid and protein sequences.

ACKNOWLEDGMENTS

I am grateful to Dr. P. M. Catling, now of the Biosystematics
Research Institute, Agriculture Canada, Ottawa, for permission to
use his *Spiranthes* data and to Ms. K. Pryer, Mr. J. Helie
and Mr. G. Ben-Tchavtchavadze for assistance in the preparation
of the illustrations. I am indebted to Mr. L. P. Lefkovitch and
the Statistics and Engineering Research Institute of Agriculture
Canada for some of the computer programs used and to Dr. F. J.
1f, State University of New York, Stony Brook, for others.
Some of the research described was carried out while I was a
research scientist with Agriculture Canada and their support and,
more recently, that of the Natural Sciences and Engineering
Research Council (NSERC) of Canada are acknowledged.

REFERENCES

Anderson, E. 1936. The species problem in *Iris*. *Ann. Mo.*
    Bot. Gard. 23: 457–501.
Anderson, E. 1949. *Introgressive Hybridization*. Wiley,
    New York.
Anderson, E. 1953. Introgressive hybridisation. *Biol. Rev.*
    28: 280–307.
Boivin, B. 1967. Enumération des plantes du Canada. III –
    Herbidées, 1 partie: Digitae: Dimerae, Liberae. *Naturaliste*

Can. 93: 583–646.

Boivin. B. 1968. Flora of the Prairie Provinces. Part II – Digitae, Dimerae, Liberae. *Phytologia* 16: 219–339.

Briggs, D. and Walters, S.M. 1969. *Plant Variation and Evolution.* McGraw-Hill, New York & Toronto.

Burkhill, I.H. 1895. On the variations in number of stamens and carpels. *J. Linn. Soc. (Bot.)* 31: 216–245.

Camp, W.H. 1943. The herbarium in modern systematics. *Am. Nat.* 77: 322–344.

Camp, W.H. and Gilly, C.L. 1943. The structure and origin of species. *Brittonia* 4: 323–385.

Catling, P.M. and Cruise, J.E. 1974. *Spiranthes casei,* a new species from northeastern North America. *Rhodora* 76: 526–536.

Davidson, J.F. 1947. The polygonal graph for simultaneous portrayal of several variables in population analysis. *Madrono* 9: 105–110.

Davis, P.H. and Heywood, V.H. 1963. *Principles of Angiosperm Taxonomy.* Oliver & Boyd, Edinburgh & London.

Duncan, T. and Baum, B.R. 1981. Numerical phenetics: its uses in botanical systematics. *Annu. Rev. Ecol. Syst.* 12: 387–404.

Dunn, G. and Everitt, B.S. 1982. *An Introduction to Mathematical Taxonomy.* Cambridge Univ. Press, Cambridge.

Farris, J.S. 1977. On the phenetic approach to vertebrate classification. In *Major Patterns in Vertebrate Classification* (M.K. Hecht, B.M. Hecht and P.C. Goody, eds.). Plenum, New York, pp. 823–850.

Felsenstein, J. 1982. Numerical methods for inferring evolutionary trees. *Quart. Rev. Biol.* 57: 379–404.

Felsenstein, J. 1983a. *Package for Inferring Phylogenies.* Author, Seattle.

Felsenstein, J. 1983b. *Numerical Taxonomy: Proceedings of a NATO Advanced Studies Institute.* Springer, Berlin.

Gillett, J.B. 1960. *Indigofera hirsuta* L. and *I. astragalina* DC. *Kew Bull.* 14: 290–295.

Gordon, A.D. 1981. *Classification.* Chapman and Hall, London.

Heiser, C.B. and Pickersgill, B. 1969. Names for the cultivated *Capsicum* species (Solanaceae). *Taxon* 18: 277–283.

Heslop-Harrison, J. 1953. *New Concepts in Flowering Plant Taxonomy.* Heinemann, London.

Hutchinson, A.H. 1936. The polygonal presentation of polyphase phenomena. *Trans. Roy. Soc. Can.,* ser. 3, sect. V. 30: 19–26.

Jardine, N. and Sibson, R. 1971. *Mathematical Taxonomy.* Wiley, London.

Lee, A. 1902. Dr. Ludwig on variation and correlation in plants. *Biometrika* 1: 316–319.

Löve, A. and Löve, D. 1956. Chromosomes and taxonomy of eastern North American *Polygonum*. *Can. J. Bot.* 34: 501–521.

Legendre, L. and Legendre, P. 1983. *Numerical Ecology.* Elsevier, Amsterdam.

Ludwig, F. 1901. Variationsstatistische Probleme und Materialen.

Biometrika 1: 11-29.

McNeill, J. 1978. *Silene alba* and *S. dioica* in North America and the generic delimitation of *Lychnis, Melandrium,* and *Silene* (Caryophyllaceae). *Can. J. Bot.* 56: 297-308.

McNeill, J. 1980. Purposeful phenetics. *Syst. Zool.* 28: 465-482.

McNeill, J. 1981. The taxonomy and distribution in eastern Canada of *Polygonum arenastrum* (4x=40) and *P. monspeliense̲ (6x=60)*, introduced members of the *P. aviculare complex.* *Can. J. Bot.* 59: 2744-2751.

McNeill, J. 1983. Taximetrics to-day. In *Current Topics in Plant Taxonomy* (V.H. Heywood, ed.). Academic Press, London.

Mertens, T.R. and Raven, P.H. 1965. Taxonomy of *Polygonum,* section *Polygonum (Avicularia)* in North America. Madrono 18: 85-92.

Neff, N.A. and Marcus, L.F. 1980. *A Survey of Multivariate Methods for Systematics. Privately published, New York (for Am. Soc. Mammalog.).*

Pickersgill, B., Heiser, C.B. and McNeill, J. 1979. Numerical taxonomic studies on variation and domestication in some species of Capsicum. In *The Biology and Taxonomy of the Solanaceae* (J.G. Hawkes, R.N. Lester and A.D. Skelding, eds.). Academic Press, London. pp. 679-700.

Riley, H.P. 1938. A character analysis of colonies of *Iris fulva, Iris hexagona* var. *giganticaerulea* and natural hybrids. *Am. J. Bot.* 25: 727-738.

Simpson, R.C. and Catling, P.M. 1978. *Spiranthes lacera* var. lacera X *S. romanzoffiana,* a new natural hybrid orchid from Ontario. *Can. Field-Nat.* 92: 350-358.

Sneath, P.H.A. and Sokal, R.R. 1973. *Numerical Taxonomy.* Freeman, San Francisco.

Sokal, R.R. and Sneath, P.H.A. 1963. *Principles of Numerical Taxonomy.* Freeman, San Francisco.

Swofford, D.L. and Selander, R.B. 1981. BIOSYS-1: a FORTRAN program for the comprehensive analysis of electrophoretic data in population genetics and systematics. *J. Hered.* 72: 281- 283.

Wolf, S.J. and McNeill, J. 1984. A revision of *Polygonum* sect. *Polygonum* in Canada. in prep.

Problems of Hybridity in the Cladistics of *Crataegus* (Rosaceae)

J. B. Phipps

Department of Plant Science
University of Western Ontario
London, Ontario, Canada

INTRODUCTION

Systematics has been through many phases: α-taxonomy, the grand phylogenetic schemes for major taxa, biosystematics, numerical taxonomy, cladistics. Indeed, in the 1980's all these branches with their distinctive methodologies and aims continue to flourish and many investigators cross and combine areas. It is thus fitting that biosystematics, with its roots in the 20's and 30's, its name coined by Camp and Gilly (1943) and its aim being to identify the natural biotic units, be linked with the dynamically active field of cladistics for both possess an essentially evolutionary orientation.

In my studies of *Crataegus*, a large, north-temperate genus of some 155 broadly-defined species (Phipps 1983) I have perforce had to play out all the roles of a taxonomist. The genus has been revised for Ontario (Phipps and Muniyamma 1980) and the revision for the Vascular Flora of the South-Eastern United States is underway. The large amount of field and herbarium work involved brings one up sharply against the different type of variation patterns found among the different species or species complexes. While there appears to be a considerable number of sexual diploids (eg. North American *C. punctata* Jacq., with clinical variation, and the promiscuous *C. monogyna* Jacq., native to Europe), there are also polyploid and apomictic complexes with sharply but finely dissected local population variation (e.g. *C. crus-galli* sens lat., *C. pruinosa* sens. lat., *C.* § *Rotundifoliae*), as initially indicated by the studies of Rickett (1936,37) and in much more detail by Sinnott and Phipps (1983) and Dickinson and Phipps (1983). Apomixis, long suspected and noted by Palmer (1932) from the unpublished incomplete experiments of Sax, has been proven by us cytologically (Muniyamma and Phipps 1979a;

Dickinson 1983; Smith unpubl.) and much experimental work on reproductive behavior has now been carried out.

Ploidy is extensively documented (e.g. Muniyamma and Phipps 1979b; Gladkova 1968; and others) with counts for some half of all *Crataegus* species showing somewhat similar frequencies of $2x$, $3x$, and $4x$ cytotypes ($x = 17$). Autoploidy presumably exists in apomictic complexes where very similar, though distinct morphs are often $3x$ and $4x$ (Dickinson 1983; Smith unpubl.). Apomictic complexes often have fuzzy edges probably partly due to hybridisation with other species complexes (e.g. Dickinson 1983), and sometimes by links to species from other series. Consequently the α -taxonomy is not 'clean', because 'species' as such are often difficult to define.

Considering the difficulties in species delimitation, it is no surprise, therefore, that relatively little attempt has been made to develop understandings of inter-serial or inter-sectional relationships. The 20-25 series of Loudon (1838), Schneider (1906), Rehder (1940), Rusanov (1965) and others are basically similar, but as mere lists of series or sections, they do not indicate interrelationships. Cinovskis' reticulate graph (1971) is so complex as to confuse, while El-Gazaar's treatment (1980), creating two subgenera *Crataegus* and *Americanae*, is essentially erroneous as I have discussed elsewhere (Phipps 1983).

It is nonetheless an essential preliminary, however, to have a map of the genus showing phenetic interrelationships and my numerical taxonomic studies both on 49 varied morphological characters and 11 leaf characters indicate that some 5 or 6 major groupings stand out, like the species, not always fully clearly, some with odd outliers which may or may not be subsequently placeable as intermediates. To a remarkable extent these major groups parallel biogeographic patterns (Phipps 1983) and this is supportive that *Crataegus* is not totally intractable taxonomically.

On the basis of known hybrids, apomixis, polyploidy, the particular modes of variation, together with a difficult taxonomy, the question must therefore be posed - to what extent has hybridity entered into the evolution of *Crataegus*? - and answers to this lie in combining the disciplines of biosystematics and cladistics with some exact morphometric analyses.

EVIDENCE FOR THE EXTENT OF HYBRIDITY IN *CRATAEGUS* EVOLUTION

What is the evidence for hybridisation in *Crataegus*?

a) The subfamily Maloideae has its origin in a polyploid hybrid cross $(x = 8 + x = 9) = (x = 17)$ as is widely and plausibly believed (Challice 1981). Polyploidy (entirely $3x$, $4x$) is widespread in the subfamily and there is no *a priori* reason not to expect further alloploids besides the subfamilial ancestor.

b) Numerous intergeneric hybrids with other Maloidae (Phipps 1983) have been reported.

c) Several extant hybrid complexes have been demonstrated in *Crataegus*: e.g. *C. monogyna* × *C. laevigata* (Byatt 1975, 1976); *C. monogyna* × *C. douglasii* (Love and Feigen 1979).

d) Documented horticulturally-generated hybrids e.g. *C. prunifolia (C. crus-galli* × *C. macracantha), C. lavallei (C. mexicana* × *C. crus-galli, C. 'Toba' (C. monogyna* derivative) etc. are known.

e) Numerous taxa, ranging in frequency from sporadic individuals e.g. *C. disperma, C. grandis*, to whole series, e.g. *Brainedianeae, Triflorae, Dilatatae*, and others, are "Gestalt" intermediates. Proof, however, of their hybrid origin is lacking at this time. Likewise the gigantism of Sect. *Coccineae* and its putative autoploidy represents its main difference from Sect. *Tenuifoliae*.

f) Aggressive, vigorous, weedy, polyploid, apomictic complexes commonly manifest high levels of hybridity.

In fact, we have some proven hybrids, ample evidence of the potential for hybridisation, a few morphometrically investigated intermediates and a considerable number of 'Gestalt' intermediates requiring classification. It is, therefore, entirely probable that hybrid taxa, possibly including very distinct ones, exist quite widely in *Crataegus* and it is pertinent to seek these out.

In order to estimate the extent to which hybridity has been involved in the main lines of *Crataegus* evolution, one must therefore consider (a) characteristics of different hybrid and

TABLE I. *Generalised characteristics of hybrid types*

	Characters size	Characters of form	1 or few gene characters	Totally new character states
Normal Hybrid	intermediate (within parental range)	intermediate (within parental range)	resemble only one parent	no
Alloploid	larger	intermediate (within parental range)		yes
Autoploid	larger	same as parent		

polyploid situations and (b) how hybrids are incorporated into cladograms.

HYBRID CHARACTERISTICS

In cladograms incorporating hybridity, such as those reported by V. Funk (this conference), the hybrid may be identified on the basis of the apparent synapomorphy sum of two ancestors. In some illustrated examples, hybridity is proposed even though total parsimony was only very slightly increased. While this approach represents a straightforward algorithm it is clear that more complex criteria may be required in complex cases. Consider three main possibilities (see Table I):

   a) Hybrid may be 'simple' (i.e. F_1 and modal derivatives with no ploidy change).

Here it may be anticipated that criteria of morphological intermediacy would satisfy inference very strongly. Experimental investigation is, in principle, simple.

   b) 'Hybrid' may be autoploid.

Here some 'gigantism' is often indicative, with little other change. Autoploids may be very difficult to detect algorithmically in cladograms as they can have the potential to engander wholescale character trend reversal. Also they often possess strong similarity to another species.

   c) Hybrid may be alloploid or segmental alloploid.

This case is also complicated. As Gottlieb (1976) argues persuasively, genic interaction may modify some characters such

that they represent "new" states thus complicating the phenetic
identification of alloploid ancestors. Nevertheless many
characters of form prove to be intermediate between parents.

Certain general comments may also be noted, that hold true
irrespective of the type of hybrid (see Table II). Character
states symplesiomorphic for both ancestors should usually also be
found in the hybrid (occasionally not, and most likely in
alloploids; or, also, if character wrongly coded). Finally, it
must be noted that taxa of ancient hybrid origin may no longer
perfectly fit the expected criteria due to continued evolution.
This argues for setting a tolerance zone (≃ confidence band)
around the algorithmic threshold expectation for hybridity.

PROBLEMS CONCERNING THE PROOF OF HYBRIDITY

The processes leading to proof may be long and complicated as the
outline below indicates:

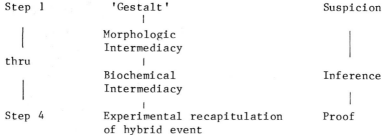

We are well aware that our analysis will often not go beyond
inference, due to the large number of possible instances.
However, if there were a known frequency of inference leading to
proof, then the reliability of inference would have a probability
level.

Cladograms such as we are dealing with, reflecting the course
of evolution of a large genus since the early or mid-Tertiary,
must of necessity be concerned with main lines and distinctive
taxa. Thus, the hybrid must be wide enough for it to be
recognized as a clearly defined taxon, and, in the simple case we
must postulate selection around the F_1 or the intermediate mode in
the early history of the hybrid.

Biochemical tests may therefore be helpful. Phenolics
extracted under standardized conditions are valuable, due to the
quite large number of easily returnable characters. Isozyme
analysis is very useful but tends to generate relatively few
characters for the price. As with morphometric analysis, some
intermediate, or additive, outcome is required as supporting the
hybrid hypothesis. Horticulturalists have generated a series of
hybrid entities (some true-breeding) of known parentage. These
could be tested by biochemical methods to validate a methodology
for general use in putative cases. If effective, it would become
the basis for an extensive assay in *Crataegus*.

IDEAL CHARACTER

SUITES FOR HYBRID

(3 STATE CHARACTERS 0,1,2)

A. NO NEW PLOIDY

Char. No	1	2	3	4	5	6	7	8	9	10	11	12	13	14	15	ORDER
HYB	1	0	.5	0	0	1.5	.5	1.5	2	0	.5	1	1	1.5	0	11.5
OTU2	2	0	1	0	0	1	0	2	2	0	1	1	2	2	0	14
OTU1	0	1	0	0	0	2	1	1	2	0	0	1	0	1	0	9

B. ALLOPLOID

Char. No	1	2	3	4	5	6	7	8	9	10	11	12	13	14	15	ORDER
HYB	2	1	1	0	0	1.5	.5	1.5	4	2	.5	2	2	3	0	21
OTU2	2	0	1	0	0	1	0	2	2	2	1	1	2	2	0	14
OTU1	0	1	0	0	0	2	1	1	2	0	0	1	0	1	0	9

C. Autoploid

HYB	0	2	0	0	0	2	1	1	4	0	0	2	0	2	0	14
OTU2	0	1	0	0	0	2	1	1	2	0	0	1	0	1	0	9
Char. No	1	2	3	4	5	6	7	8	9	10	11	12	13	14	15	

'SYNERGISTIC' CHARACTER UNDERLINED IN BOTH PLOIDS

FIG. 1. *Formalisation of criteria for hybridity: ideal character suites.*

TABLE II. *Some morphometric rules for detecting hybridity*

PRIMITIVE CHARACTER STATES

   1) Hybrid has all found in both ancestors
   2) Hybrid has no other characters in the primitive state

CHARACTERS OF SHAPE AND FORM

     Usually intermediate

CHARACTERS OF SIZE

     Where no new ploidy - intermediacy
     Where ploidy   - increase frequent

CONSIDERATION OF ORDER (GRADISTIC LEVEL)

     With no new ploidy - intermediate between parents
     With new ploidy   - generally greater than either
                         parent

THE THREE GENERAL STRATEGIES FOR HYBRID PLACEMENT IN CLADOGRAMS

From a procedural point-of-view, a cladistical analysis of a genus such as *Crataegus* may logically be approached in one of three ways (see Table III) (e.g. Wagner 1983).
   Is there a natural advantage to one of these approaches? Clearly, this depends on whether reticulation in a cladistic tree

TABLE III. *Strategical options in cladogenesis with hybridity*

---

   1. Identify and remove hybrids
      Make cladogram
      Re-insert hybrids

   2. Ignore hybridity
      Use preferred cladistic algorithm
      Revise tree on extrinsic bases

   3. Incorporate algorithmic rules for putative hybrid
      situations
      Do cladogram direct
      Flag putative hybrids and check

POSSIBLE HYBRID SITUATIONS

(PARENTS ARE L, M, H = HYBRIDS)

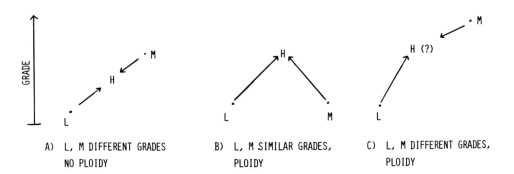

FIG. 2. *Formalisation of criteria for hybridity: affects on grade.*

is, in reality, common enough to be concerned about, and whether, if it is, it would affect the choice of tree-building systems. It also depends on the accuracy with which hybrids may be identified. On this basis Herb Wagner has used method 1 with ferns as addressed earlier.

While there are theoretical reasons for preferring one or other of these approaches, there are also practical ones. Firstly, does one know all the hybrids? Second, what is the proof of hybridity? Recognizing that in a large genus like *Crataegus* the definitive proof of hybridity in all inferred cases would take much time to demonstrate, a heuristic approach is appropriate.

ALGORITHMIC CRITERIA FOR HYBRIDITY IN THE CLADOGRAM

In a large and old genus like *Crataegus*, dating back perhaps some 50 My., narrow hybrids are not sought, both due to problems in detection and due to their relative lack of cladistic significance. To help handle the diversity of hybrid types and some of the 'looseness' inherent in their detection, the program PRIMO3 has been developed. Its data base consists of (in this case) 77 OTUs (including one known hybrid) × 23 polarised characters, derived from outgroup comparisons with other Maloideae. Then an agglomerative fusion approach uniting OTU's/HTU's in descending order of synapomorphic relationship is executed. The program basically has two options:

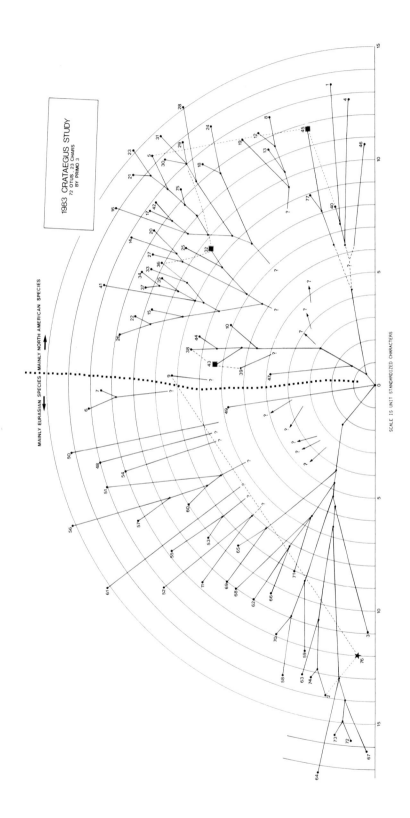

1983 CRATAEGUS STUDY
72 OTUS 23 CHARS
BY PRIMO 3

MAINLY EURASIAN SPECIES ● MAINLY NORTH AMERICAN SPECIES

SCALE IS UNIT STANDARDIZED CHARACTERS

   a) proceed to completion assuming no hybridity,
   b) use a stopping rule where hybridity is suspected or major
      isolation demonstrated - for instance, where the Manhattan
      distance exceeds a set threshold between the basal member of
      a clade (as defined at that point in the program) and its
      nearest outgroup neighbor as described in Phipps (1983). An
      appropriate value of this threshold appears to be about one-
      third the maximum Manhattan distance possible in the study.
   When the critical threshold is reached, the search for putative
hybridity is executed. With formalisation of the criteria
discussed earlier e.g. Figs. 1 and 2, the program searches for
pais of putative ancestors and estimates levels of acceptability.
Where putative hybrids are relatively few they might be built
directly into the cladogram. Where, however, putative hybridity
and extreme isolation appear more frequent, it is more pointful to
set out each hybrid option with pertinent data for separate study
and potential verification by known methods.
   Therefore, quite elaborate tests are required, and these are
built into PRIMO3.

RESULTS

The cladogram for 77 OTU's (including one known hybrid as a
standard) is shown in Fig. 3. It may be compared with Felsenstein
compatability taken to one level (Fig. 4) which is broadly
similar. Other cladistic algorithms executed did not provide
readily interpretable results.
   Four putative hybrid situations are set out in Tables II-V.
These are selected purely for illustrative purposes and were not
necessarily found by the algorithm. Fig. 6 illustrates
distinguishing features of the latter three cases which are
considered in less detail.
   The first case considered, that of *C. monogyna* × *C. punctata*, is
selected to verify the methodology, as this is a certain hybrid
situation. The parental and hybrid characteristics are
pictorially shown in Fig. 3 while the overall intermediacy of the
hybrid is shown in the PCA, Fig. 4. Table IV reveals only one
character (no. 13) outside the algorithmic standards and indicates
a loss of pigmentation in another color; since this character is
variable in *C. punctata* the result in the hybrid has little
significance.
   Table V displays the results for the very rare species *C.*
triflora the geographic range of which is almost exactly

FIG. 3. *Cladogram run with Primo 3. OTU number same as in
Phipps (1983). Certain hybrids shown (definite indicated by star;
putative by square). Long internodes disconnected deliberately
due to ambiguities. Dashed lines between hybrid and ancestors.
For synapomophies of clades see Phipps (1983, fig. 8). Branch-
swapping done to improve parsimony.*

1983 CRATAEGUS CLADOGRAM

37 BINARY CHARACTERS

FELSENSTEIN COMPATABILITY

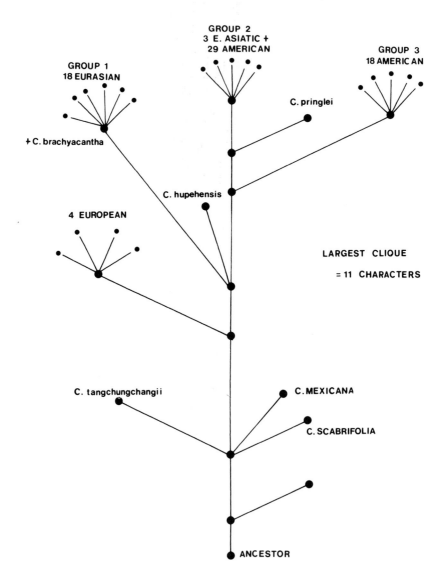

FIG. 4. *Felsenstein compatability method for 37 binary characters stopped at first clique. Only method tried generating strong resemblances to Primo 3.*

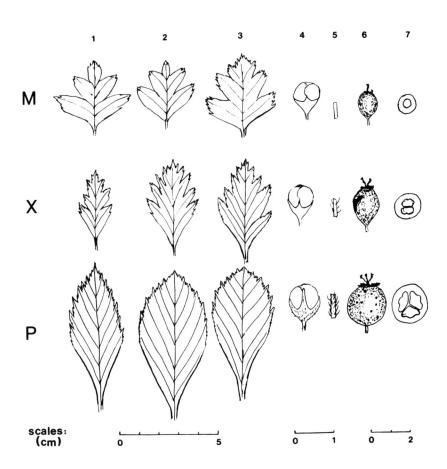

FIG. 5. *Characteristics of* C. monogyna *Jacq.*, C Jacq. and its hybrid in Ontario. Row M - C. monogyna; *row X - hybrid; row P - C.* punctata. *Columns 1-3, characteristic short shoot foliage (Col. 1 - Fansher Rd., Lambton Co.; col. 2 - Ancaster; col. 3 - Usborn Twp., Perth Co.); col. 4 - flower buds (note calyx shape and pubescence); col. 5 - pedicels; col. 6 - fruit (filaments omitted); col. 7 fruit cross-section showing pyrenes.*

sympatric to that of its putative parents *C. mollis* and *C. flava*. Illustrations of the three species may be found in Fig. 7. Two features stand out. Firstly, the algorithmic standard of intermediacy is very high. In the nonconforming characters, no. 16 (fruit size) is morphometrically intermediate but this cannot conform to the divergent fruit size evolution as encoded for the

TABLE IV. *Analysis of actual hybrid* C. monogyna × C. punctata[a]

	1	2	3	4	5	6	7	8	9	10	11	12	13	14	15	16	17	18	19	20	21	22	23	
MON X PUNCT	.5	1	1	.33	0	1	1	0	0	0	1	1	0	0	1	.5	.66	1	0	0	-1	1	1	8[b]
C. monogyna	.5	1	1	.66	0	0	1	0	0	-1	1	1	1	0	.5	1	.66	1	0	0	-1	1	.33	7[b]
C. punctata	.5	1	1	.33	0	1	0	0	0	0	1	.5	1	0	.5	0	.33	1	0	0	1	0		11[b]
Zero in all 3[c]					x			x	x					x					x	x				
Zero in hybrid but not in either parent[c]													y											
Within parental (range)[d]	x	x	x	x							x	x					x	x			x		x	
Values larger than either parent[d]															x									
Values smaller than either parent[d]																	x							

| Characters | 1 | 2 | 3 | 4 | 5 | 6 | 7 | 8 | 9 | 10 | 11 | 12 | 13 | 14 | 15 | 16 | 17 | 18 | 19 | 20 | 21 | 22 | 23 | |

[a]In this example all the Manhattan distances are large and hybrid threshold search was set at 3.5. Although this is a definite hybrid, of certain parentage, note that a few of the characteristics lie outside the expected range. $MD_{(m,p)}$ = 6.5, $MD_{(x,m)}$ = 4.7, $MD_{(x,p)}$ = 4.8.

[b]Primitive, [c]Primitive characters, [d]Characters of hybrid nonzero in both parents. x, indicates prediction validated; y, lack of congruence.

TABLE V. *Putative hybrid analysis for C. triflora (Sect. Triflorae)*[*]

	1	2	3	4	5	6	7	8	9	10	11	12	13	14	15	16	17	18	19	20	21	22	23	
TRIFLORA	0	.75	1	.33	0	1	0	0	0	1	0	.5	0	0	.5	0	.66	0	0	-1	0	1	0	13[b]
C. flava	0	.75	1	1	0	1	0	0	0	0	1	.5	1	1	.5	0	0	0	1	-1	1	1	0	11[b]
C. mollis	0	.75	1	.33	.5	1	0	0	0	0	0	.5	0	.5	.5	.66	0	0	1	1	1	1	.33	11[b]
Zero in all 3[c]	x						x	x					x					x	x					
Zero in hybrid, but not in either parent[c]																								
Within parental (range)[d]		x	x	x		x			x	x	x	x		x	x		x							
Values larger than either parent[d]																	x			y		x	x	
Values smaller than either parent[d]																x								
Characters	1	2	3	4	5	6	7	8	9	10	11	12	13	14	15	16	17	18	19	20	21	22	23	

[a] In this example, *C. triflora* is generally considered very marginal to its most obvious neighbors due to the possession of certain unusual characters. These unusual characters appear to be derived from the *C. flava* complex, widely separated from *C. mollis*. MD$_{(f,m)}$ = 8.2, MD$_{(t,f)}$ = 7.3, MD$_{(t,m)}$ = 3.8.

[b] Primitive, [c] Primitive characters, [d] Characters of hybrid nonzero in both parents.

x, indicates prediction validated; y, lack of congruence.

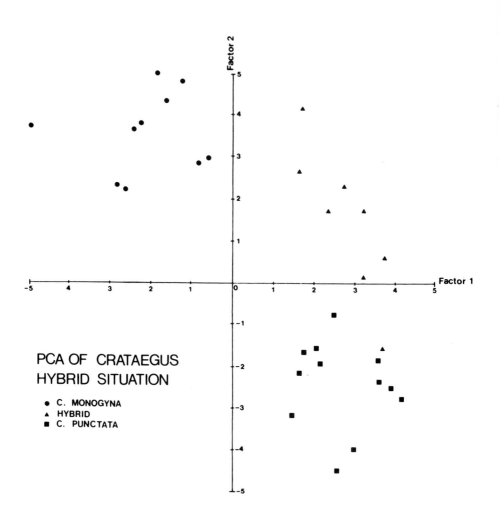

FIG. 6. *PCA of* C. monogyna *(M),* C. punctata *(P) and its
hybrid (X) from Ontario sites. 49 morphological characters.
Note position of* C. monogyna *hybrid on intercentrid axis.
69% of variance accounted for by first root.*

cladistics while character 21 (relating to leaf form) would differ
less markedly with more discriminating cladistic coding. Secondly
C. triflora is clearly much closer to *C. mollis* than to *C.
flava*. Indeed, the entire *C. flava* complex is very isolated
taxonomically but in this interpretation becomes responsible for
certain unusual apomorphies found in series *Triflorae* and
Bracteatae. This may be construed as a Holocene hybrid.

The example of *C. conspecta* (series *Dilatatae*) shown in Table VI and Fig. 7 is more ambiguous in its results. The relationship to series *Pruinosae*, most likely *C. pruinosa* var. *leiophylla*, is especially clear because of a suite of distinctive synapomorphies of the fruit, but a more exhaustive examination is required to determine if other species in series Sect. *Coccineae*, or even Sect. *Molles* represent the other parental source. All the inconsistencies are most easily explained as representing synergistic effects here operating as character reversal trends.

Table VII displays the relationship between *C. brachyacantha* (putative hybrid) and *C. viridis* and *C. douglasii* and the species concerned are illustrated in Fig. 7. This remarkable relationship was initially quite unexpected and *C. brachyacantha* has previously located in very isolated parts of various trees computed. Furthermore the nearest edges of the putative parents' existing ranges are some 2000 km apart (750 km if *C. rivularis* is treated as variety of *C. douglasii*) while *C. douglasii* ranges NW to Alaska and *C. viridis* SE to Florida. The width of the cross need not occasion surprise in view of the studies on *C. monogyna* × *C. douglasii* mentioned earlier (Love and Feigen 1979) but the allopatry and strong habitat differentiation together with its unusual thorns (expressed in char 2) suggest the possibility of a hybrid of some considerable age. *Viridis* × *saligna* and *viridis* × *rivularis* crosses would be seriously considered in an in-depth study.

Finally, such very isolated taxa as *C. cuneata* Sieb. & Zucc. (S. China), *C. hupehensis* (C. China), and *C. tangchungchangii* (W. China) provided no indications of hybrid origin, nor does *C. flava* (S.E. U.S.A.) nor *C. Sect. Aestivales* (S.E. U.S.A.) and such taxa therefore readily conform to notions of being long isolated relicts form of ancient divergence.

CONCLUSION

Bearing in mind the complex bisystematics of *Crataegus*, relatively complex algorithmic criteria were built into the cladistic search for hybridity in the program PRIMO3. One of the interests is the emerging notion of establishing the frequency of major hybridization events in *Crataegus*. By a major event we imply a wide cross with a successful outcome. Some six or seven of these are now coming to light, but this disregards the many narrower crosses attainable. The results to date are very promising, and the writer believes that suggested hybrids pulled out in this way may help in elucidating the evolutionary history of complex taxa.

ACKNOWLEDGMENTS

This work was supported by NSERC grant A-1726 to the author. The expert help of my assistants Mary MacLeod and Joanne Meininger has been indispensable.

TABLE VI. *Putative hybrid analysis for C. conspecta (Sect. Dilatatae)*[a]

Characters	1	2	3	4	5	6	7	8	9	10	11	12	13	14	15	16	17	18	19	20	21	22	23	
CONSPECTA	0	.75	1	.66	.5	1	0	0	1	0	0	.33	0	.5	.5	.66	1	0	0	0	1	0		10[b]
C. pruinosa var. leio.	.5	1	1	.5	.5	1	0	1	1	1	0	.66	0	0	.5	.66	1	.5	1	0	0	1	0	6[b]
C. pringlei	.5	.75	1	.33	.5	1	0	0	1	0	0	.66	1	.5	0	.66	.25	.5	.33	0	0	0	0	8[b]
Zero in all 3[c]									x									x	x					
Zero in hybrid, but not in either parent[c]	y																							
Within parental (range)[d]												y			y	y								
Values larger than either parent[d]			x	x	x	x		x	x	x	x		x	x			x							
Values smaller than either parent[d]																								

[a] Nearest neighbor of *C. conspecta* was OTU 25, *C. lucorum*, with $MD_{(c,luc)}$ = 2.82, but *C. lucorum* is a rare taxon of dubious nature. All chromosome counts in the series of the three species concerned are $3x$ or $4x$. $MD_{(1,pr)}$ = 6.9, $MD_{(c,pr)}$ = 6.0, $MD_{(c,1)}$ = 6.25.

[b] Primitive, [c] Primitive characters, [d] Characters of hybrid nonzero in both parents. x, indicates prediction validated; y, lack of congruence.

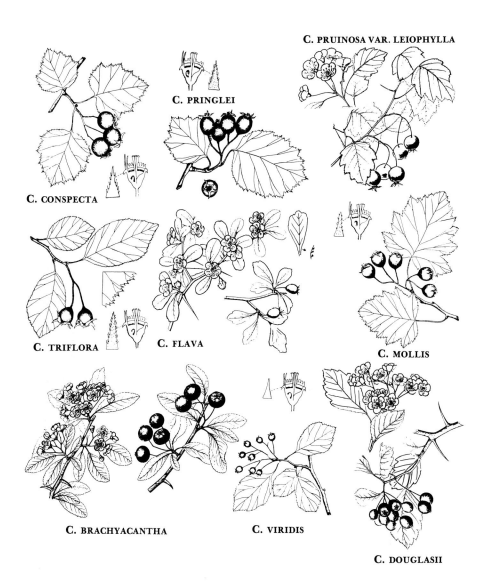

FIG. 7. *Illustrations of putative hybrid situations, adapted from Sargent (1901) and Cinovskis (1971). a) C. triflora (putative hybrid) and putative ancestors C. mollis, C. flava. b) C. conspecta (putative hybrid) and putative ancestors C. pringlei, C. pruinosa var. leiophylla. c) C. brachyacantha (putative hybrid) and putative ancestors C. viridis and C. douglasii.*

TABLE VII. *Putative hybrid analysis for* C. brachyacantha[a]

	1	2	3	4	5	6	7	8	9	10	11	12	13	14	15	16	17	18	19	20	21	22	23
BRACHYACANTHA	0	.50	1	.33	0	1	0	1	1	-1	.5	0	0	.5	.5	1	0	1	0	-1	.5	0	8[b]
C. viridis	0	.75	1	.33	0	1	0	1	1	0	0	0	0	.5	1	.33	0	0	0	1	0	0	12[b]
C. douglasii	1	.75	1	.66	0	1	0	1	1	.5	.66	.5	.5	.5	1	1	1	0	1	0	1	.33	5[b]
Zero in all 3[c]					x		x																
Zero in hybrid, but not in either parent[c]						x																	
Within parental (range)[d]	x		x	x				x	x		x	x	x	x			x		x			x	
Values larger than either parent[d]										x								x					
Values smaller than either parent[d]															x	x				x			
		y																			y		
Characters	1	2	3	4	5	6	7	8	9	10	11	12	13	14	15	16	17	18	19	20	21	22	23

[a] In this example all the Manhattan distances between pairs are very large. But *C. brachyacantha*, seemingly related to little except *C. saligna* Greene, scores well on hybridity criteria in the above table. MD(v,d) = 6.5, MD(v,b) = 5.4, MD(b,d) = 8.57.

[b] Primitive characters, [c] Primitive characters, [d] Characters of hybrid nonzero in both parents, x, indicates prediction validated; y, lack of congruence.

REFERENCES

Byatt, J.I. 1975. Hybridisation between *Crataegus monogy Jacq. and C. laevigata* (Poir.) DC. in south-eastern England. *Watsonia* 10: 253-264.

Byatt, J.I. 1976. The structure of some *Crataegus* populations in northeastern France and south-eastern Belgium. *Watsonia* 11: 105-115.

Camp, W.H. and Gilly, C.L. 1943. The structure and origin of species. *Brittonia* 4: 323-385.

Challice, J. 1981. Chemotaxonomic studies in the family Rosaceae and the evolutionary origins of the subfamily Maloideae. *Preslia* 53: 289-304.

Cinovskis, R. 1971. Boyaritschniki Pribaltici. Crataegi Baltici. Riga (in Russian).

Dickinson, T.A. 1983. A study of the *Crataegus crus-galli* L. species complex in southern Ontario. Ph.D. Thesis, Univ. Western Ontario, London, Ontario.

Dickinson, T.A. and Phipps, J.B. 1983. The *Crataegus crus-galli* L. species complex in southern Ontario. I. Reproductive behaviour comparison with *C. punctata* Jacq. *Am. J. Bot.* submitted.

El-Gazaar, A. 1980. The taxonomic significance of leaf morphology in *Crataegus* (Rosaceae). *Bot. Jahrb. Syst.* 101: 457-469.

Gladkova, V.N. 1968. Karyological studies on the genera of *Crataegus* L. and *Cotoneaster* Medik. (Maloideae) in relation to their taxonomy. *Bot. Zh.* 52: 354-356.

Gottleib, L.D. 1976. Biochemical consequences of speciation in plants. In *Molecular Evolution Symposium.* (A. Ayala, ed.). Publ. Seminar.

Loudon, J.C. 1838. *Arboretum et Fruticetum Brittanicum.* London.

Love, R. and Feigen, M. 1979. Intraspecific hybridisation between native and naturalised *Crataegus* (Rosaceae) in western Oregon. *Madrono* 25: 211-217.

Muniyamma, M. and Phipps, J.B. 1979a. Pollen meiosis and polyploidy in *Crataegus. Can. J. Genet. Cytol.* 21: 231-241.

Muniyamma, M. and Phipps, J.B. 1979b. Cytological proof of apomixis in *Crataegus Am. J. Bot.* 66: 149-155.

Palmer, E.J. 1932. *Crataegus.* problem. *J. Arnold Arbor.* 13: 342-362.

Phipps, J.B. 1983. Biogeographic, taxonomic and cladistic relationships between East Asiatic and North American *Crataegus. Ann. Mo. Bot. Gdn.* In press.

Phipps, J.B. and Muniyamma, M. 1980. A revision of *Crataegus* for Ontario. *Can. J. Bot.* 58: 1621-1699.

Rehder, E. 1940. Manual of Cultivated Trees and Shrubs Hardy in North America. 2nd ed. MacMillan, New York.

Rickett, H.W. 1936. Forms of *Crataegus pruinosa. Bot. Gaz.* 97: 780-793.

Rickett, H.W. 1937.   Forms of *Crataegus crus-galli*.   *Bot. Gaz*. 98: 609–616.

Rusanov, F.N. 1965.   Introdutsironavye boyaritschniki botanicheskogo sada an UzSSR, *In* Dendrologie Uzbekistanii (in Russian).

Schneider, C.K. 1906.   Illustriestes Handbuch des Laubholzkunde I: 766–802, 1001–1008, Fischer, Jena.

Sinnott, Q.P. and Phipps, J.B. 1983.   Variation patterns in *Crataegus* Sect. Pruinosae (Rosaceae) in southern Ontario. Syst. Bot.

Wagner, W.H. Jr. 1983.   Reticulistics: The recognition of hybrids and their role in cladistics and classification.   In *Advances in Cladistics* (N.I. Platnick and V.A. Funk, eds.).   Columbia Univ. Press 2: 63–79.

Population Biology and Biosystematics: Current Experimental Approaches

Barbara A. Schaal

Department of Biology
Washington University
St. Louis, Missouri, U.S.A.

INTRODUCTION

Population biology and systematics have had a long and close
relationship. This is particularly the case in the plant sciences
where much of population biology stems directly from
biosystematics. In fact, many of the researchers who have made
major contributions to understanding the population biology of
plants have their origins and early training in systematics. Both
fields are important anchors of evolutionary biology.
Biosystematics has traditionally been concerned with the
evolutionary relationships among taxa and with the mechanisms of
speciation and evolution. Population biology, on the other hand,
considers the ecological and genetic mechanisms operating within a
single species in an evolutionary context. There is necessarily a
great deal of overlap between the two fields; population biology
places more emphasis on evolutionary mechanisms in the short term
while biosystematics tends to emphasize evolutionary relationships
on a larger scale.

A simultaneous consideration of both populational and
biosystematic features is beneficial and often essential to
research in both fields. For example, a population analysis of
allozyme gene frequencies can aid in differentiation of closely
related taxa. Such populational studies have been useful in
resolving systematic relationships in a number of groups (e.g.
Gallez and Gottlieb 1982). Likewise, a comparison of the
population biology of two closely related taxa has been a powerful
tool in understanding the mechanisms of various population
processes. The value of the comparative approach is illustrated
by the classical studies of Sarukhan (1974) which detail the
numerical dynamics of three *Ranunculus* species.

The following is a comparison of the population biology of two

closely related annual species of lupines, *Lupinus texensis*
and *Lupinus subcarnosus*. The comparison has two purposes:
(1) to illustrate the kinds of populational differences two
closely related species can exhibit and (2) to demonstrate the
value of the comparative approach in the study of plant population
biology.

THE SPECIES

Lupinus texensis and *L. subcarnosus* are two closely related,
winter annuals endemic to portions of Texas. Both occur in
central Texas where they are edapically separated from each other.
L. texensis occurs on calcarious prairie soils or on the rocky
soils of the Edwards plateau. *L. subcarnosus* is more restricted
in range and occurs naturally only in sandy soils. Both species
are morphologically similar and can be somewhat difficult to
distinguish in the field. Morphological characters such as size,
leaf shape, pubescence, and number of flowers are highly variable
in both taxa. *Lupinus texensis* is distinguished most reliably
from *L. subcarnosus* by the lack of inflation in the wing
petals and by the whitish appearance of unopened flowers on the
raceme. Both taxa have a chromosome number of $n = 18$ (Turner
1957). Both species are predominantly outcrossing, although both
will set seed in greenhouse self-crosses at a very low frequency
(less than 1%). The species can be hybridized in the greenhouse
with very low fertility, less than 0.05% (Schaal unpublished).
Hybridization is unlikely to occur in nature because they are
edaphically separated in their distributions and because of the
low seed set of such interspecific crosses. These lupines share a
suite of bee pollinators, and both taxa show similar adaptations
for bee pollination, such as changes in banner spot color (Schaal
and Leverich 1980).

GENETIC STRUCTURE

A major area of inquiry in population biology is that of the
genetic structure of species. Gene flow, the movement or
migration of genes, is one of the most important processes which
can influence genetic structure. While the genetic consequences
of various levels of gene flow have been extensively studied on a
theoretical basis (e.g. Wright 1943; Kimura and Weiss 1964;
Felsenstein 1975; Takahata 1983) there is still relatively little
information on gene migration within and among natural populations
(Levin and Kerster 1974; Slatkin 1981; Handel 1982). Levels of
migration have a significant effect on the systematic
relationships of species. If gene flow is very restricted,
genetic differentiation can occur, both within a population and
among the populations of a species. On the other hand, if gene
migration is widespread among populations, then genetic uniformity
is expected throughout a species. Finally, any gene flow between

species can result in hybrid zones or can potentially limit
differentiation of the two species.

It is in this framework that the genetic structure of the two
species will be compared. First levels of gene flow will be
detailed. Then, within-population genetic structure will be
analyzed. Finally, inter-population variation will be compared.

We expect levels of gene flow to be similar in both species.
Floral characteristics of species are essentially the same. Both
species have flowers of the same size, shape and color. Flowers
are blue with a white banner spot which in both species changes to
red after the flower is no longer fertile. The distributions of
the number of flowers per inflorescence are broadly overlapping in
the two taxa. The species share the same pollinators, and no
significant differences are found in pollinator foraging patterns
between the two species (Schaal unpublished). It can be
reasonably assumed that levels of gene flow within populations of
the two species are very similar.

The actual pattern of gene flow was measured in an experimental
population of *Lupinus texensis* (Schaal 1980). The population
consisted of plants homozygous for one of several phosphoglucose-
isomerase alleles; plants at the center of the population were
homozygous for a different allele than were the other individuals.
Bees were allowed to forage in the population, the resulting
progeny were collected, and these plants were assayed for
genotype. Gene flow was detected by heterozygous genotypes in the
progeny. Figure 1 shows both the pollinator and gene dispersal
distributions within the population. Gene flow is clearly
restricted, although not as restricted as the pollinator foraging
distribution alone would suggest. Mean gene migration is 1.82 m
(SE = 0.10 m) whereas the mean pollinator foraging flight is
0.97 m (SE = 0.08 m). Both distributions are significantly
leptokurtic (Schaal 1980). Differences between the two curves are
due to pollen carryover.

Gene migration in plants also occurs by seed dispersal. Seeds
of *L. texensis* and *L. subcarnosus* are dispersed by explosive
dehiscence of the fruit. Seed dispersal is also restricted; the

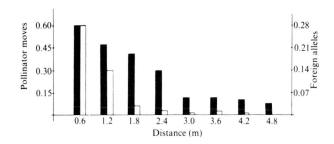

FIG. 1. *Gene flow in* Lupinus texensis. *The actual gene
flow is plotted by the solid bars. The pollinator foraging
distribution is plotted by the open bars. From Schaal (1980).*

mean dispersal distance for *L. texensis* in an experimental
population is 0.58 m (SE = 0.04 m) (Schaal 1980). Since both
pollen and seed components of gene flow are restricted, it is
expected that lupine populations would not undergo random mating.
Wright (1946) has defined the genetic neighborhood as that area
within a population which undergoes panmixia. Neighborhood size
calculated for these lupines is 95.4 plants. Lupine populations
can be very large, comprising often over 10,000 plants; one
expects such populations to consist of many genetic neighborhoods.
    Such restricted gene flow patterns, where a population is
subdivided into genetic neighborhoods, are expected to have
important genetic consequences for a species. If gene flow is
restricted, genetic subdivision of population is predicted (Wright
1969). Subdivision is manifested by significant genetic
heterogeneity within a population and often by inbreeding.
Populations of both lupine species were analyzed for genetic
subdivision. Two populations of each species were chosen for
intensive analysis. A sample of 30 plants was collected at 5 m
intervals along a 50 meter transect and allozyme frequencies for
three loci were measured. Subdivision was determined by an
analysis of F-statistics (Wright 1965). Table I shows the loci
and F-statistics values for the populations. Both species show
significant genetic substructuring as indicated by $F(st)$. $F(st)$
is the standardized genetic variance. Values of $F(st)$ can range
from 0 to 1. Zero indicates no substructure, and 1 is the maximum
possible substructuring. $F(st)$ ranges from 0.06 to 0.18 in *L.
subcarnosus* and from 0.18 to 0.2 in *L. texensis*. Thus genetic
differentiation of allozyme frequencies occurs within populations
of both species. *L. texensis* appears to be slightly more
subdivided than *L. subcarnosus*; mean $F(st)$ values are 0.16
and 0.11, respectively. Both species also show significant
deviations from Hardy-Weinberg expectations, as measured by $F(is)$.
A positive $F(is)$ value indicates heterozygote deficiency, whereas
a negative $F(is)$ value is due to heterozygotes in excess of Hardy-
Weinberg expectations. $F(is)$ values are similar for both species;
mean $F(is)$ for *L. texensis* is 0.25 and ranges from 0.20 to 0.30,
while mean $F(is)$ for *L. subcarnosus* is 0.28 and ranges from
0.24 to 0.32. Total fixation indices are about the same for the
two species. The greater $F(is)$ in *L. subcarnosus* is balanced
by the greater $F(st)$ in *L. texensis*. The analysis of F-
statistics indicates that both species have essentially the same
population substructure, although there is some heterogeneity
between species, populations, and loci. This similarity in
population substructuring of the two species is not surprising,
considering the similarity of their gene flow patterns.
    The last aspects of population structure to be considered are
the levels and organization of genetic variability within each
species. While the two species show no major differences in
population substructure or in deviation from Hardy-Weinberg
expectations, the species are quite different in their overall
levels of genetic variability. Babbel and Selander (1974) in a

study of the genetic organization of these two species, found
consistent differences in the overall level of variation within
populations of the species. In *L. texensis* the average number
of alleles per locus per population is 3.12 and the average genic
heterozygosity is 0.36, whereas in *L. subcarnosus* the average
number of alleles per locus per population is 1.84 and the average
heterozygosity is 0.09 (Babbel and Selander 1974). In addition to
differences in the within-population component of genetics
structure, levels of geographical variation also differ slightly
between the species. Mean genetic identity of populations of
L. texensis is 0.957 and mean genetic identities of populations

TABLE I. *F-statistics of* Lupinus texensis *and* L. subcarnosus

| | Lupinus subcarnosus | | |
	F(is)	F(st)	F(it)
Population 1			
ADH	.29	.06	.33
GOT	.32	.07	.37
6-PGDH	.31	.12	.39
Population 2			
ADH	.24	.18	.37
GOT	.27	.13	.36
6-PGDH	.26	.11	.34
Mean	.28	.11	.36
	Lupinus texensis		
	F(is)	F(st)	F(it)
Population 1			
ADH	.28	.19	.41
GOT	.21	.20	.37
6-PGDH	.22	.15	.34
Population 2			
ADH	.30	.13	.39
GOT	.28	.14	.38
6-PGDH	.20	.17	.34
Mean	.25	.16	.37

for *L. subcarnosus* is 0.975 (Babbel and Selander 1974).
Neither taxon shows major intraspecific genetic differentiation;
within each species, populations are fairly consistent in their
patterns and levels of genetic variability. However, the two
species are quite distinct from each other; for 80 pairwise
interspecific comparisons, mean genetic identity is 0.354 (Babbel
and Selander 1974).

There are several conclusions that can be drawn from these
genetic studies. First, levels of gene flow are similar between
the two species. This similarity in gene flow pattern is most
likely the major determinant of the within-population organization
of genetic variation, the deviation from Hardy-Weinberg
equilibrium, and genetic substructuring. The species differ
significantly in their overall levels of genetic variation,
Lupinus texensis being much more variable than *L. subcarnosus*.
This difference in overall levels of variability must relate to
determinants of genetic structure other than gene flow.

LIFE HISTORY FEATURES

Perhaps the most notable development in plant ecology in the last
two decades has been the emphasis on the numerical dynamics of
plant populations, the demographic approach to plant ecology.
Plant demography deals with age specific features of plant
species, such as the time, intensity, and duration of
reproduction, the survivorship schedule, and reproductive values.
Interest in the demographic aspects of plant ecology was
stimulated by John Harper's seminal paper, *A Darwinian Approach
to Plant Ecology* (Harper 1967). Subsequently, much has been
learned about plant population dynamics. The demographic features
of a wide range of plant species have been been chronicled. Such
a life-history approach to plant ecology has several objectives.
The first is to describe the demographic features of a wide
variety of plant types. The second is to determine how specific
demographic features are adaptive to particular habitats.
Finally, a third objective is to determine how these features have
evolved and why they differ among species.

In pursuing this ecological area of population biology, a
systematic approach can also be very useful. Closely related
species share much of their evolutionary history, yet often occupy
different environments. By comparing demographic features of such
species, much information can be gained on how specific life-
history features are adaptive. The following section will compare
the age-specific features of survivorship and reproduction in the
two lupine species. The purpose of the comparison is to
illustrate the nature of life-history divergence that can occur
between species and to relate these life-history features to the
overall biology of the species.

The basic data for any life-history study consists of the
number of individuals alive and their reproduction for each age
class during the life-cycle. The data used in the following study

were collected from greenhouse populations of the two lupine
species. Seeds from each species were germinated and the
resulting plants followed throughout their life-span. The number
of plants alive was recorded at biweekly intervals and the number
of flowers produced by each plant was tallied every two weeks.
These raw data were then used to calculate a life table.
Demographic studies can be done either in the field or in the
greenhouse, depending on the type of information desired. Field
studies yield information on the actual values for demographic
parameters, but the values that these measures take are a
reflection not only of the underlying genotype of the plant but
also of phenotypic plasticity in response to varying environments.
Greenhouse studies, on the other hand eliminate much of the
environmental component of variation so that genetic differences
between taxa can be observed. However, such studies are often
difficult to relate back to natural communities unless a great
deal of additional information is known.

Life tables for *L. texensis* and *L. subcarnosus* are given in
Tables II and III. The pattern of survivorship for the two
species is different, even in the greenhouse. Figure 2 plots the
age-specific survivorship curve for the lupines. Both species
show a Type I survivorship curve in the greenhouse, where most
mortality occurs late in the life cycle. A Type I survivorship
curve is to be expected in the greenhouse, since environmental
conditions of water, nutrients, and lack of pest pressure are
ideal. Some differences in the survivorship curves of the two
species do exist. In the early stages of seedling growth,
immediately after germination, *L. texensis* suffers significant
mortality, about 30%, whereas almost no early mortality occurs in

TABLE II. *Life Table for* Lupinus texensis

Age	Number surviv-ing	l(x)	Number of flowers	m(x)	l(x)m(x)	E(x)	v(x)
0	320	1.000	0	0	0	7.14	72.6
1	223	.697	0	0	0	8.81	104.1
2	221	.691	0	0	0	7.87	105.0
3	219	.684	0	0	0	6.94	106.1
4	214	.669	0	0	0	6.07	108.5
5	213	.666	121	0.57	0.38	5.14	110.0
6	212	.662	4046	19.08	12.63	4.12	109.1
7	211	.659	9082	43.04	28.36	3.14	90.4
8	205	.641	7290	35.56	22.79	2.20	48.7
9	118	.369	1457	12.35	4.56	2.08	22.8
10	77	.241	762	9.90	2.38	1.67	16.0
11	42	.131	440	10.48	1.37	1.21	11.3
12	9	.028	36	4.00	0.11	1.00	3.9

L. subcarnosus. Furthermore, population decline begins much earlier in *L. texensis* than in *L. subcarnosus* age class 8 as opposed to age class 12, a difference of 8 weeks. These differences in survivorship schedules are also manifested in the expectation of life for each age class, $E(x)$. $E(x)$ is the expected duration of life (in units of age class) for individuals in age class x. $E(x)$ is calculated by

$$E_{(x)} = \sum_{x \to t} \frac{1_{(t)}}{1_{(x)}}$$

where $l(x)$ is the age specific survivorship of age class x. Figure 3 plots $E(x)$ for both species. *L. subcarnosus* shows a gradual decline in $E(x)$ as the population ages. *L. texensis* however, shows at first an increase in expectation of life between germination and the first age class. After the first age class, the slope of the $E(x)$ curve is similar to that of *L. subcarnosus* until the last age clasees, where *L. texensis* death rate is less rapid than the death rate for *L. subcarnosus*. The two species show differences in survivorship (or conversely, mortality) at several stages in the life cycle, even in the uniform environment of the greenhouse. It can be reasonably concluded that the two species have become differentiated in some genetic features which affect survivorship. Such characters are probably, in part, physiological ones and reflect the different physiological optima of the species. The differences in

TABLE III. *Life Table for* Lupinus subcarnosus

Age	Number surviving	l(x)	Number of flowers	m(x)	l(x)m(x)	E(x)	v(x)
0	117	1.00	0	0	0	12.07	12.4
1	115	.983	0	0	0	11.26	12.7
2	115	.983	0	0	0	10.26	12.7
3	115	.983	0	0	0	9.25	12.7
4	115	.983	0	0	0	8.26	12.7
5	115	.983	0	0	0	7.28	12.7
6	115	.983	0	0	0	6.26	12.7
7	111	.949	7	0.1	0.06	5.45	13.1
8	110	.940	142	1.3	1.21	4.49	13.2
9	110	.940	536	4.9	4.58	3.49	11.9
10	105	.897	604	5.8	5.16	2.61	7.3
11	96	.821	128	1.3	1.09	1.76	1.7
12	54	.462	37	0.7	0.32	1.35	0.7
13	19	.162	3	0.3	0.03	1.00	0.2

survivorship curves that are observed in the greenhouse are a
reflection of the relative departure of the greenhouse environment
from the optimal conditions for each species.

Tables II and III also list age specific reproduction, m(x).
Age-specific reproduction is defined as the average number of
offspring produced per individual in an age class. In this study
the actual number of offspring, the number of seeds, was not used
since the study was interested in genetic differentiation between
the species. Seed production in these outbreeding lupines is a
strong function of the environment, for example, pollinator
service. Figure 4 plots age specific reproduction measured as
number of flowers produced for each species. Reproduction is
weighted by the survivorship value, l(x), since the l(x)m(x) value
is the true contribution to reproduction. The species are very
different in their pattern of reproduction. *L. texensis*

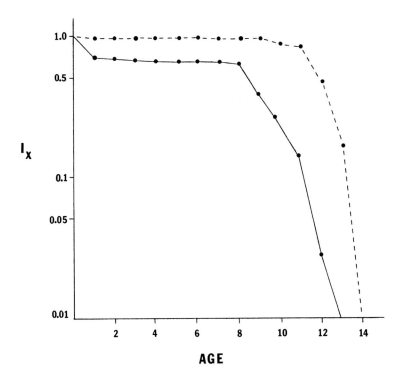

FIG. 2. *Age-specific survivorship in* Lupinus. *l(x) is the
proportion of individuals from the initial cohort alive at the
beginning of an age class. l(x) is plotted on a log scale. Each
age-class represents a two week interval. Lupinus texensis*
is represented by the solid line; *L. subcarnosus* is
represented by the dashed line.

produces many more flowers and much earlier than *L. subcarnosus*.
The net reproductive rate, R(o), is the average number of offspring produced per plant during the life cycle, and is calculated
as the sum of all the age specific l(x)m(x) values. True R(o)
values cannot be calculated from these data, since the number of
flowers rather than the number of seeds were counted. However,
the R(o) values obtained from the flowering data can be used for a
comparison of the two species. R(o) for *L. texensis* is 72.6
whereas the net reproductive rate of *L. subcarnosus* is much
smaller, 12.4. The time of maximum reproduction of the two
species differs by 8 weeks (Figure 4). As *L. texensis*
begins its reproductive decline, *L. subcarnosus* reaches peak
reproduction. The reproductive data for the two species can be
examined in another way. The age specific reproductive value,
v(x), is the age-specific contribution to the ancestory of future

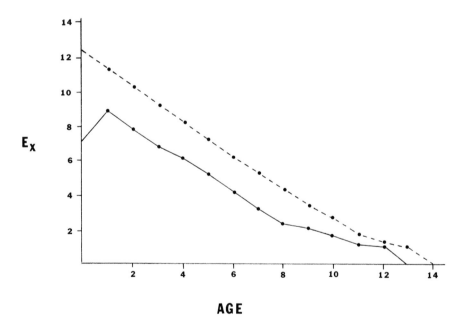

AGE

FIG. 3. *Age-specific expectation of life. E(x) is the average
expectation of life for individuals in age class x. The values of
E(x) are given in age class units. Lupinus texensis is*
the solid line; *L. subcarnosus is the dashed line.*

populations. Reproductive value is calculated by

$$V_{(x)} = \sum_{x \to t} \frac{1_{(t)} \; m_{(t)}}{1_{(x)}}$$

Figure 5 plots reproductive values for the two species. Again the
species differ, both in the magnitude and the overall shape of the
curves. *L. texensis* reaches much greater reproductive
values than does *L. subcarnosus*. Reproductive value for
L. texensis increases rapidly at first, then continues to
increase gradually until the time of first reproduction. *L.
subcarnosus*, on the other hand, maintains a high reproductive
value curve until well after reproduction is initiated. In *L.
texensis* the major contribution to the ancestry of future
populations occurs between age classes 1 and 7, whereas in *L.
subcarnosus* significant contributions are made during age
classes 0 through 10.
    The detailed analysis of reproduction indicates that the

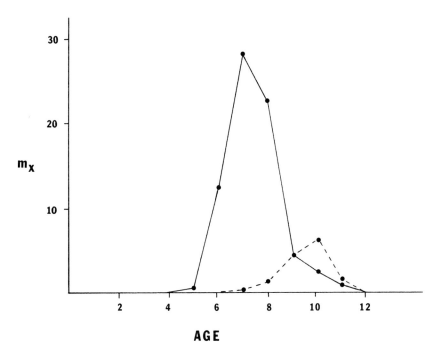

FIG. 4. *Age-specific reproduction. The age-specific
reproduction m(x) is defined as the average number of offspring
produced per individual of an age class. Here m(x) is the average
number of flowers produced per individual. Solid line is
Lupinus texensis; dashed line is L. subcarnosus.*

species are quite different in their reproductive performance. As
with the survivorship data, these differences in reproduction must
reflect in part the different physiological optima of the two
species. Under natural conditions, where each species occupies a
habitat and experiences conditions to which it is presumably
adapted, the intensity of reproduction is not so strikingly
different, although differences in reproduction persist.
Moreover, the species still possess temporal flowering differences
in natural populations. The differentiation between the two
species that is observed in the greenhouse cannot be explained
simply as differences in tolerance to greenhouse conditions. If
this were the case, we might expect that both the survivorship and
reproductive characteristics of the species would be similarly
affected. Yet, *L. subcarnosus* has the greatest survivorship,
far greater than *L. texensis*. Conversely, *L. texensis* clearly
reproduces much more than *L. subcarnosus*. Thus the differences
that are observed must be more than just different degree of
vigor in the greenhouse. The demographic differentiation of the
two species observed in the greenhouse must also be indicative of
the demographic adaptations that the two species have undergone in
response to their different environments.

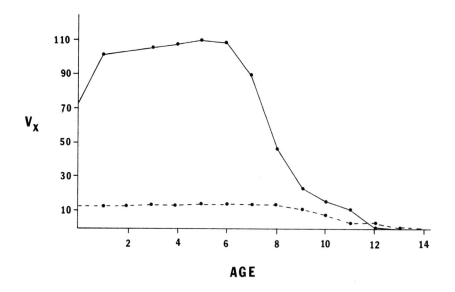

FIG. 5. *Age-specific reproductive value. Reproductive value
is the age specific contribution to future population growth.
Lupinus texensis is the solid line; L. subcarnosus is
the dashed line.*

CONCLUSIONS

The taxonomically close lupine species have a number of
similarities and, also, a number of dissimilarities in their
population biology. Morphology appears to place a constraint on
how divergent population features can become. Hence, the similar
flowers of *L. texensis* and *L. subcarnosus* result in similar
gene flow patterns which result, in turn, in very similar
population structures. Likewise, the habit, size, and number of
ovules are similar between the two species. These factors
constrain the ecological aspects of the species. The timing and
length of the flowering season must necessarily be close in the
two species, since they are both Texas winter annuals whose
growing season is limited by the same seasonal environment.

Beyond these features which are strongly affected by morphology
or habit, a great deal of divergence occurs between the species.
These differences in the population biology of the two species are
not immediately apparent. The overall genetic structures of the
species are quite dissimilar. *L. texensis* is much more
variable than is *L. subcarnosus*. The life history features
of the two species are quite different for two winter annuals of
the same region. *L. texensis* is much more reproductive than
is *L. subcarnosus*, but shows a greater early mortality.
Reproductive value is spread out over much of the flowering season
in *L. subcarnosus* while it is more restricted over time in
L. texensis. Both the demography and the overall genetic
structure of the two species are quite different.

What factors could have resulted in this divergence? Some
insight into this question can be gained from a consideration of
the habitats of the two species. *L. texensis* is a widespread
species, which at times may be considered an aggressive weed. It
occurs in many different habitats. *L. subcarnosus*, on the
other hand, is very restricted in its habitat. Not only is it
restricted to sandy soils, but it is also associated with a
particular plant community, one closely tied to gopher
disturbances. Thus one species is widespread, the other is
narrowly endemic. The divergences in population biology between
the two species is explainable in terms of the environment the
species occupy. Greater genetic variation is expected from a
species that is faced with many environments. Babbel and Selander
(1974) used this explanation of differences in ranges to account
for differences in genetic variability of the lupines. The
species also show the appropriate demographic specialization for
their different habitats. *L. texensis*, as the more "weedy"
of the two species, has the greater reproduction and the lower
survivorship curve expected for weedy plants (Pianka 1970). *L.
subcarnosus* occupies not only a very specific environment, but
also a very constant, predictable one. In such an environment one
expects high survivorship levels, and correspondingly low
reproductive levels. *L. texensis* reproduces earlier than
L. subcarnosus. Again this is consistent with its status as

a more widespread, weedy species.

Thus, the determinants of the population biology of the two species fall into two areas. First are those features which are strongly influenced by morphology and thus are similar in both species. The second major influence on the populational aspects of the two species appears to be the nature of the environment the species occupy, specifically the heterogeneity and the predictability of the environment. The ability to distinguish the morphological from the environmental influences on the population biology would be extremely difficult in a study of a single species. Only by considering and using the systematic relationships of the two species is it possible to even begin to understand the underlying causes for specific populational features.

ACKNOWLEDGMENTS

I thank W.J. Leverich for his help throughout this study. This work was supported by N.S.F. Grant DEB 79-05198.

REFERENCES

Babbel, G.R. and Selander, R.K. 1974. Genetic variability in edaphically restricted and widespread plant species. *Evolution* 28: 619-630.

Felsenstein, J. 1975. *Genetics* 81: 191-207.

Gallez, G.P. and Gottlieb, L.D. 1983. *Evolution* 36: 1158-1167.

Handel, S.N. 1982. *Am. J. Bot.* 69: 1538-1546.

Harper, J. 1967. A Darwinian approach to plant ecology. *J. Ecol.* 55: 247-270.

Kimura, M. and Weiss, G.H. 1964. *Genetics* 49: 561-576.

Leverich, W.J. 1979. Age-specific survivorship and fecundity in *Phlox drummondii*. *Am. Nat.* 113: 881-903.

Levin, D.A. and Kerster, H.W. 1974. Gene flow in seed plants. *Evol. Biol.* 7: 139-220.

Pianka, E. 1970. On r- and K- selection. *Am. Nat.* 104: 592-597.

Sarukhan, J. 1974. *J. Ecol.* 62: 151-157.

Schaal, B.A. 1980. Measurement of gene flow in *Lupinus texensis*. *Nature* 284: 450-451.

Schaal, B.A. and Leverich, W.J. 1980. Banner markings and pollination in *Lupinus texensis* (Leguminosae). Southwestern Natur. 25: 280-281.

Slatkin, M. 1981. *Genetics* 99: 323-335.

Takahata, N. 1983. *Genetics* 104: 497-512.

Turner, B.L. 1957. *Madrono* 14: 13-16.

Wright, S. 1943. Isolation by distance. *Genetics* 28: 114-138.

Wright, S. 1965. *Evolution* 19: 395-420.

Wright, S. 1969. *Evolution and Genetics of Populations*. Vol. II. Univ. Chicago Press, Chicago.

Cytogeography and Biosystematics

C. Favarger
Institut de Botanique
Université de Neuchâtel
Neuchâtel, Switzerland

INTRODUCTION

Since von Wettstein (1898) introduced the geographical-morphological method, it is known that the spatial distribution of a plant taxon is as useful in the development of its definition as is its morphology. It is even "a fortiori" in Biosystematics where at times morphology does not allow one to distinguish individual populations (cytodemes) though their areas do not coincide. Biosystematics, on the other hand, not only involves the definition of the microspecies, it is also the science of the mechanisms of evolution (Merxmüller, 1970). Now, "Evolution is a process of development which is based on space, time and shape" (Croizat, 1966, p. 24); consequently, there will always be a geographical aspect involved.

This paper will be divided into three parts; in the first part there will be a brief overview of the role of cytogeography in clarifying the structure of a complex group. Cytogeography, however, is also implicated in historical botanical geography, and therefore, in the second part of this paper, certain aspects of this symbiosis will be presented in greater detail. Finally, the third part will be concerned with demonstrating that cytogeography must be used in conjunction with other biosystematic methods if valid conclusions are to be drawn from any results obtained using this method.

Throughout this paper, the terms "cytotypes", "cytodemes" and "chromosomal races" will be used as defined in a recent article (Favarger et Küpfer, 1983). The term "chromosomal race" has been, at times, discredited (cf. Stebbins, 1980, p. 499). It seems to us to be a very useful term for distinguishing, within a species, a group of cytodemes which possess a unique geographical distribution when we cannot or will not come to a decision as to the taxonomic status which would best suit it, whatever its mode

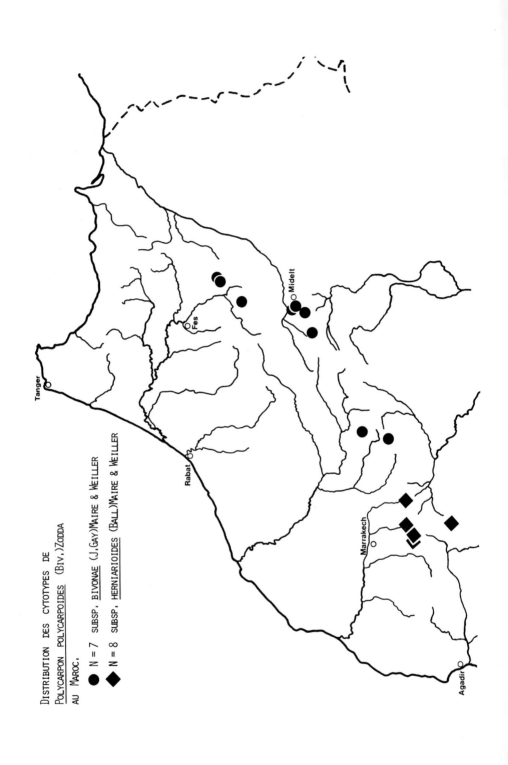

DISTRIBUTION DES CYTOTYPES DE
POLYCARPON POLYCARPOIDES (BIV.)ZODDA
AU MAROC.

● N = 7 SUBSP. BIVONAE (J.GAY)MAIRE & WEILLER

◆ N = 8 SUBSP. HERNIARIOIDES (BALL)MAIRE & WEILLER

of formation (auto- or allopolyploid in the genetic sense).

Part One: CYTOGEOGRAPHY AND STRUCTURE OF A GROUP OF RELATED TAXA

During the course of investigating the chromosome numbers of a
population of a species over its entire range, it is not rare to
encounter different cytotypes which upon further analysis are
found to be chromosomal races that are more or less allopatric.
(This is one of the first steps that a biosystematist takes.) In
the cases of di- and polyploid races, there are so many examples
that we need not dwell on this point here. (It is, nevertheless,
rare to find that a species has been studied over its entire range
using a sufficiently large sample of populations. Thanks, in
particular, to the efforts of IOPB these conditions are being more
frequently completed.) Sometimes the cytotypes have a range but
are either not easily differentiated or not at all distinguishable
morphologically: for example, the octoploid
(western Alps, south of Lautaret) and hexaploid (Alps, north of
Lautaret, Jura, etc.) cytotypes of *Leucanthemum adustum* (Koch)
Gremli (Favarger 1975a). Once the cytotaxonomist has more
opportunities - or becomes more skilled - he will discover taxa as
yet unknown. As in *Ranunculus plantagineus* All. where Küpfer
(1974) showed that there was a good correlation between the number
of chromosomes, geographic distribution, the number of stamens
relative to that of carpels and the pollen characteristics, which
allowed the author to distinguish the subsp. *occidentalis* Küpfer
(south-western Alps) from subsp. *plantagineus* (panalpine).
   Cases of intraspecific dysploidy with allopatrism of the
chromosomal races are relatively infrequent. We shall cite here
the example - recently discovered (Galland unpublished) - of
Polycarpon polycarpoides represented in the moroccan Atlas by a
western noncalcareous race of $n = 8$ and an eastern calcareous race
of $n = 7$ (Fig. 1). Maire (1963), without knowing the chromosome
number, had distinguished these races under the name of subsp.
hemiarioides, for the first subspecies and subsp.
bivonae, for the second, but it must be acknowledged that
morphological differences are weak. A genuine dysploid complex is
found in the orophilous flora of Europe for that of *Carduus
defloratus* L. (Gremaud 1981), where we find the main cytotypes
are of $2n = 18$, 20, 22 and 24. From the point of view of
botanical geography and evolution this example is very intersting.
In fact, those populations situated on the periphery of the area
are karyologically stable ($2n = 22$ or $2n = 24$), whereas
the situation is quite confusing ($2n = 18$ to $2n = 26$)
for those at the center of the area. We can therefore ask
ourselves if the evolution of this group was centrifugal ("genetic

---

FIG. 1 *Distribution of the cytotypes of* Polycarpon polycarpoides
in Morocco.

pool hypothesis", Davis and Heywood 1963) or centripetal: the
marginal populations representing the primitive types and having
been produced by hybridization of the central populations
(Favarger 1982).

The complex *Erysimum grandiflorum-australe-sylvestre* located
in the western Mediterranean, and in the Alps, Pyrenees and
Cantabrian range (Favarger 1980a), is an example of a polyploid and
dysploid complex. This difficult group offers a large range of
morphological variations, not only among the different regions of
the area but also among the populations. To this is superimposed
a clinal variation in the decreasing length of the style as we
move from North Africa (*E. grandiflorum*) to Italy (*E. sylvestre),*
across Spain and France (*E. australe* and related taxa).
In North Africa, where evolution of the complex seems quite recent
(Favarger and Galland 1982), we did not find any correlation
between morphology and chromosome number; the distribution of
cytotypes was found difficult to interpret ($2n$ = 14, 26, 32, 36
and 38). The most we can do is point out the scarcity of the
diploids which are confined to the megaatlasic region with a
disjunct area, and that the most frequent cytotype with $2n$ = 26 is
localized in Morocco. In Europe, there is, in general, a good
correlation between chromosome number and geographic location.
This would suggest that differentiation of the taxa there is
older, however the relationships with morphology are far from
always being obvious (Küpfer 1981). The existence of such
complexes poses some difficult problems to the taxonomist.
Polatschek (1974, 1979) distinguished seven species of this group
in Italy and 12 species in Spain based on morphology and cytogeography. (It should be noted that the chromosome numbers
determined by Polatschek do not always coincide with our own).
Many of these species, however, are very difficult to distinguish.
In our opinion, it would be more practical to adopt an intermediary position between the maintenance of a large collective
species, and splitting, which will lead to the creation of almost
as many species as there are mountains.

The behavior of *Erysimum* of the *grex grandiflorum-*
australe-sylvestre complex is somewhat analogous with that of
the tubered *Claytonia* from North America: *C. virginica*
(Lewis *et al.* 1967; Lewis and Semple 1977) and *C. caroliniana*
(Lewis *et al.* 1967; Gervais and Grandtner 1981). However,
morphological variation is weaker and the cytotypes are more often
sympatric in the *Claytonia*, group than those of *Erysimum*
(at least those from the North of the Mediterranean). Perhaps the
Claytonia complex has a more recent origin.

Similar phenomena have occurred in the evolution of the genus
Pulmonaria, in Europe (Sauer, 1975; Bolliger, 1982) with
varying degrees of morphological differentiation depending on the
groups (inter- or intraspecific dysploidy or polyploidy). In such
complexes it is often very difficult to know the respective roles
of dysploidy (at the diploid or polyploid level) and hybridization
between some basic cytotypes.

These examples are particularly interesting because they reveal at the microevolutionary level what may have happened in the diversification of a polybasic genus or family (Stebbins 1966; Favarger and Huynh 1980; Favarger 1981). This has been termed *pachyphyletism* by Heslop-Harrison (1958).

Part Two: CYTOGEOGRAPHY AND HISTORICAL BOTANICAL GEOGRAPHY

Applications of cytogeography to the history of the flora have been based until now almost exclusively on polyploidy. Thus, it is obvious that no conclusions would be possible in this field if polyploidy in natural populations was an easily reversible phenomenon. This problem has been discussed especially by Jones (1970) and by Stebbins (1971, 1980). According to the latter (1980, p. 496) polyhaploids have no possibility of success unless they result from recently originated polyploids which have only slightly differentiated. We support the opinions of these authors. It follows that a polyploid taxon is, with rare exceptions (there is no hard and fast rule in biology), younger than the corresponding diploid taxon(a). This gives a historical dimension to cytogeography. We will now examine (1) migrations, (2) the relative distribution of di- and polyploid chromosomal races, and (3) the age of "chromosomal races".

Migrations

If most phytogeographers have accepted Darwin's theory that a species originates from a confined area (= center of origin) and later emigrates in order to occupy unlimited territory, sometimes far from the center, then the ideas of Croizat (1966, 1968), which have gained much support, especially from zoologists, cannot be ignored. Croizat negates the idea of centers of origin and migrations and thinks that speciation occurred only by vicariance. That quite a few species from the boreal flora, remaining diploid, or possessing diploid races, were differentiated by vicariance from a common ancestor belonging to the arcto-tertiary flora, is an undeniable fact. According to the results of research done by Gervais (1973, 1981) this is the case for *Avenula sulcata* and *bromoides* (western Mediterranean), *A. compressa* (central Europe and the Balkans), *A. schelliana* (from the Ukraine to Manchuria) and *A. hookeri* (North America). (According to Löve *et al.* (1971), citing Weber's Rocky Mountain Flora (1967), this latter species is identical to *A. asiatica* of the mountains of Central Asia.) But the migrations occurring after the climatic changes at the end of the Tertiary and Quaternay periods seem difficult to contest. Polyploidy, in our opinion, brings some evidence in support of this hypothesis, since it seems certain that polyploid races or species are spreading far from their center of origin to conquer new territories. This was well demonstrated in the example of *Primula* subgenus *Farinosae* (Stern 1949) by Darlington (1973), and Knaben

(1982) speaks of the "waves of polyploids" which reached
Scandinavia in the Quaternary Period.

Many authors (e.g. Landolt 1960; Hess *et al.* 1967) admit that
the alpine orophytes of the boreal branch (Diels 1910) were
differentiated during the Tertiary Period from ancestors occupying
the old Central Asian Mountains (Angaria), which would implicate
migrations. Kress (1963a) showed that the *Primula*, section
Auricula, which are all polyploid, ($2n = 66$) descended, in all
probability, from the asiatic diploid *Nivales* ($2n = 22$)
and the east to west migration was accompanied in Europe by a gradient
decrease in number of primitive characters as well as in dysploidy
($66 \rightarrow 64 \rightarrow 62$). According to Kress (1963b), a similar reasoning
applies to the case of *Androsace* (section *Aretia*) which are
all polyploids in the mountains of Europe (Favarger 1958; Kress
1963b). Müller (1982) has also drawn the same conclusions from a
cytotaxonomic study of the section *Cyclostigma* of the genus
Gentiana; he believes that the primitive base number was $x = 5$.
Here too the "eastern" species (e.g., *G. pumila*: $n = 10$;
G. tergestina sensu Soltokovic, *G. pontica*, *G. oschtenica*:
$n = 15$) have remained closer to the primitive base number than
has *G. verna* ($n = 14$) or *G. orbicularis* ($n = 16$) which were
developed through dysploidy. It is interesting to note that no
species of *Primula* (section *Auricula*), nor of *Androsace*
(section *Aretia*), or *Gentiana* (section *Cyclostigma*),
are found in the alpine flora of Corsica. According to the so-
called messinian theory (Hsu *et al.* 1973, 1977; Bocquet
et al. 1978), during the drying up of the Mediterranean at
the end of the Miocene era and with the aid of cooling of the
climate to a certain degree, the movement of alpine species
towards the high mountains of Corsica became possible. We must
therefore conclude that the European migration of those ancestors
of asiatic origin of some of our present day alpine orophytes
happened after the messinian salinity crisis, that is, probably in
the Pliocene era.

Relative Distribution of Di- and Polyploid Races

If the appearance of polyploid cytotypes in a diploid population
no longer poses a problem today (this occurrence, it seems, arose
as a result of the production of nonreduced gametes; Harlan and de
Wet 1975; de Wet 1980; W. Lewis 1980), the same does not hold true
for that of a polyploid race which occupies a particular area. In
geographical botany there is always the difficult and
controversial problem of *the origin of the ranges*. Merxmüller
(1952) expressed it very well: "one of the most important tasks
of biosystematics consists of discovering the stages through which
a new and viable karyological arrangement leads to a well
established race" (original in German). Stebbins (1971) contributed to the clarification of this problem with his classification
of polyploid complexes. We still feel that the subject has not
been thoroughly exhausted.

Among the approximately 50 polyploid complexes of the European and North African flora studied in some depth at Neuchâtel, the majority fit into at least one of Stebbins's classifications. For example: *Leucanthemum vulgare* agg. (Villard 1970; Favarger 1975a); *Cerastium arvense* agg. (Sollner 1954; Favarger *et al.* 1979); *Arenaria ciliata* agg. (Favarger 1965; Beuret 1977); *Bupleurum ranunculoides* (Küpfer 1974); *Anthericum liliago* (Küpfer 1974); *Senecio doronicum* (Küpfer 1974); *Leucanthemopsis alpina* (Contandriopoulos and Favarger 1959; Küpfer 1974), etc. (Only the most recent results of our laboratory are cited in this review. Some of these same complexes have been studied by other authors in other regions of Europe; their results however, have not invalidated our conclusions.) In effect, at the base of each of these complexes there are many diploid taxa which are usually allopatric and more or less morphologically differentiated. The range occupied by the polyploid taxa are either intermediate between those of the corresponding diploid taxa or they overlap more or less extensively. In these complexes it is reasonable to assume that hybridization between the diploid taxa, which the vicissitudes of the history of plants (e.g., changes of climate) brought closer together, is responsible for the origin of polyploids.

On the other hand, the situation is different for about ten polyploid complexes, in the sense that (1) they stem from only one diploid taxon, (2) the diploid race and the corresponding polyploid races are more or less completely allopatric, (3) the polyploids were propagated in one or two preferential directions with at times an increasing gradient of polyploidy. This is what we call polarized distribution with allopatrism more or less complete of the chromosomal races (Hebert 1980; Favarger and Küpfer 1980).

The objection may perhaps be raised that these complexes do not offer more than two levels of chromosome numbers and that they do not include such entities as what Stebbins (1971) called "compilospecies" and Ehrendorfer (1963) "Dachsippen". It should be noted, however, that some of Stebbins's examples (*Tragopogon, Aegilops, Zauschneria*) have neither more than two levels of chromosome numbers, nor the "compilospecies" stemming from pachyphyletism as the diagram of this American author suggests (1971, p. 154).

The idea which led Stebbins to choose his examples was that a polyploid species or race always originates, according to him, following the crossing between many diploid taxa whose genomes differ more or less markedly (intervarietal autopolyploidy or allopolyploidy). In the groups about which we are speaking, on the other hand, it seems as if a specific diploid taxon produced a polyploid through autopolyploidy. This is probably not autopolyploidy in the strictest sense since in an allogamous species, all the individuals of a panmictic population are necessarily heterozygotes. This heterozygosity seems to us to be sufficient to assure that the autopolyploid will have the advantages of poly-

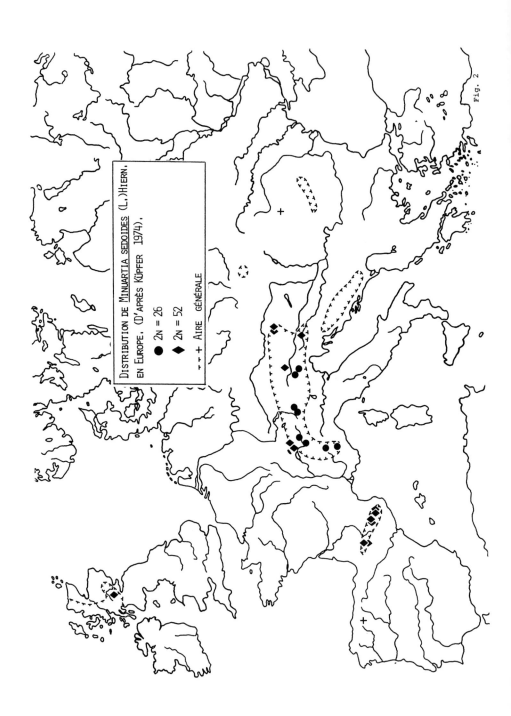

DISTRIBUTION DE MINUARTIA SEDOIDES (L.)HIERN,
EN EUROPE, (D'APRÈS KÜPFER 1974).

● 2N = 26
◆ 2N = 52
+-+ AIRE GÉNÉRALE

Fig. 2

ploidy so well described by Stebbins (1980). We can ask ourselves
if the genic differentiation existing between two allopatric
diploid races is necessarily more pronounced than between two
biotypes from a single diploid population. W. Lewis (1980) showed
also that "neoautoploids may increase biochemical diversity at the
primary enzyme level and in other ways by immediate derepression
of repressed genic expressions of the diploid. This may extend
the range of environments and habitatis in which successful
development of polyploids occurs."

One example of these *polarized polyploid complexes* (where
the evolution is centrifugal) is that of *Minuartia sedoides*
very well analyzed by Küpfer (1974) (Fig. 2). This species is
isolated systematically and is the only representative of
Subsectio Cherleria McNeill (McNeill 1962). Mattfeld (1922)
considers it as an ancient endemic of the high mountains of
Europe. Diploid in the western and southern Alps, this species
has formed a tetraploid, morphologically very close, looking
identical, to the diploid. This tetraploid race was discovered in
the northern Prealps of Switzerland and Austria (as far as
Karawanken) in the Pyrenees and in Scotland. It is impossible
that the diploid *M. sedoides* hybridized with another taxon
to form the tetraploid and there are no known morphologically
distinct geographical races of this species. Although Küpfer
(1974) suggested that karyotypically distinct geographical races
of *M. sedoides* could have hybridized during the glacial
period, we believe that this hypothesis - which remains difficult
to verify - is not necessary to explain the success of the
tetraploid because, being an allogamous species, individuals of
the same population are surely more or less heterozygotes.

Table I is based on the model of Stebbins (1971, p. 164).
Here, we have grouped examples of those "polarized" polyploid
complexes (whose allopatrism is more or less complete) which we
know in the flora of Europe and North Africa along with their main
characteristics. The case of *Oxytropis halleri* compares
excellently from a geographical point of view, with that of
Minuartia sedoides. The resemblance would be even more pronounced
if *M. sedoides* was tetraploid in the Tatras. Unfortunately,
we do not yet know the chromosome number of the populations of
this chain. In one of these complexes (*Silene ciliata*), there
exists what we will call an "escalating" polyploidy going as high
as 20X the basic number. Clearly these elevated numbers result
from crossings between chromosomal races followed at times by a
doubling of the chromosome number; but all these phenomena still
occur within the limits of the species.

In this first series of examples it is likely that polyploidy
originated at the *periphery of the range* of the diploid taxon
("Randabspaltung"). In the case of the orophilous complexes, the

FIG. 2 *Distribution of* Minuartia sedoides *in Europe.*

TABLEAU I/TABLE I.   Complexes polyploïdes polarisés avec
ploïdes/ Polarized polyploid complexes of diploid and polyploid

Complexe et références	No. de base Basic no. (x)	Degré de ploïdie Polyploid level	Distribution géographic Diploïdes
Minuartia sedoides Favarger 1962; Küpfer 1974	13	$4x$	Alpes occidentales et méridionales Western and southern Alps
Oxytropis halleri Favarger 1965; Küpfer 1974	8	$4x$	Vallées intraalpiennes à climat continental 400–1700 m

Intraalpine valleys having continental climates, 400–1700 m |
| *Helictotrichoa montanum* | 7 | $4x$ | Alpes occidentales, Massif central, Pyrénées orientales, Mts. Cantabres, Sierra Nevada, Atlas

Western Alps, Massif central, Eastern Pyrenees, Cantabrian Mts., Sierra Nevada, Atlas |
| *Ranunculus plantagineus* | 8 | $4x$ | Alpes austro-occidentales

Austro-western Alps |
| *Silene boryi* | | $4x, 6x$ | Ht. Atlas marocain, Sierra Nevada

Morrocan Atlas heights, Sierra Nevada |

allopatrisme plus ou moins complet des races diploïdes et poly-
races having more or less complete allopatrism

Geographical distribution Polyploïdes	Direction de propagation des polyploïdes Direction of dissemination of polyploids	Sympatrisme entre race $2x$ et race(s) polyploïde(s) Sympatrism between race $2x$ and its polyploid race(s)
Préalpes du Nord, Pyrénées, Ecosse Northern PreAlps, Pyrenees, Scotland	S → N puis → W et NW S → N then → W and NW	– –
Alpes septentrion- ales et orientales, à l'étage alpin. Pyrénées oriental, Ecosse, Tatras	S → N puis → W NW et → E	Sippes 2 et 4 partiellement sym- patriques aux Alpes orientales mais ne croissant pas à la même altitude
Alpine level of north- ern and western Alps, western Pyrenees, Scotland, Tatry mts.		Species $2x$ and $4x$ partially sympatric in western Alps but not crossing at the same altitude
Pyrénées centrales	E → W	–
Central Pyrenees	E → W	–
Central Pyrenees	E → W	–
Toute la chaîne des Alpes, Corse (?)	SW → NE	Léger sympatrisme probable dans les Alpes austro- occidentales.
Entire chain of the Alps, Corsica (?)	SW → NE	Slight sympatrism probable in the austro-western Alps
$4x$: Sierras espagnoles, de la Sierra Nevada à la Chaîne cantabrique $6x$: Sierra Nevada	S → N ($4x$)	Sympatrisme entre $2x$, $4x$ and $6x$ dans le sud de l'Espagne, mais $2x$ croît à des altitudes plus élevées
4 : Spanish Sierra from Sierra Nevada to the Cantabrian mt. chain		

TABLEAU I/TABLE I.   Continued.

Complex and references	No. de base	Degré de ploïdie	Distribution géographic
	Basic no. (x)	Polyploid level	Diploïdes
			Morrocan Atlas heights, Sierra Nevada
Silene ciliata Küpfer 1974	12	$4x$, $16x$, $18x$, $20x$	Centre et NW de L'Espagne
			Central and NW Spain
Silene legionensis Küpfer 1974	12	$4x$	Sud et Est de l'Espagne
			South and East Spain
Cotoneaster integerrimus Favarger 1971 and unpublished	17	$(3x)$, $4x$, $6x$	Pyrénées orientales, Alpes occidentales, Jura meridional
			Eastern Pyrenees, western Alps, and southern Jura

Geographical distribution Polyploids	Direction de propagation des polyploïdes Direction of dissemination of polyploids	Sympatrisme entre race 2x et race(s) polyploïde(s) Sympatrism between race 2x and its polyploid race(s)
4x: Spanish Sierra, from the Sierra Nevada to the Cantabrian mt. chain 6x: Sierra Nevada	S → N (4x)	Sympatrism between 2x and 6x in the south of Spain, but 2x grows at higher altitudes
Chaîne cantabrique, Pyrénées, Massif central français	S → N puis → E et NE	Sympatrisme entre 2x et 4x dans la Chaîne cantabrique
Cantabrian mt. chain, Pyrenees, Massif central (French region)	S → N then → E and NE	Sympatrism between 2x and 4x in the Cantabrian mt. chain
NW de l'Espagne. Chaîne cantabrique (Prov. de Léon)	S et E → NW	–
NW Spain. Cantabrian mt.	S and E → NW	–
Pyrénées centrales. Europe du Jura central à la Scandinavie et à la Roumanie	SW → N et E (E → W aux Pyrénées)	Alpes occidentales, dans la zone de contact entre 2x et 4x; Suède, entre 4x et 6x
Central Pyrenees. Europe, from central Jura to Scandinavia and to Rumania	SW → N and E (E → W in the Pyrenees)	Western Alps, in the contact zone between 2x and 4x; Sweden, between 4x and 6x

unique conditions created by the advance of the glaciers and the
"descent" of the alpine plants during the glacial stages are
probably responsible for that phenomenon. Then there have been
the tetraploid migrations in one or two preferential directions.
This relates to what we have said previously concerning migrations
where the phenomenon was discussed at the infrageneric level. It
is interesting to note that these polarized polyploid complexes
with allopatric races also exist in North America and the example
of *Larrea divaricata* (Yang and Lowe 1968; W. Lewis 1980) compares
very well to our European examples. Lewis (1980, p. 131) attributes this relationship to ecological factors although we feel
that there is also a historical aspect.

In another series of examples (Table II) there is likewise a
"centrifugal" distribution of polyploids, polarized on one or two
large axes, however, *diploids and their corresponding polyploids
are sympatric in the major part of the species' territory while
the polyploids occupy only one part of the area.*

As long as sympatrism is not secondary, polyploidy is not here
the result of "Randabspaltung". It must be admitted, therefore,
that the barrier of different chromosome numbers creates sufficient sexual isolation in the common territory and especially
since the diploid and corresponding polyploid grow at a short
distance from each other and in comparable plant associations
(e.g.: *Centaurium minus* 2*x* and 4*x*; Zeltner 1970 and
personal commun.).

Contrary to what seems to have happened in the *Galium incurvum*
complex (Ehrendorfer 1980, p. 48), we do not believe that other
species contributed to the origin of these "filler polyploids"
(Ehrendorfer, 1980), because the diploids and tetraploids of
Centaurium minus and *Blackstonia perfoliata* resemble each
other more than they resemble neighboring taxa. In our opinion,
this is another case of autopolyploidy. We shall speak, in this
case, of polarized polyploid complexes with partial "sorting" (The
difficult to translate German term "Entmischung" fits best here)
of the races. (Intervarietal autopolyploidy in *Centaurium minus*
and *C. tenuiflorum* is nevertheless not excluded, reminding us of
the "Stebbinsian" complexes. In fact, Zeltner (1970) distinguished two diploid subspecies in the former and two diploid
varieties in the latter.)

Finally, a third type of polyploid complex which drew our
attention is made up of diploid and polyploid cytodemes which are
almost completely sympatric, displaying quite a random
distribution within the common territory. This is the case for
diploid and tetraploid *Centaurium tenuiflorum* which are sympatric
in the greater part of the Mediterranean Basin where they sometimes grow together in the same locality (Zeltner 1970, 1978;
unpublished data; see paragraph above). This is the case, also,
for many species of the genus *Sedum*, for example, *Sedum
urvillei* and *S. litoreum* in Greece (according to t'Hart 1978) or
S. sediforme (t'Hart 1978; Hébert 1980), a species where polyploidy is further complicated by dysploidy at the diploid (15 → 16)

or polyploid level. In the latter complex and in *S. tenuifolium* there is an escalation of chromosome numbers which results undoubtedly from hybridization between the different cytotypes.

These complexes, which with Hébert (1980), we call "nonpolarized" are difficult to explain. The sympatrism of most of the *Sedum* cytotypes may be related to a breaking up of the populations of these saxicolous plants and to a very active vegetative multiplication which allows the cytotypes, even those with odd chromosome numbers, to form small populations. This explanation, however, does not fit for *Centaurium tenuiflorum*. As for the absence of polarity, this could be explained by the fact that in the Mediterranean, there were no important external factors to impose a definite geographical direction on the evolution of polyploids, as for example, that which occurred from the glaciations in Central Europe. Thus, as Hébert (1980) wrote: "in the Mediterranean region, in the Iberian and Balkan peninsulas especially, the polyploid complexes form mosaics of small populations of different chromosome races." In the *S. tenuifolium* complex there is all the same, a certain polarity, in the sense that the diploid cytotype is confined in the Ibero-Mauritanian region, while in the *S. forsterianum* complex, the octoploid race ($2n = 96$) is allopatric and more "nordic" than the others. [Between the diploid ($2n = 40$) and the tetraploid ($2n = 80$) races of *Sedum acre* (t'Hart 1971, 1978), there is a pseudo-vicariance S → N, which brings us back to the first type of polyploid complex or again to a "Stebbinsian" complex, due to the morphological differentiation of the diploids.) Be that as it may, the complexes of this third type are somewhat the despair of the cytogeographer who is unable to grasp the determining factors for the presence of a chromosomal race in a given locality and finds himself incapable of making predictions based on only a few results.

There are obviously intermediaries among these three types of complexes according to the importance of sympatrism. On the other hand, the boundary between these and the "Stebbinsian" complexes are not always clear.

The problem we have addressed in the past few pages is a very difficult one, because, as we have previously stated (Favarger 1967), the distribution of chromosomal races depends above all on suitable characters, and the biology and the history of each taxon, which makes generalizing hazardous. This is not a reason to despair. To reject all comparisons is an easy solution and one which would render as forever fanciful the idea of "panbio-geography" which Croizat (1958) dreamed of well ahead of his time. Most assuredly, the geographic behavior of "chromosomal races" is quite varied, but there are still some very striking similarities, and we share the opinion of W. Lewis (1980) who, supporting the work of Johnson and Packer (1965), wrote: "even though these data were not directed toward infraspecific polyploidy *per se*, they do lend support to the idea that, by and large, elements of the same ancestral flora respond to certain major environmental

TABLEAU II/TABLE II. *Complex polyploïdes polarisés avec triage selection of races*

Nom du complexe et références/ Complex name and references	*Blackstonia perfoliata* Zeltner 1970 et/and nonpublié/unpublished	*Cerastium gibraltaricum* Favarger *et al.* 1979
No. de base/ Basic number	$x = 10$	$x = 18$
Polyploïdie	$4x$	$4x$, $6x$
Distribution Géographique		
Diploïdes/ Diploids	Région méditerranéenne $\leqslant 43°$ latitude Mediterranean region $\leqslant$ latitude $43°$	Maroc, S de l'Espagne/ Morocco, Southern Spain
Polyploïdes/ Polyploids	Région méditerranéenne, Europe centrale, Maroc/ Mediterranean region, Central Europe, Morocco	$4x$: Maroc, S de l'Espagne, Corse/Corsica, $6x$: Sardaigne/Sardinia
Direction de propagation des polyploïdes/ Direction of dissemination of polyploids	S → N (Europe) N → S (Maroc/ Morocco)	W → E
Sympatrisme entre race $2x$ et race(s) polyploïde(s)/ Sympatrism between race $2x$ and its polyploid race(s)	Portugal, S de l'Espagne/Southern Spain, Baléares/Balearic Islands, Corse/Corsica, Sardaigne/Sardinia, Grèce/Greece	Maroc/Morocco, S de l'Espagne ($2x$ et/and $4x$)

*Ce complexe pourrait, à la rigueur, être classé parmi les sont assez différenciés(deux sous-espèces reconnues)./This complexes, since according to Zeltner (1970) the diploids are

partiel des races/Polarised polyploid complexes with partial

Amelanchier ovalis Favarger 1971; Favarger et/and Stearn 1983	*Centaurium minus* Zeltner 1970, 1980 et/and nonpublié/unpublished *
$x = 17$	$x = 10$
$4x$	$4x$
Région méditerranéenne occidentale. Alpes occidentales et méridionale, Croatie; Alpes apuanes/Eastern mediterranean region, Eastern southern Alps, Croatia; Apennine Alps	Région méditerranéenne/ Mediterranean region $\leqslant 43°$ latitude
Région méditerranéenne Pyrénées centrales, Europe occidentales → Allemagne, Italie centrale, Sicile, Crimée, Caucase/Mediterranean region, Central Pyrenees, Eastern Europe → Germany, Central Italy, Sicily, Crimean peninsula, Caucasus	Région méditerranéenne/ Mediterranean region, Europe, Asie occidentale (Turquie)/Eastern Asia (Turkey)
à peu près/approximately S → N et/and NE	S → N
Sierra de/of Cazorla(?), Pyrénées orientales/Eastern Pyrenees, S de la France/Southern France, Alpes occidentales/Eastern Alps	S de la France/Southern France, Corse/Corsica, Sardaigne/Sardinia, S de l'Italie/Southern Italy, Grèce/Greece, une hexaploïde en Iran, peut-être hybridogène/a hexaploid in Iran, perhaps a genetic hybrid

complexes stebbinsiens, car d'après Zeltner (1970) les diploïdes
complex could perhaps be classified among the Stebbinsian
differentiated (two recognized subspecies).

parameters similarly." We should, however, make an exception for
the regions which were severely glaciated, for Merxmüller (1952)
has shown that the areas of species belonging to different
elements of the flora have been dislocated and modified in the
same way. It is true that we are not speaking in this case of
polyploidy or chromosomal races.

The Age of Chromosome Races

Since the time we drew attention (Favarger 1961) to the fact that
in historical botanical geography one should not lump all polyploids together in the same "drawer", many authors have been using
such notions as paleo-, meso- and neopolyploidy. The spectrum of
relative age of a flora which we have proposed is, understandably,
far from perfect. It can certainly be modified and improved as
needed. However, it has been used successfully as such by Johnson
and Packer (1967) for the arctic flora of northwestern Alaska. In
order to better explain the case of a local flora of a temperate
region, where very definite associations exist within species
groups, Ehrendorfer (1980) created an interesting version of our
method. His version took into consideration the phytosociology
and ecology of the species. All these authors considered, as we
do, that the neopolyploids were young taxa of recent or subrecent
formation. However, over the past few years (cf. Favarger and
Küpfer 1980) we have raised new questions concerning the real age
of chromosomal races of a single species. Already Küpfer (1974),
while studying the distribution of polyploid races of *Ranunculus
plantagineus* in the Alps and Corsica, and of *Poa cenisia*
in the Alps, Corsica and in the Pyrenees, reached the conclusion
that polyploid complexes of these two species most probably dated
to the Neogene era and certainly dated prior to the Messinian era
(ca. 8 million years), unless we would like to imagine a polytopic
origin for the polyploid races followed by the extinction of the
diploid forms in Corsica (for both species) and in the Pyrenees
for *Poa cenisia*, which is quite improbable.
　　Some of the polyploid complexes of the Mediterranean flora are
no longer young. For example, the tetraploid race of *Centaurium
majus* (Zeltner 1970, 1978 and unpublished) is found on a series of
islands (Balears, Sardinia, Sicily, Crete, Rhodes, Cyprus) which
would suggest either that there was accidental transportation by
man from island to island or that the territories were once
physically connected. One race of *Sedum sediforme* ($2n = 169$–171)
is found both in Sicily and in Tunisia which would suggest that it
originated prior to the sinking of the siculo-tunisian bridge
(Hébert 1980).
　　The idea that polyploid complexes can be in existence for
millions of years is not extraordinary in itself. Stebbins (1971)
thinks that the "mature" polyploid complexes date to a period
varying between 500,000 and 10 million years ago. But the
complexes to which this American author refers are not usually
intraspecific. It does, however, seem a bit surprising that a

polyploid race, of slight or no morphological differences could persist for such long periods without evolving into a species. It is exactly this slight degree of morphological differentiation which led us to propose, after Monnier (1959), the term *neopolyploid*. Could we apply this term to a race dating from the Miocene era? Before answering this question, it would be better to make sure that the above mentioned chromosomal races (as well as others not discussed) are really that old. If we can prove that, (1) There had been no recent physical transportation either by man or animals and, (2) That no polytopic origin is involved (which is not an impossibility for chromosomal races but quite improbable when we are dealing with many separated islands), it would be proper then, to take into account the old "chromosomal races" in the analysis of the age of a flora.

In all classifications there is always the delicate problem of borderline cases. Between a mesopolyploid and a neopolyploid there may be solely the difference that the ancestral diploid has or has not survived. For example, if the subsp. *occidentalis* of *Ranunculus plantagineus* ($2x$) disappears (which may be the case in a few centuries), *R. plantagineus* would be classified as a mesopolyploid, attributing to it as an ancestor, *R. pyrenaeus*, or some other closely related diploid taxon of the Pyrenees. The same applies for cytological classification of endemic taxa. Between the ancient apoendemic, and paleoendemic, there is only one difference: whether or not the primitive diploid taxon has been preserved. Now this survival may depend on fortuitous circumstances. These two classifications both come up against the same difficulties: the speed of evolution, undoubtedly quite varied from one group to another, and the hazardous survival of the ancestral diploid taxon. At the present state of our knowledge, these difficulties appear almost insurmountable.

In the meantime, we can modify the spectrum of relative age of a given flora by including in Class 2 (elements of the Middle Ages), those chromosomal races which have conclusively proven to be old. The mesopolyploids will no longer be solely the "good species" whose most probable ancestor is another taxon of the same rank, but will also consist of chromosomal races whose origins go back at least as far as the tertiary period, and which have not evolved since.

Part 3: THE OCCURRENCE OF OTHER BIOSYSTEMATIC METHODS ON
          CYTOGEOGRAPHY

Cytogeography is not an independent discipline, it is only one of the many techniques of biosystematics. The cytogeographer should have ever-present in his mind that his job is not simply the mapping of chromosome numbers, but involves living entities which have certain biological, genetic and ecological properties. Only in this way will a person be able to draw valid conclusions in phylogeny and in historical geography. A few examples shall be sufficient to demonstrate this belief:

(1) Ockendon (1968, 1971) showed in flax of the Alps that there

were two taxa related to *Linum perenne* which he had distinguished.
The one native to the mountains of South Europe (Alps, Pyrenees,
etc.) is the subsp. *alpinum* (diploid), the other, whose range
includes southern Jura and Bavaria is the subsp. *montanum*
(tetraploid). Küpfer (1974) found two tetraploid populations of
alpine flax in the central Pyrenees and would have been tempted to
reunite them to the subsp. *montanum* by explaining their
disjunction, either by calling for a glacial migration, or a
polytopic origin. Now, if there is a polytopism of the chromosome
number, then this is not a polytopic taxon, since the population
from the central Pyrenees has tricolpate pollen like the subsp.
alpinum, and not hexaporous as is the case for the subsp.
montanum. The populations of central Pyrenees are autopolyploids of the subsp. *alpinum* while those of Jura and Bavaria
have another origin. In this example the use of palynology was
the determining factor.

(2) It was as a result of a precise study of morphology that
Küpfer (1974) also showed that the hexaploid populations of
Bupleurum ranunculoides from the Picos de Europa had a completely
different origin from the hexaploid race of central Jura and the
Prealps of Switzerland. By relying only on the mapping of
chromosomes we risk making quite erroneous conclusions.

(3) In another vein, it was the genetic studies (crossings,
meiotic observations), as well as the use of ecology and
reproduction, which allowed Urbanska-Worytkiewicz (1977) and
Urbanska-Worytkiewicz and Landolt (1978) to state precisely the
details of the origin of a neopolyploid taxon of the central Alps:
Cardamine X *Schulzii*.

(4) By using the techniques of biometry, anatomy of the
achenes, and genetics, together with very precise chromosome
mapping of cytotypes, Gremaud (1981) was able to unravel the
difficult taxonomy of *Carduus defloratus* agg. Here again, the
simple geography of chromosome numbers would have led to erroneous
conclusions.

(5) Finally, knowledge of the mode of reproduction, sexual or
apomictic, is very important. Without the discovery of apomixis,
it would be difficult to explain the cytotypes of the *Poa alpina*
complex in which there is a multiplicity of chromosome numbers,
many of which have odd numbers (17 cytotypes have been observed on
Swiss territory), and the distribution which is not only apolar
but to a large degree quite erratic (Duckert and Favarger 1983).
However, the sexual phenomenon is not entirely absent and seems to
be one factor that contributes not insignificantly to the
karyological variability of this complex (cf. Skalinska 1952;
Müntzing 1954; Duckert and Favarger 1983).

REFERENCES

Beuret, E. 1977. Contribution à l'étude de la distribution
    géographique et de la physiologie de taxons affines di- et
    polyploïdes. *Bibl. Bot.* 133: 1-80.

Bocquet, G. *et al.* 1978. The Messinian Model. A new outlook for the floristics and systematics of the Mediterranean area. *Candollea* 33: 269-287.

Bolliger, M. 1982. *Pulmonaria* in West-Europa. *Phanerogamarum Monogr.* T. VIII J. Cramer. Vaduz, 5-214.

Contandriopoulos, J. et Favarger, C. 1959. Existence de races chromosomiques chez *Chrysanthemum alpinum* L. Leur répartition dans les Alpes. *Rev. Gén. Bot.* 66: 1-17.

Croizat, L. 1958. Panbiogeography 1, 2a, 2b. Caracas.

Croizat, L. 1966. L'âge des Angiospermes en général et de quelques Angiospermes en particulier. *Andansonia* 6: 65-104.

Croizat, L. 1968. Introduction raisonnée à la biogéographie de l'Afrique. *Mem. Soc. Broter.* 20: 1-45.

Davis, P.H. and Heywood, V.H. 1963. *Principles of angiosperm taxonomy.* Edinburgh and London, 1-556.

De Wet, J.M.J. 1980. Origins of polyploids. In *Polyploidy, biological relevance* (W.H. Lewis, ed.), Plenum, N.Y. 3-51.

Diels, L. 1910. Genetische Elemente in der Flora der Alpen. *Beibl. Bot. Jahrb.* 102: 7-46.

Duckert, M.-M. et Favarger, C. 1983. Index des nombres chromosomiques des Spermatophytes de la Suisse. 1. *Poaceae.* L. Genre *Poa. Bot. Helvet.* (en cours de publication).

Ehrendorfer, F. 1963. Probleme, Methoden und Ergebnisse der experimentellen Systematik. *Planta Med.* 3: 234-251.

Ehrendorfer, F. 1980. Polyploidy and distribution. In *Polyploidy, biological relevance.* (W.H. Lewis, ed.), Plenum, N.Y. 45-58.

Favarger, C. 1958. Contribution à l'étude cytologique des genres *Androsace* et *Gregoria. Veröff. Geobot. Inst. Rübel Zürich* 33: 59-80.

Favarger, C. 1961. Sur l'emploi des nombres de chromosomes en géographie botanique historique. *Ber. Geobot. Inst. Eidg. Tech. Hochsch. Stift. Rübel Zürich* 32: 119-146.

Favarger, C. 1962. Contribution à l'étude cytologique des Genres *Minuartia* et *Arenaria. Bull. Soc. Neuchâtel. Sci. Nat.* 85: 53-81.

Favarger, C. 1965a. A striking polyploid complex in the alpine flora. *Bot. Not.* 118: 273-380.

Favarger, C. 1965b. Notes de caryologie alpine IV. *Bull. Soc. Neuchâtel. Sci. Nat.* 88: 5-60.

Favarger, C. 1967. Cytologie et distribution des plantes. *Biol. Rev.* 42: 163-206.

Favarger, C. 1971. Relations entre la flore méditerranéenne et celle des enclaves à végétation subméditerranéenne d'Europe centrale. *Boissiera* 19: 149-168.

Favarger, C. 1975. Sur quelques marguerites d'Espagne et de France (étude cytotaxonomique). *An. Inst. Bot. Cavanilles* 32(2): 1209-1243.

Favarger, C. 1980a. Un exemple de variation cytogéographique: le complexe de l'*Erysimum grandiflorum-sylvestre. An. Inst. Bot. Cavanilles* 35: 361-393.

Favarger, C. 1980b. Le nombre chromosomique des populations

alticoles d'*Erysimum* des Picos de Europa. *Bull. Soc.*
Neuchâtel. Sci. Nat. 103: 85-90.

Favarger, C. 1981. Cytotaxonomie et problèmes fondamentaux de
géographie botanique. *Mem. Soc. Biogeogr.* 11:37-49.

Favarger, C. 1982. Hybridation et systématique. Table ronde.
Museum Nat. Hist. Nat. Paris. pp. 143-159.

Favarger, C. et Galland, N. 1982. Contribution à la cytotaxonomie
des *Erysimum* vivaces d'Afrique du Nord. *Bull. Inst.*
Sci. Rabat 6: 73-87.

Favarger, C., Galland, N. et Küpfer, Ph. 1979. Recherches
cytotaxonomiques sur la flore orophile du Maroc. *Nat.*
Monspel. Sér. Bot. 29: 1-64.

Favarger, C. et Huynh, K.-L. 1980. Contribution à la
cytotaxonomie des Caryophyllacées méditerranéennes. *Bol.*
Soc. Brot. 53: 493-514.

Favarger, C. et Küpfer, Ph. 1980. Application de la cytotaxonomie
à quelques problèmes d'origine et de mise en place de la flore
méditerranéenne. *Nat. Monspel.* Emberger, 53-65.

Favarger, C. et Küpfer, Ph. 1983. Index des nombres
chromosomiques des Spermatophytes de la Suisse. Introduction,
matériel et méthodes. *Bot. Helvet.* 93: 3-7.

Favarger, C. et Stearn, W.T. 1983. Contribution à la
cytotaxonomie de l'*Amelanchier ovalis Medikus (Rosaceae)*.
Bot. J. Linn. Soc. (à l'impression).

Gervais, C. 1973. Contribution à l'étude cytologique et
taxonomique des avoines vivaces (g. *Helictotrichon* Bess.
et *Avenochloa* Holub). *Mém. Soc. Helv. Sci. Nat.* 88: 1-166.

Gervais, C. 1981. Notes sur la phylogénie des avoines vivaces
(Genres *Avenula* Dumort. et *Helictotrichon* Bess.) à
la lumière d'hybridations récentes. *Bull. Soc. Neuchâtel.*
Sci. Nat. 104: 153-166.

Gervais, C. et Grandtner, M.M. 1981. Étude cyto-écologique de
quatre populations de *Claytonia caroliniana* var.
caroliniana au Québec. *Can. J. Bot.* 59: 1685-1701.

Gremaud, M. 1981. Recherches de taxonomie expérimentale sur le
Carduus defloratus L. s.l. *(Compositae). Rev. Cytol.*
Biol. Végét. Bot. 4: 1-75; 111-171; 207-268; 341-386.

Harlan, J.R. and De Wet, J.M.J. 1975. On Ö. Winge and a prayer.
The origins of polyploidy. *Bot. Rev.* 41: 361-390.

t'Hart, H. 1971. Cytological and morphological variation in *Sedum*
acre L. in Western Europe. *Acta Bot. Neerl.* 20: 282-290.

t'Hart, H. 1978. Biosystematic studies in the Acre-group and the
Series Rupestria Berger of the genus *Sedum* L.
(Crassulaceae). Thèse, Utrecht, 1-153.

Hébert, L.-Ph. 1980. Recherches cytogéographiques et
cytotaxonomiques sur des espèces méditerranéennes du genre
Sedum L. *(Crassulaceae D.C.).* Thèse, Neuchâtel, 1-186.

Heslop-Harrison, J. 1958. The unisexual flower — a reply to
criticism. *Phytomorphology* 8: 177-184.

Hess, H.E., Landolt, E. und Hirzel, R. 1967. Flora der Schweiz.

Hsü, K.J. *et al.* 1973. Late Miocene dessication of the

Mediterranean. *Nature* 242: 240-244.

Hsü, K.J. *et al.* 1977. History of the Mediterranean salinity crisis. *Nature* 267: 399-403.

Johnson, A.W. and Packer, J.G. 1965. Polyploidy and environment in arctic Alaska. *Science* 148: 237-239.

Johnson, A.W. and Packer, J.G. 1967. Distribution, ecology and cytology of the Ogotoruk Creek flora and history of Beringia. *The Bering land bridge*. Stanford Univ. Press., 245-265.

Knaben, G.S. 1982. Om arts- og rasedannelsi i Europa under kvartaertiden. I. Endemiske arter i Nord - Atlanteren. *Blyttia* 40: 229-235.

Kress, A. 1963a. Zytotaxonomische Untersuchungen an den Primeln der Sektion*Auricula* Pax. *Oster. Bot. Ztschr.* 11D (1): 53-102.

Kress, A. 1963b. Zytotaxonomische Untersuchungen an den *Androsace-* Sippen der Sektion *Aretia* (L.) Koch. *Ber. Bayer. Bot. Ges.* 36: 33-39.

Küpfer, Ph. 1974. Recherches sur les liens de parenté entre la flore orophile des Alpes et celle des Pyrénées. *Boissiera* 23: 1-322.

Küpfer, Ph. 1981. Les processus de différenciation des taxons orophiles en Méditerranée occidentale. Acta III. Congr. Optima. *An. Jard. Bot. Madrid* 37: 321-337.

Landolt, E. 1960. Unsere Alpenflora. Verlag Schweizer Alpen Club. Zöllikon-Zürich, 1-217.

Lewis, W.H. 1980. Polyploidy in species populations. In *Polyploidy, biological relevance* (W.H. Lewis, ed.). 103-144.

Lewis, W.H., Oliver, R.L. and Suda, Y. 1967. Cytogeography of *Claytonia virginica* and its allies. *Ann. Mo. Bot. Gard.* 54: 153-171.

Lewis, W.H. and Semple, J.C. 1977. Geography of *Claytonia virginica* cytotypes. *Am. J. Bot.* 64: 1078-1082.

Löve, A., Löve, D. and Kapoor, B.M. 1971. Cytotaxonomy of a century of Rocky Mountain orophytes. *Arct. Alp. Res.* 3: 139-165.

Maire, R. 1963. Flore de l'Afrique du Nord 9, 1-300 Paris.

Mattfeld, J. 1922. Geographisch-genetische Untersuchungen über die Gattung *Minuartia* (L.) Hiern. *Repert. Spec. Nov Regni. Veget. Beih.* 15: 1-228.

McNeill, J., 1962. Taxonomic studies in the *Alsinoideae*: I. Generic and infrageneric groups. *Notes R. Bot. Gard. Edinburgh* 24: 79-155.

Merxmüller, H. 1952. Untersuchungen zur Sippengliederung und Arealbildung in den Alpen. München, 1-105.

Merxmüller, H. 1970. Provocation of biosystematics. *Taxon* 19: 140-145.

Monnier, P. 1960. Biosystématique de quelques *Spergularia* méditerranéens. *C.R. Acad. Sci.* 251: 117-119.

Müller, G. 1982. Contribution à la cytotaxonomie de la section *Cyclostigma* Griseb. du genre *Gentiana* L. *Feddes Repert.* 93: 625-722.

Müntzing, A. 1954. The cytological basis of polymorphism in

Poa alpina. Hereditas 40: 459-516.

Ockendon, D.J. 1968. Biosystematic studies in the *Linum perenne* group. *New Phytol.* 67: 787-813.

Ockendon, D.J. 1971. Taxonomy of the *Linum perenne* groupe in Europe. *Watsonia* 8: 205-235.

Polatschek, A. 1974. Systematisch-nomenklatorische Vorarbeit zur Gattung *Erysimum* in Italien. *Ann. Naturhistor. Mus. Wien* 78, 171-182.

Polatschek, A. 1979. Die Arten der Gattung *Erysimum* auf der Iberischen Halbinsel. *Ann. Naturhist. Mus. Wien* 82: 325-362.

Sauer, W. 1975. Karyo-systematische Untersuchungen an der Gattung *Pulmonaria (Boraginaceae). Bibl. Bot.* 131: 1-85.

Skalinska, M. 1952. Cyto-ecological studies in *Poa alpina* L. var. *vivipara* L. *Bull. Acad. Pol. Sci. Lett. Ser. Sci. B.* (I): 253-283.

Söllner, R. 1954. Recherches cytotaxonomiques sur le genre *Cerastium. Bull. Soc. Bot. Suisse* 64: 2221-354.

Stebbins, G.L. 1966. Chromosomal variation and evolution. *Science* 152: 1463-1469.

Stebbins, G.L. 1971. *Chromosomal evolution in higher plants.* MacMillan, London, 1-216.

Stebbins, G.L. 1980. Polyploidy in plants: unsolved problems and prospects In *Polyploidy, biological relevance* (W.H. Lewis, ed.).

Stern, F.C. 1949. Chromosome numbers and taxonomy. *Proc. Linn. Soc. London* 161: 119-129.

Urbanska-Worytkiewicz, K. 1977. Reproduction in natural triploid hybrids (2*n* = 24) between *Cardamine rivularis* Schur and *C. amara* L. *Ber. Geobot. Inst. Eidg. Tech. Hochsch. Stift. Rübel Zürich* 44: 42-8.

Urbanska-Worytkiewicz, K. et Landolt, E. 1978. Recherches démographiques et écologiques sur une population hybridogène de *Cardamine* L. *Ber. Geobot. Inst. Eidg. Tech. Hochsch. Stift. Rübel Zürich* 45: 30-53.

Villard, M. 1970. Contribution à l'étude cytotaxinomique et cytogénétique du genre *Leucanthemum.* Adans. Em. Briq. et Cav. *Bull. Soc. Bot. Suisse* 80: 96-188.

Weber, W.A. 1967. *Rocky Mountain Flora.* Univ. Colorado

Wettstein, R. von 1898. *Grundzüge der geographisch-morphologischen Methode der Pflanzensystematik.* Jena, 1-64.

Yang, T.W. and Lowe, C.H. 1968. Chromosome variation in ecotypes of *Larrea divaricata* in the North American desert. *Madrono* 19: 161-164

Zeltner, L. 1970. Recherches de biosystématique sur les genres *Blackstonia* Huds. et *Centaurium* Hill. *Bull. Soc. Neuchatel. Sci. Nat.* 93: 1-164.

Zeltner, L. 1978. Notes de cytotaxonomie sur les genres *Blackstonia* Huds. et *Centaurium* Hill en Crète. *Bull. Soc. Neuchatel. Sci. Nat.* 101: 107-117.

Zeltner, L. 1980. Contribution à la cytotaxonomie des populations iraniennes du genre *Centaurium* Hill. (Gentianacées). *Biol. Ecol. Médit.* 7: 57-62.

Biosystematics of Macaronesian Flowering Plants

Liv Borgen
Botanical Garden and Museum
University of Oslo
Oslo, Norway

INTRODUCTION

Insular floras are particularly interesting as a subject for the
study of evolution owing to their geographical delimination, which
has remained constant over a considerable period of time and
effectively inhibited the migration of species. Endemism is a
general phenomenon in all territories which are geographically
well defined, and especially islands which have a great reputation
as centres of differentiation and speciation.

The phytogeographical region Macaronesia (see Sunding 1979)
consists of five Atlantic archipelagos: the Azores, the Madeiras,
the Salvage Islands, the Canary Islands, and the Cape Verde
Islands, with an enclave in Morocco (Fig. 1). The total area of
the archipelagos is about 14,400 km^2. The number of flowering
plant species is approximately 3200, of which some 680, ca. 20%,
constitute an endemic element (Humphries 1979; Sunding 1979).

The large number of endemic species has made the Macaronesian
islands an outstanding area for studies of evolution and
speciation. As to chromosome numbers, much has been done on
endemic Macaronesian species (see Borgen 1979), but hard evidence
on the nature of speciation can only come from detailed crossing
experiments. In this contribution I will therefore concentrate on
those genera in which also extensive experimental hybridization
has been done.

The endemic Macaronesian element includes widespread endemics
as well as endemics of single archipelagos and single islands.
Some 32 genera are recognized as endemic to Macaronesia (Humphries
1979). Twenty are restricted to one archipelago, and among these,
crossing experiments have been undertaken in *Sinapidendron* Lowe
from Madeira by Harberd (1972) and Rustan (1980a).

Twelve endemic genera are represented in more than one

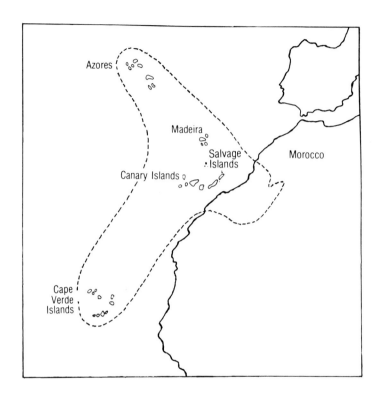

FIG. 1. *The biogeographical region of Macaronesia. After Sunding (1979).*

archipelago. *Argyranthemum* Webb ex Schulz Bip. from the Canary Islands, the Salvage Islands, and Madeira has been studied biosystematically by Humphries (1973, 1975, 1976b), Borgen (1976), and Brochmann (1983), but the crossing experiments only include species from the Canary Islands.

Most of the Macaronesian endemic species belong to more widespread genera with continental sister groups in the mainland floras of the Mediterranean area. Biosystematic studies have been undertaken in Macaronesian species of *Scrophularia* L. by Dalgaard (1979), *Diplotaxis* DC. by Rustan (1980a), *Asteriscus* (Tourn.) Mill. s. lat. by Halvorsen (1981), and *Lobularia* Desv. by Borgen (1983). Some examples from these biosystematic studies are given, and the evolutionary patterns which can be traced are discussed.

SINAPIDENDRON AND *DIPLOTAXIS*

Sinapidendron s. lat., of the Brassiceae-Brassicinae, Cruciferae (Schulz 1919), is restricted to the Macaronesian region. The

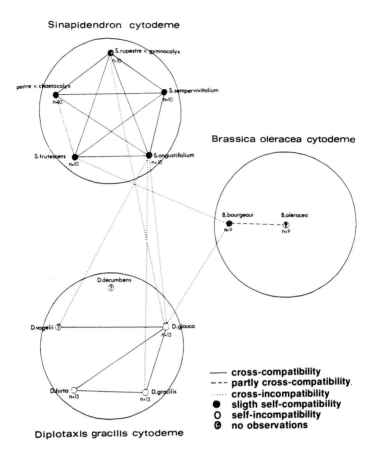

FIG. 2. *Classification of* Sinapidendron *s. lat. in cytodemes. After Rustan (1980a).*

circumscription formerly included species in Madeira, the Canary Islands, and the Cape Verde Islands. Borgen *et al.* (1979) and Rustan (1980a,b) showed that *Sinapidendron* s. str. is confined to the Madeira Islands. Rustan's (1980) classification of *Sinapidendron* s. lat., based on the cytodeme concept of Harberd (1972), is shown in Fig. 2.

In *Sinapidendron* s. str. four taxa are found in Madeira proper and one, *S. sempervivifolium* Mnzs., is restricted to the neighboring islet Deserta Grande. They grow on inaccessible cliffs and gravel slopes in both dry and humid habitats. Their distribution ranges from sea level to 1700 m alt. The species share the chromosome number $n = 10$ and are predominantly outcrossing and interfertile. Crosses with other $n = 10$ cytodemes

or species previously included in *Sinapidendron* have failed
(Harberd 1972; Rustan 1980a).

Brassica bourgeaui (Webb) O. Kuntze, a rare endemic of the
Canary Islands, was previously included in *Sinapidendron*, but
is closely related to *B. oleracea* L. and has the same chromosome
number, n = 9 (Borgen *et al.* 1979). Crosse between *B.*
bourgeaui and *B. oleracea* yielded seeds when *B. oleracea* was
the female parent, indicating at least partial interfertility.
Rustan (1980) therefore included *B. bourgeaui* in the *B. oleracea*
cytodeme.

Sinapidendron palmense (O. Kuntze) O.E. Schulz was described
from the Canary Islands, but is identical with the Mediterranean
Sinapsis pubescens L. (Rustan 1980a,b), placed in a separate
cytodeme with n = 9 (Harberd 1972).

The Cape Verdian species are shown (Rustan 1980a) to
belong to *Diplotaxis*, sect. *Catocarpum* DC. emend O.E. Schulz.
The five species are endemic, suffruticose, self-incompatible,
interfertile perennials with n = 13 (Fig. 2). They grow on steep
cliffs in humid areas and on gravel slopes in dry areas. Their
distribution ranges from sea level to 1800 m alt. Rustan (1980a)
placed these species in a *D. gracilis* cytodeme.

LOBULARIA

Lobularia, of the Alysseae, Cruciferae, is indigenous to the
coasts of W and S Europe, N Africa, the Near Orient and all the
Macaronesian archipelagos, except Madeira (Fig. 4). *L. maritima*
(L.) Desv. is a suffruticose, mainly outcrossing perennial with
n = 12 and has a Mediterranean distribution with a centre in the
west, extending to the Azores (Fig. 3). *L. libyca* (Viv.) Meisn.
is an autogamous annual with n = 11 and has its main distribution
in N Africa, with extensions to S Europe and the Canary Islands.
L. arabica (Boiss.) Muschler, also an autogamous annual, with
n = 21, has a narrow distribution in NE Africa and the Near
Orient. The endemic Macaronesian species, *L. intermedia* Webb et
Berth., *L. marginata* (Webb) Christ, and *L. palmensis* Webb ex
Christ, are suffruticose, mainly out-crossing perennials with
n = 11. *L. intermedia* ssp. *intermedia* occurs in the
Salvage Islands, the Canary Islands, and the Cape Verde Islands;
the ssp. *spathulata* (J.A. Schmidt) B. Petters. in the Cape Verde
Islands; *L. palmensis* in the western Canary Islands; and *L.*
marginata in the eastern Canary Islands and the Moroccan enclave
(Fig. 3). The crossing polygon (Fig. 4) summarizes the cross
compatibility between these taxa, based on the viability and the
fertility of the F_1- hybrids.

L. arabica and *L. libyca* are sympatric, separated from each
other by (1) initial barriers to crossability due to autogamy,
and (2) chromosomal sterility barriers. They are separated from
the suffruticose perennials by (1) initial barriers to
crossability, (2) barriers due to differences in chromosome
numbers, and (3) chromosomal or genic sterility barriers between

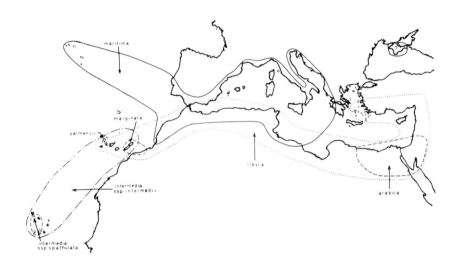

FIG. 3. *The distribution of seven taxa of* Lobularia.
The boundaries are drawn somewhat arbitrarily, especially for
L. maritima, *which occurs as a cultivation escape outside*
the area indicated, for instance in the Canary Islands, the Cape
Verde Islands, Madeira and the coasts of western Europe.

L. libyca and the Macaronesian species which share the chromosome
number *n* = 11. *L. arabica* is geographically isolated from the
perennials; *L. libyca* coexists sympatrically.

L. maritima is effectively isolated from the other species by
the differences in chromosome numbers. Due to its worldwide
introduction as an ornamental and its capacity of subsequent
spreading, it may occur adjacent to all the other species, but
its natural distribution overlaps with that of *L. libyca* only.

Between the endemic Macaronesian taxa, internal isolation
barriers are absent or weak. There are no chromosomal differences
between these taxa, and they seem to be able to exchange genes
quite freely. They are isolated or semi-isolated in nature by (1)
geographical and (2) ecological barriers to gene exchange.

L. intermedia ssp. *intermedia* is a polymorphic taxon,
represented by numerous populations in the coastal as well as the
montaneous regions, in artificial habitats, such as roadsides, as
well as in natural cliff habitats. It partly overlaps in
distribution and ecological preferences with the other
Macaronesian taxa and sometimes forms intergrades. The other
taxa, *L. intermedia* ssp. *spathulata, L. marginata,* and *L.
palmensis*, occupy disjunct areas and habitats with distinct and
different ecology. They do not intergrade with each other, but
since they all intergrade with *L. intermedia* ssp. *interme*

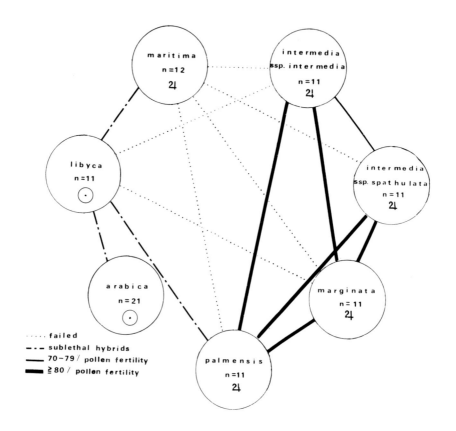

FIG. 4. *Crossing polygon based on pollen stainability and vitality of the F_1 hybrids among seven taxa of* Lobularia. *Dotted line: failed; Dash-dotted line: sublethal hybrids; narrow solid line: 70-79% pollen fertility; wide solid line: 80% or better pollen fertility.*

is doubtful if specific rank should be maintained for any of them.

SCROPHULARIA

The Macaronesian species of *Scrophularia*, of the sect. *Scrophularia*, subsect. Scorodonia, Scrophulariaceae, is subdivided by Dalgaard (1979) into three groups, on the basis of crossing experiments (Fig. 5) and comparative morphological and cytological studies.

The *scorodonia* group comprises seven taxa of suffruticose, self-compatible, predominantly out-crossing perennials with $n = 29$. *S. scorodonia* L. is widespread and occurs in the Azores, the Madeiras, W Europe, and N Africa. The other

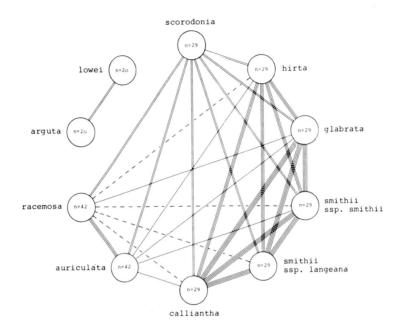

FIG. 5. *Crossing polygon summarizing hybridizations between taxa of* Scrophularia *in Macaronesia.* No line joining circles: *No capsule development or capsules very small;* Dashed line: *Plump capsules but only aborted seeds;* single solid line: *Only very few good seeds obtained; F_1 hybrids completely seed sterile; pollen fertility less than 32%;* double line: *abundant good seeds obtained; F_1 hybrids completely or highly seed sterile; pollen fertility less than 52%;* triple line: *F_1 hybrids seed fertile; pollen fertility 52-85%;* quadruple line: *F_1 hybrids seed fertile; pollen fertility 86-100%. After Dalgaard (1979).*

taxa are Macaronesian endemics: *S. hirta* Lowe in Madeira; *S. smithii* Hornem. (with three subspecies), *S. calliantha* Webb et Berth, and *S. glabrata* Ait. in the Canary Islands.

S. scorodonia is the closest relative of the endemic species, and a small amount of gene exchange may occur between *S. scorodonia* and the endemics, at least under experimental conditions.

Within the endemic Macaronesian species of the *scorodonia* group, internal barriers to gene exchange are absent or weak. Only in crosses between the Madeiran *S. hirta* and the Canarian species, pollen fertility is slightly reduced. No genetic barriers exist between the Canarian endemics. Despite their capacity of interbreeding and the sympatric occurrence of some of them, natural hybrids are not common. External geographical and ecological barriers therefore seem sufficient to maintain species integrity in the group.

The *auriculata* group consists of two self-compatible, mainly out-crossing, allopatric perennials with slightly lignified bases and $n = 42$. *S. auriculata* L. is found in the Azores, W Europe, and N Africa, whereas *S. racemosa* Lowe is endemic to Madeira. These species are separated from the other Macaronesian species by distinct barriers to gene exchange due to differences in chromosome numbers. In crosses between the two species, the hybrids have slightly reduced pollen fertility.

The *arguta* group comprises two predominantly autogamous, allopatric annuals with $n = 20$: *S. arguta* Sol. ex Ait, which is found in the Salvage Islands, the Canary Islands, the Cape Verde Islands, S Spain, N Africa and Ethiopia to Muscat; and *S. lowei* Dalgaard, endemic to Madeira. These species are separated from the other Macaronesian species by (1) initial barriers to crossability, and (2) chromosomal barriers. They are separated from each other by chromosomal or genic sterility barriers, giving rise to hybrids which die before flowering or are highly seed and pollen sterile.

ASTERISCUS

In *Asteriscus* s. lat., of the Inuleae-Inulinae, Compositae, 12 species are recognized in Macaronesia. The results of some crossing experiments made by Halvorsen (1981) on nine of these species are summarized in Fig. 6. Seven species are suffruticose perennials with $n = 7$: the three Cape Verdian endemics *A. daltonii* (Webb) Walp., *A. smithii* (Webb) Walp., and *A. vogelii* (Webb) Walp.; the three Canarian endemics *A. intermedius* (DC.) Pit. et Proust, *A. sericeus* (L.f.) DC., and *A. stenophyllus* (Link in Buch) O. Kuntze; and one Canarian-Moroccan species, *A. odorus* (Schousb.) DC. Two are Mediterranean-Canarian annuals, *A. aquaticus* (L.) Less., with $n = 7$, and *A. maritimus* (L.) Less., with $n = 6$.

The annuals *A. aquaticus* and *A. maritimus* are mainly autogamous and isolated from the others by initial barriers to crossability. Differences in chromosome numbers between *A. maritimus* and the others represent another barrier to gene exchange. Also *A. aquaticus*, which share the chromosome number $n = 7$ with the perennial species, seems to be isolated from them by chromosomal or genic barriers.

The perennial species are mainly outcrossing, the Canarian ones self-compatible and the Cape Verdian ones self-incompatible. Lack of flower formation in some of the F_1-families from crosses between Cape Verdian and Canarian species indicate partial hybrid sterility between species from the two archipelagos. There is a seasonal isolation, the Canarian species flowering one or two months earlier than the Cape Verdian ones, which may perhaps influence the flower formation in the hybrids.

Most cross-combinations of the perennial species gave vigorous and fully fertile F_1-hybrids, with regular meiosis and pollen

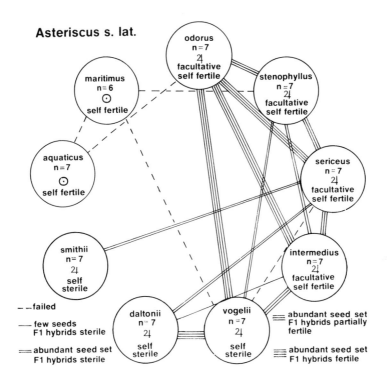

FIG. 6. *Crossing polygon summarizing hybridizations between some taxa of* Asteriscus *s. lat. in Macaronesia. Seed formation and fertility of F_1 are included. After Halvorsen (1981).*

fertility values above 50%. These results indicate a high degree of chromosomal homology and cross-compatibility between the perennial species. Natural hybrids have not been reported. Mostly geographical, ecological, and seasonal barriers seem to maintain species integrity among the Macaronesian perennial species of *Asteriscus* s. lat.

ARGYRANTHEMUM

Argyranthemum, of the Anthemideae, Compositae, comprise species (Humphries 1976a; Borgen 1980) 3 endemic to Madeira, 1 to the Salvage Islands and 19 to the Canary Islands. All species are perennial shrubs, and have the chromosome number $n = 9$, and a similar karyotype (Humphries 1975). They outbreeders, but can be self-compatible (Humphries 1976b). Five sections are recognized, based on differences in cypselas 23 morphology.

The results from Humphries' (1976b) crossing experiments on

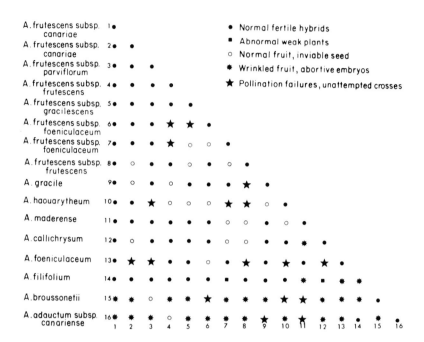

A. frutescens subsp. canariae	1 ●		●	Normal fertile hybrids
A. frutescens subsp. canariae	2 ● ●		■	Abnormal weak plants
A. frutescens subsp. parviflorum	3 ● ● ●		○	Normal fruit, inviable seed
A. frutescens subsp. frutescens	4 ● ● ● ●		✳	Wrinkled fruit, abortive embryos
A. frutescens subsp. gracilescens	5 ● ● ● ● ●		★	Pollination failures, unattempted crosses

FIG. 7. *Crossing relationships between Canary Island species of* Argyranthemum. *After Humphries (1979).*

populations from the Canary Islands are summarized in Fig. 7.
Some of his intersectional crosses failed, for instance are
between *A. broussonetii* (Pers.) Humphr. of the sect.
Sphenismelia, and species of the sect. *Argyranthemum*. My own
reciprocal crosses between *A. broussonetii* and *A.*
frutescens (L.) Sch. Bip. of the sect. *Argyranthemum*, gave fully
fertile hybrids. Mostly, infrasectional as well as intersectional
crosses are successful and give rise to normal plants with
intermediate morphology and high pollen fertility values
(Humphries 1973, 1976b; Borgen 1976, Borgen 1983). Meiosis in the
F_1 and F_2 hybrids is always normal, but a slight depression of the
overal chiasma frequency occurs (Humphries 1975).

The pairing behavior in meiosis suggests that differences
between the taxa are the result from genic changes and not from
major chromosomal rearrangements. High pollen fertility values in
the hybrids also indicate high chromosomal homology between the
species.

The isolation between the Canarian species seems to be entirely
extrinsic, due to ecological and geographical barriers. Genetic
isolation has lagged a long way behind morphological and
physiological divergences.

According to Humphries (1976b), the evolution in *Argyranthemum*

has taken place by adaptive radiation and has been influenced by
strong selection in steep ecological gradients. Adaptive trends
are found in leaf shape, habit, woodiness, and flower size. The
largest, most lignified and shrubby member of the genus is the
laurel forest species *A. broussonetii*, denoted A1 in Fig. 8.
It can attain a diameter of 6 m and a height of 2 m and has
bipinnatifid, almost sessile, leaves. Similar leaf shapes are
found in other shade or forest species (A2-A3).

From these species, Humphries (1976b) distinguishes four series
of adaptive trends. In arid south facing lowland areas, reduction
in lignification, habit, capitulum size, and leaf area results in
forms of the B and C type in Fig. 9. *A. filifolium* (Sch.
Bip.) Humphr. (C1) is the most extreme example, with a slender,
unbranched stem up to 1 m, trifid, pinnatisect, filiform leaves,
and extremely narrow capitula.

Another reduction series is found in taxa of subalpine and high
montane environments (D1-D3).

Plants of the E series have also reduced size and
inflorescences, pinnatilobed or pinnatifid leaves, but increased
succulence, and large capitula. They are found in exposed north
coastal areas of the Canary Islands and Madeira (E1-E4).

There are so far three documented examples of natural,
intersectional hybridization in *Argyranthemum*. In all three cases
the hybrids are completely viable and fertile. The hybridization
between *A. coronopifolium* (Willd.) Humphr., of the sect.
Sphenismelia, a rare endemic of Tenerife, and the widespread
A. frutescens, of the sect. *Argyranthemum*, has been analysed
by Humphries (1973, 1976b) and Brochmann (1983). The hybrid seems
to spread along the roadsides and an introgression of *frutescens*
genes has occurred to such an extent that *A. coronopifolium*
has almost disappeared in its pure form.

Borgen (1976) analysed a large hybrid swarem between *A.
adauctum* (Link) Humphr. ssp. *canariense* (Sch. Bip.) Humphr., of
the sect. *Preauxia*, and *A. filifolium*, of the sect. *Monoptera,*
on the southwestern coast of Gran Canaria, and indicated the
possibility of a similar hybrid origin for a third species in
western Gran Canaria, *A. escarrei* (Svent.) Humphr., of the sect.
Monoptera.

A third case analysed by Brochmann (data unpublished), is
between *A. frutescens*, of the sect. *Argyranthemum*, and
A. broussonetii, of the sect. *Sphenismelia*, in Tenerife
(Figs. 9-11).

A. broussonetii ssp. *broussonetii* is mainly restricted
to the Anaga peninsula (Fig. 9), and occurs in openings of the
laurel forest on tertiary basalts between 550 and 1000 m alt.

A. frutescens ssp. *frutescens* is widespread on cliffs along
the north and south coast, as a weed in disturbed areas, and may
even occur in some xerophytic scrub communities, usually between 5
and 300 m alt. It intergrades with ssp. *succulentum* Humphr. on
the northeastern coast of Tenerife. The map (Fig. 9) includes the
distribution of both subspecies.

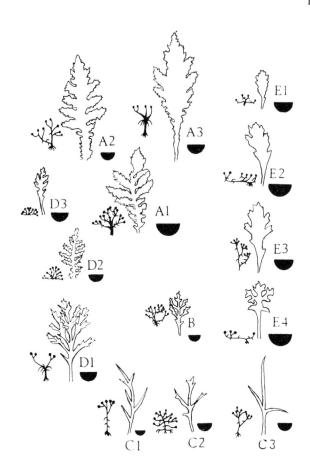

FIG. 8. *Adaptive trends in leaf shape, habit and capitulum diameter in* Argyranthemum.

SPECIES	ECOLOGY
A1 A. broussonetii	Sheltered, warm wet forest
A2 A. adauctum ssp. jacobaeifolium	
A3 A. pinnatifidum ssp. pinnatifidum	
B A. frutescens ssp. frutescens	Xerophytic zone
C1 A. filifolium	
C2 A. frutescens ssp. gracilescens	Arid, S facing slopes of xerophytic zone
C3 A. gracile	
D1 A. foeniculaceum	Low montane (xerophytic zone, sheltered ladera

On the north coast of Anaga a hybrid swarm between *A. broussonetii* ssp. *broussonetii* and *A. frutescens* ssp. *frutescens* (Fig. 9) occurs in a clearing of the vegetation locally known as Fayal-Brezal at about 500 m alt. This hybrid swarm is particularly interesting because of the proposed hybrid origin of *A. sundingii* Borgen (Borgen 1980). *A. sundingii* grows at ca. 200 m alt. on the south coast of Anaga (Fig. 9). It is intermediate between its putative parents in ecological preference as well as in morphological characters.

The swarm and populations of *frutescens, broussonetii,* and *sundingii* have been analysed by different techniques by Brochmann, and the results of an analysis based on Anderson's (1949) hybrid index method, where exact character values, ranged after Gower (1971), have been used instead of scaled values, are presented in Fig. 11.

The upper square (Fig. 11) shows the variation in different parental populations, denoted Fl-3 for *frutescens* and Bl-3 for *broussonetii*. In the lower square, the parental populations are united, and experimental F_1-hybrids are shown to be quite intermediate. The lower histogram shows the variation in the natural hybrid swarm; the S histogram the variation in *A. sundingii*, and the SC histogram the variation in the cultivated progeny of *A. sundingii*.

The variation in *A. sundingii* does not exceed that of either proposed parent, and the population of *sundingii* is intermediate between the experimental F_1-hybrids and *A. frutescens*. Due to cypsela morphology, *A. sundingii* is placed in sect. *Argyranthemum* (Borgen 1980). Brochmann's results may suggest that it has originated as a segregate in a hybrid swarm of *A. broussonetii* X *frutescens*, possibly from a generation of backcrosses to *A. frutescens*. *A. sundingii* grows in a natural cliff habitat surrounded by farm land. It is kept apart from other species of *Argyranthemum* by ecological isolation. An eventual former hybrid swarm in the area may have become extinct, and the only trace of the hybridizing activity is the isolated population of *A. sundingii*.

D2 A. adauctum *ssp.* adauctum	*Arid, high montane regions*
D3 A. tenerifae	*Submacaro-alpine zone*
E1 A. pinnatifidum *ssp.* succulentum	
E2 A. coronopifolium	*Halophytic coastal cliffs of the N facing slopes of*
E3 A. maderense	*the xerophytic zone*
E4 A. frutescens *ssp.* succulentum	

Scales: plant height/width (⊢——⊣ = 1 m); leaf length/width (⊢————⊣ = 10 cm); capitulum diameter (⊢—⊣ = 10 mm). After Humphries (1979).

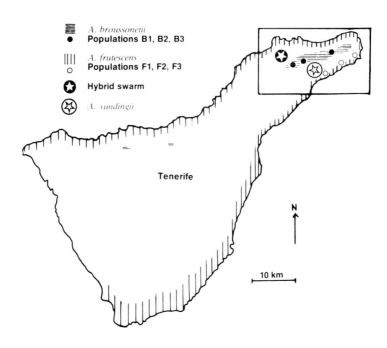

FIG. 9. *The distribution of* Argyranthemum broussonetii. A. frutescens, A. sundingii *and a hybrid swarm,* A. broussonetii X frutescens, *in Tenerife. After Brochmann unpubl.).*

CONCLUDING REMARKS

The isolation mechanisms in the genera cited are both internal and external, and different patterns of speciation and isolation can be recognized.

Between some species and species groups internal barriers are strong and involve prefertilization and postfertilization incompatibility barriers, hybrid inviability, hybrid sterility, and hybrid breakdown. These barriers are either based on differences in chromosome numbers or on chromosomal repatterning and genic changes.

1. *Chromosomal Isolation and Speciation by Means of Changes in Basic Chromosome Numbers*

Differences in basic chromosome numbers are the main barrier to gene exchange between *Diplotaxis (n = 13)* and *Sinapidendron (n = 10)*; between the *scorodonia (n = 29)*, *auriculata (n = 42)*, and *arguta (n = 20)* groups of *Scrophularia*; between *Asteriscus maritimus (n = 6)* and the other species of *Asteriscus* in Macaronesia *(n = 7)*; and between *Lobularia maritima (n = 12)* and the Macaronesian *(n = 11)* species

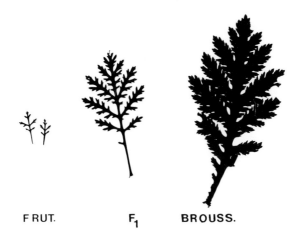

F RUT.   F₁   BROUSS.

FIG. 10. *Leaf silhouettes of* Argyranthemum broussonetii, A. frutescens *and their experimental F₁ hybrid. After Brochmann (unpubl.).*

of that genus. Aneuploid, dysploid or polyploid divergence has occasioned strong internal barriers in all these cases. The evolutionary changes involved have probably taken place a long time ago, presumably in the Tertiary or late Cretaceous period, at which time the palaeo-endemic Macaronesian flora is believed to have been established (see Sunding 1979). The different species or species groups may have some common ancestor, or can have reached the Macaronesian region at different times and originated from different mainland stocks. Morphological divergence is pronounced in all groups except *Lobularia*.

2. *Chromosomal Isolation and Speciation by Means of Chromosomal Repatterning*
Another speciation process involves a sudden differentiation of a population with respect to morphology and particularly with respect to chromosomal repatterning, with consequent reproductive isolation (see Stebbins and Major 1965; Grant 1981).

Isolation barriers between Macaronesian species with the same chromosome number may be due to chromosomal repatterning, but may also be due to genic changes. Such barriers are found between the annual species of the *Scrophularia arguta* group ($n = 20$), between the annual *Asteriscus aquaticus* ($n = 7$), and the perennial Macaronesian species of that genus ($n = 7$), and between the annual *Lobularia libyca* ($n = 11$) and the perennial endemic Macaronesian species of *Lobularia* ($n = 11$). The annual species in the examples cited are also isolated from their perennial relatives by initial barriers to crossability due

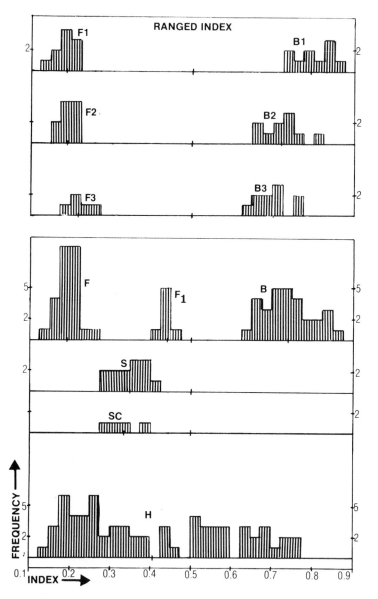

FIG. 11. *Histograms showing the frequency distribution of the hybrid index totals in populations of* Argyranthemum broussonetii, A. frutescens, A. sundingii *and the hybrid swarm* A. broussonetii × frutescens. *F:* A. frutescens *(F1-3 three populations).* B: A. broussonetii *(B1-3 three populations).* F_1: *Experimental F_1-hybrids.* S: A. sundingii. SC: *Cultivated progeny of* A. sundingii. H: *Hybrid swarm. After Brochmann (unpubl.).*

to autogamy.

According to Grant (1981) it is a prevalence of chromosomal repatterning among annual plants. These plants are often, as the Macaronesian ones, autogamous, and inbreeding may induce chromosome breakage and rearrangements, at least during certain periods in the history of a plant population (Lewis and Raven 1958; Lewis 1962).

3. *Speciation by Means of Hybridization*

A third major process of speciation involves hybridization between related, previously isolated populations, eventually followed by introgression, polyploidy, or both.

Despite their capacity of interbreeding, hybrids are not common among the Macaronesian species of the genera cited. Most examples are found in *Argyranthemum*, in which genus hybridization may have influenced species diversity. *A. sundingii* is a fairly well documented case of a species of hybrid origin. Introgression has had a swamping effect on *A. coronopifolium*, due to the extensive hybridization with *A. frutescens*, and thus representing a reversal of the primary divergence in the genus.

In other Macaronesian genera field records of rare hybrids exist, for instance in *Aeonium* and *Greenovia* (Praeger 1929, 1932; Kunkel 1972; Voggenreiter 1974) and *Echium* (Bramwell 1973), but experimental data are lacking.

Polyploidy has played a minor role in the speciation in Macaronesia. The overall frequency of polyploids is low, ca. 25% in the Canarian flora (Borgen 1979), some of the polyploids being neopolyploids or mesopolyploids, others paleopolyploids (Borgen 1969, 1979).

The *Lobularia* and *Diplotaxis* species, which have moderately high basic numbers, may have a polyploid origin (see Stebbins 1971), and the *Scrophularia* specie certainly are polyploids. Polyploidy in these cases probably is an ancient one, and the diversification within each group has taken place at the same ploidy level.

4. *Ecogeographical Isolation and Gradual Speciation*

In all the genera cited, crossing experiments have shown that internal barriers to gene exchange are absent or weak among closely related suffruticose, perennial Macaronesian species of flowering plants, i.e. the Madeiran species of *Sinapidendron*, the Canarian species of *Argyranthemum*, the Canarian species of the *scorodonia* group of *Scrophularia*, the Canarian species of *Asteriscus*, the Cape Verdian species of *Diplotaxis*, and the Macaronesian endemic species of *Lobularia*. In the biosystematic hierarchy these species are ecotypes or ecospecies, and each group constitutes a coenospecies.

The isolation mechanisms in these groups are mainly spatial and environmental. Occasionally, as in *Asteriscus*, the species are also seasonally isolated.

The fertility relationship between the species of each group

corresponds to the *Ceanothus* pattern described by Grant
(1981). The species are intercompatible, interfertile, more or
less woody shrubs or subshrubs, with a predominantly outcrossing
breeding system, and chromosomal homology. They are isolated or
semi-isolated in nature by geographical and ecological barriers.

The pattern is most pronounced within a single archipelago.
The hybrid fertility drops with increasing distance between the
parental populations, for instance between the Cape Verdian and
Canarian species of *Asteriscus* and between the Madeiran
Scrophularia hirta and the Canarian species of the *S. scorodonia*
group (see also Kruckeberg 1957).

Within each archipelago speciation has taken place by a gradual
divergence of previously similar populations, mainly by the
process of adaptive radiation in response to a wide range of
habitats under the isolated island conditions (see Stebbins and
Major 1965; Humphries 1979; Grant 1981). The endemic species of
each of these adaptively radiated groups all share the same
chromosome number and are schizoendemics according to the
classification of Favarger and Contandriopoulos (1961).

Also other genera of Macaronesian flowering plants than those
cited are schizoendemic groups of woody shrubs or subshrubs,
resulting from adaptive radiation, such as *Aeonium* (Lems 1960),
Echium (Bramwell 1975) and *Sonchus*, subgenus *Dendrosonchus*
(Bramwell 1972). The species of these genera also probably lack
internal barriers to gene exchange, are isolated by geographical
and ecological barriers, and may constitute coenospecies in the
biosystematic hierarchy.

According to Grant (1981) most trees, shrubs and perennial
species are outcrossing, whereas most groups of annuals are partly
or predominantly self-fertilizing. This correlation between
sterility barriers and life form, which may reflect a more
fundamental relationship, is found in all Macaronesian
biosystematically studied species. Biosystematic studies of
Macaronesian species have therefore not brought any essentially
new evolutionary mechanisms into the light, but have substantiated
the occurrence also in isolated island floras of some common
patterns of speciation.

REFERENCES

Anderson, E. 1949. *Introgressive Hybridization*. Wiley, New
York. 109 p.
Borgen, L. 1976. Analysis of a hybrid swarm between
Argyranthemum adauctum and *A. filifolium* in the Canary
Islands. *Norw. J. Bot.* 23: 121-137.
Borgen, L. 1979. Karyology of the Anarian Flora In *Plants and
Islands* (D. Bramwell, ed.), Academic Press, New York. pp.
329-346.
Borgen, L. 1980. A new species of *Argyranthemum* (Compositae)
from the Canary Islands. *Norw. J. Bot.* 27: 163-165.
Borgen, L. 1983. Chromosome numbers and fertility relationships

in *Lobularia*, Cruciferae. A preliminary report. *Webbia*, in press.

Borgen, L., Rustan, Ø.H. and Elven, R. 1979. *Brassica bourgeaui* (Cruciferae) in the Canary Islands. *Norw. J. Bot.* 26: 255–264.

Bramwell, D. 1972. Endemism in the flora of the Canary Islands. In *Taxonomy, Phytogeography and Evolution* (D.H. Valentine, ed.), Academic Press, New York. pp. 141–159.

Bramwell, D. 1973. Studies in the genus *Echium* from Macaronesia. *Monogr. Biol. Canar.* 4: 71–82.

Bramwell, D. 1975. Some morphological aspects of the adaptive radiation of Canary Island Echium species. *Anal. Inst. Bot. Cavanilles* 32(2): 241–254.

Brochmann, C. 1983. Hybridization and distribution of *Argyranthemum coronopifolium* (Asteraceae: Anthemideae) in the Canary Islands. *Nord. J. Bot.*, in press.

Dalgaard, S. 1979. Biosystematics of the Macaronesian species of *Scrophularia*. *Opera Bot.* 51: 1–64.

Favarger, C. and Contandriopoulos, J. 1961. Essai sur l'endemisme. *Bull. Soc. Bot. Suisse* 71: 383–408.

Gower, J.C. 1971. A general coefficient of similarity and some of its properties. *Biometrics* 27: 857–874.

Grant, V. 1981. *Plant Speciation*. Columbia Univ. Press, New York. 563 p.

Halvorsen, T. 1981. *The perennial species of* Bubonium *(Asteriscus p.p.) (Asteraceae) in the Canary Islands and the Cape Verde Islands*. Unpubl. Cand. real. Thesis, Univ. Oslo. 120 p.

Harberd, D.J. 1972. A contribution to the cytotaxonomy of *Brassica* (Cruciferae) and its allies. *J. Linn. Soc. Bot.* 65: 1–23.

Humphries, C.J. 1973. *A taxonomic study of the genus Argyranthemum*. Unpubl. Ph.D. Thesis, Univ. Reading, England. 376 p.

Humphries, C.J. 1975. Cytological studies in the Macaronesian genus *Argyranthemum* (Compositae-Anthemideae). *Bot. Not.* 128: 239–255.

Humphries, C.J. 1976a. A revision of the Macaronesian genus *Argyranthemum* Webb ex Schultz Bip. (Compositae-Anthemideae). *Bull. Br. Mus. Nat. Hist. Bot.* 5 (4): 147–240.

Humphries, C.J. 1976b. Evolution and endemism in*Argyranthemum* Webb ex Schultz Bip. (Compositae: Anthemideae). *Bot. Macaronesica* 1: 25–50.

Humphries, C.J. 1979. Endemism and evolution in Macaronesia. In *Plants and Islands* (D. Bramwell, ed.), Academic Press, New York. pp. 171–199.

Kruckeberg, A.R. 1957. Variation in fertility of hybrids between isolated populations of the serpentine species, *Streptanthus glandulosus* Hook. *Evolution* 11: 185–211.

Kunkel, G. 1972. Plantas Vasculares de Gran Canaria. *Monogr. Biol. Canar.* 3: 1–86.

Lems, K. 1960. Botanical notes on the Canary Islands. II. The evolution of plant forms, in the islands: *Aeonium*. *Ecology* 41: 1-17.

Lewis, H. 1962. Catastrophic selection as a factor in speciation. *.Evolution* 16: 257-271.

Lewis, H. and Raven, P.H. 1958. Rapid evolution in *Clarkia*. *Evolution* 12: 319-336.

Praeger, R.L. 1929. Semperviva of the Canary Islands area. *Proc. R. Irish Acad.* 38, Sect. B., 454-499.

Praeger, R.L. 1932. *An Account of the Sempervivum Group.* R. Hortic. Soc., London. 265 p.

Rustan, Ø.H. 1980a. *Biosystematic studies in* Sinapidendron s. lat. *(Brassicaceae). 1. The generic delimination.* Unpubl. Cand. scient. Thesis. Univ. Oslo. 75 p.

Rustan, Ø.H. 1980b. *Sinapidendron palmense* (Brassicaceae) is *Sinapis pubescens*. *Norw. J. Bot.* 27: 301-305.

Schulz, O.E. 1919. Cruciferae – Brassiceae. Subtribus I. Brassicinae et II. Raphanineae. In *Das Pflanzenreich* 70 (IV 105) (A. Engler, ed.), pp. 1-290.

Stebbins, G.L. 1971. *Chromosomal Evolution in Higher Plants.* Arnold, London. 216 p.

Stebbins, G.L. and Mayor, J. 1965. Endemism and speciation in the Californian Flora. *Ecol. Monogr.* 35: 1-35.

Sunding, P. 1979. Origins of the Macaronesian flora. In *Plants and Islands* (D. Bramwell, ed.), Academic Press, New York. pp. 13-40.

Voggenreiter, V. 1974. Geobotanische Untersuchungen an der Natürlichen Vegetation der Kanareninsel Tenerife (Anhang; Vergleiche mit La Palma und Gran Canaria). Als Grundlage für den Naturschutz. *Diss. Bot. 26.* Cramer, Leutershausen. 718 p.

Biosystematics of Tropical Forest Plants: A Problem of Rare Species

P. S. Ashton
The Arnold Arboretum
Harvard University
Cambridge, Massachusetts, U.S.A.

INTRODUCTION

Tropical forests continue to lure evolutionary biologists because of their often exceptional species richness. This richness poses two interrelated questions: How has it arisen, and how is it maintained?

These questions have in turn been addressed in two ways. One, the synecological, analyses species population patterns in time and space, between and within sample forest communities. The recurrence of patterns in population distributions or performance be they correlated with soil, with variation in light climate for instance within the dynamic cycle, or with the structure of the mature forest, or ultimately solely with the distribution of other species within a uniform community; or again, be there spatial interactions between juveniles and adults within species populations; all imply that selective processes are mediating performance and mortality at some stage in the life history, and thus help explain how species richness is maintained. Of special interest to the biosystematist is the detection of positive or negative association, in time or space, between taxa with close systematic affinity. Absence of recurring patterns, on the other hand, implies that the community is a random assemblage of species, whose composition is not maintained in equilibrium by environmental forces, but which represents no more than a balance between rates of immigration and extinction following island biogeography theory (Macarthur and Wilson 1967; Hubbell 1979).

The second approach is biosystematic. Using the synecological approach, coupled with systematic, including cladistic analysis, groups of allied species which are representative of the leading community-wide patterns of ecology, biogeography and plant form are identified. Knowledge of breeding systems can then establish

whether the means for gene exchange, and the maintenance of
genetic variability within populations exists, and whether genetic
variability is restricted by intrinsic factors. Direct evidence
may be sought for genetic variability and patterns of allele
distribution within and between populations, while demographic
data can provide evidence for selective mortality through the life
cycle.

The biosystematist, synthesizing the evidence of both
approaches, can thereby address both the maintenance and origin of
species richness.

At present, our knowledge of the biosystematics of rain forest
plants remains too rudimentary and fragmented to do more than
provide preliminary inferences, and to identify promising areas
for future work. In this paper I will attempt this, in summary,
by taking for my examples one region, the aseasonal tropics of
western Malesia excluding the Philippines (often called Sundaland
by the zoologists); one life form, namely trees; and recently
monographed taxa, particularly from two representative but
contrasting mature-phase rain forest tree families,
Dipterocarpaceae and Sapindaceae.

COMMUNITY STRUCTURE

The search for patterns in plant communities is complicated by the
need to predict the expected scale of the pattern before sampling.
If the sample size is substantially smaller or the sampling
duration shorter than the mean scale of pattern, it cannot be
detected. Forests of the aseasonal humid equatorial lowlands are
particularly amenable to studies of temporal patterns for, unlike
most other terrestrial ecosystems, many are apparently not subject
to those occasional climatic catastrophes which play such a large
though still little understood part in the ecology - and
biosystematics - of species in seasonal climates. Though the
great majority of rain forest plants are long-lived, growth and
mortality rates can be readily calculated in these predictable
environments through extrapolation from fairly short-term
measurements provided sample sizes are large enough, while spatial
patterns of temporal changes within communities can be inferred
from maps of horizontal stand structure.

On the other hand, the analysis of spatial patterns themselves
pose greater problems in tropical rain forest than elsewhere.
This is on account of the very large scale of pattern which is
occasioned by the species richness, and the ultimate size as well
as the range of sizes represented by the species.

At the regional level, species richness in the Far East is
strongly correlated with rainfall seasonality. Unfortunately,
complete tree species enumerations of sample plots larger than 1
ha in mixed evergreen rain forest are not available for the
seasonal regions of Asia. The total number of tree species in
seasonal lowland mixed forests south of the Cardamome Mountains in
southern Cambodia, an area of 6000 sq km, has been estimated as

less than 500 (M. Marcotte, pers. comm.), while the total number
in a similar forest, the Ma-Da forest northeast of Ho Chi Minh
City (formerly Saigon), Vietnam, ca. 150 sq km in area, is
estimated to be 350 (Pham Hoang Ho, pers. comm.). For comparison,
580 tree species exceeding 10 cm dbh were recorded in five, 2 ha
plots at Pasoh forest, in the aseasonal lowlands of peninsular
Malaysia, and the total number in that forest alone, now only of
1600 ha extent, may be expected to exceed 700. According to
Ashton (1967) there are 139 dipterocarp species in the lowland
mixed dipterocarp forests of peninsular Malaysia, only 31 in
seasonal mixed evergreen forest in adjacent Thailand. Whitmore
(1975) commented that there is a similar difference among species
in the pioneer genus *Macaranga* (Euphorbiaceae).

There is a striking increase in the number of series of
congenerics cohabiting single guilds within a community type in
the forests of the West Malesian aseasonal tropics, though the
genera concerned do generally occur in the seasonal regions to the
north as well.

There is thus a correlation between the absence of the rigors
of an annual drought and high species richness of both whole
forests and individual genera. It may be significant that many
genera in families, such as Dipterocarpaceae and Caesalpiniaceae,
which are speciose in the aseasonal tropics manifest supraannual
mast fruiting there but fruit annually in seasonal regions.

Forests on different soils are well known to vary in species
richness within a climatically uniform region (Richards 1952). In
addition, though, individual forest types characteristic of one
climate and soil vary in species richness throughout a region: 4
ha samples of mixed dipterocarp forest, on relatively uniform
leached yellow clay loams of moderate to low fertility, differ
substantially in species richness between Sri Lanka, peninsular
Malaysia and Borneo (Fig. 1). The relative poverty of these
forests in aseasonal Sri Lanka where, even formerly, they occupied
less than one-quarter of the land area of that small island, and
are separated by thousands of kilometers from those of Madagascar
and West Malesia, is to be expected according to island
biogeography theory (Macarthur and Wilson 1967). Immigration
rates and opportunities for allopatric speciation are limited,
even if extinction rates are low. But the exceptional species
richness of the sample from Northwest Borneo is surprising. The
lowland forests of peninsular Malaysia overlie an ancient land
surface, and are floristically rather uniform through most of the
country. Northwest Borneo by contrast is geomorphologically
young, and soils shallow and variable so that lowland forests
there are fragmented into ecological islands of varying but often
small area. Phytogeographical and geomorphological evidence
suggests that such islands fragment and coalesce on a time scale
of tens to hundreds of thousands of years (Ashton 1964, 1972,
1977). The species richness there is contrary to first expectations from island biogeography. One explanation could be that
periodic isolation and coalescence on these scales could allow for

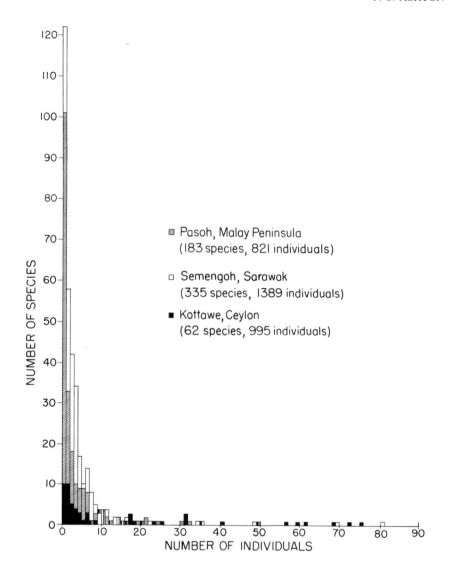

FIG. 1. *Population densities of species occurring in 3, 4 ha, rain forest samples.*

rapid allopatric speciation and immigration rates. On the other hand, though species populations may only persist within a specific soil range, chance may allow sporadic individuals to establish themselves, and perhaps reproduce at reduced fecundity, in suboptimal habitats adjacent to the permanent breeding population. Yap (1982) observed, for instance, that populations of the self-compatible androdioecious tree *Xerospermum*

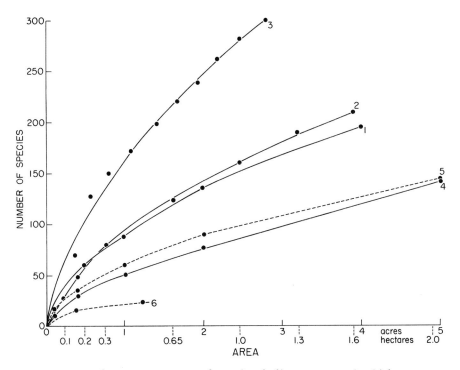

FIG. 2. *Species-area curves for mixed dipterocarp (solid lines) and heath forests (broken lines) in West Malesia. 1: Sungei Menyala Forest Reserve, Peninsular Malaysia, clay; 2: E. Borneo, clay; 3: N.W. Borneo, rhyolite; 4: N.W. Borneo, clay; 5: N.W. Borneo, podsol; 6: E. Borneo, podsol. Date from Wyatt-Smith, 1966 (1); Riswan 1982 (2,6); P.S. Ashton (unpubl. data).*

noronhianum (Sapindaceae) are concentrated on alluvium within Pasoh forest, peninsular Malaysia and that, though some trees are scattered on adjacent hills, the fecundity of those trees is lower. The first of these interpretations is compatible with selectionist views of species richness, while the second implies that chance plays a major role.

That nonequilibrium, island biogeographic, factors do play some part is implied from Table I, and Fig. 2, where the floristic richness of samples of mixed dipterocarp forest on leached yellow clay loams, and Heath forest on lowland podsols, are compared between northwest and east Borneo. Whereas the two soil types occupy extensive, if fragmented, areas in the northwest, the area of podsols in the east is small, and divided into more or less remote islands. These "oceanic" islands bearing a highly distinctive biome are either less successful in accumulating

TABLE I. *Species richness in West Malesian forest samples:*
trees exceeding 10 cm dbh

Sample area (ha)	Locality	Soil	Number of individuals	Number of species
1.6	Sungei Menyala Forest Reserve Malaya[1]	Yellow/red clay loam	1075	197
1.6	Lempake, Samarinda, E. Borneo[2]	Yellow/red clay loam	712	209
1.6	Kuala Belalong, Brunei, N.W. Borneo[3]	Yellow clay	712	125
1.6	Lambir National Park, Sarawak, N.W. Borneo[3]	Yellow sand	1025	283
0.5	Samboja, Samarinda[2]	Sand podsol	375	24
0.5	Badas, Brunei[3]	Sand podsol	369	72

[1]Wyatt-Smith 1966; [2]Riswan 1982; [3]Ashton, unpublished data.

species, or in sustaining their populations. That these
differences are real rather than a freak of sampling error is
proven by the presence, in northwest Borneo, of a substantial
endemic element on podsols. Also, another endemic element exists
there on the deep leached sands which, in that region, not only
characterize the marginal ecotones of podsolised beach terraces
and plateaux, but cover substantial areas of poorly consolidated
Mio-Pliocene sediments at several points up the coast (Ashton
1964, see fig. 59). In Fig. 3, the distribution of the
dipterocarp flora specific to these soils is indicated. Though it
is possible that east Borneo experienced greater climatic change
than the northwest during the Pleistocene, this argument cannot be
used to explain the poverty of these elements in Riau and east
peninsular Malaysia. In east Borneo I have observed the
widespread clay soil flora occupying the peripheral yellow sands
of podsolized terraces in the absence of a distinct yellow sand
tree flora there.

We know that forest composition varies climatically and edaphically. We expect patches of associated species to occur within uniform sites, representing the elements of the dynamic cycle of the forest. We also expect structurally defined guilds to occur. But are species populations randomly sorted within each of these guilds? In particular, are the population distributions of species within those congeneric series that are so well represented in West Malesian forests (Fedorov 1966) randomly associated with one another? This could occur in the presence of specific selective factors, for instance if Janzen's (1971) view is correct that seed predation limits population density. This is testable, for individuals in populations thus limited should be regularly disposed (i.e. overdispersed in the statistical sense). Seed predators may instead be genus-specific, in which case individuals of all the species of that genus should show overdispersion. What little evidence there is sugggests that seed predator specialization is usually supraspecific. If so, then lack of association between congenerics must be accepted as evidence for Hubbell's (1979) nonequilibrium hypothesis, derived from island biogeography theory, that the species composition of guilds is random.

In practice, Hubbell's hypothesis is difficult to test without biosystematic and ecophysiological study. For instance, it is well known that species of *Shorea* section *Mutica* (Dipterocarphaceae), several of which cohabit the emergent guild of mature phase Malesian mixed dipterocarp forests, differ in growth rates and, apparently, light requirements (Vincent 1961). Canopy individuals of the different species can thereby be randomly associated though their juveniles, once established, should show negative association. Also, many species exhibit clumped distributions at all stages (Poore 1968; Ashton 1969), apparently on account of poor fruit dispersal. At first sight this would appear to support the nonequilibrium hypothesis provided clumps of different species are not associated. Dipterocarps, which are always clumped, are ectotrophic mycorrhizal though, and the possibility therefore exists that conditions for early establishment are thereby most favorable within the root-run of conspecifics (Ashton 1982; W. Smits *in litt*.).

Detailed analysis of spatial and temporal variation within large uniform blocks of rain forest, such as is being undertaken by S.P. Hubbell at Barro Colorado Island, Canal Zone, are absolutely essential though, in order to define problems for further research. Such a study is not imperative in a region, such as West Malesia, where speciose tree genera dominate floristic composition (Fig. 4).

BIOSYSTEMATICS

Tropical forest tree species are, of course, defined principally on the basis of discontinuities in the pattern of morphological variation. Now that at least some groups have been farily well

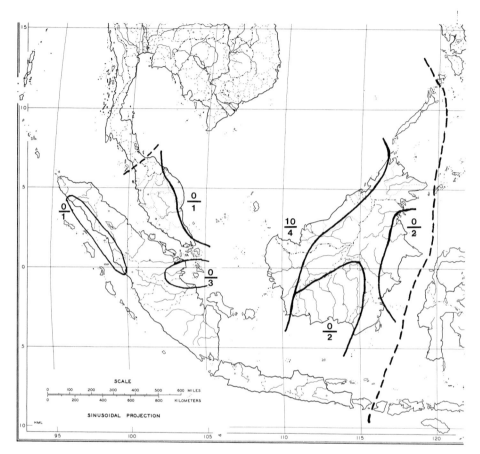

FIG. 3 (left). *Distribution of West Malesian dipterocarp species confined to podsols; endemics.*

collected, distinct patterns of geographical variation are
emerging. Within most speciose tree genera, flower morphology
varies rather little between species, though there may frequently
be variation in flower size. The differences between species are
mainly therefore in tree size, leaf, fruit and sometimes bark.
Some exceptions do exist in which the species are indistinguishable on sterile leaf characters but differ in flower or fruit:
examples include many neotropical *Eschweilera* (Lecythidaceae)
(G.H. Prance pers. comm.) and some Malesian *Polyalthia*
(Anonaceae). Species of the *sumatrana* group of *Polyalthia*,
at least, can coexist in the same forest, providing an ideal
subject for biosystematic study (S.H. Rogstad pers. comm.).
    In evergreen forest, taxa are generally morphologically uniform
within populations, though this is frequently not the case among
congenerics inhabiting savanna (see, e.g. Ashton 1982 regarding

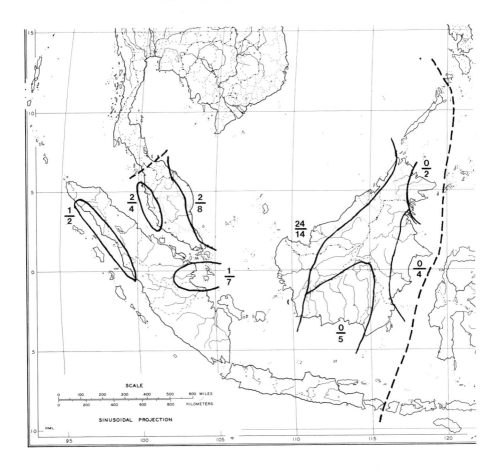

FIG. 3 (right). *Distribution of West Malesian dipterocarp species confined to deep yellow sands; wides.*

Dipterocarpaceae).
     Regional patterns of morphological variation differ in some
respects between families. Patterns of variation among the
commoner dipterocarps are rather well known, though by no means
understood. There is a matrix of widespread species which are
common throughout West Malesia excluding the Philippines. Some
are so uniform that provenance of unlabelled specimens cannot be
guessed at. Others are subdivided, generally rather clearly, into
rather constant geographical forms which I have described as
subspecies. Many are fragmented into vicariants on each of the
major land masses. Many widespread species of yellow-red inland
clay soils have edaphically differentiated coastal sibling taxa,
particularly in northwest Borneo. Isolated outcrops of distinct
lithology and soil, such as limestone, ultramaphic rock, basalt
and rhyolite can bear distinct and constant forms of widespread

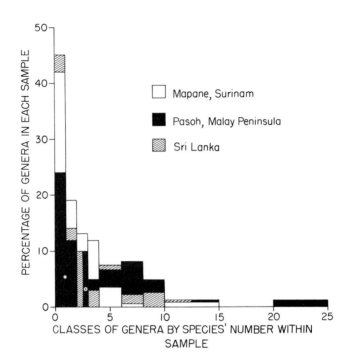

FIG. 4. *The proportion of genera of different size in three rain forests. 4 ha samples; trees above 20 cm dbh only.*

species. These forms sometimes apparently persist in very small populations, reminiscent of mountain peak endemics in *Syzygium* and *Eugenia* (Myrtaceae) in Sri Lanka (Ashton 1980) or *Calophyllum* (Clusiaceae) in peninsular Malaysia (Stevens 1980). Allozyme data are consistent with evidence from comparative morphology (Ashton and Gan, in prep.) that speciation here is allopatric, and that the most closely allied congenerics rarely occupy the same guild (e.g. *S. mareoptera* subspecies cited in Ashton, 1969; see also Fig. 5; the case of *Hopea jucunda,* Fig. 7, is to be discussed).

    Patterns among *Pometia, Allophyllus, Lepisanthes, Dimocarpus* and *Xerospermum,* Sapindaceae which have been the subject of critical treatments by Jacobs (1962) and Leenhouts (1967, 1969, 1971, 1983) respectively, differ considerably from those described among dipterocarps, but bear a family stamp of their own. Here, allopatric and ecotypic differentiation is also the rule, and more or less distinct local forms occur. Compared with dipterocarps, though, each genus contains far fewer sharp discontinuities in its overall pattern of variation. Instead, one or more very widespread taxa occur, within which a complex of more or less reticulate but stepped rather than clinal pattern of allopatric or

FIG. 5. Shorea macroptera *Dyer ssp.* bailloni *(Heim)*
Ashton, left; and ssp. macropterifolia *Ashton, right. Ssp.*
bailloni *is a main canopy tree, ssp.* macropterifolia
an emergent, but both can coexist in the same community.

ecotypic differentiation is discernable. Though two or more forms
within the species can co-occur in the same locality and even the
same forest which, in that locality, remain constantly distinct
the regional variation contains intermediate forms. An example is
in *Xerospermum noronhianum* in the peninsular Malysian part of its
distribution, where a muricate-fruited form, "wallichii", and a
tuberculate-spiny-fruited, "intermedium", often co-occur but seem
to flower in different seasons (Yap 1976). It is not that the
overall pattern of differentiation is different between the two
families, or that the totality of variation is necessarily less,
but that speciation appears not to proceed to completion in these
Sapindaceous genera.

Further research may confirm that the key to these familial
differences is in their reproductive biology. Dipterocarps bear
hermaphrodite flowers. All closed forest dipterocarp species
examined have proven to be highly, but not totally, self-
compatible (Chan 1981); most are diploid (Jong 1982). The smaller
dipterocarp genera appear to be pollinated by meliponid bees, but
the species within the big sympatric congeneric series share

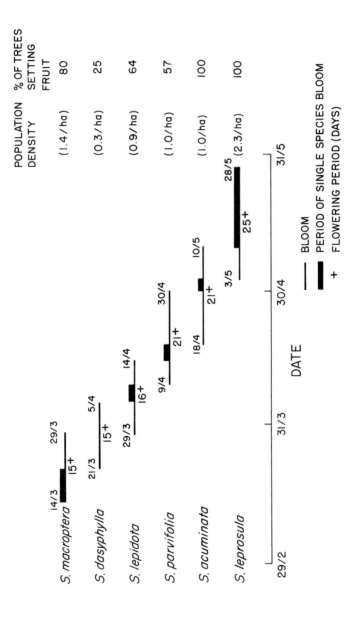

FIG. 6. *Flowering phenology and fruit production among six species of Shorea, section Mutica, tagged along a 1.6 × 0.1 km transect in Pasoh forest, peninsular Malaysia (modified after Chan and Appanah 1980).*

common pollinators which are fecund and have short life-cycles.
Species of *Shorea* section *Mutica,* for instance, are apparently
exclusively pollinated by thrips (Thysanoptera) (Chan and Appanah
1980). Flowering, which is supraannual, is highly synchronized
within species populations, but staggered among co-occuring
members of the section (Fig. 6). Hybridization among members of
the section is almost unknown. The fruit , are dispersed by
gyration occasionally assisted by gusts of wind, or they merely
drop. Both pollen and seed dispersal are therefore probably very
short range, but the means do exist for occasional transfer over
considerable distances.

Sapindaceous genera examined by Ha (1978) are monoecious
(Pometia, Allophyllus) or androdioecious *(Xerospermum,Nephelium).*
All appear to be diploid. *Pometia pinnata* and *Allophyllus
cobbe* were found to be fully self-compatible. The reproductive
biology of *Xerospermum noronhianum* has been the subject of
detailed studies by Ha (1978), Yap (1976, 1982) and Appanah
(1982). They found that "female" trees produce hermaphrodite
flowers with pollen grains of normal form and germination rates,
but the anthers do not dehisce. The stigma remains receptive for
three days. If cross-pollination does not occur, toward the end
of anthesis the innermost stamens of the hermaphrodite flower arch
over and the anthers become pressed against the stigmatic
surfaces, when the pollen sacs collapse and selfing occurs. The
pollinators of *Pometia, Nephelium* and *Xerospermum* are principally
bees, *Trigona* spp. in the lattermost. The fruit
are dispersed by birds (*Allophyllus,* perhaps *Lepisanthes*) and
primates.

The seeds of probably the majority of rain forest trees are
dispersed by vertebrates. Many of these trees are dioecious
(Ashton 1969; Bawa and Opler 1975). Nevertheless, the patterns of
differentiation found among dipterocarps seems to be more general
among mature phase genera in other families than that manifested
in Sapindaceae, although the level of local endemism among
dipterocarps is attained in few. Complex patterns of
intraspecific variation do sometimes occur, for instance in the
hermaphrodite flowered genus *Durio* (Bombacaceae). Here the
evidence from bat-pollinated *D. zibethinus,* section *Durio*
(Soepadmo and Eow 1977) and *Trigona*-pollinated *D. griffithii,*
section *Boschia* (Chan in Ha 1978) suggests high self-
incompatibility. Durian seeds are dispersed by vertebrates. It
would seem, though, that partial self-compatibility in conjunction
with animal dispersal may play, in a way yet to be explained, a
leading part in mediating the complex variation that exists in
sapindaceous genera.

THE PROBLEM OF RARE SPECIES

Uniform stands of mixed rain forests appear always to possess an
assemblage of species which are relatively abundant, sexually
mature individuals varying in population density between 0.5-3 per

ha (among trees exceeding 20 cm dbh) and comprising ca. 10% of all species, and 50% of the individuals in the stand (Fig. 1). Many, perhaps most, of these species seem to be aggregated into clumps of variable area (e.g. see Poore 1968; Chan 1980). Under uniform site conditions we can surmise that these clumps disperse and reform over time; there would not appear to be difficulties in pollen movement through these populations over time if the species are pollinated by animals. It is likely that direct interspecific competition between individuals of species in this assemblage could be sufficiently common to influence selection among them.

It is the remaining 90% of the species, which by definition exist in low, often exceptionally low, population densities that distinguish mixed lowland tropical rain forests for the biosystematist. Though rarity has many manifestations it is taxa, local or widespread but rare on account of their low population densities, which I wish now to discuss. Do the means exist to maintain fecundity among these populations, either through clumping of genetically dissimilar individuals of outbreeders, through breakdown of incompatibility systems, through the services of wide-ranging pollinators, or through vegetative reproduction? What is the size of the gamodeme (Gilmour and Gregor 1939)? Clearly, competition between trees cannot be interspecific among rare members of a species rich guild. Even if more or less host-specific seed predators and pathogens prevent population density from rising above a certain threshold what factors, other than chance, can force a species to decline to extinction? What, if any, form of natural selection can mediate survival of the progeny of those rare trees which can maintain high fecundity?

These problems have rarely been studied in the field. In view of the dominant role that graduate students are playing in addressing theoretical biological questions in rain forests, this is hardly surprising! Instead of the clusters of small plots which we have generally used up until now to examine stand characteristics, or patterns of community variation relative to site, we need to map species population distributions in substantial blocks of forest, and monitor them over time. Without such large samples, we cannot even ascertain which are rare species. The following comments are therefore merely inferences, from the inadequate and generally circumstantial observations that have arisen from other studies.

There is no indication, from the analyses of floristic variation among ca. 400 plots, varying in area between 0.2–0.6 ha, which have been sampled on a wide variety of lowland soils in northwest Borneo, that species which generally exist in low population densities are either more, or less, restricted in their soils range compared with those with generally high population densities. Some species, such as many of the gap phase species (e.g. *Parkia,* Mimosaceae; *Alstonia,* Apocynaceae; *Pterocymbium,* Sterculiaceae) which reach emergent size at maturity are constrained by their ecology always to exist in low mean population densities, and often as isolated individuals, in old

forest. It is interesting that the pollinators of species in this
guild, where known, are far-ranging; *Apis dorsata* in *Alstonia
scholaris* (pers. obs.), the bat *Eonycteris spelaea* in *Parkia*
spp. in the Far East (Start and Marshall 1976), and this is the
case in many rare emergent species of the mature phase too, e.g.
Durio section *Durio* species which are also pollinated by bats,
and *Sindora* spp. (Mimosaceae) by Xylocopid bees (Appanah 1979).
It is also interesting that no local endemism exists in these gap
phase genera throughout west Malesia, and that most species occur
throughout the region.

It may be supposed that competition for the services of
pollinators among these rare species may be particularly acute.
That it does occur is implied by Start's (1974) observation of
staggered flowering among sympatric stands of *Parkia speciosa* and
P. roxburghii, and the sequential appearance in the Smarinda market
of durians from the four wild species in eastern Borneo (pers.
obs.). In *Shorea* section *Mutica,* Chan (1981) observed that
isolated individuals of *S. leprosula* set less fruit than those
in clumps. Also, a lower proportion of individuals in a low
density population of *S. dasyphylla* set fruit than among five
other sympatric species in the same section that had higher
population densities (Fig. 6). *S. dasyphylla,* unlike the
other species, lacked a period during which its population alone
was in flower.

Clearly, the influence of isolation on fecundity among self-
incompatible species provides a mechanism for extinction. The
occurrence of scattered individuals of dioecious species,
therefore, demands explanation. Such occurs in *Garcinia*
(Clusiaceae), but Treub (1911) demonstrated that the cultivated
mangosteen, *G. mangostana,* is apomictic through adventive
embryony, while Ha (1978) inferred adventive embryony in the wild
peninsular Malaysian species *Garcinia forbesii* and *G. parvifolia,*
which he found to be associated with high ploidy levels. Kaur
(Kaur *et al.* 1978; elaborated in Jong and Kaur 1979)
demonstrated nucellar polyembryony in two dipterocarp species, and
inferred it in others on the basis of consistent triploidy or
multiple seedlings originating from a single seed. Adventive
embryony is now inferred in rain forest trees among a remarkable
number of families including Anacardiaceae, Apocynaceae,
Clusiaceae, Dipterocarpaceae, Ebenaceae, Myrtaceae and Rutaceae.
There is no indication that a higher proportion of rare than
abundant species are thus apomictic.

It is probable that apomixis is generally facultative. Kaur
(1977) found the proportion of fruit in multiple seedlings to vary
between 0-70% within a population of *Shorea macroptera.*
Ashton (1979) has pointed out that many dipterocarps, including
S. macroptera (Fig. 4), in which geographical subspeciation
is manifest appear to be facultative apomicts.

Co-occurring closely related rain forest tree taxa,
distinguished morphologically by small, often quantitative,
differences that one would consider as justification for, at most,

FIG. 7. Hopea jucunda *Thw*. ssp jucunda, *left; and* ssp. modesta
DC. which coexist in Sri Lanka forests. ssp. jucunda *is a main
canopy,* ssp. modesta *an understorey tree.*

subspecific rank do occur, but their breeding systems are unknown.
Examples exist of pairs of taxa which share identical flower,
fruit and leaf qualities, but in which there are consistent
differences in dimensions. Thus, in *Hopea jucunda*,
Dipterocarpaceae (Fig. 7), subspecies *jucunda* is a main canopy
tree in which the size of parts including, surprisingly, the leaf
is greater than the size of those of the understorey ssp.
modesta which frequently coexist with it and shares the same
geographical range. *Shorea rotundifolia* is a rare taxon which
co-occurs with the widespread *S. amplexicaulis* of Borneo,
differing solely (though distinctively) in its leaf shape (Fig. 8).
This latter pattern of rarity, which is unusual among tropical
trees, could result from occasional hybridization or mutation. I
mention this case only because it might just represent sympatric
speciation, through punctuated evolution, *in statu nascendi*
(Fedorov 1966). Some species of *Stemonoporus,* a dipterocarp genus

FIG. 8. Shorea rotundifolia *Ashton, which occurs as rare scattered groups in association with the common* S. amplexicaulis *Ashton (right) in Central Borneo.*

endemic to Sri Lanka, such as *S. canaliculatus* (Fig. 9), which has been divided into three species by Kostermans (1981-82), combine differences in habit between small single-stemmed trees (*S. canaliculatus sensu stricto*) and lax pauciflorus shrubs. The shrubby forms appear to be allopatric, albeit within the overall range, which only extends 35 km diameter, of the tree form.

The morphological constancy of even these rare forms suggests rarity of hybridization. It also suggests that sometimes, at least, selection is sufficiently generalized to allow survival of substantial random mutations, and of attributes that have lost their adaptive value over evolutionary time. Van Steenis (1969, 1978) has doggedly upheld Goldschmidts' (1933) hypothesis of "hopeful monsters" in the context of tropical rain forest plants. If support is found for Hubbell's nonequilibrium hypothesis for explaining species richness in tropical forests, then monsters may indeed be lurking in them after all, though among van Steenis' examples the rheophytes, for example (1981), differ from their putative forest ancestors in ways that appear to be clear adaptations to the rigors of their physical environment. Once again, claims that a character lacks any adaptive value must ultimately rest on negative evidence, though van Steenis' point, that adaptive value must be demonstrated before it can be accepted, is well taken. Stronger evidence, perhaps, against the existence of specific biotically defined niches lies in the abundance of extraordinarily uniform yet widespread species, in the mature phase of west Malesian forests whose overall species

FIG. 9. *Variation in* Stemonoporus canaliculatus *Thw.* sensu lato.
The taxon described as S. marginalis *Kostermans, left, is known from
one collection and, with that named* S. bullatus *Kostermans, above
right, is confined within the geographical range of the local
endemic* S. canaliculatus *sensu stricto, lower right.*

*composition varies considerably geographically. In the case of
the common and well collected dipterocarp Shorea leprosula of*
peninsular Malaysia, Sumatra and Borneo, for instance, it is
impossible to detect the provenance of individual specimens, yet
Gan *et al.* (1977) found, on allozymic as well as morphological
evidence, that individual trees are surprisingly variable within a
population.

It appears that apomixis may serve on the one hand to maintain
fecundity where pollen dispersal is inadequate, and on the other
to allow rapid spread of favorable ecotypes in heterogeneous
terrain. It is unclear to me why, if present evidence proves
correct, adventive embryony should be favored over other forms of
vegetative reproduction in rain forest trees, or over self-
compatibility. Though apomixis will not lead to gene fixation, it
will lead to lowered genetic variability within populations, and
thus may still result in increased probability of extinction in
biotically changing environments.

If tropical rain forest trees have narrow biotically defined
niches, then apomixis must be regarded as a dead end and a prelude
to extinction (Ashton 1977). Apomictic siblings of widespread
outbreeding species will diversify into transient niches and be
extinguished when the niche disappears. On the other hand, if
these rare sibling species are not niche-specific, these biotypes

might be expected to accumulate over time, though chance
extinction through island biogeographic phenomena may be expected
in all but the most homogeneous terrain (Hubbell 1979). If
Hubbell's hypothesis is correct, we should expect relatively
higher proportions of apomicts among species with low population
densities.

TOWARDS A SYNTHESIS

Though our knowledge of tropical tree biosystematics is growing,
we are still far from being able to answer whether plant species
in tropical rain forests are niche-specific, or whether random
processes play a significant part in allowing the survival of at
least some of the extraordinary number of species they contain.

It has long been recognized that more or less distinct guilds,
of unrelated species sharing a common habit, growth, vegetative
morphology, and structural role exist in rain forests. Whether or
not the mature phase of the forest is distinctly stratified
vertically, constant differences in habit, morphology and
reproductive biology exist between emergent and understorey trees
(Richards 1952; Hallé et al. 1978; Yap 1982). Ontogenesis
thus follows predictable patterns in the larger trees (Ashton
1969, 1982). Richards also noticed consistent differences in
guild characteristics on different soils, which Ashton (1964)
showed to vary continuously along edaphic gradients. We can
conclude that selection acts on these characteristics, which
include tree habit and leaf shape, uniformly on all individuals
within a guild irrespective of their systematic relationships. In
these respects, these species are ecologically complementary.

We must suspect that seed predators, and many herbivores and
pathogens, mediate tree population density at a supra-specific,
sometimes generic or even familial level, thus allowing accumulation of suites of congeneric species, within a single physical
habitat, up to a certain maximum population density. Also, there
is the possibility of biological specialization beneath the soil,
for instance in relation to mycorrhizal associations. These are
subjects in urgent need of attention.

Even if, as appears likely, groups of species share common
responses to individual selective factors, does this mean that the
total sum of responses, to all factors, is common to groups of
species? Do groups of species sharing a common habitat and
geography respond identically from flower-bud initiation to
senility to light, mycorrhizal invasion, water stress, seed
predation? Do they share the same pollinators and seed vectors?
In particular, why are some species abundant, others rare? Two
interpretations of these phenomena remain possible. Resolution
must rely upon detailed systematic and especially cladistic
knowledge of the species concerned, and biosystematic comparisons
between rare and abundant congenerics. We need to confirm whether
the means exist for gene flow; and then to directly measure the
level and pattern of genetic variability. Though comparisons of

genetic variability between populations of consistently common, and rare, species would be instructive, this cannot explain how specific natural selection is. The most amenable source of evidence is to test for positive or negative associations between species population distributions in space, and demographic and population genetic study. Here, evidence for random association can be strengthened through tests for consistency by repeated observation.

Trees have been the life form to which most attention has been given. This is not surprising on account of the accessability of trees, and hence the ease with which their spatial disposition and certain growth parameters can be recorded. Gentry (1982) has pointed out that tree species richness varies more or less independently on other life forms, such as shrubs and epiphytes, in the neo-tropics. This is not surprising, as the physical conditions which favor these life forms differ. All the same it is noteworthy that, whereas species in large tree genera generally share a common pollinator and differ principally in leaf and indumentum, those in large epiphyte genera, among Orchidaceae and Bromeliaceae for example, are often indistinguishable except when in flower and apparently attract specific pollinators. Could this imply that the role of chance in successful establishment is greater in the ephemeral and apparently rather uniform substrate of tree branches than in the forest understorey? New methods of access to the forest canopy are a high priority in evolutionary botany.

If it is eventually concluded that random processes do play a substantial part in determining the survival of plant species in tropical rain forests, it must surely be also concluded that these phenomena are universal, if less important, in other vegetation. Tropical rain forest, with its aseasonality and its immunity, in some areas a least, from unpredictable fluctuations, is clearly the most amenable biome for this research. Substantial tracts of unmodified species rich forest are essential. With the news that, in the unprecedented drought which the Asian tropics have endured this year, an estimated one million hectares of forest, already fragmented by increased cultivation and modified by logging, succumbed to fire in East Borneo alone, the urgency of our task cannot be underestimated.

REFERENCES

Appanah, S. 1979. *The ecology of insect pollintion of some tropical rain forest trees.* Ph.D. thesis, Univ. Aberdeen.
Appanah, S. 1982. *Biol. J. Linn. Soc.* 18: 11–34.
Ashton, P.S. 1964. Ecological studies in the mixed dipterocarp forests of Brunei State. *Oxford For. Mem.* 25. 75p.
Ashton, P.S. 1967. Climate versus soils in the classification of Southeast Asian tropical lowland vegetation. *J. Ecol.* 55: 67–68.
Ashton, P.S. 1969. Speciation among tropical forest trees: some

deductions in the light of recent evidence. *Biol. J. Linn. Soc.* 1: 155-196.

Ashton, P.S. 1972. The quaternary geomorphological history of western Malesia and lowland forest phytogeography. In *Transactions Second Aberdeen-Hull Symposium on Malesian Ecology* (P. and M. Ashton, eds.), Univ. Hull.

Ashton, P.S. 1977. A contribution of rain forest research to evolutionary theory. *Ann. Mo. Bot. Gard.* 64: 694-705.

Ashton, P.S. 1979. Some geographic trends in morphological variation in the Asian tropics and their possible significance. In *Tropical Botany* (K. Larsen and L.B. Holm-Neilson, eds.), Academic Press, London.

Ashton, P.S. 1980. Myrtaceae. In *A revised handbook to the flora of Ceylon* (M.D. Dassanayake, ed.), Smithsonian Institution, Washington, 1, 166-196.

Ashton, P.S. 1982. Dipterocarpaceae. In *Flora Malesiana* (C.G.G.J. van Steenis, ed.), 1, 9, 2, 237-552. Nijhoff.

Bawa, K.S. and Opler, P.A. 1975. *Evolution* 29: 167-179.

Chan, H.T. 1980. Reproductive biology of some Malaysian dipterocarps. II. *Malaysian For.* 43: 438-451.

Chan, H.T. 1981. *Malaysian For.* 44: 28-36.

Chan, H.T. and Appanah, S. 1980. *Malaysian For.* 43: 132-143.

Fedorov, A.A. 1966. *J. Ecol.* 54: 1-11.

Gan, Y.Y., Robertson, F.W., Ashton, P.S., Soepadmo, E. and Lee, D.W. 1977. *Nature, London* 269: 323-325.

Gentry, A.H. 1982. *Evol. Biol.* 15: 1-84.

Gilmour, J.S.L. and Gregor, J.W. 1939. Demes: a suggested new terminology. *Nature, London* 144: 333-334.

Goldschmidt, R. 1933. *Science* 78: 539-547.

Ha, C.O. 1978. *Embryological and cytological aspects of the reproductive biology of some understorey rain forest trees.* Ph.D. thesis, Univ. Malaya, Kuala Lumpur 189 p.

Hallé, F., Oldeman, R.A.A. and Tomlinson, P.B. 1978. *Tropical trees and forests.* Springer, Berlin. 441p.

Hubbell, S.P. 1979. *Science* 203: 1299-1309.

Jacobs, M. 1962. *Pometia* (Sapindaceae), a study in variability. *Reinwardtia* 6: 109-144.

Janzen, D.H. 1971. Seed predation by animals. *Annu. Rev. Ecol. Syst.* 2: 465-492.

Janzen, D.H. 1970. *Am. Nat.* 104: 501-528.

Jong, K. 1982. Cytotaxonomy. In Ashton, P.S. Dipterocarpaceae. In *Flora Malesiana* (C.G.G.J. Van Steenis, ed.), 1, 9, 2, 268-273, Nijhoff, The Hague.

Jong, K. and Kaur, A. 1979. A cytotaxonomic view of the Dipterocarpaceae with some comments on polyploidy and apomixis. *Am. Nat. Hist. Mus.* (Paris) B,26: 4-49.

Kaur, A. 1977. *Embryological and cytological studies in some members of the Dipterocarpaceae.* Ph.D. thesis, Univ. Aberdeen.

Kaur, A., Ha, C.O., Jong, K., Sands, V.E., Chan, H.T., Soepadmo, E. and Ashton, P.S. 1978. Apomixis may be widespread among trees of the climax rain forest. *Nature* 271: 440-441.

Kostermans, A.J.G.H. 1981-1982. *Stemonoporus* Thw.
(Dipterocarpaceae): a monograph. Part 1. 1982. *Adansonia*
3: 321-358; Part 2, 1981. *Bull. Mus. Nat. Hist.* Paris 4, 3,
373-405.
Leenhouts, P.W. 1967. *Blumea* 15: 301-358.
Leenhouts, P.W. 1969. *Blumea* 17: 33-91.
Leenhouts, P.W. 1971. *Blumea* 19: 113-131.
Leenhouts, P.S. 1983. *Blumea* 28: 389-401.
Macarthur, R.H. and Wilson, E.O. 1967. *The theory of island
biogeography*. Monogr. Populat. Biol. 1. Princeton. 203p.
Poore, M.E.D. 1968. *J. Ecol.* 56: 143-196.
Richards, P.W. 1952. *The tropical Rain Forest: An ecological
study*. Cambridge Univ. Press. 450p.
Riswan, S. 1982. *Ecological studies on primary, secondary and
experimentally cleared mixed dipterocarp and kerangas forest in
East Kalimantan, Indonesia*. Ph.D. thesis, Univ. Aberdeen.
342p.
Soepadmo, E. and Eow, B.K. 1977. The reproductive biology of
Durio zibethinus Murr. *Gard. Bull. Singapore* 29: 25-33.
Start, A.N. 1974. *The feeding biology in relation to food
sources of nectarivorous bats (Chiroptera: Macroglossinae) in
Malaysia*. Ph.D. thesis, Univ. Aberdeen. 259p.
Start, A.N. and Marshall, A.G. 1976. Nectarivorous bats as
pollinators of trees in West Malaysia. In *Variation,
breeding and conservation of tropical forest trees* (J. Burley
and E.G. Styles, eds.), Academic Press, London.
Steenis, C.G.G.J. van 1969. *Biol. J. Linn. Soc.* 1: 97-133.
Steenis, C.G.G.J. van 1978. Patio ludens and extinction in
plants. Notes, *Roy. Bot. Gard. Edinburgh* 36: 317-323.
Steenis, C.G.G.J. van 1981. *Rheophytes of the world*. Sijthoff
and Noordhoff. Alphen. 407p.
Stevens, P.F. 1980. A revision of the Old World species of
Calophyllum (Gultiferae). *J. Arnold Arbor. Harvard Univ.* 61:
2-31, 117-699.
Treub, M. 1911a. *Ann. Jard. Bot. Btzg.* 24: 1-27.
Treub, M. 1911b. *Ann. Sci. Nat. Bot.* 9: 1-17.
Vincent, A.J. 1961. A note on the growth of three meranti (LHW
Shorea) hill forest species in naturally and artifically
regenerated forest, Malaya. *Malayan For. Dept. Res. Pamphlet* 37.
Whitmore, T.C. 1975. *Tropical rain forests of the Far East*.
Clarendon, Oxford. 282p.
Wyatt-Smith, J. 1966. Ecological studies in Malayan forests. I.
Malayan For. Dept. Res. Pamphlet 52.
Yap, S.K. 1976. *The reproductive biology of some understorey
fruit tree species in the lowland dipterocarp forest of West
Malaysia*. Ph.D. thesis, Univ. Malaya. 146p.
Yap, S.K. 1982. The phenology of some fruit tree species in a
lowland dipterocarp forest. *Malaysian For.* 45: 21-35.

Biosystematics of Bryophytes: An Overview

Robert Wyatt and *Ann Stoneburner*
Department of Botany
University of Georgia
Athens, Georgia, U.S.A.

INTRODUCTION

Bryophytes have been treated traditionally as the division Bryophyta comprised of three classes: Musci (mosses), Hepaticae (liverworts), and Anthocerotae (hornworts). Crosby (1980a) estimated the relative sizes of these groups as 800 genera an 9,000 species of mosses, 400 genera and 5,500 species of liverworts, and five genera and 300 species of hornworts. Ot authors (e.g. Anderson 1980) have presented higher estimates, pointing out that the bryophyte flora of tropical regions rem largely unknown.

Some recent authors, however, have chosen to recognize two (Bold 1977) or three (Crandall-Stotler 1980) divisions. Thes authors emphasize differences among these groups and also poi their presumed polyphyletic origins. Nevertheless, the origi bryophytes is shrouded in mystery with some workers adhering the view that they are reduced vascular land plants (e.g. Miller 1974), while others maintain that bryophytes represent an evolutionary line derived independently from green algal ancestors (e.g. Fott 1974). Study of recently discovered fossil gametophytes from the Rhynie Chert has convinced Remy (1982) that bryophytes and vascular plants represent two parallel evolutionary lineages emerging from a nexus of Lower Devonian land plants. The former developed dependent sporophytes, while the latter featured independent sporophytes. This view is basically in accord with that of Steere (1969) and Crum (1976).

Certainly it is true that much of the unity of bryophytes as a group derives from their uniform possession of an alternation of generations unique among land plants. The sexual life cycle involves a dominant, free-living, haploid gametophyte alternating with a reduced, dependent, diploid sporophyte. Despite the elaborate specialization of its tissues in some groups, the

sporophyte of all bryophytes remains attached to, and nutrition-
ally dependent on, the female gametophyte throughout its lifetime.

Possession of two multicellular stages in the life cycle
creates a taxonomic dilemma without parallel among vascular
plants: should classifications be based on sporophytic characters
alone, gametophytic characters alone, or some mixture of both?
The situation is comparable to that faced by insect taxonomists,
whose classifications and phylogenies differ enormously depending
on the life history stage considered: egg, larva, nymph, or adult
(Sokal and Sneath 1963).

In general, bryophyte taxonomists have used sporophytic
characters (especially those of the peristome) to delimit groups
of generic, familial, and ordinal rank. Gametophytic characters,
on the other hand, have been used at specific or, less commonly,
generic levels. Crosby (1980b) formalized this tendency by
stating that "the most important principle of moss systematics is
that mosses with similar peristomes must be grouped together,
regardless of the morphology of their gametophytes." The logical
extreme of this view was advocated by Szweykowski (1982), who
argued that gametophytes should be regarded as equivalent to
gametes in diploid organisms and "cannot have any relevance to
taxonomy."

The rationale behind this emphasis on sporophytic characters is
simple. "Sporophytes, attached to gametophytes and thus protected
from the rigors of the environment, are less subject to natural
selection than gametophytes. For that reason, the characters of
the sporophyte ... are relatively stable." Conversely, "gametophytes, growing in more intimate contact with the environment,
respond more readily to forces of selection" (Buck and Crum 1978).
Additionally, less variability might be expected in sporophytic
characters because of the genetic "buffering effect" of heterozygosity made possible by the diploid condition.

If these arguments are correct, it should be possible to
demonstrate that, even within a species, sporophytic characters
are inherently less variable than gametophytic ones. We tested
this hypothesis by comparing 21 gametophytic characters with 14
sporophytic characters measured for five peristomate species of
the moss genus *Weissia* (Stoneburner 1981). Mean coefficients of
variation were nearly half again as large for sporophytes.
Indeed, for all five species, sporophytic characters were more
variable, contrary to expectation. In angiosperms, floral
characters are considered more stable than vegetative ones and,
hence, are more heavily relied upon in classification. This is
generally attributed to the relatively greater protection of
floral parts from weather extremes; flowers are exposed only
during the most favorable periods. In mosses, however, sporophytes are considerably more long-lived and may therefore be
subjected to a wider range of environmental conditions during
development. In fact, their elevation above the insulating mat
formed by the gametophytes may actually project them into a much
less protected, severe microclimate.

Another sort of difficulty commonly encountered in bryophyte systematics is the lack of information about sporophytic charac- ters for those taxa that have never been found with sporophytes. Gemmell's (1950) original estimate that 14.3% of the 573 mosses in the British flora never produce sporophytes has now been reduced to only about nine species that are truly sterile (Longton 1976). Nevertheless, such taxa pose real problems in developing classifi- cations. Information regarding the sporophyte of the recently described monotypic genus *Donrichardsia* (Crum and Anderson 1979), for example, would be most helpful in deciding whether its affinities lie with the Amblystegiaceae or the Brachytheciaceae (Wyatt and Stoneburner 1980). An overemphasis of peristome characters also seems likely to lead to problems, because one of the most prevalent trends in bryophyte evolution is that of reduction or even total loss of the peristome (Anderson 1980). Among the liverworts, sporophytes are much more ephemeral, and the number of species for which sporophytes are unknown is greater (Schuster 1966).

A related problem concerns what should be done when characters expressed in both generations give conflicting views of evolution- ary relationships. In the Mniaceae, for example, we have discovered some taxa in which the flavonoids produced by gameto- phytes are identical to those produced by sporophytes. There are other taxa, however, in which the flavonoid profiles differ, some- times substantially (Wyatt and Stoneburner unpublished). This evidence, therefore, calls into question conclusions from some previous flavonoid studies of mosses in which gametophytes or sporophytes or both were used to make taxonomic comparisons (e.g. Koponen and Nilsson 1977). Obviously, the only strictly valid comparisons are those involving gametophytes with gametophytes and sporophytes with sporophytes. Any other basis for comparison is likely to lead to errors of interpretation.

BIOSYSTEMATIC APPROACHES

Morphological Variation
Numerous papers in recent years have heralded the introduction of the biosystematic approach to bryology and have reviewed exhaus- tively the available literature on this subject (e.g. Stotler 1976; Smith 1978a, 1979; Longton 1979, 1982). Unfortunately, the number of reviews nearly equals the number of actual studies to be reviewed. We propose, nevertheless, to provide a critical over- view of the biosystematics of bryophytes, a field that we agree with Smith (1978a) is in its very earliest stages of development.

Suprisingly few taxonomic studies of bryophytes have employed quantitative methods to assess variability within species, degrees of overlap between species, and overall morphological similarity among species within a genus. More commonly, bryophyte systema- tists have been content to detect one or a few characters which allow pigeonholing of taxa. If a difference can be discovered, in most cases the variants are described as distinct species. Many

bryologists disavow naming any variants at infraspecific ranks
(e.g. Isoviita 1966; Andrus 1974), presumably because taxa of less
than species rank are likely to be overlooked in future work.
Certainly, it is partly as a result of such views in the past that
numerous new species were described, about half of which are today
being reduced to synonymy (Crosby 1980b). Touw's (1974) tabula-
tions of moss species named after 1801 showed that only 32% of
these were still considered valid species in 1974. Smith (1978a)
noted that of Dixon and Jameson's (1924) 290 taxa of varietal rank
in the British moss flora, his recent flora (Smith 1978b) recognized only 90, some of which may also prove of doubtful validity
upon further study.

Concepts of taxonomic ranks are a general source of confusion
in bryophyte taxonomy, just as they are in vascular plant
taxonomy. The situation is compounded in bryophytes because of
their small size. For mosses and liverworts Cronquist's (1978)
"ordinary means" must include use of a compound light microscope.
This makes comparisons of species tedious and time-consuming.
Although some bryologists pay lip service to the idea of species
as "interbreeding populations" (Stotler 1976) and mention
reproductive isolation as an important criterion for distinguishing species (e.g. Horton 1979; Longton 1982), only one actual
study of this sort has been performed (Proctor 1972). Species of
bryophytes usually are defined strictly on the basis of
morphological discontinuities.

One of the earliest studies to take a quantitative approach was
that of Lodge (1960a) on the hydrophytic moss genus *Drepanocladus*.
Using simple histograms, he discovered wide variation in leaf
size and shape, costa width, and cell length in *D. fluitans* and
D. exannulatus. Variation in each species, however, centered
around four "foci", distinguished in part by the nature of the
alar cells, a previously overlooked character that proved
extremely constant in both species. Lodge (1960b) also investigated the basis of these morphological characters by growing
apparent clones under four different sets of environmental conditions. Most of the characters were altered considerably, displaying great phenotypic plasticity, but the structure and arrangement
of the alar cells remained remarkably stable. On the basis of
this evidence, Lodge (1960b) distinguished two varieties of *D.
fluitans* and two of *D. exannulatus*. Sonesson (1966) grew
D. trichophyllus under a range of conditions and reached
conclusions similar to those of Lodge (1960a). Leaf size and
shape were plastic, while costa length, denticulation, and alar
cell structure appeared to have a strong genetic basis.

A more complex situation was encountered in a biosystematic
study of four taxa in the moss genus *Dicranum* (Briggs 1965).
Cultivation in a greenhouse showed that much of the variation in
leaf characters used to separate infraspecific taxa was due to
phenotypic plasticity, which "blurs the distinctness of particular
genotypes." Briggs (1965), therefore, decided not to give formal
taxonomic recognition to the previously designated varieties.

Hatcher (1967) grew five genetically distinct clones of the liverwort *Lophocolea heterophylla* on five identical series of media and used simple univariate statistics to analyze the patterns of variation. He discovered that the range of variation in a single clone under the influence of different environments could be as great as that among the five clones grown on any single medium. He concluded that it was impossible to distinguish between genetically-based variation and that induced by the environment under field conditions.

Koponen (1967) carried out a biometrical analysis of morphological variation in a mixed stand of the mosses *Mnium affine* and *M. medium*, the latter having been treated often as a variety of the former. Using t-tests, he found that the two taxa differed significantly in most of the characters. Because of marked differences in the structure of the teeth on the leaf margin, costa width, cell shape, and length of the decurrent base of the leaves, Koponen (1967) concluded that the taxa ought to be treated as distinct species. Szweykowski and Krzakowa (1967) measured six characters for each of 47 samples of the liverwort *Pleuroclada albescens* from the Tatry Mountains of Poland. Using the "dendrite method", they clustered the samples on the basis of overall morphological similarity and found two distinct groups that coincided perfectly with var. *albescens* and var. *islandica*. Newton (1968) found, however, that the two described varieties of the moss *Tortula muralis* intergraded to such an extent that about 16% of small diploid plants could not be distinguished from haploids. Although factor analysis applied to 64 populations measured for nine characters showed clear separation of the chromosome races, Newton (1968) decided that there was no simple morphological basis on which to recognize a discontinuity in the variation pattern. Rahman (1972) obtained nearly the same result applying principal components analysis to 86 samples of *Sphagnum subsecundum* measured for 23 characters. Although this technique and a discriminant analysis both showed a small region of overlap, Rahman (1972) advocated continued recognition of var. *inundatum* and var. *auriculatum*. Cultivation of three samples under various conditions showed that the morphological characters were plastic, but all except one sample of var. *inundatum* remained within the boundary of their original cluster.

Wigh (1972) cultivated material of seven species of the moss genus *Mnium* under uniform conditions. He found considerable plasticity in a number of characters, including the degree of decurrency of the leaves, a very important taxonomic character according to the concepts of Koponen (1968). Species differed in their degree of plasticity for the same characters. Wigh (1975) also grew material of the mosses *Brachythecium rutabulum* and *B. rivulare* in nine different environments. He noted that gametophytic characters were more plastic than sporophytic ones and that variations in humidity gave rise to more extensive changes than did variations in light intensity. Different

characters responded differently to environmental factors, and
some proved relatively stable, including leaf shape, leaf
denticulation, seta papillation, and size and shape of capsules
and peristomes. Smith and Hill's (1975) investigation of the moss
Ulota crispula showed it to be merely a small form of *U. crispa*.
Their analysis also supported reduction of *U. bruchii* to a
variety of *U. crispa*. The degree of distinctiveness of these taxa
was assessed by correspondence analysis (Hill and Smith 1976), a
multivariate form of principal components analysis developed for
use with discrete characters. Hill (1976) examined variation in
10 characters commonly used to distinguish *Sphagnum capillaceum*
and *S. rubellum*. Principal components analysis did not support
the existence of discrete taxa, prompting Hill (1976) to reject
retention of *S. rubellum* even as a variety.

Lewis and Smith (1977) analyzed variation in natural populations of the moss genus *Pohlia* that regularly produce
axillary bulbils, concluding that seven species exist Growth under
uniform conditions demonstrated the stability of bulbil morphology
with respect to environment. The most variable taxon, *P.
proligera*, could not be clearly subdivided along the lines
recommended by previous authors. In the moss genus *Campylopus*,
Florschütz-de Waard and Worrell-Schets (1980) cultivated plants
under uniform conditions and found that presence of pseudostereids
is environmentally determined. This convinced them that two
segregates, *C. leucognodes* and *C. argyrocaulon*, should
be considered synonymous with *C. subconcolor*. Earlier, using
using similar methods, Corley (1976) had concluded that although
C. fragilis, C. pyriformis, C. paradoxus, C. setifolius, and
C. shawii "remained quite distinct, much of the infraspecific
variation was reduced, or disappeared."

Smith and Whitehouse (1978) cultured numerous collections of
mosses from the *Bryum bicolor* complex on sterilized soil under
uniform conditions. Combining these observations with other data,
they concluded that four taxa exist in Great Britain. Unfortunately, quantitative results of their culturing experiments were
not presented. Steel (1978) made numerous collections of the
liverworts *Lophocolea bidentata* and *L. cuspidata* to use in
culturing experiments and reciprocal transplants. He varied pH,
light, and humidity and found great phenotypic plasticity in leaf
length, leaf width, and cell width. Nevertheless, the differences
between the two species were retained, or even accentuated, when
they were grown under identical conditions. The taxa also
remained distinct in the eciprocal transplants.

Lane (1981) studied variation in selected problematical taxa of
Sphagnum, emphasizing the large range of variation possible
within single plants. On this basis he questioned the validity of
the two varieties of *S. imbricatum*. Culturing under uniform
conditions showed, on the other hand, that the differences between
the varieties of *S. macrophyllum* were stable. Reciprocal
transplants and common garden cultivation demonstrated that

variants of *S. lescurii* that resemble *S. cyclophyllum*
could be distinguished from the latter. Based partly on Agnew's
(1958) cultivation experiments, which showed that the teeth on the
leaf margin of *S. trinitense* disappeared when grown in
exposed situations, Lane (1981) synonymized that taxon with *S.
cuspidatum*. In a very comprehensive study of intraspecific
variation, Longton (1981) grew 24 clones of *Bryum argenteum* under
a range of environmental conditions. The clones proved morphologically distinct in protonemal morphology, shoot arrangement,
and leaf morphology. Some differences observed in field-collected
material, such as length of the leaf apiculus, were maintained in
culture, while other characters, such as stem length, leaf length,
and nerve excurrence, were altered.

Stoneburner and Wyatt (1983) analyzed variation in populations
of *Weissia* from the southwestern United States. Using correlation
coefficients and analyses of variance, they identified morphological characters that were most useful in distinguishing among
taxa. The resulting data set of 32 characters was subjected to
cluster analysis, principal components analysis, factor analysis,
and discriminant analysis. It was concluded that the ten taxa
previously recognized in the genus should be reduced to only seven
distinct, natural species.

Chemical Variation

Bryophytes are unexpectedly rich in chemical variation. Indeed,
Giannasi (1978) concluded that bryophytes are as evolutionarily
advanced physiologically as flowering plants. A recent compendium
on the phytochemistry of mosses (Spencer 1980) listed 277
compounds isolated from 166 taxa. Markham and Porter (1978), in a
thorough review of secondary metabolites of bryophytes, pointed
out that their diverse, and often unique, chemistry can provide
useful, additional taxonomic characters. They further noted,
however, a paucity of information on secondary products of mosses
as compared to liverworts.

Studies of flavonoid and ligin chemistry suggest a phylogenetic
link between bryophytes and vascular plants, but fatty acids of
bryophytes are more similar to those of algae (Markham and Porter
1978). There are marked biochemical differences between mosses
and liverworts (Markham and Porter 1978; Asakawa 1982), and
terpenoid studies indicate that liverworts are more similar to
algae and higher plants, while mosses are closer to ferns (Asakawa
1982). Considering the resources available and the potential
usefulness of such information, the lack of biosystematic studies
utilizing chemical evidence is surprising. A recent review of the
chemotaxonomy of bryophytes (Suire and Asakawa 1979) indicated
that most of the work presently being done involves surveys for
the occurrence of chemical compounds, rather than studies aimed at
particular problems in bryophyte systematics.

Flavonoids appear to be widely distributed in both mosses and
liverworts, although only the liverworts have been sampled

adequately, and no flavonoids have yet been detected in a hornwort
(Markham and Porter 1978). Work on mosses has been directed
primarily toward examination of the natural product chemistry of
unrelated species (e.g. Nilsson 1967, 1970; Osterdahl 1979).
Using paper chromatography, McClure and Miller (1967) screened 70
species of mosses and discovered that 34 produced flavonoids.
They also noted that flavonoids were more common in acrocarpous
than in pleurocarpous species. The only truly taxonomic study of
mosses is Koponen and Nilsson's (1977) investigation of species
pairs in the Mniaceae. Recognizing that morphological data were
ambiguous with respect to progenitor-derivative relationships
between diploid and tetraploid species, Koponen and Nilsson (1977)
used flavonoid evidence to test their hypotheses concerning
appropriate pairs. Unfortunately, their chemosystematic evidence
was fraught with problems. In addition to a number of methodolog-
ical oversights, they failed to present most of their evidence,
and the data presented were ambiguous with respect to likely
progenitors (Wyatt 1983).

   There have been more biosystematic investigations of flavonoid
patterns in species complexes of liverworts. Markham and Porter
(1974) showed that two varieties of *Marchantia polymorpha* were
better treated as ecological races. Chemical races with distinctive
geographical ranges also were discovered in *Concocephalum conicum*
(Markham *et al.* 1976b; Szweykowski and Krzakowa 1 Porter
1981), and two problematical species of *Hymenophyton* were
shown to be readily distinguishable biochemically (Markham *et
al.* 1976a). *Marchantia berteroana* showed major changes in
flavonoid biosynthesis that accompanied the formation of antheri-
diophores and archegoniophores (Markham and Porter 1978).
Flavonoid studies also showed that *Carrpos sphaerocarpos*, a minute
liverwort discovered in 1956 on the Central Australian saltpans,
belongs to the Marchantialean mainstream and should no longer be
allied with the primitive genus *Corsinia* (Markham 1980).

   Terpenoids are ubiquitous in bryophytes, but again liverworts
have been studied in greater detail. Sesquiterpenoids are highly
concentrated in the cellular oil bodies of liverworts and Markham
and Porter (1978) suggested that they may function in deterring
herbivory. The majority of studies have not had biosystematic
implications, although several studies corroborate taxonomic
distinctions based on morphological evidence (Suire and Asakawa
1979). The similarity of sesquiterpenes in the morphologically
similar *Porella vernicosa* and *P. macroloba* was used by Suire
and Asakawa (1979) to support the inclusion of these two taxa in a
single species. Because a major biochemical dichotomy in the
genus *Frullania* is not accompanied by morphological differences,
Suire and Asakawa (1979) did not support recognition of a segregate genus. They did, however, suggest changes in the taxonomy of
certain species based on sesquiterpene patterns. Stotler (1976)
indicated that sesquiterpenes were not stable and, therefore,
could not provide reliable taxonomic information. This was
disputed by Suire and Asakawa (1979) who cited a study comparing

plants of two species collected in southwestern France over a one-year period which showed that terpenoid patterns are consistent. Other studies comparing terpenoid chemistry have centered on establishing phylogenetic relationships among the higher categories of liverworts and among liverworts, vascular plants, and algae (Asakawa 1982).

Fatty acid composition of the lipids of mosses is being evaluated for potential use in taxonomy (Anderson *et al.* 1974). The work thus far is largely of a survey nature, but Anderson *et al.* (1974) have suggested that the synthesis of short chain fatty acids, characteristic of xeric mosses, is an adaptation to drought.

Cytological Variation

Cytological studies have produced some of the most valuable insights into the relationships between species and the structure of populations of bryophytes. Two recent reviews by Smith (1978a) and Newton (1979) discussed various aspects of the cytogenetics of bryophytes, while Anderson (1980) provided a thorough review of the subject in mosses and Berrie (1963), in liverworts. Chromosomes of bryophytes are generally smaller than those of flowering plants. Fully condensed meiotic chromosomes of mosses range from 1-2 μm in length (Anderson and Lemmon 1972). Chromosomes of liverworts tend to be somewhat larger (Smith and Ramsey 1982).

Among liverworts and hornworts, there is little variation in basic chromosome numbers and relatively few cases of polyploidy (Berrie 1963; Smith 1978a). For about 75% of the species the basic chromosome number is $n = 9$, and only about 15% of the taxa are polyploid (Berrie 1963). For the hornworts, counts of $n = 5$ and 6 generally are reported (Smith 1978a). Some authors suggest that present-day liverworts represent ancient polyploids derived from taxa with a base number of $x = 4$ or 5 (Schuster 1966). Others believe that the base number in hepatics is $x = 8-10$ (Smith 1978a; Crosby 1980a). The diversity of chromosome numbers seen among mosses contrasts sharply with the uniformity of liverworts, and polyploidy apparently has played a considerable role in the evolution of the group. For *Sphagnum* the majority of counts are $n = 19$ or 38 and for *Andreaea*, 10 or 11.

Within the majority of mosses, however, the haploid chromosome number varies considerably within and among the three major peristomate groups. Mosses with sporophytes bearing a single articulated peristome (Haplolepideae) have a high frequency of the numbers $n = 12-16$; those with a double articulated peristome (Diplolepideae) have high frequencies of the numbers $n = 6$, 10, and 11; and those with a solid, nonarticulated peristome (Nematodonteae) generally have numbers of 7, 8, 9, and 14 (Smith 1978a; Anderson 1980). For the majority of orders and families of mosses, a base number of $x = 6$ or 7 is most probable with primary, possibly ancient, polyploidy varying from 17-98%, and secondary polyploidy, stemming from haploid numbers of $n = 10-15$, ranging from 0-68%.

Heterochromatin, differentially condensed chromatin that results in structural differences between otherwise identical chromosomes, was first described in a liverwort by Heitz (1928). Newton (1977, 1980) used Giemsa C-banding techniques to identify heterochromatin as a marker in a cytotaxonomic evaluation of relationships among species in the liverwort genus *Pellia*. Heterochromatin also has been studied in association with sex chromosomes. Sex chromosomes in plants were first described in the liverwort *Sphaerocarpos* by Allen (1917), who established a relationship between the heteromorphic bivalent and sex determination. All subsequent studies of other taxa have presented only circumstantial, and sometimes controversial, evidence of sex chromosomes (Smith 1978a; Anderson 1980; Ramsay and Berrie 1982). Ono (1970) reported structural sex chromosomes in dioicous species of the Mniaceae and Polytrichaceae. Ramsay (1966) observed heterochromatic, dimorphic bivalents in three dioicous species of *Macromitrium* in which dwarf males are epiphytic on female plants and the spores are of two distinct size classes.

Very small, supernumerary chromosomes called "m-chromosomes" are commonly found in bryophytes and were first described in a liverwort (Showalter 1921). They may vary in size and number within a species, are generally heterochromatic, and undergo heterochromatic, and undergo premature disjunction in meiosis (Smith 1978a; Anderson 1980). Anderson and Lemmon (1972), in a study demonstrating that the distribution of m-chromosome populations of *W. controversa* is nonrandom, could not detect any phenotypic differences among such populations from the Coastal Plain, Piedmont, and Mountains of the southeastern United States. Wigh (1973) also found a nonrandom distribution of m-chromosomes of *Brachythecium glareosum* in Sweden. Smith and Ramsay (1982), noting that in Britain monoicous species with m-chromosomes are more common than those without, suggested that m-chromosomes may increase genetic variability in monoicous species in a manner similar to B-chromosomes of flowering plants. Alternatively, because all reports of m-chromosomes in Britain are from polyploids, they postulated that m-chromosomes may reduce pairing between homoeologous chromosomes in new autopolyploids. While the significance of m-chromosomes is not understood, their value as cytological markers in population studies is apparent. From reciprocal transplants of populations with differing m-chromosomes, Anderson and Lemmon (1974) showed that for monoicous *Weissia controversa,* the average distance travelled by sperm is 12.3 mm. Aneuploidy, which is more common in mosses than in liverworts (Smith 1978a; Crosby 1980), is seen in the standard chromosome complement as well as in m-chromosomes. Aneuploid races within a number of species show distinctive geographical ranges (Smith 1978a). Although Wigh (1975) did not observe aneuploidy in Scandinavian material of *Brachythecium rutabulum,* he accepted its reported occurrence elsewhere because of earlier work in which irradiation of sporophytes resulted in a stable aneuploid series.

Polyploidy in bryophytes, whether interspecific or intraspecific, is considered to have originated through autopolyploidy
(Smith 1978a; Anderson 1980). Apospory is generally suggested as
the mechanism leading to autopolyploidy, partly because diploid
gametophytes can be produced relatively easily from sporophytic
tissue in culture (Crum 1976; Anderson 1980). On the other hand,
it is difficult to produce diploid gametophytes aposporously from
sporophytic tissue of liverworts, partly because of a lack of
chlorophyll and the evanescent nature of the sporophyte. It has
been done successfully, however, in a few genera, including
Marchantia (Burgeff 1937). Smith (1978a) pointed out that there
is one doubtful record of apospory in naturally occurring populations and suggested that diplospory, which has been observed in
nature, is likely to be more important in the production of polyploids. Wyatt and Anderson (1983) argued, however, that Smith
underestimated the importance of apospory. They pointed out that
the common occurrence of browsed sporophytes and likelihood of
their subsequent burial in soil and litter could lead to rapid
regeneration of diploid protonema in the field. They suggested
that the high frequency of polyploidy in acrocarpous mosses that
occur on disturbed substrates further supports this view.

There has been only one report of allopolyploidy in bryophytes.
Khanna (1960) believed *Weissia exserta* (n = 26) to be an
amphidiploid derived from a cross between *W. (Astomum) crispa*
(n = 13) and *W. controversa* (n = 13) based on its intermediate
cytology, morphology and ecology. None of the irregularities in
spore formation commonly seen in other *Weissia-Astomum* hybrids
were detected, and Khanna (1960) believed the spores to be fully
viable. Unfortunately, viability was not tested, and the hybrid
has not been synthesized. Anderson and Lemmon (1972) and Anderson
(1980) questioned the allopolyploid nature of *W. exserta*, noting
that the evidence is entirely circumstantial.

In most naturally occurring autopolyploids, there is a notable
absence of *gigas* effects and meiosis is regular (Anderson 1980).
Experimentally induced polyploidy may result in increased cell
sizes, but over successive generations the *gigas* effects
diminish and eventually disappear (Wettstein and Straub 1942).
Meiotic irregularities also occur in induced autopolyploids as
demonstrated by the pioneer work of Marchal and Marchal (1911).
These induced polyploids, however, demonstrate the capacity to
return rapidly to normal levels of fertility (Wettstein and Straub
1942). More recent studies of polyploids in the moss *Tortula
muralis* (Newton 1968) add support from natural populations to
earlier experimental work. Although Newton did not find any
meiotic irregularities in the polyploid populations, they did fall
into two gametophytic size classes, one more robust and indicative
of recent origin.

A number of bryophyte genera have been reported to have
"species pairs", in which a tetraploid, monoicous species has a
diploid, dioicous progenitor (Lowry 1948; Yano 1957a,b; Schuster

1966. Unless stated otherwise, ploidy levels refer to the sporophyte. Therefore, the gametophyte of a tetraploid species
contains a diploid number of chromosomes; the gametophyte of a
diploid species contains a haploid number of chromosomes.) The
moss genus *Mnium*, for example, is thought to contain at least four
species pairs (Lowry 1948; Koponen and Nilsson 1977; Wyatt 1983).
Species pairs have also been studied in *Atrichum* (Lowry
1954) and *Fissidens* (Smith and Newton 1968). Evidence in most
studies is circumstantial, based on morphological similarities
between diploids and presumed tetraploid derivatives. Morphologically, members of a species pair are most easily differentiated
by the changes in sexuality that accompany the doubled chromosome
number. There are discrepancies in the taxonomic treatment of the
diploid and tetraploid derivatives. Generally each is given
specific rank; in *Mnium cuspidatum*, however, both are
included in a single taxon.

Intraspecific polyploidy also occurs commonly in mosses.
According to Anderson (1980), intraspecific polyploidy exists in
more than 10% of the 1,000 or so moss species that have been
investigated cytologically. Morphologically the polyploids are
usually indistinguishable from their counterparts, unless the
progenitor is dioicous and chromosome doubling results in a change
in sexuality to the monoicous condition. This occurs in about 75%
of the cases (Wyatt and Anderson 1983). There is considerable
disagreement over the taxonomic disposition of cytological races.
If there are no differences in morphology or sexuality, the
tendency has been to recognize a single taxon (e.g. Newton 1968).
There are those, however, who argue that each ploidy level
deserves recognition as a separate, reproductively isolated
species (Lazarenko 1967; Lazarenko and Lesnyak 1977). Smith
(1978a) also stressed that whether they are morphologically
distinguishable or not, cytological races cannot interbreed.
Wyatt and Anderson (1983) argued that fixed rules would be
difficult to apply and recommended leaving such decisions to the
discretion of individual researchers.

It does not appear that ploidy level in bryophytes is correlated with latitudinal variation or other geographical factors
(Newton 1980). Smith (1978a) suggested that there is insufficient
information on the chromosome numbers of tropical mosses to allow
any conclusions about latitudinal correlations. Geographical
comparisons, however, within well-studied regions fail to show any
relationship between polyploidy and distribution of bryophytes.

Genetic Variation
Considering how frequent and widespread hybridization is believed
to be in vascular plants, it is amazing that there are so few
reports of interspecific crosses in mosses (Anderson 1980) and
none in liverworts (Crundwell 1970). Undoubtedly, this is due in
part to spatial isolation of natural populations. Gamete
dispersal distances are very short (Wyatt 1982) so that even when
two species have similar ecological tolerance ranges, they seldom

occur sufficiently close together to interbreed. As Anderson
(1980) has noted, most cases of hybridizaton involve bryophyte
taxa that grow in disturbed habitats where spores and gametophores
can become intermixed.

Another reason that reports of hybrid mosses may be rare is
that the products of interspecific crosses may not be recognized.
Nearly all of the well-documented hybrids are intergeneric and
involve one parent with cleistocarpous capsules and another with
stegocarpous capsules. Hybrids between such different plants are
easily recognized. Gametophytes derived from hybrid sporophytes
are possible and would be expected to show genetic segregation.
Such hybrid gametophytes have been reported, but Anderson (1980)
rejects all of these as based on insufficient evidence.
Nevertheless, it seems likely that hybridization is more widespread in mosses than is currently believed but that it is often
difficult to detect morphologically (Smith 1979).

The possibility of intergeneric hybrids in the Funariaceae
between *Physcomitrium* and *Funaria* and between *Aphanorrhegma*
and *Physcomitrium* was raised by Bayrhoffer (1849) and Britton
(1895), respectively. Andrews (1918) postulated that hybrid
sporophytes found on *Physcomitrella patens* had been fathered
by *Physcomitrium pyriforme*. The first artificial cross of two
bryophytes was performed by Wettstein (1923) between *P. pyriforme*
and *Funaria hygrometrica*. He later produced sporophytes by
crossing *P. patens* X *Physcomitrium eurystomum* and *P. patens* X
F. hygrometrica (Wettstein 1924, 1928, 1940). Pettet (1964)
reported natural hybrids between *P. patens* and *P. pyriforme*
and between *P. patens* and *Physcomitrium sphaericum* (including
the reciprocal). There are numerous reports of intergeneric
hybrids in the Pottiaceae between cleistocarpous *Astomum* species
and species of stegocarpous *Weissia* (Anderson and Lemmon 1972).
Similar intergeneric hybrids occur in the Ditrichaceae between
stegocarpous *Ditrichum* and cleistocarpous *Pleuridium* (Anderson
and Snider 1982).

In most of these cases, intergeneric hybrid sporophytes
produced only inviable spores. The rare instances in which
fertile spores were observed in artificial crosses in the
Funariaceae were attributed by Wettstein (1924) to an unusual
meiotic segregation of entire maternal and paternal genomes.
Nevertheless, only the maternal genome was compatible within the
maternal cytoplasm (Anderson 1980). *Weissia-Astomum* hybrids also
showed high sterility despite apparently normal meiotic divisions
(Anderson and Lemmon 1972). On the other hand, *Ditrichum-
Pleuridium* hybrids displayed complete chromosomal incompatibility,
forming 52 univalents in meiosis (Anderson and Snider 1982). It
is apparent, therefore, that post-fertilization barriers to
hybridization exist in bryophytes and that these, like those of
vascular plants, may be either chromosomal or genic in origin
(Anderson 1980).

Surprisingly few attempts have been made to cross bryophytes
for systematic or other purposes. Since the classic work of

Wettstein (1923, 1924, 1928, 1940) on moss species of the
Funariaceae and Burgeff (1943) on the liverwort *Marchantia*, there
have been virtually no attempts to perform artificial crosses
(Anderson 1974). A notable exception is the work of Proctor
(1972) who found that the liverwort *Riella americana* is comprised
of at least three reproductively isolated geographical races. In
addition, a number of workers have successfully crossed the moss
Physcomitrella patens in monospecific cultures to assess the
genetic basis of biochemical and morphological mutations (Engel,
1968; Ashton and Cove 1977; Polley 1978). Although their main
purposes were not biosystematic, the success of such studies
demonstrates the potential use of monosporic cultures and artificial crosses as tools in evaluating the degree of concordance
between the traditional morphological species concept of bryologists and the reproductive (or "biological") species concept.
Such data could also be used to determine directly the genetic
bases of taxonomically important morphological characters in
bryophytes.

Unexpectedly high levels of genetic variability have been
detected in electrophoretic studies of both mosses and liverworts
(Wyatt, 1982). Early work demonstrated that starch and acrylamide
gel electrophoresis can be applied to bryophytes. Subsequent
studies demonstrated polymorphism within species of both mosses
(Cummins and Wyatt 1981) and liverworts (Krzakowa and Szweykowski
1979). Biosystematic use of such data is illustrated by Krzakowa
(1977) and Krzakowa and Szweykowski's (1977a,b) discovery that
populations of the taxonomically difficult species *Pellia
epiphylla, P. borealis, P. endiviifolia,* and *P. neesiana* show
species-specific banding patterns for peroxidase. In *P.
asplenioides*, however, polymorphism exists, and the species can
be differentiated into several distinct geographical races that
differ significantly from each other in frequencies of six peroxidase phenotypes. In *Conocephalum conicum*, two different
different enzyme phenotypes were shown to correspond perfectly
with previously recognized morphological races in Poland
(Szweykowski and Bobowicz 1979; Szweykowski and Krzakowa 1979;
Szweykowski *et al.* 1980). These results are somewhat at variance
with other studies by Gliddon (1980) in Great Britain and Yamazaki
(1981) in Japan. Gliddon (1980) calculated a theoretical value of
0.026 for heterozygosity in *C. conicum* and two other thallose
liverworts; Yamazaki (1981) detected actual heterozygosity in
this species measured at an average of 0.167 over 11 loci.

CONCLUSION

There is suprising paucity of biosystematic studies of mosses and
liverworts. Nearly all authors are in agreement on this point.
Smith (1978a), for example, puts it very bluntly: "bryophyte
biosystematics is in its infancy." Why should this be so, when
over the same 40-year period experimental approaches to vascular
plant taxonomy have blossomed?

One of the chief reasons commonly advanced is that there are comparatively few workers in the field (Stotler 1976; Crosby 1980b). This is presumably because "the systematics of mosses is notoriously difficult, and few botanists possess the expertise to unravel the taxonomic complexities" (Taylor 1980). To gain insight into the relative dearth of bryophyte taxonomists, we determined the average number of species per active worker for both bryophytes and vascular plants (Table I). To our surprise, on a per species basis, bryophyte taxonomists outnumbered vascular plant taxonomists by about 2:1! There are obvious potential sources of error in these calculations. The number of vascular plant taxonomists is probably an underestimate, since many practicing taxonomists may not belong to the American Society of Plant Taxonomists. The number of bryophyte taxonomists may be overestimated, as many of the individuals listed in the Directory do not devote their energies entirely to this field. On the other hand, the numbers of species of bryophytes is more likely to be an overestimate than is the number of vascular plants (Touw 1974; Crosby 1980b). Lane (1983) suggested that the low genus:species ratio for bryophytes implies over-splitting. The fact that there are relatively greater numbers of bryologists in North America than any other part of the world except, perhaps, Japan and Great Britain, should also apply to vascular plant taxonomists. It appears, therefore, that the lack of biosystematic studies of bryophytes is not easily explained on the basis of comparatively few researchers.

Crundwell (1970), in a very engaging discussion of infraspecific taxonomy of bryophytes, suggested a number of alternative explanations for the lack of biosystematic studies in this group. He noted that "they are the wrong size." The plants are small, and many critical distinctions can be made only by examination of microscopic characters. This further complicates the process of comparing material with authentic herbarium specimens and makes

TABLE I. *Numbers of species of bryophytes and vascular plants and numbers of taxonomists working on these groups in North America*

	Species	Taxonomists	Ratio
Mosses	1,170[1]		
Liverworts and Hornworts	526[2]		
Total Bryophytes	1,696	162[3]	10.47
Vascular Plants	21,921[4]	938[5]	23.02

[1]From Crum et al. (1973).
[2]From Stotler and Crandall-Stotler (1977).
[3]From Gradstein (1979).
[4]From Kartesz and Bell (1980).
[5]From Tomb (1980).

quantitative approaches more tedious. Crundwell (1970) also
emphasized that "there is no money in them," and it is true that
bryophytes have few economic uses. They do, however, constitute
an important component of the vegetation in some regions, and at
least some organizations providing support for botanical research
do not use economic criteria in awarding grants. Nevertheless, it
is true that much of the earlier work in bryophyte taxonomy was
done by amateurs, who lacked the requisite training and facilities
to pursue experimental work.

Another problem related to their small size is that bryophytes,
unlike vascular plants, are often collected in an immature or
unhealthy state. Crum and Anderson (1981) noted that bryologists
"feel compelled to work over barren collections, forgetting that a
sterile mat of *Brachythecium* may be no more distinctive and no
more interesting than a tuft of grass without spikelets." The
rarity of sexual processes in some bryophyte species, which
reproduce largely by asexual means, complicates identification and
renders crossing work impossible. Crundwell (1970) likens this
taxonomic situation to that of apomictic vascular plants with all
of their attendant problems. Aberrant growth induced by environmental stimuli is less likely to be recognized at the time of
collection of a bryophyte. Therefore, some of the monstrosities
avoided by vascular plant collectors often find their way into the
bryophyte herbarium.

There is a general feeling among bryologists that morphological
characters, especially those of the gametophyte, are extremely
plastic (e.g. Crundwell 1970; Smith 1978a). This has led a
number of authors to the conclusion that such extreme variants
induced by the environment should be given taxonomic recognition
by informal names (e.g. the *modificatio* concept of Buch 1928).
This approach has been endorsed by, among others, Crundwell (1970),
who contrasts the *modificatio*, a variant with no genetic basis,
with the *forma*, a minor variant with a genetic basis. Use of
such nomenclature, albeit informal, is likely to lead to confusion
and an overly cumbersome terminology. Lodge (1960b), for example,
designated those plants he studied with ovate-lanceolate, falcate,
green leaves as *Drepanocladus exannulatus* var. *exannulatus*
mod. *latifolia-falcifolia-viridus*." Since different
environmental stimuli may produce the same phenotype (Basile and
Basile 1983), it is unclear exactly what is to be gained by naming
such variants. It should also be apparent that there is no end to
the possible phenotypic differences that might be included in the
modificatio designation.

Another reason frequently cited for a lack of biosystematic
data is that "bryophytes are difficult to cultivate" (Crundwell
1970). While this may be true for taxa that are restricted to
unusual environments or substrates, many bryophytes, especially
thallose liverworts and acrocarpous mosses that grow on soil, are
readily cultured. Smith (1978a, 1979) argued forcibly that
bryophytes are generally rather easily grown and cited numerous
physiological and genetic, as well as taxonomic, studies that made

use of material raised in culture. Genecological studies also have employed uniform garden experiments and cultivation under a range of conditions (Longton 1974, 1979) with successful results, despite some problems with pleurocarpous taxa (e.g. *Pleurozium schreberi*: Longton and Greene 1979).

Smith (1978a) suggested that "where cultivation is difficult or impossible it is feasible to use the alternative approach of comparing species growing in mixed stands." Unfortunately, the use of these "mixed collections" has become widespread among bryologists as a short-cut for more rigorous study of the genetic basis of morphological characters (Wyatt *et al.* 1982). The reasoning behind this approach is that closely related but morphologically distinct plants that grow intermixed are subjected to the same environment. Therefore, the two morphs must be genetically different and should be recognized as distinct taxa (Isoviita 1966). The method is simple to use and has been extended from mixed stands in the field to mixed specimens in herbarium packets. This method, however, ignores the possibility of microenvironmental variation and assumes that virtually no genetic variation exists in bryophyte populations. Wyatt *et al.* (1982) urged strongly that this method be rejected and suggested that its appealing simplicity may have inhibited the use of other, more rigorous experimental techniques in bryophyte systematics.

In conclusion, we are less inclined to be laudatory of the recent progress of bryophyte systematics than was Longton (1982), who noted "major advances in biosystematic studies of bryophytes." Nevertheless, we do agree that the field is wide open and that "it is high time that more experimental rather than descriptive work is carried out" (Smith 1979).

ACKNOWLEDGMENTS

We thank Lewis E. Anderson for help in obtaining literature.

REFERENCES

Agnew S. 1958. *A Study in the Experimental Taxonomy of Some British* Sphagna *(Section Cuspidata) with Observations on Their Ecology.* Ph.D. Thesis, Univ. Wales.

Allen, C.E. 1917. A chromosome difference correlated with sex differences in *Sphaerocarpos*. *Science* 46: 466-467,

Anderson, L.E. 1974. Bryology 1947-1972. *Ann. Mo. Bot. Gard.* 61: 56-85.

Anderson, L.E. 1980. Cytology and reproductive biology of mosses. In *The Mosses of North America*. (R.J. Taylor and A.E. Leviton, eds.), Pacific Div., AAAS, San Francisco.

Anderson, L.E. and Lemmon, B.E. 1972. Cytological studies of natural hybrids between species of the moss genera, *Astomum* and *Weissia*. *Ann. Mo. Bot. Gard.* 59: 382-416.

Anderson, L.E. and Lemmon, B.E. 1974. Gene flow distances in the moss, *Weissia controversa* Hedw. *J. Hattori Bot. Lab.* 38: 67-90.

Anderson, L.E. and Snider, J.A. 1982. Cytological and genetic barriers in mosses. *J. Hattori Bot. Lab.* 52: 241-254.

Anderson, W.H., Hawkins, J.M., Gellerman, J.L. and Schlenk, H. 1974. Fatty acid composition as criterion in taxonomy of mosses. *J. Hattori Bot. Lab.* 38: 99-103.

Andrews, A.L. 1918. A new hybrid in *Physcomitrium*. *Torreya* 18: 52-54.

Andrus, R.E. 1974. *The Sphagna of New York State.* Ph.D. Thesis, State Univ. N.Y., Coll. Environ. Sci. For., Syracuse, N.Y.

Asakawa, Y. 1982. Terpenoids and aromatic compounds as chemosystematic indicators in the Hepaticae and Anthocerotae. *J. Hattori Bot. Lab.* 53: 283-293.

Ashton, N.W. and Cove, D.J. 1977. The isolation and preliminary characterisation of auxotrophic and analogue resistant mutants of the moss, *Physcomitrella patens*. *Mol. Gen. Genet.* 154: 87-95.

Avise, J.C. 1975. Systematic value of electrophoretic data. *Syst. Zool.* 23: 465-481.

Basile, D.V. and Basile, M.R. 1983. Desuppression of leaf primordia of *Plagiochila arctica* (Hepaticae) by ethylene antagonists. *Science* 220: 1051-1053.

Bayrhoffer, J. 1849. Übersicht der Moose, Lebermoose und Flechten des Taunus. *Jahrb. Nassau. Ver. Naturk.* 5: 101.

Berrie, G.K. 1963. Cytology and phylogeny of liverworts. *Evolution* 17: 347-357.

Bold, H.C. 1977. *The Plant Kingdom.* Prentice-Hall, Englewood Cliffs, N.J.

Briggs, D. 1965. Experimental taxonomy of some British species of the genus *Dicranum*. *New Phytol.* 64: 366-386.

Britton, E.G. 1895. Contributions to American Bryology. IX. *Bull. Torrey Bot. Club* 22: 62-68.

Buch, H. 1928. Die Scapanien Nordeuropas und Sibiriens. --2. Systematischer Teil. *Soc. Sci. Fenn. Comm. Biol.* 3: 1-177.

Buck, W.R. and Crum, H. 1978. A re-interpretation of the Fabroniaceae with notes on selected genera. *J. Hattori Bot. Lab.* 44: 347-369.

Burgeff, H. 1937. Über Polyploidie bei *Marchantia*. *Z. Indukt. Abstamm. Vererbungsl.* 73: 394-403.

Burgeff, H. 1943. *Genetische Studien an* Marchantia. Fischer, Jena.

Corley, M.F.V. 1976. The taxonomy of *Campylopus pyriformis* (Schultz) Brid. and related species. *J. Bryol.* 9: 193-212.

Crandall-Stotler, B. 1980. Morphogenetic designs and a theory of bryophyte origins and divergence. *BioScience* 30: 580-585.

Cronquist, A. 1978. Once again, what is a species? In *Biosystematics in Agriculture.* Allanheld and Osmun, Montclair, N.J.

Crosby, M.R. 1980a. Polyploidy in bryophytes with special emphasis on mosses. In *Polyploidy: Biological Relevance* (W.H. Lewis, ed.), Plenum Press, New York.

Crosby, M.R. 1980b. The diversity and relationships of mosses. In *The Mosses of North America* (R.J. Taylor and A.E. Leviton, eds.), Pacific Div., AAAS, San Francisco.

Crum, H. 1976. *Mosses of the Great Lakes Forest*. Rev. ed. Univ. Michigan Herb., Ann Arbor, Mich.

Crum, H. and Anderson, L.E. 1979. *Donrichardsia*, a new genus of Amblystegiaceae (Musci). *Fieldiana Bot.* 1: 1-8.

Crum, H. and Anderson, L.E. 1981. *Mosses of Eastern North America*. Columbia Univ. Press, New York.

Crum, H., Steere, W.C. and Anderson, L.E. 1973. A new list of mosses of North America north of Mexico. *Bryologist* 76: 85-130.

Crundwell, A.C. 1970. Infraspecific categories in Bryophyta. *Biol. J. Linn. Soc.* 2: 221-224.

Cummins, H. and Wyatt, R. 1981. Genetic variability in natural populations of the moss *Atrichum angustatum*. *Bryologist* 84: 30-38.

Dixon, H.N. and Jameson, H.G. 1924. *The Student's Handbook of British Mosses*. Sumfield, Eastbourne.

Engel, P.P. 1968. The induction of biochemical and morphological mutants in the moss *Physcomitrella patens*. *Am. J. Bot.* 55: 438-446.

Florschütz-de Waard, J. and Worrell-Schets, M. 1980. Studies on Colombian cryptogams. VII. Culture studies on the taxonomic relevance of costal anatomy in the *Campylopus leucognodes - subconcolor* complex and in *Campylopus pittieri*. *Proc. Nederlandse Akad. Wetenschappen, Ser. C.* 83: 37-45.

Fott, B. 1974. The phylogeny of eucaryotic algae. *Taxon* 23: 449-461.

Gemmell, A.R. 1950. Studies in the Bryophyta. I. The influence of sexual mechanism in varietal production and distribution of British Musci. *New Phytol.* 49: 64-71.

Giannasi, D.E. 1978. Systematic aspects of flavonoid biosynthesis and evolution. *Bot. Rev.* 44: 399-429.

Gliddon, C.J. 1980. Studies on the populational biology of four species of thallose liverwort. *Bull. Br. Bryol. Soc.* 36: 14.

Gottlieb, L.D. 1971. Gel electrophoresis: new approach to the study of evolution. *BioScience* 21: 939-943.

Gradstein, S.R. 1979. *Directory of Bryologists and Bryological Research*. IBPTN, Bohn, Scheltema, and Holkena, Utrecht.

Hatcher, R.E. 1967. Experimental studies of variation in Hepaticae. I. Induced variation in *Lophocolea heterophylla*. *Brittonia* 19: 178-201.

Heitz, E. 1928. Das Heterochromatin der Moose. I. *Jahr. Wiss. Bot.* 69: 762-818.

Hill, M.O. 1976. A critical assessment of the distinction between *Sphagnum capillaceum* (Weiss) Schrank and *S. rubellum* Wils. in Britain. *J. Bryol.* 9: 185-191.

Hill, M.O. and Smith, A.J.E. 1976. Principal components analysis of taxonomic data with multi-state discrete characters. *Taxon* 25: 249-255.

Horton, D.G. 1979. *Eucalypta vittiana* sp. nov. and *E.
flowersiana* sp. nov. from North America. *Bryologist* 82: 368-381.
Isoviita, P. 1966. Studies on *Sphagnum* L. I.
Nomenclatural revision of the European taxa. *Ann. Bot. Fenn.*
3: 199-264.
Kartesz, J.T. and Bell, C.R. 1980. A summary of the taxa in the
vascular flora of the United States, Canada and Greenland. *Am.
J. Bot.* 67: 1495-1500.
Khanna, K.R. 1960. Studies on natural hybridization in the genus
Weissia. *Bryologist* 63: 1-16.
Koponen, T. 1967. Biometrical analysis of a mixed stand of *Mnium
affine* Funck and *M. medium*. B.S.G. *Ann. Bot. Fenn.*
4: 67-73.
Koponen, T. 1968. Generic revision of Mniaceae Mitt. (Bryophyta).
Ann. Bot. Fenn. 5: 117-151.
Koponen, T. and Nilsson, E. 1977. Flavonoid patterns and species
pairs in *Plagiomnium* and *Rhizomnium* (Mniaceae). *Bryophyt.
Bibliogr.* 13: 411-425.
Krzakowa, M. 1977. Isozymes as markers of inter- and
intraspecific differentiation in hepatics. *Bryophyt. Bibliogr.*
13: 427-434.
Krzakowa, M. and Szweykowski, J. 1977a. Peroxidases as taxonomic
characters in two critical *Pellia* taxa (Hepaticae, Pelliaceae).
Bull. Acad. Pol. Sci. Sér. Sci. Biol. 25: 203-204.
Krzakowa, M. and Szweykowski, J. 1977b. Peroxidases as taxonomic
characters. II. *Plagiochila asplenioides* (L.) Dum. *sensu*
Grolle (= *P. maior* S. Arnell) and *Plagiochi porelloides*
(= *P. asplenioides* aucti, non Grolle; Hepat Plagiochilaceae).
Bull. Soc. Sci. Lett. Poznan, Sér. D. 17: 33-36.
Krzakowa, M. and Szweykowski, J. 1979. Isozyme polymorphism in
natural populations of a liverwort, *Plagiochila asplenioides*.
Genetics 93: 711-719.
Lane, D.M. 1981. Variation in certain taxa of *Sphagnum* from
the Atlantic Coastal Plain. *J. Hattori Bot. Lab.* 49: 169-245.
Lane, D.M. 1983. A quantitative study of *The Mosses of North
America*. In *Essays on the Biology of Mosses* (R.A. Zander and
M.R. Crosby, eds.), Mo. Bot. Gard., St. Louis.
Lazarenko, A.S. 1967. Polyploidy in the evolution of Musci
[English summary]. *Tsitol. Genet.* 1: 15-26.
Lazarenko, A.S. and Lesnyak E.N. 1977. On chromosome races of the
moss *Atrichum undulatum* (Hedw.) Brid. in the west of the U.S.S.R.
[English Summary], *Ukr. Bot. Zh.* 34: 383-388.
Lewis, K. and Smith, A.J.E. 1977. Studies on some bulbiferous
species of *Pohlia* section *Pohliella*, I. Experimental
investigations. *J. Bryol.* 9: 539-556.
Lodge, E. 1960a. Studies of variation in British material of
Drepanocladus fluitans and *Drepanocladus exannulatus*. I. An
analysis of the variation. *Sven. Bot. Tidskr.* 54: 368-386.
Lodge, E. 1960b. Ibid. II. An experimental study of the variation.
Sven. Bot. Tidskr. 54: 387-393.

Longton, R.E. 1974. Genecological differentiation in bryophytes.
    J. Hattori Bot. Lab. 38: 49-65.
Longton, R.E. 1976. Reproductive biology and evolutionary
    potential in bryophytes. *J. Hattori Bot. Lab.* 41: 205-223.
Longton, R.E. 1979. Climatic adaptation of bryophytes in relation
    to systematics. In *Bryophyte Systematics* (G.C.S. Clarke
    and J.G. Duckett, eds.). Academic Press, London.
Longton, R.E. 1981. Inter-population variation in morphology and
    physiology in the cosmopolitan moss *Bryum argenteum* Hedw. *J.
    Bryol.* 11: 501-520.
Longton, R.E. 1982. The biosystematic approach to bryology.
    J. Hattori Bot. Lab. 53: 1-19.
Longton, R.E. and Greene, S.W. 1979. Experimental studies of
    growth and reproduction in the moss *Pleurozium schreberi* (Brid.)
    Mitt. *J. Bryol.* 10: 321-338.
Lowry, R.J. 1948. A cytotaxonomic study of the genus *Mnium*. *Mem.
    Torrey Bot. Club* 20: 1-42.
Lowry, R.J. 1954. Chromosome numbers and relationships in the
    genus *Atrichum* in North America. *Am. J. Bot.* 41: 410-414.
Marchal, E. and Marchal, E. 1911. Aposporie et sexualité chez les
    mousses. III. *Bull. Acad. R. Sci. Belg.* 1911: 750-778.
McClure, J.W. and Miller, H.A. 1967. Moss chemotaxonomy. A
    survey for flavonoids and the taxonomic implications. *Nova
    Hedwigia* 14: 111-125.
Markham, K.R. 1980. Phytochemical relationships of *Carrpos* with
    Corsinia and other Marchantialean genera. *Biochem. Syst.
    Ecol.* 8: 11-15.
Markham, K.R. and Porter, L.J. 1974. Flavonoids of the liverwort
    Marchantia polymorpha. *Phytochemistry* 13: 1937-1942.
Markham, K.R. and Porter, L.J. 1978. Chemical constituents of the
    bryophytes. *Prog. Phytochem.* 5: 181-289.
Markham, K.R., Porter, L.J., Campbell, E.O., Chapin, J. and
    Bouillant, M.-L. 1976a. Phytochemical support for the
    existence of two species in the genus *Hymenophyton*.
    Phytochemistry 15: 1517-1521.
Markham, K.R., Porter, L.J., Mues, R., Zinsmeister, H.D. and
    Brehm, B.G. 1976b. Flavonoid variation in the liverwort
    Conocephalum conicum: evidence for geographic races.
    Phytochemistry 15: 147-150.
Miller, H.A. 1974. Rhyniophytina, alternation of generations and
    the evolution of bryophytes. *J. Hattori Bot. Lab.* 38: 161-168.
Newton, M.E. 1968. Cyto-taxonomy of *Tortula muralis* Hedw. in
    Britain. *Trans. Br. Bryol. Soc.* 5: 523-535.
Newton, M.E. 1977. Heterochromatin as a cyto-taxonomic character
    in liverworts: *Pellia, Riccardia* and *Cryptothallus*. *J.
    Bryol.* 9: 327-342.
Newton, M.E. 1979. Chromosome morphology and bryophyte
    systematics. In *Bryophyte Systematics* (G.C.S. Clarke and
    J.G. Duckett eds.), Academic Press, London.
Newton, M.E. 1980. Chromosome studies in some Antarctic and sub-
    Antarctic bryophytes. *Br. Antarct. Surv. Bull.* 50: 77-86.

Nilsson, E. 1967. Moss pigments. 7. Preliminary investigations of the violet pigments in *Sphagnum nemoreum*. *Acta Chem. Scand.* 21: 1942-1951.

Nilsson, E. 1970. Moss pigments. 9. Scutellarein glucoside in *Bryum weigelii*. *Ark. Kemi*. 31: 475-480.

Ono, K. 1970. Karyological studies on Mniaceae and Polytrichaceae, with special reference to the structural sex-chromosomes. III. *J. Sci. Hiroshima Univ., Ser. B., Div.* 2 13: 167-221.

Österdahl, B.-G. 1979. Chemical studies on bryophytes. Isolation and identification of flavones and flavone glycosides. *Acta Univ. Ups.* 516: 3-55.

Pettet, A. 1964. Hybrid sporophytes in the Funariaceae. *Trans. Br. Bryol. Soc.* 4: 642-648.

Polley, L.D. 1978. Purification and characterization of 3-dehydroquinate hydrolase and shikimate oxidoreductase: evidence for a bifunctional enzyme. *Biochim. Biophys. Acta* 526: 259-266.

Porter, L.J. 1981. Geographic races of *Conocephalum* (Marchantiales) as defined by flavonoid chemistry. *Taxon* 30: 739-748.

Proctor, V.W. 1972. The genus *Riella* in North and South America: distribution, culture, and reproductive isolation. *Bryologist* 75: 281-289.

Rahman, S.M.A. 1972. Taxonomic investigations on some British Sphagna. I. *Sphagnum subsecundum sensu lato*. *J. Bryol.* 7: 169-179.

Ramsay, H.P. 1966. Sex chromosomes in *Macromitrium*. *Bryologist* 69: 293-311.

Ramsay, H.P. and Berrie, G.K. 1982. Sex determination in bryophytes. *J. Hattori Bot. Lab.* 52: 225-274.

Remy, W. 1982. Lower Devonian gametophytes: relation to the phylogeny of land plants. *Science* 215: 1625-1627.

Schuster, R.M. 1966. *The Hepaticae and Anthocerotae of North America*. Columbia Univ. Press, New York.

Showalter, A.M. 1921. Chromosomes of *Conocephalum conicum*. *Bot. Gaz.* 72: 245-249.

Smith, A.J.E. 1978a. Cytogenetics, biosystematics and evolution in the Bryophyta. *Adv. Bot. Res.* 6: 195-276.

Smith, A.J.E. 1978b. *The Moss Flora of Britain and Ireland*. Cambridge Univ. Press.

Smith, A.J.E. 1979. Towards an experimental approach to bryophyte taxonomy. In *Bryophyte Systematics* (G.C.S. Clarke and J.G. Duckett, eds.), Academic Press, London.

Smith, A.J.E. and Hill, M.O. 1975. A taxonomic investigation of *Ulota bruchii* Hornsch. *ex*. Brid., *U. crispa* (Hedw.) Brid. and *U. crispula* Brid. I. European materia *J. Bryol.* 8: 423-433.

Smith, A.J.E. and Newton, M.E. 1968. Chromosome studies on some British and Irish mosses. *Trans. Brit. Bryol. Soc.* 5: 463-522.

Smith, A.J.E. and Ramsey, H.P. 1982. Sex, cytology and frequency of bryophytes in the British Isles. *J. Hattori Bot. Lab.* 52: 275-281.

Smith, A.J.E. and Whitehouse, H.K.K. 1978. An account of the British species of the *Bryum bicolor* complex including *B. dunense* sp. nov. *J. Bryol.* 10: 29-47.

Sokal, R.R. and Sneath, P.H.A. 1963. *Principles of Numerical Taxonomy.* Freeman, San Francisco.

Sonesson, M. 1966. On *Drepanocladus trichophyllus* in the Torneträsk area. *Bot. Not.* 119: 379-400.

Spencer, K.C. 1980. Chemical constituents of the Musci. *Phytochem. Bull.* 13: 46-63.

Steel, D.T. 1978. The taxonomy of *Lophocolea bidentata* (L.) Dum. and *L. cuspidata* (Nees) Limpr. *J. Bryol.* 10: 49-59.

Steere, W.C. 1969. New thoughts on old theories of alternation of generations. In *Current Topics in Plant Science* (J. Gunckel ed.), Academic Press, New York.

Stoneburner, A. 1981. *Variation and Taxonomy of* Weissia *(Musci: Pottiaceae) in the Southeastern United States.* Ph.D. Thesis, Duke Univ., Durham, N.C.

Stoneburner, A. and Wyatt, R. 1983. Variation and taxonomy of *Weissia* in the southwestern United States. I. Biometrical analyses. In *Essays on the Biology of Mosses* (R.H. Zander and M.R. Crosby, eds.), Mo. Bot. Gard., St. Louis.

Stotler, R.E. 1976. The biosystematic approach in the study of the Hepaticae. *J. Hattori Bot. Lab.* 41: 37-46.

Stotler, R.E. and Crandall-Stotler, B. 1977. A checklist of the liverworts and hornworts of North America. *Bryologist* 80: 405-428.

Suire, C. and Asakawa, Y. 1979. Chemotaxonomy of bryophytes: a survey. In *Bryophyte Systematics* (G.C.S. Clarke and J.G. Duckett, eds.), Academic Press, London.

Szweykowski, J. 1982. Genetic differentiation of liverwort populations and its significance for bryotaxonomy and bryogeography. *J. Hattori Bot. Lab.* 53: 21-28.

Szweykowski, J. and Bobowicz, M.A. 1979. Morphological variation of *Conocephalum conicum* (L.) Dum. (Hepaticae, Marchantiales) in Poland. *Bull. Pol. Acad. Sci., Ser. D.* 27: 21-35.

Szweykowski, J. and Krzakowa, M. 1967. Variation in *Pleuroclada albescens* (Hook.) Spruce (Hepaticae, Hygrobiellaceae) in the Tatry Mts. II. The dendrite method. *Bull. Soc. Amis Sci. Lett. Poznan Ser. D.* 8: 97-102.

Szweykowski, J. and Krzakowa, M. 1979. Variation of four enzyme systems in Polish populations of *Conocephalum conicum* (L.) Dum. (Hepaticae, Marchantiales). *Bull. Acad. Pol. Sci. Sér. Sci. Biol.* 27: 37-41.

Szweykowski, J. and Krzakowa, M. 1980. Variation in phenolic compounds content in Polish populations of the liverwort *Conocephalum conicum* (L.) Dum. (Hepaticae, Concephalaceae). *Bull. Acad. Pol. Sci. Sér. Sci. Biol.* 28: 299-303.

Szweykowski, J., Odrzykoski, I. and Zielinski, R. 1980. Further data on the geographic distribution of two genetically different forms of the liverwort *Conocephalum conicum* (L.) Dum.: the sympatric and allopatric regions. *Bull. Acad. Pol. Sci. Sér. Sci. Biol.* 28: 437–449.

Taylor, R.J. 1980. Preface. In *The Mosses of North America* (R.J. Taylor and A.E. Leviton, eds.), Pacific Div. AAAS, San Francisco.

Tomb, S. 1980. *Membership List of the American Society of Plant Taxonomists.* ASPT, Manhattan, Kansas.

Touw, A. 1974. Some notes on taxonomic and floristic research on exotic mosses. *J. Hattori Bot. Lab.* 38: 123–128.

Wettstein, F. von. 1923. Kreuzungsversuche mit multiploiden Moosrassen. *Biol. Zentralbl.* 43: 71–82.

Wettstein, F. von. 1924. Kreuzungsversuche mit multiploiden Moosrassen. *Biol. Zentralbl.* 44: 145–168.

Wettstein, F. von. 1928. Morphologie und Physiologie des Formwechsels der Moose auf genetische Grundlage. *Bibl. Genet.* 10: 1–216.

Wettstein, F. von. 1940. Experimentelle untersuchungen zum Artbildungs–problem. II. *Ber. Dtsch. Bot. Ges.* 58: 374–388.

Wettstein, F. von, and Straub, J. 1942. Experimentelle Untersuchungen zum Artbildungs–problem. *Z. Indukt. Abstamm. Vererbungsl.* 80: 271–280.

Wigh, K. 1972. Cytotaxonomical and modification studies in some Scandinavian mosses. *Lindbergia* 1: 130–152.

Wigh, K. 1973. Accessory chromosomes in some mosses. *Hereditas* 74: 211–224.

Wigh, K. 1975. Scandinavian species of the genus *Brachythecium* (Bryophyta), I. Modification and biometric studies in the *B. rutabulum-B. rivulare* complex. *Bot. Not.* 128: 463–475.

Wyatt, R. 1982. Population ecology of bryophytes. *J. Hattori Bot. Lab.* 52: 179–198.

Wyatt, R. 1983. Chemosystematics of the Mniaceae. I. Identities of Lowry's species pairs. In *Essays on the Biology of Mosses* (R.A. Zander and M.R. Crosby, eds.), Mo. Bot. Gard., St. Louis.

Wyatt, R. and Anderson, L.E. 1983. Breeding systems of bryophytes. In *The Experimental Biology of Bryophytes* (A.F. Dyer, ed.), Academic Press, London.

Wyatt, R. and Stoneburner, A. 1980. Distribution and phenetic affinities of *Donrichardsia*, an endemic moss from the Edwards Plateau of Texas. *Bryologist* 83: 512–520.

Wyatt, R., Lane, D.M. and Stoneburner, A. 1982. The misuse of mixed collections in bryophyte taxonomy. *Taxon* 31: 698–704.

Yamazaki, T. 1981. Genic variabilities in natural population of haploid plant, *Conocephalum conicum.* I. The amount of heterozygosity. *Jpn. J. Genet.* 56: 373–383.

Yano, K. 1957a. Cytological studies on Japanese mosses. II. Hypnobryales. *Mem. Takada Branch, Niigata Univ.* 1: 85–127.

Yano, K. 1957b. Ibid. III. Isobryales, Polytrichinales. *Mem. Takada Branch, Niigata Univ.* 1: 129–159.

Biosystematic Studies on Pteridophytes in Canada: Progress and Problems

D. M. Britton
Department of Botany
University of Guelph
Guelph, Ontario, Canada

Although the theme of this International Symposium is "Plant Biosystematics: *40* Years Later", most pteridologists would stress that this is *33* Years Later because of our debt to Irene Manton and her book of 1950. Here she clearly delineated among others:

1. Aceto-carmine squashes of spore mother cells to arrive at the gametic chromosome number.
2. Pairing analysis of chromosomes in hybrids.
3. Controlled or artificial crosses.
4. Artificial production of apogamous offspring, i.e. polyhaploids and the analysis of chromosome pairing in the offspring.
5. Somatic chromosome numbers from root tips.

Manton also suggested that photographic evidence of the results together with explanatory diagrams should be presented to allow others to judge the evidence on which the decisions were made. We are also fortunate to have two good review articles from Manton's students to help us keep abreast of developments since 1950 (Lovis 1977; Walker 1979).

So, where are we today? It can be said as a starting point that most of the species in Canada have a chromosome number determination. Using the book by Cody and Britton (1984) as a basis, the Canadian pteridophyte flora has 15 families, 35 genera, 141 species, plus an additional 23 subspecies and varieties. There are only 8 species for which we have no chromosome number determinations (Table I). Also, if you are a nationalist, then there are another 27 species which are lacking determinations from Canadian material (Table II). I will not spell out here the authors for the chromosome determinations of the 106 species

TABLE I. *Species without, or uncertain, chromosome numbers*

Isoetes hieroglyphica A.A. Eaton
Selaginella oregana D.C. Eaton
Selaginella wallacei Hieron.
Mecodium wrightii (van den Bosch) Copeland
Cheilanthes gracillima D.C. Eaton
Woodwardia fimbriata J.E. Smith
Marsilea vestita Hook. & Grev.
Azolla mexicana Presl

---

studied from Canadian materials. Instead, I will merely mention
ten references (Table III) which are post Löve *et al.* (1977)
– the *Cytotaxonomical Atlas* – a most useful work.

So, progress has been made in 33 years, and some genera have
received more work than others, by Canadians and non-Canadians
alike. For example, *Dryopteris* with 11 species (10 eastern)
has received its share of attention from afar, first by Manton
(1950), then by her student S. Walker (1961) and then by his
student Gibby (1977; Gibby and Walker 1977; Gibby *et al.*
1977; Gibby *et al.* 1978). In North America there have been
extensive studies by Wagner (1970) and Britton (1974). The basic
information can be shown in a genome chart (Fig. 1).

A solid page in the *Cytotaxonomical Atlas* (Löve *et al.* 1977)
attests to a species such as *Dryopteris goldiana* having
$2n = 82$ and $x = 41$. There have been no aneuploids reported, and
this unusual basic chromosome number is well documented. There is
of course, the tantalizing missing genome.

We have a similar scheme for *Polystichum* (Fig. 2) which has

TABLE II. *Species with chromosome number determinations from
non-Canadian material*

Lycopodium annotinum	*Adiantum capillus-veneris*
L. alpinum	*Woodsia obtusa*
L. selago	*Polystichum californicum*
Equisetum pratense	*P. lemmonii*
E. telmateia	*P. kruckebergii*
Botrychium oneidense	*P. scopulinum*
B. rugulosum	*Dryopteris arguta*
B. crenulatum	*Thelypteris simulata*
B. simplex	*Asplenium platyneuron*
B. boreale	*Asplenium ruta-muraria*
Schizaea pusilla	*Camptosorus rhizophyllus*
Cheilanthes feei	*Marsilea quadrifolia*
Aspidotis densa	*Azolla caroliniana*
Pityrogramma triangularis	

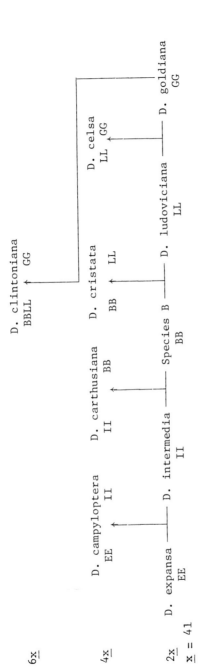

FIG. 1. *Evolutionary scheme for* Dryopteris, *after Walker* (1961), *Wagner* (1970), *Gibby* (1977), *Gibby and Walker* (1977).

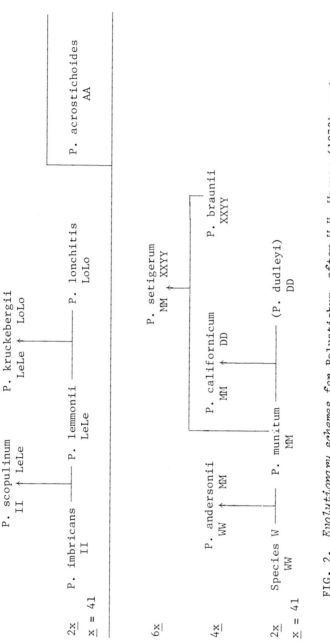

FIG. 2. *Evolutionary schemes for* Polystichum, *after W.H. Wagner* (1973), and D.H. Wagner (1979).

TABLE III. *References for chromosome number determinations post 1977 (Cytotaxonomical Atlas)*

Beitel and Wagner (1982)
Britton, In Cody and Britton (1984)
Britton, Ceska and Ceska, In Cody and Britton (1984)
Cody and Mulligan (1982)
Hersey and Britton (1981)
Kott (1981)
Kott and Britton (1980)
Mulligan and Cody (1979)
Sarvela *et al.* (1981)
Wagner, D.H. (1979)

---

11 species in our flora also (10 western). Again, $x = 41$, and again no aneuploids, and again a key missing genome.

A similar scheme can be presented for our Canadian Polypodiums (Fig. 3). Our problems are small compared to those encountered in California. It should be noted that in all three cases (Figs. 1-3) we have genome analysis and we consider that we are dealing with pillar complexes or derived allopolyploids.

These stories are probably "old hat" to you, so I will not dwell on them, although I should point out that they have not solved all our problems. Examples would be *D. campyloptera* vs. *D. expansa*, eastern *expansa* vs. western *expansa*, *Polystichum andersonii* vs. *braunii* vs. *setigerum*, *P. lemmonii* vs. *P. mohrioides*, etc. However, let me move to more controversial ground. Table IV shows some cases of "intraspecific ploidy." These cases are fertile ground for biosystematics! Here we need chromosomal analysis of natural hybrids, artificial hybrids, chemotaxonomy, SEM of spores, phytogeography — all the tools of the trade to solve such questions as: Are the derived polyploids autos, segmentals or allos? What are their origins? To some, I am sure "the war is over" and the "peace treaty has been signed" and I am late! But others are not yet convinced, or perhaps have paid little attention to some of the problems. More can be done, and is being done on some of these interesting situations. Putting on "judgmental glasses," let me briefly indicate how some of these situations are handled at this time.

Case 1. Wagner (1966) and Sarvela (1978) have treated the diploid as ssp. *disjunctum* and the tetraploid as ssp. *dryopteris*. We should know more about the Asian species, e.g. *G. continentale*.

2. There are two diploid ssp. recognized in Europe and the tetraploid is called ssp. *quadrivalens*. We are not sure what we have in North America (Moran 1982).

3. The diploid is in Japan, our Northern Beech Fern is an apogamous triploid (Mulligan *et al.* 1972) and Mulligan and Cody (1979) buried a hybrid of this with *P. hexagonoptera*

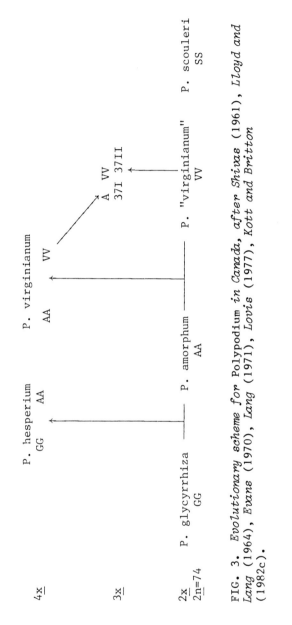

FIG. 3. *Evolutionary scheme for* Polypodium *in Canada, after Shivas (1961), Lloyd and Lang (1964), Evans (1970), Lang (1971), Lovis (1977), Kott and Britton (1982c).*

TABLE IV. *"Intraspecific" polyploidy*

Taxon	Basic number x	Ploidy levels		
		$2x$	$3x$	$4x$
Gymnocarpium dryopteris	40	x		x
Asplenium trichomanes	36	x		x
Phegopteris connectilis	30	x	x	x
Polypodium virginianum	37	x	x	x
Woodsia oregana	38	x		x
Pellaea glabella	29	x		x
Cryptogramma crispa	30	x		x
Pityrogramma triangularis	30	x		x
Phyllitis scolopendrium	36	x		x
Lycopodium lucidulum	67	x		x

in the former.

4. The relationship of 4*x* *P. virginianum* to 2*x* *P. virginianum* requires more experimental work in spite of the suggestion of Lloyd and Lang (1964) that this is another example of allopolyploidy.

5. I would consider the tetraploid to be a species, *Woodsia cathcartiana* but again more experimental work is required and the relationship with *W. mexicana* should be clarified.

6. The names are available – *P. pumila* for the diploid and *P. glabella* for the eastern apogamous tetraploid, *P. suksdorfiana* for the western apogamous tetraploid. The monographer of the group felt this obscured the obvious relationships of the three and preferred the classical "var." designation, viz. var. *occidentalis*, var. *glabella* and var. *simplex*.

7. Our species is a basic diploid, whereas the European *Cryptogramma crispa* is a tetraploid. Some experimental work is required which encompasses spp. *sitchensis*.

8. *Pityrogramma* has received quite a lot of attention in California, but our populations at the extremes of their range have not been studied.

9. Our *Phyllitis* is a tetraploid as are the ones in Japan. The European ones are diploid. In spite of a Ph.D. by one of Manton's students, the relationships are far from clear. Löve has referred the tetraploids to *P. japonicus* whereas some Asplenium workers in Europe feel with autoploids that ssp. are more reasonable. The argument is now being reiterated by those who want to place genera such as *Camptosorus, Phyllitis* and *Ceterach* in *Asplenium*.

10. It is exciting to have a basic diploid entity in this group with a lower chromosome number. You will have to ask

Beitel and Wagner how they will handle the taxonomy.

The above schemes are all very Mantonian. They are her methods and our results are sometimes limited by not doing all the things she suggested. I understand from the "grape-vine" that we in North America have been criticized for *not* using two of Manton's methods - (1) Artificial crosses, and (2) Raising apogamous offspring in culture. I should attempt to answer that "rap". The first problem is funding. Ferns are not corn or barley! The second is the time required to bring for example a *Polypodium* to fruiting (seven years), against the so-called normal time for a Ph.D. The third reason is undoubtedly field work. Most taxonomists greatly enjoy field work and come back laden down with collections that need label data, *Carex* and Violets that have to be keyed out, etc. It all takes time and effort. Manton was not one to deal with mass collections. It is sadly axiomatic in this sort of work, that if you are in the field having a great time looking at lots of variation and becoming familiar with living plants, then you are not in the lab grinding things up with the molecular biologists, or looking down the microscope doing cytology.

What other methods have we exploited? We have found the chemical evidence worthwhile in many cases. The path leads from the flavonoids of Smith and Levine (1963) on *Asplenium*, to the flavonoids of *Gymnocarpium* (Pryer *et al.* 1983) to the phloro-glucinols of *Dryopteris* (Widén *et al.* 1983). Von Euw *et al.* (1980) have written a readable account of the phloroglucinol story and stressed the positive aspects of the findings. For example, chemical analysis was a tool to determine if there were 5, 6 or 7 species in Kenya. Secondly, because hybrids and allopolyploids usually have additive patterns of phloroglucinols, this has allowed critical taxa to be clarified. The third example, I found particularly intriguing. This was when a 140 year old herbarium sheet was used as the source of a rhizome to compare the results with living material.

I will now put on my critical glasses and give you two examples of disappointments. It would seem that the phloroglucinols are at their best with such taxa as *Dryopteris marginalis*. In this species one finds a large array of distinctive compounds. This is matched by the species having a very distinctive morphology. Since the species does not seem to be involved in the evolution of further species, we are unable to make use of this interesting information. Conversely, when one would like a great number of phloroglucinol markers for taxa such as *D. expansa* and *D. campyloptera,* the evidence is disappointing. In fact at times, the phloroglucinols are absent! Widén *et al.* (1978) indicate that there are three main chemotypes of *D. expansa*: (1) In Europe and western North America, (2) Eastern North America and, (3) Asia. One might even use this evidence in favor of species status for each, until one reads further that the Asian taxon reaches into Fennoscandia and there is such great variation in Fennoscandia that one cannot recognize distinct chemotypes.

There are also complications with *Dryopteris campyloptera*. The cytological evidence suggests that it is derived from *D. intermedia* and *D. expansa* (Gibby 1977). This view is not supported by the phloroglucinol analysis, which shows the suppression of compounds from *D. intermedia*. Of course, we could also point to the "suppression" of a morphological character such as indusial glands as well. If we now examine the phloroglucinols from Gibby's artificial hybrid (Gibby *et al.* 1978), we find that there seems to be no contribution from *D. expansa*. Accordingly, the artificial diploid hybrid and the putative derived allotetraploid are very different chemically.

The phloroglucinols are also of little help with the numerous hybrids of *Dryopteris celsa* and *D. clintoniana*, e.g. *celsa* × *goldiana, clintonia* × *goldiana,* and *celsa* × *clintonia* etc. Why this is so, can be shown easily, if one reverts to genomes (Table V). The method is unsatisfactory to compare GGLL vs. GG(L); GGLLBB vs. GGLL(B); and GG(L)(B) vs. GG(L), etc. (Widen *et al.* 1975). In reality, this whole group is rather poor in both the numbers and quantity of phloroglucinols.

The second example is a strange twist to the direction that I thought we were going! I had always considered that when one had completed the genome analysis together with the phloroglucinol evidence, that if two taxa were different, e.g. *D. expansa* (EE) vs. *D. campyloptera* (EEII), that we were quite right to insist that each is a separate and distinct species, even if there are problems in identification. The other side of the coin I had also tacitly accepted. That is, if two taxa had the same genomes and the same chemistry they should be the same species. One should be as happy merging two entities as creating two entities from one!

Gibby and Walker (1977; Gibby *et al.* 1978; von Euw *et al.* 1980) have now shown that *Dryopteris intermedia, D. azorica* and *D. maderensis* have the same genomes and all three are nearly identical in their phloroglucinols, and yet all three are to be recognized as separate species. Von Euw *et al.* (1980) say "they are geographically widely separated," and that they are morphologically still similar, but distinct, and adapted to different ecological conditions. This also allows recognition of two derived allotetraploids, *D. crispifolia* derived from *aemula* × *azorica* and *D. guanchica* from *aemula* × *maderensis,* which again will have the same genomes but again are geographically separated and adapted to different ecological conditions (Gibby *et al.* 1977 and 1978). Of particular interest to North American botanists is the relationship of the European *Dryopteris dilatata s.str.* with our *D. campyloptera*. Again, they have the same genomes (Gibby 1977) and essentially the same phloroglucinols (von Euw *et al.* 1980), but the authors say, "perhaps originated separately at different places and different times," and are therefore maintained as separate species.

There are other chemical systems that can be exploited for the analysis of our pteridophytes. Ferns such as *Pityrogramma* with

TABLE V. *Phloroglucinol composition in the* Dryopteris goldiana–celsa–clintoniana *complex*

Taxon	Genomes	"Albaspidins" 1	2	3	Para-asp.	Desasp.	Trisdes.	Flav. Acid
D. goldiana	GG	(+)	+++	++	+	--	--	+++
D. celsa	GGLL	+/++	++	+	+/++	--	--	+++
D. clintoniana	BBGGLL	++	+++	+	-/+	++	+	+++
celsa × goldiana	GG(L)	++	+++	++	+	--	--	+++
clintoniana × gold.	GG(L)(B)	++	+++	++	+	+	(+)	+++
celsa × clintoniana	GGLL(B)	++	+++	+	+/++	+	+	+++

its interesting indument have shown promise (Smith 1980;
Wollenweber *et al.* 1981). I am sure there are other systems
as yet untapped.

Other workers are not as interested in species differences as
they are in characterizing larger groups, e.g. genera or even
families and orders (Cooper-Driver 1980; Wallace *et al.* 1982).
Recently, Widén *et al.* (1983) have discussed the occurrence of the
phloroglucinols on a broad basis from the flowering plants, e.g.
Myrtus, Mallotus, Hypericum, down to the "corner" that they
occur in the Pteridophytes and discuss whether this is a family,
or subfamily occurrence.

Some of the genera and species have been singularly resistant
to study by the Mantonian techniques, but are once again under
study. Work is underway on *Athyrium* (Schneller) and also on
Cystopteris by Haufler. A promising new approach by the latter
is isozyme analysis (Soltis *et al.* 1980; Gottlieb 1981; Gastony
and Gottlieb 1982). This should help to delineate more clearly
the various taxonomic entities, or at least give us valuable
information on population biology. At the same time, I foresee
difficulties for a future pteridologist in Ontario. At the
present time, if you were to ask me how many species of
Cystopteris there are, I would reply that we have the wide-
ranging tetraploid, *C. fragilis,* the ubiquitous *Thuja* swamp
diploid, *C. bulbifera* and the rare northern tetraploid *C. montana.*

In short, three species with one so rare that for most field
people it is two. We have recently found the diploid *C. protrusa*
in southern Ontario and I consider *C. laurentiana,* a hexaploid
derived from *C. bulbifera* X *fragilis,* a good species, so we now
have five. If *C. fragilis* var. *mackayii* (tetraploid) is
distinctive enough to be recognized as a species, *C. tenuis*
(Lellinger 1981) we can add one, and some are impressed with *C.
fragilis* var. *dickieana (Woodsia*-like spores), that is another.
So our new total is seven. Now you know these biosystematic
pteridologists! They like to study hybrids. How many interspecific hybrids can we have with 7 species? The first can
hybridize with 6 others, the second with 5 others, etc. - 6+5+4+3+
2+1, i.e. 21 interspecific hybrids! Then of course, there are
still others, and one is already with us. One can have an
unreduced egg plus a normal sperm to yield an autoploid. Shed a
tear for the classical taxonomist!

I had the same schizophrenic feelings after reading Bruce's
thesis (1975) from Michigan. On the one hand excitement and
enjoyment at the cytological implications of his work, and on the
other a sinking feeling for the pragmatic identifier or amateur
field companion. Here we have the innocuous little Bog Clubmoss,
L. inundatum in Michigan just west of Ontario, shown to
consist of at least six taxa - a diploid (n = 78), three
variously derived tetraploids and two triploid hybrids. Moving
east in Canada to the Maritimes, Bruce (1975) says "In Nova Scotia
for example there are *ten* taxa." So you can see that Cody
and Britton (1984) are most conservative in suggesting a mere two!

Seriously, how do we handle this? We are faced with collections
without strobili, collections with immature spores, collections
not made in the Fall. There is precious little for the herbarium
worker to arrive at a logical conclusion other than "*L. inundatum*
complex" - and complex it is indeed.

Bruce (1975) is a good reference to note that the 13 species of
Lycopodium in Canada (Cody and Britton 1984) have received
considerable attention from Wagner's school in Michigan. From
Wilce (1965), and Beitel (1979) on the *lucidulum-selago*
complex, biosystematics has been to the fore. The studies have
embraced chromosomes, SEM of spores, gametophytic studies, modern
morphological and anatomical studies and phytogeography. We can
say the Lycopods have had a new rennaissance. At the same time,
an objective observer (by definition a non-pteridologist) might
notice a strong trend towards segregation - one species to many,
e.g. *L. inundatum;* one genus to many, e.g. *Diphasiastrum,*
Lepidotis, Lycopodiella, etc. and even one family to two, e.g.
the Huperziaceae for the *selago* alliance. It was refreshing that
after Hickey (1977) segregated *L. dendroideum* from *L. obscurum*
and its variety *isophyllum,* that these three taxa were not
separable by the flavonoid system utilized by Fusiak (1982).
Recent work by Takamiya and Tanaka (1982) suggests that the *L.*
clavatum alliance will be the next group to show complexities.

As a cytogeneticist, I should not let the Lycopods go by
without mentioning the peculiar situation in *L. complanatum,*
digitatum and *tristachyum* and their so-called hybrids (Hersey
and Britton 1981). Why in this group, if these are indeed interspecific hybrids, do we have such complete chromosome pairing?
Could Linnaeus have been right and they are all one species? Or,
is chromosome pairing in this group quite unlike pairing in the
higher ferns or even in *Equisetum*? It is not possible to
invoke pairing of homeologous sets (Conant and Cooper-Driver 1980)
when dealing with a basic haploid number such as 23. These 23
must be pairing with 23 homologues. Subsets seem unlikely as the
best division would be 11 plus 12 and this would still not allow
complete bivalent formation. It is too bad that we are unable to
sow the spores (they do not germinate), artificially hybridize and
raise the progeny as the followers of Manton suggest we might!

The Isoetaceae has recently received attention. Boom (1979),
Kott (1980), Britton, Ceska and Ceska (unpubl.) and Carl Taylor
(1982) have shown that biosystematic methods - chromosome numbers
(Kott and Britton 1980), SEM of spores (Kott 1980), flavonoids
(Kott and Britton 1982a), and biological criteria (Kott and
Britton 1982b) are helpful in delineating the species in this
group. Taylor in Wisconsin, is culturing and crossing species and
is using isozyme analysis techniques. Kott (1980) has shown that
spore sizes are most useful for extrapolation of ploidy level.
The Canadian species are listed in Table VI, but as yet no genome
analysis has been performed. Problems which demand attention are
the relationships of *Isoetes echinospore, muricata* and *asiatica;*
I. macrospora vs. *I. luacustris; I. howellii* vs. *I. melanopoda.*

TABLE VI. *Canadian species of* Isoetes *and their chromosome numbers*

Species	2 n
Eastern	
I. echinospora Dur.	22
I. eatonii Dodg	22
I. riparia Engelm.	44
I. tuckermanii A. Br.	44
I. acadiensis Kott	44
I. macrospora Dur.	110
Western	
I. nuttallii A. Br.	22
I. howellii Engelm.	22
I. bolanderi Engelm.	22
I. echinospora Dur.	22
I. maritima Underw.	44
I. occidentalis Henderson	66

Do we have hybrids in nature? Are some of the taxa apomictic?
     The genus *Equisetum* appears to be uniform cytologically
(n = 108), although few workers have attained Manton's degree
of technical accuracy for this genus. There seem to be few
experts in the world, though the names Hauke (1978), Duckett
(1979) and Page (1972) are exceptions. There is an amazing array
of hybrids in the lab (Duckett 1979) and in the field in Britain
(Duckett and Page 1975).
     The SEM of spores, glands (Pryer 1981) and hairs has shown the
surface of these with magnificent clarity. At times we see
biological unity and at others diversity. For example, the spores
of the spinulose *Dryopteris* (Britton 1972a) are quite
different from those of the *D. clintoniana* group (Britton
1972b). There is much variation in the spores of *Lycopodium*
species (Wilce 1972), but disappointing uniformity in the spores
of *L. complanatum, L. digitatum* and *L. tristachyum* (Hersey
1978). Also, there is not as much diversity as one might wish in
Gymnocarpium if one is interested in discriminating between ssp.
disjunctum and ssp. *dryopteris,* or between *G. jessoense*
ssp. *parvulum* and *G. robertianum* (Pryer and Britton 1983).
     Modern molecular methods which compare for example the
nucleotide sequences of 5S ribosomal RNA (Anderson *et al.* 1982)
on a Phycomycete, or papers such as Fox *et al.* (1980) which discuss
the evolution of the prokaryotes would seem to be source works for
pteridologists. The technology is available to directly compare
the genetic apparatus of different species. Another approach

which avoids "reading out" all the DNA of a genome (Stein *et al.*
1979) is to compare various chloroplast DNAs.

Let me return very briefly to my theme of "Progress and
Problems" in connection with the Canadian Pteridophyte flora.
Progress is evident in the species that are recognized in the
flora. The list is gradually being "tidied up". From what I have
said, you will realize that *Lycopodium* will require further
revision. *Selaginella* requires some work on *S. densa,* and *S.
sibirica.* Our list for *Botrychium* is out-of-date. For the rest
of the list, I would consider that one could expect further work
on the taxa in Table III, as well as work which is going forward
on *Woodsia, Cystopteris* and *Athyrium.*

Under problems, I would consider that the most untidy parts of
any descriptive flora must be "subspecies" and "varieties". The
seven subspecies and 16 varieties in Cody and Britton (1984) are
no exception. I know that I share with some of you the fervent
wish that we could start again and have more meaningful subspecies
and varieties. But this is not reality. Ferns are replete with
entities such as the Ostrich Fern, *Matteuccia struthiopteris* var.
pensylvanica. Or, is our plant a separate species? *Isoetes*
var. or ssp. *muricata* or is it *I. muricata? Pteridium
aquilinum* var. *latiusculum, Thelypteris palustris* var.
pubescens, etc. Hopefully, with more objective analysis, based
on more work, these will become clearer. This cytogeneticist
believes that we do not need any more name shifting without more
evidence.

I am an optimist that new technologies, new techniques and
bright ideas will be helpful in arriving at our goal of a more
natural classification (Britton 1974). After reviewing Tryon and
Tryon (1982) which certainly is biosystematics, with emphasis on
morphology, phytogeography, cytology, SEM of spores and some
chemistry, that most workers consider this is the logical
approach. At the same time, the major problem with the ultimate
synthesis is when the evidence is quite contradictory, e.g. plants
that look alike but we insist are different, chromosome numbers
which should be the same but are different, spore types which are
unexpectedly divergent or unexpectedly similar, chemistry which is
unexpected. Different workers "mind set" or basic prejudices then
come into play. Which evidence will be paramount? Do not look at
me for an answer! I am a disciple of Manton and a trained
cytogeneticist, so my mind is made up!

SUMMARY

The pteridophyte flora of Canada is presented as 35 genera, 141
species and 23 additional subspecific taxa. Only eight species
lack chromosome number determinations, whereas 106 have determinations from Canadian material, and 27 have determinations from
material collected outside Canada. Allopolyploidy and genome
analysis are well documented for *Dryopteris, Polystichum* and
Polypodium. Chemotaxonomic studies have been done on *Asplenium,*

Dryopteris, Gymnocarpium and *Isoetes*. SEM studies of spores have been useful for *Dryopteris, Gymnocarpium, Phegopteris, Lycopodium* and *Isoetes*. Problems have arisen concerning the recognition of autoploidy vs. segmental alloploidy vs. genomic alloploidy. This creates a problem for the recognition of biological species (segregate taxa) vs. pragmatic species complexes. Biosystematic studies have clarified species concepts, but they have greatly complicated species recognition.

REFERENCES

Anderson, J., Andresini, W. and Delihas, N. 1982. On the phylogeny of *Phycomyces blakesleeanus*. Nucleotide sequence of 5S ribosomal RNA. *J. Biol. Chem.* 257: 9114-9118.

Beitel, J.M. 1979. Clubmosses (*Lycopodium*) in North America. *Fiddlehead Forum* 6: 1-8.

Beitel, J.M. and Wagner, F.S. 1982. The chromosomes of *Lycopodium lucidulum*. *Am. Fern J.* 72: 33-35.

Boom, B.M. 1979. Systematic studies of the genus *Isoetes* in the southeastern United States. *M.Sc. Thesis*, Univ. Tennessee, Knoxville.

Britton, D.M. 1972a. Spore ornamentation in the *Dryopteris spinulosa* complex. *Can. J. Bot.* 50: 1617-1621.

Britton, D.M. 1972b. The spores of *Dryopteris clintoniana* and its relatives. *Can. J. Bot.* 50: 2027-2029.

Britton, D.M. 1974. The significance of chromosome numbers in ferns. *Ann. Missouri Bot. Gard.* 61: 310-317.

Bruce, J.G. 1975. Systematics and morphology of subgenus *Lepidotis* of the genus *Lycopodium* (Lycopodiaceae). *Ph.D. Thesis*, Univ. Michigan, Ann Arbor.

Cody, W.J. and Britton, D.M. 1984. *Fern and Fern Allies of Canada*. *Canada Agric.*, in press.

Cody, W.J. and Mulligan, G.A. 1982. Chromosome numbers of some Canadian ferns and fern allies. *Nat. Can.* 109: 273-275.

Conant, D.S. and Cooper-Driver, G. 1980. Autogamous allohomoploidy in *Alsophila* and *Nephelea* (Cyatheaceae): A new hypothesis for speciation in homoploid homosporous ferns. *Am. J. Bot.* 67: 1269-1288.

Cooper-Driver, G.A. 1980. The role of flavonoids and related compounds in fern systematics. *Bull. Torrey Bot. Club.* 107: 116-127.

Duckett, J.G. 1979. An experimental study of the reproductive biology and hybridization in the European and North American species of *Equisetum*. *Bot. J. Linn. Soc.* 79: 205-229.

Duckett, J.G. and Page, C.N. 1975. *Equisetum*. In *Hybridization and the flora of the British Isles*, C.A. Stace (ed.), pp. 99-103, Academic Press, London.

Evans, A.M. 1970. A review of systematic studies of the pteridophytes of the Southern Apalachians. In *The distributional history of the Biota of the Southern Appalachians. Part II: Flora. Res. Div. Mon.* 2. Virginia

Polytechnic Inst., Blackburg, Va. pp. 117–146.

Fox, G.E. *et al.* 1980. The phylogeny of prokaryotes. *Science* 209: 457–463.

Fusiak, F. 1982. Flavonoids of the *Lycopodium obscurum* complex of North America. *Bot. Soc. Am., Misc. Publ.* 162: 76. Abstr.

Gastony, G.J. and Gottlieb, L.D. 1982. Evidence of genetic heterozygosity in a homosporous fern. *Am. J. Bot.* 69: 636–639.

Gibby, M. 1977. The origin of *Dryopteris campyloptera*. *Can. J. Bot.* 55: 1419–1428.

Gibby, M., Jermy, A.C., Rasbach, H., Rasbach, K., Reichstein, T. and Vida, G. 1977. The genus *Dryopteris* in the Canary Islands and Azores and the description of two new tetraploid species. *Bot. J. Linn. Soc.* 74: 251–277.

Gibby, M. and Walker, S. 1977. Further cytogenetic studies and reappraisal of the diploid ancestry in the *Dryopteris cathusiana* complex. *Fern Gaz.* 11: 315–324.

Gibby, M., Widen, C.-J. and Widen, H.K. 1978. Cytogenetic and phytochemical investigations in hybrids of Macaronesian *Dryopteris* (Pteridophyta; Aspidiaceae). *Plant Syst. Evol.* 130: 235–252.

Gottlieb, L.D. 1981. Electrophoretic evidence and plant populations. *Prog. Phytochem.* 7: 1–45.

Hauke, R.L. 1978. A taxonomic monograph of *Equisetum* subgenus *Equisetum*. *Nova Hedwigia* 30: 385–455.

Hersey, R.E. 1978. A study of variation in three species of *Lycopodium* L. (Section *Complanata* Vict.) in Ontario. *M.Sc. Thesis*, Univ. Guelph, Guelph, Ont.

Hersey, R.E. and Britton, D.M. 1981. A cytological study of three species and a hybrid taxon of *Lycopodium* (Section *Complanata*) in Ontario. *Can. J. Genet. Cytol.* 23: 497–504.

Hickey, R.J. 1977. The *Lycopodium obscurum* complex in North America. *Am. Fern. J.* 67: 45–48.

Kott, L.S. 1980. The taxonomy and biology of the genus *Isoetes* L. in northeastern North America. *Ph.D. Thesis*, Univ. Guelph, Guelph, Ont.

Kott, L.S. 1981. *Isoetes acadiensis*, a new species from eastern North America. *Can. J. Bot.* 59: 2592–2594.

Kott, L. and Britton, D.M. 1980. Chromosome numbers for *Isoetes* in northeastern North America. *Can.J.Bot.* 58: 980–984.

Kott, L.S. and Britton, D.M. 1982a. Comparison of chromatographic spot patterns of some North American *Isoetes* species. *Am. Fern J.* 72: 15–18.

Kott, L.S. and Britton, D.M. 1982b. A comparative study of spore germination of some *Isoetes* species of northeastern North America. *Can. J. Bot.* 60: 1679–1687.

Kott, L.S. and Britton, D.M. 1982c. A comparative study of sporophyte morphology of the three cytotypes of *Polypodium virginianum* in Ontario. *Can. J. Bot.* 60: 1360–1370.

Lang, F.A. 1971. The *Polypodium vulgare* complex in the

Pacific Northwest. *Madrono* 21: 235–254.

Lellinger, D.B. 1981. Notes on North American ferns. *Am. Fern J.* 71: 90–94.

Lloyd, R.M. and Lang, F.A. 1964. The *Polypodium vulgare* complex in North America. *Br. Fern Gaz.* 9: 168–177.

Löve, A., Löve D. and Pichi Sermolli, R.E.G. 1977. *Cytotaxonomical atlas of the Pteridophyta.* J. Kramer, Vaduz, Germany.

Lovis, J.D. 1977. Evolutionary patterns and processes in ferns. *Adv. Bot. Res.* 4: 229–415.

Manton, I. 1950. *Problems of cytology and evolution in the Pteridophyta.* Cambridge, England.

Moran, R.C. 1982. The *Asplenium trichomanes* complex in the United States and adjacent Canada. *Am. Fern J.* 72: 5–11.

Mulligan, G.A. and Cody, W.J. 1979. Chromosome numbers in Canadian *Phegopteris. Can. J. Bot.* 57: 1815–1819.

Mulligan, G.A., Cinq-Mars, L. and Cody, W.J. 1972. Natural interspecific hybridization between sexual and apogamous species of the beech fern genus *Phegopteris* Fee. *Can. J. Bot.* 50: 1295–1300.

Page, C.N. 1972. An assessment of inter-specific relationships in *Equisetum* subgenus *Equisetum. New Phytol.* 71: 355–368.

Pryer, K.M. 1981. Systematic studies in the genus *Gymnocarpium* Newm. in North America. *M.Sc. Thesis,* Univ. Guelph, Guelph, Ont.

Pryer, K.M. and Britton, D.M. 1983. Spore studies in the genus *Gymnocarpium. Can. J. Bot.* 61: 377–388.

Pryer, K.M., Britton, D.M. and McNeill, J. 1983. A numerical analysis of chromatographic profiles in North American taxa of the fern genus *Gymnocarpium. Can. J. Bot.* in press.

Sarvela, J. 1978. A synopsis of the fern genus *Gymnocarpium. Ann. Bot. Fenn.* 15: 101–106.

Sarvela, J. Britton, D.M. and Pryer, K. 1981. Studies on the *Gymnocarpium robertianum* complex in North America. *Rhodora* 83: 421–431.

Shivas, M.G. 1961. Contributions to the cytology and taxonomy of species of *Polypodium* in Europe and America. I. Cytology. *J. Linn. Soc.* London, Bot. 58: 13–25.

Smith, D.M. 1980. Flavonoid analysis of the *Pityrogramma triangularis* complex. *Bull. Torrey Bot. Club* 107: 134–145.

Smith, D.M. and Levin, D.A. 1963. A chromatographic study of reticulate evolution in the Appalachian *Asplenium* complex. *Am. J. Bot.* 50: 952–958.

Soltis, D.E., Haufler, C.H. and Gastony, G.J. 1980. Detecting enzyme variation in the fern genus *Bommeria:* an analysis of methodology. *Syst. Bot.* 5: 30–38.

Stein, D.B., Thompson, W.F. and Belford, H.S. 1979. Studies on DNA sequences in the Osmundaceae. *J. Molec. Evol.* 13: 215–232.

Takamiya, M. and Tanaka, R. 1982. Polyploid cytotypes and their habitat preferences in *Lycopodium clavatum. Bot. Mag.* 95: 419–434.

Taylor, W.C. 1982. Crossing studies with *Isoetes* from northeastern United States. *Bot. Soc. Am. Misc. Publ.* 162: 77.
Abstr.

Tryon, R.M. and Tryon, A.F. 1982. *Ferns and Allied Plants with
Special Reference to Tropical America.* Springer-Verlag, New
York.

von Euw, J., Lounasma, J., Reichstein, T. and Widen, C.-J. 1980.
Chemotaxonomy in *Dryopteris* and related fern genera.
Review and evaluation of analytical methods. *Studia Geobot.*
1: 275-311.

Wagner, D.H. 1979. Systematics of *Polystichum* in western North
America north of Mexico. *Pteridologia* 1: 1-64.

Wagner, W.H. 1966. New data on North American oak ferns,
Gymnocarpium. *Rhodora* 68: 121-138.

Wagner, W.H. 1970. Evolution of *Dryopteris* in relation to
the Appalachians. *In* Distributional History of the Biota
of the Southern Appalachians, Part II: Flora. *Res. Div.
Monogr*. 2. Va. Polytech. Inst. State Univ., Blacksburg,
Virginia, pp. 147-192.

Wagner, W.H. 1973. Reticulation of Holly Ferns (Polystichum) in
the western United States and adjacent Canada. *Am. Fern
J.* 63: 99-115.

Walker, S. 1961. Cytogenetic studies in the *Dryopteris
spinulosa* complex II. *Am. J. Bot.* 48: 607-614.

Walker, T.G. 1979. The cytogenetics of ferns. In *The
experimental biology of ferns*. (A.F. Dryer, ed.). Academic
Press, London.

Wallace, J.W., Markham, K.R., Giannasi, D.E., Mickel, J.T., Yopp,
D.L., Gomez, L.D., Pittillo, J.D. and Soeder, R. 1982. A survey
for 1,3,6,7-tetrahydroxy-C-glycosylxanthones emphasizing the
"primitive" leptosporangiate ferns and their allies. *Am. J.
Bot*. 69: 356-362.

Widén, C.-J., Britton, D.M., Wagner, W.H. and Wagner, F.S. 1975.
Chemotaxonomic studies on hybrids of *Dryopteris* in eastern
North America. *Can. J. Bot.* 53: 1554-1567.

Widén, C.-J., Widen, H.K. and Gibby, M. 1978. Chemotaxonomic
studies of synthesised hybrids of the *Dryopteris carthusiana*
complex. *Biochem. Syst. Ecol.* 6: 5-9.

Widén, C.-J., Sarvela, J. and Britton, D.M. 1983. On the location
and distribution of phloroglucinols (filicin) in ferns. New
results and review of the literature. *Ann. Bot. Fenn.* in press.

Wilce, J.H. 1972. Lycopod spores, I. General spore patterns and
the generic segregates of *Lycopodium*. *Am. Fern. J.* 62: 65-79.

Wilce, J.H. 1965. Section *Complanata* of the genus *Lycopodium*.
Nova Hedwigia 19: 1-233.

Wollenweber, E., Smith, D.M. and Reeves, T. 1981. Flavonoid
patterns and chemical races in the California cloak-fern,
Notholaena californica. *Proc. Int. Bioflavonoid Symp.*,
Munich, pp. 221-226.

Biosystematics and Medicine

Walter H. Lewis
Department of Biology
Washington University
St. Louis, Missouri, U.S.A.

INTRODUCTION

Populations of species growing in different environments throughout their geographic ranges are subject to different selections
and are usually genetically as well as physiologically,
chemically, and morphologically different. Such inherent
diversity of plant species is well recognized by biosystematists,
but it is not fully appreciated by some researchers in
biomedicine. In fact, most pharmacognosists stress the importance
of environmental variables, such as temperature and nutrients, on
plant growth and secondary metabolite production and have often
assumed genetic influences as nonvariable (Solomon and Crane
1970). This assumption essentially ignores genetic polymorphism
as perhaps the greatest asset of species, and rejects the effects
of natural selection on this diversity and genetically-regulated
biosynthesis during ontogeny. Environmental influences
superimposed on this framework are important, but they should not
overshadow paramount effects of genetic determinants.
  Chemical polymorphism is a significant variable of plants as
new compounds and high concentrations of well-known ones are
needed for therapeutic use. Examples of known or suspected
genetic, chromosomal, and genomic regulation of this diversity is
the focus of this paper. Included also are brief discussions of
the significance of conserving and classifying such diversity for
long-range and efficient utilization of plants for our well-
being.

DIVERSITY

1. *Genetic Polymorphism of Species*
The genetic regulation of secondary metabolic pathways is not
understood for the majority of medicinal plants. Indeed the

genetics of most are undoubtedly highly complex and their
elucidation will be difficult and time-consuming, but a few
examples exist that show the genetic basis of secondary compounds
and its diverse expression in biotypes of a species throughout its
range. The classic example is that of cyanogenic polymorphism of
Trifolium repens (white clover) examined by Daday (1954a,b,
1958, 1965) throughout the world, where it has escaped from its
native range in Europe and western Asia. Cyanogenic and
acyanogenic characteristics are regulated by two genes: the
synthesis of the cyanogenic glycoside, lotaustralin, from amino
acid precursors by alleles of the gene *Ac*, and hydrolysis of the
cyanogenic glycoside, linamarase, by alleles of the independently
inherited gene *Li* (Fig. 1). Only plants that possess at least
one dominant allele of both genes are cyanogenic. By selective
breeding, it is possible to obtain the four homozygous genotypes
AcAcLiLi, AcAclili, acacLiLi, and *acaclili,* only the first of
which is cyanogenic by releasing HCN when crushed (Conn 1979).
These and their heterozygotes occur in Europe where adaptation to
environmental gradients of latitude (Fig. 2) and altitude (Fig. 3)
have given rise to clines from southern and western Europe where
the dominant genetic cyanogenic plants predominate, to northern
and eastern Europe where the recessive genetic acyanogenic ones
are typical, and from cyanogenic low altitude populations to
acyanogenic high altitude populations (Daday 1954a). Even in
North America, following post-Columbian introductions, selection
has occurred along a similar latitudinal cline (Daday 1958).
Natural selections correlate to some extent with temperature, for
linamarase is activated by cold and the HCN liberated causes an
irreversible inhibition of the respiratory system of the *AcLi*
genotype resulting in tissue death (Daday 1965).
     Genetically-controlled cyanogenesis is also recognized among
other species of *Trifolium,* as well as for *Lotus, Manihot,* and
Sorghum, giving rise to both sweet (acyanogenic) and bitter
(cyanogenic) phenotypes. In some species of the same genus,
however, HCN formation is absent (Grant and Sidhu 1967), which can
be confounding for the biosystematist using chemical data for
purposes of classification, but it does not necessarily mean that
the biosynthetic pathway is absent, but rather that an active
degradative pathway is present.
     A second example of direct implication in human medicine is the
selection and hybridization of high and low alkaloid-producing

FIG. 1. *Genetics of cyanogenesis.*

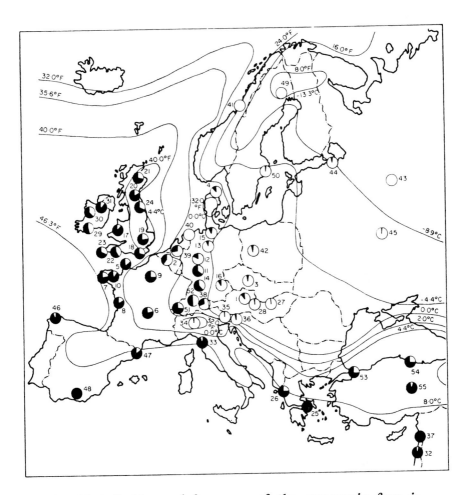

FIG. 2. *Distribution and frequency of the cyanogenic form in European and Middle Eastern wild populations of* Trifolium repens (black section = frequency of cyanogenic form, white section = frequency of acyanogenic forms; — January isotherm). (From Daday 1954a and Jones 1973).

cultivars of *Papaver somniferum* (opium poppy) by Nyman and associates in Sweden. Searching for an opium poppy having reduced morphine production, Nyman and Hall (1974) found plants with a morphine content of only 1-2% in opium and 0.02-0.03% in dry capsules. These selections were made from among a cultivar where morphine averaged 21% in opium and 0.3-0.6% in capsules. They also hybridized *P. somniferum* with *P. orientale* (Oriental poppy) and found about 4% morphine content in opium and 0.3% in dry capsules among the F_1. However, up to 10 times as much opium

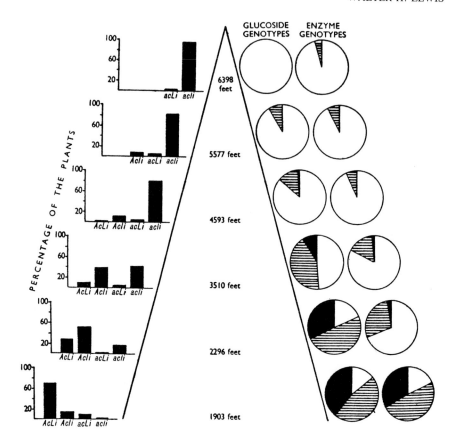

FIG. 3. *Phenotypic and genotypic frequencies in wild populations of* Trifolium repens *from different altitudes of the Central European Alps* (phenotypes on left: *AcLi* = glycoside and enzyme, *Acli* = glycoside only, *acLi* = enzyme only, *acli* = neither glycoside nor enzyme; genotypes on right: black section = dominant homozygotes, lined section = heterozygotes, white section = recessive homozygotes). (From Daday 1954b).

could be collected from the hybrids as from *P. somniferum.* Later, Nyman and Hall (1976) and Nyman (1980a) isolated a spontaneous mutant with red latex, low morphine production, and high thebaine content. Thebaine normally ranges from 2–5% in opium (Table I), but the selected plant had 24% thebaine and only 0.5% morphine. Total morphine alkaloid content was within the normal range for the species, but this mutant probably altered normal enzymatic activity along the biosynthetic pathway from thebaine (→ neopinone → codeinone → codeine →) to morphine, or perhaps earlier, resulting in high thebaine and low morphine production. Other alkaloids, such as the phtalideisoquinolines (noscapine, narcotoline) and

TABLE I. *Selected mutants with low productivity of secondary metabolites, % of dry wt* (from Nyman and Hall, 1976, and courtesy of the publishers of *Hereditas*).

Species	Vernacular name	Compound	Normal content	Mutant content	Organ	Gene
Brassica napus cv. Bronowski	rapeseed	total glucosinolates	5.0	0.2	seed	--
Lupinus albus	white lupine	total alkaloids	1.7	0.06[1]	seed	s, r[3]
L. angustifolius cv. incundus	European blue lupine	total alkaloids	1.1	0.001	seed	s, r
L. luteus cv. liber	European yellow lupine	total alkaloids	0.9	0.02[1]	seed	s, r
Nicotiana tabacum cv. pauper	tobacco	nicotine	1.5	0.08	leaf	s, r
Papaver somniferum cv. soma	opium poppy	morphine (traces of codeine, thebaine)	0.3	0.03	capsule[2]	s, r[4]

[1]Intensive breeding programs whereby toxic quinolizidines have been reduced further to below 0.01% among 'sweet' cultivars (Hudson 1979).
[2]Normal ranges of morphinanes in opium from latex are morphine 20–25%, codeine 2–5%, and thebaine 2–5%, while ranges in mutant (Sv 74204) are morphine 1–2%, codeine 0.1–0.3%, and thebaine 0.1–0.3%. No other alkaloid was detected in the mutant cultivar.
[3]Single, recessive
[4]However, the isoquinoline alkaloids s.l. of *P. somniferum* appear under polygenic control (Nyman 1980a).

benzylisoquinolines (papaverine), are present in considerable
quantities in the opium poppy and they may markedly influence
morphine content: when the phtalideisoquinolines are low or
absent, presumably because of a block in the biosynthetic steps,
morphine content is high. This interaction has useful
implications, for as the authors suggest it should be possible to
increase yields of already existing high morphine selections by
breeding for low content of phtalideisoquinolines, and alternately
by breeding for high phtalideisoquinoline content to reduce
morphine content. This might be important in areas where opium
poppy is grown chiefly for seed purposes and where it is the
source of morphine used in illicit heroin production. Finally, by
superimposing infraspecific crossings on mutants already selected
for low or high morphine content, Nyman (1980b) produced even more
extreme morphine-containing plants in the F_2, some having only
traces of the alkaloid and others with capsule content as high as
1.27%.

Other examples of reduction in secondary metabolite content,
usually involving a recessive gene, are given in Table I. They
further emphasize the inherent variability of chemical races of
plants that through selection and hybridization can be important
sources of greater or lesser concentrations of compounds that may
be pertinent to our health.

2. *Chromosomal and Genomic Polymorphism of Species*

In addition to single or polygenes, do chromosomal and genomic
mutations contribute to medically-useful diversity? There are few
examples involving aneuploidy, but as described by Mechler and
Haun (1981) alkaloid content among trisomics of *Datura stramonium*
(Jimson weed) may vary markedly. Compared to hyoscyamine and
scopolamine leaf content of nontrisomics ($2n$ = 24) significant
divergence was found among three trisomic strains with reduced
alkaloid content (only 60–65% of $2n$ = 24) and seven trisomics
with significantly (136%–143%) or more commonly highly significant
(155%–227%) increases in alkaloid content. Because all plants
were strictly cultivated under similar conditions and because
leaves were harvested at the same developmental stage, Mechler and
Haun concluded that differences in alkaloid content may be caused
by the additional chromosome with tropane biosynthesis influenced
by the particular chromosome duplicated.

Even greater influence on secondary metabolite diversity can be
anticipated from both spontaneous and induced genomic mutations
within species. Infraspecific polyploids are very common in
nature (Lewis 1980) and they represent an extremely important
reservoir of well-adapted, often heterozygous, biotypes that have
been rarely utilized in producing significant cultivars either for
use in agriculture (but see Bingham 1980) or medicine. Yet, early
research by Rowson (1944) showed emphatically that leaves of
colchicine-induced tetraploids of *Datura stramonium* and its purple
form (*D. tatula*) have approximately double the total alkaloid
content of diploid plants, while the proportions of individual

alkaloids remain unchanged. In roots of tetraploids the alkaloids were increased up to threefold over those of corresponding diploids (Jackson and Rowson 1953). Likewise, Rowson (1945) found alkaloids of leaves and flowering tops of autotetraploid *Atropa belladonna* increased by 153.6%, the average increase being 93% compared to diploids. Selection of the highest yielding plants and their cultivation could have a major impact on the quantity of hyoscyamine and scopolamine produced per hectare (these anticholinergic alkaloids are among the most widely used naturally-occurring compounds in human and veterinary medicine), yet selecting for high yielding chemical races seems never to have been attempted, at least in the West. Recently in eastern Europe, however, Dzhurmansk and Yankulov (1978, 1981) and Yakulov and Dzhurmanski (1979, 1982) have shown greater concentrations of atropine and scopolamine (6-104%) among induced tetraploids of *Datura ferox, D. inoxia, D. metel* (as *D. fastuosa*), and *D. stramonium* (as *D. inermis*), noting the advantage of bringing the polyploid forms into cultivation for greater alkaloid production.

Other alkaloids of induced tetraploids of medicinal interest, such as those of *Catharanthus roseus* (Madagascar periwinkle), greatly exceed the production of their diploid progenitors (Gogitidze and Laptev 1981). Indeed, alkaloid concentration is generally increased in polyploids; as they are among the most important secondary compounds used therapeutically, much greater research effort should be directed to the selection of these high-yielding polyploids. Vitamin C, certain crude fat extracts, and other compounds of interest in medicine are also found at higher levels among polyploids (Lewis 1980), but one group of secondary metabolites, notably the glycosides, may concentrate in reduced amounts. Thus, Kennedy (1978) found that *Digitalis lanata* × *D. grandiflora* amphiploids induced by colchicine treatment contained digoxin at lower levels than *D. lanata*. In this instance the allotetraploid would be of less value than the current commercial source of the cardioactive glycoside from the diploid *D. lanata*.

Apart from quantitative chemical differences between diploids and polyploids, there may be important qualitative diversities within single or closely allied species. For example, the cytotypes of *Urginea indica* (Indian squill) possess various steroidal glycosides (Table II): the $2x$ and $6x$ races with proscillaridin A and scillaren A, the $3x$ having these glycosides plus scilliphaeoside, and the $4x$ with all three in addition to anhydroscilliphaeosidin (an aglycone). The aglycones of the first two glycosides are also found in *U. maritima* (European squill) and both possess digitalis-like cardiotonic properties. It is important medically to understand the diversity of these cytotypes, which may vary by two of four known active principles, and it is also important biosystematically to know that such diversity is infraspecific and correlated with the ploidy.

In a parallel example naturally-occurring triploids of *Achillea millefolium* form azulenogenic substances over a

TABLE II. *Distribution of bufadienolides[1] in bulbs of Indian populations of Urginea indica (Liliaceae) cytotypes (+ present, – absent) (from Jha and Sen 1981 and courtesy of the publishers of Phytochemistry)*

Cytotype (2n=20)	Number of populations	Proscilla- ridin A	Scillaren A	Scillipha- eoside	Anhydro- scilliphaeosidin
2x	5	+	+	–	–
3x	2	+	+	+	–
4x	3	+	+	+	+
6x	2	+	+	–	–

[1]Approximate concentrations (% dry wt) of proscillaridin A, scillaren A, scilliphaeoside, and ahydroscilliphaeosidin in bulbs were 0.18, 0.10, 0.10, and 0.07, respectively.

complete growing season, yet phenotypically identical hexaploids
show no azulenes during a similar period in any plant organ
(Spurna *et al.* 1970). This difference is important
medically for azulenes may inhibit plant growth and they may also
be cytotoxic to animals; presence or absence of these
sesquiterpenoids in races of the same ssp. *millefolium* is
fundamental to an understanding of its growth and development as
well as of its toxicity in relation to human and grazing animal
health.

I have described chemical differences over broad geographic
regions of a species, and both quantitative and qualitative
differences among polyploid and diploid biotypes. It is also
known that within a restricted region polyploids may successfully
colonize areas and habitats different from the diploids, and are
thus adapted to different moisture and temperature regimes and
soil types (Lewis 1980). Over more local ranges of a species
then, it is possible to find ecological segregation of ploidy
types and, as secondary metabolites are often metabolized
differently among cytotypes, there can be little doubt that
chemical assays of diploids in one habitat and polyploids in
another could show striking differences.

Short-term evolution of many plants is dominated by polyploidy,
for the mutational process may increase biochemical diversity at
the primary enzyme level and in other ways. It may allow the
immediate expression of derepressed enzymes with different
properties from those of the diploids that not only could extend
the range of environments and habitats in which normal and
successful development can occur (Lewis 1980), but could result in
qualitative and quantitative diversity of secondary metabolic
products. This resource needs greater utilization in biomedicine.

3. *Ontogenic Polymorphism*

One wonders why Ayurvedic (Hindu) physicians practicing
thousands of years ago followed a set of regulations governing the
collection of plants for greatest efficacy in treating various
illnesses (Lewis and Elvin-Lewis 1979). Indeed Theophrastos
mentioned that in the 4th century B.C. certain herbs were
gathered at night, others by day, and some before the sun struck
them (Robinson 1974). Was this simply magic associated with
ancient cures or were these early practitioners recognizing
individual plant differences of value in their practices?
Seemingly these herb collectors were realizing differences in the
turnover and accumulation of specific compounds, and yet it was
not many years ago that biochemists considered alkaloids and many
other secondary metabolites inert metabolic end products. Now,
however, these compounds are considered dynamic products
fluctuating in both total concentration and in rate of turnover,
and as such are more in harmony with the emperical selective
practices of ancient herbal medicine.

A. *Annual Change*

Alkaloid changes during ontogeny are striking in *Catharanthus roseus*, the periwinkle containing the antineoplastics vincristine and vinblastine. Of several dozen indole alkaloids isolated from the species virtually no alkaloid is present in the seeds. They first appear during germination and by three weeks are present throughout the plant; they gradually disappear almost completely and finally reappear in about eight weeks (Mothes *et al.* 1965). In the opium poppy, *Papaver somniferum*, the seeds are likewise essentially free from alkaloids, but on germination the seedlings give rise to narcotine in three days and to codeine, morphine, and papaverine when the seedlings are about 7 cm high (Wealth of India 1966). Total alkaloid content slowly increases until flowering when there is a sharp increase lasting until the floral leaves fall (Wealth of India 1966).

The alkaloids of poison hemlock, *Conium maculatum*, also vary with development (Fairbairn and Suwal 1961). During flowering little alkaloid is present, although γ-coniceine predominates over coniine, but by the fourth week (a dry season) and the fifth week (a wet season) the content of coniine in the fruit increases dramatically to about 80 µg/fruit and 130 µg/fruit, while the γ-coniceine content is only about 10 µg/fruit and 5 µg/fruit in these two years. Thus, an overall change from γ-coniceine to coniine is associated with rapid fruit development, but at later stages of fruit maturation coniine content falls to only about 11 µg/fruit after eight weeks (wet season) and there is a slight increase in the γ-coniceine content, indicating a possible reconversion of coniine to γ-coniceine.

Concentrations of the three major alkaloids of *Baptisia leucophaea* also vary according to growth and development. Anagyrine increases from a preflowering concentration of 11% content of total alkaloids, 28% when plants are flowering, to 56% when plants are fruiting. Methylcytisine parallels this increase with maturation, though at more modest concentrations (1%, 4%, 17%), but cytosine concentration decreases with maturity, from 88% of total alkaloid content at preflowering stage, 64% when flowering, to only 26% during the fruiting condition (Cranmer and Turner 1967).

There are many other well-documented examples of ontogenic changes in alkaloid content that generally follow the principle of rapid increase at the time of cell enlargement and vacuolization followed by a slow decline in concentration during senescence.

B. *Diurnal Fluctuations*

Active metabolism of these alkaloids can be further demonstrated by measuring fluctuations during one day. Continuing with the analysis of poison hemlock, Fairbairn and Suwal (1961) showed that amounts of coniine and γ-coniceine vary considerably during the day and that an increase in one corresponds to decrease in the other (Table III). As already noted, large amounts of coniine accumulate in the fruit by

TABLE III. *Amounts of coniine and γ-coniceine from* Conium
maculatum *(µg/fruit during four-hourly samples in one day of*
week five following flowering (from Fairbairn and Suwal 1961,
and courtesy of the publishers of *Phytochemistry*).

Sample	Coniine	γ-Coniceine
4 am	226	0
8 am	130	2
noon	174	9
4 pm	8	21
8 pm	200	0
midnight	213	0

the fifth week following flowering (130–226 µg/fruit of
coniine), except around 4 pm when the amount drops to an
exceedingly low 8 µg/fruit. At that time the amount of
-coniceine rises to a maximum of 21 µg/fruit. Surely the
Grecian executioners appreciated this periodicity and
collected their samples of poison hemlock accordingly!
     Fluctuations also occur daily during latex collection from
capsules of *Papaver somniferum* two to three weeks after
petal fall. As reported by Fairbairn and Wassel (1964), the
opiate alkaloids morphine, codeine, and thebaine vary widely,
with morphine accumulating rapidly in the forenoon and
dropping dramatically around noon with a complementary rise
of both codeine and thebaine (Fig. 4). This pattern could
reflect the biosynthetic pathway from thebaine to codeine to
morphine, but as with coniine, morphine is always
predominant, so that more morphine is made than can be
accounted for by conversion of the thebaine and codeine that
disappear (Robinson 1974). However, Fairbairn and Wassel (1964)
suggest that morphine is converted into a nonalkaloidal molecule
which remains in the latex or passes into some other tissue of the
capsule. Likewise, leaves of medically-important *Datura* spp.
collected in the early morning contain more alkaloids than those
harvested in the evening (Wealth of India 1952).

C. Organ Divergence
In *Papaver somniferum* harvested in India about two weeks after
petal fall total morphine content varies as follows (% dry wt):
whole plant, 0.09–0.18; leaves, 0.06–0.07; stems, 0.01–0.03; and
capsules, 0.23–0.41 (Wealth of India 1966). In fact from the
beginning of flowering until this time about 75% of total

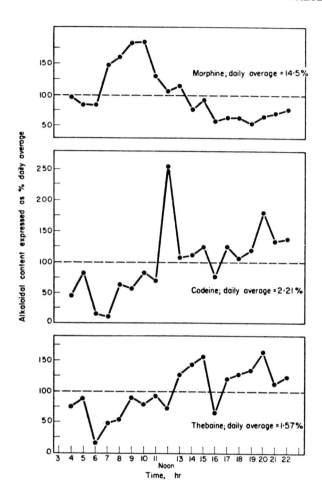

FIG. 4. *Diurnal fluctuations of opiate alkaloids of* Papaver somniferum. (From Fairbairn and Wassel 1964).

alkaloids are accumulated, with morphine, narcotoline, and papaverine preferentially accumulating in the capsules, codeine maximally concentrating in the roots, and thebaine higher in vegetative parts compared to capsules. But, as emphasized by Hughes and Genest (1973), these results are only valid for particular cultivars of *P. somniferum,* for others especially when grown under different climatic and cultural conditions may have divergent alkaloidal spectra. It must be emphasized further that the site of highest concentration may not always be the site of alkaloid formation. For example, the root synthesizes morphine at night and this is then transported to aboveground parts where it accumulates, usually in cell vacuoles, until about midday. The root is also the site of major alkaloid formation in *Atropa*, *Datura*, *Hyoscyamus*, and *Nicotiana* (Hughes and Genest 1973), but in contrast major sites of alkaloid production in *Lupinus albus* are in aboveground parts. Moreover, total alkaloid content increases during the day, but diminishes at night with the appearance of alkaloids in the roots seemingly due to a downward transport (Hughes and Genest 1973).

Even these statements are over-simplified, for variation in alkaloid production is found within the same organ. Morphine content in the opium poppy varies with the number and position of the capsules on the plant and also the mode of lancing (Wealth of India 1966). For example, it is maximal at first lancing and then rapidly decreases in successive lancings. The upper leaves and branches of *Datura* are richer in alkaloids than are those nearer the base of the plant (Wealth of India 1952).

D. *Environmental and Cultural Factors*

Commercial collection of tropane alkaloids from *Datura* spp. is closely governed by external factors. For example, total alkaloid content is considerably less after a rainy period than after clear weather. Indeed, the difference is so marked that collecting is done only following a period of clear days (Wealth of India 1952). Leaves of *Hyoscyamus niger* grown at higher altitudes (above 1524 m) are richer in tropanes than those of plants grown in the plains of India (Wealth of India 1959). Fertilizers high in nitrogen tend to increase yields of alkaloids (Hughes and Genest 1973). In general, factors favoring growth and development favor alkaloid formation (Robinson 1974, 1981).

Leaves dried in shade contain more alkaloids on the average than those dried in the sun, and leaves allowed to dry on the plant often contain more alkaloids than those dried after removal. Alkaloid yield of leaves is also increased when floral buds are removed (Wealth of India 1952). Apparently methods of harvest, light, and photoperiod are all important factors relative to secondary compound yields.

Tissue and cell-suspension cultures are being used as tools for studying secondary metabolic biosynthesis and also with the hope of producing commercially-valuable compounds (Robinson 1981). Media and culture conditions, often including growth regulators, need be optimal for maximum yield of specific compounds. However, if high yields are to be maintained in cell cultures, then individual plants within the species producing these compounds should be selected "because high-yielding individuals are likely to give high-yielding cell cultures" (Dougall 1979). Should frequencies of high-yielding clones decline with increased passaging of cultures, then high-yielding clones can be selected and propagated. Nutritional and plant growth substances alone are only "quick fixes" to enhance yields; if yields are to be maintained at high levels the genetic determinants for high production must be present in these cultures. Until this seemingly simple fact is fully appreciated, industrial exploitation of these techniques will have only limited success.

CONSERVATION OF DIVERSITY

From the above examples it is obvious that biosystematists must
know more about species of known or potential medicinal relevance.
How many neglected taxa exist of applied importance? They are
neglected not just for chemical assays, but all too often for
cytogenetics, population dynamics, floral biology, breeding
systems, and other basic biological properties. It behooves us to
study species of known economic worth whenever possible, for when
everything is equal, why not study a taxon that is going to be
extirpated because of overzealous use by humans or habitat
depletion, or that is expected to possess active principles of
human value? A good example has been the neglect of *Panax
quinquefolium* (American ginseng), a highly sought species
because of reputed medicinal value of "adoptogenic" steroidal
saponins concentrated in the roots. At one time, the distribution
of the species was essentially continuous throughout eastern North
America, but now it is rare over large areas and in others exists
only in small isolated populations even in protected zones (Lewis
and Elvin-Lewis 1979). An important consideration here, as
always, concerns the loss of biotypes from the many extirpated
populations that might have been valuable in an understanding of
both its biosystematics and significance to health. Virtually
nothing is known of population dynamics until recently (Lewis and
Zenger 1982, 1983) and if extirpation continues the biodynamic
principles that presumably affect health may never be fully known.
To this example may be added the poorly known genus *Dioscorea*
(yams), the roots of which yield the aglycone diosgenin used as
the semisynthetic base for most human hormones produced. Plants
are collected in the wild in Mexico and elsewhere and although
culture has begun it does not keep pace with the demand and a
number of imperfectly known species may be threatened with
extinction. This is always a danger if sudden medical value
followed by exploitation of a species is not integrated with
natural history, cultural, and other studies relative to its
survival.

PHYLOGENETIC CLASSIFICATION OF DIVERSITY

Biosystematists by the nature of their experimental research in
cytogenetics, breeding, and population dynamics concentrate their
studies at the species level or often on populations or
individuals of a single species. Nevertheless, major
contributions are also made to phylogenetic classifications of
higher taxa and this research based largely on morphological,
cytological, and chemical characteristics and relationships, is or
ought to be of concern to biomedical researchers. Too often in
the past, conclusions of systematists were given scant
consideration, and broad and expensive surveys of higher plants

for medical purposes were conducted without consideration of relationships and classifications.

It is important for biomedical researchers to recognize relationships when they coincide with a natural classification of plants based on all known data, and to initiate new or improved ideas within the framework of an existing classification (Lewis and Elvin-Lewis 1977, 1979). For example, a hypothetically useful compound to treat heart disease is found in the Apocynaceae. An informed researcher who wants to find plants with similar or related compounds would know immediately to examine members of the Asclepiadaceae. Unless unlimited resources and time are available, the researcher would be unwise to expend too much effort sampling for such compounds purely at random among thousands of plants belonging to several hundred angiosperm families. We know, of course, that similar compounds may be synthesized by essentially similar biosynthetic pathways in species belonging to very dissimilar families; the existence of such examples of parallel evolution in no way reflects true relationships, but simply independent origins of compounds. We know, also that there are instances of different pathways or mechanisms in plants that lead to the same compound, analogous occurrences that offer no information regarding phylogenetic relationship. Excellent examples of chemical constituents restricted to various plants and how these may be of value in improving classification are outlined by Gibbs (1974). Similar chemical structures are likely to be found in plants of a given systematic grouping, and this realization has considerable practical value when searching for medically-significant compounds.

SUMMARY

Interests of biosystematics and medicine converge at two major points, one involving diversity and conservation of species and populations, and the other regarding phylogenetic relationships of higher taxa. Plant diversity is fundamentally under genetic regulation and although environmental influences are important in selecting for and modifying secondary metabolites and other compounds of value in medicine, they should not overshadow the paramount effects of genetic determinants. Because genetic influences have been considered nonvariables in much medically-related research with plants, conventional cytogenetic techniques, such as selection and hybridization, have been under-utilized in the search for new products and better yields of known compounds, and the results have been decidedly equivocal. In the future biosystematic research must impact and influence these programs, while also contributing to a better understanding of phylogentic classification so necessary for the efficient choice of plants and compounds of potential importance to health.

REFERENCES

Bingham, E.T. 1980. Maximizing heterozygosity in autopolyploids.
In *Polyploidy: Biological Relevance* (W.H. Lewis, ed.),
Plenum Press, New York. pp. 471–489.

Conn, E.E. 1979. Cyanide and cyanogenic glycosides. In
Herbivores: Their Interaction with Secondary Plant Metabolites
(G.A. Rosenthal and D.H. Janzen, eds.), Academic Press, New
York. pp. 387–412.

Cranmer, M.F. and Turner, B.L. 1967. Systematic significance of
lupine alkaloids with particular reference to *Baptisia*
(Leguminosae). *Evolution* 21: 508–517.

Daday, H. 1954a. Gene frequencies in wild populations of
Trifolium repens. I. Distribution by latitude. *Heredity*
8: 61–78.

Daday, H. 1954b. Gene frequencies in wild populations of
Trifolium repens. II. Distribution by altitude. *Heredity*
8: 377–384.

Daday, H. 1958. Gene frequencies in wild populations of
Trifolium repens. L. III. World distribution. *Heredity*
12: 169–184.

Daday, H. 1965. Gene frequencies in wild populations of
Trifolium repens. L. IV. Mechanism of natural selection.
Heredity 20: 355–36.

Dougall, D.K. 1979. Factors affecting the yields of secondary
products in plant tissue cultures. In *Plant Cell and Tissue
Culture Principles and Applications* (W.R. Sharp, P.O. Larsen,
E.F. Paddock, V. Raghavan, eds.) Ohio State Univ. Press,
Columbus. pp. 727–743.

Dzhurmanski, G. and Yankulov, J. 1978. Experimentally obtained
autotetraploid thorn-apple and yellow poppy. *Genet. Sel.*
11: 5–16.

Dzhurmanski, G. and Yankulov, J. 1981. Changes in herb yield and
alkaloid content of diploid and tetraploid forms of *Datura
inoxia* following various dates of seeding. *Rastenievud. Nauki*
18: 53–63.

Fairbairn, J.W. and Suwal, P.N. 1961. The alkaloids of hemlock
(*Conium malculatum* L.)—II. Evidence for a rapid turnover
of the major alkaloids. *Phytochemistry* 1: 38–46.

Fairbairn, J.W. and Wassel, G. 1964. The alkaloids of *Papaver
somniferum* L.—I. Evidence for a rapid turnover of the major
alkaloids. *Phytochemistry* 3: 253–258.

Gibbs, R.D. 1974. *Chemotaxonomy of Flowering Plants*, Vol. 1,
Constituents. McGill-Queens' Univ. Press, Montreal. 680 p.

Gogitidze, T.R. and Laptex, Y.P. 1981. Comparative evaluation of
experimental polyploids of *Catharanthus roseus* and their
parent forms. *Genetika* 17: 563–564.

Grant, W.F. and Sidhu, B.S. 1967. Basic chromosome number,
cyanogenetic glucoside variation, and geographic distribution of
Lotus species. *Can. J. Bot.* 45: 639–647.

Hudson, B.J.F. 1979. The nutritional quality of lupin seed. *Qual. Plant.* 29: 245-251.

Hughes, D.W. and Genest, K. 1973. Alkaloids. In *Phytochemistry, Vol. 2, Organic Metabolites* (L.P. Miller, ed.), Van Nostrand Reinhold, New York. pp. 118-170.

Jackson, B.P. and Rowson, J.M. 1953. Alakloid biogenesis in tetraploid stramonium. *J. Pharm. Pharmacol.* 5: 778-793.

Jha, S. and Sen, S. 1981. Bufadienolides in different chromosomal races of Indian squill. *Phytochemistry* 20: 524-526.

Jones, D.A. 1973. Co-evolution and cyanogenesis. In *Taxonomy and Ecology* (V.H. Heywood, ed.), Academic Press, New York. pp. 213-242.

Kennedy, A.J. 1978. Cytology and digoxin production in hybrids between *Digitalis lanata* and *Digitalis grandiflora*. *Euphytica* 27: 267-272.

Lewis, W.H. 1980. Polyploidy in species population. In: *Polyploidy: Biological Relevance* (W.H. Lewis, ed.), Plenum Press, New York. pp. 103-144.

Lewis, W.H. and Elvin-Lewis, M.P.F. 1977. *Medical Botany: Plants Affecting Man's Health*. Wiley-Interscience, New York. 515 p.

Lewis, W.H. and Elvin-Lewis, M.P.F. 1979. Systematic botany and medicine. In *Systematic Botany, Plant Utilization and Biosphere Conservation* (I. Hedberg, ed.), Almqvist & Wiksell, Stockholm. pp. 24-31.

Lewis, W.H. and Zenger, V.E. 1982. Population dynamics of the American ginseng *Panax quinquefolium* (Araliaceae). *Am. J. Bot.* 69: 1483-1490.

Lewis, W.H. and Zenger, V.E. 1983. Breeding systems and fecundity in the American ginseng *Panax quinquefolium* (Araliaceae). *Am. J. Bot.* 70: 466-468.

Mechler, E. and Haun, N. 1981. Alkaloid content in trisomic mutants of *Datura stramonium*. *Planta Med.* 42: 102-103.

Mothes, K.I., Richter, I., Stolle, K. and Groger, D. 1965. Physiologische Bedingungen der Alkaloid-synthese bei *Catharanthus roseus* G. Don. *Naturwissenschaften* 52: 431.

Nyman, U. and Hall, O. 1974. Some varieties of *Papaver somniferum* L. with changed morphinane alkaloid content. *Hereditas* 84: 69-76.

Nyman, U. and Hall, O. 1976. Morphine content variation in *Papaver somniferum* L. as affected by the presence of some isoquinoline alkaloids. *Hereditas* 88: 17-26.

Nyman, U. 1980a. Selection for high thebaine/low morphine content in *Papaver somniferum* L. *Hereditas* 93: 121-124.

Nyman, U. 1980b. Alkaloid content in the F_1 and F_2 generations obtained from crosses between different chemoprovarieties in *Papaver somniferum* L. *Hereditas* 93: 115-119.

Robinson, T. 1974. Metabolism and function of alkaloids in plants. *Science* 184: 430-435.

Robinson, T. 1981. *The Biochemistry of Alkaloids*, ed. 2. Springer-Verlag, New York. 225 p.

Rowson, J.M. 1944. Increased alkaloidal contents of induced polyploids of *Datura*. *Nature* 154: 81–82.

Rowson, J.M. 1945. Increased alkaloidal contents of induced polyploids of *Datura*, *Atropa* and *Hyoscyamus*. *Quart. J. Pharm. Pharmacol*. 18: 175–193.

Solomon, M.J. and Crane, F.A. 1970. Influences of heredity and environment on alkaloidal phenotypes in Solanaceae. *J. Pharm. Sci*. 57: 1670–1672.

Spurna, V., Plchova, S. and Karpfel, Z. 1970. Study of some biotypes in the genus *Achillea*. *Naturwissenschaften* 57: 196–197.

Wealth of India--Raw Materials. 1952. *Datura*, (3: 14–19); 1959 *Hyoscyamus*, (5: 151–154); 1966 *Papaver* (7: 231–248). Counc. Sci. Ind. Res., New Delhi.

Yankulov, J. and Dzhurmanski, G. 1979. Morphogenetic alkaloid concentration variability in tetraploid thorn-apple *Datura godronii* cultivar minka. *Genet. Sel*. 12: 416–422.

Yakulov, J.K. and Dzhurmanski, G.T. 1982. Morphogenetic variability of the concentration of scopolamine and atropine in tetraploid form of *Datura inoxia*. *Fitologiya* 20: 42–49.

Modes of Evolution in Plants under Domestication

Daniel Zohary
Department of Botany
The Hebrew University
Jerusalem, Israel

INTRODUCTION

Variation patterns and variation ranges in cultivated plants are
frequently very different from those encountered in their wild
relatives. Crops present us with divergent patterns which can be
very confusing to botanists familiar with classification of wild
plant groups. As succinctly stated by Harlan (1975) morphological
divergence among *genetically related breeds* in crops is
frequently of a different order of magnitude compared with that
found in the wild. This is true for many vegetative traits as
well as for floral and fruit structures which in the wild
relatives stay conservative over ranges of species, or even ranges
of genera or families. Taxonomic treatment of crops can be
difficult and confusing. When attempted on the basis of
conventional morphological criteria it often leads to excessive
splitting, and to species delimitation which is utterly
incompatible with the cytogenetic information available in
numerous crops.

Some of the divergence encountered in plants under
domestication is obviously the outcome of deliberate, conscious
selection by the cultivator. The impact of such type of selection
is very clear when a given crop is being bred for several uses as
the case is in cabbage, beet or maize. More significant still are
the automatic, unconscious selection pressures caused by the
transfer of plants from their native wild environments into new
and contrasting systems of cultivation, the use of different
methods of maintenance, and the various ways of utilization of
crops. The present paper aims at sketching the main determinants
of unconscious selection in plants under domestication and at the
valuation of the constrasting modes of evolution they bring about.
It stresses the fact that these agronomic practices determine, to
a large extent, the genetic systems and the patterns of divergence

encountered in plants under domestication. For these reasons
these determinants can serve as important guidelines in
biosystematic treatments of crop plants.

REPRODUCTIVE SYSTEMS IN CULTIVATED PLANTS

Plants under cultivation differ markedly from wild plants in the
ways by which they are reproduced and maintained. [For background
data on various crops, consult the various chapters in Simmonds
(1976) book.] In the various families of the flowering plants
allogamy (cross-fertilization) is the principal genetic system.
Most species build outbreeding populations and disperse themselves
by seed. Other systems, such as self-pollination, apomixis or
vegetative propagation operate as well, and are even common in
some genera and families. Yet, compared to allogamy, their
overall weight is small. In contrast, in plants under
domestication sexual reproduction by means of cross-pollination is
relatively rare. Most crops are maintained by either one of the
following two systems: (a) self-pollination, and (b) vegetative
propagation. In contrast to allogamy both these systems are
effective in bringing about immediate "fixation" and isolation of
desired genotypes.
     (a) *Self-pollination,* or more exactly almost full self-
pollination, is the principal mating system operating in grain
crops, and in many vegetables. Consequently, variation in these
crops is structured in the form of numerous, distinct true
breeding lines, which in traditional agriculture formed aggregates
of land races. Almost all the 60-70 cereals, pulses and other
grain crops important in world food production are predominantly
selfed. Only a few (such as maize, rye, buckwheat, scarlet runner
bean, broad bean) stand as exceptions to this rule and are cross-
pollinated. Now that the wild progenitors of the majority of the
grain crops are already satisfactorily identified, we know that as
a rule the wild ancestors of self-pollinated crops are also
selfers. In other words, self-pollination in grain crops is not
the outcome of selection under domestication. It is rather a
"preadaptation" of the wild ancestor which considerably enhances
its chances to be successfully taken into cultivation. After all,
the majority of plant species in the various geographic areas of
plant domestication are outcrossers, not selfers.
     b) *Vegetative propagation* is the second, widely adopted
means to "fix" and maintain desired genotypes under domestication.
This method prevails in the cultivated fruit trees, root and tuber
crops as well as in numerous ornamentals. In contrast to self-
pollination, vegetative propagation is a new genetic system
introduced by the grower. The wild relatives of most vegetatively
propagated crops reproduce in nature, exclusively or almost
exclusively, from seed. Practically all are allogamous plants and
in many of them cross-pollination is determined in the wild either
by self-incompatibility or by sex-determination (dioecy). In
other words, in domestication of fruit trees, root and tuber crops

and many ornamental plants there is first of all a shift from
sexual reproduction and panmixis (in the wild) to vegetative
propagation of clones (under domestication). Also here one is
faced with "preadaptation" to domestication. This time in terms
of how well wild stocks lend themselves to vegetative propagation.
The history of Old World fruit trees (Zohary and Spiegel-Roy 1975)
indicates that wild fruit bearing plants which could be easily
vegetatively propagated had a premium over those that were not.
The "classical" fruit trees (olive, grape, fig, date palm) which
root easily from twigs or suckers were part of Old World food
production already in the 4th millennium B.C. Rosaceous fruit
trees (apple, pear, plum) which do not lend themselves to rooting
were taken into cultivation much later – apparently only after the
invention of grafting (1st millennium B.C.).

The realization that in cultivated crops we are faced
principally with cases of self-pollination and shifts from
panmixis to vegetative propagation have also critical implications
in crop plant classification. Taxonomists dealing with crops
should always bear in mind that in treating variation found in
crops they are first of all concerned with the contrasting
complexities of (a) aggregates of self-pollinated annuals and (b)
agamic complexes.

METHODS OF MAINTENANCE AND THEIR IMPACT

Two main methods are employed by the cultivator to maintain plants
under domestication: (a) *planting of seed* and (b)
vegetative propagation. The choice between these two
agronomic practices is also the choice between two contrasting
patterns of selection and modes of evolution under domestication.

With very few exceptions (such as nucellar seed in *Citrus*
or mango) planting of seed means sexual reproduction. Cultivated
plants maintained by seed (the bulk of the grain crops, numerous
vegetables and truck crops, some ornamentals) undergo a
recombination-and-selection cycle every planting. In other words,
such crops have had, under domestication, numerous (hundreds or
thousands) generations of selection. They have been continuously
molded into either (a) clusters of inbred lines in predominantly
self-pollinated crops (the large majority of grain crops), or (b)
distinct cultivated races in cross-pollinated crops (few grain
crops, some vegetables). In numerous sexually reproducing crops
the results of such continuous selection are indeed striking.
Under domestication these crops diverged considerably from their
wild progenitors. They are now distinguished from them by complex
syndromes of both morphological and physiological traits.

Vegetatively propagated crops (most fruit trees, root and tuber
crops, few vegetables, many ornamentals) have had an entirely
different history of selection. "Cultivars" in these crops are
not true races but just clonal replications of "exceptional
individuals" which are as a rule highly heterozygous. They have
been picked up by the cultivator from variable panmictic wild

populations, and later in domestication also from among
segregating progeny of spontaneous or artificial crosses between
cultivated × *wild* or *cultivated* × *cultivated* individuals. In
terms of selection, domestication of vegetatively propagated crops
is largely a single step operation. With the exception of rare
somatic mutations, selection is completed the moment the clone is
picked up. In traditional agriculture the turnover of clones was
apparently slow, particularly in fruit trees. Appreciated
genotypes were maintained for long periods of time. Thus,
vegetatively propagated crops underwent, in cultivation, only few
recombination-and-selection cycles. In sharp contrast to sexually
reproducing cultivated plants, their cultivars do not represent
true breeding races, i.e. populations which have been continuously
molded in response to new and changing environments, but only
clones which as a rule are also highly heterozygous and segregate
widely when progeny tested. Significantly, the large majority of
the segregating progeny are not only economically worthless, but
often regress towards the mean found in spontaneous populations,
showing striking resemblance to the wild forms.

THE CHOICE OF THE PLANT'S PART AND ITS IMPACT

Different crops are grown for different parts of the plant's body.
Some cultivated plants are raised for their *vegetative* parts
(roots, corms, leaves, stems, etc.); in others the *reproductive*
parts (flowers, fruits, seed) constitute the agricultural
products. Also, the choice of the organ leads automatically to
the operation of different and contrasting selection pressures,
particularly in regard to the reproductive system of the crop.
    When crops are *grown for their seed* (or at least when they
are reproduced by seed), they stay under constant stabilizing
selection which keeps their reproduction system intact. *Grain
crops* provide us with the most rigid cases of such normalizing
selection. Yields in these crops depend decisively on the proper
development of flowers and fruits, normal chromosome pairing and
full fertility. Deviants are weeded out automatically and the
reproduction system is kept in balance. It is no wonder, that
among cultivated plants grain crops are the most conservative in
this regard. They are characterized by strictly balanced
chromosome systems and show very little chromosome divergence
under domestication. With very few exceptions, (such as addition
hexaploidy in bread wheat) the chromosome complements in the
cultivated varieties of grain crops are identical to those found
in their wild progenitors.
    In seed propagated plants *grown for their fruits* the
pressure of stabilizing selection to maintain seed fertility and
seed yield is many times somewhat laxer. Thus *seed propagated
vegetables* grown for their multi-seed fruits (i.e. tomato,
eggplant, cucumber, melon) are also automatically selected for
retention of seed production. But since their fleshy fruits are
at premium, wider variation in seed set occurs in these crops.

Drastic changes in seed fertility (as well as in the chromosome system) can be tolerated when plants are *grown for their fruits* but are *maintained by vegetative propagation*. Crops in this group (the bulk of fruit trees) do not depend, under domestication, on seed fertility for their propagation; yet have to maintain the basic reproductive elements to assure the development of fruits. Moreover, in fruit crops under cultivation there is often an advantage for types in which the size and number of stones or pips has been reduced, or for the production of seedless fleshy fruits. Several solutions for the problem how to reduce seed fertility without harming fruit set evolved automatically in cultivation. They include the establishment of polyploid clones, some of them meiotically unbalanced types (e.g. tetraploidy in some apples, or triploidy in some pears or in bananas), or the incorporation of mutations inducing parthenocarpy (e.g. bananas, common fig, some pears).

In crops *grown for their vegetative parts* one is confronted with unavoidable antagonism between production of vegetative parts and development of reproductive organs. In such cases continuous selection pressure is exerted on the crop to allot more of its resources to build desired vegetative parts. In many vegetables (cabbage, lettuce, beet, carrot, celery, etc.) the development of generative parts is automatically buffered by the practice of *growing the vegetables from seed*. However, these crops are continuously selected against early bolting. Consequently they many times manifest attenuation of their vegetative growth phase. Some have been molded under cultivation to have biannual growth habit, the first year devoted to vegetative production, while initiation of flowering of seed set occurs a year later.

Crops *grown for their vegetative parts* and *maintained by vegetative propagation* exhibit the most drastic disruption of their reproductive systems and the most variable and bizarre chromosomal situations among cultivated plants. Because of their mode of propagation, the pressures exerted on such crops to increase vegetative output are rarely counterbalanced by selection to retain normal sexual reproductive functions. Tropical root and tuber crops provide us with outstanding examples for this mode of evolution. Cultivated clones of cassava, yam, sweet potato, and sugarcane frequently show drastic reduction of flowering. In some cultivated clones flowering ceases altogether, or almost altogether. When flowers develop they are frequently sterile or semi-sterile. Also chromosomally, many of these crops are highly polymorphic and frequently contain clones with different levels of polyploidy or variable, high aneuploid chromosome numbers. Clones may also contain $3x$, $5x$ or even higher polyploid levels of meiotically unbalanced chromosome complements. Thus in the yams, *Dioscorea alata* shows all chromosomal levels between $3x$ and $8x$; while in *D. esculenta* $4x$, $6x$, $9x$ and $10x$ forms are known. Sugarcanes confront us with even a more complex chromosomal picture. Cultivated clones in this crop are all highly polyploid and frequently aneuploid. Modern forms range

from $2n = 100$ to $2n = 125$ chromosomes. Older cultivars vary
from $2n = 80$ to $2n = 124$ (Simmonds 1976). Another feature of
sugarcane as well as many other vegetatively propagated crops
grown for their vegetative parts, is the frequent origin of clones
by means of distant interspecific hybridization. Since they do
not have to pass through the sieve of sexual reproduction, such
sterile or almost sterile and meiotically largely unbalanced
hybrids are easily retained under cultivation.

Many ornamental plants *grown for their flowers* and
vegetatively propagated, show close similarity to root and
tuber crops in their mode of evolution under domestication. Also
in clones of such plants (e.g. Hyacinths, Narcissus, Tulips) one
finds frequent cases of sterility, initiation of polyploidy and
unbalanced chromosomal situations. Many times cultivars of
vegetatively propagated ornamental plants are of hybrid origin,
and represent derivatives of interspecific crosses. The
ornamental plants differ, of course, from the previous group by
the fact that they are continuously selected by the grower for
effective production of flowers. But being vegetatively
propagated, these plants tolerate drastic changes in flower
structure, high levels of sterility and variable post-flowering
breakdowns. In fact, unbalanced meiotic chromosomal situations
and sterility in ornamental plants are frequently automatically
selected for. Without pollination and fertilization flowers
frequently keep longer.

THE IMPACT OF REAPING AND SOWING

Traditonal grain agriculture is based on the practice of
sowing the crop in the tilled field, *reaping* the reproductive
parts in the mature plants and *threshing out* the grains.
Basically this practice did not change much in modern, mechanized
agriculture. Reaping and sowing automatically initiate selection
towards the following changes in plants grown for their grain
(Harlan *et al.* 1973), setting them apart from their wild
progenitors:

First, there is an automatic selection for the retainment of
mature seed on the mother plant, i.e., for the breakdown of the
wild mode of seed dispersal. Most conspicuous is the shift from
shattering spikes or panicles in wild cereals to the nonshattering
condition in their cultivated counterparts, and the evolvement of
nondehiscent pods and capsules in legumes and other grain crops.
Retainment of seed on the mature plants means also a drastic
relaxation of the pressure of normalizing selection which under
wild conditions keeps the elaborate structure and function of the
seed dispersal units in strict order.

Second, the various mechanisms of inhibition of seed
germination, characteristic to wild plants, are selected against.
This results in breakdown of wild-type regulation of germination
in place and time.

Third, there is a stress to develop forms with erect habit, and to reduce the growth of side tillers or branches. In cases of biennial or perennial wild stocks, there is also an automatic selection for a shift to annuality. Finally, because grain yields are important to the grower, unconscious selection operates also to increase the number of fertile ovules in the reproductive parts of the crops (e.g. addition of fertile flowers to wheat spikelets, or development of fertile lateral spikelets in barley).

ORCHARD CULTIVATION AND ITS CONSEQUENCES

As already stated, most fruit trees under cultivation are derived from allogamic wild progenitors in which cross-pollination is maintained either by self-incompatibility or by dioecy. Because of this background, the shift from sexual reproduction (in the wild) to planting of vegetatively propagated clones (in the orchard) introduces serious limitations on fruiting. Cultivation of a single self-incompatible clone, or alternatively a female clone (or clones), would not bring about fruit set. Several agronomic devices assuring fruit set in the orchard have been empirically adopted. They are accompanied by unconscious selection for several types of mutations that partially resolved the restrictions set by self-incompatibility and sex determination (Zohary and Spiegel-Roy 1975).

In *self-incompatible fruit trees* (e.g. apple, pear, olive) planters very likely had realized, already at the start of horticulture, that to obtain satisfactory fruit set it is necessary to plant together two or more synchronously flowering clones. The traditional cultivation of such fruit trees is based on mixed planting. Two additional solutions are the outcome of automatic selection. In some fruit trees we find mutations that brokedown self-incompatibility (e.g. some almonds, $4x$ cherries); or at least rendered this system "leaky" (e.g. some olives, plums, apples and pears). In still other varieties of fruit trees derived from self-incompatible stocks, pollination was rendered unnecessary for fruit set by incorporation of mutations confering parthenocarpy (e.g. some pears, pineapple).

In *dioecious fruit trees* under domestication, fruit set in the plantation is safeguarded in parallel ways. In several crops, some male individuals are planted together with the female clones (e.g. pistachio, papaya); or natural pollination is replaced by artificial pollination (e.g. date palm, Smyrna type figs). Also in dioecious fruit crops unconscious selection added two solutions: A genetic shift from dioecy in the wild to hermaphroditism in cultivation (e.g. grape); or the replacement of pollination by parthenocarpy (e.g. common fig, orynth type grapes).

CONCLUSIONS

This paper stresses the fact that evolution of plants under domestication was not a uniform process. Different crops (or groups of crops) have had different and contrasting wild backgrounds and have been molded under cultivation in entirely different ways.

The examination reveals that variation and evolution in crops have largely been determined by the following agronomic practices:

(1) The mode of maintenance of the crop by the grower: whether by planting of seed (sexual reproduction) or vegetative propagation.

(2) The parts of the crop exploited: whether vegetative (roots, corms, leaves, etc.) or reproductive (flowers, fruits, seed). These alternative practices lead automatically to contrasting types of selection, and resulted in widely different modes of divergence under domestication.

Evolution under domestication is based on genetic systems that are rather exceptional in the flowering plants. In seed crops self-pollination is the rule, both for the cultigens and for their wild progenitors. In tuber crops and fruit trees domestication brought about first of all a shift from cross-pollination to vegetative propagation. Biotaxonomists dealing with cultivated crops should therefore be aware of the fact that they are largely concerned with the complexity of: (a) Aggregates of self-pollinated annuals (in most grain crops and in many vegetables) and (b) The bizarre morphological and cytological variation characteristic of agamic complexes (in fruit trees, root and tuber crops, many ornamentals). "Cultivars" in each category have totally different genetic make up; they have evolved in widely different ways!

REFERENCES

Harlan, J.R. 1975. *Crops and Man*. Am. Soc. Agron. Madison, Wisconsin. pp. 107–110.

Harlan, J.R., deWet, J.M.J. and Price, E.G. 1973. Comparative evolution of cereals. *Evolution* 27: 311–325.

Simmonds, N.W. 1976. *Evolution of Crop Plants*. Longmans, London, New York.

Zohary, D. and Spiegel-Roy, P. Beginnings of fruit growing in the Old World. *Science* 187: 319–327.

Zea—A Biosystematical Odyssey

Hugh H. Iltis
Herbarium and Department of Botany
University of Wisconsin
Madison, Wisconsin, U.S.A.

John F. Doebley
Department of Genetics
North Carolina State University
Raleigh, North Carolina, U.S.A.

A. INTRODUCTION

Charles Gilly (1911-1970), onetime plant taxonomist at the New York Botanical Garden and coauthor with Wendell H. Camp of the classic *Biosystematics* of 1943, which this symposium has been convened to honor, was the first to seriously study teosinte systematics (All wild species of *Zea*, formerly placed in the genus *Euchlaena*, are commonly referred to as "teosinte".). This occurred in the mid-1940's, shortly after he returned to Iowa and its State College at Ames, and became associated with I.E. Melhus, who for long had been interested in the agronomy of this grass, which, as the closest relative of maize, was and is a center of evolutionary controversy. Though Gilly influenced later classifications (Table I), his contribution remains unacknowledged. This is not surprising, since he never published on teosinte save for a short abstract in which he hints at his taxonomic difficulties: "An examination of available herbarium specimens reveals that geographical distribution and range of variability within this group of plants are considerably greater than is indicated in the existing literature" (Gilly and Melhus 1946).

Why did he not publish more? Part of the reason was inherent in his subject. Maize, teosinte, and the related *Tripsacum* are all large, complex grasses, difficult and time-consuming to collect, hence woefully underrepresented in herbaria. "Making an accurate and complete record of a *Tripsacum* plant on an ordinary herbarium sheet is like attempting to stable a camel in a dog kennel" (Cutler and Anderson 1941). Teosintes look so much like maize that it takes an experienced collector even to find them, woven as these weedy plants so often are into the very

TABLE I. *Comparative taxonomy of Teosinte and Maize, the genus Zea**

Gilly (Unpublished thesis proposal and specimen annotations, 1948)	Wilkes (Teosinte, The Closest Relative of Maize, 1967)	Iltis and Doebley (1980) Doebley and Iltis (1980) (Taxonomy of *Zea* I and II)
Zea	*Zea*	*Zea*
Zea mays	Section *Zea*	Section *Zea*
	Zea mays	*Zea mays*
		ssp. *mays*
Euchlaena	Section *Euchlaena*	ssp. *mexicana*
Euchlaena mexicana	*Zea mexicana*	
(Incl. in Durango Phase?)	Nobogame Race	Nobogame Race
Durango Phase	(Incl. in Central Plateau Race)	(Incl. in Central Plateau Race)
Bajio Phase	Central Plateau Race	Central Plateau Race
Barranca Phase	(Incl. in Central Plateau Race?)	(Incl. in Central Plateau Race)
(Guadalajara)		
Chalco Phase	Chalco Race	Chalco Race
		ssp. *parviglumis*
Guerrero Phase	Balsas Race	var. *parviglumis*
Huehuetenango Phase	Huehuetenango Race	var. *huehuetenangensis*
		Section *Luxuriantes*
Jutiapa Phase	Guatemala Race	*Zea luxurians*
Euchlaena perennis	*Zea perennis*	*Zea perennis*
		Zea diploperennis

* Modified from Iltis and Doebley 1980.

fabric of a maize field. Thus, Gilly had hardly any herbarium
material to work with, perhaps no more than 20 collections
totaling 40 specimens. In addition, teosinte taxa are so
perplexingly similar that to tell them apart requires large
population samples, the kind of "mass collections" dear to the
hearts of Edgar Anderson (1949, 1953), Norman Fassett (1941) and,
of course, their close friend Camp. But these mass collections
had not yet been made.

Lastly, to grow teosinte as far north as Iowa is difficult, and
at best results in abnormal, strongly tillered plants (Wilkes
1967, Figs. 18–24). Inexplicably, we have not seen a single
herbarium specimen of teosinte collected by Gilly or Melhus, in
either the Herbarium of Iowa State University or any other, even
though Melhus explored for teosinte in Guatemala every year
between 1944 and 1953 and discovered many new stations (Melhus and
Chamberlain 1953). But unvouchered by specimens, most of these
remain systematically unknown (Iltis and Kolterman 1984). Thus
Gilly's classification was constructed on flimsy grounds.
Nevertheless, gifted taxonomist that he was, he put future workers
on the right track, even though they knew of his taxonomy only
from annotations, written on herbarium sheets in a neat,
calligraphic hand. Curiously, when he later collaborated with a
Mexican maize specialist on a distribution map of teosinte, he
ignored his own classification. Melhus must not have been pleased
with it either, nor with Gilly for that matter, because in his
subsequent teosinte publications neither was ever mentioned.

Today, 40 years later, the biosystematics of teosinte and maize
are well understood, and allow us as marvelous a view of evolution
as one may wish. For, next to *Drosophila*, there is hardly
another genus that has ever been studied as intensively, or with
so much success, as *Zea*.

B. MAIZE AND TEOSINTE AS OBJECTS OF CLASSIFICATION

Zea biosystematics has followed three themes: (1) the
gradual nomenclatural movement of the wild teosintes closer to
cultivates *Zea mays*, the two originally classified into
separate genera placed into unrelated tribes, but ending up as
species of the same genus, and eventually, *pro parte*, as
subspecies of the same species; (2) the search for a sound
classification, to tell us which taxon might be ancestral to
maize; and lastly, (3) as the central problem underlying *Zea*
systematics, the morphological evolution of that magnificently
monstrous enigmatic anomaly, the ear of maize.

I. THE EARLY HISTORY OF *ZEA*, A MONOTYPE WITHOUT RELATIVES

Maize was first noted by Europeans on November 5, 1492, when a
delegation, sent to the center of the island which now is called
Cuba to pay homage to the Great Khan of China by Columbus, who in
his geographic misconception believed that this emperor resided

there in a splendid city, returned disappointed from a poor
village without any of the hoped-for riches, but instead with
several ears of a peculiar grain called "maiz." Soon, maize was
pictured in the great herbals of the 16th century, starting with a
magnificent plate in Fuchs' *De Historia Stirpium* of 1542,
eventually to be christened *Zea mays* by Linnaeus in his
Species Plantarum of 1753 (Finan 1950). Controversy
surrounded maize from the start. Most botanists believed it to be
native to the Old World, hence Ruellius' and Fuchs' use of
"Turcicum Frumentum" (as much misnamed, and probably for the same
reasons, as our native American Thanksgiving "turkey"!), while
others favored its origin in the New World.

II. THE GENERIC PROBLEM - ENTER TEOSINTE, "GODS GRAIN"

Thus, as late as the early 1800's, people debated unendingly the
geographic origin of maize. It should therefore have been a
sensation when seeds of a teosinte, sent in 1832 from Mexico and
planted in Goettingen, grew into an unknown maize-like grass.
But, named *Euchlaena mexicana* and misplaced into the
Olyrineae, the similarity of this new genus to maize was never
even mentioned. Furthermore, in 1829, Auguste de Saint-Hilaire
had received from Brasil a peculiar maize ear with greatly
enlarged floral bracts enclosing each grain, which he named *Zea
mays* var. *tunicata;* this, he said, was wild maize, and South
America its original home. Though the peculiarities of this "pod
corn" are now known to be due to an atavistic mutation on
chromosome 4, the misconception that it played a role in the
origin of maize has now persisted for over 150 years in both
textbooks and monographs (Grant 1977; Mangelsdorf 1983).
    In 1849, teosinte was again introduced to Europe, but this
time, under the generic name *Reana*, correctly placed next to
Zea. Later, Ascherson (1875, 1876) compared the recently
introduced Guatemalan *Reana luxurians* with *Euchlaena mexicana*
and concluded not only that they were congeneric but that
Euchlaena was obviously related to and intermediate between
maize and *Tripsacum*, a small New World genus centered in
Mexico (DeWet *et al.* 1976, 1981, 1982; Doebley 1983b). At
long last, the relationship of maize to teosinte and its origin in
the New World, if not in Mexico was established. But dreams of
Asiatic ancestry continued to bedevil some biosystematists far
into the 20th century (Anderson 1945; Stonor and Anderson 1949).
(See Wilkes 1967 and Mangelsdorf 1974 for a detailed historical
account.)
    The first and foremost biosystematic fact concerning maize and
teosinte is this: they hybridize with the greatest of ease and,
despite immense structural differences in the female
inflorescences, produce small, intermediate, 4-rowed ears of
exquisite symmetrical design. But not until 1891 were these first
described, and, named *Zea canina* (n.v. "maiz de coyote,"
meaning wild maize), were then pronounced to be the long sought

ancestor of maize. But the Mexican botanist José Segura soon set
the record straight: "*Zea canina* Watson is [nothing] but the
result of hybridization of 'asese' [i.e. teosinte] with maize"
(Harshberger 1896). Ever since, *Euchlaena* and *Zea* have remained
linked in all evolutionary discussions of maize. Nevertheless,
Euchlaena continued to be accepted as an independent genus.

In the 1920's, all annual types of teosinte, as well as maize,
were shown to be diploid and to have 20 chromosomes. Only
Euchlaena perennis, a recently discovered primitive perennial
from Jalisco, Mexico, had $2n = 40$ (Fig. 1; Table I). Crosses
of maize with Mexican teosinte were often 100% fertile, those with
Guatemalan teosinte (*E. luxurians*) less so. Nevertheless, even
E. perennis could be crossed with artificially produced
tetraploid maize (Emerson and Beadle 1930, 1932). The
unencumbered hybridization of maize and teosinte and their
cytogenetic similarities had an important taxonomic consequence:
in 1942, Reeves and Mangelsdorf formally merged teosinte into *Zea*,
this now consisting of *Zea mays*, *Z. mexicana* and *Z. perennis*.

This logical disposition, while accepted by biosystematists
(Clausen *et al.* 1945; Rollins 1953; Stebbins 1956), was
uniformly rejected by agronomists (Sprague 1959; and, of all
people, Gilly and Melhus 1946), maize geneticists (Randolph 1955;
Weatherwax 1955; Anderson 1952, p. 160), and the two foremost
American agrostologists, Hitchcock and Chase (1950). This
explains in part why two previous generic mergers of teosinte and
maize, Kuntze's of 1904 and, independently, Kozh's of 1939 (see
Wilkes 1967, p. 21) were ignored, so much so that when Reeves and
Mangelsdorf made the same decision, they were unaware of either.
Evidently – genetics and easy hybridization be damned – "common
sense" dictated that teosinte, with its slender, 6- to 12-grained
ears which *look* like those of *Tripsacum*, must obviously
belong to a different genus than maize which, with its gigantic
multigrained ears, looks like nothing else.

Yet the close *genetic* affinity of teosinte and maize was
already recognized by Harshberger (1896): "that teosinte and
maize can be crossed and a fertile progeny results shows that the
two plants are united by the close and intimate bonds of kinship."
Similarly, Clausen, Keck and Hiesey (1945, p. 136) suggested
that, on cytogenetic evidence, "annual teosinte and corn should be
biosystematically regarded as members of one ecospecies."
Identical comments by Darlington (1956) and Miranda (1966)
eventually culminated in the formal "lumping" of all annual
teosintes into *Zea mays* as subspecies: *Z. mays mexicana*
(Iltis, in Galinat 1971, p. 450) and *Z. m. luxurians* (Iltis 1972).
Annual teosintes, as *species* different from maize, finally
ceased to exist.

III. THE TAXA OF TEOSINTE – ARE THEY HARD TO TELL APART!

The taxonomy of teosinte developed gradually, with *mexicana*
described in 1832, *luxurians* in 1872 and *perennis* in

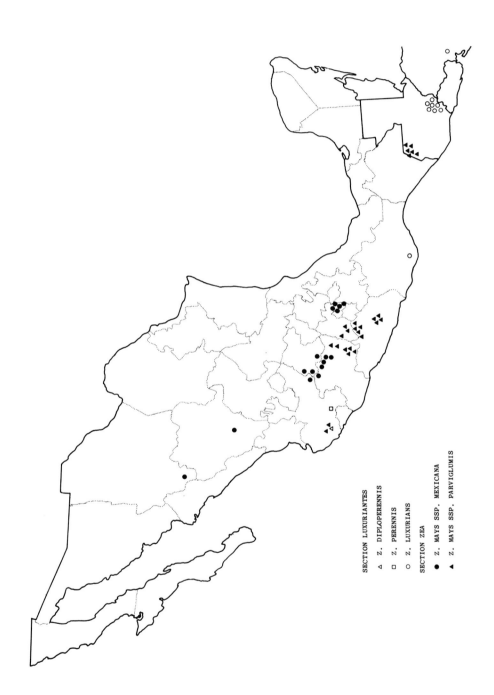

SECTION LUXURIANTES

△ Z. DIPLOPERENNIS

□ Z. PERENNIS

○ Z. LUXURIANS

SECTION ZEA

● Z. MAYS SSP. MEXICANA

▲ Z. MAYS SSP. PARVIGLUMIS

1922. But these and other epithets (cf. Iltis and Doebley 1980)
were rarely used by geneticists, who usually labelled their plants
with the name of the region or town where originally collected:
thus "Chalco," "Durango," or "Guatemala" ("Florida") teosinte.
Other informal names soon appeared: "Huehuetenango" (Kempton and
Popenoe 1937) and "Guerrero" teosinte (Gilly, in herbarium
annotations, ca. 1948), both referring to extensive and truly wild
populations, their surprisingly late discovery finally
demonstrating that teosinte is not always a weed of maize fields.
In any case, each kind was usually looked on simply as a local
phase of a taxonomic mess, collectively called "teosinte," with
insufficient differentiation to justify formal recognition.
Comparative spikelet structure remained unstudied. Hence, even
the most distinctive teosinte, the $2n = 40$ perennial, was
sometimes questioned as to its species status (Mangelsdorf and
Reeves 1939; Iltis 1972) and erroneously interpreted as an
autotetraploid of the annual *mexicana* (Randolph, in Clausen
et al. 1945), this at that time the only species known from
which it could be derived (but see Mangelsdorf 1947, p. 175, and
Galinat 1971, who predicted a diploid perennial). Though Collins
(1921, 1930) noted that many of these teosintes were
morphologically distinct, to the geneticists – and they were
mostly geneticists then, who worked with maize and teosinte –
these differences seemed trivial. To them, only interfertility
and crossing-over, not glume size or fruit-shape, were of
importance.

Yet, by the 1940's, the time was ripe for a modern teosinte
classification. That Gilly's turned out as well as it did was due
to his geographic intuition as a taxonomist, and to the soon
apparent realization that teosinte ranges were in fact all
allopatric. Thus, Gilly was the first to differentiate "Guerrero"
("Balsas") teosinte. At the same time, however, his intuition
misled him too, for his "Barranca" and "Durango phases" (cf.
Doebley 1983c) are in no way distinct from his "Bajio phase"
("Central Plateau" teosinte, *Z. m. mexicana*), demonstrating
that his classification was at best quite preliminary (Table I).

Any classification must ultimately rest on classical taxonomy,
on detailed field work and ample herbarium material, even if
augmented by modern techniques. For teosinte, this was
accomplished a decade later by Mangelsdorf's student Garrison
Wilkes (1967), an ethnobotanist-geneticist, who energetically
explored all over Mexico and Guatemala. His book-length monograph

---

FIG. 1. *Distribution of native populations of the genus*
Zea. *In* Zea *mays ssp.* mexicana, *the two northern stations
represent the Nobogame and Durango populations respectively, the
two southern clusters Races Central Plateau (western) and Chalco
(eastern). In* Zea *mays ssp.* parviglumis, *the southern Mexican
cluster represents var.* parviglumis, *and the Guatemalan one var.*
huehuetenangensis *(From Doebley and Iltis 1980, p. 989).*

is a major landmark in teosinte research. It described under
Zea mexicana six "races" of annual teosinte, but neither
named them formally nor ranked them hierarchically. Both Gilly
and Wilkes produced teosinte distribution maps [the first,
coauthored with Hernandez-X., stuck almost as an afterthought and
without comment in the middle of *The Races of Maize in
Mexico* (Wellhausen *et al.* 1951, 1952)], in which all
stations were mapped with identical symbols. While insufficient
data prevented the authors of the first map from knowing the
detailed ranges of Gilly's "phases," Wilkes (1967) was able to
show in his text that each of his "races" had not only its own
morphological tendencies but also its own allopatric range.

To the taxonomist, this was a most significant discovery. It
demonstrated clearly that "teosinte" is a good *genus*
composed of distinct wild taxa, which must have evolved their
ranges in geographic isolation (Fig. 1). This in turn implied
that the taxa were relatively ancient, at least of Pleistocene
origins, and not recent fortuitous hybrids of a "wild" *Zea mays*
with *Tripsacum* (Mangelsdorf and Reeves 1939), nor pre-Columbian
introgressants with cultivated *Zea mays* (Wilkes 1967). Lastly,
maize must have evolved recently out of this more ancient group.

IV. THE SUBGENERIC DIVISIONS OF *ZEA* - OR "COMMON SENSE"
     WILL LEAD YOU ASTRAY

While Kuntze (1904) was perceptive in combining maize and teosinte
into one genus, his division of *Zea* into sect. *Zea (Zea mays)*
and sect. *Euchlaena (Z. mexicana)* was unfortunate, if understandable. By separating the cultigen, with its gigantic,
polystichous ears and tassel spikes, from the wild teosinte, with
its slender distichous ears and tassel spikes, he utilized
differences so striking, so seemingly fundamental, that they
forcefully misled searchers for the origin of maize for decades
(Fig. 3). Thus Wilkes (1967) accepted this division, placing all
teosintes into sect. *Euchlaena* and only the cultigen into
sect. *Zea*.

But to segregate any cultigen into a major subdivision of its
own creates a real taxonomic dilemma (cf. Doebley and Iltis 1980,
p. 982). Maize, after all, is the epitomy of an human artifact:
"All the unique peculiarities of corn are concentrated in the
structure of the female inflorescence, the corn [ear], and all can
be easily interpreted as the result of human selection for human
needs for more food: for greater quantity and for greater, more
efficient harvestability" (Iltis 1971). Maize is an *obligatory*
cultigen quite unable to survive outside of agriculture, its seeds
permanently attached to the cob and permanently imprisoned by
husks. In contrast to the ears, the tassels of maize and teosinte
are nearly indistinguishable. Anderson (1944, 1951) and his
student Alava (1952) pioneered their use in maize race
classification. They also suggested that they might be useful in
uncovering relationships with the wild ancestors. Expanding on

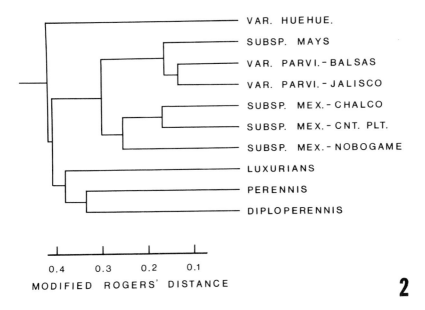

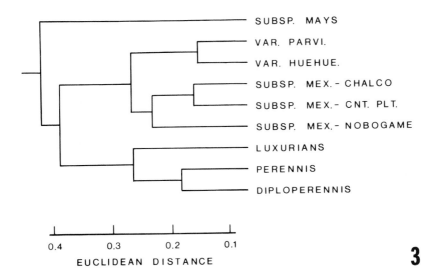

FIG. 2. *Average linkage cluster analysis of species and races of* Zea *(cf. Table I), using modified Rogers' distances, based on isozyme data (ex Doebley* et al. *1984).* FIG. 3. *Average linkage cluster analysis of species and races of* Zea *(cf. Table I) based on tassel morphology (ex Doebley* et al. *1984).*

this theme, one of us (Iltis 1974) suggested that, since human
selection focused on the edible ears, these could hardly be useful
as indicators of true phyletic relationships of maize to teosinte,
unlike the inedible tassels which, not directly subject to human
selection, survived domestication relatively unchanged. Intensive
and systematic studies of *Zea* male inflorescences (i.e. tassels)
soon showed that the genus can indeed be split along new and more
natural lines (Doebley and Iltis 1980; Doebley 1983a):

> First of all, despite its astoundingly massive female
> inflorescence, maize is here not segregated into a section
> all its own; rather, it is closely and properly aligned with
> those wild teosinte populations from which, presumably, it
> evolved into cultivation some 8,000 years ago. Secondly,
> despite its annual habit, the eastern Guatemalan *Z. luxurians*
> does not remain aligned with the other annuals, from which it
> differs in many important characters of the tassel; instead it
> is placed with the two perennial species of western Mexico, its
> morphologically close though geographically distant relatives.
> This new classification finally allows botanists and
> anthropologists to place the domestication of *Zea mays* and
> its anthropogenic antecedents in a morphologically realistic,
> phylogenetically coherent, and geographically localized
> conceptual framework (Doebley and Iltis 1980, p. 986).

Thus we described a new section (Fig. 1; Table I),
Luxuriantes, which comprised *Z. luxurians* and the two perennials
from Jalisco, Mexico, *Z. diploperennis* and *Z. perennis* (Kato
1979; Guzman 1982). The old sect. *Zea* still contained only
one species, *Zea mays*, this now however including next to
the cultivated ssp. *mays* also all Mexican and *western*
Guatemalan annual teosintes: the large-grained ssp. *mexicana*
of the Mexican plateau [with the three races "Chalco", "Central
Plateau" and "Nobogame" as recognized by Wilkes (1967)], and the
newly-named, small-grained ssp. *parviglumis* of mid-elevations
on the southern slopes [with the "Balsas" ("Guerrero") and
"Huehuetenango" teosintes of previous workers recognized as
varieties]. By placing ssp. *mays* into a position equivalent
to ssp. *mexicana* and *parviglumis*, its close relationship to
these wild taxa was emphasized. This now allowed speculation as
to its ancestry: exactly where and from what did *Zea mays*
evolve? (See note 1)

C. CURRENT RESEARCH ON ZEA BIOSYSTEMATICS

During the past five years, a number of genetic and biochemical
studies have allowed a still closer examination of evolutionary
relationships within *Zea*. This work has been carried out in
several laboratories, including those headed by Goodman and
Stuber and Timothy and Levings at North Carolina State University,
and DeWet and Harlan at the University of Illinois. Contributions
in cytology have come from McClintock at the Carnegie Institution

of Washington at Cold Spring Harbor, New York, Galinat at the University of Massachusetts, and Kato at the Colegio de Postgraduados at Chapingo, Mexico. Below we will review the bearing of this work on the taxonomy of *Zea* as defined by morphology (Fig. 3; Table I) and on the origin of maize itself.

On the basis of morphology, Doebley and Iltis (1980) split the genus into sections *Luxuriantes* and *Zea*, a division now confirmed by several independent lines of research. First, it has been shown that the DNAs of mitochondria and chloroplasts differ between the sections but are less divergent within each (Timothy *et al.* 1979, 1983; Sederoff *et al.* 1981). Second, isozyme diversity in *Zea* demonstrated that sections *Luxuriantes* and *Zea* differ markedly in their isozyme constitution, although *Z. mays* var. *huehuetenangensis* is somewhat unique and not clearly associated with either (Fig. 2; Doebley *et al.* 1984). Third, Smith and Lester (1980) demonstrated that seed proteins in sect. *Zea* are electrophoretically similar to one another but clearly distinct from those of *Z. luxurians* in sect. *Luxuriantes*.

On a morphological basis, we recognized three distinct species within sect. *Luxuriantes*. That *Z. luxurians* should be treated as a species (Bird 1978) and not as a geographic subspecies, as one of us once hastily presumed (Iltis 1972), is now biochemically confirmed. Its cytoplasmic genome is distinct from all others (Timothy *et al.* 1979) and its isozymes differ dramatically as well (Fig. 2; Doebley *et al.* 1984; Smith *et al.* 1984). The two perennials in sect. *Luxuriantes* are more closely related to one another, but clearly deserve separate species status. In our revision, we treated the diploid *Z. diploperennis* and the tetraploid *Z. perennis* as distinct species because of differences in habit, rhizome structure and adaptation, as well as geography (Iltis *et al.* 1979). Timothy and coworkers (1983) have now shown that the two perennials have distinct, though similar, mitochondrial DNAs. Doebley *et al.* (1984) found the two perennials to be similar, though distinct in isozyme patterns. Kato (pers. comm. to Goodman 1981) found the chromosome knob constitution to differ as well. Thus suggestions (Galinat and Pasupuleti 1982) that *Z. diploperennis* be submerged as a subspecies of *Z. perennis* seem now unjustified.

In sect. *Zea*, we recognized the single species *Z. mays*, with three subspecies:

(1) Subspecies *parviglumis* included both the middle altitude, small-grained Mexican annual teosinte (Balsas) and the morphologically similar populations from western Guatemala (Huehuetenango). Because of their geographic isolation and minor morphological and ecological differences, we kept these separate, as var. *parviglumis* and var. *huehuetenangensis*.

(2) Subspecies *mexicana* included the large-grained, upland Mexican annual teosintes which, in this thoroughly utilized landscape, occur almost exclusively as weeds in maize fields, where locally they produce abundant F_1 hybrids with maize and occasional backcrosses.

(3) Finally, the treatment of ssp. *mays*, the cultigen, as conspecific with the wild subspecies emphasized the Mexican annual teosintes as the ancestors of maize.

Because ssp. *parviglumis* rarely hybridizes with maize (see note 2) and has more delicate tassels, while ssp. *mexicana* commonly crosses with maize and is robust and very maize-like, one of us felt rather strongly that this subspecies must be THE ancestor of maize (Doebley and Iltis 1980; Iltis and Doebley 1980). The proximity of subsp. *mexicana* to the extensive pre-Columbian ruins in the Valley of Mexico no doubt also helped to encourage this opinion. The more cautious collaborator (Doebley 1983a) preferred to reserve judgement, a wise decision, as we shall see.

Current genetic and biochemical studies have greatly clarified evolutionary relationships within sect. *Zea*. While they confirm our morphological relationships in some ways, they also reveal dramatically the shortcomings of depending solely on morphology. One most important result concerns *Z. mays* var. *huehuetenangensis*. Fruitcase and spikelet morphology and seed protein studies showed it to be very similar to var. *parviglumis*. However, the genetic data reveal it to be quite distinct. Isoenzymatically it is distinct from all other taxa of *Zea* (Fig. 2), though closest to var. *parviglumis* (Doebley *et al.* 1984). The isozyme data also suggest that it is to some extent intermediate between the two sections, a relationship supported by chromosome knob studies (Longley 1941a,b; Kato 1976). This relationship is hard to understand, since in morphology it is practically indistinguishable from var. *parviglumis* (Fig. 3). Nevertheless, these data suggest that var. *huehuetenangensis* might well deserve recognition at a higher taxonomic rank.

The genetic and biochemical data have had their greatest impact on our understanding of the relationship between Mexican annual teosinte and maize. Many authors agreed that var. *parviglumis* is a "pure" wild teosinte, with little or no introgression from maize (Wilkes 1967, 1977; Iltis and Doebley 1980). Concurrently, many authors believed that ssp. *mexicana* (especially race Chalco) is either highly contaminated with maize or is its direct ancestor (Wilkes 1967, 1977; Bird 1978; Doebley and Iltis 1980; Iltis and Doebley 1980). These opinions, based mostly on morphology, have been challenged by isozyme studies (Doebley *et al.* 1984; Smith *et al.* 1984): ssp. *mexicana* is hardly if at all affected by maize introgression; furthermore, it is unlikely to have been the direct ancestor of maize. In sharp contrast, var. *parviglumis* is isoenzymatically indistinguishable from maize, and hence would seem to be a most likely candidate for the direct ancestor of maize. This conclusion is strengthened by cytological studies which show that abnormal chromosome 10 (type 1), occasionally found in maize, occurs in teosinte only in var. *parviglumis* (Kato 1976; McClintock *et al.* 1981).

In retrospect, others have previously singled out var.

parviglumis as ancestral to maize. Miranda (1966)
championed the semi-tropical Balsas River basin at middle
altitudes (ca. 1000 m), where wild beans and extensive stands of
this teosinte occur together, as the region of co-domestication.
Barbara McClintock also suggested this relationship in a remark at
the Harvard University Corn Conference of 1972. Iltis, who
attended, was then unable to accept this, for both morphology and
lack of hybridization (see note 2) did not suggest such a
relationship. [Curiously, in her recent book with Kato and
Blumenschein (1981), McClintock does not specify which of the
Mexican annual teosintes gave rise to maize.]

We may now hope that these demonstrations of the intimate
relationship between maize and var. *parviglumis* will
encourage an intensive re-exploration of the Balsas River basin,
and lead archaeologists to a new round of excavations and
discoveries as dramatic and significant as those at Tehuacán
(MacNeish 1981).

D. FROM TEOSINTE TO MAIZE - THE MORPHOLOGICAL CONUNDRUM

Throughout the development of *Zea* systematics, there
reverberated a bewildering theme, the question of questions of
maize evolution, the phylogenetic problem that underlied all
others, and was more deeply interesting than any other: *the
origin of the ear of maize*. What structure was ancestral?
What genetic mechanisms were involved? What forces produced this
monstrous, polystichous (radially symmetrical) ear within the
nearly 1000 species of strictly distichous (bilaterally
symmetrical) Andropogoneae? And what role might humans have had
in this most dramatic of all morphological revolutions?

Almost to a person, maize specialists struggling with this
problem had always sensed that there was something very special
about the maize ear, something mysterious, even miraculous, unique
and inexplicable (Weatherwax 1955, p. 10). Starting with Schuman
in 1904, most concluded that (a) maize evolution was extremely
rapid; (b) all characters evolved together as a syndrome; and (c)
a once useless grass (teosinte) *all at once* became useful.
But Collins (1930 p. 206) objected:

"The multiplicity of gene differences between [maize and
teosinte] seems...a serious objection to the 'origin by gross
mutation' hypothesis...It might be urged that the mutation was of
another kind, involving a simultaneous change in a large number of
genes. There is no parallel of this type of mutation in maize or
anywhere else...and to explain the origin of maize in this way
is...to fall back upon special creation."

Mangelsdorf (1947, p. 191) expressed this puzzlement best:

The differences between teosinte and maize are complex both
morphologically and genetically and it does not seem possible
that maize could have been derived from teosinte *during
domestication* [sic] by any genetic mechanism now known. If

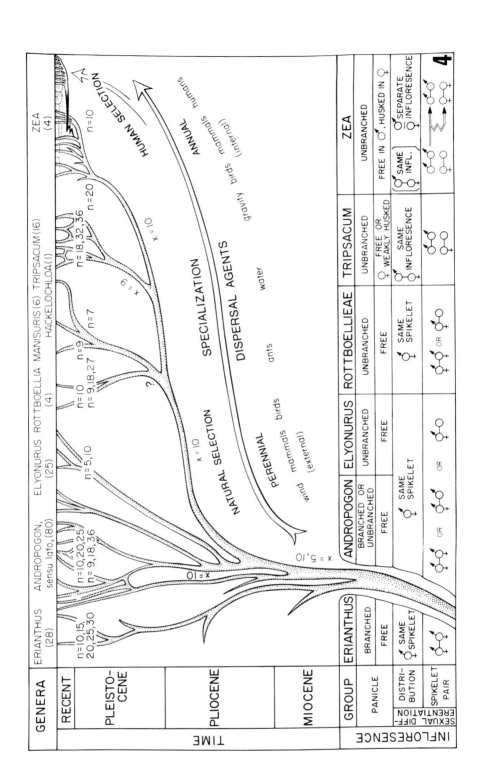

maize has originated from teosinte it represents the widest departure of a cultivated plant from its wild ancestor which still comes within man's purview. One must indeed allow a considerable period of time for its accomplishment or one must assume that cataclysmic changes, of a nature unknown, have been involved.

Unfortunately, he rejected that clairvoyant insight in the very next sentence!

a. THE TRIPARTITE HYPOTHESIS: RIGHT STRUCTURE, WRONG PARENT

Since maize is a cultigen, work on its evolution fell largely to geneticists and agronomists. These usually lacked opportunity to study teosinte in its native habitat, where it grows the way evolution "meant" it to grow (a crucial consideration!), and thus often misinterpreted its morphology. Taxonomists did not do much better. Ascherson (1880) suggested by way of an illustration of a feminized "tassel" (reproduced in Iltis 1983a) that polystichy resulted from *fusion* of four or more distichous inflorescences, a misinterpretation which, like *fasciation*, was frequently entertained from then on for decades to explain the maize ear. Hybridization was used as well, maize said to have arisen from a cross of teosinte with some unknown grass (Iltis 1911; Collins 1912, 1921, 1930), or with another, previously domesticated teosinte (Harshberger 1896). Others favored differential sterilization of perfect-flowered panicles, forgetting that teosinte was already monoecious (Montgomery 1906; Goebel 1910; East 1913; Stebbins 1982). As Kempton (1937) pointed out, nearly all evaded polystichy, the key problem, by simply pushing its origin back in time: maize evolved from an *already* polystichious "wild" species, now extinct, or from an *already* polystichous tassel spike *of maize* (sic).

This nonsolution to the problem was evidently convenient for the "Hypothesis of Remote Common Ancestry", namely that *Zea, Euchlaena* [teosinte], and *Tripsacum* represent three independently evolved parallel lines (Weatherwax 1918, 1935, 1955; Randolph 1955). The "Tripartite Hypothesis" of Mangelsdorf and Reeves (1939) similarly proposed that (1) "wild maize" was a "pod corn" and already polystichous; (2) *Tripsacum* hybridized with maize to produce "Guatemala teosinte," and this in turn with maize to give rise to the other annual teosintes; and (3) many modern maize races are the result of introgression of teosinte and/or *Tripsacum* into maize. A "hybrid" itself of various earlier bizarre phylogenies (e.g. Schiemann 1932) with one of Anderson's more far-fetched insights (Mangelsdorf and Reeves 1939, p. 172; Mangelsdorf 1983, p. 221), this theory ignored the origin of polystichy in "wild" maize, as well as the near impossibility of

FIG. 4. *The phylogeny of the genus* Zea.

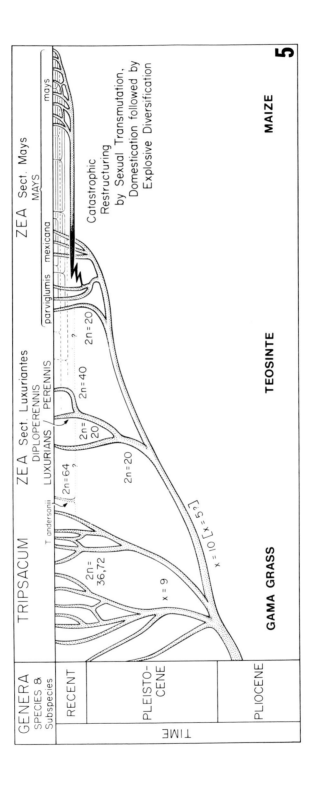

FIG. 5. *The phylogeny of Zea mays.*

crossing *Tripsacum* (2*n* = 36, 72, 90, 108) with *Zea*
(2*n* = 29, 40) (but see note 3). To its credit, the Tripartite
Hypothesis did stress the well-known homology of the polystichous
central spike of the maize tassel to the maize ear, but without
expanding this relationship to teosinte, which, because it was
thought to be a hybrid of *Tripsacum* and maize, *had* to be rejected
as a maize ancestor. Ironically, others rejected teosinte because
its central tassel spike was *not* polystichous (Kempton, 1937).
In short, this hypothesis identified, in a hazy sort of way, the
right structure (the central tassel spike), but choose the wrong
parent (the imaginary polystichous "wild maize"). This may
explain the reconstruction of an ancestral "wild maize" by Galinat
(Mangelsdorf 1958, p. 1318, and 1974, p. 180; Mangelsdorf *et
al.* 1964, p. 542), in which a small, 8-rowed ear, borne
terminally on the unbranched main stem(!), is tipped by a tassel
spike. Based on archaeological remains and morphological
abnormalities, it is essentially correct, but mostly for the wrong
reasons.

Eventually, Mangelsdorf (1974) himself had to abandon
Tripsacum as one of the parents of teosinte, only to again
champion "hybridity" and "wild maize" in the recently proposed but
taxonomically equally impossible "Wilkes" hypothesis, in which
polystichous "wild maize" crossed with *Zea diploperennis* to
give rise to the annual teosintes, and these in turn with (wild)
maize to yield the modern races of maize (Wilkes 1979; Mangelsdorf
et al. 1981; MacNeish 1982; Mangelsdorf 1983).

b. THE TEOSINTE HYPOTHESIS: RIGHT PARENT, WRONG STRUCTURE

The Tripartite Hypothesis was immediately opposed by Beadle
(1939), Langham (1940) and Longley (1941), all of whom on genetic
grounds favored teosinte as the ancestor of maize. But they were
not concerned as much with the *morphological* steps needed to
produce polystichy, though that was, and still is, the basic
question! For teosinte and maize are practically indistinguishable - except for the ears! Cover the ears, and it takes a
specialist to tell them apart. Maize is teosinte - domesticated!
On this all True Believers in the Teosinte Hypothesis agree. But
compare a massive 24-rowed, 1000-grained ear of maize to a
slender, 2-rowed, 10-grained ear of teosinte - and be astonished.
How could such a monstrous elephant evolve from such a fragile
mouse? (cf. Fig. 7)

While geneticists searched for the genes differentiating maize
and teosinte, Mangelsdorf (1947, 1974) and Galinat (1971, 1978)
defended, correctly as it turns out, a complex multi-factorial
inheritance instead of the five or six major Mendelian genes
postulated by Beadle (1972, 1978, 1980). Meanwhile, morphological
studies by Weatherwax (1935, 1955) and Galinat (1956, 1959)
derived by homology the teosinte "cupulate fruit case" [and later
(1970, 1975), the cupule of maize; but see Iltis 1983b, Table I]
from an ancient reduction series involving several genera of

Andropogoneae. Already in 1919, however, Collins not only
illustrated the uninterrupted series of hybrid intermediates by
which the ears of teosinte are connected to those of maize, but
showed how, by condensation and twisting, polystichy was derived
from distichy [He used a distichous tassel (!), but, as he stated,
for the sake of illustration only]. Nevertheless, Collins was
unconvinced that teosinte was ancestral. Thus, with one exception
(Beadle 1939), the Teosinte Hypothesis was not seriously
considered again until the 1969 Illinois Corn Conference, where it
was vigorously defended by Beadle (1972) on genetic grounds, and
by Iltis (1971, 1972) on morphological grounds. Since then widely
accepted (Galinat 1970, 1971, 1975, 1977, 1978; Kato 1976; DeWet
et al. 1972; Harlan 1975; Harlan *et al.* 1973; Flannery
1973; Beadle 1978, 1980), the Teosinte Hypothesis included the
following propositions:

　　1. *Condensation of the teosinte ear* by horizontal
"slippage" of rachids in the apical meristem early in ontogeny,
leading first to the formation of a dense but still 2-ranked ear,
and then, with still greater condensation, to the spontaneous
twisting of the inflorescence into a polystichous 4- or more-
ranked maize ear according to principles of optimal packing, when
"...a threshold in torsion created by a basal gradient of twisting
from condensation is suddenly relaxed by slippage into a higher
order of ranking." (Galinat 1975, p. 323; cf. Iltis 1983b, Fig.
3). That Collins' (1919) brilliant concepts of "condensation" and
"twisting" must be correct, cannot be denied. And that the
uninterrupted series of hybrid intermediates do indeed illustrate
this hypothetical transformation quite plausibly (even if
misleadingly) has often been noted (Weatherwax 1923);

　　2. *Reactivation of the suppressed spikelet* in the female
rachids of teosinte (each with one grain), which doubled the grain
number (to two) in those of maize, this said to be analogous to
the reactivation of the sterile spikelets of a triplet in barley
(*Hordeum*);

　　3. *Increase in apical dominance* coupled to the multiplication
of rachids, leading to suppression of all ears but one per branch,
to increases in row length and row number as well as seed size,
and, concurrently, by loss of abscission layers, to nonshattering
ears and tassels; and finally,

　　4. *Clustering of ears* preceeded by loss of branching as
the first step in maize domestication (Flannery 1973; Galinat
1975). But such "unbranched" plants, common in any teosinte
population, simply represent the response of a monopodial annual
to crowding and shading (Iltis 1983b).

　　All along, the proponents of the Teosinte Hypothesis
(including, alas, even Iltis in 1971) failed to consider the
position of the maize ear with respect to the teosinte ears.
This most important morphological fact is the real crux of maize
evolution: while *maize ears are always terminal on a stout
primary branch, ears of teosinte are always lateral to the primary
branch* (they are of course terminal, but on secondary,

tertiary, etc. branchlets). In turn, any well-developed primary
teosinte branch is always terminated by a tassel (Fig 6). Thus,
maize ears could not have come from teosinte ears, unless one
assumes unlikely morphological acrobatics. In sum, the Teosinte
Hypothesis recognized the right parent (annual teosinte), but
derived the maize ear from the wrong structure (the teosinte ear).

There are other objections to the Teosinte Hypothesis (Brown
1978): (a) the total absence of the hard teosinte fruit cases
concurrent with the oldest archaeological maize, these to be
expected in quantity had they been collected for food; (b) the
absence of intermediates (other than hybrids) between maize and
teosinte today or in the archaeological record; and (c) the
improbability of using the stone-hard teosinte fruit-cases for
food in the first place (Kempton 1937; Mangelsdorf 1983; Iltis, in
Crosswhite 1982, p. 197).

Useful in drawing attention to the derivation of maize from
teosinte, the Teosinte Hypothesis must nevertheless now be
modified, and its homologies reinterpreted. In fact, homology was
at the heart of the problem all along: "Hybrids between maize and
teosinte will always exhibit suggestive series" wrote Weatherwax
in 1923, "but, until we are more sure of the homologies between
these two genera, it is futile to expect much information from the
hybrids, for they will be speaking in a language that we cannot
understand."

c. THE CATASTROPHIC SEXUAL TRANSMUTATION THEORY (CSTT) - OLD WINE
   IN A NEW BOTTLE

Several facts of maize morphology have been well known for years:
   1. *The maize ear and the central spike of the maize tassel
are homologous* (Kellerman 1896; Montgomery 1906; Iltis 1911;
Weatherwax 1918, 1935; Kempton 1937, p. 397). Both are
polystichous and governed by identical genes (Anderson 1944;
Mangelsdorf 1945; Anderson and Brown 1948), and the central spike
of the tassel is frequently feminized into an ear by abnormal
environments (Schaffner 1927, 1930, 1935; Richey and Sprague 1932;
Heslop-Harrison 1961) or by diseases such as corn smut, *Ustilago
maydis* (Iltis 1911); in other words,
   2. *The maize ear is a feminized tassel reduced to its
terminal spike.* Starting with Boccone (in Robert Morison's
Icones of 1674), abnormal tassels have often been
illustrated (see reproductions in Iltis 1983a), and usually
consist of a polystichous ear subtended by several distichous,
male or female branches. Such abnormalities thus show:
   3. *The occurrence of polystichy and distichy in the same
inflorescence.* Though this is also obvious in every maize
tassel, its significance has apparently never been pointed out
before. The central spike, by always blooming first, assumes (if
female) a powerful nutritional priority (Iltis 1983b; Allen and
Starr 1982). Polystichy, therefore, is a genetically fixed
position effect related to apical dominance, and not due to a gene

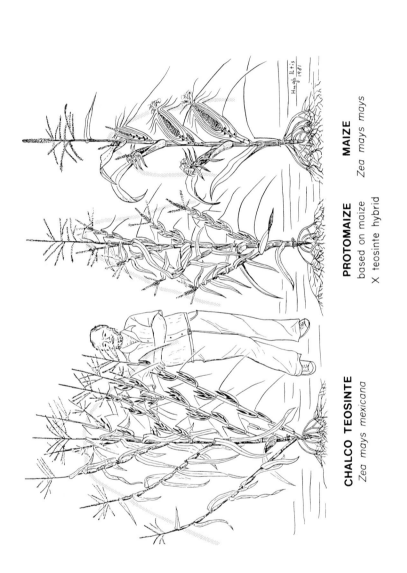

CHALCO TEOSINTE
Zea mays mexicana

PROTOMAIZE
based on maize
X teosinte hybrid

MAIZE
Zea mays mays

for polystichy as such. Since distichy vs. polystichy is one main difference between teosinte and maize (spikelets single vs. double is the other), many have searched for its Mendelian alleles for years, without success. This is hardly surprising, since, in fact, they do not as such exist.

The new synthesis assembled here from the facts cited above is called the *Catastrophic Sexual Transmutation Theory* (CSTT), and derives the maize ear not from the teosinte ear but from the teosinte *tassel* (Iltis 1979, 1983a, b; Allen and Iltis 1980; Allen and Starr 1982).

Annual teosintes are strongly monopodial, each node of the main stem producing one primary lateral branch (Fig. 6). Each primary lateral branch terminates in a tassel, as does the main stem. The many small ears, on the other hand, are all borne *laterally* to the primary branches or the main stem, at the end of clustered secondary, tertiary, etc. branchlets. A teosinte plant, then, has an outer "male zone" of tassels and an inner "female zone" of ears. The terminal male tassels bloom much in advance of the ears on the branchlets of lower rank, which follow in hierarchical sequence (Allen and Iltis 1980; Allen and Starr 1982). When, for whatever reasons, the primary branches became shortened, their tassels moved into the female hormonal zone, and their staminate spikelets changed sex and started to produce grain. The CSTT proposes then that *the maize ear is the transformed, feminized central spike of the tassel terminating each of the primary lateral branches of the ancestral annual teosinte.*

Branch-tassels in teosinte are developmentally dominant because of their apical position. Hence, once past the threshold of male sexuality, these now increasingly female "tassels" progressively suppressed the many female teosinte ears on the later-blooming

---

FIG. 6. *THE ORIGIN OF THE MAIZE EAR BY CATASTROPHIC SEXUAL TRANSMUTATION. The contraction of the teosinte branch internodes was coupled to a shift of their terminal male inflorescences (tassels) into the hormonal zone of female expression and to the suppression of the lateral female inflorescences of teosinte. The narrow line of shading indicates the threshold zone below which only female inflorescences form. Habit sketches are shown on the left side of each plant, diagrams of internode patterns etc., on the right-hand side. Note increase in apical dominance associated with the increased feminization of the apical inflorescences of the primary lateral branches. The maize plant here illustrated is in an early hypothetical stage of domestication, in which presumably the lower ears were the larger (Based in part on Galinat's splendid drawing in Mangelsdorf 1958, Fig. 9). In modern maize that situation is reversed. Recent work indicates that the direct ancestor of maize was probably ssp. parviglumis, rather than ssp. mexicana (Doebley 1983). After Iltis (1983b). Copyright 1983, AAAS, with permission.*

lateral branchlets beneath them, concurrently drawing the abundant nutrients of the whole branch system into themselves (Kellerman 1895). In short, what at the end of each branch had been a nutrient undemanding male "governor", shunting nutrients even-handedly to the many small female ears beneath it, suddenly evolved into a nutrient-demanding, apically dominant, female "dictator", arrogating all nutrients to itself. The consequent catastrophic changes in resource allocation throughout the plant thus eventually transformed each of the distichous, slender, male tassel spikes into an increasingly condensed, enlarged and efficient female nutrient sink.

Because of its terminal position, and the already great number of rachis segments, as well as its soft-glumed and free male spikelets, each tassel spike (Fig. 7d-3) was eminently preadapted to turn into a soft-glumed, free-grained and multi-grained maize ear. The often postulated mutation of the Teosinte Hypothesis to multiply the teosinte ear rachids is not needed: unlike the 6 to 12-grained teosinte ear (i.e., with 6 to 12 rachids), each tassel spike *already* has 40 or more rachids and is basically indeterminate (Bonnet 1948). Similarly, the mutation reactivating the pedicellate spikelet of each spikelet pair, which is suppressed in the teosinte ear (Fig. 7a-c), is not needed either: each rachid of the teosinte tassel *already* had two fully developed, fertile spikelets ready and waiting to be feminized (Fig. 7d-e; Montgomery 1906, Fig. 5). Thus facile comparisons (Iltis 1971) to spikelet reactivation in barley (*Hordeum*) were based on a false analogy. Finally, as seen in the mutation "tassel-seed", any grains formed on a teosinte tassel were free from the start, making them attractive to human selection.

In truth, there are no *essential* differences between maize and teosinte, only structural ones related to sexual expression and apical dominance. In other words, the basic differences between teosinte and maize are found not between their ears, but between the specialized teosinte ears on the one hand and the primitive teosinte tassels on the other, both on the same plant, both controlled by the same genes. Since these primitive attributes (paired spikelets; distichous spikes; soft glumes) are basic to all Andropogoneae, we can assume them to be well-buffered genetically by a collection of polygenes (Mangelsdorf 1947; Galinat 1971, 1978) which, accumulated over tens of millions of years, can no longer be identified individually.

Once the tassels crossed the sexual threshold, the feminization of their central spike invited nutrient overloading and condensation, and hence polystichy. Concurrently, or subsequently, the lateral tassel branches, tassel branching-space, and tassel peduncle as well as all the many teosinte ears became suppressed by indirect human selection aimed at concentrating all resources in the edible ears.

At what point did the presumably automatic consequences of the CST end and human selection take over? We do not know. Evidently, increased condensation in the ear was coadaptively linked (as effect or cause?) to a shortening of the branch internodes into a stout "shank", and this then to the transfer of the once widely spaced branch leaves (i.e., sheaths) to their new function as overlapping, ear-protecting "husks". Other innovations were selected indirectly, such as nonshattering in the tassel and polystichy in its spike, these due to human selection aimed originally at the target structure, the maize ear. Reduction of ears to two or three per plant, and synchronization of grain maturation within the ear, within the plant, and finally, race by race, within each field and region, soon followed (Harlan et al. 1973). Local introgression by teosinte in Mexico and Guatemala is said to have aided racial diversification (Wellhausen et al. 1951, 1952; Wilkes 1977), but that this was not necessary is proven by the racial explosion in Peruvian maize, where teosinte has never grown. Introgression of maize into teosinte has often been postulated, but its supposed effects appear to be due to phenotypic convergence (Doebley 1984).

Nothing in the CSTT is contradicted by either maize genetics or morphology. The CSTT is not, in fact, a radically new idea. Kellerman (1895), in her classic one page *tour-de-force*, outlined the effects of feminizing a central spike of *maize*, leading to apical dominance. In his brilliant 1906 paper, Montgomery set the stage for his flat statement of 1913 (p.18, reproduced in Iltis 1983a), that the teosinte tassel spike gave rise to the maize ear. Yet his text is so muddled that it has never been quoted, nor even cited, by anyone. The Mexican agronomist Torres (1938) was dealt a similar fate, and for similar reasons.

The CST could have been initiated repeatedly, and then within many individuals of a population, as a kind of evolutionary experiment, only to produce nonviable abnormalities, the naked seeds quickly consumed by birds or rodents. But once accidentally placed into the lap of a developing Mexican agriculture 7500 or more years ago, the same abnormalities led to macroevolutionary consequences of almost unbelievable dimensions. Threshhold selection (Mayr 1963, 1970) by the accumulation of many minor canalizing genes (Waddington 1957) quickly buffered the epigenesis of this monstrosity, which then increased its prevalence within the population and was thus saved for posterity (Iltis 1983b).

The CST may involve shifting interactions of male-inducing gibberellins produced by the seedling leaves and female-inducing cytokinins produced by the seedling roots (Chailakhyan and Khryanin 1980). The CSTT resolves beautifully the many paradoxes of archaeology (soft glumes, distichous ears, and a slender, flexible rachis in the oldest Tehuacán maize; absence of teosinte grains in cave deposits), morphology (terminal maize ears from lateral teosinte ears; male "tail" on maize ears), and genetics (absence of major genetic differences between maize and teosinte).

But as to its causes, we can only guess. Two possibilities should be considered. The first is the genetic assimilation of

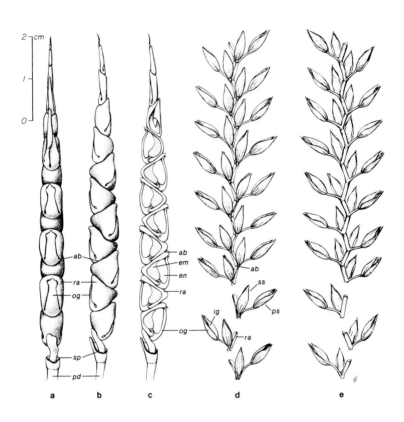

FIG. 7. *Female (ears) and male (tassel) branches of* Zea
mays *subsp.* mexicana, *race Chalco, the annual teosinte of
the Valley of Mexico: (a) side view of ear (left is back); (b)
front (abaxial) view of ear; (c) longitudinal section of b; (d)
front (abaxial) and (e) back (adaxial) view of tassel spike or
branch. In a - c (female inflorescence): ra = rachid (cupule);
og = outer glume (which, together with ra, forms the cupulate
fruitcase); em = embryo; en = endosperm; pd = peduncle;
sp = spathe scar (most of the spathe removed); ab = abscission
layer; papery inner glume, lemmas and paleas of both suppressed
and grain-forming spikelets are not shown; in a, the eighth rachid
from the base shows also a non-suppressed pedicellate spikelet, a
rare abnormality. In d - e (male inflorescence): ra = rachid;
og = outer glume; ig = inner glume; ss = sessile spikelet;
ps = pedicellate spikelet; ab = abscission layer; lowest three
rachids with their spikelets are shown disarticulated. Female
ears from Ixtapaluca (Iltis and Doebley 10b), tassels from
5.5 km N of Los Reyes (Iltis et al. 769); all drawings (by
Lucy Taylor) to same scale; note that 11 female rachids equal
about 17 male rachids, hence feminization of the latter will*

previously present, genetically determined, morpho-physiological
potentials waiting to be expressed if given the appropriate
abnormal conditions. Switches of sexual expression in cultivated
Zea are easily triggered experimentally by short days and/or
cold nights, the frequency and intensity ranging from low to high
depending on the cultivar used, the condition therefore clearly
under some genetic control (Richey and Sprague 1933; Heslop-
Harrison 1961). Hormone-releasing or -stimulating pathogens, such
as stunt viruses, viroids or corn smut (*Ustilago maydis*)
have analogous effects (Iltis 1911; cf. Crosswhite 1982). This
hypothesis is attractive because the putative mutations changing
teosinte to maize have not been found. Alternatively, the CST
could have been initiated by a mutation, either early in ontogeny,
shortening the branch internodes and placing the tassel into the
zone of female expression, or late in ontogeny, activating the
vestigial gynoecia in the male spikelets to produce "tassel-seed".
In either case, the specific genes have not been identified,
despite the immense amount of information available on *Zea*
inheritance. It will be one of the great challenges of future
work to uncover the exact genetic basis of this transmutation, and
experimentally produce maize from teosinte.

The CSTT demonstrates that tremendous morphological revolutions
can and do occur in evolution, given appropriately pre-adapted
organisms, with hardly any genetic changes and in the shortest
conceivable time. Though comparable catastrophic transmutations
must be extremely rare, they may have played an analogous
role in the evolution of cones and the cone-like flowers of the
primitive angiosperms (e.g. *Magnolia*), a mystery for now
well over a century.

The CSTT does not deny the commonly accepted tenets of
evolution by natural selection, the more or less gradual
modification of species over time (even sexual threshold selection
must have been in the Darwinian mode), but it does support, as an
extreme example, the occurrence of rapid and basic structural
change which may or may not be followed by long periods of stasis.
Analogous concepts of catastrophic morphological revolutions
include Goldschmidt's (1940) "hopeful monsters", D. Dwight Davis'
"The Panda's Thumb" (Gould 1980), and the ideas championed by
D'Arcy Thompson (1943) in *Of Growth and Form*, the final
paragraphs of which (pp. 1094-1095) seem an appropriate way to end
this biosystematic odyssey:

[To] seek for stepping stones across the gaps between [one basic
morphological type and another] is to seek in vain, for
ever...This is no argument against the theory of evolutionary

---

result, automatically, in condensation and deflection of grains
away from the rachids. It should be noted that ssp.
parviglumis *is nearly identical in structure, but slightly
smaller. From Iltis (1983b). Copyright 1983, AAAS, with
permission.*

descent. It merely states that formal resemblance...ceases in
certain other cases to serve us, because under certain
circumstances it ceases to exist. Our geometrical analogies
weigh heavily against Darwin's conception of endless small
continuous variations; they help to show that discountinuous
variations are a natural thing, that [macro-] "mutations"--or
sudden changes, greater or less--are bound to have taken place,
and new "types" to have arisen, now and then.

ACKNOWLEDGMENTS

Support by NSF Grant DEB 80-22772, Pioneer Hi-bred International
Inc., Johnston, Iowa, Harris Seed Company, Rochester, New York and
the E.K. and O.N. Allen Herbarium Fund, University of Wisconsin-
Madison is gratefully acknowledged. Publication costs supported
in part by Asgrow Seed Company (Upjohn Co.), Kalamazoo, Michigan.
We thank Lucy Taylor for her artistic assistance, Ted Cochrane for
help with Figs. 4 and 5, and Duane Kolterman for a critical
reading of the manuscript. Finally, the first author wishes to
acknowledge Hugo Iltis (1882-1952), who called his son's attention
to sexual teratologies in maize, both in publications (1910, 1911)
and in conversations during botanical excursions through the
Virginia countryside. In a very real sense, the present paper is
a belated harvest of these seminal observations.

NOTES

1. Many earlier workers also believed that maize evolved from
teosinte, so Miranda Colin, Darlington, Longley, Langham, Beadle,
Torres Barusta, Emerson, Blaringham, Montgomery, Harshberger,
Vinson and Ascherson. But their documentation was often
insufficient (what with teosinte then in many ways an unknown
entity), and, with the opposition of practically the whole maize
establishment to contend with (e.g. Anderson, Collins, Galinat,
Kempton, Mangelsdorf, Randolph, Reeves, Weatherwax), their views
did not prevail. In addition, bizarre hypotheses continued to
obfuscate the issues, then as now, and by their closely-reasoned
inner logic often overwhelmed even the most obvious facts (Wilkes
1979; Mangelsdorf *et al.* 1981; Stebbins 1982; MacNeish
1982; Mangelsdorf 1983). What is it about the "Origin of Maize"
controversy that makes otherwise competent botanists toss all
caution to the wind?
2. No one has done a detailed genetic study of why some
populations of teosinte hybridize with maize and others do not.
On the Mexican Plateau, near Mexico City or Morelia, for example,
hybridization of maize and ssp. *mexicana* is common (Wilkes
1967, 1977; pers. observ.). Similarly, ssp. *parviglumis*
var. *huehuetenangensis* hybridizes readily at San Antonio
Huista, Guatemala, where fully 8% of the plants in a maize field
were F_1 hybrids (Iltis, pers. observ., 1976; cf. Wilkes 1967,
1977). In contrast, during seven separate visits to the var.

parviglumis stands west of Teloloapan, Guerrero, we found
only four F₁ hybrids among several thousand plants. We can now
report a recently (1982) discovered stand of var.
parviglumis where hybridization does commonly occur.
Southeast of Zenzontla, at the eastern base of the Sierra de
Manantlán, Jalisco, Mexico, between La Lima (1450 m) and El
Rodeo (1200 m), we found extensive hybridizing teosinte stands,
both on the edges of, and within, the maize fields, as well as in
open moist *Ficus - Quercus* forests. Evidently,
hybridization in var. *parviglumis* depends on the genetic
nature of the local population and the local ecology.
3. There now does appear to be one remarkable, naturally-formed,
putative *Tripsacum* × *Zea* hybrid, appropriately named *T.*
andersonii Gray, a completely sterile, vegetatively
reproduced taxon with 64 chromosomes (apparently with 36 + 18 from
Tripsacum, 10 from *Zea*) which, as Guatemala Grass, is
often cultivated in Latin America for forage and for erosion
control (DeWet *et al.* 1983).

REFERENCES

NOTE: Due to space limitations, and the immensity of the maize
literature, many well-known, older references cited in the text
are not listed here but may be found in Mangelsdorf (1974),
Mangelsdorf and Reeves (1939), Galinat (1971, 1975), Wilkes
(1967), and in the cited papers of the authors (esp. Doebley
1983a and Iltis 1983b).

Allen, T.F.H. and Iltis, H.H. 1980. Overconnected collapse to
    higher levels: Urban and agricultural origins, a case study.
    In *Systems Science and Science* (G.H. Banathy, ed.).
    Proc. 24th Annu. N. A. Meet., Soc. Gen. Systems Res.).
    Louisville, Ky., pp. 96-103.
Allen, T.F.H. and Starr, T.B. 1982. *Hierarchy. Perspectives
    for Ecological Complexity*. Univ. Chicago Press, Chicago.
Anderson, E. 1944. Homologies of the ear and tassels in *Zea
    mays*. *Ann. Mo. Bot. Gard.*. 31: 325-342.
Anderson, E. 1952. *Plants, Man and Life*. Little, Brown,
    Boston.
Beadle, G.W. 1939. Teosinte and the origin of maize. *J.
    Hered.* 30: 245-247.
Beadle, G.W. 1980. The ancestry of corn. *Sci. Am.* 242:
    112-119, 162. (Note: due to editorial rearrangements, the
    figures on p. 116 were totally mixed-up and do not reflect
    Beadle's conceptions.)
Brown, W.L. 1978. Introductory remarks to the session on
    evolution. In: *Maize Breeding and Genetics* (D.B. Walden,
    ed.). Wiley, New York, pp. 87-91.
Clausen, J., Keck, D.D. and Hiesey, W.M. 1945. *Experimental
    Studies on the Nature of Species II*. Carnegie Inst.
    Washington Publ. 564, pp. 135-137.

Chailakhyan, M. Kh. and Khryanin, V.N. 1980. Hormonal regulation
of sex expression in plants. In: *Plant Growth Substances*
(F. Skoog, ed.). Springer, Berlin, pp. 331-334.

Collins, G.N. 1919. Structure of the maize ear as indicated in
Zea-Euchlaena hybrids. *J. Agric. Res. (Washington,
D.C.)* 17: 127-135.

Crosswhite, F.C. 1982. Corn (*Zea mays*) in relation to wild
relatives (Editorial summary, unsigned). *Desert Plants* 3:
193-202.

DeWet, J.M.J., Fletcher, G.B., Hilu, K.W. and Harlan, J.R. 1983.
Origin of *Tripsacum andersonii* (Gramineae). *Am. J. Bot.*
68: 269-276.

Doebley, J.F. 1983a. The maize and teosinte male inflorescence:
A numerical taxonomic study. *Ann. Mo. Bot. Gard.* 70: 32-70.

Doebley, J.F. 1983b. The taxonomy and evolution of *Tripsacum*
and teosinte, the closest relatives of maize. In *Proc. Int.
Maize Virus Disease Colloquium and Workshop* (D.T. Gordon et
al., eds.). Wooster, Ohio: Ohio Agric. Res. Devel. Center. In
press.

Doebley, J.F. 1983c. A brief note on the rediscovery of Durango
teosinte. *Maize Genet. Coop. Newslett.* 57: 127-128.

Doebley, J.F. 1984. Maize introgression into teosinte--a
reappraisal. *Ann. Mo. Bot. Gard.* 71: In press.

Doebley, J.F. and Iltis, H.H. 1980. Taxonomy of *Zea* I.
Subgeneric classification with key to taxa. *Am. J. Bot.*
67: 982-993.

Doebley, J.F., Goodman, M.M. and Stuber, C.W. 1984. Isoenzymatic
variation in *Zea* (Gramineae). *Syst. Bot.* In press.

Galinat, W.C. 1971. The origin of maize. *Annu. Rev. Genet.*
5: 447-478.

Galinat, W.C. 1975. The evolutionary emergence of maize.
Bull. Torr. Bot. Club 102: 313-324.

Galinat, W.C. 1978. The inheritance of some traits essential to
maize and teosinte. In *Maize Breeding and Genetics* (D.B.
Walden, ed.). Wiley, New York, pp. 93-111.

Gilly, C.L. and Melhus, I.E. 1946. Distribution and
variability in teosinte. *Am. J. Bot.* Suppl. 33: 235. Abstr.

Guzmán M., R. 1982. *El Teosinte en Jalisco: su Distribucion
y Ecologia.* Univ. de Guadalajara (Escuela de Agric.),
Zapopán, Mexico. Thesis.

Harlan, J.R., DeWet, J.M.J. and Price, E.J. Comparative evolution
of cereals. *Evolution* 27: 311-325.

Heslop-Harrison, J. 1961. The experimental control of sexuality
and inflorescence structure in *Zea mays* L. *Proc. Linn.
Soc. London* 172: 108-123.

Iltis, H. 1911. Über einige bei *Zea mays* L. beobachtete
Atavismen, ihre Verursachung durch den Maisbrand, *Ustilago
maydis* D.C. (Corda) und über die Stellung der Gattung
Zea im System. *Z. Indukt. Abstamm. Vererbungsl.* 5: 38-57.

Iltis, H.H. 1971. The maize mystique--A reappraisal of the origin
of corn. Corn Conf., Univ. Ill., Urbana, 1969, and Univ. Iowa,

Ames, 1970. Botany Dept., Univ. Wisconsin, Madison, Photo-offset. 4 pp. Abstr.

Iltis, H.H. 1972. The taxonomy of *Zea* (Gramineae). *Phytologia* 23: 248-249.

Iltis, H.H. 1974. Morphological-systematic studies of the male inflorescence of teosinte. N.S.F. grant proposal. Unpublished. Univ. Wisc. Herbarium, Madison.

Iltis, H.H. 1983a. The catastrophic sexual transmutation theory (CSTT): From the teosinte tassel spike to the ear of corn. *Maize Genet. Coop. Newslett.* 57: 81-91.

Iltis, H.H. 1983b. From teosinte to maize: the catastrophic sexual transmutation. *Science* 222: 886-894.

Iltis, H.H. and Doebley, J.F. 1980. Taxonomy of *Zea* (Gramineae) II. Subspecific categories in the *Zea mays* complex and a generic synopsis. *Am. J. Bot.* 67: 994-1004.

Iltis, H.H. and Doebley, J.F., Guzmán M., R. and Pazy, B. 1979. *Zea diploperennis* (Gramineae): A new teosinte from Mexico. *Science* 203: 186-188.

Iltis, H.H. and Kolterman, D.A. 1984. Need for accurate documentation of germ plasm: the case of the lost Guatemalan teosintes (*Zea*, Gramineae). *Econ. Bot.* In press.

Kato Y., A. 1976. Cytological studies of maize (*Zea mays* L.) and teosinte (*Zea mexicana* (Schrader) Kuntze) in relation to their origin and evolution. *Univ. Mass. Agric. Exp. Stn. Bull.* No. 635.

Kellerman, W.A. 1895. Primitive corn. *Meehan's Monthly* 5: 44, 53.

Kempton, J.H. 1937. Maize--Our heritage from the Indian. *Smithson. Inst. Annu. Rep. 1937*: 385-408.

MacNeish, R.S. 1981. Tehuacan's accomplishments. In *Suppl. to the Handbook of Middle American Indians* 1: *Archeology* (J.A. Sabloff, ed.). Univ. Texas Press, Austin, pp. 31-47.

Mangelsdorf, P.C. 1974. *Corn: its Origin, Evolution and Improvement*. Belknap, Harvard Univ., Cambridge, Mass.

Mangelsdorf, P.C. 1983. The mystery of corn: new perspectives. *Proc. Am. Philos. Soc.* 127: 215-246.

Mangelsdorf, P.C. and Reeves, R.G. 1939. The origin of Indian corn and its relatives. *Texas Agric. Exp. Stn. Bull.* 574: 1-315.

Mangelsdorf, P.C., Roberts, L.M. and Rogers, J.S. 1981. The probable origin of annual teosintes. *Bussey Inst., Harvard Univ.* 10: 39-68.

McClintock, B., Kato Y., A. and Blumenschein, A. 1981. *Chromosome Constitution of Races of Maize*. Colegio de Postgraduados, Chapingo, Mexico.

Miranda C., S. 1966. Discussion sobre el origen y la evolución del maíz. *Memorias Segunda Congreso Nac. Fitogenética*. Monterey, N.L., Mexico, Colegio de Postgraduados, Chapingo, Mexico, pp. 233-251.

Montgomery, E.G. 1906. What is an ear of corn? *Pop. Sci. Mon.* 68: 55-62.

Montgomery, E.G. 1913. *The Corn Crops*. Macmillan, New York.

Pasupuleti, C.V. and Galinat, W.C. 1982. *Zea diploperennis* I. Its chromosomes and comparative cytology. *J. Hered.* 73: 168–170.

Richey, F.D. and Sprague, G.F. 1932. Some factors affecting the reversal of sex expression in the tassels of maize. *Am. Nat.* 66: 433–443.

Schaffner, J.H. 1930. Sex reversal and the experimental production of neutral tassels in *Zea mays*. *Bot. Gaz. (Chicago)* 90: 279–298.

Sederoff, R.R., Levings, C.S., Timothy, D.H. and Hu, W.W.L. 1981. Evolution of DNA sequence organization in mitochondrial genomes of *Zea*. *Proc. Natl. Acad. U.S.A.* 78: 5953–5957.

Stebbins, G.L., Jr. 1982. *From Darwin to DNA, Molecules to Humanity*. Freeman, San Francisco.

Smith, J.S.C., Goodman, M.M. and Stuber, C.W. 1984. Variation within teosinte III. Numerical analysis of allozyme data. *Econ. Bot.* In press.

Smith, J.S.C. and Lester, R.N. 1980. Biochemical systematics and evolution of *Zea, Tripsacum* and related genera. *Econ. Bot.* 34: 201–218.

Timothy, D.H., Levings, C.S., Pring, D.R., Conde, M.F. and Kermicle, J.L. 1979. Organelle DNA variation and systematic relationships in the genus *Zea*: Teosinte. *Proc. Natl. Acad. Sci.* U.S.A. 76: 4220–4224.

Timothy, D.H., Levings, C.S., Hu, W.W.L. and Goodman, M.M. 1983. Plasmid-like mitochondrial DNAs in diploperennial teosinte. *Maydica*. In press.

Weatherwax, P. 1955. Structure and development of reproductive organs. In *Corn and Corn Improvement* (G.F. Sprague, ed.). Academic Press, New York, pp. 89–121.

Wilkes, H.G. 1967. *Teosinte: The Closest Relative of Maize*. The Bussey Inst., Harvard Univ., Cambridge, Mass.

Wilkes, H.G. 1977. Hybridization of maize and teosinte in Mexico and Guatemala and the improvement of maize. *Econ. Bot.* 31: 254–293.

Wilkes, H.G. 1979. Mexico and Central America as a centre for the origin of agriculture and the evolution of maize. *Crop Improv.* (India) 6: 1–18.

Biosystematics and Hybridization in Horticultural Plants

Willem A. Brandenburg
Department of Taxonomy of Cultivated Plants and Weeds
The Agricultural University
Wageningen, The Netherlands

INTRODUCTION

Camp and Gilly (1943) defined biosystematy. Lawrence (1951) renamed the discipline biosystematics. Biosystematists attempt to describe, as exactly as possible, natural relations between populations. In 1952, Anderson stated: "Most modern taxonomists do next to nothing with cultivated plants; many deliberately avoid studying or even collecting them. As a result the scientific botanical name affixed to most cultivated plants becomes just an elaborate way of saying I do not know." By now, taxonomy of cultivated plants has been recognized as a separate discipline with its own problems with regard to classification and nomenclature (Baum 1981; Brandenburg 1982; de Wet 1981; Hanelt 1972, 1973; Harlan and de Wet 1971; Hawkes 1980, 1981; Lwrence 1963; Mansfeld 1953, 1954; Danert 1962).

Heiser (1949) stated with regard to hybridization that the term hybrid is hard to define. According to Dansereau (1940) taxonomical, cytological and genetical hybrids have to be distinguished. Taxonomic hybrids are the result of crosses between plants of different taxa. In taxonomic literature, it is remarkable that hardly any distinction is made between natural and artificial hybrids.

Plants, grown under horticultural conditions, are called horticultural plants. Compared with other kinds of cultivated plants, like agricultural and forestry plants, they are grown on relatively small areas, and they are labor intensive. Horticultural plants are grown for many different reasons (vegetables, fruits, ornamentals and industrial plants).

Dealing with implications of biosystematics on hybridization in horticultural plants, the historical background of horticultural plants, the stage of domestication of horticultural plants, and

the classification and nomenclature of cultivated plants are to be
treated in this subsequential order.

HISTORICAL BACKGROUND OF HORTICULTURAL PLANTS

Nowadays, many horticultural plants are grown, some of them on a
worldwide scale, others only in certain regions. Each
horticultural crop has its own history depending on where, when,
how, and by whom it has been recognized being useful for mankind.
Knowledge about the history of horticultural plants means that
hypotheses on biosystematic relationships between horticultural
plants and their allies can be shaded. As a consequence,
effective strategies for studying these relationships may be
formulated.

     Banga (1963) studied the history of cultivated carrots in
Western Europe. He made use of herbals, paintings and old
agricultural books. He concluded that our current carotene carrot
originated in the Netherlands during the 17th Century. The carrot
populations from which the carotene carrot has been derived were
introduced by the Moors from Afghanistan and Turkey into Spain
during the 13th Century. From there they dispersed over Europe.
Banga neglected, however, the possibility of introgression from
wild or weedy carrots into cultivated carrots (Brandenburg 1981b).
Prehistoric evidence, indicating that centuries ago carrots were
growing around human settlements (Van Zeist 1968, 1980), and the
remarkable variation between forage carrot cultivars, have been
reasons for us to study biosystematic relationships between wild,
weedy and cultivated carrots in the Netherlands. Ornamentals have
been introduced as populations or sometimes even as individual
plants. Described at the place of introduction, they were
considered to be distinct species. They are, in fact, a part of
the total variation of a species or just a cultivar as Brandenburg
and Van de Vooren (1982) have shown for cultivated *Clematis*.
Knowing more details of particular introductions may be a good
starting point in determining the nature of an introduced taxon.
From research on crop history can be learned that the initial
purpose for bringing plants into cultivation, can be different
from our current use. In the late Middle Ages, European *Clematis*
species were exclusively appreciated as medicinal plants (Nylandt
1682), whereas nowadays *Clematis* is a well-known ornamental. Many
present-day ornamentals, vegetables and fruits were brought into
cultivation because of their medicinal or aromatic characters [see
herbals of e.g. Dodoens (1554) and Fuchs (1543)], rather than
because of their current use. Chang (1970) stated that the first
cultivated plants were medicinal and aromatic plants. Data from
historic and ethnobotanic research often fit this view remarkably
well (Sauer 1969).

STATE OF DOMESTICATION OF HORTICULTURAL PLANTS

According to de Wet (1981), "Plant domestication refers to changes
in adaptation that insure total fitness in habitats especially
prepared by man for his cultigens." Domestication makes plants
dependent on man for survival. They may lose the ability of
generative or vegetative propagation under natural circumstances
(de Wet and Harlan 1975).

Considering the wide range of horticultural plants, the
following questions are to be dealt with:

What, in general, is the effect of domestication on cultivated
plants?

In which stages of domestication are horticultural plants?

How are the effects of domestication taxonomically to be
interpreted?

Schwanitz (1967) has described special trends by which cultivated
plants are more or less characterized: Giant growth, and
especially allometric growth of those plant parts desired by man;
increased number of plant parts desired by man; loss of
morphological characters unfavourable for man; loss of chemical
characters unfavourable for man; loss of mechanical defenses (e.g.
spines); adaptation of habitus to anthropogene circumstances.

The more pronounced one or more of these trends are expressed,
the more domesticated is the cultivated plant. Differences
between horticultural plants of minor importance and their wild or
weedy allies are often insignificant. Mostly, such horticultural
plants have not yet lost their ability to survive under natural
circumstances. Concerning important horticultural plants,
however, domestication has often proceeded to the extent, that
only through biosystematic investigations their relationships with
their wild or weedy allies may be revealed. It is difficult to
decide upon their ancestors.

When studying the domestication of cultivated lettuce
(*Lactuca sativa*) the important question is to find out if the
present-day lettuce once originated from natural populations of
L. sativa or from populations of *L. serriola,* known as the wild
(or prickly) lettuce. Before this question can be solved,
however, biosystematic studies must show if *L. sativa* and
L. serriola are two separate species or, in fact, just one
biological species (Ferakova 1977; Lindquist 1960b; Prince and
Carter 1977; Thompson 1941, 1943; Weges 1983).

Similar questions have arisen with respect to the cultivated
cucumber (*Cucumis sativus*). Again the status of wild and
weedy allies is subject to controversy (Boos 1959; Filov 1964,
1967; Gabaev 1933; Jeffrey 1980; Meshcherov and Zal'kaln 1964;
Royle 1839; Van Leeuwen 1979; Van Leeuwen and Den Nijs 1980;
Whitaker 1973). The crucial point in these and similar cases is
how to interpret the effects of domestication in a uniform way.
Therefore, it is important to know the way in which domestication
proceeds. This depends on: (a) the duration and locality of human

influence; (b) the kind of human influence (direct or indirect)
(Schwanitz 1967); (c) the way in which plants are used by man
(Anderson 1952; Schwanitz 1967).

Lettuce and cucumber are examples of vegetables which have been
cultivated for a long time. Because they originated in different
regions (lettuce in the Middle East; cucumber in India, Himalayan-
area); Zeven and De Wet 1982), the ratio of direct/indirect human
influence may have been different for both. In addition to that,
their dispersal by man is interacted with totally different human
migrations.

Arctic bramble (*Rubus arcticus*) is a good example of
present-day domestication (Tammisola 1982). Its infraspecific
variation is now being studied in order to evaluate its potential
as a fruit crop.

CLASSIFICATION OF CULTIVATED PLANTS

In biosystematics, the biological species has to be taken as a
starting-point (Camp and Gilly 1943; Camp 1951; Harlan and de Wet
1971; Hawkes 1980, 1981; Löve 1964; Vavilov 1940; Whyte 1963).
In my opinion, the ideal situation would be achieved, if the
biological species and the taxonomical species are identical,
although the opposite is often asserted (Darlington 1940; Van
Steenis 1957). Grant (1971) surveyed different species concepts
and added the evolutionary species concept to the taxonomic and
the biological species concept (after Simpson 1961).

An evolutionary species is considered to be a unit, which is
changing by means of evolutionary processes over long time
periods. However, the biological species, being a natural unit,
under natural circumstances, of interbreeding individuals
(Dobzhansky 1937), is basic with regard to the classification of
cultivated plants (Harlan and de Wet 1971; Harlan 1975; Vavilov
1940).

A fundamental problem arises, when dealing with cultigens like
Cucumis sativus and *Lactuca sativa*. The biological species
concept of Dobzhansky cannot be fully applied, because these
species do not grow under natural conditions. Regarding cultigen
species, this means that there will be a considerable gap between
the concepts of the taxonomic and biological species. This may be
demonstrated by the fact that in cultivation many interspecific
hybrids are produced and commercially grown, whereas their
parental species have been perfectly isolated under natural
conditions.

With regard to the biological species, ecological variation
patterns, caused by adaptation to differentiating habitats are
worthwhile studying (Clausen *et al.* 1940; Clausen 1951;
Stebbins 1950; Valentine 1975) as these patterns cannot be
similarly interpreted for cultigens. This can be learned from the
many infraspecific classifications for cultivated plants, based on
agro-ecological criteria, as formulated by Vavilov (1926). These
classifications are highly artificial (e.g. *Brassica oleracea*

- Bailey 1922, 1930; Helm 1963; Schulz 1936; see also Snögerup 1980, p. 126), whereas many special ranks were necessary to cover the total variation (Jeffrey 1968; Jirasek 1966; Mansfeld 1953, 1954; Parker 1978).

Harlan and de Wet (1971) proposed the gene-pool concept. The primary gene-pool is to be considered a biological species, partly consisting of cultivated plants. The secondary gene-pool is formed by allied species, from which genes can be transferred by means of introgressive hybridization (Anderson 1949; Heiser 1949). From the tertiary to the primary gene-pool, gene transfer is only possible with the use of advanced breeding techniques.

Looking for possibilities to broaden the genetic base of cultivated plants, plant breeders have evaluated interspecific relationships between cultivated plants and their allies (Hawkes 1980; Zeven and Van Harten 1979). In this respect, Hogenboom (1973, 1975, 1979a,b, 1983a,b) introduced the term incongruity in order to approach mechanisms of interpopulational divergence. Although interpopulational relations are subject to biosystematic research, up to now, incongruity has hardly been used in biosystematics. According to Hogenboom, in pistil-pollen relationships, at least two mechanisms for nonfunctioning occur: (1) incompatibility, preventing or disturbing the functioning of the relationships, although the potential for functioning of both pollen and pistil is complete; (2) incongruity, standing for nonfunctioning due to incompleteness of the relationship; genic systems of both partners do not fully fit together. In an incongruent combination the penetration capacity of the pollen is too restricted compared with the barrier capacity of the pistil, for whatever reason. Incongruity is coherent with evolutionary divergence between populations, whereas incompatibility is a result of a positive selection pressure within populations, favoring genes that prevent self-fertilization. Nowadays, breeding programs have been developed which make use of incongruity in order to obtain F_1 hybrids (Hogenboom 1979b).

Although the concept of incongruity has proved to be useful, there are still controversies with respect to the need for distinction between incompatibility and incongruity (Hermsen and Sawicka 1979; Hogenboom 1979a; Pandey 1979). Working with *Nicotiana* Pandey (1979) stated, that S-gene polymorphism has been developed by evolutionary processes, thus leading in the first place to interspecific incompatibility, and in the second place to infraspecific incompatibility. Comparing Hogenboom's concept of incongruity with Pandey's hypothesis, Hermsen and Sawicka (1979) came to the conclusion that interspecific relationships are more simply and more widely explained by means of incongruity, than by means of complex S-gene polymorphism. Although some effects like pollen-tube growth inhibition can be caused by both incompatibility as well as incongruity, many phenomena (pistil length vs. maximum pollen-tube growth, or pollen unable to stick on alien stigma lobes) may only be explained by the concept of incongruity.

At the Department of Taxonomy of Cultivated Plants and Weeds,

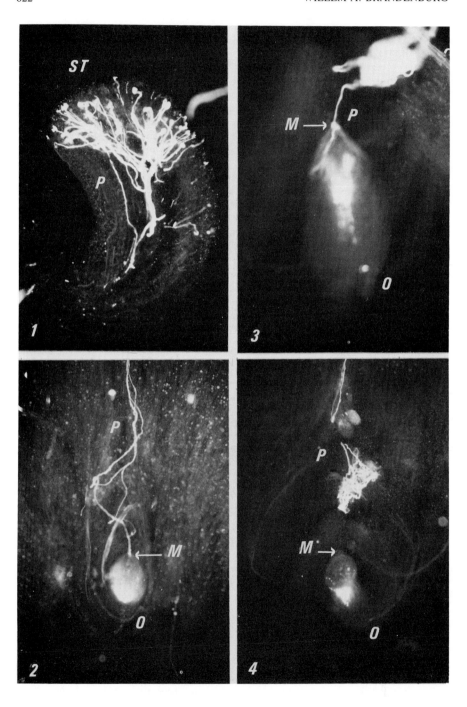

we are carrying out a diallel cross experiment between *Celmatis*
species, in order to determine the degree of relationship between
species from different sections and series. By doing this, we try
to establish a biosystematic base for the classification of
cultivated *Clematis* (Brandenburg 1981a). The results of this
experiment will explain, together with additional botanical data,
the nature and structure of *Clematis* species. This is made
possible by using the concept of incongruity. Besides seed set
resulting from interspecific crosses, cytogenetic analysis of
hybrids, the pistil-pollen interaction between partners,
agamospermy, and flower biology are being studied.

The concept of incongruity is forcing us to interpret all data
in the light of evolutionary divergence, whereas S-gene
polymorphism can never explain the coherence of these characters,
unless a large degree of pleiotropy is supposed. Consequently, by
having an approach to determine the degree of relationship, it may
be possible that *Clematis* species, up to now considered to be
distinct, have to be united into one biological species.

Similar programs are carried out with *Cucumis* (Dane-Kloene,
1976; Deakin *et al.* 1971; Den Nijs and Oost 1980; Kho *et al.*
1980; Kroon *et al.* 1979; Robinson and Kowalewski 1978) and
Lactuca (Eenink *et al.* 1982; Lindquist 1960a; Thompson 1941,
1943).

It is to be stressed that, besides genetic criteria, natural
dispersal patterns and ecological adaptation should also be taken
into account, as can be learned from biosystematic research on
Lycopersicon (Rick 1950; Rick *et al.* 1977).

Considering the systematic position of cultivated plants with
respect to their allies, a careful examination of the nature of
weedy populations needs to be done (de Wet 1981; de Wet and Harlan
1975; Harlan 1965, 1975). Depending on variation and distribution
patterns of natural populations three different evolutionary
relationships between cultivated plants and weedy relatives can be
distinguished (Pickersgill 1981): (1) weedy populations are
ancestral to the cultivated plant; (2) weedy populations are
derived from the cultivated plant; (3) cultivated plants and weedy
populations diverge simultaneously from a common ancestral
population. Interacting with introgressive hybridization
(Anderson 1952) these types of relationships are important for the

---

FIG. 1. Clematis montana *'Alba': Incompatibility reaction in
the stigma after self-pollination*. FIG. 2. C. chrysocoma X C.
montana: *congruent pistil-pollen relationship between two
Clematis species*. FIG. 3. C. montana X C. chrysocoma: *The
reciprocal of Fig. 2; also shows a congruent pistil-pollen
relationship*. FIG. 4. *An incongruent pistil-pollen
relationship between two different subpopulations of* Clematis
montana. *0 = ovule, P = pollen tubes, ST = stigma, M = micropylar
end of the ovule. Fluoresence technique modified after Kho and
Baër (1968)*.

classification of cultivated plants. Their impact on classification, however, depends on whether cultivated plants and wild or
weedy relatives are conspecific or not. In the conspecific
situation, the first two types of relationships occur. The crop-
weed complex of *Capsicum annuum* (Pickersgill 1971, 1977) is
an example in which cultivated plants originated from weedy
populations. The opposite situation has been shown by Small
(1978, 1984) for *Daucus carota*. In these and similar cases
classification of infraspecific variation needs re-examination,
because different views are obviously conflicting (Brandenburg
1983), e.g. with regard to the rank of subpsecies: different
subspecies for wild, weedy and cultivated populations (Harlan and
de Wet 1971); subspecies as an interbreeding, morphologically
more or less distinct, but geographically distinct part of a
species (Camp and Gilly 1943; Meikle 1957; Fuchs 1958).

    In the case where cultivated plants and their allies are not
conspecific, evolutionary relationships, in which cultivated
plants and weedy populations diverge simultaneously from a common
ancestral natural population, are of special importance
(Pickersgill 1981). Two different situations are then to be
distinguished: (1) domesticated populations are distinct from wild
and weedy populations, whereas wild and weedy populations are more
or less similar in adaptation and phenotype; (2) wild populations
are distinct from weedy and domesticated populations, whereas
weedy and domesticated populations show similarities in ecological
pattern and phenotype. In the first case a great range of
variation within and between wild and weedy populations will occur
because of introgressive hybridization. The second case may give
rise to mimetic weeds, whether from cultivated populations, from
hybridization between cultivated and wild populations or from
abandoned cultivars (de Wet 1981; Evans and Weir 1981; Ford-Lloyd
and Williams 1975).

    Referring to horticultural plants, a somewhat paradoxical
situation has to be faced in order to establish a sound
classification system for cultivated plants: one crop to be
assigned to more than one species (pumpkins - *Cucurbita
spp.*), vs. several crops to be assigned to only one species
(cabbages, kales, cauliflower, kohlrabi, Brussels sprouts, etc. -
Brassica oleracea). Therefore, two different starting points
have to be chosen for the classification of cultivated plants
(Brandenburg 1982; Brandenburg and Oost 1982; Brandenburg *et al.*
1982; Duyvendak *et al.* 1981: the biological species, as well as
the cultivar, "denoting an assemblage of cultivated plants which
is clearly distinguished by any characters (morphological,
physiological, cytological or others) and which, when reproduced
(sexually or asexually), retains its distinguishing characters"
(International Code of Nomenclature for Cultivated Plants, 1980).

    With regard to infraspecific variation to be assigned to
cultivated plants, there is no need to split up this part of
infraspecific variation in different, hierarchically ordered taxa,
as between cultivated plants of the same botanical taxon popula-

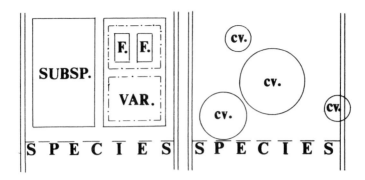

FIG. 5 (left). *Closed classification under the ICBN. Hierarchial order of ranks.* FIG. 6 (right). *Open classification of cultivars, independent of botanical classification.*

tions of wild or weedy plants may occur. On the contrary, it is necessary to group cultivars properly (Baum 1981), thus making an open classification system in which all categories not only need to be described, but also circumscribed (Brandenburg 1982).

NOMENCLATURE OF CULTIVATED PLANTS

There is an apparent need for an unequivocal nomenclature of cultivated plants, which can be easily applied in economic and legal affairs, and which, on the other hand, reflects biological affinities (Schneider 1982). Two Codes regulate the nomenclature of cultivated plants: (1) ICBN: International Code of Botanical Nomenclature (1978); (2) ICNCP: International Code of Nomenclature for Cultivated Plants (1980). The latter Code is mainly dealing with nomenclature, registration and even classification of cultivars (Brickell 1982). However, according to the ICBN, art. 28, note 1, in cases both Codes may be followed, authors are given free choice as to which Code they will apply in nomenclature of cultivated plants. Particularly, in cases of cultivated inter-specific or intergeneric hybrids, similar taxonomic treatments cannot always be compared due to the use of different nomenclature systems [e.g. *Lilium* Midcentury Hybrids (collective name in modern language - ICNCP); *Astilbe* X *arendsii* (collective name in Latin - ICBN).] Because it is difficult to define the term hybrid in taxonomy (Darlington 1937; Pactl 1952; Yeo 1981), the nomenclature of hybrid plants is subject to many controversial opinions (Grassl 1963; Little 1960; Rowley 1964; Yeo 1981). A proposal to rewrite the hybrid appendix of the ICBN entirely has been submitted by Yeo (1981) to the Sydney Botanical Congress. However, neither in this proposal, nor in any other proposal concerning nomenclature of hybrid plants, any distinction has been made in naming hybrids, which have arisen in cultivation, and wild or weedy hybrids. Because the nature of hybrid cultivated plants

is in general completely different from the nature of wild or
weedy hybrid plants, this difference should be reflected in their
nomenclature.

   The ICNCP is meant to regulate the nomenclature of cultivated
plants. Besides this, the ICNCP also tends to regulate classification and registration of cultivated plants. It therefore seems
logical, that the ICNCP (and the ICNCP only) should contain the
nomenclature rules for hybrid *cultivated* plants. Naming of *wild*
and *weedy* hybrids will remain subject to the ICBN. In this
respect, the link between the two Codes should be more clearly
stated in order to avoid confusing situations in nomenclature. In
future, more efforts have to be made in close international
cooperation to develop a uniform classification system, connected
with unequivocal nomenclature rules regarding cultivated plants.

ACKNOWLEDGMENTS

The author expresses his appreciation to Dr. Ir. N.G. Hogenboom,
Ir. E.H. Oost and Mr. J.G. van de Vooren for their helpful remarks
and stimulating discussions; to Ir. M.P.H. van Loosdrecht for
supplying Figures 1-4; to Mrs. M. de Geus for drawing Figures 5
and 6; and to Mrs. A. la Bastide and Miss C. den Hartog who did
the typing.

REFERENCES

Anderson, E. 1949. *Introgressive Hybridization*. Wiley, New York.
Anderson, E. 1952. *Plants, Man and Life*. Little Brown,
   Boston.
Bailey, L.H. 1922. The cultivated Brassicas I. *Gentes Herbarum*
   1(2): 51-108.
Bailey, L.H. 1930. The cultivated Brassicas II. *Gentes Herbarum*
   2(5): 211-267.
Banga, O. 1963. *Western Carotene Carrot, Main Types and Origin*.
   Tjeenk Willink, Zwolle.
Baum, B.R. 1981. Taxonomy of the infraspecific variability of
   cultivated plants. *Kulturpflanze* 29: 209-239.
Boos, G.V. 1959. On the classification of the cucumber. *Tr.
   Prikl. Bot. Genet. Sel.* 32: 80-86 (Russ.)
Brandenburg, W.A. 1981a. Historical background and taxonomy of
   cultivated large-flowered Clematis in Europe. *Kulturpflanze*
   29: 321-323.
Brandenburg, W.A. 1981b. Possible relationships between wild and
   cultivated carrots (*Daucus carota* L.) in the Netherlands.
   Kulturpflanze 29: 369-375.
Brandenburg, W.A. 1982. Taxonomy of cultivated plants. 21st Int.
   Hort. Congr. Vol. II, 1967. Abstr.
Brandenburg, W.A. 1983. Taxonomy of cultivated plants with regard
   to breeding value in accessions. *Genetika* Suppl. In press.
Brandenburg, W.A. and Oost, E.H. 1982. Cultivar registration and
   problems in cultivar grouping. 21st Int. Hort. Congr. Vol. II,

1971. Abstr.

Brandenburg, W.A., Oost, E.H. and Van de Vooren, J.G. 1982. Taxonomic aspects of the germplasm conservation of cross pollinated cultivated plants. In *Seed Regeneration in Cross-Pollinated Species*. (E. Porceddu and G. Jenkins, eds.), Balkema, Rotterdam. pp. 33-41.

Brandenburg, W.A. and Van de Vooren, J.G. 1982. Taxonomy and history of large-flowered cultivated Clematis in Europe. 21st Int. Hort. Congr. Vol. II, 1976. Abstr.

Brickell, C.D. 1982. The International Code of Nomenclature for Cultivated Plants and the importance of international registration of cultivar names. Proc. 21st Int. Hort. Congr. 2: 1077-1088.

Camp, W.H. 1951. Biosystematy. *Brittonia* 7: 113-127.

Camp, W.H. and Gilly, C.L. 1943. The structure and origin of species. *Brittonia* 4: 323-385.

Chang, K.C. 1970. The beginnings of agriculture in the Far East. *Antiquity* 44: 175-185.

Clausen, J. 1951. Stages in the Evolution of Plant Species. Cornell Univ., Ithica, N.Y.

Clausen, J., Keck, D.D. and Hiesey, W.M. 1940. Experimental studies on the nature of species. I. The effect of varied environment on western North American plants. Carnegie Inst. Washington Publ. 520.

Dane-Kloene, F. 1976. Evolutionary studies in the genus *Cucumis*. Thesis, Colorado Univ.

Danert, S. 1962. Uber Gliederungsprobleme bei Kulturpflanzen. *Kulturpflanze* 10: 350-358.

Dansereau, P. 1940. Etudes sur les hybrides de Cistes. *Ann. Epiphyt. Phytogenet.* 6: 7-26.

Darlington, C.D. 1937. What is a hybrid? *J. Hered.* 28: 308.

Darlington, C.D. 1940. Taxonomic species and genetic systems. In The New Systematics. (J. Huxley, ed.), Clarendon, Oxford. pp. 137-160.

Deakin, J.R., Bohn, G.W. and Whitaker, T.W. 1971. Interspecific hybridization in the genus *Cucumis*. *Econ. Bot.* 25: 195-211.

Den Jijs, A.P.M. and Oost, E.H. 1980. Effect of mentor pollen on pistil pollen incongruities among species of *Cucumis* L. *Euphytica* 29: 267-272.

De Wet, J.M.J. 1981. Species concepts and systematics of domesticated cereals. *Kulturpflanze* 29: 177-198.

De Wet, J.M.J. and Harlan, J.R. 1975. Weeds and domesticates: Evolution in the man-made habitat. *Econ. Bot.* 29: 99-107.

Dobzhansky, Th. 1937. *Genetics and the Origin of Species*. Columbia Univ. Press, New York.

Dodoens, R. 1554. *Cruydeboeck*. Antwerpen.

Duyvendak, R., Luesink, B. and Vos, H. 1981. Delimitation of taxa and cultivars of red fescue (*Festuca rubra* L. *sensu lato*). *Rasen-Turf-Gazon* 3: 53-62.

Eenink, A.H., Groenwold, R. and Dieleman, F.H. 1982. Resistance of lettuce (*Lactuca*) to the leaf aphid *Nasonovia ribes-*

nigri 1. Transfer of resistance from *L. virosa* to *L. sativa*
by interspecific crosses and selectioning of resistant breeding
lines. *Euphytica* 31: 291-300.

Evans, E. and Weir, J. 1981. The evolution of weed beet in sugar
beet crops. *Kulturpflanze* 29: 301-310.

Feráková, V. 1977. The genus *Lactuca* in Europe. Univ.
Komenshého, Bratislava.

Filov, A.I. 1964. The wild parent of the cucumber. *Bull.*
Glav. Bot. Sad. 52: 105-1061.

Filov, A.I. 1967. The origin of the cucumber (*Cucumis sativus* L.)
and its ecological evolution. *Bull. Glav. Bot. Sad.*
66: 31-37.

Ford-Lloyd, B.V. and Williams, J.T. 1975. A revision of Beta
section Vulgares (Chenopodiaceae), with new light on the origin
of cultivated beets. *Bot. J. Linn. Soc.* 71: 89-102.

Fuchs, H.P. 1958. Historische Bemerkungen zum Begriff der
Subspezies. *Taxon* 7: 44-52.

Fuchs, L. 1543. *Neu Kreuterbuch.* Basel.

Gabaev, S.G. 1933. Systematische Untersuchungen an Gurkenarten
und Varietäten. *Angew. Bot.* 15: 290-307. Grant, V. 1971.
Plant Speciation. Columbia Univ. Press, New York.

Grassl, C.O. 1963. Proposals for modernizing the international
rules of nomenclature for hybrids. *Taxon* 12: 337-347.

Hanelt, P. 1972. Die infraspezifische Variabilität von *Vicia*
faba L. und ihre Gliederung. *Kulturpflanze* 20: 75-128.

Hanelt, P. 1973. Merkmalsvariabilität bei *Vicia faba* L.I.
Künstliche morfologische Systeme bei Kulturpflanzen-Arten.
Kulturpflanze 21: 55-60.

Harlan, J.R. 1965. The possible role of weed races in the
evolution of cultivated plants. *Euphytica* 14: 173-176.

Harlan, J.R. 1975. *Crops and Man.* Crop Sci. Soc. Am.,
Madison, Wis.

Harlan, J.R. and De Wet, J.M.J. 1971. Towards a rational taxonomy
of cultivated plants. *Taxon* 20: 509-517.

Hawkes, J.G. 1980. The taxonomy of cultivated plants and its
importance in plant breeding research. In *Perspectives in*
World Agriculture. CAB, Farnham Royal, pp. 49-66.

Hawkes, J.G. 1981. Biosystematic studies of cultivated plants as
an aid to breeding research and plant breeding.
Kulturpflanze 29: 327-336.

Heiser, C.B. 1949. Natural hybridization with particular
reference to introgression. *Bot. Rev.* 15: 645-687.

Helm, J. 1963. Morphologisch-taxonomische Gliederung der
Kultursippen von *Brassica oleracea* L. *Kulturpflanze* 11: 92-210.

Hermsen, J.G.Th. and Sawicka, E. 1979. Incompatibility and
incongruity in tuberbearing *Solanum* species. In *The Biology*
and Taxonomy of the Solanaceae. (J.G. Hawkes, R.N. Lester and
A.D. Skelding, eds.). Academic Press, London, pp. 445-453.

Hogenboom, N.G. 1973. A model for incongruity in intimate partner
relationships. *Euphytica* 22: 219-233.

Hogenboom, N.G. 1975. Incompatibility and incongruity: two

different mechanisms for the non-functioning of intimate partner relationships. *Proc. R. Soc. London Ser. B.* 188: 361-375.

Hogenboom, N.G. 1979a. Incompatibility and incongruity in *Lycopersicon* In *The Biology and Taxonomy of the Solanaceae* (J.G. Hawkes, R.N. Lester and A.D. Skelding, eds.). Academic Press, London. pp. 435-444.

Hogenboom, N.G. 1979b. Exploitation of incongruity, a new tool for hybrid seed production. In *Broadening the Genetic Base of Crops* (A.C. Zeven and A.M. Van Harten, eds.) Pudoc, Wageningen. pp. 299-309.

Hogenboom, N.G. 1983a. Bridging a gap between related fields of research: Pistil-pollen relationships and the distinction between incompatibility and incongruity in nonfunctioning host-parasite relationships. *Phytopathology* 73: 381-383.

Hogenboom, N.G. 1983b. Incongruity: Nonfunctioning of intercellular and intracellular partner relationships through nonmatching information. In *Cellular Interactions* (J. Heslop-Harrison and H.F. Linskens, eds.). *Encyclopedia of Plant Physiology*, New Series, Springer Verlag. In press.

International Code of Botanical Nomenclature 1978. *Regnum Veg.* 97.

International Code of Nomenclature for Cultivated Plants 1980. *Regnum Veg.* 104.

Jeffrey, C. 1968. Systematic categories for cultivated plants. *Taxon* 17: 109-114.

Jeffrey, C. 1980. Further notes on Cucurbitaceae V. The Cucurbitaceae of the Indian subcontinent. *Kew Bull.* 34: 789-809.

Jirasek, V. 1966. The systematics of cultivated plants and their taxonomic categories. *Preslia* 38: 267-284.

Kho, Y.O. and Baër, J. 1968. Observing pollen tubes by means of fluorescence. *Euphytica* 17: 298-302.

Kho, Y.O., Den Nijs, A.P.M. and Franken, J. 1980. Interspecific hybridization in *Cucumis* L. II. The crossability of species, an investigation of in vivo pollen tube growth and seed set. *Euphytica* 29: 661-672.

Kroon, G.H., Custers, J.B.M., Kho, Y.O., Den Nijs, A.P.M. and Varekamp, H.Q. 1979. Interspecific hybridization in Cucumis L. I. Need for genetic variation, biosystematic relations and possibilities to overcome crossability barriers. *Euphytica* 28: 723-728.

Lawrence, G.H.M. 1951. *Taxonomy of Vascular Plants*. MacMillan, New York.

Lawrence, G.H.M. 1963. The taxonomy of cultivated plants. In *Vistas in Botany* (W.B. Turrill, ed.). Pergamon, Oxford. 2: 199-214.

Lindquist, K. 1960a. Cytogenetic studies in the serriola group of *Lactuca*. *Hereditas* 46: 75-151.

Lindquist, K. 1960b. On the origin of cultivated lettuce. *Hereditas* 46: 319-350.

Little, E.L. 1960. Designating hybrid forest trees. *Taxon*

9: 225–231.

Löve, A. 1964. The biological species concept and its
evolutionary structure. *Taxon* 13: 33–45.

Mansfeld, R. 1953. Zur allgemeinen Systematik der Kulturpflanzen
I. *Kulturpflanze* 1: 138–155.

Mansfeld, R. 1954. Zur allgemeinen Systematik der Kulturpflanzen
II. *Kulturpflanze* 2: 130–142.

Meikle, R.D. 1957. What is the subspecies? *Taxon* 6:   102–105.

Meshcherov, E.T. and Zal'kaln, A.A. 1964. The problem of the
ancestors and relations of *Cucumis sativus* L. *Nausch Tr.
Maikopophyt. Sta.* VIR 1: 62–67 (Russ.).

Nylandt, P. 1682. *De Nederlandtse Herbarius of Kruidt-boek.*
Amsterdam.

Paclt, J. 1952. Hybrids and taxonomy. *Taxon* 1: 117–118.

Pandey, K.K. 1979. The genus *Nicotiana*: evolution of
incompatibility in flowering plants. In *The Biology and
Taxonomy of the Solanaceae* (J.G. Hawkes, R.N. Lester and A.D.
Skelding, eds.). Academic Press, London. pp. 421–434.

Parker, P.F. 1978. The classification of crop plants. In
Essays in Plant Taxonomy (H.E. Street, ed.). Academic Press,
London. pp. 97–124.

Pickersgill, B. 1971. Relationships between weedy and cultivated
forms in some species of chili peppers (genus *Capsicum*).
Evolution 25: 683–691.

Pickersgill, B. 1977. Chromosomes and evolution in

Capsicum. In Capsicum 77 (E. Packard, ed.). 3me Congr.
EUCARPIA, Montfavet-Avignon. pp. 27–37.

Pickersgill, B. 1981. Biosystematics of crop-weed complexes.
Kulturpflanze 29: 377–388.

Prince, S.D. and Carter, R.N. 1977. Prickly lettuce (*Lactuca
serriola* L.) in Britain. *Watsonia* 11: 331–338.

Rick, C.M. 1950. Pollination relations of Lycopersicon esculentum
in native and foreign regions. *Evolution* 4: 110–122.

Rick, C.M., Fobes, J.F. and Holle, M. 1977. Genetic Variation in
Lycopersicon pimpinellifolium: Evidence of Evolutionary Change
in Mating Systems. *Pl. Syst. Evol.* 127: 139–170.

Robinson, R.W. and Kowalewski, E. 1978. Interspecific
hybridization of Cucumis. CGC Rep. 1: 40.

Rowley, G.D. 1964. The naming of hybrids (2). *Taxon* 13: 64–65.

Royle, J.F. 1839. *Illustrations of the Botany and Other Branches
of Natural History of the Himalayan Mountains and of the Flora
of Cashmere.* London.

Sauer, C.O. 1969. *Seeds, Spades, Hearths and Herds. 2nd ed.*
MIT Press, Cambridge, Mass.

Schneider, F. 1982. Nomenclature and common names. 21st Int.
Hort. Congr. 1968, 2: Abstr.

Schulz, O.E. 1936. Brassica. In *Die Naturliche
Pflanzenfamilien* (A. Engler and K. Prantl eds.). Leipzich.
17: 321–330.

Simpson, G.G. 1961. *Principles of Animal Taxonomy*.
Columbia Univ. Press, New York.

Small, E. 1978. A numerical taxonomic analysis of the *Daucus carota* complex. *Can. J. Bot.* 56: 248-276.

Small, E. 1984. Hybridization in the domesticated-weed-wild complex. In *Plant Biosystematics* (W.F. Grant ed.).
Academic Press, Toronto.

Snogerup, S. 1980. The wild forms of the *Brassica oleracea* group (2n = 18) and their possible relations to the cultivated ones. In *Brassica Crops and Wild Allies* (S. Tsunoda, K. Hinata and C. Gomez-Campo, eds.). Japan Sci. Soc. Press, Tokyo. pp. 121-132.

Stebbins, G.L. 1950. *Variation and Evolution in Plants*.
Columbia Univ. Press, New York.

Tammisola, J. 1982. Principles of genetic sampling and conservation in perennial species - A case study in arctic bramble (*Rubus arcticus* L.). In *Seed Regeneration in Cross-Pollinated Plants*. (E. Porceddu and G. Jenkins, eds.). Balkema, Rotterdam. pp. 165-189.

Thompson, R.C. 1941. Interspecific genetic relationships in *Lactuca*. *J. Agr. Res.* 63: 91-107.

Thompson, R.C. 1943. Further studies on interspecific genetic relationships in *Lactuca*. J. Agric. Res. 66: 41-48.

Valentine, D.H. 1975. The taxonomic treatment of polymorphic variation. *Watsonia* 10: 385-390.

Van Leeuwen, L. 1979. Een bijdrage tot de identificatie van wilde Cucumis L. species. (Stud. Res. Rep.). Vakgroep Taxonomie van Cultuurgewassen en -begeleiders, Landbouwhogeschool, Wageningen.

Van Leeuwen, L. and Den Nijs, A.P.M. 1980. Problems with the identification of Cucumis L. taxa. CGC Rep. 3: 55-60.

Van Steenis, C.G.G.J. 1958. Specific and infra specific delimitation, *Flora Malesiana,* ser. 1, 5(3): 167-234.

Van Zeist, W. 1968-1970. Prehistoric and early historic food plants in the Netherlands. *Palaeohistoria* 14: 41-173.

Van Zeist, W. 1980. Prehistorische Cultuurplanten. *Intermediair* 16(3): 35-47.

Vavilov, N.I. 1926. Studies on the origin of cultivated plants. *Bull. Appl. Bot.* 26(2): 248 p.

Vavilov, N.I. 1940. The new systematics of cultivated plants. In *The New Systematics* (J. Huxley, ed.). Clarendon Press, Oxford. pp. 549-566.

Weges, R. 1983. Taxonomisch onderzoek aan cultuursla en haar wilde verwanten. (Stud. Res. Rep.). Vakgroep Taxonomie van Cultuurgewassen en -begeleiders, Landbouwhogeschool, Wageningen.

Whitaker, T.W. 1973. Cucurbits of India. Proc. Trop. Reg. Am. Soc. Hort. Sci. 17: 255-259.

Whyte, R.O. 1963. The evolution and domestication. *Span* 6: 6-10.

Yeo, P.F. 1981. Proposal no. 164 to the Sydney Congress affecting the rules of nomenclature for hybrids. *Taxon* 30: 260-267.

Zeven, A.C. and de Wet, J.M.J. 1982. *Dictionary of Cultivated Plants and their Regions of Diversity - Excluding Most Ornamentals, Forest Trees and Lower Plants.* 2nd ed. Pudoc, Wageningen.

Zeven, A.C. and Van Harten, A.M. 1979. *Broadening the Genetic Base of Crops.* Pudoc, Wageningen.

Biosystematics and Conservation

David Bramwell

Jardin Botanico "Viera y Clavijo"
Las palmas de Gran Canaria
Canary Islands, Spain

INTRODUCTION

Rather than attempt to use a narrow definition of Biosystematics
for the purpose of this review, I prefer to permit myself the
luxury of placing the widest possible interpretation on the term,
allowing for both those who consider the classification of the
end-products of evolution to be the ultimate goal of
biosystematics and those who are more preoccupied with the
processes of microevolution and the origin of plant diversity.
Mostly, however, I intend to concern myself with the interface
between the information obtained by biosystematists for whatever
purpose and its relevance to the needs of conservationists.

It has been estimated that the number of species threatened
with extinction by other than natural processes is in the region
of 20,000. It is difficult, however, to get a realistic idea of
the accuracy of this estimate simply because for vast areas of the
world we do not have sufficient data with which to work and it has
been said that most of the World's Floras are so badly known that
is is almost impossible to assess their conservation requirements
at species level. As Heywood (1971) stated "we can scarcely
conserve what we do not know."

In the same paper on the conservation of the European flora,
which must be one of the better known floras, Heywood suggests
that of the 16-19000 species found in Europe and in the wider
Medditerranean region, probably less than a couple of thousand
have ever been studied in any detail, that is sampled
cytologically on a population basis rather than as single
individuals or studied experimentally or biosystematically and
even when North America is included, well-studied species on a
world scale can only number a few thousand. If we consider only
the threatened species of the world, those 20,000, there really

must be no more than a few hundred that have ever been studied
biosystematically and most of them have scarcely been studied at
all. For many of the known species there is only an original
description and in some cases a little more information in local
basic floras or floristic catalogues. So despite over 200 years
of intensive exploration and study of the world's plants we still
have an extraordinarily long path to tread even before we know
what grows where, without even worrying about why or how?

Coming back to my observation that it is difficult to get even
an approximate figure for the number of threatened plant species
on the planet I am sure that the estimate of 20,000 originally
produced by Melville is very low indeed. I base this observation
on the fact that Given (1976) warns that some 350 species are
endangered in New Zealand alone, my own estimate for the Canary
Islands is of a similar magnitude and Woolliams (1976) has
suggested that of a native flora of some 2734 species in the
Hawaiian archipelago 1318 are in some way endangered, that is
almost 50%. If we take Degener's estimate of 20,000 taxonomic
species for the Hawaiian flora the archipelago alone must have
between 8 and 10,000 endangered plant species. Even without
Degener's figure New Zealand, the Canaries and Hawaii have between
them over 2000 threatened species and I do not really believe that
over 10% of the World's threatened plant species occur in these
three small island groups alone. What these figures do, however,
demonstrate, is the critical situation in which many of the
world's insular floras find themselves and they emphasize the need
for intensive study of insular enedemics in their natural
evolutionary laboratories before it is too late. Recently some
excellent work has been carried out in this direction and one only
has to look at the publications of Humphries (1975) and Borgen
(1976) on *Argyranthemum* in Macaronesia, Strid (1972) on
Nigella on the Mediterranean islands and of course the late
George Gillet (1972) and now Ganders (1984) on *Bidens* and
other groups on Hawaii to see that biosystematists have almost
unlimited opportunities for carrying out most stimulating research
on island floras and work which can at the same time provide
important data for conservation. The work of Gillett on
hybridization in *Bidens* in Hawaii, as well as raising
various taxonomic problems, suggests that Degener's estimate of
species is a considerable exaggeration and that conservationists
and taxonomists will not, fortunately, have to cope with 20,000
Hawaiian endemics.

According to Raven (1976) two thirds of the world's plant
species are tropical (probably about 150,000 species) and, in view
of the present rate of destruction of the vegetation of tropical
regions, we must pose the questions: What proportion of these
species will still be with us by the end of this Century? and, How
many of them will disappear without us even having been aware of
their existence?

The problems of conservation are, therefore, complex, and the
task of conserving almost certainly well beyond the limits of our

present capabilities and we may well not even be able to complete
the catalogue of plant species before losing a considerable
proportion of them. Major habitat destruction is, of course, the
biggest and least controllable threat to the plant kingdom as we
know it and it is accelerating daily. But even this rather
pessimistic view of our ability to save endangered species (I see
no reason whatsoever for being optimistic about the situation)
does not mean that there is not a moral obligation to attempt it
and taxonomists and biosystematists obviously have a most
important role to play in conservation.

TAXONOMIC DATA AND CONSERVATION

The requirement for information in conservation concerns the
taxonòmist/biosystematist at two levels. The first is the need,
in Prance's words, to complete the inventory (Prance 1983). This
implies a great deal more basic exploration and alpha-taxonomy,
the preparation of floristic lists for poorly known areas, surveys
of neglected areas and so on, and does not initially require the
sophistication of biosystematic studies. As Given (1976) points
out, the primary requirement even in areas where floras have at
least been catalogued (this excludes vast areas of the tropics and
subtropics) is for an inventory with an assessment of the
conservation status and necessities of the species.
Unfortunately, as Heywood has noted, the number of working
taxonomists in an area is usually inversely proportional to the
number of species occurring there and Heslop-Harrison (1976) said
at a Conservation Symposium a few years ago "I sometimes feel
despair when I see grant applications proposing more and more
intensive work on the smaller and smaller problems of European and
North American floras, whether from physiologists, or indeed
taxonomists." I can only add that a great international effort
and rethink is necessary if we are ever to have enough
systematists working on this enormous task.
    At the second level when threatened and endangered species have
been identified we usually find that the published data are so
out-of-date as to be useless for conservation purposes and there
is a need for more modern information so that appropriate
conservation measures can be taken. It is not enough just to
produce a computerized list of endangered plants of, for example,
the island of Madeira and then consider the task to have been
completed, and it is at this level where the research carried out
and information produced by biosystematists can come into its
own.

BIOSYSTEMATIC RESEARCH AND CONSERVATION

A very high proportion of biosystematical research projects must
include within the groups studied a number of rare or threatened
species of interest to the conservationist. As the biosystematist
is generally concerned with populations of more or less similar

individuals rather than the development and evolution of whole
communities he is more involved in what can be rather loosely
termed autecology rather than synecology and the information
obtained about the plants studied is often within the scope of
requirements suggested by Heslop-Harrison when he recently
referred to the need for data on, for example, reproductive
physiology as an aid to conservation (Heslop-Harrison 1976), data
such as those on the effects of environmental conditions on
flowering; on breeding systems, pollination mechanisms,
incompatibility, pollen fertility, pollen vectors, etc.; and seed
biology information such as seed-set and viability, dispersal,
conditions for germination and establishment and so on. It is
interesting to enquire how much of this information is actually
obtained by biosystematic researchers and then lost because it is
not considered relevant to the publication of the taxonomic or
biosystematic results of the work carried out or is not included
for publication because page costs of journals do not permit the
so-called luxury of publication of such data. Robert Jenkins
(1981) recently drew attention to the fact that "the best
information is often locked away in the minds of the professional
scientific community" and in this same vein a warning against the
taxonomist discarding information, especially cytological,
genetical or biochemical, if it does not correlate with data on
morphological variation has been given by Hawkes (1978) as in this
way valuable information on the genetic diversity and reproductive
physiology of rare species may be permanently lost to the
conservationist. Crovello (1981) has also drawn attention to this
problem when discussing "literature information gaps" with respect
to the literature's role in providing rare plant data and he lists
a series of difficulties for data retrieval and stresses the need
for the preservation of what he terms "supporting data" that is,
those data not included in a publication but accumulated and used
in preparation of the summary information or conclusions presented
in the publication and may include valuable field and laboratory
notes, photographs, etc., which should be made available if
possible. Such information could be published in conservation
journals or simply passed on to organizations such as the
Botanical Gardens Conservation Coordinating Body or the
Conservation Monitoring Units of IUCN (Int. Union for the
Conservation of Nature) providing that the person producing the
information is aware of its potential conservation value.

Our knowledge of the biology of most species is rudimentary and
of the world's floras those known to the level of, for example,
the *Biological Flora of the British Isles* must be a very
tiny fraction of one percent. Fortunately at least a little more
attention has been given to special fields such as the wild
relatives of crop plants and Hawkes (1978) recently, in a review
of the subject, drew attention to the fact that Vavilov's
recognition of the need for biosystematic-type studies in order to
define and describe the genetic diversity of the wild relatives of
crops has led to a certain concentration of effort in this field

and this can be seen, for example, in the work of Harberd on
Brassica relatives (Harberd 1972; Harberd and McArthur
1980); Hawkes and collaborators on potatoes (Hawkes and Hjerting
1969) and Zohary on wheats and other cereals (Zohary 1971) and
Iltis on corn (1981; Iltis and Doebley 1984). Hawkes further
emphasizes the need for the experimental taxonomist to enter into
the field of wild genetic resource study while the ancestors and
precursors of most of our crop species still survive and to
provide good basic work on experimental and evolutionary taxonomy
of cultivated species and their relatives in order to aid both
plant breeders trying to improve cultivars and conservationists
hoping to preserve the wild forms and prevent loss of potentially
useful material by genetic erosion or extinction. A stimulating
starting point for anyone considering this field of research may
be found in Frankel and Soule's recent book *Conservation and
Evolution* (1981).

Outside the field of cultivars and their wild relatives there
remains, of course, an enormous area of almost unknown or at best
understudied rare and threatened species. Indeed, the
biosystematist or taxonomist must, in many cases, be the only
person who, as a result of his studies can orientate
conservationists as to which species merit their special interest,
what are the factors contributing to rareness and so on, and also
in some cases even whether taxa previously considered to be rare
local endemics, for example, are worth recognising taxonomically
or not.

As previously noted, much of the information available to us
from the literature or from the herbarium is old, and no use at
all to conservationists. It is the biosystematist carrying out
modern studies who has the necessary data to indicate that a
particular species has become severely restricted in its
distribution over a period of time and who should be able to tell
us which areas to consider for "in situ" conservation and where
"last ditch" rescue operations might be needed. Biosystematists
can also help conservation by providing important data on
populations of rare plants in their natural habitats, information
which can help to form the scientific basis of management policy
for the establishment and running of biosphere reserves. Also, as
biosystematists have to work with living plants in cultivation,
they must solve the problems of how to get them to flower, etc.
Such information derived from cultivation experiments and, indeed,
transplant and controlled environment experiments, is extremely
useful to, for example, specialist botanic gardens trying to
maintain reserve collections of rare and endangered species, a
task becoming more and more important at least for holding
disappearing species while their potential value is assessed,
especially if we do not succeed in obtaining the urgently need "in
situ" reserves for such plants. Experimental taxonomists can
often, at the same time determine whether or not a rare species is
reproductively vulnerable through delicate pollinator
relationships, low pollen fertility or poor seed-set as shown by

the work of Macior on *Pedicularis* (1978, 1980), and reverting
to the point I made before about publication, in many cases
information is lost through not being published or is at best
hidden away in almost inaccessible doctoral theses. Studies of
relationships between population size and variation, the effects
of decreasing numbers on genetic erosion, effects of drift and
sudden drops in population size are important as the information
obtained can be vital in deciding conservation strategies in
individual cases and the biosystematic models provided by the
studies of Aegean plants carried out by the Lund school can, for
example, be very useful tools for conservationists working in such
fields (Strid 1972).

Frankel and Soulé (1981) list a series of factors which
contribute to the extinction of biological species:

1. *Biotic Factors:* a) competiton; b) predation; c)
parasitism and disease.

2. *Isolation*

3. *Habitat Alteration:* a) slow geological change; b)
climatic change; c) catastrophe; d) man.

Of these the most important is, of course, habitat alteration
by Man. As Frankel and Soulé suggest the survivor species are
those able to overcome extinction factors by adaptation and those
species unable to adapt become extinct. Biosystematists are well
placed by the nature of their scientific background to make a
major contribution to the study of the response of species to
biotic factors and the influence of Man, as biosystematics is
concerned with the genetics and evolution of populations and for
conservation purposes we also need to be concerned with the same
subject usually for very small populations.

Up to now I have been suggesting that the information obtained
by biosystematists during their normal course of study is also,
somewhat incidentally, useful to conservationists even though it
has not been gathered specifically with conservation purposes in
mind. Now I am proposing that biosystematists should enter
directly into the conservation field and positively apply their
methodology to endangered species, to small populations of
threatened plants, and so on, as data on such species will be
invaluable if we are ever to save, and then manage adequately,
even a small fraction of such plants. We urgently need for
deciding conservation strategies, studies of tropical species
especially as the structure of populations in tropical forests
tends to be considerably different from those of temperate
regions. As Ashton (1984) has demonstrated, the rare species of
tropical forests tend to occur as individual plants scattered over
a wide area and their population biology and "in situ"
conservation pose particular problems.

One must sympathize with Heslop-Harrison's despair about the
plants and regions we study so intensively at the present time. A
recent paper by Cronk (1980) on the extinction of the species on
Ascension Island serves to bring this message home and illustrates
at the same time the magnitude of our ignorance about vast

proportions of the world's flora. For example, about *Oldenlandia adscensionis* Cronk states that it shows little resemblance to any other species of its genus and that "nothing is known about its breeding biology" and that it is now almost certainly extinct. On *Sporobolus durus*, Cronk notes "This is a distinct species and its affinities are not clear" and that it is most probably extinct due to competition from alien grass species. I mention both these examples because they are species from biosystematically interesting genera, both without clear affinities and both on the point of disappearing without us knowing anything about them other than the information provided with their original discovery and description last century.

The idea of having to study such groups is often rather daunting for the young research worker who might feel safer studying *Ononis* on the sand-dunes of Western Europe but one only has to read the papers of Paul Berry, for example on South America *Fuchsia* (Berry, 1982) to see what can be achieved under what must have been extremely difficult circumstances. A conservationist needing to take decisions on rare species in this case would have an excellent data base on which to work.

CONCLUSIONS

I have stressed during the course of this paper the need for biological information on threatened and rare plants for conservation purposes and the role a biosystematist can play in obtaining it. This is because I see no other group of people so well placed to provide this service. Taxonomists in general are not always well considered by some of their so-called progressive colleagues, though biosystematists do not perhaps suffer as much as those more classically orientated, but within the major framework of conservation there is a tremendous need for taxonomic and biosystematic work both at the alpha and much more advanced levels. There is a need for a positive approach to studies of tropical species, insular and Mediterranean floras and other such regions with high proportions of endangered plants and wild genetic resources and these should be given priority over already relatively well-known areas of study.

The publication or at least the making available of biological information even though it may not be directly relevant to the results or end-product of a biosystematic study, can be important for conservation purposes as at least it will prevent duplication of effort, as well as the loss of painstakingly gathered data which may otherwise be considered unimportant.

Scientists as a community need to make their views on conservation more strongly felt and biosystematists, by working in the conservation field on threatened, rare species, crop-plant relatives, etc. and by giving the maximum publicity to their work and results, can make a major contribution to saving at least part of the plant resources of the planet Earth for the needs of future generations.

REFERENCES

Ashton, P.S. 1984. Biosystematics of tropical forest plants: A
    problem of rare species. In *Plant Biosystematics* (W.F.
    Grant, ed.). Academic Press, Toronto.
Berry, P.E. 1982. the systematics of *Fuchsia* Sect. *Fuchsia*
    (Onagraceae). *Ann. Mo. Bot. Gard.* 69: 1-198.
Borgen, L. 1976. Analysis of a hybrid swarm between
    Argyranthemum adauctum and *A. filifolium* in the Canary Islands.
    Norw. J. Bot. 23: 121-137.
Cronk, Q.C.B. 1980. Extinction and survival in the endemic
    vascular flora of Ascension Island. *Biol. Conser.* 17:
    207-220.
Crovello, T.J. 1981. The literature as a rare plant information
    resource. In *Rare Plant Conservation: Geographical Data
    Organization* (L.E. Morse and M.S. Henifin, eds.). N.Y. Bot.
    Gard. pp. 83-94.
Frankel, O.H. and Soulé, M.E. 1981. *Conservation and
    Evolution.* Cambridge Univ. Press, Cambridge. pp. 327.
Ganders, F.R. 1984. The role of hybridization in the evolution of
    Bidens on the Hawaiian Islands. In *Plant Biosystematics*
    (W.F. Grant, ed.). Academic Press, Toronto.
Gillett, G.W. 1972. The role of hybridization in the evolution of
    the Hawaiian flora. In *Taxonomy, Phytogeography and
    Evolution* (D.H. Valentine, ed.), Academic Press, New York.
    pp. 205-219.
Given, D.R. 1976. A register of rare and endangered indigenous
    plants of New Zealand. *N. Z. J. Bot.* 14: 135-149.
Harberd, D.J. 1972. A contribution to the cytotaxonomy of
    Brassica (Cruciferae) and its allies. *Bot. J. Linn. Soc.*
    65: 1-23.
Harberd, D.J. and McArthur, E.D. 1980. Meiotic analysis of some
    species and genus hybrids in the Brassiceae. In *Brassica
    Crops and Wild Allies* (Isunoda *et al.*, eds.), Japan
    Sci. Soc. Press, Tokyo. pp. 65-87.
Hawkes, J. 1978. The taxonomist's role in the conservation of
    genetic diversity. In *Essays in Plant Taxonomy* (H.E.
    Street, ed.), Academic Press, New York. pp. 125-142.
Hawkes, J.G. and Hjerting, J.P. 1969. *The Potatoes of
    Argentina, Brazil, Paraguay and Uruguay, A Biosystematic
    Study.* Oxford Univ. Press, Oxford. pp. 525.
Heslop-Harrison, J. 1976. Reproductive physiology. In
    Conservation of Threatened Plants (J.B. Simmons *et al.* eds.),
    Plenum Press, New York. pp. 199-205.
Heywood, V.H. 1971. Preservation of the European flora: The
    taxonomist's role. *Bull. Jard. Bot. Nat. Belg.* 41: 153-166.
Humphries, C.J. 1975. Cytological studies in the Macaronesian
    genus *Argyranthemum* (Compositae: Anthemideae). *Bot. Not.*
    128: 239-255.

Iltis, H.H. 1981. The catastrophic sexual transmutation theory (CSTT) from the teosinte tassel spike to the ear of corn. 13th Int. Bot. Cong., Sydney, Australia. Mimeog. 16 pp.

Iltis, H.H. and Doebley, J.F. 1984. *Zea* - A biosystematical odyssey. In *Plant Biosystematics* (W.F. Grant, ed.), Academic Press, Toronto.

Jenkins, R.E. 1981. Rare plant conservation through elements of diversity information. In *Rare Plant Conservation: Geographical Data Organization* (L.E. Morse and M.S. Henifin, eds.), N.Y. Bot. Gard. pp. 33-40.

Macior, L.W. 1978. The pollination ecology and endemic adaptation of *Pedicularis furbishiae* S. Wats. *Bull. Torrey Bot. Club* 105: 268-277.

Macior, L.W. 1980. Population ecology of the Furbish Lousewort, *Pedicularis furbishiae* S. Wats. *Rhodora* 82: 105-111.

Prance, G.T. 1983. Completing the inventory. In *Current Topics in Plant Taxonomy* (V.H. Heywood and D.M. Moore, eds.), Academic Press, London. In press.

Raven, P.H. 1976. Ethics and attitudes. In *Conservation of Threatened Plants* (J.B. Simmons *et al.* eds.), Plenum press, New York. pp. 155-179.

Strid, A. 1972. Some evolutionary and phytogeographical problems in the Aegean. In *Taxonomy, Phytogeography and Evolution* (D.H. Valentine, ed.), Academic press, New York. pp. 289-300.

Woolliams, K.R. 1976. The propagation of Hawaiian endangered species. In *Conservation of Threatened Plants* (J.B. Simmons, ed.). Plenum Press, New York. pp. 73-83.

Zohary, D. 1971. Origin of south-west Asiatic cereals; wheat, barley, oats and rye. In *Plant Life in South-West Asia* (R. H. Davis *et al.*, eds.). *Bot. Soc. Edinburgh. pp. 235-263.*

A Comparison of Taxonomic Methods in Biosystematics

Warren H. Wagner, Jr.
Department of Botany
The University of Michigan
Ann Arbor, Michigan, U.S.A.

INTRODUCTION

After a number of decades of contributions from biosystematics, we still have the problem of expressing them taxonomically. Perhaps biosystematic data should not be expressed at all, and we should merely report our findings as such and leave classification alone. In my discussion below, I shall confine myself to botanical taxonomy, rather than include the somewhat different problems of zoology. And I shall deal with the inter-relationships between traditional taxonomic methodology and biosystematics, and attempt to conclude with an over-all philosophy of how we can handle biosystematic data in classification. In such a short article as this, I cannot hope to cover the vast amount of publication that bears upon this subject, so I shall only hit some of the high spots.

There are actually two ways of expressing biosystematic results in categories, namely by the methods of regular taxonomy, used by all botanists the world around, and by *ad hoc* methods developed by the biosystematists themselves. Ideally any taxonomic categories should be hierarchical, proceeding in successive ranks, each subordinate to the one above it; to use the popular term, they should be "nested." The categories should be clearly defined, and thus easy to use. As we all know, the latter is hard to achieve, even in the simple hierarchy of traditional taxonomy. Our standard categories of every-day classification at the levels that can possibly pertain to biosystematics are *genus, species, subspecies, variety,* and *form.* They are, of course, nicely hierarchical, but they are not clearly defined. Their application is a matter of custom, and may change through time.

So that my points will be clear, I shall define taxonomy here as classification in the traditional sense, emphasizing the Code of Nomenclature and the standard categories. Biosystematics is the study of reproductive characteristics, involving such fields as genecology and experimental taxonomy. All analytical researchers who discover new comparative characters wish to have their results embodied in the universal taxonomy of botany. Thus the plant anatomist with newly discovered information from the stele or from the vascular pattern of the flower attempts to tie in his data with classification, and in some cases to modify it. The biosystematist is no exception. However, the biosystematist has tended to go outside the traditional methods and to set up systems to embody biosystematic data alone. The biosystematist may have two systems -- the classical taxonomic one, and a distinct one of biosystematic categories.

In standard taxonomy, the purpose of giving a name to a taxonomic group is not to indicate its characters or history, but to supply a means of referring to it, and to indicate its rank. The International Code aims for a precise and simple system of ranks and rules for correct names. The full set of ranks of concern here are as follows: *genus, subgenus, sectio, subsectio, series, subseries, species, subspecies, varietas, subvarietas, forma,* and *subforma.* The actual taxonomic use of these categories has been a matter of custom largely, although there is still much variation on their use by different authors. However, this is appropriate -- nomenclaturists do not tell us how to use our categories; they tell us only what is the ranking of the correct choice of names. The system has been enormously successful, at least to the extent of its global usage, and the fact that our entire information system is based upon it. To be sure, it has been criticized from time to time, but it is our only standard, our only *modus operandi.*

Many systems of biosystematic categories have been proposed over the last sixty years. The ones originated by Turreson (1922a,b) have received much publicity but little actual use. The best examples are in Clausen, Keck, and Hiesey (1940) in their *Experimental Studies on the Nature of Species.* The terms *Coenospecies, Ecospecies,* and *Ecotype,* are surely hierarchical, being based largely upon degrees of crossability. However, their definitions are subtle, and there are no clear-cut boundaries. Their relationship to taxonomic categories is chaotic (*coenospecies* -- "more than one taxonomic species ... sometimes corresponds to a taxonomic section or a genus;" *ecospecies* -- "may or may not correspond to the Linnaean taxonomic species;" *ecotype* -- "if morphogically distinguishable, we classify them as taxonomic subspecies." *op. cit.*).

The system proposed by Camp and Gilly (1943) has no doubt received more attention than any other biosystematic approach. Their goal was admirable: "(1) to delimit the natural biotic units and (2) to apply to these units a system of nomenclature adequate to the task of conveying precise information regarding their defined limits, relationships, variability, and dynamic structure." Camp and Gilly believed that "... the species is a kind of population." "Genetic structure of a species is what

makes it a recognizable unit" was a principle that they realized
could not be applied, for it would result in millions of species,
an absurd and appalling possibility. Their system was actually
very simple: *homogeneon* (homogeneous, interfertile),
phenon (intersterile segments), *parageneon* (some aberrant
segments), *dysploideon* (aneuploid series), *euploideon* (polyploid
segments), *alloploideon* (allopolyploid), *micton* (swarms of
interfertile species), *rheogameon* (interfertile subspecies),
cleistogameon (at least part cleistogamous), *apogameon* (both
sexual and apomictic), *agameon* (only apomictic), and *heterogameon*
(found in certain oenotheras). It is immediately evident, of
course, that this system is not hierarchical, nor are the
categories really clearly defined, as there are clearly numerous
overlaps possible between them. The main value of the Camp and
Gilly system, in my opinion, is pedagogical; by separating out
these different categories, it emphasizes the kinds of
populations. I do not know of any taxonomists who have actually
used these categories in their classifications. I do not know of
indicated, although I can imagine some such method as we use today
with hybrids; instead of using the X sign between the genus and
the epithet, one could use *phn* for *phenon,* *pag* for *parageneon,*
and so on.

A rather similar effort was the so-called "Deme Terminology"
promulgated by Gilmour (1960). The categories, however,
emphasized somewhat different features. The system "... arose
because of the multitude of categories which had been proposed for
the units of experimental taxonomy (alternatively known as
biosystematy and genecology)." A "deme" is any group of
individuals of a specific taxon, and the categories were
topodeme (associated with a specific geographical area),
ecodeme (habitat), *phenodeme* (phenotype), *genodeme*
(genotype), *plastodeme* (mod *gamodeme* (in position to interbreed),
autodeme (mainly autogamous), *endodeme* (mainly inbreeding),
agamodeme (mainly apomictic), *clinodeme* (part of a cline).
Overlaps can be expressed by combined forms such as genoecodeme,
hologamodeme, and so on. Some of these categories are more   or
less like those of Camp and Gilly; for example, apogameon is
equivalent to agamodeme, and rheogameon is rather similar to
clinodeme. Again, these terms have not been used in formal
taxonomy, although presumably they could be interpolating their
abbreviations between the genus and species epithet (or between
the species and infraspecies, substituting the deme abbreviation
for subspecies or variety). [I might point out that the deme
terminology can be applied to other characters beside
genecological, e.g. *morphodeme* (a group characterized by a
distinct morphological character or character-complex),
histodeme (tissue), and *chemodeme* (distinctive chemistry).]

Verne Grant (1971, 1981) took over the word species from
traditional taxonomy and proceeded to define it in six different
ways: *taxonomic species* (morphological species, phenetic species),
biological species (genetical species), *microspecies*

(agamospecies, apomict), *successional species* (palaeospecies),
biosystematic species (sterility barriers, ecospecies,
coenospecies), and *semispecies* (two different senses: a
partially differentiated intermediate stages of divergence,
"neither good races nor good species," and products of natural
interspecific hybridization where the breeding barriers are less
than 100% effective and where the distinctions between the
original species become blurred, making them semispecies).   It is
not likely that this system will influence the traditional
taxonomic methods.   In addition to these categories, Grant uses
still others, including *evolutionary species*, which is a
combination of his biological, micro-, and successional species.
He favors the presumptuously labelled biological species concept
based on reproductively isolated systems of breeding populations.
He clearly does *not* favor the taxonomic species, which he
regards as subjective and in many cases out of concordance with
the natural units.   The biological species concept has been
rejected by most botanists and has been strongly criticized (e.g.,
by Sokal and Crovello 1970).   One problem with Grant's categories
is that there may be combined in one taxonomic species different
components which are inseparable except by biosystematic
characters.   Also, they are not as clearly defined as might be
assumed.   Grant himself recognizes that there are all degrees of
interfertility (e.g. 1981, p. 74).   Accordingly the biological
species concept is just as subjective in its application as the
taxonomic species.
  I personally believe that the taxonomic species, as it is
actually used by taxonomists today, embodies much of what Grant
says it does not.   In the light of present usage, I would define
*species as a convenient taxonomic category that defines a unit
of organismic diversity in a given time frame and composed of
individual organisms that resemble one another in all or most of
their structural and functional characters, that reproduce true by
any means, sexual or asexual, and constitute a distinct
phylogenetic line that differs consistently and persistently
from populations of other species in gaps in character state
combinations including geographical, ecological, physiological,
morphological, anatomical, cytological, chemical, and genetic, the
character states of number and kind ordinarily used for species
discrimination in the same and related genera, and if partially or
wholly sympatric and coexistent with related species in the same
habitats, unable to cross or, if able to cross, able to maintain
the special distinctions.*   I realize that this is a cumbersome
definition (for which I am indebted in part to various authors,
especially the well known taxonomist, Cronquist 1978).   I want,
however, in this definition to get away from the simplistic
definitions that rely on only one or a few criteria, and I also
want to define it as it is actually used by most authors today.
By listing all of the different types of characters that are used,
we stress that taxonomic species are not merely morphological, no
matter how important morphological characters are.   I submit that

the vast majority of species recognized by monographers, at least in North America and Europe, conform to this definition, perhaps 95% or higher.

What are the contributions of biosystematics to the taxonomic species concept? These come mainly from intersterility criteria. However, as we all know, and as pointed out by many authors (e.g. Gilmour 1960), in plants one often finds that intersterility barriers "cut right across a grouping according to morphological similarity." The number of examples of species that include sexual and asexual segments, or species that contain diploid and polyploid segments is now very large and continuing to increase. I believe that pre-biosystematic taxonomists instinctively used intersterility as a criterion to aid them in assessing species distinctions. What could be more obvious than a "natural common garden experiment" in which a community of species belonging to the same genus co-occur and display their differences, even when these are subtle, to maximum advantage. Our own studies of grapeferns and moonworts, *Botrychium*, have shown genus communities to be of great value in discerning subtly differentiated taxa (Wagner and Wagner 1983), but long ago these species-associations were noticed, e.g. by R.J. Webb (as quoted by Hopkins 1910) who wrote "The most interesting fact revealed by my studies of botrychia is the remarkable clannishness of the species. They seem to love each other's society, and where one grows the others also do." The point is that the species grew side-by-side and did not hybridize and they keep their distinctive morphology.

Let us take two taxonomic species and examine what different biosystematic situations may prevail between them. Species A may be able to cross with species B, may occasionally cross with B, or may not be able to cross with B. If a hybrid between them is formed, it may be entirely fertile, partially fertile, and completely sterile. Species A may be diploid and species B diploid, or species B may be tetraploid, or both may be tetraploid. Species A may have $n = 12$ and species B may have $n = 12$, 12 being the base number for the genus, or species A may have $n = 12$, and B $n = 13$, or both species may be aneuploids, A with $n = 11$ and B $n = 13$. Species A may be an outbreeder and species B likewise, or species A may be an outbreeder, and B an inbreeder, or species A and species B both inbreeders. Species A and B may both be sexual, or one sexual, the other apomictic, or both species may be apomictic. If all other data support the idea that A and B are separate species, there is no reason why the biosystematic data should change this concept. In some of the situations cited here, the biosystematic data actually support the concept that A and B are separate species, but in others they seem to oppose the idea. The point is that the biosystematic data should be included as no different from any other source of comparative data, as listed in the definition of species given above. Some of these situations may seem incredible to zoologists working with higher animals, but plants are noted for these conditions, and botanists have come to accept them.

The subspecies and variety are both usually interpreted as
geographically distinctive elements of a species, in which the
taxa are not only distinct in distribution pattern but in
morphological characters as well (Camp and Gilly 1943; Fernald
1950). Zoologists tend to use the concept of subspecies entirely
(without any special designation, merely a trinomial: *genus,
species, subspecies* names in a row), and botanists tend to use
the concept of variety (three names, as in Zoology, but
interpolating the abbreviation "var." before the third name).
Many botanists feel that the distinctions between subspecies and
variety are too subtle to justify having two categories. I
believe, however, that having two categories at this level, but
with the subspecies a higher category than the variety, allows
greater flexibility in taxonomic expression, and as will be seen
below may fit better into certain applications of biosystematics
than having just one.

I have the impression that both categories are being used less
and less by taxonomists, compared to the thirties and the forties,
at least in the United States. Fernald's *Gray's Manual of
Botany* published in 1950 had out of 8340 taxa including 1205
forms and named hybrids 1612 varieties not including the typical,
19.3% of the total. However, many of his varieties have proved to
be good species, others are hybrids, others are extremes of
clines, and some are true geographical varieties. Some
taxonomists feel that taxa described as varieties or subspecies
are lost or ignored. Various guides to nomenclature do not even
include them (e.g., *Index Kewensis, Index Filicum*).

The suggestion is not new that these infraspecific categories
could be used to accommodate biosystematic data, even if the
varieties or subspecies co-exist in the same habitats. Confining
themselves to the term subspecies, Camp and Gilly (1943) noted
that in general "the subspecies would be separated from other
segments of the species by definable differences in both
morphology and distribution." They then suggested that one could
just as well use subspecies for taxa separated by ploidal
differences or sexual-apomictic differences, regardless of whether
there were geographical differences. One may question whether it
is justified to use differences in ability to interbreed in place
of geographical differentiation. I have considered this at
length, and I have come to the conclusion that both geographical
and interfertility differences provide barriers to interbreeding,
and therefore one can consider subspecies to be morphologically
differentiated members of a species that have become separated by
either geographical or breeding barriers (or both). If we retain
the category of variety, it should constitute a more minor
segregate, in which the morphological differentiation is less.
Good examples of such usage by biosystematists of the category of
subspecies are found in the maidenhair-spleenwort complex, viz.
Asplenium trichomanes L. ssp. *trichomanes, A. t.* ssp.
inexpectans Lovis, and *A. t.* ssp. *quadrivalens* D.E. Meyer

(See Lovis 1977 and references). Studies of chromosome behavior
in intersubspecific hybrids, as well as ploidal levels, are in
this case supported by minor morphological differences.

The category of form (*forma*) is probably the most trivial
of the standard categories, and the one which is most likely to
disappear from usage by competent taxonomists. The category is
used primarily for individual variants, sporadic specimens that
show distinctive features, often involving only one character.
One of the best known form types in flowering plants is the
albinic state of the corolla; often blue-flowered or red-flowered
species especially throw off individual white-flowered specimens.
Forms may be genetically or environmentally produced. There is no
reason that the category of form cannot be utilized for
biosystematically differentiated individuals, such as occasional
triploids in a normally diploid species. In ferns, some of the
forms are spectacular, such as the much-divided forms that
entirely lack reproductive structures (sori) such as *Polypodium
virginianum* L. f. *bipinnatifidum* Fernald and *Asplenium
platyneuron* (L.) Oakes f. *hortonae* (Davenp.) L.B. Smith. In these,
not only are the plants well marked morphologically but they have
no means of interbreeding at all, there being no sporangia or
spores produced; they can only reproduce by rhizome growth.
Without doubt they are genetically fixed. On the contrary, in
Onoclea sensibilis L. f. *obtusilobata* (Schkuhr) Gilbert,
only the fertile fronds are effected; they are intermediate
between the ordinary trophophylls and the sporophylls. They
produce, so far as is known, normal spores for sexual
reproduction, but the form is actually due to damage to the plant
-- fire, mowing, and other phenomena (Beitel *et al.* 1981).
Most taxonomists eschew the use of the form category because of
its very minor significance in populations, and because many or
most of them are due either to single mutations or to
environmental influences. Nevertheless, the category is there,
and can be used by the taxonomist (and biosystematist) if deemed
useful or necessary. I do not think, however, that the form
category should ever be applied to single organs of a plant, as in
the case of "f. *obtusilobata*" given above. If, for example,
a single leaf or single flower should become polyploidized, while
the rest of the plant remains normal, no formal designation should
be made. One reason that the category form is recognized is that
the plant is so distinctive that it will inevitably attract
attention, as in the case of the fern forms noted above; the
pteridologist Wherry with little doubt recognized them because of
this (Wherry 1961). Also, from a biological standpoint, forms may
be very instructive; for example, even "f. *obtusilobata*" may
give us clues to morphogenetic processes. How a simple mutation
can create such a bizarre plant as "f. *hortonae*" has considerable
developmental implications.

Perhaps the largest problem of taxonomic usage centers around
the nomenclature of hybrids. With the growth of biosystematics
over the years, it has become more and more apparent that

hybridization between species is a prominent factor in influencing
the pattern of diversity. In North American pteridophytes alone
over 20% of the taxa are believed to have originated through
hybridization. Because of the problems with the various uses of
the word hybrid, nomenclaturists at the most recent international
botanical congress (Sydney 1981) have decided to use the root
notho-(bastard) for taxa at all levels. Thus we have
nothogenus (intergeneric hybrid), *nothospecies* (interspecific),
nothosubspecies (intersubspecific), *nothovariety* (intervarietal).
In all cases, it is permissible to use the multiplication sign
(X) in the name to indicate reticulate origin.

   Some biologists have asked whether if two "genera" hybridize
they should not be placed in the same genus. This is actually a
good example of the use of biosystematic data in taxonomy. The
point is that if all other data support generic separation,
crossability alone is not considered sufficient to counteract that
separation. The orchids provide an excellent example, in which,
apparently, purely mechanical factors play a large role in
separating the gene pools of genera. Horticulturists have
produced a bewildering array of nothogenera, in some cases
involving several genera together. There are all degrees of
intersterility in plants and intersterility factors may be
independent of other divergences, so we must use all of the data
in making our generic evaluations. Using sterility factors alone
is comparable to "single character taxonomy" now assiduously
avoided by most systematists. As to how to express nothogenera,
the recommendation is to create hybrid names (e.g. *Dryopteris* X
Polystichum = X*Dryopolystichum*) by collapsing the formula.

   At the species level, nothotaxa have been expressed with great
inconsistency. Some authors use formulae, others make the hybrids
varieties of one or the other parent, others as a variety
coordinate with the two parental species treated as varieties, and
others as separate nothospecies (Wagner 1963, 1969, 1983). The
genus of woodferns, *Dryopteris*, provides a good example in
North America, where the confusion produced by the Fernaldian
school at Harvard University was straightened out through the
pioneering field studies of R.C. Benedict and E.T. Wherry and the
biosystematic studies of D.M. Britton, S. Walker, and W.H. Wagner.
The urge to use the varietal category led to some wild
inconsistencies in rank, for example:

   Dryopteris cristata X *D. spinulosa* var. *spinulosa*
   = *D.* X*uliginosa*
   Dryopteris cristata X *D. spinulosa* var. *intermedia*
   = *D.* X*boottii*

Thus with one parent constant, the others "varieties" of the same
species, the resultant nothotaxa became raised to the rank of
nothospecies.

   With nothosubspecies and nothovarieties the problem is the same
as with holosubspecies and holovarieties. As described above,
ordinary usage of nonhybrid subspecies and varieties is somewhat
controversial, with some authors eliminating one or the other

category. Whether or not an occasion would arise to use nothoform
seems questionable, but one cannot anticipate taxonomic needs of a
particular situation that may arise. Presumably the category is
available and can be used if needed.

The Code, of course, allows us to use the names in any way we
feel suitable to our goals. With regard to nothotaxa, taxonomists
have emphasized whether the hybrid is sterile or not. If the
hybrid is sterile, the hybrid sign, X , is generally used, either
in a formula (*Rosa alba* X *rubra*) or as a prefix to the epithet
in a binomial (*Rosa* X*rosea*). But if the nothotaxon is "fertile,"
i.e., reproduces itself and forms populations, the X is
eliminated and the binomial is used. This is a good example of
custom in taxonomy –– the idea of hybrid became associated with
sterility or nonreproducibility. But it became customary to give
reproductive nothospecies binomials, even when their reproduction
was totally apomictic and the population made up of sterile
individuals. This, in particular, is an area where the
biosystematist becomes important in making taxonomic judgements.

Because of the great variability of reproductive conditions in
nothospecies, *I propose that* all *taxa of reticulate origin
be designated with the* X *sign.* This tells us something about
the nature of the plant, and in particular shows that the taxon is
dictyophyletic and therefore must be removed in making character
analyses (Wagner 1983). In cladistics, a taxon may be
monophylectic, paraphyletic, or *polyphyletic,* all patterns
that depend upon divergence. But hybrids, by their intermediate
nature, possess semidivergences (or hemiapomorphies) and thus
blunt and modify the true divergent patterns.

Another custom in expressing nothospecies is dependent upon
abundance of the taxon in question. If rare, one usually uses the
taxonomic formula only; if common, one uses a binomial. However,
one cannot, of course, keep taxonomists from naming sterile,
isolated hybrids, and that is exactly what has happened in certain
groups. For example, in ferns, it is standard practice to
designate even such occasional hybrids with binomials.

To conclude, what taxonomic methods can we suggest for use by
biosystematists? Clearly all of the multitude of biosystematic
data cannot be expressed in standard classification. Furthermore,
it is necessary to coordinate the data of biosystematics with the
rest of the evidence in making decisions. This is nicely
illustrated by the methodology of numerical taxonomy, in which all
sources of information are sought and utilized. The numerical
taxonomists, with their stress on *all* biological attributes
derived from *all* disciplines may, with some justification,
contend that it is they who are the real biosystematists; the
researchers who call themselves biosystematists are actually
specialists (Wagner 1963). Whatever one's viewpoint on the
significance of numerical taxonomy or phenetics, we have to admit
that no other approach has so forcefully caused us to consider a
broad biological approach to taxonomy, including evidence from the
entire organismic spectrum.

The biosystematist can contribute to cladistics by aiding in the detection and analysis of hybrids. All hybrids should be removed from the study collection by picking out intermediates and checking for their hybrid nature (Wagner 1983). Valid cladistics depends on the determination of all true divergences (apomorphies, homoplasies, parallelisms, convergences, and reversals) so that the primitive ground plan can be estimated, and the divergence directions and distances can be plotted (Wagner 1980). In order to develop correct character polarities using in-group and out-group studies, the blunted semidivergences must be removed, and only after the holotaxa have been correlated into a phylogenetic system can the nothotaxa be placed. The taxa which arise by reticulation are expressed as network connections between the parental taxa on the cladogram.

Some polarities involved in biosystematic data are given as follows:

Character	Primitive	Derived
Sexuality	Sexual	Apomictic
Breeding system	Outcrossing	Inbreeding
Interfertility	Interfertile	Intersterile
Polyploidy	Diploidy	Polyploidy
Aneuploidy	Euploidy	Aneuploidy

Some of these are probably irreversible in most cases, such as apomixis, intersterility, and polyploidy. In most study groups (e.g. a genus or subgenus), these are rarely true apomorphies in the sense of uniquely derived states; most of them arise again and again by homoplasious evolution. Indeed this is one of the problems with much of biosystematic data — the tendency, for example, for apomixis to arise repeatedly in a genus, or intersterility between species, or polyploidy. We do not really know whether the seeming parallelisms are not actually convergences. Different genetic factors may produce the same specialized condition. Intersterility between one pair of species may be caused one way, and between another pair another way.

How can biosystematic information be utilized in classification and choice of categories? I propose that we use the broadest definition of species, such as the one given above that utilizes the whole array of evidence, whether biosystematic or nonbiosystematic. We should also keep in mind species standards that are customary in a given group, although in most cases this is not necessary at the species level; generally speaking the species level is fairly obvious in plants in my opinion, and morphology, even by itself, is a strong indicator of other differentiation in other character sets.

In making use of infraspecific categories for biosystematic situations authors can simply use one of the categories and eliminate the other, i.e., subspecies or variety. If both are used, the subspecies would be expected to have greater overall differences than the variety. If an hierarchy of categories is called for, one could, for example, have under one species, two subspecies, and under each of the latter two varieties. The categories, whatever the case, should have more than biosystematic definitions -- they should be recognizable by ordinary means such as morphological, physiological, or ecological, whatever are the characters used within the given taxonomic group. The category of form may also be used, if deemed useful taxonomically, even if the taxon is known only by solitary individuals.

For hybrids there is always the problem of introgressants that have backcrossed so extensively with one or both of the parents that the progeny have become indistinguishable from the parents. In that case, I recommend simply using the name of the parent; taxonomically the introgressant is essentially the same as the parent, and genetically the parental heredity is dominant. For the usual, more intermediate hybrids, I recommend usage of the hybrid (X) sign in all cases, whether sterile or fertile, sexual or apomictic, whether a nothogenus, nothospecies, nothosubspecies, nothovariety, or nothoform. Neither ability to reproduce, sexuality, or rank should have anything to do with it. As to whether to give formula names or binomials, this is, of course, once again a matter of taxonomic taste and is up to the author. If formula names are given, they should probably be confined to rare and sporadic hybrids, although even these may be given binomials if species, or other appropriate nonformula names if other ranks.

What should be done if there are no other demonstrable differences between populations other than biosystematic ones -- for example, essentially identical populations, one sexual, one apomictic, or one outbreeding, one inbreeding, or one diploid, one tetraploid? In ferns one of the common conditions observed by biosystematists is the occurrence of sterile diploids and fertile tetraploids in the same nothospecies (e.g. *Polystichum, Asplenium*). These may coexist in the same habitat, but they are inseparable by ordinary means, requiring instead microscopic investigation of chromosomes and spores. In all such cases, I suggest that no formal naming by conventional taxonomic categories should be made, but rather that the differences be expressed by adding parentheses after the name, for example, *Polystichum californicum* Maxon (sterile diploid) vs. *P. californicum* Maxon (fertile tetraploid); or *Pellaea glabella* Mett. ($2x$ sexual) vs. *P. glabella* Mett. ($4x$ apogamous).

The overall conclusion of this review is that biosystematic data must take its place along with all other taxonomic data in making classifications. The *ad hoc* categories of workers like Turreson, Clausen, Keck, Hiesey, Camp, Gilly, Grant, and Gilmour will never achieve widespread acceptance in taxonomy. The traditional methods are so ingrained now that they will probably never be displaced, so that we can only adapt them to our needs. The hierarchy of categories is set, and only customs in their

usage can change. There is no reason why the epithets themselves
can not be biosystematic. We already use such ecological epithets
as *aestivalis, amphibia, autumnale, campestris, sylvaticum,
pumicola,* and such chemical epithets as *aromatica, citrata,
camphorata, foetida, tinctoria.* Indeed, taxonomists have
already used biosystematic ideas in their nomenclature such as
clandestina, dioica, fertilis, hybrida, polygama. The late
Edgar T. Wherry broke new ground when he designated a well known
nothospecies of eastern North America *Dryopteris* X*triploidea.*
In addition to their contributions to our knowledge of general
reproductive biology of plants, biosystematists have contributed
valuable new characters to augment our systematic methods and
fostered construction of a meaningful and well founded taxonomy.

REFERENCES

Beitel, J.M., Wagner, W.H., Jr. and Walter, K.S. 1981. Unusual
    frond development in Sensitive Fern, *Onoclea sensibilis* L.
    Am. Midl. Nat. 105: 396–400.
Camp, W.H. and Gilly, G.L. 1943. *Brittonia* 4: 323–385.
Clausen, J., Keck, D.D. and Hiesey, W.M. 1940. Carnegie Inst.
    Washington Publ. No. 520. 452 pp.
Cronquist, A. 1978. Once again, What is a species. Beltsville
    Symp. Agric. Res. 2. Biosystematics in Agriculture: 3–20.
Fernald, M.L. 1950. *Gray's Manual of Botany.* 8th ed.
Gilmour, J.S.L. 1960. The deme terminology. *Scott. Plant
    Breed. Stn. Rep.* pp. 99–105.
Hopkins, S.N. 1910. *Am. Fern J.* 1: 3–6.
Grant, V. 1971, 1981. *Plant Speciation.* 1st Ed., 2nd Ed.
    Columbia Univ. Press, N.Y.
Lovis, J.D. 1977. Evolutionary patterns and processes in the
    ferns. In *Advances in Botanical Research* (R.D. Preston
    and L.W. Woolhouse, eds.). Academic Press, London. 4: 229–440.
Sokal, R.R. and Crovello, T.J. 1970. The biological species
    concept. *Am. Nat.* 14: 127–153.
Turreson, G. 1922a. *Hereditas* 3: 100–113.
Turreson, G. 1922b. *Hereditas* 3: 211–250.
Wagner, W.H., Jr. 1963. Biosystematics and taxonomic categories
    in lower vascular plants. *Regnum Veg.* 27: 63–71.
Wagner, W.H., Jr. 1969. The role and taxonomic treatment of
    hybrids. *BioScience* 19: 785–789.
Wagner, W.H., Jr. 1980. Origin and philosophy of the groundplan-
    divergence method of cladistics. *Syst. Bot.* 3: 173–193.
Wagner, W.H., Jr. 1983. Reticulistics: the recognition of
    hybrids and their role in cladistics and classification. In
    Advances in Cladistics (N.I. Platnick and V.A. Funk,
    eds.). Columbia Univ. Press, N.Y. 2: 63–79.
Wagner, W.H., Jr. and Wagner, F.S. 1983. Genus communities as a
    systematic tool in the study of New World *Botrychium*
    (Ophioglossaceae). *Taxon* 32: 51–63.
Wherry, E.T. 1961. *The Fern Guide.* Doubleday, N.Y.

Observations on IOPB 1983 and
Notes on the Discussions among Participants

John C. Semple
Department of Biology
University of Waterloo
Waterloo, Ontario, Canada

INTRODUCTION

The symposium included four full days of paper presentations
covering all aspects of plant biosystematics as well as many hours
of informal discussions among the participants. Two areas of
systematics, in this participant's opinion, generated considerable
discussion and concern during the symposium. These were (1)
recent discoveries on chromosome structure and the effect of these
on chromosome pairing in early stages of meiosis, and (2) the
large number of techniques that are now considered standard in any
comprehensive biosystematic study. In this final chapter I will
show how the two general themes demonstrate that plant
biosystematics is a very difficult discipline of biology to
master, but also one that can be a satisfying long term career.
The entire conference demonstrated this generalization.
Biosystematics can never fall by the wayside, since it constantly
assimilates new methodologies and subdisciplines, combining them
with traditional morphological studies to allow ever expanding
understanding of an ever increasing number of plant taxa. Not all
of the symposium participants, it should be noted, were involved
with phylogenetic studies, but all were involved in biosystematics
research that provided data necessary to the construction of such
phylogenies.

The history of the study of the small herbaceous spring
perennials *Claytonia virginica* and *C. caroliniana* (Portulacaceae)
is indicative of how biosystematists have taken new discoveries in
the field of biology in general and utilized the techniques to
increase our understanding of the evolution of a group of
populations. The Spring Beauties are together an example of a
species complex that has repeatedly been important in the history
of modern biosystematics. *Claytonia* is well known for its

Copyright © 1984 by Academic Press Canada

complex cytology revealed during the course of more than 20 years
of study by W.H. Lewis and a number of co-workers. The flavonoid
chemistry has been studied during the last 10 years. Now
Claytonia is the study organism in an investigation of variation
in rDNA base sequences presented by J.J. Doyle, R.N. Beachy and
W.H. Lewis during the symposium. The work involved the newest
techniques of recombinant DNA procedures, i.e., Biotechnology.
Analysis of variation in restriction endonuclease maps of 5S and
18-25S ribosome DNA provided data that supported conclusions about
the phylogeny of the species based on flavonoid data and data on
relatively limited morphological variation. The research
demonstrates that "old" biosystematic problems can be viewed in
new ways.

Research on *Claytonia* proceeded at different rates, but
continued because new techniques allowed more thorough analyses to
be conducted. The same can be said about many of the
subdisciplines of cytotaxonomy and cytogenetics. Not so very long
ago, studying chromosome pairing was a "new" aspect of plant
biosystematics. The presence or absence of pairing of chromosomes
during meiosis in experimentally produced hybrids was thought to
be a strong indication of the degree of relatedness between the
parental taxa. Subsequent studies indicated that the absence of
pairing could not be viewed as evidence for the absence of
homology and an indication that two taxa were not closely related.
Data presented during the symposium by R.C. Jackson, M.S. Bennett
and H. Rees suggest that it may no longer be wise to see even
complete pairing during pachytene as evidence of strong homology.
When chromosomes of different lengths can line up during early
meiosis with no apparent difficulty and without a long unpaired
portion on the longer chromosome, then it becomes necessary to
reevaluate traditional ideas about the meanings of the terms
"homologous" and "homeologous". The apparent cause for some
pairing is the presence of similar highly repetitive DNA sequences
rather than the DNA encoding the basic information that determines
the physiological and morphological properties of a plant.

The existence of large but varying amounts of repetitive DNA in
eukaryotic chromosomes raises questions about the usefulness of
traditional karyotype studies done for decades by plant
biosystematists. In his paper K. Jones noted the continued need
for such work, but with some changes in the way the data are
interpreted. The "natural karyotype" that Bennett discussed is an
example of a new method of interpretation, which can reveal
considerable information about karyotype homologies and allows
predictions about which homologue arms are most likely to undergo
considerable change. This work comes at a time when some
researchers were beginning to view karyotype work as unproductive,
which it certainly is not in many groups of plants. Thus, the
history of karyotype work is indicative of biosystematic studies
as a whole. At a time when some researchers were thinking of the
discipline as "out-dated", new techniques and new modes of
analysis came along and revived research in a quiescent area and

made the whole discipline more active.

The paper presented by R.K. Vickery, the first morning of the symposium, provided an overview of what can be done, during a long term program devoted to thoroughly investigating many different aspects of the same group of plants, in this case species of the genus *Mimulus* (Scrophulariaceae). By comparing results from morphological, chemical, cytological and breeding studies a general pattern of phylogenetic relationships was constructed. Nonetheless, some of the individual studies gave contradictory results.

Vickery's paper raised a set of questions that I asked other speakers and many participants. How can modern plant biosystematists cope with the difficulties of learning, mastering, and applying a dozen different highly specialized research skills? If we can't learn all that we would like to, can we expect our graduate students to learn it all? Should teachers of plant biosystematics require graduate students to accumulate cytological, chemical, breeding, ecological and morphological data in sufficient quantities to justify multivariate analyses, before a degree is awarded? What is a minimum number of skills that should be brought to bear on a particular systematics problem? Is collaboration among specialists becoming a necessity? Given that plant biosystematics is not passé, what area should new students specialize in?

Each specialist at the conference suggested in one way or another that their subdiscipline was worthy of study. The quality of the research presented in the other chapters in this book serves as evidence that such opinions are justified. Nearly all the specialists, however, agreed that any approach by itself could not reveal the true phylogeny of a group of species. Also, each felt that their special area of research had something to contribute, since a best approximation of the history of a phylad could be obtained only by combining the results of a number of different studies. The need for several programs of analysis is now accepted in phenetic and cladistic studies. So too is the need to combine data from chemical, cytological and morphological studies now viewed as the best way to achieve meaningful results in biosystematics.

A consensus was apparent among conference participants – collaboration among specialists was becoming a necessity. Gottlieb offered the concept of a joint project for a group of graduate students, such as is frequently done in physics laboratories and many biochemistry laboratories. Not only would a "critical mass" of interacting graduate students be ideal, but they also would all be working on resolving the phylogeny of the same large group of species, rather than working independently on different and often unrelated genera. Such an environment would ensure that a graduate student in plant biosystematics is exposed to a number of subdisciplines, but also that he or she is involved with the final merging of all the data into a single logical interpretation of the phylogeny of the plant taxa. Such an approach could raise problems for the supervisor, when it comes time for the team of graduate students to seek employment in the open market. Teachers could find themselves in the regrettable position of having to recommend one qualified student from among several competing for the same job. While this is true at times now, each student has a different story to tell. A group project would only allow each student to relate a different aspect of the same story with the same conclusion.

For those already employed, collaboration offers the opportunity to discover more about a group of organisms and thus potentially develop a more accurate history of the group. Participants at the symposium saw many advantages to joint research efforts. Collaboration between laboratory oriented and field oriented specialists would increase the number of samples available for laboratory analysis. Detailed laboratory investigations would increase the value of observations made in the field. Collaboration among three or more research teams would result in a great reduction in the time required to complete a comprehensive study of a group of plant taxa. Thus, collaboration will mean greater efficiency and the resolution of more systematics problems. Collaboration will also mean that a specialist interested in a large group of taxa can concentrate his or her efforts on keeping abreast of new techniques within his or her area of expertise and still be able to direct a multidisciplined investigation of the group that interests him or her. Of note was the discovery that many participants at the symposium were already involved in collaborative efforts.

One additional axiom of modern plant biosystematics was also readily apparent during all of the conference. Regardless of what subdiscipline a student chooses as a specialty, *only careful, well thought-out research will bring results that clarify our understanding of plant evolution.* This applies to both cultivars and undomesticated taxa; both groups still need far more study than the present number of plant biosystematists can undertake. Biosystematics has progressed greatly since Camp and Gilly proposed the term "biosystematy". All participants were in full agreement that the next forty years should be equally productive and will include the addition of more techniques to a long list of standard methods in plant biosystematics.

Index

A

Acacia, 279
Acalypha, 279
Acanthoxanthium, 4
Acanthus, 279
Achillea, 4
  millefolium, 567
Aciphylla, 255
  monroi, 257
Aconitum, 276, 279
Actinotus novae-zelandiae, 251, 252, 254
Adenophora, 145
Adesmia boronioides, 379
Adiantum capillus-veneris, 544
Adoxa, 98, 101
Aegilops, 102, 201, 459
  umbellulata, 50, 61, 62
Aegopodium podagraria, 196
Aeonium, 493, 494
Aeschynomene schimperi, 379
Afgekia sericea, 378
Agamic complexes, 246
Agamospermy, 212, 225
Age-specific, reproduction, 447
  survivorship, 447
Agropyron, 102
Alcohol dehydrogenase, 12
Aldina, 385
Alepis, 122, 123
  flavida, 124
Alexa, 387
Algae, 325
Alkaloids, 562

Allium, 30, 89, 91, 99, 102, 104
  azureum, 91
  cepa, 91, 105, 234
  dicipience, 91
  flaxum, 104, 105
  fistulosum, 91
  grayi, 225
  porrum, 68
  stamineum, 104, 105
Allogamy, 580
Allophyllus, 506
  cobbe, 509
Allozymes, 6, 9, 11, 171, 442
Alnus, 361
Aloe, 29
Alstonia, 510
  scholaris, 511
Amelanchier ovalis, 469
Amicia zygomeris, 379
Amorpha, 383
  canescens, 379
Amphiploidy, *Campanula,* 146
Amphipterygium, 2
Amsinckia, 4
Amyema, 122, 123, 125, 129, 131, 135
  species, 124, 126–128
Amylotheca, 123
Anacyclus, 6, 101, 104, 105
Anatomy, Dipsacaceae, 313
Andreaea, 527
Andropogon, 201
Androsace, 458

Anemone, 30, 101
Aneuploidy, *Campanula,* 143
　mistletoes, 120, 129
Animal pollination, 271
Anisotome, 254
　filifolia, 255, 257
Antennaria,
　alpina, 215, 216, 217
　anaphaloides, 213, 214
　dioica, 215, 216
　pulcherrima, 213, 214
　stolonifera, 215, 216, 217
Anthericum liliago, 459
Anthocyanins, 361
Anthurium, 280
Antidaphne, 120
Aphanorrhegma, 531
Apis dorsata, 511
Apium graveolens, 583
　prostratum, 251, 253
Apomixis, 237–248
　Crataegus, 419
　Garcinia, 511
Apoplanesia, 383
Aquilegia, 278, 279
Arceuthobium, 133
Arctostaphylos uva-ursi, 364, 365, 366, 373
Arenaria ciliata, 459
Argyranthemum, 478, 485, 493, 634
　adauctum, 487–489
　broussonetii, 486–492
　coronopifolium, 487, 489, 493
　escarrei, 487
　filifolium, 487, 488
　foeniculaceum, 488
　frutescens, 486–490, 492, 493
　gracile, 488
　maderense, 489
　pinnatifidum, 488, 489
　sundingii, 489, 490, 492, 493
　tenerifae, 489
Armeria, 4
Arnica, 365
　cernua, 372
　cordifolia, 366–368, 371, 373, 374
　discoidea, 371
　gracilis, 367, 370, 371, 374
　latifolia, 367, 368, 370, 373, 374
　nevadensis, 373
　spathulata, 371, 372
　venosa, 371, 372
　viscosa, 372–374
Artocarpus, 283

Asexual reproduction, 238, 580
Aspidotis densa, 544
Asplenium, 549, 550, 556
　platyneuron, 544, 649
　ruta-muraria, 544
　trichomanes, 549, 648
Aster, 353
Asteriscus, 478, 485, 493, 494
　aquaticus, 484, 491
　daltonii, 484
　intermedius, 484
　maritimus, 484, 490
　odorus, 484
　sericeus, 484
　smithii, 484
　stenophyllus, 484
　vogelii, 484
Astilbe × *arendsii,* 625
Astomum, 529, 531
Astragalus, 277
Asyneuma, 143
Ateleia arsenii, 378
Athyrium, 553, 556
Atrichum, 530
Atriplex, 8
Atropa, 573
　belladonna, 567
Avena, 102
　barbata, 6
Avenula,
　asiatica, 457
　bromoides, 457
　compressa, 457
　hookeri, 457
　schelliana, 457
　sulcata, 457
Azolla,
　caroliniana, 544
　mexicana, 544

B

Baphiopsis parviflora, 378
Baptisia, 7
　leucophaea, 570
Bee pollination, *Lupinus,* 440
Belairea, 387
Belliolum, 278
Beloperone, 279
Benthamina, 122, 123
Beta, 101, 199
　vulgaris, 230, 583

Betula, 5
Bidens, 273, 279, 634
  amplectens, 185, 187, 188, 190
  asymmetrica, 185, 188, 189
  campylotheca, 182, 185–190, 192
  cervicata, 185, 189–193
  coartata, 187, 188
  conjuncta, 185, 189, 190, 192
  cosmoides, 180–183, 185, 191
  ctenophylla, 189
  cuneata, 189
  distans, 187
  evolution of, 179
  fecunda, 187
  forbesii, 185, 188, 190–192
  fulvescens, 189
  gracilis, 190
  hawaiensis, 185, 187, 188, 190
  hillebrandiana, 180–182, 185, 186, 188,
    190–192
  macrocarpa, 185, 188, 189
  mauiensis, 180–182, 185, 188–191, 193
  menziesii, 180–182, 185, 188–190, 192, 193
  micrantha, 182, 185, 187–190, 192
  molokaiensis, 185, 188, 189, 191, 193
  pilosa, 188
  populifolia, 185, 188
  sandvicensis, 183, 185–192
  skottsbergii, 187, 188
  torta, 182, 185–190
  valida, 185, 191
  waianensis, 188
  wiebkei, 185, 188
Biogeography, 497
  Campanula, 141
  Dipsacaceae, 310
Biological species concept, 159–168
Biomass, 211
Biometric index, 402
Biosystematic categories, 643
Blackstonia perfoliata, 466, 468
Boehmeria, 244
Boisduvalia, 352
Bothriochloa, 242
Botrychium, 556, 647
  boreale, 544
  crenulatum, 544
  oneidense, 544
  rugulosum, 544
  simplex, 544
Boykinia, 101
Brachycodon, 142, 145
Brachycome, 35

Brachythecium, 534
  glareosum, 528
  rivulare, 523
  rutabulum, 523, 528
Brassica, 240, 278, 637
  bourgeaui, 480
  napus, 565
  oleracea, 480, 583, 620
Breeding systems, evolution of, 249–270
Bromus, 4
Brya, 379
Bryaspis lupulina, 379
Bryophytes, 519–542
Bryum,
  argenteum, 525
  bicolor, 524
Bubbia, 278
Bupleurum ranunculoides, 459, 472

C

Calamagrostis, 4
Caledia captiva, 29
Calophyllum, 506
Calopogonium, 383
Calorific values, 211
Caltha, 276
Calycadenia, 5
Calyophus, 352
Camissonia, 352
Camoensia, 387
  brevicalyx, 382, 388
  maxima, 382, 388
  scandens, 382, 388
Camp and Gilly, terminology, 645
Campanula, 4, 278
  aizoides, 148
  aizoon, 148
  alliariaefolium, 142
  angustifolia, 148
  bordesiana, 143
  cashmeriana, 142
  colorata, 142
  columnaris, 148
  creutzburgii, 147
  dichotoma, 149
  drabifolia, 147
  edulis, 141, 143
  erinus, 141, 147
  fastigiata, 141
  filicaulis, 143, 145
  glomerata, 142
  hawkinsiana, 149

involucrata, 148
keniensis, 141
lactiflora, 147
lusitanica, 145, 147
macrostyla, 146
mollis, 145
olympica, 146
patula, 143, 146
persicaefolia, 147, 149, 170
primulifolia, 147
psylostachya, 148
rapunculus, 146
reuterana, 145
rotundifolia, 146, 147
scutellata, 146
sparsa, 146
spathulata, 146
trachelium, 142
Campsis, 280
Camptosorus, 549
rhizophyllus, 544
Campylopus,
argyrocaulon, 524
fragilis, 524
leucognodes, 524
paradoxus, 524
pyriformis, 524
setifolius, 524
shawii, 524
subconcolor, 524
Cannabis sativa, 204, 205, 206
Canonical variates analysis, 407
Capsicum, 7
annuum, 406, 624
baccata, 406
chinense, 207, 406
frutescens, 206, 406
pubescens, 406
Cardamine, 472
amara, 220, 221, 223
insueta, 220–223
pratensis, 45
rivularis, 220, 221, 223
schulzii, 220, 222, 223
Carduus defloratus, 455, 472
Carex, 3, 550
Carrpos sphaerocarpos, 526
Cassia, 282
Castanospermum, 387
australe, 378, 391, 392
Castilleja, 279
Catastrophic sexual transmutation, 607
Catharanthus roseus, 567, 570

Ceanothus, 494
Centaurium
majus, 470
minus, 466, 469
tenuiflorum, 466
Centella uniflora, 252
Cephalanthera, 103
Cephalaria, 309, 311, 314, 318
Cerastium
arvense, 459
gibraltaricum, 468
Cestrum, 98
Ceterach, 549
Chapmannia floridiana, 379
Characters, derived, 652
primitive, 652
Cheilanthes,
feei, 544
gracillima, 544
Chemical polymorphism, 561
Chemosystematics, 7, 9, 11, 359
bryophytes, 525
ferns, 551
Chenopodium incognitum, 355
Chiasma, 69, 92
Chirocalyx, 389
Chloroplast DNA, 322
Chorthippus, 72
brunneus, 76, 77
Chromosome
aneuploidy, 129
banding, 27, 30, 43, 97–116
bryophytes, 528
mistletoes, 122
evolution, mistletoes, 117–140
homology, 89
isolation, 490
numbers, basic changes in, 490
Campanula, 143
mistletoes, 117
pairing, 27, 67, 69, 92
polymorphism, 566
races, 453, 470
repatterning, 491
symmetry, 89
Chromosomes,
B, 34, 79, 81, 137
chiasma frequency, 81, 92
DNA content, 27, 44, 123, 129
interchanges, 31
Isotoma, 170, 172
mistletoes, 122
Chrysanthemum leucanthemum, 297

Circaea, 352
Citrus, 581
Cladistics, 652
  Crataegus, 41, 425
Clarkia, 5, 164, 349, 352
  biloba, 345, 354
  franciscana, 345, 354
  lingulata, 345, 354
  pulchella, 351
  rostrata, 350
  rubicunda, 345, 354
  xantiana, 345–348
Classifications, 643
Claytonia, 5, 321, 334, 339, 656
  caroliniana, 326–338, 456, 655
  virginica, 326–338, 456, 655
  rDNA, 331
Clematis, 618
  chrysocoma, 623
  montana, 623
Clinopodium, 4
Coleogyne, 276
Coleus, 277
Collomia linearis, 297
Community structure, 498
Compilospecies, 459
Conium maculatum, 570, 571
Conocephalum conicum, 526, 532
Conservation, 633–641
Constraints, on breeding systems, 249–270
Convolvulus, 278
Coreopsis, 5
  nucensis, 345
  nuecensoides, 345
Coriandrum, 99
Corsinia, 526
Cotoneaster integerrimus, 464
Cranocarpus mezii, 379
Crataegus, 274
  brachyacantha, 433, 435
  conspecta, 433–435
  crus-galli, 417–419
  cuneata, 433
  disperma, 419
  douglasii, 419, 433, 435, 436
  flava, 429, 431, 432, 435
  grandis, 419
  hupehensis, 433
  laevigata, 419
  lavallei, 419
  lucorum, 434
  macracantha, 419
  mexicana, 419

  mollis, 429, 431, 435
  monogyna, 417–419, 427, 430, 432
  pringlei, 434, 435
  prunifolia, 419
  pruinosa, 417, 418, 433–435
  punctata, 417, 418, 427, 429, 430, 432
  rivularis, 433
  saligna, 433
  tangchungchangii, 433
  'toba', 419
  triflora, 431, 432, 435
  viridis, 433, 435, 436
Crepis, 3, 35, 102, 365
  occidentalis, 238
Crocus speciosus, 29
Crop-weed complex, 196
Cryptantha, 4
Cryptogramma crispa, 549
Cucumis, 623
  melo, 582
  sativus, 582, 619, 620
Cucurbita, 8, 353, 624
Cultivated plants, 617
  nomenclature, 624
  × wild plants, 582
Cyanogenesis, 562
Cycads, 325
Cyclostigma, 458
Cymbispatha, 32
Cynodon dactylon, 205
Cypripedium debile, 98, 103
Cystopteris, 556
  bulbifera, 553
  fragilis, 553
  laurentiana, 553
  montana, 553
  protrusa, 553
  tenuis, 553
Cytochrome C, 9
Cytodemes, 453
Cytogeography, 453–476
  mistletoes, 121
Cytology, 25–39
Cytotypes, 453

D

Dachsippen, 459
Dactylis, 201
Dalea obovata, 378
Datura, 278, 571, 573
  fastuosa, 567
  ferox, 567

inermis, 567
inoxia, 567
metel, 567
stramonium, 566, 567
tatula, 566
Daucus carota, 199, 583, 624
glochidiatus, 251, 253
Decaisnina, 123, 129
Delphinium, 276, 279
Deme, terminology, 645
Demography, *Lupinus,* 444
Dendrophthoe, 122, 123, 129
glabrescens, 124
Dendrophthora, 131
Desmodium lespedesioides, 379
Dichanthium, 242
Dicranum, 522
Dictyostelium, 322
Digitalis,
grandiflora, 567
lanata, 567
Dimocarpus, 506
Dioecy, fruit trees, 585
mistletoes, 133, 135
Dioscorea, 574
alata, 583
esculenta, 583
polyploidy in, 583
Diphasiastrum, 554
Diphysa racemosa, 379
Diplatia, 122, 123
furcata, 124
Diplotaxis, 478, 490, 493
gracilis, 480
Dipsacus, 307, 309, 310, 318
Diptera, 278, 282
Discriminant analysis, 411
Discussion on symposium, 655–658
Ditrichum, 531
Diurnal changes, 570
DNA, 14
C values, 44, 46, 54
hybridization, 8, 9, 11, 12
nuclear variation, 87
rDNA, *Claytonia,* 321–341
Dodecatheon, 282
Domesticated, plants, evolution of, 579–586
weeds, 195–210
Donrichardsia, 521
Drepanocladus,
exannulatus, 522, 534
fluitans, 522
trichophyllus, 522

Drimys, 278
Drosophila, 12, 14, 45, 322, 325, 589
Dryopteris, 545, 550, 556, 557
aemula, 551
arguta, 544
azorica, 551
campyloptera, 545, 547, 550, 551
carthusiana, 545
celsa, 545, 551, 552
clintoniana, 545, 551, 552, 555
crispifolia, 551
cristata, 545, 650
dilatata, 551
expansa, 545, 547, 550, 551
filix-mas, 545
fragrans, 545
goldiana, 544, 545, 551, 552
guanchica, 551
intermedia, 545, 551
ludoviciana, 545
maderensis, 551
marginalis, 545, 550
spinulosa, 650
× *boottii,* 650
× *Dryopopolystichum,* 650
× *Polystichum,* 650
× *triploidea,* 654
× *uliginosa,* 650
Duria, 283
Durio, 511
griffithii, 509
zibethinus, 509
Dysploidy, *Campanula,* 143
Dipsacaceae, 311
Erysimum grandiflorum, 456

E

Echeveria, 5
Echium, 493
Ecological, genetics, 211
species concept, 166
Elymus, 102
Endangered species, 633
Endemism, 477
Bidens, 186
Campanula, 149
Energy budget, 211
Eonycteris spelaea, 511
Epilobium, 165, 179
Equisetum, 554, 555
pratense, 544
telmateia, 544

Eremolepis, 120
Errazurizia,
    rotundata, 383
Eryngium vesiculosum, 251, 253
Erysimum, 278
    australe, 456
    grandiflorum-australe-sylvestre complex, 456
    sylvestre, 456
Erythraster, 389
Erythrina, 280, 383, 389
    breviflora, 390
    coralloides, 381
    costaricensis, 390
    schimpffii, 391
    variegata, 379, 390, 391
Eschweilera, 504
Eubrachion, 120
Eucalyptus, 361
Euchlaena, 587–613
    luxurians, 591
    mexicana, 588, 590
    perennis, 588
Eugenia, 506
Eupatorium, 279
Evolution, *Bidens,* 179–194
    breeding systems, 249–270
    Campanula, 141–158
    chromosomal, mistletoes, 117–140
    of domesticates, 579–586
    species concept, 166
Experimental taxonomy, 3
Eysenhardtia, 383

F

Fagopyrum esculentum, 580
Fasciation, 601
Ferns, 325, 543–560
Fertilization, cross, 229
    self, 229
Festuca, 89, 95
    drymeja, 75, 90, 92, 94
    scariosus, 75, 90, 92, 94
Ficus, 277, 281, 613
    carica, 581
Fissidens, 530
Flavonoids, 361, 364
    bryophytes, 525
Foeniculum, 99
Fossil genes, 14
Freycinetia, 281
Fritillaria, 98, 103
Frullania, 526
Fuchsia, 285, 352, 639
Funaria hygrometrica, 531

G

Galeopsis tetrahit, 3
Galium, 6
    incurvum, 466
Garcinia,
    forbesii, 511
    mangostana, 511
    parvifolia, 511
Gasteria, 29
Gaura, 352
    demareei, 345
    longiflora, 345
Geissaspis, 383
    cristata, 380
Gene, duplication, 349
    flow, 440
    maps, 46
    sequencing, 322
Genetic, diversity, 211
    polymorphism, 561
Genome, characters of, 42
    molecular, 46
    polymorphism, 566
    variation, mistletoes, 123
Gentiana, 258, 267, 278
    amabilis, 260
    antarctica, 259, 260, 262, 268
    antipoda, 259, 261, 262, 268
    astonii, 261
    corymbifera, 259, 260, 262
    divisa, 259, 261
    gibbsii, 260, 268
    grisebachii, 260
    lineata, 259, 260, 262
    montana, 260
    orbicularis, 458
    oschtenica, 458
    pontica, 458
    pumila, 458
    saxosa, 250, 261
    serotina, 250, 261
    tergestina, 458
    townsonii, 260
    verna, 458
Geranium, 2
Germplasm transfer, 199
Gibasis, 30, 31, 32, 102
    karwinskyana, 30
    pulchella, 31
    venustula, 31
Gilia, 5, 164, 279
Gingidia, 254
    decipiens, 257
    enysii, 255, 257

flabellata, 257
   montana, 256
   trifoliolata, 256
Globularia, 279
Glycine max, 99, 101, 322, 325, 334, 338
Gongylocarpus, 352
Gossypium,
   barbadense, 197
   hirsutum, 197, 362
Greenovia, 493
Gutierrezia sarothrae, 6
Gymnocarpium, 550, 557
   dryopteris, 549
   jessoense, 555
   robertianum, 555
Gymnosperms, 325

H

Haplopappus, 35
Harmomegathy, pollen, 389
Hauya, 352
Haworthia, 29
Haynaldia, 102
Hebe, 265
Helianthus
   annus, 199
   giganteus, 78
   mollis, 78
Heliconia, 4
Helictotrichos montanum, 462
Hepatica, 101
Heterogaura, 352
Heterosis, 171, 230, 243, 441
Heterostyly, 266
Heuchera, 101
Hieracium, 238
Holocarpha, 164
Homo sapiens, 30
Hopea jucunda, 506, 512
Hordeum, 6, 98, 99, 102, 104, 240, 608
   bulbosum, 50, 53, 58, 59, 63, 77, 79
   vulgare, 50, 53, 54, 58, 59, 61–63, 77, 79
Hornworts, 519
Horticultural plants, 617–632
Houttuynia, 246
Humulus lupulus, 202
Hyacinthus, 584
Hybrid, index, 402
   terminology, 625, 649
Hybridization, 165, 196, 493
   apomixis, 241
   extinction by, 206

progenitor domesticates, 197
Hybrids, chromosome pairing, 76
   DNA content, 92
   Bidens, 179, 186
   Campanula, 146
   Crataegus, 417–433
   Isotoma, 169
Hydrocotyle, 255
   americana, 251, 252, 258
   elongata, 252, 258
   novae-zelandiae, 251, 252
Hydrophyllum, 279
Hymenophyton, 526
Hyoscyamus,
   niger, 573
Hypericum, 4, 553

I

ICBN, 625
ICNCP, 625
Ileostylus, 122, 123
   micranthus, 124
Illicium, 278
Impatiens, 279
Inbreeding, 170, 229
Incompatibility, 229–235, 621
Incongruity, 621
Indigofera, 385
Inocarpus, 385
Introgression, 653
Ipomoea batatas, 583
Ipomopsis, 278
Iris, 103
   fulva, 402, 404
   hexagona, 402, 404
Isoetes, 557
   acadiensis, 555
   asiatica, 554
   bolanderi, 555
   eatonii, 555
   echinospore, 554, 555
   hieroglyphica, 544
   howellii, 554, 555
   luacustris, 554
   macrospora, 554, 555
   maritima, 555
   melanopoda, 554
   muricata, 554, 556
   nutallii, 555
   occidentalis, 555
   riparia, 555
   tuckermanii, 555

Isolating mechanisms, 164, 191
Isotoma, 32
  anaethifolia, 173
  axillaris, 173
  hybridity in, 170
  petraea, 169–177
Isozymes, 343–357, 412
  Bidens, 182
  Zea, 597
Iteropary, 212

J

Jasione, 143
Jepsonia, 4
Juniperus, 8

K

Karyotype, 28
  natural, 49
  nomenclature, 47
Kentranthus, 279
Kinagusa, 98
Knautia, 309-311

L

Lactuca, 623
  sativa, 583, 619, 620
  serriola, 619
Laevenworthia, 265
Larrea, 6, 77
  divaricata, 78, 466
  tridentata, 78
Lasioglossum, 170
Lasthenia, 4, 7
Lathyrus, 89, 91
  angulatus, 90, 93
  articulatus, 90, 93
  hirsutus, 90, 93
  latifolius, 387
  odoratus, 207
  tingitanus, 90, 93
  vernus, 387
Leapoldia, 103
Leavenworthia, 5
Ledum, 277
Legousia, 146
Leitneria, 2
Lemna,
  gibba, 218, 219, 220
  turionifera, 218, 219

Lens, 387
Lepidoceras, 120
Lepidotis, 554
Lepisanthes, 506, 509
Lesquerella, 5
Leucanthemopsis alpina, 459
Leucanthemum,
  adustum, 455
  vulgare, 459
Liatris, 279
Life tables, *Lupinus*, 445
Ligaria, 120
Lignocarpa
  baxterae, 256
  carnosula, 255, 256, 259
  diversifolia, 254, 257, 259
Lilaeopsis novae-zelandiae, 253
Lilium, 103, 625
Linaria, 279
Linum perenne, 472
Liverworts, 519
Lobularia, 478, 482, 491, 493
  arabica, 480, 481
  intermedia, 480, 481
  libyca, 480, 481, 491
  marginata, 480, 481
  maritima, 480, 481, 490
  palmensis, 480, 481
Lolium, 89, 102
  perenne, 75, 79–82, 90
  temulentum, 75, 79, 80, 82, 90
Lophocolea,
  bidentata, 524
  cuspidata, 524
  heterophylla, 523
Loranthus, 117
Lotus, 5, 562
  collinus, 379
  pedunculatus, 101, 104
Ludwigia, 352
Lupinus,
  albus, 565
  angustifolius, 565
  luteus, 565
  subcarnosus, 440–451
  texensis, 440–451
Luzula, 4
Lychnis,
  alba, 408
Lycopersicon, 14, 623
  chielewskii, 345
  esculentum, 234, 582
  parviflorum, 345

peruvianum, 232
pimpinnellifolium, 199
Lycopodiella, 554
Lycopodium, 556, 557
  alpinum, 544
  annotinum, 544
  clavatum, 554
  complanatum, 554, 555
  dendroideum, 554
  digitatum, 554, 555
  inundatum, 553, 554
  lucidulum, 549, 554
  obscurum, 554
  selago, 544, 554
  tristachyum, 554, 555
Lysiana, 122, 123, 129
  murrayi, 124

M

Macadamia, 283
Macaranga, 499
Macaronesia, 477–496
Machaeranthera, 353
Macromitrium, 528
Macrosolen, 122, 123
  cochinchinensis, 124
Macrotyloma, 381
  axillaris, 383
Madiinae, 4
Magnolia, 278, 611
Mallotus, 553
Malvaviscus, 268
Manihot, 562
  esculentum, 583
Marchantia, 529, 532
  berteroana, 526
  polymorpha, 526
Marsilea,
  quadrifolia, 544
  vestita, 544
Matteuccia struthiopteris, 556
Mecodium wrightii, 544
Mecopus nidulans, 379
Medicago,
  falcata, 200, 201, 202
  sativa, 200, 201, 202
Medicinal plants, 561–578
Melandrium, 4
Melilotus, 205
Mertensia, 278
Micropteryx, 389
Microscopy, SEM, TEM, Papilionoideae, 377

Microseris douglasii, 30
Migrations, 457
Millettia,
  australis, 378
  theuszii, 388
  usaramensis, 388
Mimosa, 279
Mimulus, 4, 279, 657
  cardinalis, 11, 13
  eastwoodiae, 11
  lewisii, 9, 11, 13
  nelsonii, 11
  rupestris, 9, 11, 13
  verbenaceus, 11
Minuartia, 146
  sedoides, 461, 462
Mistletoes, chromosomes, 117–140
Mitella, 101
Mnium,
  affine, 523
  cuspidatum, 530
  medium, 523
Molecular genetics, 14
Morus, 362
Mosses, 519
Muelleranthus trifoliatus, 379
Muellerina, 122, 123
  eucalyptoides, 124
Multivariate analysis, 402
Muntingia calabura, 250, 265, 266, 267
Musa sapientum, 583
Muscari, 278
Musci, 519
Mutation, 14, 83, 196, 240, 243, 351, 585
Mutica, 503
Myosotis, 278
Myrtus, 553

N

Narcisus, 2, 584
Nectars, 276, 281
Neopolyploidy, 461
Nephelium, 509
Nicotiana, 3, 101, 231, 573, 621
  glauca, 278
  rustica, 197
  tabacum, 197, 565
Nigella, 101, 634
Nivales, 458
Nomenclature, cultivated plants, 624
Nothospecies, 650
Notothixos, 119, 131

Nuclear DNA variation, 87
Numerical taxonomy, 7, 9, 11, 395–415
Nuytsia, 122, 123
  floribunda, 124

O

Observations on symposium, 655–658
Oenothera, 3, 35, 147, 170, 173, 278, 350, 352
  lamarkiana, 175
Oldenlandia adscencionis, 639
Olea europaea, 581
Onoclea sensibilis, 649
Ononis, 639
  repens, 300–303
  spinosa, 300–303
Ophrys, 280
Opuntia, 5
Ordination analysis, 407
Oreomyrrhis colensoi, 253
Ormosia henryii, 385
Ornamentals, 581, 617
Oryza glaberrima, 204
Ovule number, 251
Oxytropis, 277
  albiflora, 403
  halleri, 461, 462

P

Pachyphyletism, 457, 459
Pachyrhizus, 383
Palaeoecology, 272
Palynology, Dipsacaceae, 313
Panax quinquefolium, 574
Panicum maximum, 240, 242
Papaver,
  orientale, 563
  somniferum, 563, 564, 565, 570–572
Paphiopedilum, 33
Parahebe linifolia, 268
Paris, 97, 98, 103
Parkia, 510
  roxburghii, 511
  speciosa, 511
Parthenium, 373
Parthenocarpy, 583
Parthenogenesis, 240
Partula, 182
Pattern recognition, 406
Pedicularis, 266, 277, 638
Pellia, 528
  asplenioides, 532

  borealis, 532
  endiviifolia, 532
  epiphylla, 532
  glabella, 549, 653
  neesiana, 532
  suksdorfiana, 549
Penstemon, 3, 279, 280
Peraxilla, 123
Petunia, 101
Phacelia, 279
Phaseolus, 101, 381
  vulgaris, 580
Phegopteris, 557
  connectilis, 549
Phenetics, 160, 304
  classification, 397
Phenolics, 360
Phenotypic plasticity, *see* Plasticity
Phlox, 278, 285, 374
  drummondii, 8
Phoenix dactylifera, 207, 581
Phoradendron, 117, 133, 137
Phyllanthus, 146
Phyllitis, 4
  japonicus, 549
  scolopendrium, 549
Phylogenetic constraint, 249
Phylogeny, *Campanula*, 143
  Dipsacaceae, 316
Physcomitrella patens, 531–2
Physcomitrium, 531
  eurystomum, 531
  patens, 531
  pyriforme, 531
  sphaericum, 531
Phyteuma, 143, 279
Phytochemistry, 359–376, 525
Picea, 125
Pictorialized scatter diagram, 403
Pimpinella, 99
Pinus
  muricata, 5
  resinosa, 100, 101
Pisum, 101, 387
Pityrogramma, 551
  triangularis, 544, 549
Plantago, 99
Plasticity, adaptive, 296
  compensation, 294
  developmental, 294
  environmental, 294
  phenotypic, 212, 293–306
  tolerance, 294

Plectranthus, 277
Pleuridium, 531
Pleuroclada albescens, 523
Pleurozium schreberi, 535
Poa, 4, 244, 274
    cenisia, 470
Pohlia proligera, 524
Poiretia latifolia, 379
Polemonium, 406
    pulcherrimum, 400
    viscosum, 281
Pollen,
    ovule ratio, 250
    Dipsacaceae, 312
    Papilionoideae, 377–394
Pollination, 271–292
    self, 580
Pollinators, 271, 281, 509
Polyalthia, 504
Polycarpon polycarpoides, 455
Polygonal graph, 400
Polygonum,
    achoreum, 411, 412
    amphibium, 293
    cascadense, 297
Polyploidy, 32, 197, 241, 457, 566
    bryophytes, 527
    Campanula, 143
    centrifugal, 466
    Claytonia, 321
    Crataegus, 420
    Dipsacaceae, 311
    Dioscorea, 583
    escalating, 461
    ferns, 574
    Garcenia, 511
    isozymes, 352
    mistletoes, 129
    neo, 461, 470, 471
    polarized, 461, 466
    nonpolarized, 467
Polypodium, 556
    amorphum, 548
    glycyrrhiza, 548
    hesperium, 548
    scouleri, 548
    virginianum, 549, 649
Polystichum, 544, 545, 550, 556
    acrostichoides, 546
    andersonii, 546-7
    braunii, 546-7
    californicum, 544, 546, 653
    glabella, 549

imbricans, 546
kruckebergii, 544, 546
lemmonii, 544, 546, 547
lonchitis, 546
mohrioides, 547
pumila, 549
scopulinum, 544, 546
setigerum, 546-7
suksdorfiana, 549
virginianum, 549
Pometia, 506
    pinnata, 509
Population biology, 439–452
Populus, 361
Porella,
    macroloba, 526
    vernicosa, 526
Poskea, 279
Potentilla, 4
    arenaria, 244
    argentea, 241, 242, 243, 244
    calabra, 241, 242
    collina, 244
    croceolata, 244
    gracilis, 238
    neglecta, 241, 242
    tabernaemontani, 244, 245
Primula, 5, 278, 457, 458
    vulgaris, 362
Prosopsis,
    alba, 78
    ruscifolia, 78
Protein sequencing, 9, 14
Prunella, 4
    vulgaris, 300
Prunus domestica, 581
Pseudo-compatibility, 229–235
    pseudo-self-compatibility, 229–235
Pseudowintera, 278
Psoralea,
    macrostachya, 379
    pubescens, 380–81
Psorothamnus schottii, 380
Pteridium aquilinum, 556
Pteridophytes, 543–560
Pterocephalus, 307, 319
    centennii, 314
    diandrus, 314
    intermedius, 314
Pterocymbium, 510
Pterostylis, 278
Pulmonaria, 456
Punctuated equilibrium, 166

Pycnocomon, 314, 318
Pyrus,
    communis, 581
    malus, 581

Q

Quercus, 165, 613

R

Radiation, adaptive, 179, 494
Rafflesia, 278
Rain forest, 498
Randabspaltung, 461, 466
Ranunculus, 250, 267, 268, 439
    acaulis, 264
    aquatilis, 300
    auricomus, 239
    flammula, 300
    glabrifolius, 263, 264
    insignis, 263
    limosella, 263, 264
    macropus, 264
    plantagineus, 455, 462, 470, 471
    pyrenaeus, 471
    rivularis, 263, 264
Raphanus, 199
Rare species, 497–518
Reana,
    luxurians, 590
Reproduction, asexual, 211, 580
    sexual, 211
Reproductive methods, 211–228, 266, 273
    domesticated plants, 579
Resource allocation, 211
Restriction enzymes, 14
Rhoeo, 31
Rhus, 279
Ribosomal genes, *Claytonia*, 321
Riella americana, 532
Robertsonian changes, 33, 95, 122
Rosa, 5, 99
    × *rosea*, 651
    × *rubra*, 651
Rubus, 238
    arcticus, 620
Rudbeckia, 285
    hirta, 362
Rutiflorae, 2

S

Saccharum officinarum, 583

Salix, 165
Salvia, 99, 101, 279
Saxifraga oppositifolia, 274
Scabiosa, 307, 309, 313, 314, 318, 319
    palaestina, 311
Scaevola, 179
Scandia, 254
    geniculata, 256
    rosifolia, 256
Schizaea pusilla, 544
Schizeilema,
    cockaynei, 252
    exiguum, 252
    haastii, 251, 252
Schulzii, 472
Scilla, 29, 30, 44, 99, 103, 104
    autumnalis, 29, 30
    bifola, 105
    drunensis, 105
    vindobonensis, 105
Scrophularia, 478, 482, 483, 490
    arguta, 484, 491
    auriculata, 484
    calliantha, 483
    glabrata, 483
    hirta, 483, 494
    lowei, 484
    racemosa, 484
    scorodonia, 482, 483, 494
    smithii, 483
Secale, 8, 98, 102, 104
    africanum, 50, 59
    cereale, 47, 61, 580
Sedum, 365
    acre, 467
    forsterianum, 467
    litoreum, 466
    sediforme, 466, 470
    tenuifolium, 467
    urvillei, 466
Seed, dispersal, 509
    dormancy, 213
    germination, 218
Selaginella,
    densa, 556
    oregana, 544
    sibirica, 556
    wallacei, 544
Self, incompatibility, 229, 585
    pollination, 229, 265, 580
SEM, ferns, 557
    Papilionoideae, 377
Senecio doronicum, 459

Shorea, 503, 508, 509
  amplexicaulis, 512, 513
  dasyphylla, 511
  leprosula, 511, 514
  macroptera, 507, 511
  rotundifolia, 512, 513
Silene
  alba, 408
  boryi, 462
  ciliata, 461, 464
  dioica, 408, 410
  laciniata, 278
  latifolia, 408, 410, 411
  legionensis, 464
  pratensis, 408
Sinapidendron, 477–480, 490
  palmense, 480
  sempervivifolium, 479
Sinapsis pubescens, 480
Sindora, 511
Sogerianthe, 123
  sessiliflora, 124
Solanum, 282
  acaula, 199
  ajanhuiri, 199
  chaupa, 197
  curtilobum, 197
  juzepczukii, 197, 199
  megistacrolobum, 199
  melongena, 582
  oplocense, 199
  pennellii, 14
  sparsipilum, 199
  stenotomum, 197, 199
  sucrense, 199
  tuberosum, 197, 199
Solenobia, 245
Solidago, 284
  sempervirens, 297
  virgaurea, 395, 398
Sonchus, 494
Sophora,
  davidii, 391
  microphylla, 391
Sorghum, 199, 562
Spartina,
  alterniflora, 197
  anglica, 197
  maritima, 197
Speciation, 95, 183, 490, 493
  gradual, 493
Species,
  chromosome pairing, 76

  concept, 159–168, 183
  definition, 646
  entities, 3
Sphaerocarpos, 528
Sphagnum, 527
  capillaceum, 524
  cuspidatum, 525
  cyclophyllum, 525
  imbricatum, 524
  lescusii, 525
  macrophyllum, 524
  rubellum, 524
  subsecundum, 523
  trinitense, 525
Spiranthes,
  casei, 407, 408
  cernua, 407, 408
  lacera, 407, 408
  romanzoffiana, 407, 408
Spirodela polyrrhiza, 217, 218
Sporobolus durus, 639
Stamen number, 250
Stapelia, 278
Stebbinsian complexes, 467, 469
Stebbins's Rule, 229, 232, 235
Stemonoporus, 512
  bullatus, 514
  canaliculatus, 513, 514
  marginalis, 514
Stenosiphon, 352
Stephanomeria, 5
  exigua, 345
  malheusensis, 345
Succisa, 311, 313, 314, 318
Succisella, 311, 313, 314, 318
Sullivania, 101
Swartzia, 387
  fistuloides, 385
  leptopetala, 380
  madagascariensis, 385
Symphyandra, 145
Symphytum, 6, 204
Synaptonemal complex, 94
Synecology, 497
Syzygium,
  mareoptera, 506

T

Taraxacum, 238, 244
Taxillus, 122
Taxonomic methods, 643–654
TEM, Papilionoideae, 377

Templetonia
  aculeata, 389
  egena, 382, 387–389
  hookeri, 389
  incana, 387
  retusa, 382, 387–389
  stenophylla, 389
  sulcata, 389
Teosinte, 587, 599
Terminalia, 279
Terminology, Camp and Gilly, 645
  deme, 645
  Grant, 645
  international code, 644
Tetramolopium, 355
Thalictrum, 275
Thelypteris,
  palustris, 556
  simulata, 544
Theobroma, 283
Thuja, 553
Tiarella, 101
Tipuana speciosa, 378
Tolmiea, 101
Tortula muralis, 523, 529
Trachelium, 279
Tragopogon, 5, 459
  dubius, 197
  mirus, 197
  miscellus, 197
  porrifolius, 197
  pratensis, 197
Transmutation, 605
Tremastelma, 313, 318
Trichostema, 277, 279
Trifolium, 283
  repens, 562
Trigona, 509
Trillium, 6, 97, 99, 103, 104
  erectum, 107, 109
  grandiflorum, 98, 107, 109
  ovatum, 105–107
Tripsacum, 587–613
  andersonii, 613
Trisomy, *Campanula*, 147
Triticum, 8, 99, 102, 104, 201, 325
  aestivum, 43, 48, 75, 197
  tauschi, 197
  turgidum, 197
Tropaeolum, 279
Tropical forest, 497–518
Tulbaghia, 98, 103
Tulipa, 103, 584

Tupeia, 120, 123
Turions, 218

U

Ulex, 391
Ulota,
  bruchii, 524
  crispula, 524
Urginea,
  indica, 567, 568
  maritima, 567
Ustilago maydis, 611

V

Vaccinium, 3, 278
Verbascum, 279
Veronica, 280
Vicia, 89
  bithyrica, 91
  cracca, 387
  disperma, 91
  faba, 46, 87, 98, 99, 101, 123, 387, 580
  hirsuta, 91
  monantha, 87
  narbonensis, 387
  peregrina, 387
  villosa, 91
Vigna, 381, 383
Viola, 3, 5, 279
Viscum, 31, 117, 119, 131, 133, 134, 136, 138
  album, 137
  combretiocola, 135
  hildebrandtii, 135
Vitex, 279
Vitis, 6
  vinifera, 581
Vivipary, 221

W

Wahlenbergia, 147
Weeds, 623
  progenitor domesticates, 195
Weissia, 520, 535, 531
  controversa, 528–9
  crispa, 529
  exserta, 529
Wild vs. domesticated plants, 579, 582
Williamsoniella, 273
Woodsia, 553, 556
  cathcartiana, 549

mexicana, 549
obtusa, 544
oregana, 549
Woodwardia fimbriata, 544

X

Xerospermum, 506
noronhianum, 500, 507, 509

Z

Zauschneria, 279, 459
Zea, 587–616
canina, 590
diploperennis, 588
luxurians, 588
mays, 27, 48, 54, 57, 75, 100, 102, 233,
234, 325, 580, 587–613
perennis, 588
origin of, 587
polystichy, 605
Zygogynum, 278
Zygotic lethals, 175